Prealgebra 2e

Senior Contributing Authors
Lynn Marecek, Santa Ana College
MaryAnne Anthony-Smith, Formerly of Santa Ana College
Andrea Honeycutt Mathis, Northeast Mississippi Community College

Rice University
6100 Main Street MS-375
Houston, Texas 77005

Changes to original document:
Non-numbered pages prior to the "Table of Contents" page were removed. The beginning blank pages, the previous title page, this page and final blank pages have been added. Chapter title pictures were trimmed to fit inside of printing area. For the paperback version, color images and text were converted to grayscale.

Download the original document for free at:
https://openstax.org/details/books/prealgebra-2e

Table of Contents

PREFACE

Welcome to *Prealgebra 2e*, an OpenStax resource. This textbook was written to increase student access to high-quality learning materials, maintaining highest standards of academic rigor at little to no cost.

About OpenStax

OpenStax is a nonprofit based at Rice University, and it's our mission to improve student access to education. Our first openly licensed college textbook was published in 2012, and our library has since scaled to over 35 books for college and AP® Courses used by hundreds of thousands of students. OpenStax Tutor, our low-cost personalized learning tool, is being piloted in college courses throughout the country. Through our partnerships with philanthropic foundations and our alliance with other educational resource organizations, OpenStax is breaking down the most common barriers to learning and empowering students and instructors to succeed.

About OpenStax Resources

Customization

Prealgebra 2e is licensed under a Creative Commons Attribution 4.0 International (CC BY) license, which means that you can distribute, remix, and build upon the content, as long as you provide attribution to OpenStax and its content contributors.

Because our books are openly licensed, you are free to use the entire book or pick and choose the sections that are most relevant to the needs of your course. Feel free to remix the content by assigning your students certain chapters and sections in your syllabus, in the order that you prefer. You can even provide a direct link in your syllabus to the sections in the web view of your book.

Instructors also have the option of creating a customized version of their OpenStax book. The custom version can be made available to students in low-cost print or digital form through their campus bookstore. Visit your book page on openstax.org for more information.

Art attribution in *Prealgebra 2e*

In *Prealgebra 2e*, most art contains attribution to its title, creator or rights holder, host platform, and license within the caption. For art that is openly licensed, anyone may reuse the art as long as they provide the same attribution to its original source. Some art has been provided through permissions and should only be used with the attribution or limitations provided in the credit.

Errata

All OpenStax textbooks undergo a rigorous review process. However, like any professional-grade textbook, errors sometimes occur. Since our books are web based, we can make updates periodically when deemed pedagogically necessary. If you have a correction to suggest, submit it through the link on your book page on openstax.org. Subject matter experts review all errata suggestions. OpenStax is committed to remaining transparent about all updates, so you will also find a list of past errata changes on your book page on openstax.org.

Format

You can access this textbook for free in web view or PDF through openstax.org, and for a low cost in print.

About *Prealgebra 2e*

Prealgebra 2e is designed to meet scope and sequence requirements for a one-semester prealgebra course. The text introduces the fundamental concepts of algebra while addressing the needs of students with diverse backgrounds and learning styles. Each topic builds upon previously developed material to demonstrate the cohesiveness and structure of mathematics.

Students who are taking basic mathematics and prealgebra classes in college present a unique set of challenges. Many students in these classes have been unsuccessful in their prior math classes. They may think they know some math, but their core knowledge is full of holes. Furthermore, these students need to learn much more than the course content. They need to learn study skills, time management, and how to deal with math anxiety. Some students lack basic reading and arithmetic skills. The organization of *Prealgebra* makes it easy to adapt the book to suit a variety of course syllabi.

Coverage and Scope

Prealgebra 2e takes a student-support approach in its presentation of the content. The beginning, in particular, is presented as a sequence of small steps so that students gain confidence in their ability to succeed in the course. The order of topics was carefully planned to emphasize the logical progression throughout the course and to facilitate a thorough understanding of each concept. As new ideas are presented, they are explicitly related to previous topics.

Chapter 1: Whole Numbers
Each of the four basic operations with whole numbers—addition, subtraction, multiplication, and division—is modeled and explained. As each operation is covered, discussions of algebraic notation and operation signs, translation of algebraic expressions into word phrases, and the use the operation in applications are included.

Chapter 2: The Language of Algebra
Mathematical vocabulary as it applies to the whole numbers is presented. The use of variables, which distinguishes algebra from arithmetic, is introduced early in the chapter, and the development of and practice with arithmetic concepts use variables as well as numeric expressions. In addition, the difference between expressions and equations is discussed, word problems are introduced, and the process for solving one-step equations is modeled.

Chapter 3: Integers
While introducing the basic operations with negative numbers, students continue to practice simplifying, evaluating, and translating algebraic expressions. The Division Property of Equality is introduced and used to solve one-step equations.

Chapter 4: Fractions
Fraction circles and bars are used to help make fractions real and to develop operations on them. Students continue simplifying and evaluating algebraic expressions with fractions, and learn to use the Multiplication Property of Equality to solve equations involving fractions.

Chapter 5: Decimals
Basic operations with decimals are presented, as well as methods for converting fractions to decimals and vice versa. Averages and probability, unit rates and unit prices, and square roots are included to provide opportunities to use and round decimals.

Chapter 6: Percents
Conversions among percents, fractions, and decimals are explored. Applications of percent include calculating sales tax, commission, and simple interest. Proportions and solving percent equations as proportions are addressed as well.

Chapter 7: The Properties of Real Numbers
The properties of real numbers are introduced and applied as a culmination of the work done thus far, and to prepare students for the upcoming chapters on equations, polynomials, and graphing.

Chapter 8: Solving Linear Equations
A gradual build-up to solving multi-step equations is presented. Problems involve solving equations with constants on both sides, variables on both sides, variables and constants on both sides, and fraction and decimal coefficients.

Chapter 9: Math Models and Geometry
The chapter begins with opportunities to solve "traditional" number, coin, and mixture problems. Geometry sections cover the properties of triangles, rectangles, trapezoids, circles, irregular figures, the Pythagorean Theorem, and volumes and surface areas of solids. Distance-rate-time problems and formulas are included as well.

Chapter 10: Polynomials
Adding and subtracting polynomials is presented as an extension of prior work on combining like terms. Integer exponents are defined and then applied to scientific notation. The chapter concludes with a brief introduction to factoring polynomials.

Chapter 11: Graphs
This chapter is placed last so that all of the algebra with one variable is completed before working with linear equations in two variables. Examples progress from plotting points to graphing lines by making a table of solutions to an equation. Properties of vertical and horizontal lines and intercepts are included. Graphing linear equations at the end of the course gives students a good opportunity to review evaluating expressions and solving equations.

All chapters are broken down into multiple sections, the titles of which can be viewed in the Table of Contents.

Changes to the Second Edition

The *Prealgebra 2e* revision focused on mathematical clarity and accuracy. Every Example, Try-It, Section Exercise, Review Exercise, and Practice Test item was reviewed by multiple faculty experts, and then verified by authors. This intensive effort resulted in hundreds of changes to the text, problem language, answers, instructor solutions, and graphics.

However, OpenStax and our authors are aware of the difficulties posed by shifting problem and exercise numbers when textbooks are revised. In an effort to make the transition to the 2nd edition as seamless as possible, we have minimized any shifting of exercise numbers. For example, instead of deleting or adding problems where necessary, we replaced problems in order to keep the numbering intact. As a result, in nearly all chapters, there will be no shifting of exercise numbers; in the chapters where shifting does occur, it will be minor. Faculty and course coordinators should be able to use the new edition in a straightforward manner.

Also, to increase convenience, answers to the Be Prepared Exercises will now appear in the regular solutions manuals, rather than as a separate resource.

A detailed transition guide is available as an instructor resource at openstax.org.

Pedagogical Foundation and Features

Learning Objectives

Each chapter is divided into multiple sections (or modules), each of which is organized around a set of learning objectives. The learning objectives are listed explicitly at the beginning of each section and are the focal point of every instructional element.

Narrative text

Narrative text is used to introduce key concepts, terms, and definitions, to provide real-world context, and to provide transitions between topics and examples. An informal voice was used to make the content accessible to students.

Throughout this book, we rely on a few basic conventions to highlight the most important ideas:

Key terms are boldfaced, typically when first introduced and/or when formally defined.

Key concepts and definitions are called out in a blue box for easy reference.

Examples

Each learning objective is supported by one or more worked examples, which demonstrate the problem-solving approaches that students must master. Typically, we include multiple Examples for each learning objective in order to model different approaches to the same type of problem, or to introduce similar problems of increasing complexity.

All Examples follow a simple two- or three-part format. First, we pose a problem or question. Next, we demonstrate the Solution, spelling out the steps along the way. Finally (for select Examples), we show students how to check the solution. Most examples are written in a two-column format, with explanation on the left and math on the right to mimic the way that instructors "talk through" examples as they write on the board in class.

Figures

Prealgebra 2e contains many figures and illustrations. Art throughout the text adheres to a clear, understated style, drawing the eye to the most important information in each figure while minimizing visual distractions.

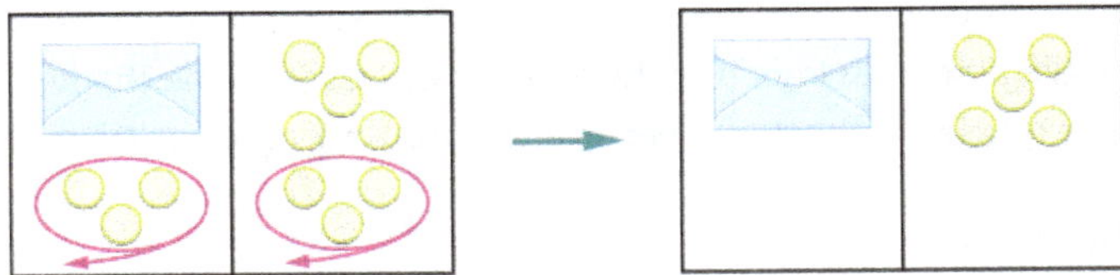

Supporting Features

Four small but important features serve to support Examples:

Be Prepared!

Each section, beginning with Section 1.2, starts with a few "Be Prepared!" exercises so that students can determine if they have mastered the prerequisite skills for the section. Reference is made to specific Examples from previous sections so students who need further review can easily find explanations. Answers to these exercises can be found in the supplemental resources that accompany this title.

How To

A "How To" is a list of steps necessary to solve a certain type of problem. A "How To" typically precedes an Example.

Try It

A "Try It" exercise immediately follows an Example, providing the student with an immediate opportunity to solve a similar problem. In the PDF and the Web View version of the text, answers to the Try It exercises are located in the Answer Key.

Media

The "Media" icon appears at the conclusion of each section, just prior to the Section Exercises. This icon marks a list of links to online video tutorials that reinforce the concepts and skills introduced in the section.

Disclaimer: While we have selected tutorials that closely align to our learning objectives, we did not produce these tutorials, nor were they specifically produced or tailored to accompany *Prealgebra 2e*.

Section Exercises

Each section of every chapter concludes with a well-rounded set of exercises that can be assigned as homework or used selectively for guided practice. Exercise sets are named Practice Makes Perfect to encourage completion of homework assignments.

Exercises correlate to the learning objectives. This facilitates assignment of personalized study plans based on individual student needs.

Exercises are carefully sequenced to promote building of skills.

Values for constants and coefficients were chosen to practice and reinforce arithmetic facts.

Even and odd-numbered exercises are paired.

Exercises parallel and extend the text examples and use the same instructions as the examples to help students easily recognize the connection.

Applications are drawn from many everyday experiences, as well as those traditionally found in college math texts.

Everyday Math highlights practical situations using the math concepts from that particular section.

Writing Exercises are included in every Exercise Set to encourage conceptual understanding, critical thinking, and literacy.

Chapter Review Features

The end of each chapter includes a review of the most important takeaways, as well as additional practice problems that students can use to prepare for exams.

Key Terms provides a formal definition for each bold-faced term in the chapter.

Key Concepts summarizes the most important ideas introduced in each section, linking back to the relevant Example(s) in case students need to review.

Chapter Review Exercises includes practice problems that recall the most important concepts from each section.

Practice Test includes additional problems assessing the most important learning objectives from the chapter.

Answer Key includes the answers to all Try It exercises and every other exercise from the Section Exercises, Chapter Review Exercises, and Practice Test.

Additional Resources

Student and Instructor Resources

We've compiled additional resources for both students and instructors, including Getting Started Guides, manipulative mathematics worksheets, Links to Literacy assignments, and an answer key. Instructor resources require a verified instructor account, which can be requested on your openstax.org log-in. Take advantage of these resources to supplement your OpenStax book.

Partner Resources

OpenStax Partners are our allies in the mission to make high-quality learning materials affordable and accessible to students and instructors everywhere. Their tools integrate seamlessly with our OpenStax titles at a low cost. To access the partner resources for your text, visit your book page on openstax.org.

About the Authors

Senior Contributing Authors

Lynn Marecek and MaryAnne Anthony-Smith taught mathematics at Santa Ana College for many years and have worked together on several projects aimed at improving student learning in developmental math courses. They are the authors of *Strategies for Success: Study Skills for the College Math Student,* published by Pearson HigherEd.

Lynn Marecek, Santa Ana College

MaryAnne Anthony-Smith, Santa Ana College

Andrea Honeycutt Mathis, Northeast Mississippi Community College

Reviewers

Tony Ayers, Collin College Preston Ridge Campus
David Behrman, Somerset Community College
Brandie Biddy, Cecil College
Bryan Blount, Kentucky Wesleyan College
Steven Boettcher, Estrella Mountain Community College
Kimberlyn Brooks, Cuyahoga Community College
Pamela Burleson, Lone Star College University Park
Tamara Carter, Texas A&M University
Phil Clark, Scottsdale Community College

Christina Cornejo, Erie Community College
Denise Cutler, Bay de Noc Community College
Richard Darnell, Eastern Wyoming College
Robert Diaz, Fullerton College
Karen Dillon, Thomas Nelson Community College
Valeree Falduto, Palm Beach State
Bryan Faulkner, Ferrum College
David French, Tidewater Community College
Stephanie Gable, Columbus State University
Heather Gallacher, Cleveland State University
Rachel Gross, Towson University
Dianne Hendrickson, Becker College
Linda Hunt, Shawnee State University
Betty Ivory, Cuyahoga Community College
Joanne Kendall, Lone Star College System
Kevin Kennedy, Athens Technical College
Stephanie Krehl, Mid-South Community College
Allyn Leon, Imperial Valley College
Gerald LePage, Bristol Community College
Laurie Lindstrom, Bay de Noc Community College
Jonathan Lopez, Niagara University
Yixia Lu, South Suburban College
Mikal McDowell, Cedar Valley College
Kim McHale, Columbia College of Missouri
Allen Miller, Northeast Lakeview College
Michelle Moravec, Baylor University TX/McLennan Community College
Jennifer Nohai-Seaman, Housatonic Community College
Rick Norwood, East Tennessee State University
Linda Padilla, Joliet Junior College
Kelly Proffitt, Patrick Henry Community College
Teresa Richards, Butte-Glenn Community College
Christian Roldan-Johnson, College of Lake County Community College
Patricia C. Rome, Delgado Community College, City Park Campus
Kegan Samuel, Naugatuck Valley Community College
Bruny Santiago, Tarrant College Southeast Campus
Sutandra Sarkar, Georgia State University
Richard Sgarlotti, Bay Mills Community College
Chuang Shao, Rose State College
Carla VanDeSande, Arizona State University
Shannon Vinson, Wake Technical Community College
Maryam Vulis, Norwalk Community College
Toby Wagner, Chemeketa Community College
Libby Watts, Tidewater Community College
Becky Wheelock, San Diego City College

1 WHOLE NUMBERS

Figure 1.1 Purchasing pounds of fruit at a fruit market requires a basic understanding of numbers. (credit: Dr. Karl-Heinz Hochhaus, Wikimedia Commons)

Chapter Outline

Introduction

Even though counting is first taught at a young age, mastering mathematics, which is the study of numbers, requires constant attention. If it has been a while since you have studied math, it can be helpful to review basic topics. In this chapter, we will focus on numbers used for counting as well as four arithmetic operations—addition, subtraction, multiplication, and division. We will also discuss some vocabulary that we will use throughout this book.

1.1 Introduction to Whole Numbers

Learning Objectives

By the end of this section, you will be able to:

- Identify counting numbers and whole numbers
- Model whole numbers
- Identify the place value of a digit
- Use place value to name whole numbers
- Use place value to write whole numbers
- Round whole numbers

Identify Counting Numbers and Whole Numbers

Learning algebra is similar to learning a language. You start with a basic vocabulary and then add to it as you go along. You need to practice often until the vocabulary becomes easy to you. The more you use the vocabulary, the more familiar it becomes.

Algebra uses numbers and symbols to represent words and ideas. Let's look at the numbers first. The most basic numbers used in algebra are those we use to count objects: and so on. These are called the **counting numbers**.

The notation "…" is called an ellipsis, which is another way to show "and so on", or that the pattern continues endlessly. Counting numbers are also called natural numbers.

MANIPULATIVE MATHEMATICS

Doing the Manipulative Mathematics activity Number Line-Part 1 will help you develop a better understanding of the counting numbers and the whole numbers.

Counting Numbers

The counting numbers start with and continue.

Counting numbers and whole numbers can be visualized on a **number line** as shown in Figure 1.2.

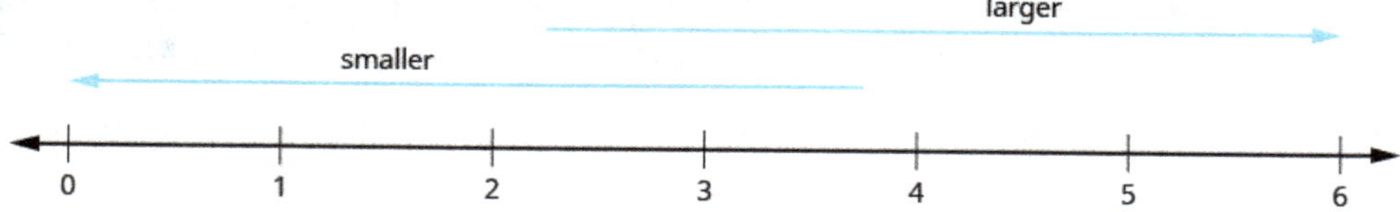

Figure 1.2 The numbers on the number line increase from left to right, and decrease from right to left.

The point labeled is called the **origin**. The points are equally spaced to the right of and labeled with the counting numbers. When a number is paired with a point, it is called the **coordinate** of the point.

The discovery of the number zero was a big step in the history of mathematics. Including zero with the counting numbers gives a new set of numbers called the **whole numbers**.

Whole Numbers

The whole numbers are the counting numbers and zero.

We stopped at when listing the first few counting numbers and whole numbers. We could have written more numbers if they were needed to make the patterns clear.

EXAMPLE 1.1

Which of the following are counting numbers? whole numbers?

—

Solution

The counting numbers start at so is not a counting number. The numbers are all counting numbers.

Whole numbers are counting numbers and The numbers are whole numbers.

The numbers — and are neither counting numbers nor whole numbers. We will discuss these numbers later.

TRY IT : : 1.1 Which of the following are counting numbers whole numbers?

—

> **TRY IT : :** 1.2 Which of the following are counting numbers whole numbers?

—

Model Whole Numbers

Our number system is called a **place value system** because the value of a digit depends on its position, or place, in a number. The number has a different value than the number Even though they use the same digits, their value is different because of the different placement of the and the and the

Money gives us a familiar model of place value. Suppose a wallet contains three bills, seven bills, and four bills. The amounts are summarized in Figure 1.3. How much money is in the wallet?

Three \$100 bills	Seven \$10 bills	Four \$1 bills
3 × \$100	7 × \$10	4 × \$1
\$300	\$70	\$4

Figure 1.3

Find the total value of each kind of bill, and then add to find the total. The wallet contains

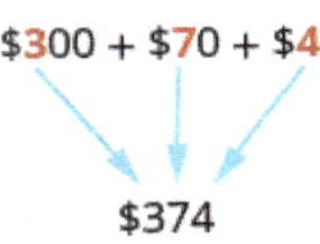
$300 + $70 + $4
$374

Base-10 blocks provide another way to model place value, as shown in Figure 1.4. The blocks can be used to represent hundreds, tens, and ones. Notice that the tens rod is made up of ones, and the hundreds square is made of tens, or ones.

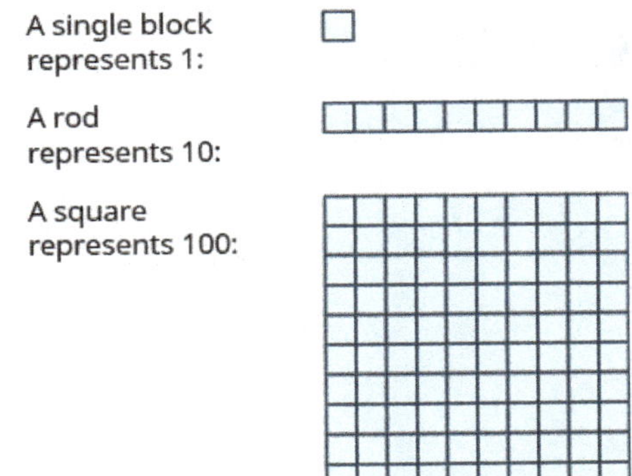
A single block represents 1:
A rod represents 10:
A square represents 100:

Figure 1.4

Figure 1.5 shows the number modeled with blocks.

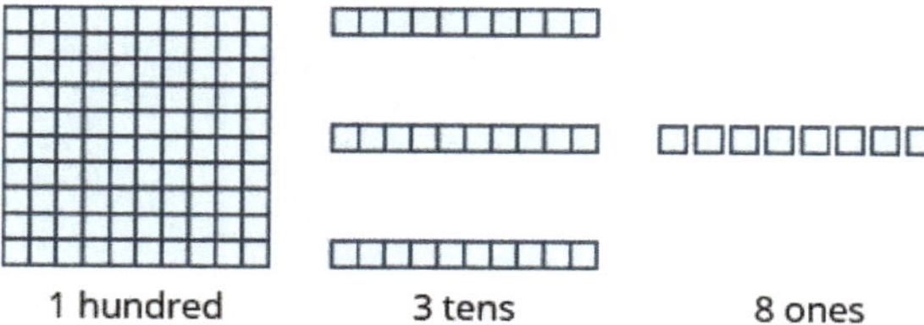

Figure 1.5 We use place value notation to show the value of the number

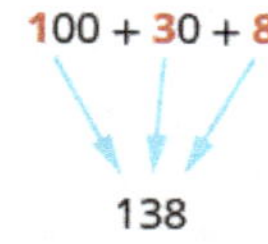

Digit	Place value	Number	Value	Total value
	hundreds			
	tens			
	ones			

EXAMPLE 1.2

Use place value notation to find the value of the number modeled by the blocks shown.

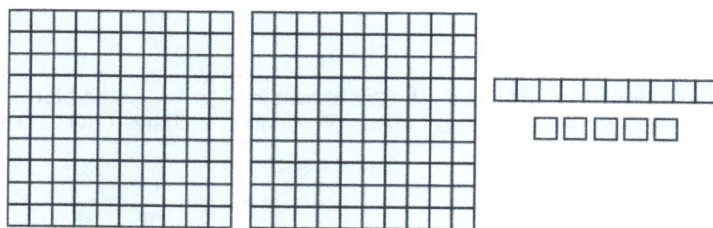

Solution

There are hundreds squares, which is

There is tens rod, which is

There are ones blocks, which is

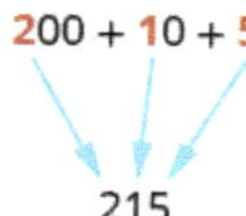

Digit	Place value	Number	Value	Total value
	hundreds			
	tens			
	ones			

The blocks model the number

 **TRY IT : :** 1.3

Use place value notation to find the value of the number modeled by the blocks shown.

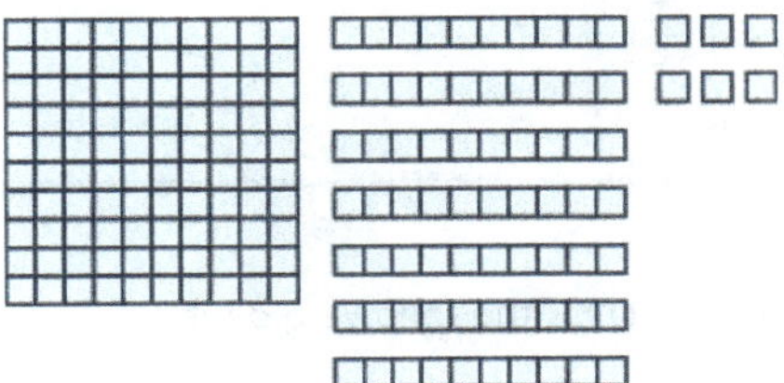

TRY IT : : 1.4

Use place value notation to find the value of the number modeled by the blocks shown.

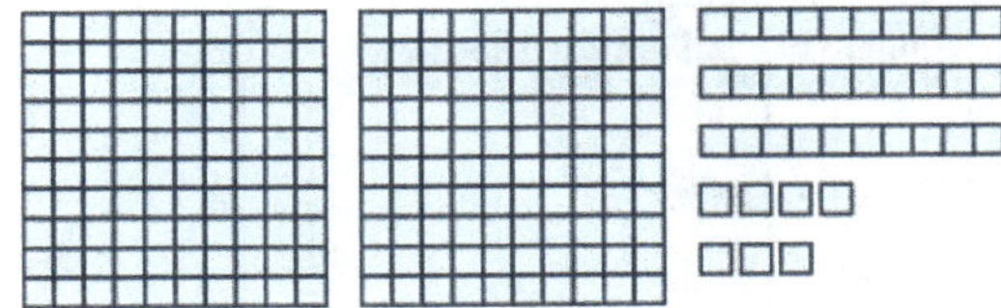

 **MANIPULATIVE MATHEMATICS**

Doing the Manipulative Mathematics activity "Model Whole Numbers" will help you develop a better understanding of place value of whole numbers.

Identify the Place Value of a Digit

By looking at money and blocks, we saw that each place in a number has a different value. A place value chart is a useful way to summarize this information. The place values are separated into groups of three, called periods. The periods are *ones, thousands, millions, billions, trillions*, and so on. In a written number, commas separate the periods.

Just as with the blocks, where the value of the tens rod is ten times the value of the ones block and the value of the hundreds square is ten times the tens rod, the value of each place in the place-value chart is ten times the value of the place to the right of it.

Figure 1.6 shows how the number is written in a place value chart.

Place Value														
Trillions			Billions			Millions			Thousands			Ones		
Hundred trillions	Ten trillions	Trillions	Hundred billions	Ten billions	Billions	Hundred millions	Ten millions	Millions	Hundred thousands	Ten thousands	Thousands	Hundreds	Tens	Ones
								5	2	7	8	1	9	4

Figure 1.6

- The digit is in the millions place. Its value is

- The digit is in the hundred thousands place. Its value is
- The digit is in the ten thousands place. Its value is
- The digit is in the thousands place. Its value is
- The digit is in the hundreds place. Its value is
- The digit is in the tens place. Its value is
- The digit is in the ones place. Its value is

EXAMPLE 1.3

In the number find the place value of each of the following digits:

Solution

Write the number in a place value chart, starting at the right.

Trillions			Billions			Millions			Thousands			Ones		
Hundred trillions	Ten trillions	Trillions	Hundred billions	Ten billions	Billions	Hundred millions	Ten millions	Millions	Hundred thousands	Ten thousands	Thousands	Hundreds	Tens	Ones
							6	3	4	0	7	2	1	8

The is in the thousands place.

The is in the ten thousands place.

The is in the tens place.

The is in the ten millions place.

The is in the millions place.

TRY IT : : 1.5 For each number, find the place value of digits listed:

TRY IT : : 1.6 For each number, find the place value of digits listed:

Use Place Value to Name Whole Numbers

When you write a check, you write out the number in words as well as in digits. To write a number in words, write the number in each period followed by the name of the period without the 's' at the end. Start with the digit at the left, which has the largest place value. The commas separate the periods, so wherever there is a comma in the number, write a comma between the words. The ones period, which has the smallest place value, is not named.

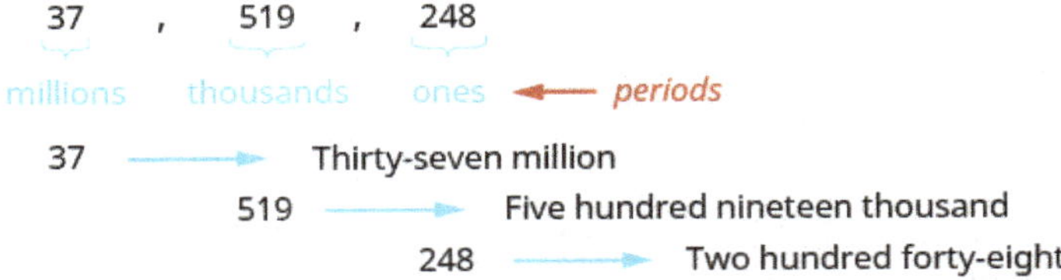

So the number is written thirty-seven million, five hundred nineteen thousand, two hundred forty-eight.

Notice that the word *and* is not used when naming a whole number.

HOW TO : : NAME A WHOLE NUMBER IN WORDS.

Step 1. Starting at the digit on the left, name the number in each period, followed by the period name. Do not include the period name for the ones.

Step 2. Use commas in the number to separate the periods.

EXAMPLE 1.4

Name the number in words.

Solution

Begin with the leftmost digit, which is 8. It is in the trillions place.	eight trillion
The next period to the right is billions.	one hundred sixty-five billion
The next period to the right is millions.	four hundred thirty-two million
The next period to the right is thousands.	ninety-eight thousand
The rightmost period shows the ones.	seven hundred ten

8 , 165 , 432 , 098 , 710
trillions billions millions thousands ones

8 → Eight trillion,
165 → One hundred sixty-five billion,
432 → Four hundred thirty-two million,
098 → Ninety-eight thousand,
710 → Seven hundred ten

Putting all of the words together, we write as eight trillion, one hundred sixty-five billion, four hundred thirty-two million, ninety-eight thousand, seven hundred ten.

TRY IT : : 1.7 Name each number in words:

TRY IT : : 1.8 Name each number in words:

EXAMPLE 1.5

A student conducted research and found that the number of mobile phone users in the United States during one month in was Name that number in words.

Solution

Identify the periods associated with the number.

327 , 577 , 529
millions thousands ones

Name the number in each period, followed by the period name. Put the commas in to separate the periods.

Millions period: three hundred twenty-seven million

Thousands period: five hundred seventy-seven thousand

Ones period: five hundred twenty-nine

So the number of mobile phone users in the Unites States during the month of April was three hundred twenty-seven million, five hundred seventy-seven thousand, five hundred twenty-nine.

TRY IT : : 1.9 The population in a country is Name that number.

TRY IT : : 1.10 One year is seconds. Name that number.

Use Place Value to Write Whole Numbers

We will now reverse the process and write a number given in words as digits.

HOW TO : : USE PLACE VALUE TO WRITE A WHOLE NUMBER.

Step 1. Identify the words that indicate periods. (Remember the ones period is never named.)

Step 2. Draw three blanks to indicate the number of places needed in each period. Separate the periods by commas.

Step 3. Name the number in each period and place the digits in the correct place value position.

EXAMPLE 1.6

Write the following numbers using digits.

fifty-three million, four hundred one thousand, seven hundred forty-two

nine billion, two hundred forty-six million, seventy-three thousand, one hundred eighty-nine

Solution

Identify the words that indicate periods.

Except for the first period, all other periods must have three places. Draw three blanks to indicate the number of places needed in each period. Separate the periods by commas.

Then write the digits in each period.

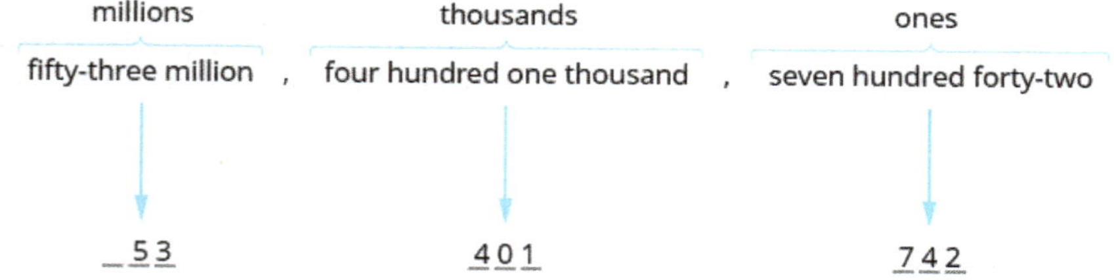

Put the numbers together, including the commas. The number is

Identify the words that indicate periods.

Except for the first period, all other periods must have three places. Draw three blanks to indicate the number of places needed in each period. Separate the periods by commas.

Then write the digits in each period.

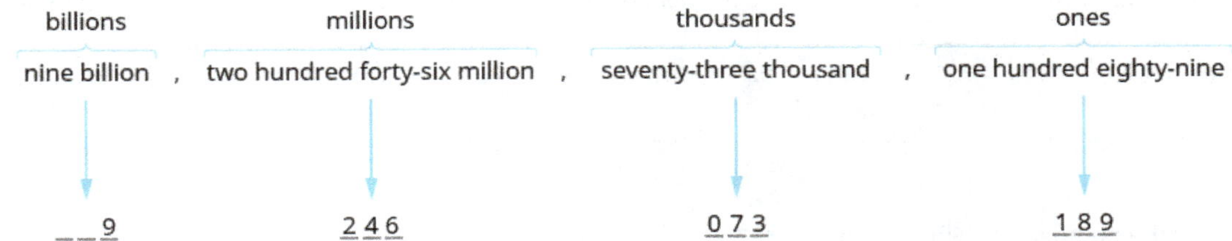

The number is

Notice that in part , a zero was needed as a place-holder in the hundred thousands place. Be sure to write zeros as needed to make sure that each period, except possibly the first, has three places.

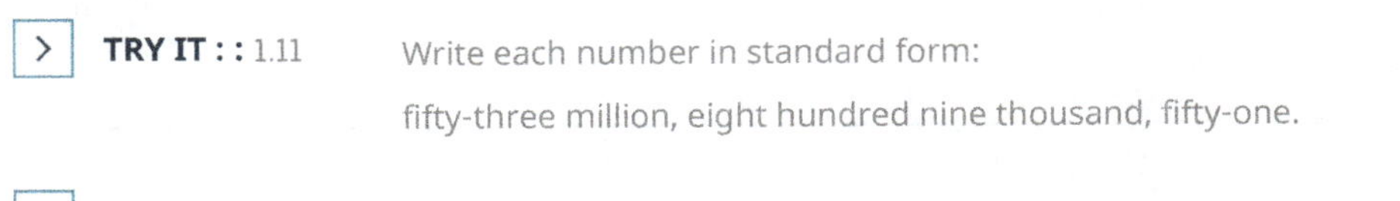

TRY IT : : 1.11 Write each number in standard form:

fifty-three million, eight hundred nine thousand, fifty-one.

TRY IT : : 1.12 Write each number in standard form:

two billion, twenty-two million, seven hundred fourteen thousand, four hundred sixty-six.

EXAMPLE 1.7

A state budget was about billion. Write the budget in standard form.

Solution

Identify the periods. In this case, only two digits are given and they are in the billions period. To write the entire number, write zeros for all of the other periods.

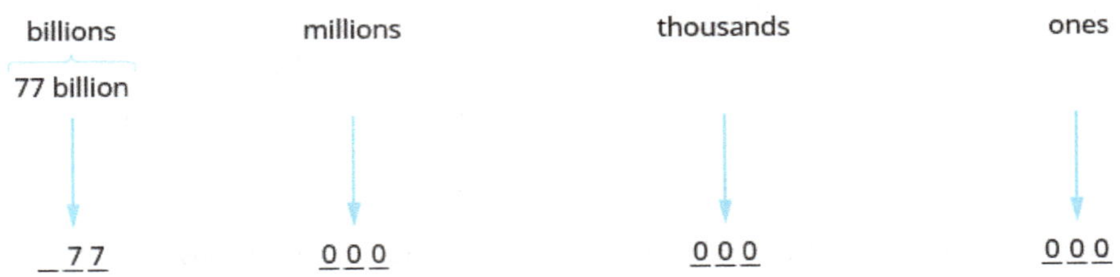

So the budget was about

TRY IT : : 1.13 Write each number in standard form:

The closest distance from Earth to Mars is about million miles.

TRY IT : : 1.14 Write each number in standard form:

The total weight of an aircraft carrier is million pounds.

Round Whole Numbers

In the U.S. Census Bureau reported the population of the state of New York as people. It might be enough to say that the population is approximately million. The word *approximately* means that million is not the exact population, but is close to the exact value.

The process of approximating a number is called **rounding**. Numbers are rounded to a specific place value depending on how much accuracy is needed. million was achieved by rounding to the millions place. Had we rounded to the one hundred thousands place, we would have as a result. Had we rounded to the ten thousands place, we would have as a result, and so on. The place value to which we round to depends on how we need to use the number.

Using the number line can help you visualize and understand the rounding process. Look at the number line in Figure 1.7. Suppose we want to round the number to the nearest ten. Is closer to or on the number line?

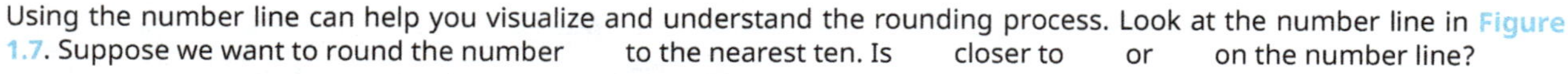

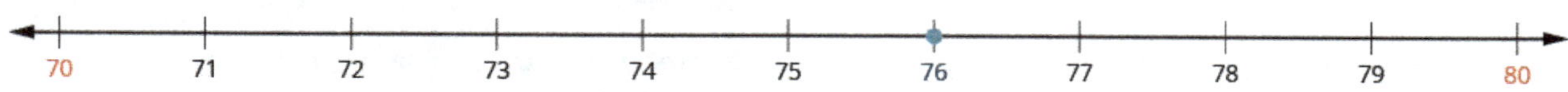

Figure 1.7 We can see that is closer to than to So rounded to the nearest ten is

Now consider the number Find in Figure 1.8.

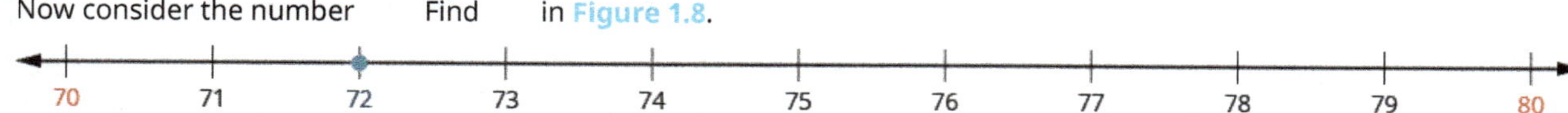

Figure 1.8 We can see that is closer to so rounded to the nearest ten is

How do we round to the nearest ten. Find in Figure 1.9.

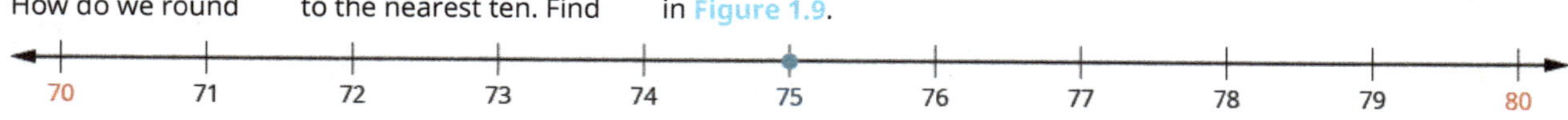

Figure 1.9 The number is exactly midway between and

So that everyone rounds the same way in cases like this, mathematicians have agreed to round to the higher number, So, rounded to the nearest ten is

Now that we have looked at this process on the number line, we can introduce a more general procedure. To round a number to a specific place, look at the number to the right of that place. If the number is less than round down. If it is greater than or equal to round up.

So, for example, to round to the nearest ten, we look at the digit in the ones place.

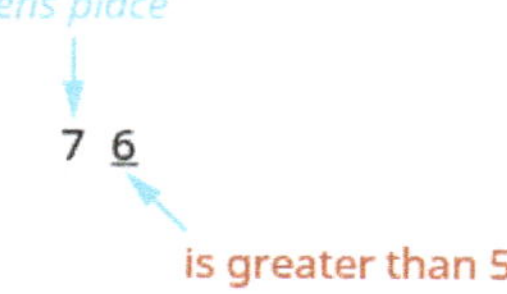

The digit in the ones place is a Because is greater than or equal to we increase the digit in the tens place by one. So the in the tens place becomes an Now, replace any digits to the right of the with zeros. So, rounds to

76 rounded to the nearest ten is 80.

Let's look again at rounding to the nearest Again, we look to the ones place.

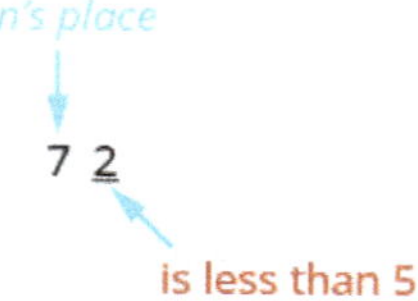

The digit in the ones place is Because is less than we keep the digit in the tens place the same and replace the digits to the right of it with zero. So rounded to the nearest ten is

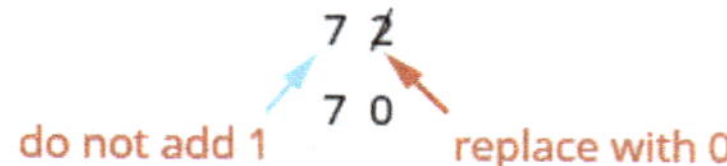

HOW TO : : ROUND A WHOLE NUMBER TO A SPECIFIC PLACE VALUE.

Step 1. Locate the given place value. All digits to the left of that place value do not change.

Step 2. Underline the digit to the right of the given place value.

Step 3. Determine if this digit is greater than or equal to

- Yes—add to the digit in the given place value.
- No—do not change the digit in the given place value.

Step 4. Replace all digits to the right of the given place value with zeros.

EXAMPLE 1.8

Round to the nearest ten.

Solution

	tens place
Locate the tens place.	843
Underline the digit to the right of the tens place.	84<u>3</u>
Since 3 is less than 5, do not change the digit in the tens place.	84<u>3</u>
Replace all digits to the right of the tens place with zeros.	84<u>0</u>
	Rounding 843 to the nearest ten gives 840.

TRY IT : : 1.15 Round to the nearest ten:

TRY IT : : 1.16 Round to the nearest ten:

EXAMPLE 1.9

Round each number to the nearest hundred:

Solution

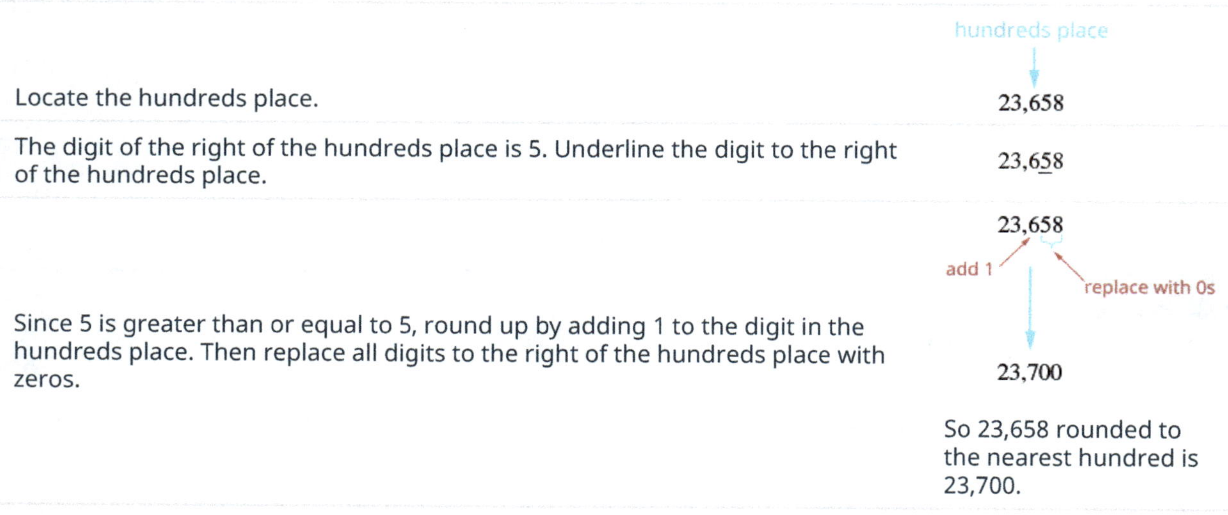
Locate the hundreds place.
hundreds place
23,658
The digit of the right of the hundreds place is 5. Underline the digit to the right of the hundreds place.
23,658
Since 5 is greater than or equal to 5, round up by adding 1 to the digit in the hundreds place. Then replace all digits to the right of the hundreds place with zeros.
23,658
add 1
replace with 0s
23,700
So 23,658 rounded to the nearest hundred is 23,700.

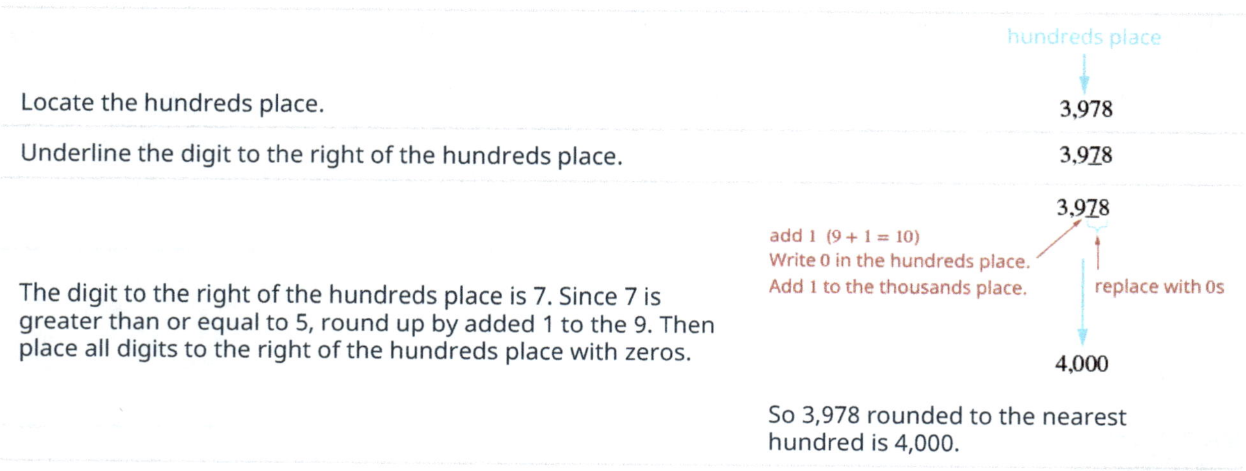
Locate the hundreds place.
hundreds place
3,978
Underline the digit to the right of the hundreds place.
3,978
The digit to the right of the hundreds place is 7. Since 7 is greater than or equal to 5, round up by added 1 to the 9. Then place all digits to the right of the hundreds place with zeros.
3,978
add 1 (9 + 1 = 10)
Write 0 in the hundreds place.
Add 1 to the thousands place.
replace with 0s
4,000
So 3,978 rounded to the nearest hundred is 4,000.

TRY IT : : 1.17 Round to the nearest hundred:

TRY IT : : 1.18 Round to the nearest hundred:

EXAMPLE 1.10

Round each number to the nearest thousand:

Solution

Locate the thousands place. Underline the digit to the right of the thousands place.	thousands place 147,032
The digit to the right of the thousands place is 0. Since 0 is less than 5, we do not change the digit in the thousands place.	147,032
We then replace all digits to the right of the thousands pace with zeros.	147,000 So 147,032 rounded to the nearest thousand is 147,000.

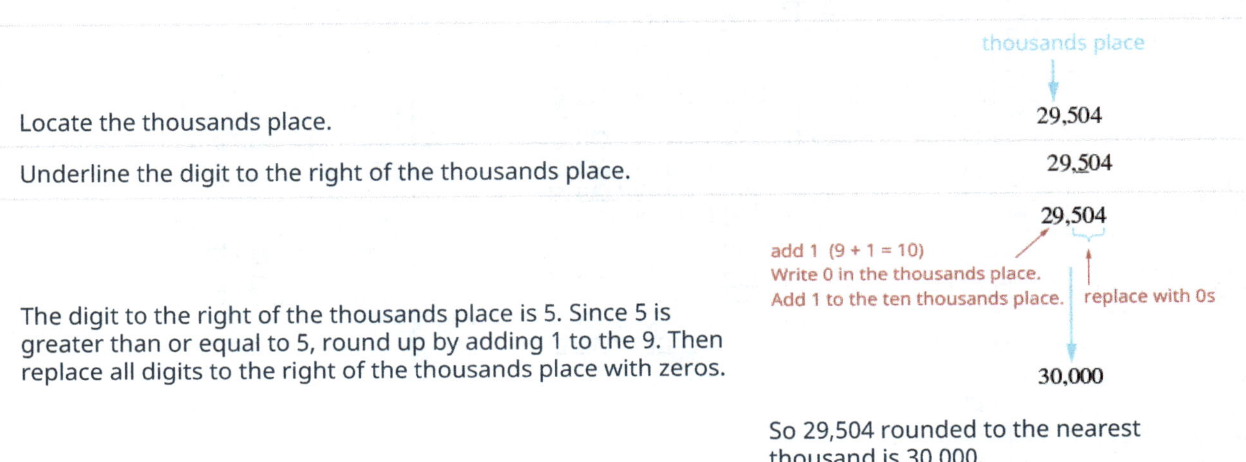

Locate the thousands place.	thousands place 29,504
Underline the digit to the right of the thousands place.	29,504
The digit to the right of the thousands place is 5. Since 5 is greater than or equal to 5, round up by adding 1 to the 9. Then replace all digits to the right of the thousands place with zeros.	29,504 add 1 (9 + 1 = 10) Write 0 in the thousands place. Add 1 to the ten thousands place. replace with 0s 30,000 So 29,504 rounded to the nearest thousand is 30,000.

Notice that in part , when we add thousand to the thousands, the total is thousands. We regroup this as ten thousand and thousands. We add the ten thousand to the ten thousands and put a in the thousands place.

TRY IT : : 1.19 Round to the nearest thousand:

TRY IT : : 1.20 Round to the nearest thousand:

MEDIA : : ACCESS ADDITIONAL ONLINE RESOURCES

- Determine Place Value (http://www.openstax.org/l/24detplaceval)
- Write a Whole Number in Digits from Words (http://www.openstax.org/l/24numdigword)

1.1 EXERCISES

Practice Makes Perfect

Identify Counting Numbers and Whole Numbers

In the following exercises, determine which of the following numbers are counting numbers whole numbers.

1. —

2. —

3. —

4. —

Model Whole Numbers

In the following exercises, use place value notation to find the value of the number modeled by the blocks.

5.

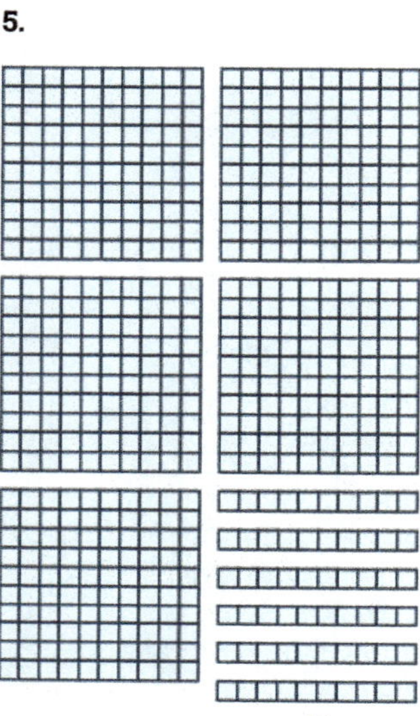

6.

7.

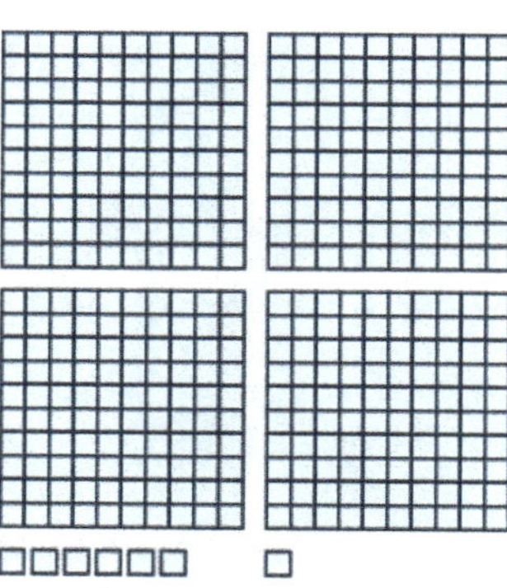

8.

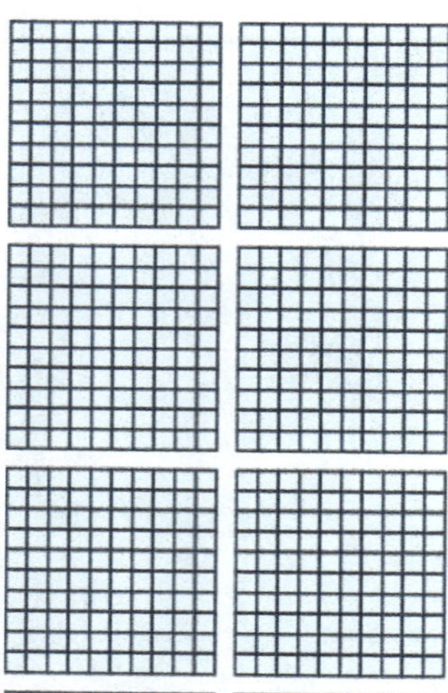

Identify the Place Value of a Digit

In the following exercises, find the place value of the given digits.

9.

9 6 0 7
5

10.

9 3 2 8
7

11.

8 6 4 7
0

12.

8 4 2 6
7

Use Place Value to Name Whole Numbers

In the following exercises, name each number in words.

13.

14.

15.

16.

17.

18.

19.

20.

21. The height of Mount Ranier is feet.

22. The height of Mount Adams is feet.

23. Seventy years is hours.

24. One year is minutes.

25. The U.S. Census estimate of the population of Miami-Dade county was

26. The population of Chicago was

27. There are projected to be college and university students in the US in five years.

28. About twelve years ago there were registered automobiles in California.

29. The population of China is expected to reach in

30. The population of India is estimated at as of July

Use Place Value to Write Whole Numbers

In the following exercises, write each number as a whole number using digits.

31. four hundred twelve

32. two hundred fifty-three

33. thirty-five thousand, nine hundred seventy-five

34. sixty-one thousand, four hundred fifteen

35. eleven million, forty-four thousand, one hundred sixty-seven

36. eighteen million, one hundred two thousand, seven hundred eighty-three

37. three billion, two hundred twenty-six million, five hundred twelve thousand, seventeen

38. eleven billion, four hundred seventy-one million, thirty-six thousand, one hundred six

39. The population of the world was estimated to be seven billion, one hundred seventy-three million people.

40. The age of the solar system is estimated to be four billion, five hundred sixty-eight million years.

41. Lake Tahoe has a capacity of thirty-nine trillion gallons of water.

42. The federal government budget was three trillion, five hundred billion dollars.

Round Whole Numbers

In the following exercises, round to the indicated place value.

43. Round to the nearest ten:

44. Round to the nearest ten:

45. Round to the nearest hundred:

46. Round to the nearest hundred:

47. Round to the nearest ten:

48. Round to the nearest thousand:

49. Round to the nearest hundred:

50. Round to the nearest thousand:

Everyday Math

51. Writing a Check Jorge bought a car for He paid for the car with a check. Write the purchase price in words.

52. Writing a Check Marissa's kitchen remodeling cost She wrote a check to the contractor. Write the amount paid in words.

53. Buying a Car Jorge bought a car for Round the price to the nearest:

ten dollars hundred dollars

thousand dollars ten-thousand dollars

54. Remodeling a Kitchen Marissa's kitchen remodeling cost Round the cost to the nearest:

ten dollars hundred dollars

thousand dollars ten-thousand dollars

55. Population The population of China was in Round the population to the nearest:

billion people hundred-million people

million people

56. Astronomy The average distance between Earth and the sun is kilometers. Round the distance to the nearest:

hundred-million kilometers

ten-million kilometers million kilometers

Writing Exercises

57. In your own words, explain the difference between the counting numbers and the whole numbers.

58. Give an example from your everyday life where it helps to round numbers.

Self Check

After completing the exercises, use this checklist to evaluate your mastery of the objectives of this section.

I can...	Confidently	With some help	No-I don't get it!
identify counting numbers and whole numbers.			
model whole numbers.			
identify the place value of a digit.			
use place value to name whole numbers.			
use place value to write whole numbers.			
round whole numbers.			

If most of your checks were...

...confidently. Congratulations! You have achieved the objectives in this section. Reflect on the study skills you used so that you

can continue to use them. What did you do to become confident of your ability to do these things? Be specific.

...with some help. This must be addressed quickly because topics you do not master become potholes in your road to success. In math, every topic builds upon previous work. It is important to make sure you have a strong foundation before you move on. Whom can you ask for help? Your fellow classmates and instructor are good resources. Is there a place on campus where math tutors are available? Can your study skills be improved?

...no—I don't get it! This is a warning sign and you must not ignore it. You should get help right away or you will quickly be overwhelmed. See your instructor as soon as you can to discuss your situation. Together you can come up with a plan to get you the help you need.

1.2 Add Whole Numbers

Learning Objectives

By the end of this section, you will be able to:

- Use addition notation
- Model addition of whole numbers
- Add whole numbers without models
- Translate word phrases to math notation
- Add whole numbers in applications

BE PREPARED : : 1.1 Before you get started, take this readiness quiz.

What is the number modeled by the blocks?

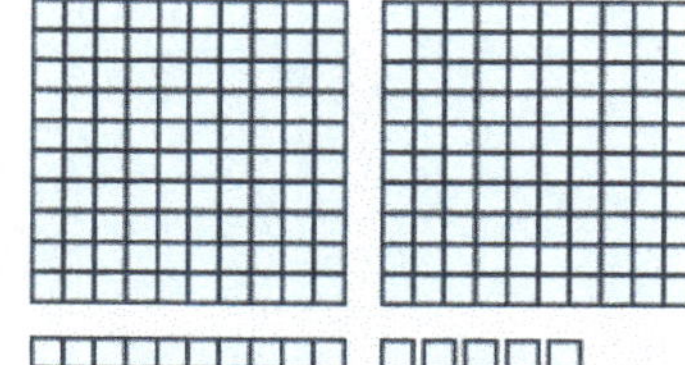

If you missed this problem, review Exercise 1.0.

BE PREPARED : : 1.2 Write the number three hundred forty-two thousand six using digits?
If you missed this problem, review Exercise 1.0.

Use Addition Notation

A college student has a part-time job. Last week he worked hours on Monday and hours on Friday. To find the total number of hours he worked last week, he added and

The operation of addition combines numbers to get a **sum**. The notation we use to find the sum of and is:

We read this as *three plus four* and the result is the sum of three and four. The numbers and are called the addends. A math statement that includes numbers and operations is called an expression.

Addition Notation

To describe addition, we can use symbols and words.

Operation	Notation	Expression	Read as	Result
Addition			three plus four	the sum of and

EXAMPLE 1.11

Translate from math notation to words:

Solution

The expression consists of a plus symbol connecting the addends 7 and 1. We read this as *seven plus one*. The result is *the sum of seven and one*.

The expression consists of a plus symbol connecting the addends 12 and 14. We read this as *twelve plus fourteen*. The result is the *sum of twelve and fourteen*.

TRY IT : : 1.21 Translate from math notation to words:

TRY IT : : 1.22 Translate from math notation to words:

Model Addition of Whole Numbers

Addition is really just counting. We will model addition with blocks. Remember, a block represents and a rod represents Let's start by modeling the addition expression we just considered,

Each addend is less than so we can use ones blocks.

We start by modeling the first number with 3 blocks.	□□□ 3
Then we model the second number with 4 blocks.	□□□ □□□□ 3 4
Count the total number of blocks.	□□□□□□□ 7

There are blocks in all. We use an equal sign to show the sum. A math sentence that shows that two expressions are equal is called an equation. We have shown that.

MANIPULATIVE MATHEMATICS

Doing the Manipulative Mathematics activity "Model Addition of Whole Numbers" will help you develop a better understanding of adding whole numbers.

EXAMPLE 1.12

Model the addition

Solution

means the sum of and

Each addend is less than 10, so we can use ones blocks.

Model the first number with 2 blocks.	2
Model the second number with 6 blocks.	2 6
Count the total number of blocks	8 There are blocks in all, so

> **TRY IT : : 1.23** Model:

> **TRY IT : : 1.24** Model:

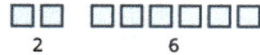

When the result is or more ones blocks, we will exchange the blocks for one rod.

EXAMPLE 1.13

Model the addition

Solution

means the sum of and

Each addend is less than 10, se we can use ones blocks.	
Model the first number with 5 blocks.	5
Model the second number with 8 blocks.	5 8
Count the result. There are more than 10 blocks so we exchange 10 ones blocks for 1 tens rod.	
Now we have 1 ten and 3 ones, which is 13.	$5 + 8 = 13$

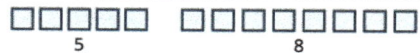

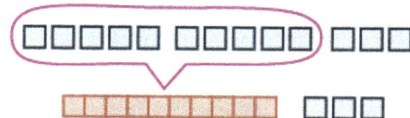

Notice that we can describe the models as ones blocks and tens rods, or we can simply say *ones* and *tens*. From now on, we will use the shorter version but keep in mind that they mean the same thing.

> **TRY IT : : 1.25** Model the addition:

> **TRY IT : : 1.26** Model the addition:

Next we will model adding two digit numbers.

EXAMPLE 1.14

Model the addition:

Solution

means the sum of 17 and 26.

Model the 17.	1 ten and 7 ones	
Model the 26.	2 tens and 6 ones	
Combine.	3 tens and 13 ones	
Exchange 10 ones for 1 ten.	4 tens and 3 ones	
We have shown that		

> **TRY IT : : 1.27** Model each addition:

> **TRY IT : : 1.28** Model each addition:

Add Whole Numbers Without Models

Now that we have used models to add numbers, we can move on to adding without models. Before we do that, make sure you know all the one digit addition facts. You will need to use these number facts when you add larger numbers.

Imagine filling in Table 1.1 by adding each row number along the left side to each column number across the top. Make sure that you get each sum shown. If you have trouble, model it. It is important that you memorize any number facts you do not already know so that you can quickly and reliably use the number facts when you add larger numbers.

+	0	1	2	3	4	5	6	7	8	9
0	0	1	2	3	4	5	6	7	8	9
1	1	2	3	4	5	6	7	8	9	10
2	2	3	4	5	6	7	8	9	10	11
3	3	4	5	6	7	8	9	10	11	12
4	4	5	6	7	8	9	10	11	12	13
5	5	6	7	8	9	10	11	12	13	14
6	6	7	8	9	10	11	12	13	14	15
7	7	8	9	10	11	12	13	14	15	16
8	8	9	10	11	12	13	14	15	16	17
9	9	10	11	12	13	14	15	16	17	18

Table 1.1

Did you notice what happens when you add zero to a number? The sum of any number and zero is the number itself. We call this the Identity Property of Addition. Zero is called the additive identity.

Identity Property of Addition

The sum of any number and is the number.

EXAMPLE 1.15

Find each sum:

Solution

The first addend is zero. The sum of any number and zero is the number.

The second addend is zero. The sum of any number and zero is the number.

TRY IT : : 1.29 Find each sum:

TRY IT : : 1.30 Find each sum:

Look at the pairs of sums.

Notice that when the order of the addends is reversed, the sum does not change. This property is called the Commutative Property of Addition, which states that changing the order of the addends does not change their sum.

Commutative Property of Addition

Changing the order of the addends and does not change their sum.

EXAMPLE 1.16

Add:

Solution

Add.

Add.

Did you notice that changing the order of the addends did not change their sum? We could have immediately known the sum from part just by recognizing that the addends were the same as in part , but in the reverse order. As a result, both sums are the same.

> **TRY IT : :** 1.31 Add: and

> **TRY IT : :** 1.32 Add: and

EXAMPLE 1.17

Add:

Solution

To add numbers with more than one digit, it is often easier to write the numbers vertically in columns.

Write the numbers so the ones and tens digits line up vertically.

Then add the digits in each place value.
Add the ones:
Add the tens:

> **TRY IT : :** 1.33 Add:

> **TRY IT : :** 1.34 Add:

In the previous example, the sum of the ones and the sum of the tens were both less than But what happens if the sum is or more? Let's use our model to find out. Figure 1.10 shows the addition of and again.

Figure 1.10

When we add the ones, we get ones. Because we have more than ones, we can exchange of the ones for ten. Now we have tens and ones. Without using the model, we show this as a small red above the digits in the tens place.

When the sum in a place value column is greater than we carry over to the next column to the left. Carrying is the same as regrouping by exchanging. For example, ones for ten or tens for hundred.

HOW TO : : ADD WHOLE NUMBERS.

Step 1. Write the numbers so each place value lines up vertically.

Step 2. Add the digits in each place value. Work from right to left starting with the ones place. If a sum in a place value is more than carry to the next place value.

Step 3. Continue adding each place value from right to left, adding each place value and carrying if needed.

EXAMPLE 1.18

Add:

Solution

Write the numbers so the digits line up vertically.

Add the digits in each place.
Add the ones:

Write the in the ones place in the sum.
Add the ten to the tens place.

Now add the tens:
Write the 11 in the sum.

TRY IT : : 1.35 Add:

TRY IT : : 1.36 Add:

EXAMPLE 1.19

Add:

Solution

Write the numbers so the digits line up vertically.	$\begin{array}{r} 324 \\ +\ 586 \\ \hline \end{array}$
Add the digits in each place value. Add the ones: Write the in the ones place in the sum and carry the ten to the tens place.	$\begin{array}{r} 3\overset{1}{2}4 \\ +\ 586 \\ \hline 0 \end{array}$
Add the tens: Write the in the tens place in the sum and carry the hundred to the hundreds	$\begin{array}{r} \overset{1}{3}\overset{1}{2}4 \\ +\ 586 \\ \hline 10 \end{array}$
Add the hundreds: Write the in the hundreds place.	$\begin{array}{r} \overset{1}{3}\overset{1}{2}4 \\ +\ 586 \\ \hline 910 \end{array}$

TRY IT : : 1.37 Add:

TRY IT : : 1.38 Add:

EXAMPLE 1.20

Add:

Solution

Write the numbers so the digits line up vertically.

Add the digits in each place value.

Add the ones:
Write the in the ones place of the sum and carry the ten to the tens place.

Add the tens:
Write the in the tens place and carry the hundred to the hundreds place.

Add the hundreds:
Write the in the hundreds place and carry the thousand to the thousands place.

Add the thousands .
Write the in the thousands place of the sum.

When the addends have different numbers of digits, be careful to line up the corresponding place values starting with the ones and moving toward the left.

TRY IT : : 1.39 Add:

TRY IT : : 1.40 Add:

EXAMPLE 1.21

Add:

Solution

Write the numbers so the place values line up vertically.

Add the digits in each place value.

Add the ones:
Write the in the ones place of the sum and carry the to the tens place.

Add the tens:
Write the in the tens place and carry the to the hundreds place.

Add the hundreds:
Write the in the hundreds place and carry the to the thousands place.

Add the thousands .
Write the in the thousands place and carry the to the ten thousands place.

Add the ten-thousands .
Write the in the ten thousands place in the sum.

This example had three addends. We can add any number of addends using the same process as long as we are careful to line up the place values correctly.

TRY IT : : 1.41 Add:

TRY IT : : 1.42 Add:

Translate Word Phrases to Math Notation

Earlier in this section, we translated math notation into words. Now we'll reverse the process. We'll translate word phrases into math notation. Some of the word phrases that indicate addition are listed in Table 1.2.

Operation	Words	Example	Expression
Addition	plus sum increased by more than total of added to	plus the sum of and increased by more than the total of and added to	

Table 1.2

EXAMPLE 1.22

Translate and simplify: the sum of and

Solution

The word *sum* tells us to add. The words *of* and tell us the addends.

	The sum of and
Translate.	
Add.	
	The sum of and is

TRY IT : : 1.43 Translate and simplify: the sum of and

TRY IT : : 1.44 Translate and simplify: the sum of and

EXAMPLE 1.23

Translate and simplify: increased by

Solution

The words *increased by* tell us to add. The numbers given are the addends.

	increased by
Translate.	
Add.	
	So increased by is

TRY IT : : 1.45 Translate and simplify: increased by

TRY IT : : 1.46 Translate and simplify: increased by

Add Whole Numbers in Applications

Now that we have practiced adding whole numbers, let's use what we've learned to solve real-world problems. We'll start by outlining a plan. First, we need to read the problem to determine what we are looking for. Then we write a word phrase that gives the information to find it. Next we translate the word phrase into math notation and then simplify. Finally, we write a sentence to answer the question.

EXAMPLE 1.24

Hao earned grades of on the five tests of the semester. What is the total number of points he earned on the five tests?

Solution

We are asked to find the total number of points on the tests.

Write a phrase.	the sum of points on the tests
Translate to math notation.	
Then we simplify by adding.	
Since there are several numbers, we will write them vertically.	
Write a sentence to answer the question.	Hao earned a total of 432 points.

Notice that we added *points*, so the sum is *points*. It is important to include the appropriate units in all answers to applications problems.

TRY IT : : 1.47

Mark is training for a bicycle race. Last week he rode miles on Monday, miles on Wednesday, miles on Friday, miles on Saturday, and miles on Sunday. What is the total number of miles he rode last week?

TRY IT : : 1.48

Lincoln Middle School has three grades. The number of students in each grade is What is the total number of students?

Some application problems involve shapes. For example, a person might need to know the distance around a garden to put up a fence or around a picture to frame it. The perimeter is the distance around a geometric figure. The perimeter of a figure is the sum of the lengths of its sides.

EXAMPLE 1.25

Find the perimeter of the patio shown.

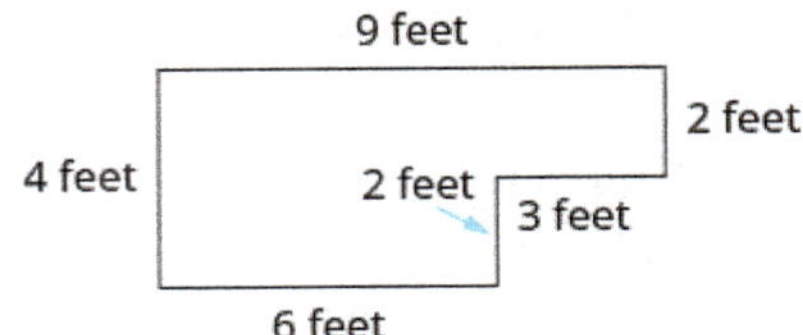

Solution

We are asked to find the perimeter.	
Write a phrase.	the sum of the sides
Translate to math notation.	
Simplify by adding.	
Write a sentence to answer the question.	
We added feet, so the sum is feet.	The perimeter of the patio is feet.

TRY IT : : 1.49 Find the perimeter of each figure. All lengths are in inches.

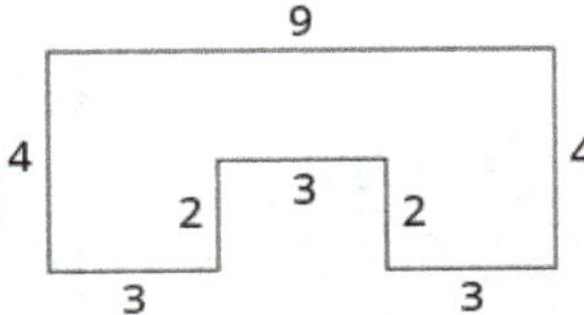

TRY IT : : 1.50 Find the perimeter of each figure. All lengths are in inches.

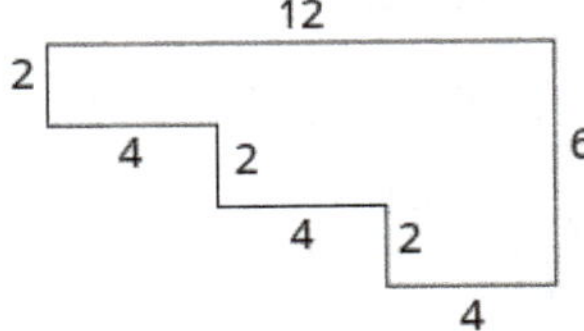

MEDIA : : ACCESS ADDITIONAL ONLINE RESOURCES

- Adding Two-Digit Numbers with base-10 blocks (http://www.openstax.org/l/24add2blocks)
- Adding Three-Digit Numbers with base-10 blocks (http://www.openstax.org/l/24add3blocks)
- Adding Whole Numbers (http://www.openstax.org/l/24addwhlnumb)

1.2 EXERCISES

Practice Makes Perfect

Use Addition Notation

In the following exercises, translate the following from math expressions to words.

59.

60.

61.

62.

63.

64.

Model Addition of Whole Numbers

In the following exercises, model the addition.

65.

66.

67.

68.

69.

70.

71.

72.

Add Whole Numbers

In the following exercises, fill in the missing values in each chart.

73.

+	0	1	2	3	4	5	6	7	8	9
0	0	1	2		4	5	6	7		9
1	1	2	3	4			7	8	9	
2		3	4	5	6		8			11
3	3		5		7	8		10		12
4	4	5			8	9		11	12	
5	5	6	7	8			11		13	
6	6	7	8		10			13		15
7			9	10		12			15	16
8	8	9		11			14		16	
9	9	10	11		13	14			17	

74.

+	0	1	2	3	4	5	6	7	8	9
0	0	1	2	3	4		6		8	9
1	1	2	3		5	6		8		10
2	2		4		6	7		9	10	
3		4		6			9		11	
4	4	5	6	7			10	11		13
5	5	6		8	9		11	12	13	
6			8	9			12	13		15
7	7	8		10		12			15	16
8	8	9	10		12		14		16	17
9			11	12	13			16		

75.

+	3	4	5	6	7	8	9
6							
7							
8							
9							

76.

+	6	7	8	9
3				
4				
5				
6				
7				
8				
9				

77.

+	5	6	7	8	9
5					
6					
7					
8					
9					

78.

+	6	7	8	9
6				
7				
8				
9				

In the following exercises, add.

79.

80.

81.

82.

83.

84.

85.

86.

87.

88.

89.

90.

91.

92.

93.

94.

95.

96.

97.

98.

99.

100.

101.

102.

103.

104.

105.

106.

Translate Word Phrases to Math Notation

In the following exercises, translate each phrase into math notation and then simplify.

107. the sum of and

108. the sum of and

109. the sum of and

110. the sum of and

111. increased by

112. increased by

113. more than

114. more than

115. the total of and

116. the total of and

117. added to

118. added to

Add Whole Numbers in Applications

In the following exercises, solve the problem.

119. Home remodeling Sophia remodeled her kitchen and bought a new range, microwave, and dishwasher. The range cost the microwave cost and the dishwasher cost What was the total cost of these three appliances?

120. Sports equipment Aiden bought a baseball bat, helmet, and glove. The bat cost the helmet cost and the glove cost What was the total cost of Aiden's sports equipment?

121. Bike riding Ethan rode his bike miles on Monday, miles on Tuesday, miles on Wednesday, miles on Friday, and miles on Saturday. What was the total number of miles Ethan rode?

122. Business Chloe has a flower shop. Last week she made floral arrangements on Monday, on Tuesday, on Wednesday, on Thursday, and on Friday. What was the total number of floral arrangements Chloe made?

123. Apartment size Jackson lives in a room apartment. The number of square feet in each room is and What is the total number of square feet in all rooms?

124. Weight Seven men rented a fishing boat. The weights of the men were and pounds. What was the total weight of the seven men?

125. Salary Last year Natalie's salary was Two years ago, her salary was and three years ago it was What is the total amount of Natalie's salary for the past three years?

126. Home sales Emma is a realtor. Last month, she sold three houses. The selling prices of the houses were and What was the total of the three selling prices?

In the following exercises, find the perimeter of each figure.

127.

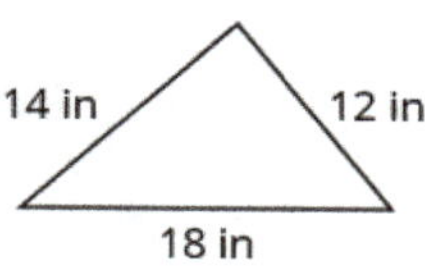

128.

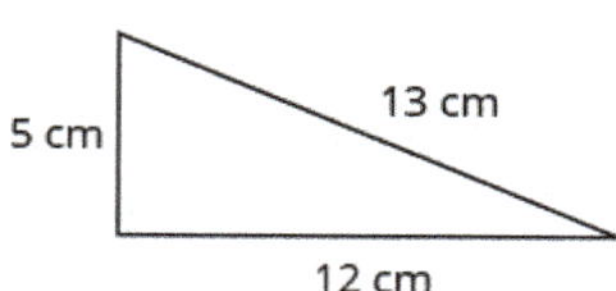

129.

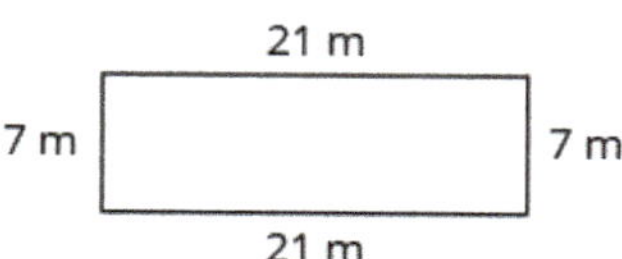

130.

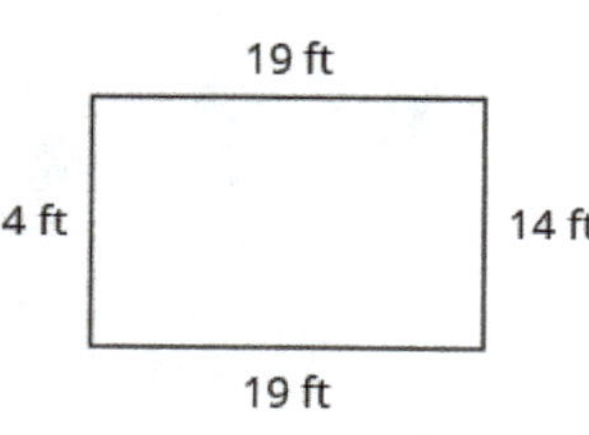

131.

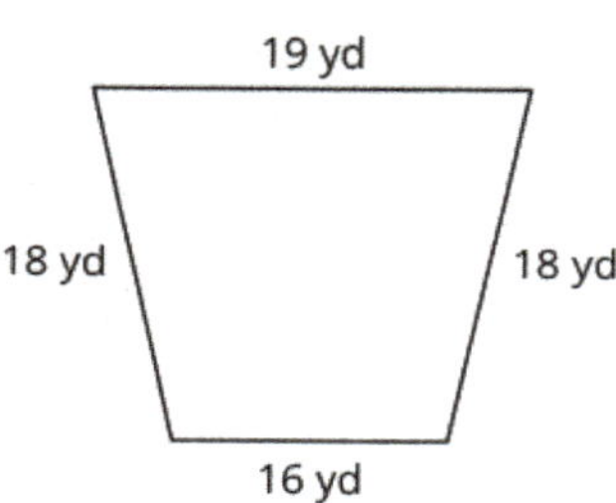

132.

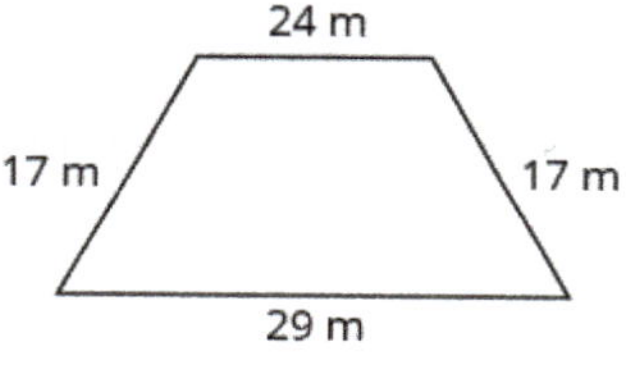

133.

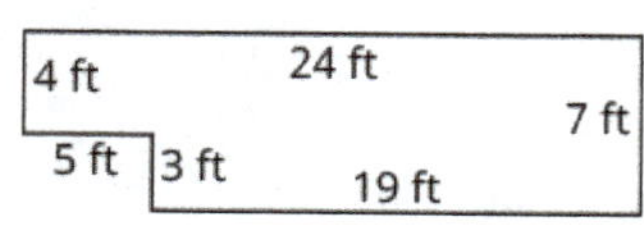

134.

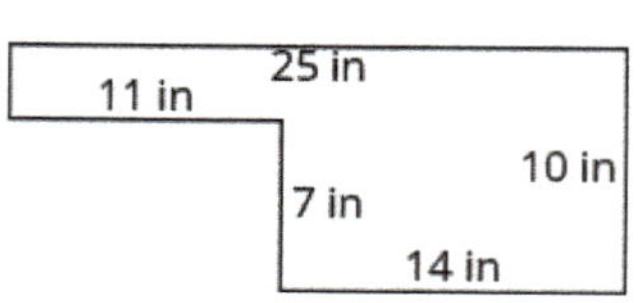

Everyday Math

135. Calories Paulette had a grilled chicken salad, ranch dressing, and a drink for lunch. On the restaurant's nutrition chart, she saw that each item had the following number of calories:

Grilled chicken salad – calories

Ranch dressing – calories

drink – calories

What was the total number of calories of Paulette's lunch?

136. Calories Fred had a grilled chicken sandwich, a small order of fries, and a chocolate shake for dinner. The restaurant's nutrition chart lists the following calories for each item:

Grilled chicken sandwich – calories

Small fries – calories

chocolate shake – calories

What was the total number of calories of Fred's dinner?

137. Test scores A students needs a total of points on five tests to pass a course. The student scored Did the student pass the course?

138. Elevators The maximum weight capacity of an elevator is pounds. Six men are in the elevator. Their weights are pounds. Is the total weight below the elevator's maximum capacity?

Writing Exercises

139. How confident do you feel about your knowledge of the addition facts? If you are not fully confident, what will you do to improve your skills?

140. How have you used models to help you learn the addition facts?

Self Check

After completing the exercises, use this checklist to evaluate your mastery of the objectives of this section.

I can...	Confidently	With some help	No-I don't get it!
use addition notation.			
model addition of whole numbers.			
add whole numbers without models.			
translate word phrases to math notation.			
add whole numbers in applications.			

After reviewing this checklist, what will you do to become confident for all objectives?

1.3 Subtract Whole Numbers

Learning Objectives

By the end of this section, you will be able to:

- Use subtraction notation
- Model subtraction of whole numbers
- Subtract whole numbers
- Translate word phrases to math notation
- Subtract whole numbers in applications

BE PREPARED : : 1.3 Before you get started, take this readiness quiz.
Model using base-ten blocks.
If you missed this problem, review Exercise 1.58.

BE PREPARED : : 1.4 Add:
If you missed this problem, review Exercise 1.58.

Use Subtraction Notation

Suppose there are seven bananas in a bowl. Elana uses three of them to make a smoothie. How many bananas are left in the bowl? To answer the question, we subtract three from seven. When we subtract, we take one number away from another to find the **difference**. The notation we use to subtract from is

We read as *seven minus three* and the result is *the difference of seven and three*.

Subtraction Notation

To describe subtraction, we can use symbols and words.

Operation	Notation	Expression	Read as	Result
Subtraction			seven minus three	the difference of and

EXAMPLE 1.26

Translate from math notation to words: .

Solution

We read this as *eight minus one*. The result is *the difference of eight and one*.

We read this as *twenty-six minus fourteen*. The resuilt is *the difference of twenty-six and fourteen*.

TRY IT : : 1.51 Translate from math notation to words:

TRY IT : : 1.52 Translate from math notation to words:

Model Subtraction of Whole Numbers

A model can help us visualize the process of subtraction much as it did with addition. Again, we will use blocks. Remember a block represents 1 and a rod represents 10. Let's start by modeling the subtraction expression we just considered,

We start by modeling the first number, 7.	7
Now take away the second number, 3. We'll circle 3 blocks to show that we are taking them away.	
Count the number of blocks remaining.	
There are 4 ones blocks left.	We have shown that .

MANIPULATIVE MATHEMATICS

Doing the Manipulative Mathematics activity Model Subtraction of Whole Numbers will help you develop a better understanding of subtracting whole numbers.

EXAMPLE 1.27

Model the subtraction:

Solution

means the difference of 8 and 2.	
Model the first, 8.	8
Take away the second number, 2.	
Count the number of blocks remaining.	
There are 6 ones blocks left.	We have shown that .

TRY IT : : 1.53 Model:

TRY IT : : 1.54 Model:

EXAMPLE 1.28

Model the subtraction:

Solution

Model the first number, 13. We use 1 ten and 3 ones.	
Take away the second number, 8. However, there are not 8 ones, so we will exchange the 1 ten for 10 ones.	 10 3
Now we can take away 8 ones.	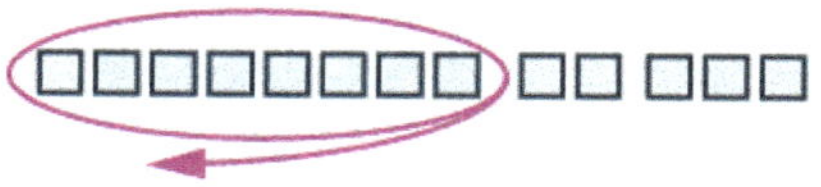
Count the blocks remaining.	
There are five ones left.	We have shown that .

As we did with addition, we can describe the models as ones blocks and tens rods, or we can simply say ones and tens.

> **TRY IT : :** 1.55 Model the subtraction:

> **TRY IT : :** 1.56 Model the subtraction:

EXAMPLE 1.29

Model the subtraction:

Solution

Because means take away we begin by modeling the

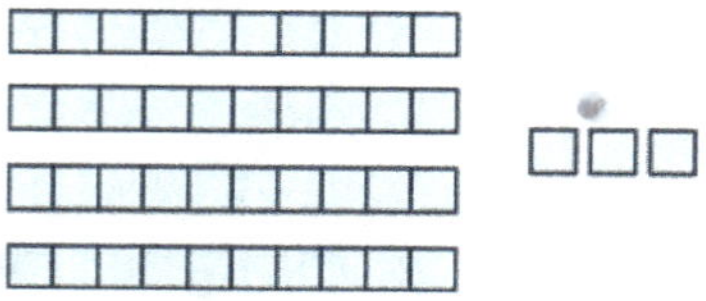

Now, we need to take away which is tens and ones. We cannot take away ones from ones. So, we exchange ten for ones.

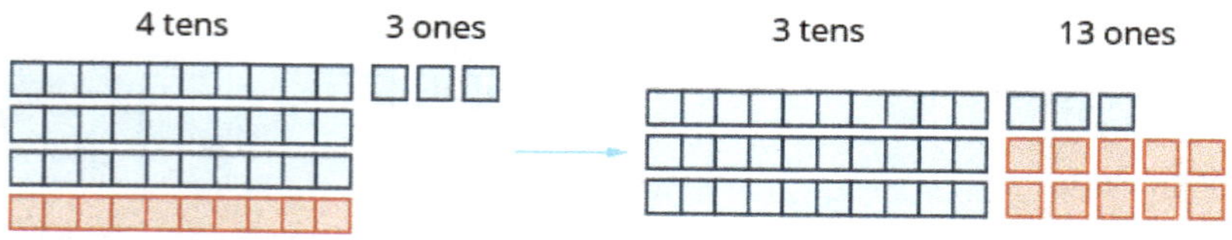

4 tens 3 ones 3 tens 13 ones

Now we can take away tens and ones.

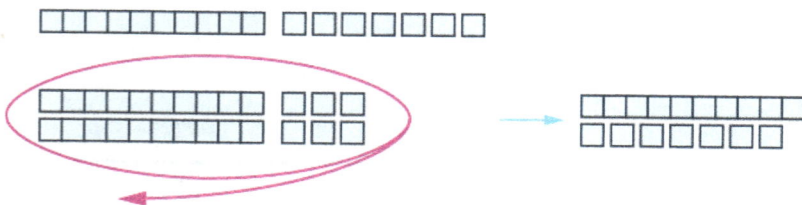

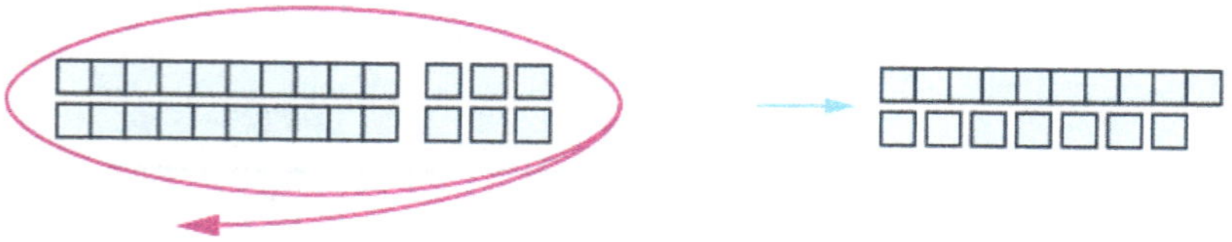

Count the number of blocks remaining. There is ten and ones, which is

TRY IT : : 1.57 Model the subtraction:

TRY IT : : 1.58 Model the subtraction:

Subtract Whole Numbers

Addition and subtraction are inverse operations. Addition undoes subtraction, and subtraction undoes addition.

We know because Knowing all the addition number facts will help with subtraction. Then we can check subtraction by adding. In the examples above, our subtractions can be checked by addition.

EXAMPLE 1.30

Subtract and then check by adding:

Solution

Subtract 7 from 9.

Check with addition.

Subtract 3 from 8.

Check with addition.

TRY IT : : 1.59 Subtract and then check by adding:

TRY IT : : 1.60 Subtract and then check by adding:

To subtract numbers with more than one digit, it is usually easier to write the numbers vertically in columns just as we did for addition. Align the digits by place value, and then subtract each column starting with the ones and then working to the left.

EXAMPLE 1.31

Subtract and then check by adding:

Solution

Write the numbers so the ones and tens digits line up vertically.

Subtract the digits in each place value.

Subtract the ones:
Subtract the tens:

Check using addition.

Our answer is correct.

TRY IT : : 1.61 Subtract and then check by adding:

TRY IT : : 1.62 Subtract and then check by adding:

When we modeled subtracting from we exchanged ten for ones. When we do this without the model, we say we borrow from the tens place and add to the ones place.

HOW TO : : FIND THE DIFFERENCE OF WHOLE NUMBERS.

Step 1. Write the numbers so each place value lines up vertically.

Step 2. Subtract the digits in each place value. Work from right to left starting with the ones place. If the digit on top is less than the digit below, borrow as needed.

Step 3. Continue subtracting each place value from right to left, borrowing if needed.

Step 4. Check by adding.

EXAMPLE 1.32

Subtract:

Solution

Write the numbers so each place value lines up vertically.	$\begin{array}{r} 43 \\ -26 \\ \hline \end{array}$
Subtract the ones. We cannot subtract 6 from 3, so we borrow 1 ten. This makes 3 tens and 13 ones. We write these numbers above each place and cross out the original digits.	$\begin{array}{r} \overset{3}{\cancel{4}}\overset{13}{\cancel{3}} \\ -26 \\ \hline \end{array}$
Now we can subtract the ones. We write the 7 in the ones place in the difference.	$\begin{array}{r} \overset{3}{\cancel{4}}\overset{13}{\cancel{3}} \\ -\ 26 \\ \hline 7 \end{array}$
Now we subtract the tens. We write the 1 in the tens place in the difference.	$\begin{array}{r} \overset{3}{\cancel{4}}\overset{13}{\cancel{3}} \\ -\ 26 \\ \hline 17 \end{array}$

Check by adding.

$$\begin{array}{r} 17 \\ +26 \\ \hline 43\checkmark \end{array}$$

Our answer is correct.

TRY IT : : 1.63 Subtract and then check by adding:

TRY IT : : 1.64 Subtract and then check by adding:

EXAMPLE 1.33

Subtract and then check by adding:

Solution

Write the numbers so each place value lines up vertically.	$\begin{array}{r} 207 \\ -\ 64 \\ \hline \end{array}$
Subtract the ones. Write the 3 in the ones place in the difference. Write the 3 in the ones place in the difference.	$\begin{array}{r} 207 \\ -\ 64 \\ \hline 3 \end{array}$
Subtract the tens. We cannot subtract 6 from 0 so we borrow 1 hundred and add 10 tens to the 0 tens we had. This makes a total of 10 tens. We write 10 above the tens place and cross out the 0. Then we cross out the 2 in the hundreds place and write 1 above it.	$\begin{array}{r} {}^{1\ 10} \\ \not{2}\not{0}7 \\ -\ 64 \\ \hline 3 \end{array}$
Now we subtract the tens. We write the 4 in the tens place in the difference.	$\begin{array}{r} {}^{1\ 10} \\ \not{2}\not{0}7 \\ -\ 64 \\ \hline 43 \end{array}$
Finally, subtract the hundreds. There is no digit in the hundreds place in the bottom number so we can imagine a 0 in that place. Since we write 1 in the hundreds place in the difference.	$\begin{array}{r} {}^{1\ 10} \\ \not{2}\not{0}7 \\ -\ 64 \\ \hline 143 \end{array}$

Check by adding.

$$\begin{array}{r} {}^{1} \\ 143 \\ +\ 64 \\ \hline 207 \checkmark \end{array}$$

Our answer is correct.

TRY IT : : 1.65 Subtract and then check by adding:

TRY IT : : 1.66 Subtract and then check by adding:

EXAMPLE 1.34

Subtract and then check by adding:

Solution

Write the numbers so each place value lines up vertically.

$$\begin{array}{r} 910 \\ -\ 586 \\ \hline \end{array}$$

Subtract the ones. We cannot subtract 6 from 0, so we borrow 1 ten and add 10 ones to the 10 ones we had. This makes 10 ones. We write a 0 above the tens place and cross out the 1. We write the 10 above the ones place and cross out the 0. Now we can subtract the ones.

$$\begin{array}{r} 9\overset{0}{\cancel{1}}\overset{10}{\cancel{0}} \\ -\ 586 \\ \hline \end{array}$$

Write the 4 in the ones place of the difference.

$$\begin{array}{r} 9\overset{0}{\cancel{1}}\overset{10}{\cancel{0}} \\ -\ 586 \\ \hline 4 \end{array}$$

Subtract the tens. We cannot subtract 8 from 0, so we borrow 1 hundred and add 10 tens to the 0 tens we had, which gives us 10 tens. Write 8 above the hundreds place and cross out the 9. Write 10 above the tens place.

$$\begin{array}{r} \overset{8}{\cancel{9}}\overset{10}{\cancel{1}}\overset{10}{\cancel{0}} \\ -\ 586 \\ \hline 4 \end{array}$$

Now we can subtract the tens.

$$\begin{array}{r} \overset{8}{\cancel{9}}\overset{10}{\cancel{1}}\overset{10}{\cancel{0}} \\ -\ 586 \\ \hline 24 \end{array}$$

Subtract the hundreds place. Write the 3 in the hundreds place in the difference.

$$\begin{array}{r} \overset{8}{\cancel{9}}\overset{10}{\cancel{1}}\overset{10}{\cancel{0}} \\ -\ 586 \\ \hline 324 \end{array}$$

Check by adding.

$$\begin{array}{r} \overset{1}{3}\overset{1}{2}4 \\ +\ 586 \\ \hline 910\ \checkmark \end{array}$$

Our answer is correct.

> TRY IT : : 1.67 Subtract and then check by adding:

> TRY IT : : 1.68 Subtract and then check by adding:

EXAMPLE 1.35

Subtract and then check by adding:

Solution

Write the numbers so each place values line up vertically.	$\begin{array}{r} 2,162 \\ -\ 479 \\ \hline \end{array}$
Subtract the ones. Since we cannot subtract 9 from 2, borrow 1 ten and add 10 ones to the 2 ones to make 12 ones. Write 5 above the tens place and cross out the 6. Write 12 above the ones place and cross out the 2.	$\begin{array}{r} {\scriptstyle 5\ 12} \\ 2,1\cancel{6}\cancel{2} \\ -\ 479 \\ \hline \end{array}$
Now we can subtract the ones.	
Write 3 in the ones place in the difference.	$\begin{array}{r} {\scriptstyle 5\ 12} \\ 2,1\cancel{6}\cancel{2} \\ -\ 479 \\ \hline 3 \end{array}$
Subtract the tens. Since we cannot subtract 7 from 5, borrow 1 hundred and add 10 tens to the 5 tens to make 15 tens. Write 0 above the hundreds place and cross out the 1. Write 15 above the tens place.	$\begin{array}{r} {\scriptstyle 15} \\ {\scriptstyle 0\ \cancel{5}\ 12} \\ 2,\cancel{1}\cancel{6}\cancel{2} \\ -\ 479 \\ \hline 3 \end{array}$
Now we can subtract the tens.	
Write 8 in the tens place in the difference.	$\begin{array}{r} {\scriptstyle 0\ 15\ 12} \\ 2,\cancel{1}\cancel{6}\cancel{2} \\ -\ 479 \\ \hline 83 \end{array}$
Now we can subtract the hundreds.	$\begin{array}{r} {\scriptstyle 10} \\ {\scriptstyle 1\ \ \cancel{0}\ 15\ 12} \\ \cancel{2},\cancel{1}\cancel{6}\cancel{2} \\ -\ 479 \\ \hline 83 \end{array}$
Write 6 in the hundreds place in the difference.	$\begin{array}{r} {\scriptstyle 1\ \ 10\ 15\ 12} \\ \cancel{2},\cancel{1}\cancel{6}\cancel{2} \\ -\ 479 \\ \hline 683 \end{array}$
Subtract the thousands. There is no digit in the thousands place of the bottom number, so we imagine a 0. Write 1 in the thousands place of the difference.	$\begin{array}{r} {\scriptstyle 1\ \ 10\ \cancel{15}\ \cancel{12}} \\ \cancel{2},\cancel{1}\cancel{6}\cancel{2} \\ -\ 479 \\ \hline 1,683 \end{array}$
Check by adding.	

Our answer is correct.

> TRY IT : : 1.69 Subtract and then check by adding:

> TRY IT : : 1.70 Subtract and then check by adding:

Translate Word Phrases to Math Notation

As with addition, word phrases can tell us to operate on two numbers using subtraction. To translate from a word phrase to math notation, we look for key words that indicate subtraction. Some of the words that indicate subtraction are listed in Table 1.3.

Operation	Word Phrase	Example	Expression
Subtraction	minus	minus	
	difference	the difference of and	
	decreased by	decreased by	
	less than	less than	
	subtracted from	subtracted from	

Table 1.3

EXAMPLE 1.36

Translate and then simplify:

the difference of and subtract from

Solution

The word *difference* tells us to subtract the two numbers. The numbers stay in the same order as in the phrase.

	the difference of 13 and 8
Translate.	
Simplify.	5

The words *subtract from* tells us to take the second number away from the first. We must be careful to get the order correct.

	subtract 24 from 43
Translate.	
Simplify.	19

TRY IT : : 1.71 Translate and simplify:

the difference of and subtract from

TRY IT : : 1.72 Translate and simplify:

decreased by less than

Subtract Whole Numbers in Applications

To solve applications with subtraction, we will use the same plan that we used with addition. First, we need to determine what we are asked to find. Then we write a phrase that gives the information to find it. We translate the phrase into math notation and then simplify to get the answer. Finally, we write a sentence to answer the question, using the appropriate units.

EXAMPLE 1.37

The temperature in Chicago one morning was degrees Fahrenheit. A cold front arrived and by noon the temperature was degrees Fahrenheit. What was the difference between the temperature in the morning and the temperature at noon?

Solution

We are asked to find the difference between the morning temperature and the noon temperature.

Write a phrase.	the difference of 73 and 27
Translate to math notation. *Difference* tells us to subtract.	
Then we do the subtraction.	$\begin{array}{r} \overset{6\ 13}{\not{7}\not{3}} \\ -\ 27 \\ \hline 46 \end{array}$
Write a sentence to answer the question.	The difference in temperatures was 46 degrees Fahrenheit.

TRY IT : : 1.73

The high temperature on in Boston was degrees Fahrenheit, and the low temperature was degrees Fahrenheit. What was the difference between the high and low temperatures?

TRY IT : : 1.74

The weather forecast for June in St Louis predicts a high temperature of degrees Fahrenheit and a low of degrees Fahrenheit. What is the difference between the predicted high and low temperatures?

EXAMPLE 1.38

A washing machine is on sale for Its regular price is What is the difference between the regular price and the sale price?

Solution

We are asked to find the difference between the regular price and the sale price.

Write a phrase.	the difference between 588 and 399
Translate to math notation.	
Subtract.	$\begin{array}{r} \overset{4\ 17\ 18}{\not{5}\not{8}\not{8}} \\ -\ 399 \\ \hline 189 \end{array}$
Write a sentence to answer the question.	The difference between the regular price and the sale price is $189.

TRY IT : : 1.75

A television set is on sale for Its regular price is What is the difference between the regular price and the sale price?

TRY IT : : 1.76

A patio set is on sale for Its regular price is What is the difference between the regular price and the sale price?

MEDIA : : ACCESS ADDITIONAL ONLINE RESOURCES

- Model subtraction of two-digit whole numbers (http://www.openstax.org/l/24sub2dignum)
- Model subtraction of three-digit whole numbers (http://www.openstax.org/l/24sub3dignum)
- Subtract Whole Numbers (http://www.openstax.org/l/24subwholenum)

1.3 EXERCISES

Practice Makes Perfect

Use Subtraction Notation

In the following exercises, translate from math notation to words.

141.

142.

143.

144.

145.

146.

Model Subtraction of Whole Numbers

In the following exercises, model the subtraction.

147.

148.

149.

150.

151.

152.

153.

154.

155.

156.

157.

158.

Subtract Whole Numbers

In the following exercises, subtract and then check by adding.

159.

160.

161.

162.

163.

164.

165.

166.

167.

168.

169.

170.

171.

172.

173.

174.

175.

176.

177.

178.

179.

180.

181.

182.

Translate Word Phrases to Algebraic Expressions

In the following exercises, translate and simplify.

183. The difference of and

184. The difference of and

185. The difference of and

186. The difference of and

187. Subtract from

188. Subtract from

189. Subtract from

190. Subtract from

191. decreased by

192. decreased by

193. decreased by

194. decreased by

195. less than

196. less than

197. less than

198. less than

Mixed Practice

In the following exercises, simplify.

199.

200.

201.

202.

203.

204.

205.

206.

In the following exercises, translate and simplify.

207. Seventy-five more than thirty-five

208. Sixty more than ninety-three

209. less than

210. less than

211. The difference of and

212. The difference of and

Subtract Whole Numbers in Applications

In the following exercises, solve.

213. Temperature The high temperature on June in Las Vegas was degrees and the low temperature was degrees. What was the difference between the high and low temperatures?

214. Temperature The high temperature on June in Phoenix was degrees and the low was degrees. What was the difference between the high and low temperatures?

215. Class size Olivia's third grade class has children. Last year, her second grade class had children. What is the difference between the number of children in Olivia's third grade class and her second grade class?

216. Class size There are students in the school band and in the school orchestra. What is the difference between the number of students in the band and the orchestra?

217. Shopping A mountain bike is on sale for Its regular price is What is the difference between the regular price and the sale price?

218. Shopping A mattress set is on sale for Its regular price is What is the difference between the regular price and the sale price?

219. Savings John wants to buy a laptop that costs He has in his savings account. How much more does he need to save in order to buy the laptop?

220. Banking Mason had in his checking account. He spent How much money does he have left?

Everyday Math

221. Road trip Noah was driving from Philadelphia to Cincinnati, a distance of miles. He drove miles, stopped for gas, and then drove another miles before lunch. How many more miles did he have to travel?

222. Test Scores Sara needs points to pass her course. She scored on her first four tests. How many more points does Sara need to pass the course?

Writing Exercises

223. Explain how subtraction and addition are related.

224. How does knowing addition facts help you to subtract numbers?

Self Check

After completing the exercises, use this checklist to evaluate your mastery of the objectives of this section.

I can...	Confidently	With some help	No-I don't get it!
use subtraction notation.			
model subtraction of whole numbers.			
subtract whole numbers.			
translate word phrases to math notation.			
subtract whole numbers in applications.			

What does this checklist tell you about your mastery of this section? What steps will you take to improve?

1.4 Multiply Whole Numbers

Learning Objectives

By the end of this section, you will be able to:

- Use multiplication notation
- Model multiplication of whole numbers
- Multiply whole numbers
- Translate word phrases to math notation
- Multiply whole numbers in applications

BE PREPARED : : 1.5 Before you get started, take this readiness quiz.

Add:

If you missed this problem, review Exercise 1.58.

BE PREPARED : : 1.6 Subtract:

If you missed this problem, review Example 1.33.

Use Multiplication Notation

Suppose you were asked to count all these pennies shown in Figure 1.11.

Figure 1.11

Would you count the pennies individually? Or would you count the number of pennies in each row and add that number times.

Multiplication is a way to represent repeated addition. So instead of adding three times, we could write a multiplication expression.

We call each number being multiplied a factor and the result the **product**. We read as *three times eight*, and the result as *the product of three and eight*.

There are several symbols that represent multiplication. These include the symbol as well as the dot, , and parentheses

Operation Symbols for Multiplication

To describe multiplication, we can use symbols and words.

Operation	Notation	Expression	Read as	Result

EXAMPLE 1.39

Translate from math notation to words:

Solution

We read this as *seven times six* and the result is *the product of seven and six*.

We read this as *twelve times fourteen* and the result is *the product of twelve and fourteen*.

We read this as *six times thirteen* and the result is *the product of six and thirteen*.

TRY IT : : 1.77 Translate from math notation to words:

TRY IT : : 1.78 Translate from math notation to words:

Model Multiplication of Whole Numbers

There are many ways to model multiplication. Unlike in the previous sections where we used blocks, here we will use counters to help us understand the meaning of multiplication. A counter is any object that can be used for counting. We will use round blue counters.

EXAMPLE 1.40

Model:

Solution

To model the product we'll start with a row of counters.

The other factor is so we'll make rows of counters.

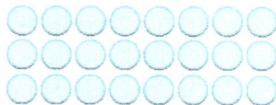

Now we can count the result. There are counters in all.

If you look at the counters sideways, you'll see that we could have also made rows of counters. The product would have been the same. We'll get back to this idea later.

TRY IT : : 1.79 Model each multiplication:

 **TRY IT : :** 1.80 Model each multiplication:

Multiply Whole Numbers

In order to multiply without using models, you need to know all the one digit multiplication facts. Make sure you know them fluently before proceeding in this section.

Table 1.4 shows the multiplication facts. Each box shows the product of the number down the left column and the number across the top row. If you are unsure about a product, model it. It is important that you memorize any number facts you do not already know so you will be ready to multiply larger numbers.

×	0	1	2	3	4	5	6	7	8	9
0	0	0	0	0	0	0	0	0	0	0
1	0	1	2	3	4	5	6	7	8	9
2	0	2	4	6	8	10	12	14	16	18
3	0	3	6	9	12	15	18	21	24	27
4	0	4	8	12	16	20	24	28	32	36
5	0	5	10	15	20	25	30	35	40	45
6	0	6	12	18	24	30	36	42	48	54
7	0	7	14	21	28	35	42	49	56	63
8	0	8	16	24	32	40	48	56	64	72
9	0	9	18	27	36	45	54	63	72	81

Table 1.4

What happens when you multiply a number by zero? You can see that the product of any number and zero is zero. This is called the Multiplication Property of Zero.

Multiplication Property of Zero

The product of any number and is

EXAMPLE 1.41

Multiply:

Solution

The product of any number and zero is zero.

Multiplying by zero results in zero.

> **TRY IT : : 1.81** Find each product:

> **TRY IT : : 1.82** Find each product:

What happens when you multiply a number by one? Multiplying a number by one does not change its value. We call this fact the Identity Property of Multiplication, and is called the multiplicative identity.

Identity Property of Multiplication

The product of any number and is the number.

EXAMPLE 1.42

Multiply:

Solution

The product of any number and one is the number.

Multiplying by one does not change the value.

> **TRY IT : : 1.83** Find each product:

> **TRY IT : : 1.84** Find each product:

Earlier in this chapter, we learned that the Commutative Property of Addition states that changing the order of addition does not change the sum. We saw that is the same as

Is this also true for multiplication? Let's look at a few pairs of factors.

When the order of the factors is reversed, the product does not change. This is called the Commutative Property of Multiplication.

Commutative Property of Multiplication

Changing the order of the factors does not change their product.

EXAMPLE 1.43

Multiply:

Solution

Multiply.

Multiply.

Changing the order of the factors does not change the product.

TRY IT : : 1.85 Multiply:

TRY IT : : 1.86 Multiply:

To multiply numbers with more than one digit, it is usually easier to write the numbers vertically in columns just as we did for addition and subtraction.

We start by multiplying by

We write the in the ones place of the product. We carry the tens by writing above the tens place.

$$\begin{array}{r} {}^{2}\\ 27 \\ \times 3 \\ \hline 1 \end{array}$$

Here are the 2 tens in 21.

Here is the 1 one in 21.

Then we multiply the by the and add the above the tens place to the product. So and Write the in the tens place of the product.

$$\begin{array}{r} {}^{2}\\ 27 \\ \times 3 \\ \hline 81 \end{array}$$

This comes from 3×2 plus the 2 we carried.

The product is

When we multiply two numbers with a different number of digits, it's usually easier to write the smaller number on the bottom. You could write it the other way, too, but this way is easier to work with.

EXAMPLE 1.44

Multiply:

Solution

Write the numbers so the digits and line up vertically.

Multiply by the digit in the ones place of

Write in the ones place of the product and carry the tens.

Multiply by the digit in the tens place of .
Add the tens we carried. .

Write the in the tens place of the product.

> **TRY IT : : 1.87** Multiply:

> **TRY IT : : 1.88** Multiply:

EXAMPLE 1.45

Multiply:

Solution

Write the numbers so the digits and line up vertically.

Multiply by the digit in the ones place of

Write the in the ones place of the product and carry the to the tens place.Multiply by the digit in the tens place of .

Add the tens we carried to get .
Write the in the tens place of the product and carry the 4 to the hundreds place.

Multiply by the digit in the hundreds place of
Add the hundreds we carried to get
Write the in the hundreds place of the product and the to the thousands place.

> **TRY IT : : 1.89** Multiply:

 **TRY IT : :** 1.90 Multiply:

When we multiply by a number with two or more digits, we multiply by each of the digits separately, working from right to left. Each separate product of the digits is called a partial product. When we write partial products, we must make sure to line up the place values.

HOW TO : : MULTIPLY TWO WHOLE NUMBERS TO FIND THE PRODUCT.

Step 1. Write the numbers so each place value lines up vertically.

Step 2. Multiply the digits in each place value.

- Work from right to left, starting with the ones place in the bottom number.
  - Multiply the bottom number by the ones digit in the top number, then by the tens digit, and so on.
  - If a product in a place value is more than carry to the next place value.
  - Write the partial products, lining up the digits in the place values with the numbers above.
- Repeat for the tens place in the bottom number, the hundreds place, and so on.
- Insert a zero as a placeholder with each additional partial product.

Step 3. Add the partial products.

EXAMPLE 1.46

Multiply:

Solution

Write the numbers so each place lines up vertically.	$\begin{array}{r} 62 \\ \times 87 \\ \hline \end{array}$
Start by multiplying 7 by 62. Multiply 7 by the digit in the ones place of 62. Write the 4 in the ones place of the product and carry the 1 to the tens place.	$\begin{array}{r} ^{1} \\ 62 \\ \times 87 \\ \hline 4 \end{array}$
Multiply 7 by the digit in the tens place of 62. Add the 1 ten we carried. . Write the 3 in the tens place of the product and the 4 in the hundreds place.	$\begin{array}{r} ^{1} \\ 62 \\ \times 87 \\ \hline 434 \end{array}$
The first partial product is 434.	
Now, write a 0 under the 4 in the ones place of the next partial product as a placeholder since we now multiply the digit in the tens place of 87 by 62. Multiply 8 by the digit in the ones place of 62. Write the 6 in the next place of the product, which is the tens place. Carry the 1 to the tens place.	$\begin{array}{r} ^{1} \\ \not{1} \\ 62 \\ \times 87 \\ \hline 434 \\ 60 \end{array}$
Multiply 8 by 6, the digit in the tens place of 62, then add the 1 ten we carried to get 49. Write the 9 in the hundreds place of the product and the 4 in the thousands place.	$\begin{array}{r} ^{1} \\ \not{1} \\ 62 \\ \times 87 \\ \hline 434 \\ 4960 \end{array}$
The second partial product is 4960. Add the partial products.	$\begin{array}{r} ^{1} \\ \not{1} \\ 62 \\ \times 87 \\ \hline 434 \\ 4960 \\ \hline 5394 \end{array}$

The product is

TRY IT : : 1.91 Multiply:

TRY IT : : 1.92 Multiply:

EXAMPLE 1.47

Multiply:

Solution

When we multiplied times the product was Notice that has one zero, and we put one zero after to get the product. When we multiplied times the product was Notice that has two zeros and we put two zeros after to get the product.

Do you see the pattern? If we multiplied times which has four zeros, we would put four zeros after to get the product

> **TRY IT : : 1.93** Multiply:

> **TRY IT : : 1.94** Multiply:

EXAMPLE 1.48

Multiply:

Solution

There are three digits in the factors so there will be partial products. We do not have to write the as a placeholder as long as we write each partial product in the correct place.

	354
	× 438
Multiply 8(354)	2832
Multiply 3(354)	1062
Multiply 4(354)	1416
Add the partial products	155,052

> **TRY IT : : 1.95** Multiply:

> **TRY IT : : 1.96** Multiply:

EXAMPLE 1.49

Multiply:

Solution

There should be partial products. The second partial product will be the result of multiplying by

Multiply 1(896)	896
	× 201
	896
Multiply 0(896)	000
Multiply 200(896)	1792
Add the partial products	180,096

Notice that the second partial product of all zeros doesn't really affect the result. We can place a zero as a placeholder in the tens place and then proceed directly to multiplying by the in the hundreds place, as shown.

Multiply by but insert only one zero as a placeholder in the tens place. Multiply by putting the from the in the hundreds place.

> **TRY IT : : 1.97** Multiply:

> **TRY IT : : 1.98** Multiply:

When there are three or more factors, we multiply the first two and then multiply their product by the next factor. For example:

to multiply
first multiply
then multiply .

Translate Word Phrases to Math Notation

Earlier in this section, we translated math notation into words. Now we'll reverse the process and translate word phrases into math notation. Some of the words that indicate multiplication are given in Table 1.5.

Operation	Word Phrase	Example	Expression
Multiplication	times product twice	times the product of and twice	

Table 1.5

EXAMPLE 1.50

Translate and simplify: the product of and

Solution

The word *product* tells us to multiply. The words *of* *and* tell us the two factors.

	the product of 12 and 27
Translate.	
Multiply.	

> **TRY IT : : 1.99** Translate and simplify the product of and

> **TRY IT : : 1.100** Translate and simplify the product of and

EXAMPLE 1.51

Translate and simplify: twice two hundred eleven.

 **Solution**

The word *twice* tells us to multiply by

	twice two hundred eleven
Translate.	2(211)
Multiply.	422

> **TRY IT : : 1.101** Translate and simplify: twice one hundred sixty-seven.

> **TRY IT : : 1.102** Translate and simplify: twice two hundred fifty-eight.

Multiply Whole Numbers in Applications

We will use the same strategy we used previously to solve applications of multiplication. First, we need to determine what we are looking for. Then we write a phrase that gives the information to find it. We then translate the phrase into math notation and simplify to get the answer. Finally, we write a sentence to answer the question.

EXAMPLE 1.52

Humberto bought sheets of stamps. Each sheet had stamps. How many stamps did Humberto buy?

 **Solution**

We are asked to find the total number of stamps.

Write a phrase for the total.	the product of 4 and 20
Translate to math notation.	
Multiply.	$\begin{array}{r} 20 \\ \times 4 \\ \hline 80 \end{array}$
Write a sentence to answer the question.	Humberto bought 80 stamps.

TRY IT : : 1.103

Valia donated water for the snack bar at her son's baseball game. She brought cases of water bottles. Each case had water bottles. How many water bottles did Valia donate?

TRY IT : : 1.104

Vanessa brought packs of hot dogs to a family reunion. Each pack has hot dogs. How many hot dogs did Vanessa bring?

EXAMPLE 1.53

When Rena cooks rice, she uses twice as much water as rice. How much water does she need to cook cups of rice?

Solution

We are asked to find how much water Rena needs.

Write as a phrase.	twice as much as 4 cups
Translate to math notation.	
Multiply to simplify.	8
Write a sentence to answer the question.	Rena needs 8 cups of water for cups of rice.

TRY IT : : 1.105

Erin is planning her flower garden. She wants to plant twice as many dahlias as sunflowers. If she plants 14 sunflowers, how many dahlias does she need?

TRY IT : : 1.106

A college choir has twice as many women as men. There are 18 men in the choir. How many women are in the choir?

EXAMPLE 1.54

Van is planning to build a patio. He will have rows of tiles, with tiles in each row. How many tiles does he need for the patio?

Solution

We are asked to find the total number of tiles.

Write a phrase.	the product of 8 and 14
Translate to math notation.	
Multiply to simplify.	
Write a sentence to answer the question.	Van needs 112 tiles for his patio.

> **TRY IT : :** 1.107
>
> Jane is tiling her living room floor. She will need 16 rows of tile, with 20 tiles in each row. How many tiles does she need for the living room floor?

> **TRY IT : :** 1.108
>
> Yousef is putting shingles on his garage roof. He will need 24 rows of shingles, with 45 shingles in each row. How many shingles does he need for the garage roof?

If we want to know the size of a wall that needs to be painted or a floor that needs to be carpeted, we will need to find its **area**. The area is a measure of the amount of surface that is covered by the shape. Area is measured in square units. We often use square inches, square feet, square centimeters, or square miles to measure area. A square centimeter is a square that is one centimeter (cm.) on a side. A square inch is a square that is one inch on each side, and so on.

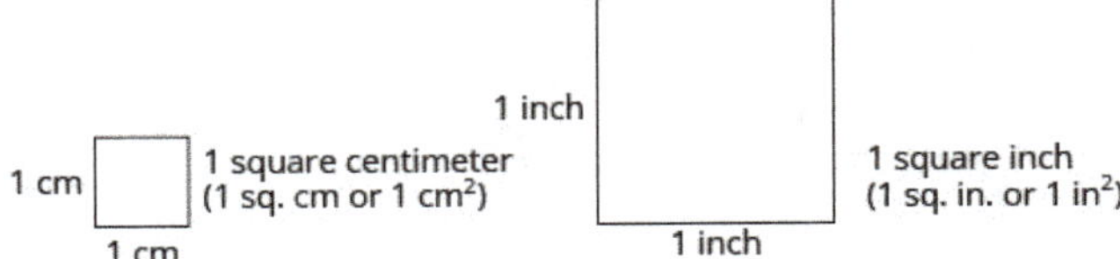

For a rectangular figure, the area is the product of the length and the width. Figure 1.12 shows a rectangular rug with a length of feet and a width of feet. Each square is foot wide by foot long, or square foot. The rug is made of squares. The area of the rug is square feet.

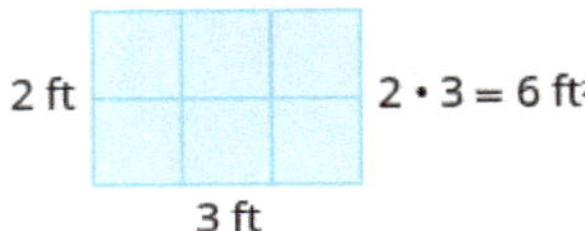

Figure 1.12 The area of a rectangle is the product of its length and its width, or square feet.

EXAMPLE 1.55

Jen's kitchen ceiling is a rectangle that measures 9 feet long by 12 feet wide. What is the area of Jen's kitchen ceiling?

Solution

We are asked to find the area of the kitchen ceiling.

Write a phrase for the area.	the product of 9 and 12
Translate to math notation.	
Multiply.	
Answer with a sentence.	The area of Jen's kitchen ceiling is 108 square feet.

> **TRY IT : :** 1.109
>
> Zoila bought a rectangular rug. The rug is 8 feet long by 5 feet wide. What is the area of the rug?

> **TRY IT : :** 1.110 Rene's driveway is a rectangle 45 feet long by 20 feet wide. What is the area of the driveway?

MEDIA : : ACCESS ADDITIONAL ONLINE RESOURCES

- Multiplying Whole Numbers (http://www.openstax.org/l/24multwhlnum)
- Multiplication with Partial Products (http://www.openstax.org/l/24multpartprod)
- Example of Multiplying by Whole Numbers (http://www.openstax.org/l/24examplemultnm)

1.4 EXERCISES

Practice Makes Perfect

Use Multiplication Notation

In the following exercises, translate from math notation to words.

225.

226.

227.

228.

229.

230.

231.

232.

Model Multiplication of Whole Numbers

In the following exercises, model the multiplication.

233.

234.

235.

236.

Multiply Whole Numbers

In the following exercises, fill in the missing values in each chart.

237.

×	0	1	2	3	4	5	6	7	8	9
0	0	0	0		0	0	0	0		0
1	0	1	2	3			6	7	8	
2		2	4	6	8		12			18
3	0		6		12	15		21		27
4	0	4			16	20		28	32	
5	0	5	10	15			30		40	
6	0	6	12		24			42		54
7			14	21		35			56	63
8	0	8		24			48		64	
9	0	9	18		36	45			72	

238.

×	0	1	2	3	4	5	6	7	8	9
0	0	0	0	0	0		0		0	0
1	0	1	2		4	5		7		9
2	0		4		8	10		14	16	
3		3		9			18		24	
4	0	4	8	12			24	28		36
5	0	5		15	20		30	35	40	
6			12	18			36	42		54
7	0	7		21		35			56	63
8	0	8	16		32		48		64	72
9			18	27	36			63		

239.

×	3	4	5	6	7	8	9
4							
5							
6							
7							
8							
9							

240.

×	4	5	6	7	8	9
3						
4						
5						
6						
7						
8						
9						

241.

×	3	4	5	6	7	8	9
6							
7							
8							
9							

242.

×	6	7	8	9
3				
4				
5				
6				
7				
8				
9				

243.

×	5	6	7	8	9
5					
6					
7					
8					
9					

244.

×	6	7	8	9
6				
7				
8				
9				

In the following exercises, multiply.

245. **246.** **247.**

248. **249.** **250.**

251. **252.** **253.**

254. **255.** **256.**

257. **258.** **259.**

260. **261.** **262.**

263. **264.** **265.**

266. **267.** **268.**

269. **270.** **271.**

272. **273.** **274.**

275. **276.** **277.**

278. **279.** **280.**

281. **282.** **283.**

284. **285.** **286.**

287. **288.**

Translate Word Phrases to Math Notation

In the following exercises, translate and simplify.

289. the product of and

290. the product of and

291. fifty-one times sixty-seven

292. forty-eight times seventy-one

293. twice

294. twice

295. ten times three hundred seventy-five

296. ten times two hundred fifty-five

Mixed Practice

In the following exercises, simplify.

297.

298.

299.

300.

301.

302.

303.

304.

305.

306.

307.

308.

In the following exercises, translate and simplify.

309. the difference of 50 and 18

310. the difference of 90 and 66

311. twice 35

312. twice 140

313. 20 more than 980

314. 65 more than 325

315. the product of 12 and 875

316. the product of 15 and 905

317. subtract 74 from 89

318. subtract 45 from 99

319. the sum of 3,075 and 95

320. the sum of 6,308 and 724

321. 366 less than 814

322. 388 less than 925

Multiply Whole Numbers in Applications

In the following exercises, solve.

323. Party supplies Tim brought 9 six-packs of soda to a club party. How many cans of soda did Tim bring?

324. Sewing Kanisha is making a quilt. She bought 6 cards of buttons. Each card had four buttons on it. How many buttons did Kanisha buy?

325. Field trip Seven school busses let off their students in front of a museum in Washington, DC. Each school bus had 44 students. How many students were there?

326. Gardening Kathryn bought 8 flats of impatiens for her flower bed. Each flat has 24 flowers. How many flowers did Kathryn buy?

327. Charity Rey donated 15 twelve-packs of t-shirts to a homeless shelter. How many t-shirts did he donate?

328. School There are 28 classrooms at Anna C. Scott elementary school. Each classroom has 26 student desks. What is the total number of student desks?

329. Recipe Stephanie is making punch for a party. The recipe calls for twice as much fruit juice as club soda. If she uses 10 cups of club soda, how much fruit juice should she use?

330. Gardening Hiroko is putting in a vegetable garden. He wants to have twice as many lettuce plants as tomato plants. If he buys 12 tomato plants, how many lettuce plants should he get?

331. Government The United States Senate has twice as many senators as there are states in the United States. There are 50 states. How many senators are there in the United States Senate?

332. Recipe Andrea is making potato salad for a buffet luncheon. The recipe says the number of servings of potato salad will be twice the number of pounds of potatoes. If she buys 30 pounds of potatoes, how many servings of potato salad will there be?

333. Painting Jane is painting one wall of her living room. The wall is rectangular, 13 feet wide by 9 feet high. What is the area of the wall?

334. Home décor Shawnte bought a rug for the hall of her apartment. The rug is 3 feet wide by 18 feet long. What is the area of the rug?

335. Room size The meeting room in a senior center is rectangular, with length 42 feet and width 34 feet. What is the area of the meeting room?

336. Gardening June has a vegetable garden in her yard. The garden is rectangular, with length 23 feet and width 28 feet. What is the area of the garden?

337. NCAA basketball According to NCAA regulations, the dimensions of a rectangular basketball court must be 94 feet by 50 feet. What is the area of the basketball court?

338. NCAA football According to NCAA regulations, the dimensions of a rectangular football field must be 360 feet by 160 feet. What is the area of the football field?

Everyday Math

339. Stock market Javier owns 300 shares of stock in one company. On Tuesday, the stock price rose per share. How much money did Javier's portfolio gain?

340. Salary Carlton got a raise in each paycheck. He gets paid 24 times a year. How much higher is his new annual salary?

Writing Exercises

341. How confident do you feel about your knowledge of the multiplication facts? If you are not fully confident, what will you do to improve your skills?

342. How have you used models to help you learn the multiplication facts?

Self Check

After completing the exercises, use this checklist to evaluate your mastery of the objectives of this section.

I can...	Confidently	With some help	No-I don't get it!
use multiplication notation.			
model multiplication of whole numbers.			
multiply whole numbers.			
translate word phrases to math notation.			
multiply whole numbers in applications.			

On a scale of 1–10, how would you rate your mastery of this section in light of your responses on the checklist? How can you improve this?

1.5 Divide Whole Numbers

Learning Objectives

By the end of this section, you will be able to:

- Use division notation
- Model division of whole numbers
- Divide whole numbers
- Translate word phrases to math notation
- Divide whole numbers in applications

BE PREPARED : : 1.7 Before you get started, take this readiness quiz.

Multiply:
If you missed this problem, review Example 1.44.

BE PREPARED : : 1.8 Subtract:
If you missed this problem, review Example 1.32

BE PREPARED : : 1.9 Multiply:
If you missed this problem, review Example 1.45.

Use Division Notation

So far we have explored addition, subtraction, and multiplication. Now let's consider division. Suppose you have the cookies in Figure 1.13 and want to package them in bags with cookies in each bag. How many bags would we need?

Figure 1.13

You might put cookies in first bag, in the second bag, and so on until you run out of cookies. Doing it this way, you would fill bags.

In other words, starting with the cookies, you would take away, or subtract, cookies at a time. Division is a way to represent repeated subtraction just as multiplication represents repeated addition.

Instead of subtracting repeatedly, we can write

We read this as *twelve divided by four* and the result is the **quotient** of and The quotient is because we can subtract from exactly times. We call the number being divided the **dividend** and the number dividing it the **divisor**. In this case, the dividend is and the divisor is

In the past you may have used the notation ⟌, but this division also can be written as — In each case the is the dividend and the is the divisor.

Operation Symbols for Division

To represent and describe division, we can use symbols and words.

Operation	Notation	Expression	Read as	Result
	— ⟌	— ⟌		

Division is performed on two numbers at a time. When translating from math notation to English words, or English words to math notation, look for the words *of* and *and* to identify the numbers.

EXAMPLE 1.56

Translate from math notation to words.

— ⟌

Solution

We read this as *sixty-four divided by eight* and the result is *the quotient of sixty-four and eight*.

We read this as *forty-two divided by seven* and the result is *the quotient of forty-two and seven*.

We read this as *twenty-eight divided by four* and the result is *the quotient of twenty-eight and four*.

TRY IT : : 1.111 Translate from math notation to words:

— ⟌

TRY IT : : 1.112 Translate from math notation to words:

— ⟌

Model Division of Whole Numbers

As we did with multiplication, we will model division using counters. The operation of division helps us organize items into equal groups as we start with the number of items in the dividend and subtract the number in the divisor repeatedly.

MANIPULATIVE MATHEMATICS

Doing the Manipulative Mathematics activity Model Division of Whole Numbers will help you develop a better understanding of dividing whole numbers.

EXAMPLE 1.57

Model the division:

Solution

To find the quotient we want to know how many groups of are in

Model the dividend. Start with counters.

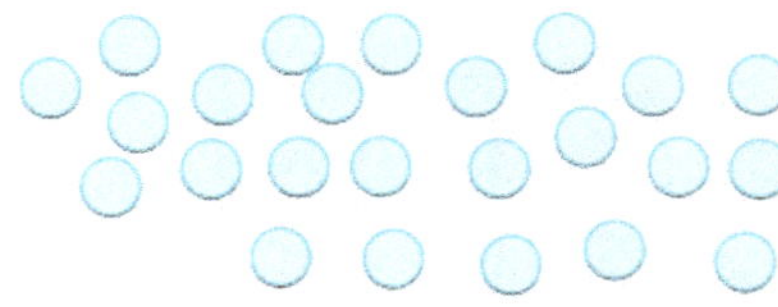

The divisor tell us the number of counters we want in each group. Form groups of counters.

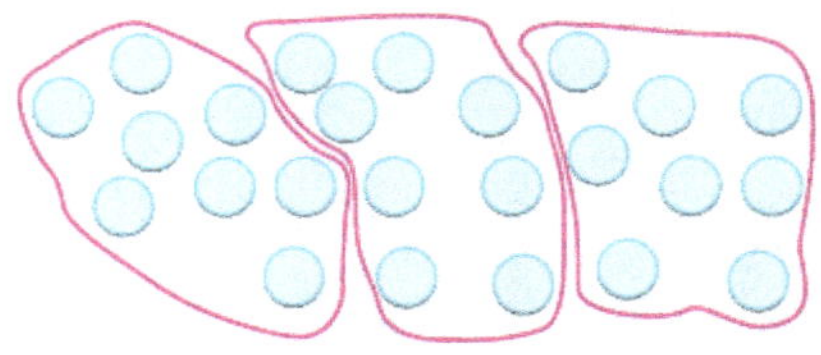

Count the number of groups. There are groups.

 **TRY IT : :** 1.113 Model:

TRY IT : : 1.114 Model:

Divide Whole Numbers

We said that addition and subtraction are inverse operations because one undoes the other. Similarly, division is the inverse operation of multiplication. We know because Knowing all the multiplication number facts is very important when doing division.

We check our answer to division by multiplying the quotient by the divisor to determine if it equals the dividend. In Example 1.57, we know is correct because

EXAMPLE 1.58

Divide. Then check by multiplying. 

 **Solution**

Divide 42 by 6.
Check by multiplying.

—
Divide 72 by 9.
Check by multiplying.

)
Divide 63 by 7.
Check by multiplying.

> **TRY IT : : 1.115** Divide. Then check by multiplying:
>
> —

> **TRY IT : : 1.116** Divide. Then check by multiplying:
>
> —)

What is the quotient when you divide a number by itself?

—

Dividing any number by itself produces a quotient of Also, any number divided by produces a quotient of the number. These two ideas are stated in the Division Properties of One.

Division Properties of One

Any number (except 0) divided by itself is one.	
Any number divided by one is the same number.	

Table 1.6

EXAMPLE 1.59

Divide. Then check by multiplying:

— ⟌

Solution

A number divided by itself is 1.

Check by multiplying.

—

A number divided by 1 equals itself.

Check by multiplying.

⟌

A number divided by 1 equals itself.

Check by multiplying.

TRY IT : : 1.117 Divide. Then check by multiplying:

—

TRY IT : : 1.118 Divide. Then check by multiplying:

— ⟌

Suppose we have and want to divide it among people. How much would each person get? Each person would get Zero divided by any number is

Now suppose that we want to divide by That means we would want to find a number that we multiply by to get This cannot happen because times any number is Division by zero is said to be *undefined*.

These two ideas make up the Division Properties of Zero.

Division Properties of Zero

Zero divided by any number is 0.	
Dividing a number by zero is undefined.	undefined

Table 1.7

Another way to explain why division by zero is undefined is to remember that division is really repeated subtraction. How many times can we take away from Because subtracting will never change the total, we will never get an answer. So we cannot divide a number by

EXAMPLE 1.60

Divide. Check by multiplying:

Solution

Zero divided by any number is zero.	
Check by multiplying.	
Division by zero is undefined.	undefined

TRY IT : : 1.119 Divide. Then check by multiplying:

TRY IT : : 1.120 Divide. Then check by multiplying:

When the divisor or the dividend has more than one digit, it is usually easier to use the $\overline{)}$ notation. This process is called long division. Let's work through the process by dividing by

Divide the first digit of the dividend, 7, by the divisor, 3.	
The divisor 3 can go into 7 two times since . Write the 2 above the 7 in the quotient.	$\begin{array}{r} 2 \\ 3\overline{)78} \end{array}$
Multiply the 2 in the quotient by 3 and write the product, 6, under the 7.	$\begin{array}{r} 2 \\ 3\overline{)78} \\ \underline{6} \end{array}$
Subtract that product from the first digit in the dividend. Subtract . Write the difference, 1, under the first digit in the dividend.	$\begin{array}{r} 2 \\ 3\overline{)78} \\ \underline{6} \\ 1 \end{array}$
Bring down the next digit of the dividend. Bring down the 8.	$\begin{array}{r} 2 \\ 3\overline{)78} \\ \underline{6} \\ 18 \end{array}$
Divide 18 by the divisor, 3. The divisor 3 goes into 18 six times.	$\begin{array}{r} 26 \\ 3\overline{)78} \\ \underline{6} \\ 18 \end{array}$
Write 6 in the quotient above the 8.	
Multiply the 6 in the quotient by the divisor and write the product, 18, under the dividend. Subtract 18 from 18.	$\begin{array}{r} 26 \\ 3\overline{)78} \\ \underline{6} \\ 18 \\ \underline{18} \\ 0 \end{array}$

We would repeat the process until there are no more digits in the dividend to bring down. In this problem, there are no more digits to bring down, so the division is finished.

Check by multiplying the quotient times the divisor to get the dividend. Multiply to make sure that product equals the dividend,

It does, so our answer is correct.

HOW TO : : DIVIDE WHOLE NUMBERS.

Step 1. Divide the first digit of the dividend by the divisor.
If the divisor is larger than the first digit of the dividend, divide the first two digits of the dividend by the divisor, and so on.

Step 2. Write the quotient above the dividend.

Step 3. Multiply the quotient by the divisor and write the product under the dividend.

Step 4. Subtract that product from the dividend.

Step 5. Bring down the next digit of the dividend.

Step 6. Repeat from Step 1 until there are no more digits in the dividend to bring down.

Step 7. Check by multiplying the quotient times the divisor.

EXAMPLE 1.61

Divide Check by multiplying:

Solution

Let's rewrite the problem to set it up for long division.	$4\overline{)2596}$
Divide the first digit of the dividend, 2, by the divisor, 4.	$4\overline{)2596}$
Since 4 does not go into 2, we use the first two digits of the dividend and divide 25 by 4. The divisor 4 goes into 25 six times.	
We write the 6 in the quotient above the 5.	$\begin{array}{r} 6 \\ 4\overline{)2596} \end{array}$
Multiply the 6 in the quotient by the divisor 4 and write the product, 24, under the first two digits in the dividend.	$\begin{array}{r} 6 \\ 4\overline{)2596} \\ \underline{24} \end{array}$
Subtract that product from the first two digits in the dividend. Subtract . Write the difference, 1, under the second digit in the dividend.	$\begin{array}{r} 6 \\ 4\overline{)2596} \\ \underline{24} \\ 1 \end{array}$
Now bring down the 9 and repeat these steps. There are 4 fours in 19. Write the 4 over the 9. Multiply the 4 by 4 and subtract this product from 19.	$\begin{array}{r} 64 \\ 4\overline{)2596} \\ \underline{24} \\ 19 \\ \underline{16} \\ 3 \end{array}$
Bring down the 6 and repeat these steps. There are 9 fours in 36. Write the 9 over the 6. Multiply the 9 by 4 and subtract this product from 36.	$\begin{array}{r} 649 \\ 4\overline{)2596} \\ \underline{24} \\ 19 \\ \underline{16} \\ 36 \\ \underline{36} \\ 0 \end{array}$
So .	
Check by multiplying. $\begin{array}{r} \overset{1\,3}{649} \\ \underline{\times\quad 4} \\ 2{,}596\checkmark \end{array}$	

It equals the dividend, so our answer is correct.

TRY IT : : 1.121 Divide. Then check by multiplying:

TRY IT : : 1.122 Divide. Then check by multiplying:

EXAMPLE 1.62

Divide Check by multiplying:

Solution

Let's rewrite the problem to set it up for long division.	$6\overline{)4506}$
First we try to divide 6 into 4.	$6\overline{)4506}$
Since that won't work, we try 6 into 45. There are 7 sixes in 45. We write the 7 over the 5.	$\begin{array}{r} 7 \\ 6\overline{)4506} \end{array}$
Multiply the 7 by 6 and subtract this product from 45.	$\begin{array}{r} 7 \\ 6\overline{)4506} \\ \underline{42} \\ 3 \end{array}$
Now bring down the 0 and repeat these steps. There are 5 sixes in 30. Write the 5 over the 0. Multiply the 5 by 6 and subtract this product from 30.	$\begin{array}{r} 75 \\ 6\overline{)4506} \\ \underline{42} \\ 30 \\ \underline{30} \\ 0 \end{array}$
Now bring down the 6 and repeat these steps. There is 1 six in 6. Write the 1 over the 6. Multiply 1 by 6 and subtract this product from 6.	$\begin{array}{r} 751 \\ 6\overline{)4506} \\ \underline{42} \\ 30 \\ \underline{30} \\ 06 \\ \underline{6} \\ 0 \end{array}$
Check by multiplying. $\begin{array}{r} \overset{3}{7}51 \\ \underline{\times \quad 6} \\ 4{,}506 \checkmark \end{array}$	

It equals the dividend, so our answer is correct.

TRY IT : : 1.123 Divide. Then check by multiplying:

TRY IT : : 1.124 Divide. Then check by multiplying:

EXAMPLE 1.63

Divide Check by multiplying.

Solution

Let's rewrite the problem to set it up for long division.

$$9\overline{)7263}$$

First we try to divide 9 into 7.

$$9\overline{)7263}$$

Since that won't work, we try 9 into 72. There are 8 nines in 72.
We write the 8 over the 2.

$$\begin{array}{r} 8 \\ 9\overline{)7263} \end{array}$$

Multiply the 8 by 9 and subtract this product from 72.

$$\begin{array}{r} 8 \\ 9\overline{)7263} \\ \underline{72} \\ 0 \end{array}$$

Now bring down the 6 and repeat these steps. There are 0 nines in 6.
Write the 0 over the 6. Multiply the 0 by 9 and subtract this product from 6.

$$\begin{array}{r} 80 \\ 9\overline{)7263} \\ \underline{72} \\ 06 \\ \underline{0} \\ 6 \end{array}$$

Now bring down the 3 and repeat these steps. There are 7 nines in 63. Write the 7 over the 3.
Multiply the 7 by 9 and subtract this product from 63.

$$\begin{array}{r} 807 \\ 9\overline{)7263} \\ \underline{72} \\ 06 \\ \underline{0} \\ 63 \\ \underline{63} \\ 0 \end{array}$$

Check by multiplying.

$$\begin{array}{r} \overset{6}{8}07 \\ \times\quad 9 \\ \hline 7{,}263\ \checkmark \end{array}$$

It equals the dividend, so our answer is correct.

> TRY IT : : 1.125 Divide. Then check by multiplying:

> TRY IT : : 1.126 Divide. Then check by multiplying:

So far all the division problems have worked out evenly. For example, if we had cookies and wanted to make bags of cookies, we would have bags. But what if there were cookies and we wanted to make bags of Start with the cookies as shown in Figure 1.14.

Figure 1.14

Try to put the cookies in groups of eight as in Figure 1.15.

Figure 1.15

There are groups of eight cookies, and cookies left over. We call the cookies that are left over the remainder and show it by writing R4 next to the (The R stands for remainder.)

To check this division we multiply times to get and then add the remainder of

EXAMPLE 1.64

Divide Check by multiplying.

Solution

Let's rewrite the problem to set it up for long division.	4)1439
First we try to divide 4 into 1. Since that won't work, we try 4 into 14. There are 3 fours in 14. We write the 3 over the 4.	3 4)1439
Multiply the 3 by 4 and subtract this product from 14.	3 4)1439 12 2
Now bring down the 3 and repeat these steps. There are 5 fours in 23. Write the 5 over the 3. Multiply the 5 by 4 and subtract this product from 23.	35 4)1439 12↓ 23 20 3
Now bring down the 9 and repeat these steps. There are 9 fours in 39. Write the 9 over the 9. Multiply the 9 by 4 and subtract this product from 39. There are no more numbers to bring down, so we are done. The remainder is 3.	359R3 4)1439 12 23 20↓ 39 36 3

Check by multiplying.

359 (carries 2 3)	quotient
× 4	divisor
1,436	
+ 3	remainder
1,439 ✓	

So is with a remainder of Our answer is correct.

> **TRY IT : : 1.127** Divide. Then check by multiplying:

> **TRY IT : : 1.128** Divide. Then check by multiplying:

EXAMPLE 1.65

Divide and then check by multiplying:

Solution

Let's rewrite the problem to set it up for long division.	$\overline{)\quad}$
First we try to divide 13 into 1. Since that won't work, we try 13 into 14. There is 1 thirteen in 14. We write the 1 over the 4.	$13\overline{)1461}$ with 1 over the 4
Multiply the 1 by 13 and subtract this product from 14.	$13\overline{)1461}$, quotient 1; 14 − 13 = 1
Now bring down the 6 and repeat these steps. There is 1 thirteen in 16. Write the 1 over the 6. Multiply the 1 by 13 and subtract this product from 16.	$13\overline{)1461}$, quotient 11; 14 − 13 = 1, bring down 6: 16 − 13 = 3
Now bring down the 1 and repeat these steps. There are 2 thirteens in 31. Write the 2 over the 1. Multiply the 2 by 13 and subtract this product from 31. There are no more numbers to bring down, so we are done. The remainder is 5. is 112 with a remainder of 5.	$13\overline{)1461}$, quotient 112R5; 14 − 13 = 1, 16 − 13 = 3, 31 − 26 = 5

Check by multiplying.

$$\begin{array}{rl} 112 & \text{quotient} \\ \times\ 13 & \text{divisor} \\ \hline 336 & \\ 1{,}120 & \\ +\ \ 5 & \text{remainder} \\ \hline 1{,}461 & \checkmark \end{array}$$

Our answer is correct.

> **TRY IT : : 1.129** Divide. Then check by multiplying:

> **TRY IT : : 1.130** Divide. Then check by multiplying:

EXAMPLE 1.66

Divide and check by multiplying:

Solution

Let's rewrite the problem to set it up for long division.

First we try to divide 241 into 7. Since that won't work, we try 241 into 74. That still won't work, so we try 241 into 745. Since 2 divides into 7 three times, we try 3. Since , we write the 3 over the 5 in 745.
Note that 4 would be too large because , which is greater than 745.

Multiply the 3 by 241 and subtract this product from 745.

$$\begin{array}{r} 3 \\ 241\overline{)74521} \\ \underline{723} \\ 22 \end{array}$$

Now bring down the 2 and repeat these steps. 241 does not divide into 222. We write a 0 over the 2 as a placeholder and then continue.

$$\begin{array}{r} 30 \\ 241\overline{)74521} \\ \underline{723} \\ 222 \end{array}$$

Now bring down the 1 and repeat these steps. Try 9. Since , we write the 9 over the 1. Multiply the 9 by 241 and subtract this product from 2,221.

$$\begin{array}{r} 309\text{ R}52 \\ 241\overline{)74521}\phantom{\text{ R}52} \\ \underline{723}\phantom{\text{ R}52} \\ 2221\phantom{\text{ R}52} \\ \underline{2169}\phantom{\text{ R}52} \\ 52\phantom{\text{ R}52} \end{array}$$

There are no more numbers to bring down, so we are finished. The remainder is 52. So
is 309 with a remainder of 52.

Check by multiplying.

$$\begin{array}{rl} \overset{3}{3}09 & \text{quotient} \\ \times\ 241 & \text{divisor} \\ \hline 309 & \\ 12{,}360 & \\ 61{,}800 & \\ \hline 72{,}469 & \\ +\ \ 52 & \text{remainder} \\ \hline 74{,}521 & \checkmark \end{array}$$

Sometimes it might not be obvious how many times the divisor goes into digits of the dividend. We will have to guess and check numbers to find the greatest number that goes into the digits without exceeding them.

TRY IT : : 1.131 Divide. Then check by multiplying:

TRY IT : : 1.132 Divide. Then check by multiplying:

Translate Word Phrases to Math Notation

Earlier in this section, we translated math notation for division into words. Now we'll translate word phrases into math notation. Some of the words that indicate division are given in Table 1.8.

Operation	Word Phrase	Example	Expression
Division	divided by quotient of divided into	divided by the quotient of and divided into	— ⟌

Table 1.8

EXAMPLE 1.67

Translate and simplify: the quotient of and

Solution

The word *quotient* tells us to divide.

We could just as correctly have translated *the quotient of* *and* using the notation

⟌ —

> **TRY IT : : 1.133** Translate and simplify: the quotient of and

> **TRY IT : : 1.134** Translate and simplify: the quotient of and

Divide Whole Numbers in Applications

We will use the same strategy we used in previous sections to solve applications. First, we determine what we are looking for. Then we write a phrase that gives the information to find it. We then translate the phrase into math notation and simplify it to get the answer. Finally, we write a sentence to answer the question.

EXAMPLE 1.68

Cecelia bought a box of oatmeal at the big box store. She wants to divide the ounces of oatmeal into servings. She will put each serving into a plastic bag so she can take one bag to work each day. How many servings will she get from the big box?

Solution

We are asked to find the how many servings she will get from the big box.

Write a phrase.	160 ounces divided by 8 ounces
Translate to math notation.	
Simplify by dividing.	
Write a sentence to answer the question.	Cecelia will get 20 servings from the big box.

> **TRY IT : : 1.135**
>
> Marcus is setting out animal crackers for snacks at the preschool. He wants to put crackers in each cup. One box of animal crackers contains crackers. How many cups can he fill from one box of crackers?

TRY IT : : 1.136

Andrea is making bows for the girls in her dance class to wear at the recital. Each bow takes feet of ribbon, and feet of ribbon are on one spool. How many bows can Andrea make from one spool of ribbon?

MEDIA : : ACCESS ADDITIONAL ONLINE RESOURCES

- Dividing Whole Numbers (http://www.openstax.org/l/24divwhlnum)
- Dividing Whole Numbers No Remainder (http://www.openstax.org/l/24divnumnorem)
- Dividing Whole Numbers With Remainder (http://www.openstax.org/l/24divnumwrem)

1.5 EXERCISES

Practice Makes Perfect

Use Division Notation

In the following exercises, translate from math notation to words.

343.

344. —

345. —

346. $\overline{)\quad}$

347.

348. —

349. $\overline{)\quad}$

350.

Model Division of Whole Numbers

In the following exercises, model the division.

351.

352.

353. —

354. —

355. $\overline{)\quad}$

356. $\overline{)\quad}$

357.

358.

Divide Whole Numbers

In the following exercises, divide. Then check by multiplying.

359.

360.

361. —

362. —

363. $\overline{)\quad}$

364. $\overline{)\quad}$

365. —

366. —

367.

368. $\overline{)\quad}$

369. —

370.

371. $\overline{)\quad}$

372. $\overline{)\quad}$

373.

374.

375. —

376. —

377.

378.

379.

380.

381. —

382. —

383. —

384. —

385. $\overline{)\quad}$

386. $\overline{)\quad}$

387.

388.

389.

390.

391.

392.

393.

394.

395.

396.

397.

398.

399.

400.

401.

402.

403.

404.

405.

406.

407.

408.

409.

410.

411.

412.

413.

414.

415.

416.

417.

418.

419.

420.

421.

422.

Mixed Practice

In the following exercises, simplify.

423.

424.

425.

426.

427.

428.

429.

430.

Translate Word Phrases to Algebraic Expressions

In the following exercises, translate and simplify.

431. the quotient of and

432. the quotient of and

433. the quotient of and

434. the quotient of and

Divide Whole Numbers in Applications

In the following exercises, solve.

435. Trail mix Ric bought ounces of trail mix. He wants to divide it into small bags, with ounces of trail mix in each bag. How many bags can Ric fill?

436. Crackers Evie bought a ounce box of crackers. She wants to divide it into bags with ounces of crackers in each bag. How many bags can Evie fill?

437. Astronomy class There are students in an astronomy class. The professor assigns them into groups of How many groups of students are there?

438. Flower shop Melissa's flower shop got a shipment of roses. She wants to make bouquets of roses each. How many bouquets can Melissa make?

439. Baking One roll of plastic wrap is feet long. Marta uses feet of plastic wrap to wrap each cake she bakes. How many cakes can she wrap from one roll?

440. Dental floss One package of dental floss is feet long. Brian uses feet of dental floss every day. How many days will one package of dental floss last Brian?

Mixed Practice

In the following exercises, solve.

441. Miles per gallon Susana's hybrid car gets miles per gallon. Her son's truck gets miles per gallon. What is the difference in miles per gallon between Susana's car and her son's truck?

442. Distance Mayra lives miles from her mother's house and miles from her mother-in-law's house. How much farther is Mayra from her mother-in-law's house than from her mother's house?

443. Field trip The students in a Geology class will go on a field trip, using the college's vans. Each van can hold students. How many vans will they need for the field trip?

444. Potting soil Aki bought a ounce bag of potting soil. How many ounce pots can he fill from the bag?

445. Hiking Bill hiked miles on the first day of his backpacking trip, miles the second day, miles the third day, and miles the fourth day. What is the total number of miles Bill hiked?

446. Reading Last night Emily read pages in her Business textbook, pages in her History text, pages in her Psychology text, and pages in her math text. What is the total number of pages Emily read?

447. Patients LaVonne treats patients each day in her dental office. Last week she worked days. How many patients did she treat last week?

448. Scouts There are boys in Dave's scout troop. At summer camp, each boy earned merit badges. What was the total number of merit badges earned by Dave's scout troop at summer camp?

Writing Exercises

449. Contact lenses Jenna puts in a new pair of contact lenses every days. How many pairs of contact lenses does she need for days?

450. Cat food One bag of cat food feeds Lara's cat for days. How many bags of cat food does Lara need for days?

Everyday Math

451. Explain how you use the multiplication facts to help with division.

452. Oswaldo divided by and said his answer was with a remainder of How can you check to make sure he is correct?

Self Check

After completing the exercises, use this checklist to evaluate your mastery of the objectives of this section.

I can...	Confidently	With some help	No-I don't get it!
use division notation.			
model division of whole numbers.			
divide whole numbers.			
translate word phrases to algebraic expressions.			
divide whole numbers in applications.			

Overall, after looking at the checklist, do you think you are well-prepared for the next Chapter? Why or why not?

CHAPTER 1 REVIEW

KEY TERMS

coordinate A number paired with a point on a number line is called the coordinate of the point.

counting numbers The counting numbers are the numbers 1, 2, 3, ….

difference The difference is the result of subtracting two or more numbers.

dividend When dividing two numbers, the dividend is the number being divided.

divisor When dividing two numbers, the divisor is the number dividing the dividend.

number line A number line is used to visualize numbers. The numbers on the number line get larger as they go from left to right, and smaller as they go from right to left.

origin The origin is the point labeled 0 on a number line.

place value system Our number system is called a place value system because the value of a digit depends on its position, or place, in a number.

product The product is the result of multiplying two or more numbers.

quotient The quotient is the result of dividing two numbers.

rounding The process of approximating a number is called rounding.

sum The sum is the result of adding two or more numbers.

whole numbers The whole numbers are the numbers 0, 1, 2, 3, ….

KEY CONCEPTS

1.1 Introduction to Whole Numbers

Place Value														
Trillions			Billions			Millions			Thousands			Ones		
Hundred trillions	Ten trillions	Trillions	Hundred billions	Ten billions	Billions	Hundred millions	Ten millions	Millions	Hundred thousands	Ten thousands	Thousands	Hundreds	Tens	Ones
								5	2	7	8	1	9	4

Figure 1.16

- **Name a whole number in words.**

  Step 1. Starting at the digit on the left, name the number in each period, followed by the period name. Do not include the period name for the ones.

  Step 2. Use commas in the number to separate the periods.

- **Use place value to write a whole number.**

  Step 1. Identify the words that indicate periods. (Remember the ones period is never named.)

  Step 2. Draw three blanks to indicate the number of places needed in each period.

  Step 3. Name the number in each period and place the digits in the correct place value position.

- **Round a whole number to a specific place value.**

  Step 1. Locate the given place value. All digits to the left of that place value do not change.

  Step 2. Underline the digit to the right of the given place value.

  Step 3. Determine if this digit is greater than or equal to 5. If yes—add 1 to the digit in the given place value. If no—do not change the digit in the given place value.

Step 4. Replace all digits to the right of the given place value with zeros.

1.2 Add Whole Numbers

- **Addition Notation** To describe addition, we can use symbols and words.

Operation	Notation	Expression	Read as	Result
Addition			three plus four	the sum of and

- **Identity Property of Addition**
  - The sum of any number and is the number.
- **Commutative Property of Addition**
  - Changing the order of the addends and does not change their sum. .
- **Add whole numbers.**

Step 1. Write the numbers so each place value lines up vertically.

Step 2. Add the digits in each place value. Work from right to left starting with the ones place. If a sum in a place value is more than 9, carry to the next place value.

Step 3. Continue adding each place value from right to left, adding each place value and carrying if needed.

1.3 Subtract Whole Numbers

Operation	Notation	Expression	Read as	Result
Subtraction			seven minus three	the difference of and

- **Subtract whole numbers.**

Step 1. Write the numbers so each place value lines up vertically.

Step 2. Subtract the digits in each place value. Work from right to left starting with the ones place. If the digit on top is less than the digit below, borrow as needed.

Step 3. Continue subtracting each place value from right to left, borrowing if needed.

Step 4. Check by adding.

1.4 Multiply Whole Numbers

Operation	Notation	Expression	Read as	Result

- **Multiplication Property of Zero**
  - The product of any number and 0 is 0.
- **Identity Property of Multiplication**
  - The product of any number and 1 is the number.
- **Commutative Property of Multiplication**
  - Changing the order of the factors does not change their product.

- **Multiply two whole numbers to find the product.**

  Step 1. Write the numbers so each place value lines up vertically.

  Step 2. Multiply the digits in each place value.

  Step 3. Work from right to left, starting with the ones place in the bottom number.

  Step 4. Multiply the bottom number by the ones digit in the top number, then by the tens digit, and so on.

  Step 5. If a product in a place value is more than 9, carry to the next place value.

  Step 6. Write the partial products, lining up the digits in the place values with the numbers above. Repeat for the tens place in the bottom number, the hundreds place, and so on.

  Step 7. Insert a zero as a placeholder with each additional partial product.

  Step 8. Add the partial products.

1.5 Divide Whole Numbers

Operation	Notation	Expression	Read as	Result
	— ⟌	— ⟌		

- **Division Properties of One**
  - Any number (except 0) divided by itself is one.
  - Any number divided by one is the same number.
- **Division Properties of Zero**
  - Zero divided by any number is 0.
  - Dividing a number by zero is undefined. undefined
- **Divide whole numbers.**

  Step 1. Divide the first digit of the dividend by the divisor.
  If the divisor is larger than the first digit of the dividend, divide the first two digits of the dividend by the divisor, and so on.

  Step 2. Write the quotient above the dividend.

  Step 3. Multiply the quotient by the divisor and write the product under the dividend.

  Step 4. Subtract that product from the dividend.

  Step 5. Bring down the next digit of the dividend.

  Step 6. Repeat from Step 1 until there are no more digits in the dividend to bring down.

  Step 7. Check by multiplying the quotient times the divisor.

REVIEW EXERCISES

1.1 Introduction to Whole Numbers

Identify Counting Numbers and Whole Numbers

In the following exercises, determine which of the following are (a) counting numbers (b) whole numbers.

453.

454.

455.

456.

Model Whole Numbers

In the following exercises, model each number using blocks and then show its value using place value notation.

457. 258

458. 104

Identify the Place Value of a Digit

In the following exercises, find the place value of the given digits.

459.

460.

Use Place Value to Name Whole Numbers

In the following exercises, name each number in words.

461.

462.

463.

464.

Use Place Value to Write Whole Numbers

In the following exercises, write as a whole number using digits.

465. six hundred two

466. fifteen thousand, two hundred fifty-three

467. three hundred forty million, nine hundred twelve thousand, sixty-one

468. two billion, four hundred ninety-two million, seven hundred eleven thousand, two

Round Whole Numbers

In the following exercises, round to the nearest ten.

469.

470.

471.

472.

In the following exercises, round to the nearest hundred.

473.

474.

475.

476.

1.2 Add Whole Numbers

Use Addition Notation

In the following exercises, translate the following from math notation to words.

477.

478.

479.

480.

Model Addition of Whole Numbers

In the following exercises, model the addition.

481.

482.

Add Whole Numbers

In the following exercises, fill in the missing values in each chart.

483.

+	0	1	2	3	4	5	6	7	8	9
0	0	1		3	4		6	7		9
1	1	2	3	4			7	8	9	
2		3	4	5	6	7	8		10	11
3	3		5		7	8		10		12
4	4	5			8	9			12	
5	5		7	8			11		13	
6	6	7	8		10			13		15
7			9			12	13		15	16
8	8	9		11			14		16	
9	9	10	11		13	14			17	

484.

+	3	4	5	6	7	8	9
6							
7							
8							
9							

In the following exercises, add.

485.

486.

487.

488.

489.

490.

491.

492.

493.

494.

Translate Word Phrases to Math Notation

In the following exercises, translate each phrase into math notation and then simplify.

495. the sum of and

496. increased by

497. more than

498. total of and

Add Whole Numbers in Applications

In the following exercises, solve.

499. Shopping for an interview Nathan bought a new shirt, tie, and slacks to wear to a job interview. The shirt cost the tie cost and the slacks cost What was Nathan's total cost?

500. Running Jackson ran miles on Monday, miles on Tuesday, mile on Wednesday, miles on Thursday, and miles on Friday. What was the total number of miles Jackson ran?

In the following exercises, find the perimeter of each figure.

501.

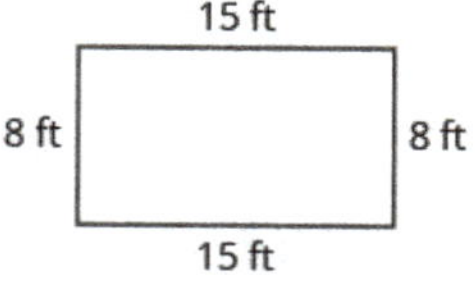

502.

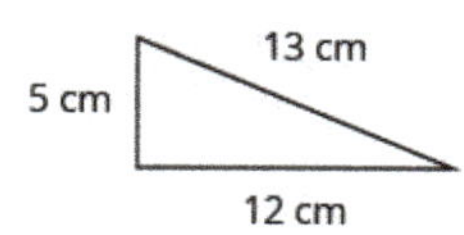

1.3 Subtract Whole Numbers

Use Subtraction Notation

In the following exercises, translate the following from math notation to words.

503.

504.

505.

506.

Model Subtraction of Whole Numbers

In the following exercises, model the subtraction.

507.

508.

Subtract Whole Numbers

In the following exercises, subtract and then check by adding.

509.

510.

511.

512.

513.

514.

515.

516.

517.

518.

519.

520.

Translate Word Phrases to Math Notation

In the following exercises, translate and simplify.

521. the difference of nineteen and thirteen

522. subtract sixty-five from one hundred

523. seventy-four decreased by eight

524. twenty-three less than forty-one

Subtract Whole Numbers in Applications

In the following exercises, solve.

525. Temperature The high temperature in Peoria one day was degrees Fahrenheit and the low temperature was degrees Fahrenheit. What was the difference between the high and low temperatures?

526. Savings Lynn wants to go on a cruise that costs She has in her vacation savings account. How much more does she need to save in order to pay for the cruise?

1.4 Multiply Whole Numbers

Use Multiplication Notation

In the following exercises, translate from math notation to words.

527.

528.

529.

530.

Model Multiplication of Whole Numbers

In the following exercises, model the multiplication.

531.

532.

Multiply Whole Numbers

In the following exercises, fill in the missing values in each chart.

533.

×	0	1	2	3	4	5	6	7	8	9
0	0	0	0	0	0		0		0	0
1	0	1	2		4	5	6	7		9
2	0		4		8	10		14	16	
3		3		9			18		24	
4	0	4		12			24			36
5	0	5	10		20		30	35	40	45
6			12	18			36	42		54
7	0	7		21		35			56	63
8	0	8	16		32		48		64	
9			18	27	36			63	72	

534.

×	3	4	5	6	7	8	9
6							
7							
8							
9							

In the following exercises, multiply.

535.

536.

537.

538.

539.

540.

541.

542.

543.

544.

545.

546.

547.

548.

Translate Word Phrases to Math Notation

In the following exercises, translate and simplify.

549. the product of and

550. ninety-four times thirty-three

551. twice

552. ten times two hundred sixty-four

Multiply Whole Numbers in Applications

In the following exercises, solve.

553. Gardening Geniece bought packs of marigolds to plant in her yard. Each pack has flowers. How many marigolds did Geniece buy?

554. Cooking Ratika is making rice for a dinner party. The number of cups of water is twice the number of cups of rice. If Ratika plans to use cups of rice, how many cups of water does she need?

555. Multiplex There are twelve theaters at the multiplex and each theater has seats. What is the total number of seats at the multiplex?

556. Roofing Lewis needs to put new shingles on his roof. The roof is a rectangle, feet by feet. What is the area of the roof?

1.5 Divide Whole Numbers

Use Division Notation

Translate from math notation to words.

557.

558.

559. —

560. $\overline{)}$

Model Division of Whole Numbers

In the following exercises, model.

561.

562. $\overline{)}$

Divide Whole Numbers

In the following exercises, divide. Then check by multiplying.

563.

564. —

565.

566. $\overline{)}$

567. —

568.

569.

570. —

571.

572. $\overline{)}$

573. —

574.

Translate Word Phrases to Math Notation

In the following exercises, translate and simplify.

575. the quotient of and

576. the quotient of and

Divide Whole Numbers in Applications

In the following exercises, solve.

577. **Ribbon** One spool of ribbon is feet. Lizbeth uses feet of ribbon for each gift basket that she wraps. How many gift baskets can Lizbeth wrap from one spool of ribbon?

578. **Juice** One carton of fruit juice is ounces. How many ounce cups can Shayla fill from one carton of juice?

PRACTICE TEST

579. Determine which of the following numbers are

counting numbers

whole numbers.

580. Find the place value of the given digits in the number

581. Write each number as a whole number using digits.

six hundred thirteen

fifty-five thousand two hundred eight

582. Round to the nearest hundred.

Simplify.

583.

584.

585.

586.

587.

588.

589.

590.

591. —

592.

593.

594.

595.

596.

597.

598. —

599.

600.

601.

602.

Translate each phrase to math notation and then simplify.

603. The sum of and

604. The product of and

605. The difference of and

606. The quotient of and

607. Twice

608. more than

609. less than

In the following exercises, solve.

610. LaVelle buys a jumbo bag of candies to make favor bags for her son's party. If she wants to make bags, how many candies should she put in each bag?

611. Last month, Stan's take-home pay was and his expenses were How much of his take-home pay did Stan have left after he paid his expenses?

612. Each class at Greenville School has children enrolled. The school has classes. How many children are enrolled at Greenville School?

613. Clayton walked blocks to his mother's house, blocks to the gym, and blocks to the grocery store before walking the last blocks home. What was the total number of blocks that Clayton walked?

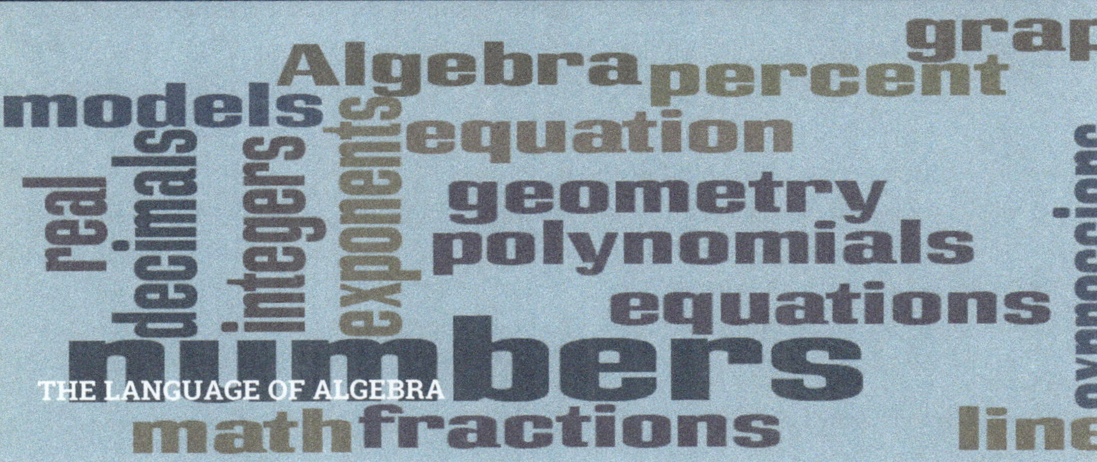

Figure 2.1 Algebra has a language of its own. The picture shows just some of the words you may see and use in your study of Prealgebra.

Chapter Outline

Introduction

You may not realize it, but you already use algebra every day. Perhaps you figure out how much to tip a server in a restaurant. Maybe you calculate the amount of change you should get when you pay for something. It could even be when you compare batting averages of your favorite players. You can describe the algebra you use in specific words, and follow an orderly process. In this chapter, you will explore the words used to describe algebra and start on your path to solving algebraic problems easily, both in class and in your everyday life.

2.1 Use the Language of Algebra

Learning Objectives

By the end of this section, you will be able to:

- Use variables and algebraic symbols
- Identify expressions and equations
- Simplify expressions with exponents
- Simplify expressions using the order of operations

BE PREPARED : : 2.1 Before you get started, take this readiness quiz.

If you missed this problem, review Example 1.19.

BE PREPARED : : 2.2

If you missed this problem, review Example 1.48.

BE PREPARED : : 2.3

If you missed this problem, review Example 1.64.

Use Variables and Algebraic Symbols

Greg and Alex have the same birthday, but they were born in different years. This year Greg is years old and Alex is so Alex is years older than Greg. When Greg was Alex was When Greg is Alex will be No matter what Greg's age is, Alex's age will always be years more, right?

In the language of algebra, we say that Greg's age and Alex's age are variable and the three is a constant. The ages change, or vary, so age is a variable. The years between them always stays the same, so the age difference is the constant.

In algebra, letters of the alphabet are used to represent variables. Suppose we call Greg's age Then we could use to represent Alex's age. See Table 2.1.

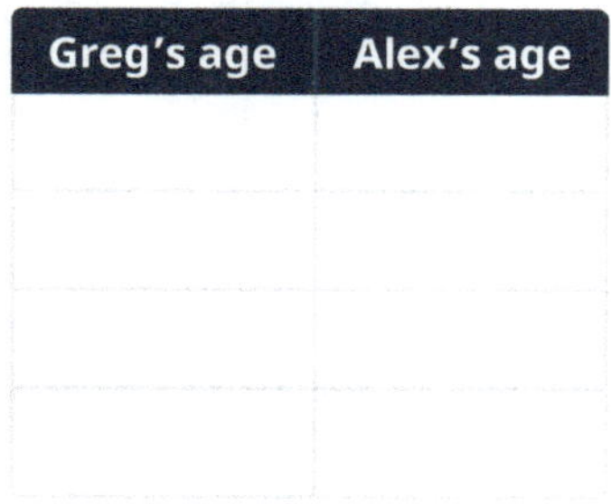

Greg's age	Alex's age

Table 2.1

Letters are used to represent variables. Letters often used for variables are

Variables and Constants

A variable is a letter that represents a number or quantity whose value may change.

A constant is a number whose value always stays the same.

To write algebraically, we need some symbols as well as numbers and variables. There are several types of symbols we will be using. In Whole Numbers, we introduced the symbols for the four basic arithmetic operations: addition, subtraction, multiplication, and division. We will summarize them here, along with words we use for the operations and the result.

Operation	Notation	Say:	The result is...
Addition			the sum of and
Subtraction			the difference of and
Multiplication			The product of and
Division	— Γ	divided by	The quotient of and

In algebra, the cross symbol, is not used to show multiplication because that symbol may cause confusion. Does mean (three times) or (three times)? To make it clear, use • or parentheses for multiplication.

We perform these operations on two numbers. When translating from symbolic form to words, or from words to symbolic form, pay attention to the words *of* or *and* to help you find the numbers.

The *sum* ***of*** ***and*** means add plus which we write as

The *difference* **of** **and** means subtract minus which we write as
The *product* **of** **and** means multiply times which we can write as
The *quotient* **of** **and** means divide by which we can write as

EXAMPLE 2.1

Translate from algebra to words:

Solution

12 plus 14
the sum of twelve and fourteen

30 times 5
the product of thirty and five

64 divided by 8
the quotient of sixty-four and eight

minus
the difference of and

TRY IT : : 2.1 Translate from algebra to words.

TRY IT : : 2.2 Translate from algebra to words.

When two quantities have the same value, we say they are equal and connect them with an *equal sign*.

Equality Symbol

The symbol is called the equal sign.

An inequality is used in algebra to compare two quantities that may have different values. The number line can help you understand inequalities. Remember that on the number line the numbers get larger as they go from left to right. So if we know that is greater than it means that is to the right of on the number line. We use the symbols and for inequalities.

Inequality

is read is less than

is to the left of on the number line

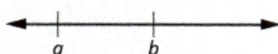

is read is greater than

is to the right of on the number line

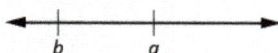

The expressions can be read from left-to-right or right-to-left, though in English we usually read from left-to-right. In general,

When we write an inequality symbol with a line under it, such as it means or We read this is less than or equal to Also, if we put a slash through an equal sign, it means not equal.

We summarize the symbols of equality and inequality in Table 2.2.

Algebraic Notation	Say
	is equal to
	is not equal to
	is less than
	is greater than
	is less than or equal to
	is greater than or equal to

Table 2.2

Symbols and

The symbols and each have a smaller side and a larger side.

smaller side larger side

larger side smaller side

The smaller side of the symbol faces the smaller number and the larger faces the larger number.

EXAMPLE 2.2

Translate from algebra to words:

Solution

20 is less than or equal to 35

11 is not equal to 15 minus 3

9 is greater than 10 divided by 2

plus 2 is less than 10

TRY IT : : 2.3 Translate from algebra to words.

TRY IT : : 2.4 Translate from algebra to words.

EXAMPLE 2.3

The information in Figure 2.2 compares the fuel economy in miles-per-gallon (mpg) of several cars. Write the appropriate symbol in each expression to compare the fuel economy of the cars.

Car	Prius	Mini Cooper	Toyota Corolla	Versa	Honda Fit
Fuel economy (mpg)	48	27	28	26	27

Figure 2.2 (credit: modification of work by Bernard Goldbach, Wikimedia Commons)

MPG of Prius____ MPG of Mini Cooper

MPG of Versa____ MPG of Fit

MPG of Mini Cooper____ MPG of Fit

MPG of Corolla____ MPG of Versa

MPG of Corolla____ MPG of Prius

Solution

	MPG of Prius___MPG of Mini Cooper
Find the values in the chart.	48___27
Compare.	48 > 27
	MPG of Prius > MPG of Mini Cooper

	MPG of Versa___MPG of Fit
Find the values in the chart.	26___27
Compare.	26 < 27
	MPG of Versa < MPG of Fit

	MPG of Mini Cooper___MPG of Fit
Find the values in the chart.	27___27
Compare.	27 = 27
	MPG of Mini Cooper = MPG of Fit

	MPG of Corolla___MPG of Versa
Find the values in the chart.	28___26
Compare.	28 > 26
	MPG of Corolla > MPG of Versa

	MPG of Corolla___MPG of Prius
Find the values in the chart.	28___48
Compare.	28 < 48
	MPG of Corolla < MPG of Prius

TRY IT : : 2.5 Use Figure 2.2 to fill in the appropriate

MPG of Prius____MPG of Versa MPG of Mini Cooper____ MPG of Corolla

TRY IT : : 2.6 Use Figure 2.2 to fill in the appropriate

MPG of Fit____ MPG of Prius MPG of Corolla ____ MPG of Fit

Grouping symbols in algebra are much like the commas, colons, and other punctuation marks in written language. They indicate which expressions are to be kept together and separate from other expressions. Table 2.3 lists three of the most commonly used grouping symbols in algebra.

Common Grouping Symbols	
parentheses	
brackets	
braces	

Table 2.3

Here are some examples of expressions that include grouping symbols. We will simplify expressions like these later in this section.

Identify Expressions and Equations

What is the difference in English between a phrase and a sentence? A phrase expresses a single thought that is incomplete by itself, but a sentence makes a complete statement. "Running very fast" is a phrase, but "The football player was running very fast" is a sentence. A sentence has a subject and a verb.

In algebra, we have *expressions* and *equations*. An expression is like a phrase. Here are some examples of expressions and how they relate to word phrases:

Expression	Words	Phrase
		the sum of three and five
	minus one	the difference of and one
		the product of six and seven
—	divided by	the quotient of and

Notice that the phrases do not form a complete sentence because the phrase does not have a verb. An equation is two expressions linked with an equal sign. When you read the words the symbols represent in an equation, you have a complete sentence in English. The equal sign gives the verb. Here are some examples of equations:

Equation	Sentence
	The sum of three and five is equal to eight.
	minus one equals fourteen.
	The product of six and seven is equal to forty-two.
	is equal to fifty-three.
	plus nine is equal to two minus three.

Expressions and Equations

An **expression** is a number, a variable, or a combination of numbers and variables and operation symbols.

An **equation** is made up of two expressions connected by an equal sign.

EXAMPLE 2.4

Determine if each is an expression or an equation:

Solution

This is an equation—two expressions are connected with an equal sign.

This is an expression—no equal sign.

This is an expression—no equal sign.

This is an equation—two expressions are connected with an equal sign.

TRY IT : : 2.7 Determine if each is an expression or an equation:

TRY IT : : 2.8 Determine if each is an expression or an equation:

Simplify Expressions with Exponents

To simplify a numerical expression means to do all the math possible. For example, to simplify we'd first multiply to get and then add the to get A good habit to develop is to work down the page, writing each step of the process below the previous step. The example just described would look like this:

Suppose we have the expression We could write this more compactly using exponential notation. Exponential notation is used in algebra to represent a quantity multiplied by itself several times. We write as and as In expressions such as the is called the **base** and the is called the exponent. The exponent tells us how many factors of the base we have to multiply.

base → 2^3 ← exponent

We say is in exponential notation and is in expanded notation.

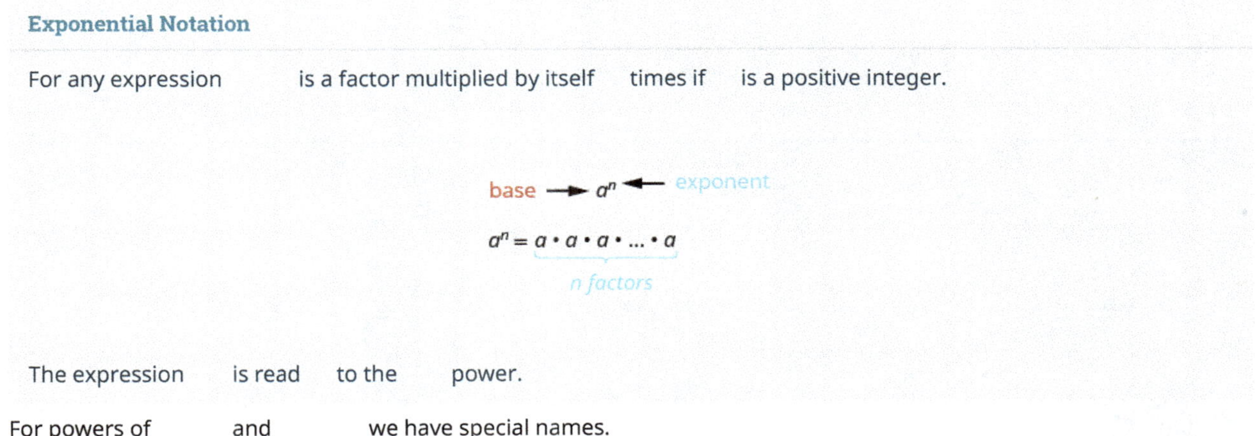

Exponential Notation

For any expression is a factor multiplied by itself times if is a positive integer.

The expression is read to the power.

For powers of and we have special names.

Table 2.4 lists some examples of expressions written in exponential notation.

Exponential Notation	In Words
	to the second power, or squared
	to the third power, or cubed
	to the fourth power
	to the fifth power

Table 2.4

EXAMPLE 2.5

Write each expression in exponential form:

Solution

The base 16 is a factor 7 times.

The base 9 is a factor 5 times.

The base is a factor 4 times.

The base is a factor 8 times.

TRY IT : : 2.9 Write each expression in exponential form:

TRY IT : : 2.10 Write each expression in exponential form:

EXAMPLE 2.6

Write each exponential expression in expanded form:

Solution

The base is and the exponent is so means

The base is and the exponent is so means

TRY IT : : 2.11 Write each exponential expression in expanded form:

TRY IT : : 2.12 Write each exponential expression in expanded form:

To simplify an exponential expression without using a calculator, we write it in expanded form and then multiply the factors.

EXAMPLE 2.7

Simplify:

Solution

Expand the expression.
Multiply left to right.
Multiply.

TRY IT : : 2.13 Simplify:

TRY IT : : 2.14 Simplify:

Simplify Expressions Using the Order of Operations

We've introduced most of the symbols and notation used in algebra, but now we need to clarify the order of operations. Otherwise, expressions may have different meanings, and they may result in different values.

For example, consider the expression:

Imagine the confusion that could result if every problem had several different correct answers. The same expression should give the same result. So mathematicians established some guidelines called the order of operations, which outlines the order in which parts of an expression must be simplified.

Order of Operations

When simplifying mathematical expressions perform the operations in the following order:

1. **P**arentheses and other Grouping Symbols
   - Simplify all expressions inside the parentheses or other grouping symbols, working on the innermost parentheses first.
2. **E**xponents
   - Simplify all expressions with exponents.
3. **M**ultiplication and **D**ivision
   - Perform all multiplication and division in order from left to right. These operations have equal priority.
4. **A**ddition and **S**ubtraction
   - Perform all addition and subtraction in order from left to right. These operations have equal priority.

Students often ask, "How will I remember the order?" Here is a way to help you remember: Take the first letter of each key word and substitute the silly phrase. **P**lease **E**xcuse **M**y **D**ear **A**unt **S**ally.

Order of Operations	
Please	**P**arentheses
Excuse	**E**xponents
My **D**ear	**M**ultiplication and **D**ivision
Aunt **S**ally	**A**ddition and **S**ubtraction

It's good that '**M**y **D**ear' goes together, as this reminds us that **m**ultiplication and **d**ivision have equal priority. We do not always do multiplication before division or always do division before multiplication. We do them in order from left to right.

Similarly, '**A**unt **S**ally' goes together and so reminds us that **a**ddition and **s**ubtraction also have equal priority and we do them in order from left to right.

MANIPULATIVE MATHEMATICS

Doing the Manipulative Mathematics activity Game of 24 will give you practice using the order of operations.

EXAMPLE 2.8

Simplify the expressions:

Solution

	$4 + 3 \cdot 7$
Are there any **p**arentheses? No.	
Are there any **e**xponents? No.	
Is there any **m**ultiplication or **d**ivision? Yes.	
Multiply first.	$4 + 3 \cdot 7$
Add.	$4 + 21$
	25

	$(4 + 3) \cdot 7$
Are there any **p**arentheses? Yes.	$(4 + 3) \cdot 7$
Simplify inside the parentheses.	$(7)7$
Are there any **e**xponents? No.	
Is there any **m**ultiplication or **d**ivision? Yes.	
Multiply.	49

TRY IT : : 2.15 Simplify the expressions:

TRY IT : : 2.16 Simplify the expressions:

EXAMPLE 2.9

Simplify:

Solution

	$18 \div 9 \cdot 2$
Are there any **p**arentheses? No.	
Are there any **e**xponents? No.	
Is there any **m**ultiplication or **d**ivision? Yes.	
Multiply and divide from left to right. Divide.	$2 \cdot 2$
Multiply.	4

	$18 \cdot 9 \div 2$
Are there any **p**arentheses? No.	
Are there any **e**xponents? No.	
Is there any **m**ultiplication or **d**ivision? Yes.	
Multiply and divide from left to right.	
Multiply.	$162 \div 2$
Divide.	81

TRY IT : : 2.17 Simplify:

TRY IT : : 2.18 Simplify:

EXAMPLE 2.10

Simplify:

Solution

	$18 \div 6 + 4(5 - 2)$
Parentheses? Yes, subtract first.	$18 \div 6 + 4(3)$
Exponents? No.	
Multiplication or division? Yes.	
Divide first because we multiply and divide left to right.	$3 + 4(3)$
Any other multiplication or division? Yes.	
Multiply.	$3 + 12$
Any other multiplication or division? No.	
Any addition or subtraction? Yes.	15

TRY IT : : 2.19 Simplify:

TRY IT : : 2.20 Simplify:

When there are multiple grouping symbols, we simplify the innermost parentheses first and work outward.

EXAMPLE 2.11

Solution

	$5 + 2^3 + 3[6 - 3(4 - 2)]$
Are there any parentheses (or other grouping symbol)? Yes.	
Focus on the parentheses that are inside the brackets.	$5 + 2^3 + 3[6 - 3(4 - 2)]$
Subtract.	$5 + 2^3 + 3[6 - 3(2)]$
Continue inside the brackets and multiply.	$5 + 2^3 + 3[6 - 6]$
Continue inside the brackets and subtract.	$5 + 2^3 + 3[0]$
The expression inside the brackets requires no further simplification.	
Are there any exponents? Yes.	
Simplify exponents.	$5 + 2^3 + 3[0]$
Is there any multiplication or division? Yes.	
Multiply.	$5 + 8 + 3[0]$
Is there any addition or subtraction? Yes.	
Add.	$5 + 8 + 0$
Add.	$13 + 0$
	13

TRY IT : : 2.21 Simplify:

TRY IT : : 2.22 Simplify:

EXAMPLE 2.12

Simplify:

Solution

	$2^3 + 3^4 \div 3 - 5^2$
If an expression has several exponents, they may be simplified in the same step.	
Simplify exponents.	$2^3 + 3^4 \div 3 - 5^2$
Divide.	$8 + 81 \div 3 - 25$
Add.	$8 + 27 - 25$
Subtract.	$35 - 25$
	10

TRY IT : : 2.23 Simplify:

TRY IT : : 2.24 Simplify:

MEDIA : : ACCESS ADDITIONAL ONLINE RESOURCES

- Order of Operations (http://openstaxcollege.org/l/24orderoperate)
- Order of Operations – The Basics (http://openstaxcollege.org/l/24orderbasic)
- Ex: Evaluate an Expression Using the Order of Operations (http://openstaxcollege.org/l/24Evalexpress)
- Example 3: Evaluate an Expression Using The Order of Operations (http://openstaxcollege.org/l/24evalexpress3)

2.1 EXERCISES

Practice Makes Perfect

Use Variables and Algebraic Symbols

In the following exercises, translate from algebraic notation to words.

1. **2.** **3.**

4. **5.** **6.**

7. **8.** **9.**

10. **11.** **12.**

13. **14.** **15.**

16. **17.** **18.**

19. **20.** **21.**

22.

Identify Expressions and Equations

In the following exercises, determine if each is an expression or an equation.

23. **24.** **25.**

26. **27.** **28.**

29. **30.**

Simplify Expressions with Exponents

In the following exercises, write in exponential form.

31. **32.** **33.**

34.

In the following exercises, write in expanded form.

35. **36.** **37.**

38.

Simplify Expressions Using the Order of Operations

In the following exercises, simplify.

39. **40.** **41.**

42. **43.** **44.**

45. **46.** **47.**

48. **49.** **50.**

51. **52.** **53.**

54. **55.** **56.**

57. **58.** **59.**

60. **61.** **62.**

63. **64.**

Everyday Math

65. Basketball In the 2014 NBA playoffs, the San Antonio Spurs beat the Miami Heat. The table below shows the heights of the starters on each team. Use this table to fill in the appropriate symbol

Spurs	Height	Heat	Height
Tim Duncan		Rashard Lewis	
Boris Diaw		LeBron James	
Kawhi Leonard		Chris Bosh	
Tony Parker		Dwyane Wade	
Danny Green		Ray Allen	

Height of Tim Duncan___Height of Rashard Lewis

Height of Boris Diaw___Height of LeBron James

Height of Kawhi Leonard___Height of Chris Bosh

Height of Tony Parker___Height of Dwyane Wade

Height of Danny Green___Height of Ray Allen

66. Elevation In Colorado there are more than mountains with an elevation of over The table shows the ten tallest. Use this table to fill in the appropriate inequality symbol.

Mountain	Elevation
Mt. Elbert	
Mt. Massive	
Mt. Harvard	
Blanca Peak	
La Plata Peak	
Uncompahgre Peak	
Crestone Peak	
Mt. Lincoln	
Grays Peak	
Mt. Antero	

Elevation of La Plata Peak___Elevation of Mt. Antero

Elevation of Blanca Peak___Elevation of Mt. Elbert

Elevation of Gray's Peak___Elevation of Mt. Lincoln

Elevation of Mt. Massive___Elevation of Crestone Peak

Elevation of Mt. Harvard___Elevation of Uncompahgre Peak

Writing Exercises

67. Explain the difference between an expression and an equation.

68. Why is it important to use the order of operations to simplify an expression?

Self Check

After completing the exercises, use this checklist to evaluate your mastery of the objectives of this section.

I can...	Confidently	With some help	No-I don't get it!
use variables and algebraic symbols.			
identify expressions and equations.			
simplify expressions with exponents.			
simplify expressions using the order of operations.			

If most of your checks were:

...confidently. Congratulations! You have achieved the objectives in this section. Reflect on the study skills you used so that you can continue to use them. What did you do to become confident of your ability to do these things? Be specific.

...with some help. This must be addressed quickly because topics you do not master become potholes in your road to success. In math, every topic builds upon previous work. It is important to make sure you have a strong foundation before you move on. Whom can you ask for help? Your fellow classmates and instructor are good resources. Is there a place on campus where math tutors are available? Can your study skills be improved?

...no—I don't get it! This is a warning sign and you must not ignore it. You should get help right away or you will quickly be overwhelmed. See your instructor as soon as you can to discuss your situation. Together you can come up with a plan to get you the help you need.

2.2 Evaluate, Simplify, and Translate Expressions

Learning Objectives

By the end of this section, you will be able to:

- Evaluate algebraic expressions
- Identify terms, coefficients, and like terms
- Simplify expressions by combining like terms
- Translate word phrases to algebraic expressions

BE PREPARED : : 2.4

Before you get started, take this readiness quiz.

Is an expression or an equation?

If you missed this problem, review Example 2.4.

BE PREPARED : : 2.5

Simplify

If you missed this problem, review Example 2.7.

BE PREPARED : : 2.6

Simplify

If you missed this problem, review Example 2.8.

Evaluate Algebraic Expressions

In the last section, we simplified expressions using the order of operations. In this section, we'll evaluate expressions—again following the order of operations.

To **evaluate** an algebraic expression means to find the value of the expression when the variable is replaced by a given number. To evaluate an expression, we substitute the given number for the variable in the expression and then simplify the expression using the order of operations.

EXAMPLE 2.13

Evaluate when

Solution

To evaluate, substitute for in the expression, and then simplify.

	$x + 7$
Substitute.	$3 + 7$
Add.	10

When the expression has a value of .

To evaluate, substitute for in the expression, and then simplify.

	$x+7$
Substitute.	$12+7$
Add.	19

When the expression has a value of

Notice that we got different results for parts and even though we started with the same expression. This is because the values used for were different. When we evaluate an expression, the value varies depending on the value used for the variable.

TRY IT : : 2.25 Evaluate:

TRY IT : : 2.26 Evaluate:

EXAMPLE 2.14

Evaluate

Solution

Remember means times so means times

To evaluate the expression when we substitute for and then simplify.

	$9x-2$
Substitute 5 for x.	$9\cdot5-2$
Multiply.	$45-2$
Subtract.	43

To evaluate the expression when we substitute for and then simplify.

	$9x-2$
Substitute 1 for x.	$9(1)-2$
Multiply.	$9-2$
Subtract.	7

Notice that in part that we wrote and in part we wrote Both the dot and the parentheses tell us to multiply.

TRY IT : : 2.27 Evaluate:

TRY IT : : 2.28 Evaluate:

EXAMPLE 2.15

Evaluate when

Solution

We substitute for and then simplify the expression.

	x^2
Substitute 10 for x.	10^2
Use the definition of exponent.	$10 \cdot 10$
Multiply.	100

When the expression has a value of

TRY IT : : 2.29 Evaluate:

TRY IT : : 2.30 Evaluate:

EXAMPLE 2.16

Solution

In this expression, the variable is an exponent.

	2^x
Substitute 5 for x.	2^5
Use the definition of exponent.	$2 \cdot 2 \cdot 2 \cdot 2 \cdot 2$
Multiply.	32

When the expression has a value of

TRY IT : : 2.31 Evaluate:

TRY IT : : 2.32 Evaluate:

EXAMPLE 2.17

Solution

This expression contains two variables, so we must make two substitutions.

	$3x + 4y - 6$
Substitute 10 for *x* and 2 for *y*.	$3(10) + 4(2) - 6$
Multiply.	$30 + 8 - 6$
Add and subtract left to right.	32

When and the expression has a value of

TRY IT : : 2.33 Evaluate:

TRY IT : : 2.34 Evaluate:

EXAMPLE 2.18

Solution

We need to be careful when an expression has a variable with an exponent. In this expression, means and is different from the expression which means

	$2x^2 + 3x + 8$
Substitute 4 for each *x*.	$2(4)^2 + 3(4) + 8$
Simplify .	$2(16) + 3(4) + 8$
Multiply.	$32 + 12 + 8$
Add.	52

TRY IT : : 2.35 Evaluate:

TRY IT : : 2.36 Evaluate:

Identify Terms, Coefficients, and Like Terms

Algebraic expressions are made up of *terms*. A **term** is a constant or the product of a constant and one or more variables. Some examples of terms are

The constant that multiplies the variable(s) in a term is called the **coefficient**. We can think of the coefficient as the number *in front of* the variable. The coefficient of the term is When we write the coefficient is since Table 2.5 gives the coefficients for each of the terms in the left column.

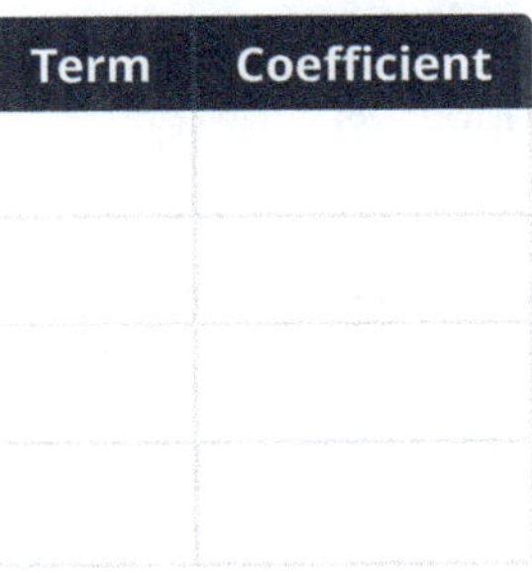

Term	Coefficient

Table 2.5

An algebraic expression may consist of one or more terms added or subtracted. In this chapter, we will only work with terms that are added together. Table 2.6 gives some examples of algebraic expressions with various numbers of terms. Notice that we include the operation before a term with it.

Expression	Terms

Table 2.6

EXAMPLE 2.19

Identify each term in the expression Then identify the coefficient of each term.

Solution

The expression has four terms. They are and

The coefficient of is

The coefficient of is

Remember that if no number is written before a variable, the coefficient is So the coefficient of is

The coefficient of a constant is the constant, so the coefficient of is

> **TRY IT : : 2.37** Identify all terms in the given expression, and their coefficients:

> **TRY IT : : 2.38** Identify all terms in the given expression, and their coefficients:

Some terms share common traits. Look at the following terms. Which ones seem to have traits in common?

Which of these terms are like terms?

- The terms and are both constant terms.
- The terms and are both terms with
- The terms and both have

Terms are called **like terms** if they have the same variables and exponents. All constant terms are also like terms. So among the terms

Like Terms

Terms that are either constants or have the same variables with the same exponents are like terms.

EXAMPLE 2.20

Identify the like terms:

Solution

Look at the variables and exponents. The expression contains and constants.

The terms and are like terms because they both have

The terms and are like terms because they both have

The terms and are like terms because they are both constants.

The term does not have any like terms in this list since no other terms have the variable raised to the power of

Look at the variables and exponents. The expression contains the terms

The terms and are like terms because they both have

The terms are like terms because they all have

The term has no like terms in the given expression because no other terms contain the two variables

> **TRY IT : : 2.39** Identify the like terms in the list or the expression:

TRY IT : : 2.40 Identify the like terms in the list or the expression:

Simplify Expressions by Combining Like Terms

We can simplify an expression by combining the like terms. What do you think would simplify to? If you thought you would be right!

We can see why this works by writing both terms as addition problems.

$$\underbrace{x+x+x}_{3x} + \underbrace{x+x+x+x+x+x}_{6x}$$

$$9x$$

Add the coefficients and keep the same variable. It doesn't matter what is. If you have of something and add more of the same thing, the result is of them. For example, oranges plus oranges is oranges. We will discuss the mathematical properties behind this later.

The expression has only two terms. When an expression contains more terms, it may be helpful to rearrange the terms so that like terms are together. The Commutative Property of Addition says that we can change the order of addends without changing the sum. So we could rearrange the following expression before combining like terms.

$$3x + 4y - 2x + 6y$$

$$\underbrace{3x - 2x} + \underbrace{4y + 6y}$$

Now it is easier to see the like terms to be combined.

HOW TO : : COMBINE LIKE TERMS.

Step 1. Identify like terms.

Step 2. Rearrange the expression so like terms are together.

Step 3. Add the coefficients of the like terms.

EXAMPLE 2.21

Simplify the expression:

Solution

	$3x + 7 + 4x + 5$
Identify the like terms.	$3x + 7 + 4x + 5$
Rearrange the expression, so the like terms are together.	$3x + 4x + 7 + 5$
Add the coefficients of the like terms.	$\underbrace{3x + 4x}_{7x} + \underbrace{7 + 5}_{12}$
The original expression is simplified to...	$7x + 12$

TRY IT : : 2.41 Simplify:

TRY IT : : 2.42 Simplify:

EXAMPLE 2.22

Simplify the expression:

Solution

	$7x^2 + 8x + x^2 + 4x$
Identify the like terms.	$7x^2 + 8x + x^2 + 4x$
Rearrange the expression so like terms are together.	$7x^2 + x^2 + 8x + 4x$
Add the coefficients of the like terms.	$8x^2 + 12x$

These are not like terms and cannot be combined. So is in simplest form.

TRY IT : : 2.43 Simplify:

TRY IT : : 2.44 Simplify:

Translate Words to Algebraic Expressions

In the previous section, we listed many operation symbols that are used in algebra, and then we translated expressions and equations into word phrases and sentences. Now we'll reverse the process and translate word phrases into algebraic expressions. The symbols and variables we've talked about will help us do that. They are summarized in Table 2.7.

Operation	Phrase	Expression
Addition	plus the sum of and increased by more than the total of and added to	
Subtraction	minus the difference of and subtracted from decreased by less than	
Multiplication	times the product of and	, , ,
Division	divided by the quotient of and the ratio of and divided into	, , —, ⌐

Table 2.7

Look closely at these phrases using the four operations:

- the sum *of* *and*
- the difference *of* *and*
- the product *of* *and*
- the quotient *of* *and*

Each phrase tells you to operate on two numbers. Look for the words ***of*** and ***and*** to find the numbers.

EXAMPLE 2.23

Translate each word phrase into an algebraic expression:

the difference of and the quotient of and

Solution

The key word is *difference*, which tells us the operation is subtraction. Look for the words *of* and *and* to find the numbers to subtract.

The key word is *quotient*, which tells us the operation is division.

This can also be written as —

TRY IT : : 2.45 Translate the given word phrase into an algebraic expression:

the difference of and the quotient of and

TRY IT : : 2.46 Translate the given word phrase into an algebraic expression:

the sum of and the product of and

How old will you be in eight years? What age is eight more years than your age now? Did you add to your present age? Eight *more than* means eight added to your present age.

How old were you seven years ago? This is seven years less than your age now. You subtract from your present age. Seven *less than* means seven subtracted from your present age.

EXAMPLE 2.24

Translate each word phrase into an algebraic expression:

Eight more than Seven less than

Solution

The key words are *more than*. They tell us the operation is addition. *More than* means "added to".

The key words are *less than*. They tell us the operation is subtraction. *Less than* means "subtracted from".

TRY IT : : 2.47 Translate each word phrase into an algebraic expression:

Eleven more than Fourteen less than

TRY IT : : 2.48 Translate each word phrase into an algebraic expression:

more than less than

EXAMPLE 2.25

Translate each word phrase into an algebraic expression:

five times the sum of and the sum of five times and

Solution

There are two operation words: *times* tells us to multiply and *sum* tells us to add. Because we are multiplying times the sum, we need parentheses around the sum of and

five times the sum of and

To take a sum, we look for the words *of* and *and* to see what is being added. Here we are taking the sum *of* five times and the sum of five times and

Notice how the use of parentheses changes the result. In part , we add first and in part , we multiply first.

TRY IT : : 2.49 Translate the word phrase into an algebraic expression:

four times the sum of and the sum of four times and

TRY IT : : 2.50 Translate the word phrase into an algebraic expression:

the difference of two times two times the difference of

Later in this course, we'll apply our skills in algebra to solving equations. We'll usually start by translating a word phrase to an algebraic expression. We'll need to be clear about what the expression will represent. We'll see how to do this in the next two examples.

EXAMPLE 2.26

The height of a rectangular window is inches less than the width. Let represent the width of the window. Write an expression for the height of the window.

Solution

Write a phrase about the height.	less than the width
Substitute for the width.	less than
Rewrite 'less than' as 'subtracted from'.	subtracted from
Translate the phrase into algebra.	

TRY IT : : 2.51

The length of a rectangle is inches less than the width. Let represent the width of the rectangle. Write an expression for the length of the rectangle.

TRY IT : : 2.52

The width of a rectangle is meters greater than the length. Let represent the length of the rectangle. Write an expression for the width of the rectangle.

EXAMPLE 2.27

Blanca has dimes and quarters in her purse. The number of dimes is less than times the number of quarters. Let represent the number of quarters. Write an expression for the number of dimes.

Solution

Write a phrase about the number of dimes.	two less than five times the number of quarters
Substitute for the number of quarters.	less than five times
Translate *times* .	less than
Translate the phrase into algebra.	

TRY IT : : 2.53

Geoffrey has dimes and quarters in his pocket. The number of dimes is seven less than six times the number of quarters. Let represent the number of quarters. Write an expression for the number of dimes.

TRY IT : : 2.54

Lauren has dimes and nickels in her purse. The number of dimes is eight more than four times the number of nickels. Let represent the number of nickels. Write an expression for the number of dimes.

MEDIA : : ACCESS ADDITIONAL ONLINE RESOURCES

- Algebraic Expression Vocabulary (http://openstaxcollege.org/l/24AlgExpvocab)

2.2 EXERCISES

Practice Makes Perfect

Evaluate Algebraic Expressions

In the following exercises, evaluate the expression for the given value.

69. 70. 71.

72. 73. 74.

75. 76. 77.

78. 79. 80.

81. 82. 83.

84. 85. 86.

87. 88.

Identify Terms, Coefficients, and Like Terms

In the following exercises, list the terms in the given expression.

89. 90. 91.

92.

In the following exercises, identify the coefficient of the given term.

93. 94. 95.

96.

In the following exercises, identify all sets of like terms.

97. 98. 99.

100.

Simplify Expressions by Combining Like Terms

In the following exercises, simplify the given expression by combining like terms.

101. 102. 103.

104. 105. 106.

107. 108. 109.

110.

111.

112.

113.

114.

115.

116.

Translate English Phrases into Algebraic Expressions

In the following exercises, translate the given word phrase into an algebraic expression.

117. The sum of 8 and 12

118. The sum of 9 and 1

119. The difference of 14 and 9

120. 8 less than 19

121. The product of 9 and 7

122. The product of 8 and 7

123. The quotient of 36 and 9

124. The quotient of 42 and 7

125. The difference of and

126. less than

127. The product of and

128. The product of and

129. The sum of and

130. The sum of and

131. The quotient of and

132. The quotient of and

133. Eight times the difference of and nine

134. Seven times the difference of and one

135. Five times the sum of and

136. Nine times five less than twice

In the following exercises, write an algebraic expression.

137. Adele bought a skirt and a blouse. The skirt cost more than the blouse. Let represent the cost of the blouse. Write an expression for the cost of the skirt.

138. Eric has rock and classical CDs in his car. The number of rock CDs is more than the number of classical CDs. Let represent the number of classical CDs. Write an expression for the number of rock CDs.

139. The number of girls in a second-grade class is less than the number of boys. Let represent the number of boys. Write an expression for the number of girls.

140. Marcella has fewer male cousins than female cousins. Let represent the number of female cousins. Write an expression for the number of boy cousins.

141. Greg has nickels and pennies in his pocket. The number of pennies is seven less than twice the number of nickels. Let represent the number of nickels. Write an expression for the number of pennies.

142. Jeannette has and bills in her wallet. The number of fives is three more than six times the number of tens. Let represent the number of tens. Write an expression for the number of fives.

Everyday Math

In the following exercises, use algebraic expressions to solve the problem.

143. Car insurance Justin's car insurance has a deductible per incident. This means that he pays and his insurance company will pay all costs beyond If Justin files a claim for how much will he pay, and how much will his insurance company pay?

144. Home insurance Pam and Armando's home insurance has a deductible per incident. This means that they pay and their insurance company will pay all costs beyond If Pam and Armando file a claim for how much will they pay, and how much will their insurance company pay?

Writing Exercises

145. Explain why "the sum of x and y" is the same as "the sum of y and x," but "the difference of x and y" is not the same as "the difference of y and x." Try substituting two random numbers for and to help you explain.

146. Explain the difference between times the sum of and and "the sum of times and

Self Check

After completing the exercises, use this checklist to evaluate your mastery of the objectives of this section.

I can...	Confidently	With some help	No-I don't get it!
evaluate algebraic expressions.			
identify terms, coefficients, and like terms.			
simplify expressions by combining like terms.			
translate word phrases to algebraic expressions.			

After reviewing this checklist, what will you do to become confident for all objectives?

2.3 Solving Equations Using the Subtraction and Addition Properties of Equality

Learning Objectives

By the end of this section, you will be able to:

- Determine whether a number is a solution of an equation
- Model the Subtraction Property of Equality
- Solve equations using the Subtraction Property of Equality
- Solve equations using the Addition Property of Equality
- Translate word phrases to algebraic equations
- Translate to an equation and solve

BE PREPARED : : 2.7 Before you get started, take this readiness quiz.

If you missed this problem, review Example 2.13.

BE PREPARED : : 2.8

If you missed this problem, review Example 2.14.

BE PREPARED : : 2.9 Translate into algebra: the difference of and

If you missed this problem, review Example 2.24.

When some people hear the word *algebra*, they think of solving equations. The applications of solving equations are limitless and extend to all careers and fields. In this section, we will begin solving equations. We will start by solving basic equations, and then as we proceed through the course we will build up our skills to cover many different forms of equations.

Determine Whether a Number is a Solution of an Equation

Solving an equation is like discovering the answer to a puzzle. An algebraic equation states that two algebraic expressions are equal. To solve an equation is to determine the values of the variable that make the equation a true statement. Any number that makes the equation true is called a **solution** of the equation. It is the answer to the puzzle!

Solution of an Equation

A **solution to an equation** is a value of a variable that makes a true statement when substituted into the equation.

The process of finding the solution to an equation is called solving the equation.

To find the solution to an equation means to find the value of the variable that makes the equation true. Can you recognize the solution of If you said you're right! We say is a solution to the equation because when we substitute for the resulting statement is true.

Since is a true statement, we know that is indeed a solution to the equation.

The symbol asks whether the left side of the equation is equal to the right side. Once we know, we can change to an equal sign or not-equal sign

HOW TO : : DETERMINE WHETHER A NUMBER IS A SOLUTION TO AN EQUATION.

Step 1. Substitute the number for the variable in the equation.

Step 2. Simplify the expressions on both sides of the equation.

Step 3. Determine whether the resulting equation is true.

- If it is true, the number is a solution.
- If it is not true, the number is not a solution.

EXAMPLE 2.28

Solution

	$6x - 17 = 16$
Substitute 5 for *x*.	$6 \cdot 5 - 17 \stackrel{?}{=} 16$
Multiply.	$30 - 17 \stackrel{?}{=} 16$
Subtract.	$13 \neq 16$

So is not a solution to the equation

TRY IT : : 2.55

TRY IT : : 2.56

EXAMPLE 2.29

Solution

Here, the variable appears on both sides of the equation. We must substitute for each

	$6y - 4 = 5y - 2$
Substitute 2 for *y*.	$6(2) - 4 \stackrel{?}{=} 5(2) - 2$
Multiply.	$12 - 4 \stackrel{?}{=} 10 - 2$
Subtract.	$8 = 8 \checkmark$

Since results in a true equation, we know that is a solution to the equation

TRY IT : : 2.57

TRY IT : : 2.58

Model the Subtraction Property of Equality

We will use a model to help you understand how the process of solving an equation is like solving a puzzle. An envelope represents the variable - since its contents are unknown - and each counter represents one.

Suppose a desk has an imaginary line dividing it in half. We place three counters and an envelope on the left side of desk, and eight counters on the right side of the desk as in Figure 2.3. Both sides of the desk have the same number of counters, but some counters are hidden in the envelope. Can you tell how many counters are in the envelope?

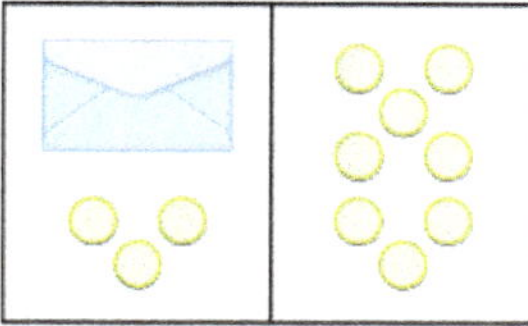

Figure 2.3

What steps are you taking in your mind to figure out how many counters are in the envelope? Perhaps you are thinking "I need to remove the counters from the left side to get the envelope by itself. Those counters on the left match with on the right, so I can take them away from both sides. That leaves five counters on the right, so there must be counters in the envelope." Figure 2.4 shows this process.

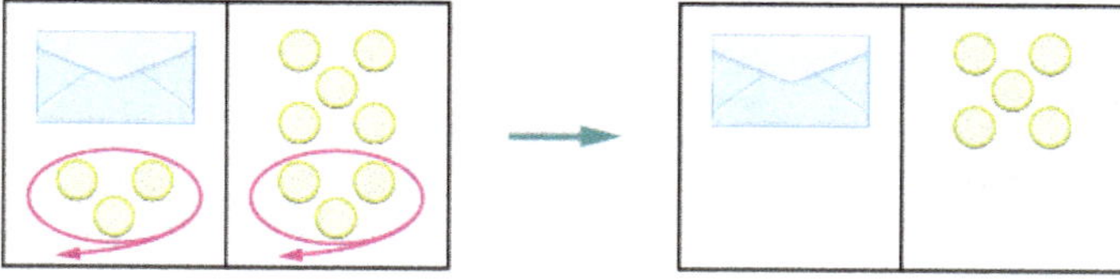

Figure 2.4

What algebraic equation is modeled by this situation? Each side of the desk represents an expression and the center line takes the place of the equal sign. We will call the contents of the envelope so the number of counters on the left side of the desk is On the right side of the desk are counters. We are told that is equal to so our equation is

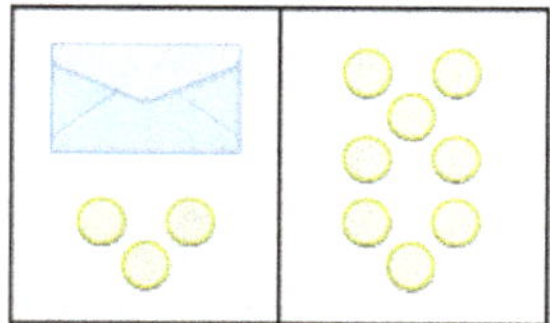

Figure 2.5

Let's write algebraically the steps we took to discover how many counters were in the envelope.

$$x+3=8$$

First, we took away three from each side.	$x+3-3=8-3$
Then we were left with five.	$x=5$

Now let's check our solution. We substitute for in the original equation and see if we get a true statement.

$$x+3=8$$

$$5+3\stackrel{?}{=}8$$

$$8=8\ \checkmark$$

Our solution is correct. Five counters in the envelope plus three more equals eight.

MANIPULATIVE MATHEMATICS

Doing the Manipulative Mathematics activity, "Subtraction Property of Equality" will help you develop a better understanding of how to solve equations by using the Subtraction Property of Equality.

EXAMPLE 2.30

Write an equation modeled by the envelopes and counters, and then solve the equation:

Solution

On the left, write for the contents of the envelope, add the counters, so we have .

On the right, there are counters.

The two sides are equal.

Solve the equation by subtracting counters from each side.

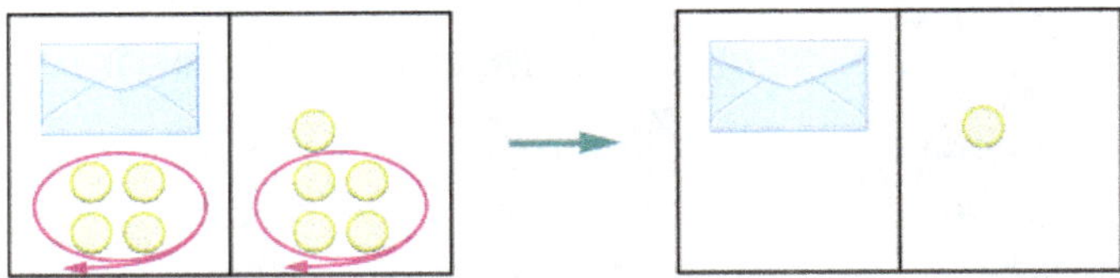

We can see that there is one counter in the envelope. This can be shown algebraically as:

$$\begin{aligned} x + 4 &= 5 \\ x + 4 - 4 &= 5 - 4 \\ x &= 1 \end{aligned}$$

Substitute for in the equation to check.

$$\begin{aligned} x + 4 &= 5 \\ 1 + 4 &\stackrel{?}{=} 5 \\ 5 &= 5 \checkmark \end{aligned}$$

Since makes the statement true, we know that is indeed a solution.

TRY IT : : 2.59 Write the equation modeled by the envelopes and counters, and then solve the equation:

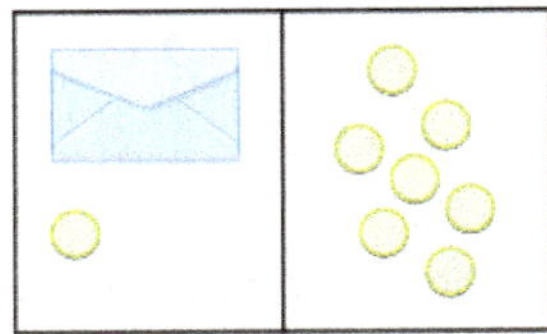

> **TRY IT : :** 2.60 Write the equation modeled by the envelopes and counters, and then solve the equation:

Solve Equations Using the Subtraction Property of Equality

Our puzzle has given us an idea of what we need to do to solve an equation. The goal is to isolate the variable by itself on one side of the equations. In the previous examples, we used the Subtraction Property of Equality, which states that when we subtract the same quantity from both sides of an equation, we still have equality.

Subtraction Property of Equality

For any numbers and if

then

Think about twin brothers Andy and Bobby. They are years old. How old was Andy years ago? He was years less than so his age was or What about Bobby's age years ago? Of course, he was also. Their ages are equal now, and subtracting the same quantity from both of them resulted in equal ages years ago.

HOW TO : : SOLVE AN EQUATION USING THE SUBTRACTION PROPERTY OF EQUALITY.

Step 1. Use the Subtraction Property of Equality to isolate the variable.
Step 2. Simplify the expressions on both sides of the equation.
Step 3. Check the solution.

EXAMPLE 2.31

Solve:

Solution

We will use the Subtraction Property of Equality to isolate

	$x+8=17$
Subtract 8 from both sides.	$x+8-8=17-8$
Simplify.	$x=9$
	$x+8=17$
	$9+8=17$
	$17=17\ \checkmark$

Since makes a true statement, we know is the solution to the equation.

TRY IT : : 2.61 Solve:

TRY IT : : 2.62 Solve:

EXAMPLE 2.32

Solve:

Solution

To solve an equation, we must always isolate the variable—it doesn't matter which side it is on. To isolate we will subtract from both sides.

	$100 = y + 74$
Subtract 74 from both sides.	$100 - 74 = y + 74 - 74$
Simplify.	$26 = y$
Substitute for to check. $100 = y + 74$ $100 \stackrel{?}{=} 26 + 74$ $100 = 100 \checkmark$	

Since makes a true statement, we have found the solution to this equation.

TRY IT : : 2.63 Solve:

TRY IT : : 2.64 Solve:

Solve Equations Using the Addition Property of Equality

In all the equations we have solved so far, a number was added to the variable on one side of the equation. We used subtraction to "undo" the addition in order to isolate the variable.

But suppose we have an equation with a number subtracted from the variable, such as We want to isolate the variable, so to "undo" the subtraction we will add the number to both sides.

We use the Addition Property of Equality, which says we can add the same number to both sides of the equation without changing the equality. Notice how it mirrors the Subtraction Property of Equality.

Addition Property of Equality

For any numbers , and , if

then

Remember the twins, Andy and Bobby? In ten years, Andy's age will still equal Bobby's age. They will both be

We can add the same number to both sides and still keep the equality.

HOW TO : : SOLVE AN EQUATION USING THE ADDITION PROPERTY OF EQUALITY.

Step 1. Use the Addition Property of Equality to isolate the variable.
Step 2. Simplify the expressions on both sides of the equation.
Step 3. Check the solution.

EXAMPLE 2.33

Solve:

Solution

We will use the Addition Property of Equality to isolate the variable.

	$x - 5 = 8$
Add 5 to both sides.	$x - 5 + 5 = 8 + 5$
Simplify.	$x = 13$
Now we can check. Let $x = 13$.	
$x - 5 = 8$	
$13 - 5 \stackrel{?}{=} 8$	
$8 = 8$ ✓	

TRY IT : : 2.65 Solve:

TRY IT : : 2.66 Solve:

EXAMPLE 2.34

Solve:

Solution

We will add to each side to isolate the variable.

	$27 = a - 16$
Add 16 to each side.	$27 + 16 = a - 16 + 16$
Simplify.	$43 = a$
Now we can check. Let $a = 43$.	$27 = a - 16$
	$27 \stackrel{?}{=} 43 - 16$
	$27 = 27 \checkmark$

The solution to is

TRY IT : : 2.67 Solve:

TRY IT : : 2.68 Solve:

Translate Word Phrases to Algebraic Equations

Remember, an equation has an equal sign between two algebraic expressions. So if we have a sentence that tells us that two phrases are equal, we can translate it into an equation. We look for clue words that mean *equals*. Some words that translate to the equal sign are:

- is equal to
- is the same as
- is
- gives
- was
- will be

It may be helpful to put a box around the *equals* word(s) in the sentence to help you focus separately on each phrase. Then translate each phrase into an expression, and write them on each side of the equal sign.

We will practice translating word sentences into algebraic equations. Some of the sentences will be basic number facts with no variables to solve for. Some sentences will translate into equations with variables. The focus right now is just to translate the words into algebra.

EXAMPLE 2.35

Translate the sentence into an algebraic equation: The sum of and is

Solution

The word *is* tells us the equal sign goes between 9 and 15.

Locate the "equals" word(s).	The sum of 6 and 9 is 15.
Write the = sign.	The sum of 6 and 9 = 15.
Translate the words to the left of the *equals* word into an algebraic expression.	$6 + 9 = \_\_\_$
Translate the words to the right of the *equals* word into an algebraic expression.	$6 + 9 = 15$

> **TRY IT : :** 2.69 Translate the sentence into an algebraic equation:
>
> The sum of and gives

> **TRY IT : :** 2.70 Translate the sentence into an algebraic equation:
>
> The sum of and is

EXAMPLE 2.36

Translate the sentence into an algebraic equation: The product of and is

Solution

The location of the word *is* tells us that the equal sign goes between 7 and 56.

Locate the "equals" word(s).	The product of 8 and 7 [is] 56.
Write the = sign.	The product of 8 and 7 = 56.
Translate the words to the left of the *equals* word into an algebraic expression.	$8 \cdot 7 = \_\_$
Translate the words to the right of the *equals* word into an algebraic expression.	$8 \cdot 7 = 56$

> **TRY IT : :** 2.71 Translate the sentence into an algebraic equation:
>
> The product of and is

> **TRY IT : :** 2.72 Translate the sentence into an algebraic equation:
>
> The product of and gives

EXAMPLE 2.37

Translate the sentence into an algebraic equation: Twice the difference of and gives

Solution

Locate the "equals" word(s).	Twice the difference of x and 3 [gives] 18.
Recognize the key words: *twice; difference of .... and ....*	*Twice* means two times.
Translate.	Twice the difference of x and 3 [gives] 18. $2 \quad (x - 3) \quad = \quad 18$

> **TRY IT : :** 2.73 Translate the given sentence into an algebraic equation:
>
> Twice the difference of and gives

> **TRY IT : :** 2.74 Translate the given sentence into an algebraic equation:
>
> Twice the difference of and gives

Translate to an Equation and Solve

Now let's practice translating sentences into algebraic equations and then solving them. We will solve the equations by using the Subtraction and Addition Properties of Equality.

EXAMPLE 2.38

Translate and solve: Three more than is equal to

Solution

	Three more than x is equal to 47.
Translate.	$x+3=47$
Subtract 3 from both sides of the equation.	$x+3-3=47-3$
Simplify.	$x=44$
We can check. Let .	$x+3=47$
	$44+3\stackrel{?}{=}47$
	$47=47\ \checkmark$

So is the solution.

TRY IT : : 2.75 Translate and solve:

Seven more than is equal to

TRY IT : : 2.76 Translate and solve:

Eleven more than is equal to

EXAMPLE 2.39

Translate and solve: The difference of and is

Solution

	The difference of y and 14 is 18.
Translate.	$y-14=18$
Add 14 to both sides.	$y-14+14=18+14$
Simplify.	$y=32$
We can check. Let .	$y-14=18$
	$32-14\stackrel{?}{=}18$
	$18=18\ \checkmark$

So is the solution.

TRY IT : : 2.77 Translate and solve:

The difference of and is equal to

TRY IT : : 2.78 Translate and solve:

The difference of and is equal to

MEDIA : : ACCESS ADDITIONAL ONLINE RESOURCES

- Solving One Step Equations By Addition and Subtraction (http://openstaxcollege.org/l/24Solveonestep)

2.3 EXERCISES

Practice Makes Perfect

Determine Whether a Number is a Solution of an Equation

In the following exercises, determine whether each given value is a solution to the equation.

147.

148.

149.

150.

151.

152.

153.

154.

155.

156.

157.

158.

Model the Subtraction Property of Equality

In the following exercises, write the equation modeled by the envelopes and counters and then solve using the subtraction property of equality.

159.

160.

161.

162.

Solve Equations using the Subtraction Property of Equality

In the following exercises, solve each equation using the subtraction property of equality.

163.

164.

165.

166.

167.

168.

169.

170.

171.

172.

173.

174.

Solve Equations using the Addition Property of Equality

In the following exercises, solve each equation using the addition property of equality.

175.

176.

177.

178.

179.

180.

181.

182.

183.

184.

185.

186.

Translate Word Phrase to Algebraic Equations

In the following exercises, translate the given sentence into an algebraic equation.

187. The sum of and is equal to

188. The sum of and is equal to

189. The difference of and is equal to

190. The difference of and is equal to

191. The product of and is equal to

192. The product of and is equal to

193. The quotient of and is equal to

194. The quotient of and is equal to

195. Twice the difference of and gives

196. Twice the difference of and gives

197. The sum of three times and is

198. The sum of eight times and is

Translate to an Equation and Solve

In the following exercises, translate the given sentence into an algebraic equation and then solve it.

199. Five more than is equal to

200. Nine more than is equal to

201. The sum of and is

202. The sum of and is

203. The difference of and is equal to

204. The difference of and is equal to

205. less than is

206. less than is

207. less than gives

208. less than gives

Everyday Math

209. Insurance Vince's car insurance has a deductible. Find the amount the insurance company will pay, for an claim by solving the equation

210. Insurance Marta's homeowner's insurance policy has a deductible. The insurance company paid to repair damages caused by a storm. Find the total cost of the storm damage, by solving the equation

211. Sale purchase Arthur bought a suit that was on sale for off. He paid for the suit. Find the original price, of the suit by solving the equation

212. Sale purchase Rita bought a sofa that was on sale for She paid a total of including sales tax. Find the amount of the sales tax, by solving the equation

Writing Exercises

213. Is a solution to the equation How do you know?

214. Write the equation in words. Then make up a word problem for this equation.

Self Check

After completing the exercises, use this checklist to evaluate your mastery of the objectives of this section.

I can...	Confidently	With some help	No-I don't get it!
determine whether a number is a solution of an equation.			
model the subtraction property of equality.			
solve equations using the subtraction property of equality.			
solve equations using the addition property of equality.			
translate word phrases to algebraic equations.			
translate to an equation and solve.			

What does this checklist tell you about your mastery of this section? What steps will you take to improve?

2.4 Find Multiples and Factors

Learning Objectives

By the end of this section, you will be able to:

- Identify multiples of numbers
- Use common divisibility tests
- Find all the factors of a number
- Identify prime and composite numbers

BE PREPARED : : 2.10 Before you get started, take this readiness quiz.
Which of the following numbers are counting numbers (natural numbers)?

If you missed this problem, review Example 1.1.

BE PREPARED : : 2.11 Find the sum of and
If you missed the problem, review Example 2.1.

Identify Multiples of Numbers

Annie is counting the shoes in her closet. The shoes are matched in pairs, so she doesn't have to count each one. She counts by twos: She has shoes in her closet.

The numbers are called multiples of Multiples of can be written as the product of a counting number and The first six multiples of are given below.

A **multiple of a number** is the product of the number and a counting number. So a multiple of would be the product of a counting number and Below are the first six multiples of

We can find the multiples of any number by continuing this process. Table 2.8 shows the multiples of through for the first twelve counting numbers.

Counting Number

Table 2.8

Multiple of a Number

A number is a multiple of if it is the product of a counting number and

Recognizing the patterns for multiples of will be helpful to you as you continue in this course.

MANIPULATIVE MATHEMATICS

Doing the Manipulative Mathematics activity "Multiples" will help you develop a better understanding of multiples.

Figure 2.6 shows the counting numbers from to Multiples of are highlighted. Do you notice a pattern?

1	2	3	4	5	6	7	8	9	10
11	12	13	14	15	16	17	18	19	20
21	22	23	24	25	26	27	28	29	30
31	32	33	34	35	36	37	38	39	40
41	42	43	44	45	46	47	48	49	50

Figure 2.6 Multiples of between and

The last digit of each highlighted number in Figure 2.6 is either This is true for the product of and any counting number. So, to tell if any number is a multiple of look at the last digit. If it is then the number is a multiple of

EXAMPLE 2.40

Determine whether each of the following is a multiple of

Solution

Is 489 a multiple of 2?	
Is the last digit 0, 2, 4, 6, or 8?	No.
	489 is not a multiple of 2.

Is 3,714 a multiple of 2?	
Is the last digit 0, 2, 4, 6, or 8?	Yes.
	3,714 is a multiple of 2.

TRY IT : : 2.79 Determine whether each number is a multiple of

TRY IT : : 2.80 Determine whether each number is a multiple of

Now let's look at multiples of Figure 2.7 highlights all of the multiples of between and What do you notice about the multiples of

1	2	3	4	5	6	7	8	9	10
11	12	13	14	15	16	17	18	19	20
21	22	23	24	25	26	27	28	29	30
31	32	33	34	35	36	37	38	39	40
41	42	43	44	45	46	47	48	49	50

Figure 2.7 Multiples of between and

All multiples of end with either or Just like we identify multiples of by looking at the last digit, we can identify multiples of by looking at the last digit.

EXAMPLE 2.41

Determine whether each of the following is a multiple of

Solution

Is 579 a multiple of 5?	
Is the last digit 5 or 0?	No.
	579 is not a multiple of 5.

Is 880 a multiple of 5?	
Is the last digit 5 or 0?	Yes.
	880 is a multiple of 5.

TRY IT : : 2.81 Determine whether each number is a multiple of

TRY IT : : 2.82 Determine whether each number is a multiple of

Figure 2.8 highlights the multiples of between and All multiples of all end with a zero.

1	2	3	4	5	6	7	8	9	10
11	12	13	14	15	16	17	18	19	20
21	22	23	24	25	26	27	28	29	30
31	32	33	34	35	36	37	38	39	40
41	42	43	44	45	46	47	48	49	50

Figure 2.8 Multiples of between and

EXAMPLE 2.42

Determine whether each of the following is a multiple of

Solution

Is 425 a multiple of 10?	
Is the last digit zero?	No.
	425 is not a multiple of 10.

Is 350 a multiple of 10?	
Is the last digit zero?	Yes.
	350 is a multiple of 10.

TRY IT : : 2.83 Determine whether each number is a multiple of

TRY IT : : 2.84 Determine whether each number is a multiple of

Figure 2.9 highlights multiples of The pattern for multiples of is not as obvious as the patterns for multiples of

1	2	3	4	5	6	7	8	9	10
11	12	13	14	15	16	17	18	19	20
21	22	23	24	25	26	27	28	29	30
31	32	33	34	35	36	37	38	39	40
41	42	43	44	45	46	47	48	49	50

Figure 2.9 Multiples of between and

Unlike the other patterns we've examined so far, this pattern does not involve the last digit. The pattern for multiples of is based on the sum of the digits. If the sum of the digits of a number is a multiple of then the number itself is a multiple of See Table 2.9.

Table 2.9

Consider the number The digits are and and their sum is Since is a multiple of we know that is also a multiple of

EXAMPLE 2.43

Determine whether each of the given numbers is a multiple of

Solution

Is a multiple of

Find the sum of the digits.	
Is 15 a multiple of 3?	Yes.
If we're not sure, we could add its digits to find out. We can check it by dividing 645 by 3.	
The quotient is 215.	

Is a multiple of

Find the sum of the digits.	
Is 16 a multiple of 3?	No.
So 10,519 is not a multiple of 3 either..	
We can check this by dividing by 10,519 by 3.	

When we divide by we do not get a counting number, so is not the product of a counting number and It is not a multiple of

TRY IT : : 2.85 Determine whether each number is a multiple of

TRY IT : : 2.86 Determine whether each number is a multiple of

Look back at the charts where you highlighted the multiples of of and of Notice that the multiples of are the numbers that are multiples of both and That is because Likewise, since the multiples of are the numbers that are multiples of both and

Use Common Divisibility Tests

Another way to say that is a multiple of is to say that is divisible by In fact, is so is Notice in Example 2.43 that is not a multiple When we divided by we did not get a counting number, so is not divisible by

Divisibility

If a number is a multiple of then we say that is divisible by

Since multiplication and division are inverse operations, the patterns of multiples that we found can be used as divisibility tests. Table 2.10 summarizes divisibility tests for some of the counting numbers between one and ten.

Divisibility Tests	
A number is divisible by	
	if the last digit is
	if the sum of the digits is divisible by
	if the last digit is or
	if divisible by both and
	if the last digit is

Table 2.10

EXAMPLE 2.44

Determine whether is divisible by

Solution

Table 2.11 applies the divisibility tests to In the far right column, we check the results of the divisibility tests by seeing if the quotient is a whole number.

Divisible by...?	Test	Divisible?	Check
	Is last digit *Yes.*	yes	
	Yes.	yes	
	Is last digit or *Yes.*	yes	
	Is last digit *Yes.*	yes	

Table 2.11

Thus, is divisible by

TRY IT : : 2.87 Determine whether the given number is divisible by

TRY IT : : 2.88 Determine whether the given number is divisible by

EXAMPLE 2.45

Determine whether is divisible by

Solution

Table 2.12 applies the divisibility tests to and tests the results by finding the quotients.

Divisible by...?	Test	Divisible?	Check
	Is last digit No.	no	
	Yes.	yes	
	Is last digit is or Yes.	yes	
	Is last digit No.	no	

Table 2.12

Thus, is divisible by and but not or

TRY IT : : 2.89 Determine whether the given number is divisible

TRY IT : : 2.90 Determine whether the given number is divisible

Find All the Factors of a Number

There are often several ways to talk about the same idea. So far, we've seen that if is a multiple of we can say that is divisible by We know that is the product of and so we can say is a multiple of and is a multiple of We can also say is divisible by and by Another way to talk about this is to say that and are factors of When we write we can say that we have factored

$$\underbrace{8 \cdot 9}_{factors} = \underbrace{72}_{product}$$

Factors

If then are factors of and is the product of

In algebra, it can be useful to determine all of the factors of a number. This is called factoring a number, and it can help us solve many kinds of problems.

MANIPULATIVE MATHEMATICS

Doing the Manipulative Mathematics activity "Model Multiplication and Factoring" will help you develop a better understanding of multiplication and factoring.

For example, suppose a choreographer is planning a dance for a ballet recital. There are dancers, and for a certain scene, the choreographer wants to arrange the dancers in groups of equal sizes on stage.

In how many ways can the dancers be put into groups of equal size? Answering this question is the same as identifying the factors of Table 2.13 summarizes the different ways that the choreographer can arrange the dancers.

Number of Groups	Dancers per Group	Total Dancers

Table 2.13

What patterns do you see in Table 2.13? Did you notice that the number of groups times the number of dancers per group is always This makes sense, since there are always dancers.

You may notice another pattern if you look carefully at the first two columns. These two columns contain the exact same set of numbers—but in reverse order. They are mirrors of one another, and in fact, both columns list all of the factors of which are:

We can find all the factors of any counting number by systematically dividing the number by each counting number, starting with If the quotient is also a counting number, then the divisor and the quotient are factors of the number. We can stop when the quotient becomes smaller than the divisor.

HOW TO : : FIND ALL THE FACTORS OF A COUNTING NUMBER.

Step 1. Divide the number by each of the counting numbers, in order, until the quotient is smaller than the divisor.

- If the quotient is a counting number, the divisor and quotient are a pair of factors.
- If the quotient is not a counting number, the divisor is not a factor.

Step 2. List all the factor pairs.

Step 3. Write all the factors in order from smallest to largest.

EXAMPLE 2.46

Find all the factors of

Solution

Divide by each of the counting numbers starting with If the quotient is a whole number, the divisor and quotient are a pair of factors.

Dividend	Divisor	Quotient	Factors
72	1	72	1, 72
72	2	36	2, 36
72	3	24	3, 24
72	4	18	4, 18
72	5	14.4	–
72	6	12	6, 12
72	7	~10.29	–
72	8	9	8, 9

The next line would have a divisor of and a quotient of The quotient would be smaller than the divisor, so we stop. If we continued, we would end up only listing the same factors again in reverse order. Listing all the factors from smallest to greatest, we have

---

> **TRY IT : :** 2.91 Find all the factors of the given number:

> **TRY IT : :** 2.92 Find all the factors of the given number:

Identify Prime and Composite Numbers

Some numbers, like have many factors. Other numbers, such as have only two factors: and the number. A number with only two factors is called a **prime number**. A number with more than two factors is called a **composite number**. The number is neither prime nor composite. It has only one factor, itself.

Prime Numbers and Composite Numbers

A prime number is a counting number greater than whose only factors are and itself.

A composite number is a counting number that is not prime.

Figure 2.10 lists the counting numbers from through along with their factors. The highlighted numbers are prime, since each has only two factors.

Number	Factors	Prime or Composite?		Number	Factor	Prime or Composite?
2	1,2	Prime		12	1,2,3,4,6,12	Composite
3	1,3	Prime		13	1,13	Prime
4	1,2,4	Composite		14	1,2,7,14	Composite
5	1,5	Prime		15	1,3,5,15	Composite
6	1,2,3,6	Composite		16	1,2,4,8,16	Composite
7	1,7	Prime		17	1,17	Prime
8	1,2,4,8	Composite		18	1,2,3,6,9,18	Composite
9	1,3,9	Composite		19	1,19	Prime
10	1,2,5,10	Composite		20	1,2,4,5,10,20	Composite
11	1,11	Prime				

Figure 2.10 Factors of the counting numbers from through with prime numbers highlighted

The prime numbers less than are There are many larger prime numbers too. In order to determine whether a number is prime or composite, we need to see if the number has any factors other than and itself. To do this, we can test each of the smaller prime numbers in order to see if it is a factor of the number. If none of the prime numbers are factors, then that number is also prime.

HOW TO : : DETERMINE IF A NUMBER IS PRIME.

Step 1. Test each of the primes, in order, to see if it is a factor of the number.

Step 2. Start with and stop when the quotient is smaller than the divisor or when a prime factor is found.

Step 3. If the number has a prime factor, then it is a composite number. If it has no prime factors, then the number is prime.

EXAMPLE 2.47

Identify each number as prime or composite:

Solution

Test each prime, in order, to see if it is a factor of , starting with as shown. We will stop when the quotient is smaller than the divisor.

Prime	Test	Factor of
	Last digit of is not	No.
	and is not divisible by	No.
	The last digit of is not or	No.
		No.
		No.

We can stop when we get to because the quotient is less than the divisor.

We did not find any prime numbers that are factors of so we know is prime.

Test each prime, in order, to see if it is a factor of

Prime	Test	Factor of
	Last digit is not	No.
	and is not divisible by	No.
	the last digit is not or	No.
		Yes.

Since is divisible by we know it is not a prime number. It is composite.

TRY IT : : 2.93 Identify the number as prime or composite:

TRY IT : : 2.94 Identify the number as prime or composite:

LINKS TO LITERACY

The Links to Literacy activities *One Hundred Hungry Ants*, *Spunky Monkeys on Parade* and *A Remainder of One* will provide you with another view of the topics covered in this section.

MEDIA : : ACCESS ADDITIONAL ONLINE RESOURCES

- Divisibility Rules (http://openstaxcollege.org/l/24Divisrules)
- Factors (http://openstaxcollege.org/l/24Factors)
- Ex 1: Determine Factors of a Number (http://openstaxcollege.org/l/24Factors1)
- Ex 2: Determine Factors of a Number (http://openstaxcollege.org/l/24Factors2)
- Ex 3: Determine Factors of a Number (http://openstaxcollege.org/l/24Factors3)

2.4 EXERCISES

Practice Makes Perfect

Identify Multiples of Numbers

In the following exercises, list all the multiples less than for the given number.

215. **216.** **217.**

218. **219.** **220.**

221. **222.** **223.**

224.

Use Common Divisibility Tests

In the following exercises, use the divisibility tests to determine whether each number is divisible by

225. **226.** **227.**

228. **229.** **230.**

231. **232.** **233.**

234. **235.** **236.**

237. **238.** **239.**

240. **241.** **242.**

Find All the Factors of a Number

In the following exercises, find all the factors of the given number.

243. **244.** **245.**

246. **247.** **248.**

249. **250.**

Identify Prime and Composite Numbers

In the following exercises, determine if the given number is prime or composite.

251. **252.** **253.**

254. **255.** **256.**

257. **258.** **259.**

260. **261.** **262.**

Everyday Math

263. Banking Frank's grandmother gave him at his high school graduation. Instead of spending it, Frank opened a bank account. Every week, he added to the account. The table shows how much money Frank had put in the account by the end of each week. Complete the table by filling in the blanks.

Weeks after graduation	Total number of dollars Frank put in the account	Simplified Total

264. Banking In March, Gina opened a Christmas club savings account at her bank. She deposited to open the account. Every week, she added to the account. The table shows how much money Gina had put in the account by the end of each week. Complete the table by filling in the blanks.

Weeks after opening the account	Total number of dollars Gina put in the account	Simplified Total

Writing Exercises

265. If a number is divisible by and by why is it also divisible by

266. What is the difference between prime numbers and composite numbers?

Self Check

After completing the exercises, use this checklist to evaluate your mastery of the objectives of this section.

I can...	Confidently	With some help	No-I don't get it!
identify multiples of numbers.			
use common divisibility tests.			
find all the factors of a number.			
identify prime and composite numbers.			

On a scale of 1–10, how would you rate your mastery of this section in light of your responses on the checklist? How can you improve this?

2.5 Prime Factorization and the Least Common Multiple

Learning Objectives

By the end of this section, you will be able to:

- Find the prime factorization of a composite number
- Find the least common multiple (LCM) of two numbers

BE PREPARED : : 2.12 Before you get started, take this readiness quiz.

Is divisible by

If you missed this problem, review Example 2.44.

BE PREPARED : : 2.13 Is prime or composite?

If you missed this problem, review Example 2.47.

Write in exponential notation.

If you missed this problem, review Example 2.5.

Find the Prime Factorization of a Composite Number

In the previous section, we found the factors of a number. Prime numbers have only two factors, the number and the prime number itself. Composite numbers have more than two factors, and every composite number can be written as a unique product of primes. This is called the **prime factorization** of a number. When we write the prime factorization of a number, we are rewriting the number as a product of primes. Finding the prime factorization of a composite number will help you later in this course.

Prime Factorization

The prime factorization of a number is the product of prime numbers that equals the number.

MANIPULATIVE MATHEMATICS

Doing the Manipulative Mathematics activity "Prime Numbers" will help you develop a better sense of prime numbers.

You may want to refer to the following list of prime numbers less than as you work through this section.

Prime Factorization Using the Factor Tree Method

One way to find the prime factorization of a number is to make a factor tree. We start by writing the number, and then writing it as the product of two factors. We write the factors below the number and connect them to the number with a small line segment—a "branch" of the factor tree.

If a factor is prime, we circle it (like a bud on a tree), and do not factor that "branch" any further. If a factor is not prime, we repeat this process, writing it as the product of two factors and adding new branches to the tree.

We continue until all the branches end with a prime. When the factor tree is complete, the circled primes give us the prime factorization.

For example, let's find the prime factorization of We can start with any factor pair such as and We write and below with branches connecting them.

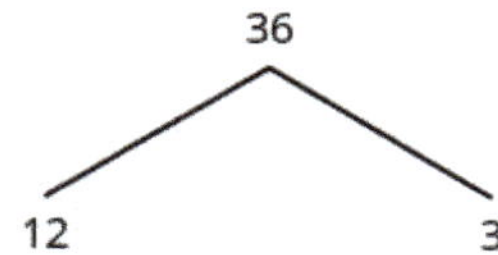

The factor is prime, so we circle it. The factor is composite, so we need to find its factors. Let's use and We write these factors on the tree under the

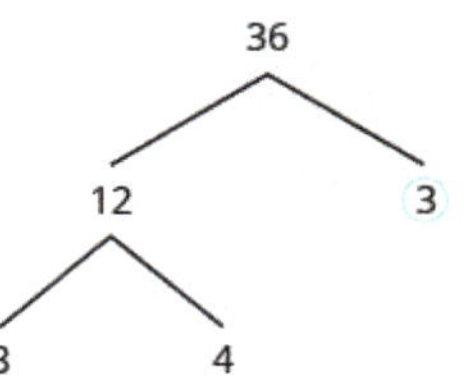

The factor is prime, so we circle it. The factor is composite, and it factors into We write these factors under the Since is prime, we circle both

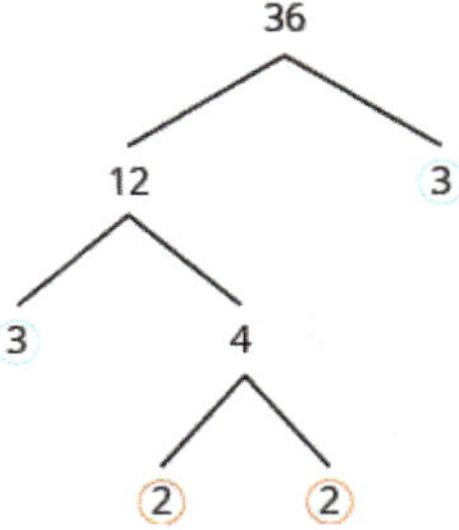

The prime factorization is the product of the circled primes. We generally write the prime factorization in order from least to greatest.

In cases like this, where some of the prime factors are repeated, we can write prime factorization in exponential form.

Note that we could have started our factor tree with any factor pair of We chose and but the same result would have been the same if we had started with and and

HOW TO : : FIND THE PRIME FACTORIZATION OF A COMPOSITE NUMBER USING THE TREE METHOD.

Step 1. Find any factor pair of the given number, and use these numbers to create two branches.

Step 2. If a factor is prime, that branch is complete. Circle the prime.

Step 3. If a factor is not prime, write it as the product of a factor pair and continue the process.

Step 4. Write the composite number as the product of all the circled primes.

EXAMPLE 2.48

Find the prime factorization of using the factor tree method.

Solution

We can start our tree using any factor pair of 48. Let's use 2 and 24.
We circle the 2 because it is prime and so that branch is complete.

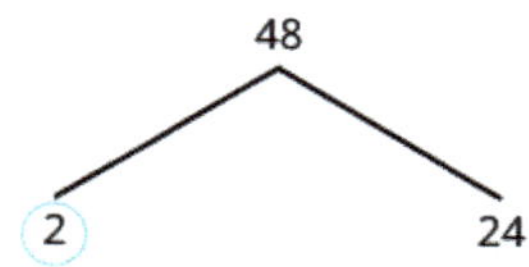

Now we will factor 24. Let's use 4 and 6.

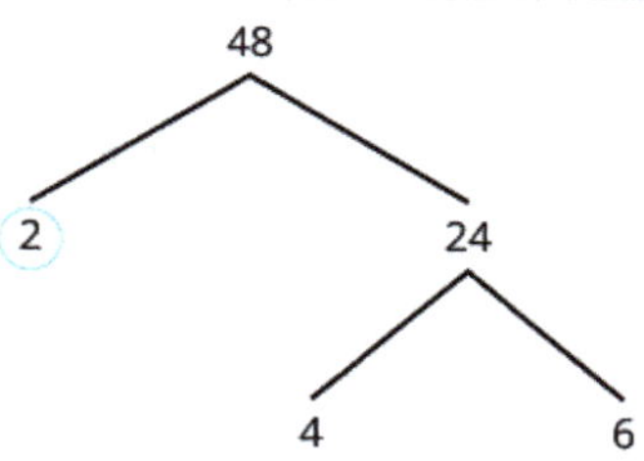

Neither factor is prime, so we do not circle either.
We factor the 4, using 2 and 2.
We factor 6, using 2 and 3.

We circle the 2s and the 3 since they are prime. Now all of the branches end in a prime.

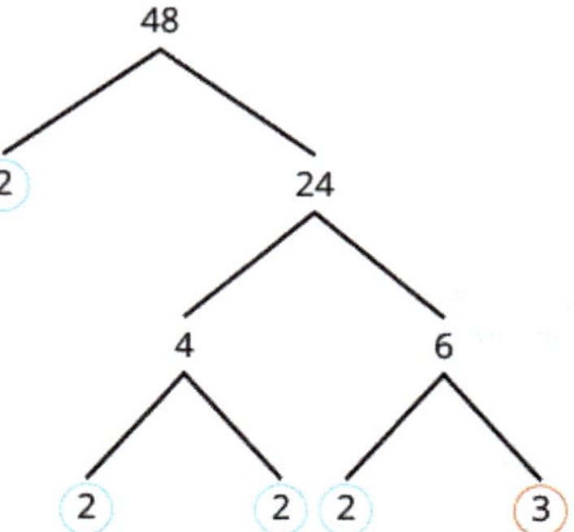

Write the product of the circled numbers.

Write in exponential form.

Check this on your own by multiplying all the factors together. The result should be

TRY IT : : 2.95 Find the prime factorization using the factor tree method:

TRY IT : : 2.96 Find the prime factorization using the factor tree method:

EXAMPLE 2.49

Find the prime factorization of 84 using the factor tree method.

Solution

We start with the factor pair 4 and 21.
Neither factor is prime so we factor them further.

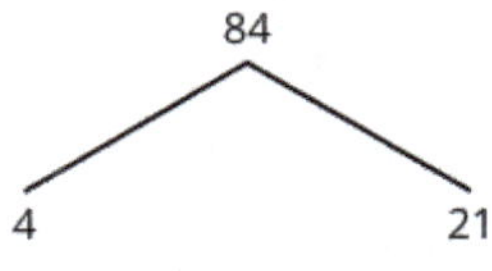

Now the factors are all prime, so we circle them.

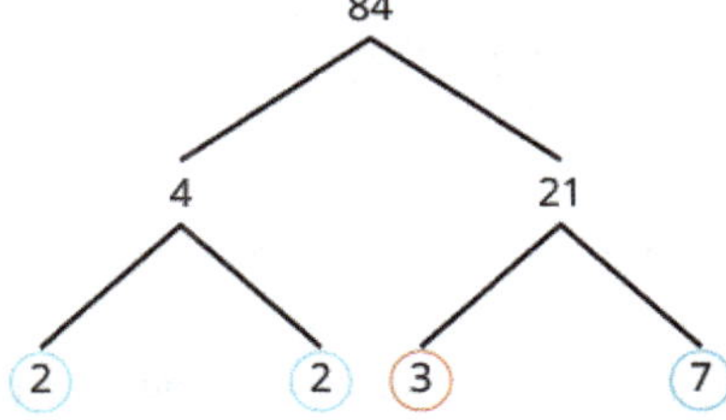

Then we write 84 as the product of all circled primes.

Draw a factor tree of

> **TRY IT : :** 2.97 Find the prime factorization using the factor tree method:

> **TRY IT : :** 2.98 Find the prime factorization using the factor tree method:

Prime Factorization Using the Ladder Method

The ladder method is another way to find the prime factors of a composite number. It leads to the same result as the factor tree method. Some people prefer the ladder method to the factor tree method, and vice versa.

To begin building the "ladder," divide the given number by its smallest prime factor. For example, to start the ladder for we divide by the smallest prime factor of

$$\begin{array}{r} 18 \\ 2\overline{)36} \end{array}$$

To add a "step" to the ladder, we continue dividing by the same prime until it no longer divides evenly.

$$\begin{array}{r} 9 \\ 2\overline{)18} \\ 2\overline{)\ 36} \end{array}$$

Then we divide by the next prime; so we divide by

$$\begin{array}{r} 3 \\ 3\overline{)9} \\ 2\overline{)18} \\ 2\overline{)\ 36} \end{array}$$

We continue dividing up the ladder in this way until the quotient is prime. Since the quotient, is prime, we stop here.

Do you see why the ladder method is sometimes called stacked division?

The prime factorization is the product of all the primes on the sides and top of the ladder.

Notice that the result is the same as we obtained with the factor tree method.

HOW TO : : FIND THE PRIME FACTORIZATION OF A COMPOSITE NUMBER USING THE LADDER METHOD.

Step 1. Divide the number by the smallest prime.

Step 2. Continue dividing by that prime until it no longer divides evenly.

Step 3. Divide by the next prime until it no longer divides evenly.

Step 4. Continue until the quotient is a prime.

Step 5. Write the composite number as the product of all the primes on the sides and top of the ladder.

EXAMPLE 2.50

Find the prime factorization of using the ladder method.

Solution

Divide the number by the smallest prime, which is 2.	$2\overline{)120}$ with quotient 60
Continue dividing by 2 until it no longer divides evenly.	$2\overline{)120}$, $2\overline{)60}$, $2\overline{)30}$, quotient 15
Divide by the next prime, 3.	$2\overline{)120}$, $2\overline{)60}$, $2\overline{)30}$, $3\overline{)15}$, quotient 5
The quotient, 5, is prime, so the ladder is complete. Write the prime factorization of 120.	

Check this yourself by multiplying the factors. The result should be

TRY IT : : 2.99 Find the prime factorization using the ladder method:

TRY IT : : 2.100 Find the prime factorization using the ladder method:

EXAMPLE 2.51

Find the prime factorization of using the ladder method.

Solution

Divide the number by the smallest prime, 2.	$2\overline{)48}$ quotient 24
Continue dividing by 2 until it no longer divides evenly.	$2\overline{)48}$, $2\overline{)24}$, $2\overline{)12}$, $2\overline{)6}$, quotient 3
The quotient, 3, is prime, so the ladder is complete. Write the prime factorization of 48.	

TRY IT : : 2.101 Find the prime factorization using the ladder method.

TRY IT : : 2.102 Find the prime factorization using the ladder method.

Find the Least Common Multiple (LCM) of Two Numbers

One of the reasons we look at multiples and primes is to use these techniques to find the least common multiple of two numbers. This will be useful when we add and subtract fractions with different denominators.

Listing Multiples Method

A common multiple of two numbers is a number that is a multiple of both numbers. Suppose we want to find common multiples of and We can list the first several multiples of each number. Then we look for multiples that are common to both lists—these are the common multiples.

We see that and appear in both lists. They are common multiples of and We would find more common multiples if we continued the list of multiples for each.

The smallest number that is a multiple of two numbers is called the **least common multiple** (LCM). So the least LCM of and is

HOW TO : : FIND THE LEAST COMMON MULTIPLE (LCM) OF TWO NUMBERS BY LISTING MULTIPLES.

Step 1. List the first several multiples of each number.

Step 2. Look for multiples common to both lists. If there are no common multiples in the lists, write out additional multiples for each number.

Step 3. Look for the smallest number that is common to both lists.

Step 4. This number is the LCM.

EXAMPLE 2.52

Find the LCM of and by listing multiples.

Solution

List the first several multiples of and of Identify the first common multiple.

The smallest number to appear on both lists is so is the least common multiple of and

Notice that is on both lists, too. It is a common multiple, but it is not the least common multiple.

> **TRY IT : :** 2.103 Find the least common multiple (LCM) of the given numbers:

> **TRY IT : :** 2.104 Find the least common multiple (LCM) of the given numbers:

Prime Factors Method

Another way to find the least common multiple of two numbers is to use their prime factors. We'll use this method to find the LCM of and

We start by finding the prime factorization of each number.

Then we write each number as a product of primes, matching primes vertically when possible.

Now we bring down the primes in each column. The LCM is the product of these factors.

$$\begin{array}{l} 12 = 2 \cdot 2 \cdot 3 \\ 18 = 2 \cdot \quad\ 3 \cdot 3 \\ \hline \text{LCM} = 2 \cdot 2 \cdot 3 \cdot 3 \\ \text{LCM} = 2 \cdot 2 \cdot 3 \cdot 3 = 36 \end{array}$$

Notice that the prime factors of and the prime factors of are included in the LCM. By matching up the common primes, each common prime factor is used only once. This ensures that is the least common multiple.

HOW TO : : FIND THE LCM USING THE PRIME FACTORS METHOD.

Step 1. Find the prime factorization of each number.
Step 2. Write each number as a product of primes, matching primes vertically when possible.
Step 3. Bring down the primes in each column.
Step 4. Multiply the factors to get the LCM.

EXAMPLE 2.53

Find the LCM of and using the prime factors method.

Solution

Write each number as a product of primes.	$15 = 3 \cdot 5$ $\quad 18 = 2 \cdot 3 \cdot 3$
Write each number as a product of primes, matching primes vertically when possible.	$15 = \quad\ 3 \cdot \quad 5$ $18 = 2 \cdot 3 \cdot 3$
Bring down the primes in each column.	$\begin{array}{l} 15 = \quad\ 3 \cdot \quad 5 \\ 18 = 2 \cdot 3 \cdot 3 \\ \hline \text{LCM} = 2 \cdot 3 \cdot 3 \cdot 5 \end{array}$
Multiply the factors to get the LCM.	The LCM of 15 and 18 is 90.

TRY IT : : 2.105 Find the LCM using the prime factors method.

TRY IT : : 2.106 Find the LCM using the prime factors method.

EXAMPLE 2.54

Find the LCM of and using the prime factors method.

Solution

Write the prime factorization of each number.	$50 = 2 \cdot 5 \cdot 5$ $100 = 2 \cdot 2 \cdot 5 \cdot 5$
Write each number as a product of primes, matching primes vertically when possible.	$50 = \quad 2 \cdot 5 \cdot 5$ $100 = 2 \cdot 2 \cdot 5 \cdot 5$
Bring down the primes in each column.	$50 = \quad 2 \cdot 5 \cdot 5$ $100 = 2 \cdot 2 \cdot 5 \cdot 5$ $\text{LCM} = 2 \cdot 2 \cdot 5 \cdot 5$
Multiply the factors to get the LCM.	The LCM of 50 and 100 is 100.

TRY IT : : 2.107 Find the LCM using the prime factors method:

TRY IT : : 2.108 Find the LCM using the prime factors method:

MEDIA : : ACCESS ADDITIONAL ONLINE RESOURCES

- Ex 1: Prime Factorization (http://openstaxcollege.org/l/24PrimeFactor1)
- Ex 2: Prime Factorization (http://openstaxcollege.org/l/24PrimeFactor2)
- Ex 3: Prime Factorization (http://openstaxcollege.org/l/24PrimeFactor3)
- Ex 1: Prime Factorization Using Stacked Division (http://openstaxcollege.org/l/24stackeddivis)
- Ex 2: Prime Factorization Using Stacked Division (http://openstaxcollege.org/l/24stackeddivis2)
- The Least Common Multiple (http://openstaxcollege.org/l/24LCM)
- Example: Determining the Least Common Multiple Using a List of Multiples (http://openstaxcollege.org/l/24LCM2)
- Example: Determining the Least Common Multiple Using Prime Factorization (http://openstaxcollege.org/l/24LCMFactor)

2.5 EXERCISES

Practice Makes Perfect

Find the Prime Factorization of a Composite Number

In the following exercises, find the prime factorization of each number using the factor tree method.

267.

268.

269.

270.

271.

272.

273.

274.

275.

276. 1560

In the following exercises, find the prime factorization of each number using the ladder method.

277.

278.

279.

280.

281.

282.

283.

284.

285.

286.

In the following exercises, find the prime factorization of each number using any method.

287.

288.

289.

290.

291.

292.

293.

294.

Find the Least Common Multiple (LCM) of Two Numbers

In the following exercises, find the least common multiple (LCM) by listing multiples.

295.

296.

297.

298.

299.

300.

301.

302.

In the following exercises, find the least common multiple (LCM) by using the prime factors method.

303.

304.

305.

306.

307.

308.

In the following exercises, find the least common multiple (LCM) using any method.

309.

310.

311.

312.

Everyday Math

313. Grocery shopping Hot dogs are sold in packages of ten, but hot dog buns come in packs of eight. What is the smallest number of hot dogs and buns that can be purchased if you want to have the same number of hot dogs and buns? (Hint: it is the LCM!)

314. Grocery shopping Paper plates are sold in packages of and party cups come in packs of What is the smallest number of plates and cups you can purchase if you want to have the same number of each? (Hint: it is the LCM!)

Writing Exercises

315. Do you prefer to find the prime factorization of a composite number by using the factor tree method or the ladder method? Why?

316. Do you prefer to find the LCM by listing multiples or by using the prime factors method? Why?

Self Check

After completing the exercises, use this checklist to evaluate your mastery of the objectives of this section.

I can...	Confidently	With some help	No-I don't get it!
find the prime factorization of a composite number.			
find the least common multiple (LCM) of two numbers.			

Overall, after looking at the checklist, do you think you are well-prepared for the next Chapter? Why or why not?

CHAPTER 2 REVIEW

KEY TERMS

coefficient The constant that multiplies the variable(s) in a term is called the coefficient.

composite number A composite number is a counting number that is not prime.

divisibility If a number is a multiple of , then we say that is divisible by .

equation An equation is made up of two expressions connected by an equal sign.

evaluate To evaluate an algebraic expression means to find the value of the expression when the variable is replaced by a given number.

expressions An expression is a number, a variable, or a combination of numbers and variables and operation symbols.

least common multiple The smallest number that is a multiple of two numbers is called the least common multiple (LCM).

like terms Terms that are either constants or have the same variables with the same exponents are like terms.

multiple of a number A number is a multiple of if it is the product of a counting number and .

prime factorization The prime factorization of a number is the product of prime numbers that equals the number.

prime number A prime number is a counting number greater than 1 whose only factors are 1 and itself.

solution of an equation A solution to an equation is a value of a variable that makes a true statement when substituted into the equation. The process of finding the solution to an equation is called solving the equation.

term A term is a constant or the product of a constant and one or more variables.

KEY CONCEPTS

2.1 Use the Language of Algebra

Operation	Notation	Say:	The result is...
Addition			the sum of and
Multiplication			The product of and
Subtraction			the difference of and
Division	— ⌐	divided by	The quotient of and

- **Equality Symbol**
  - is read as is equal to
  - The symbol is called the equal sign.
- **Inequality**
  - is read is less than
  - is to the left of on the number line

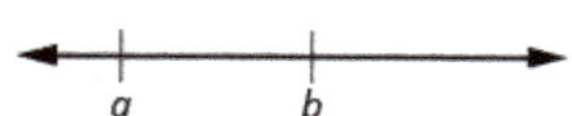

  - is read is greater than
  - is to the right of on the number line

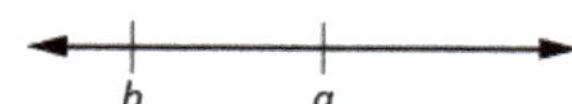

Algebraic Notation	Say
	is equal to
	is not equal to
	is less than
	is greater than
	is less than or equal to
	is greater than or equal to

Table 2.14

- **Exponential Notation**
    - For any expression is a factor multiplied by itself times, if is a positive integer.
    - means multiply factors of

base → a^n ← exponent

$$a^n = \underbrace{a \cdot a \cdot a \cdot \ldots \cdot a}_{n \text{ factors}}$$

    - The expression of is read to the power.

Order of Operations When simplifying mathematical expressions perform the operations in the following order:

- Parentheses and other Grouping Symbols: Simplify all expressions inside the parentheses or other grouping symbols, working on the innermost parentheses first.
- Exponents: Simplify all expressions with exponents.
- Multiplication and Division: Perform all multiplication and division in order from left to right. These operations have equal priority.
- Addition and Subtraction: Perform all addition and subtraction in order from left to right. These operations have equal priority.

2.2 Evaluate, Simplify, and Translate Expressions

- Combine like terms.

  Step 1. Identify like terms.

  Step 2. Rearrange the expression so like terms are together.

  Step 3. Add the coefficients of the like terms

2.3 Solving Equations Using the Subtraction and Addition Properties of Equality

- Determine whether a number is a solution to an equation.

  Step 1. Substitute the number for the variable in the equation.

  Step 2. Simplify the expressions on both sides of the equation.

  Step 3. Determine whether the resulting equation is true. If it is true, the number is a solution.

  If it is not true, the number is not a solution.
- Subtraction Property of Equality
    - For any numbers , , and ,

if	
then	

- Solve an equation using the Subtraction Property of Equality.
  Step 1. Use the Subtraction Property of Equality to isolate the variable.
  Step 2. Simplify the expressions on both sides of the equation.
  Step 3. Check the solution.
- Addition Property of Equality
  - For any numbers , , and ,

if	
then	

- Solve an equation using the Addition Property of Equality.
  Step 1. Use the Addition Property of Equality to isolate the variable.
  Step 2. Simplify the expressions on both sides of the equation.
  Step 3. Check the solution.

2.4 Find Multiples and Factors

Divisibility Tests	
A number is divisible by	
2	if the last digit is **0, 2, 4, 6,** or **8**
3	if the sum of the digits is divisible by **3**
5	if the last digit is **5** or **0**
6	if divisible by both **2** and **3**
10	if the last digit is **0**

- Factors If , then and are factors of , and is the product of and .
- Find all the factors of a counting number.
  Step 1. Divide the number by each of the counting numbers, in order, until the quotient is smaller than the divisor.
    a. If the quotient is a counting number, the divisor and quotient are a pair of factors.
    b. If the quotient is not a counting number, the divisor is not a factor.
  Step 2. List all the factor pairs.
  Step 3. Write all the factors in order from smallest to largest.
- Determine if a number is prime.
  Step 1. Test each of the primes, in order, to see if it is a factor of the number.
  Step 2. Start with 2 and stop when the quotient is smaller than the divisor or when a prime factor is found.
  Step 3. If the number has a prime factor, then it is a composite number. If it has no prime factors, then the number is prime.

2.5 Prime Factorization and the Least Common Multiple

- Find the prime factorization of a composite number using the tree method.
  Step 1. Find any factor pair of the given number, and use these numbers to create two branches.
  Step 2. If a factor is prime, that branch is complete. Circle the prime.
  Step 3. If a factor is not prime, write it as the product of a factor pair and continue the process.
  Step 4. Write the composite number as the product of all the circled primes.
- Find the prime factorization of a composite number using the ladder method.
  Step 1. Divide the number by the smallest prime.
  Step 2. Continue dividing by that prime until it no longer divides evenly.
  Step 3. Divide by the next prime until it no longer divides evenly.
  Step 4. Continue until the quotient is a prime.
  Step 5. Write the composite number as the product of all the primes on the sides and top of the ladder.
- Find the LCM by listing multiples.
  Step 1. List the first several multiples of each number.
  Step 2. Look for multiples common to both lists. If there are no common multiples in the lists, write out additional multiples for each number.
  Step 3. Look for the smallest number that is common to both lists.
  Step 4. This number is the LCM.
- Find the LCM using the prime factors method.
  Step 1. Find the prime factorization of each number.
  Step 2. Write each number as a product of primes, matching primes vertically when possible.
  Step 3. Bring down the primes in each column.
  Step 4. Multiply the factors to get the LCM.

REVIEW EXERCISES

2.1 Use the Language of Algebra

Use Variables and Algebraic Symbols

In the following exercises, translate from algebra to English.

317.

318.

319.

320.

321.

322.

323.

324.

Identify Expressions and Equations

In the following exercises, determine if each is an expression or equation.

325.

326.

327.

328.

Simplify Expressions with Exponents

In the following exercises, write in exponential form.

329.

330.

331.

332.

In the following exercises, write in expanded form.

333. **334.** **335.**

336.

In the following exercises, simplify each expression.

337. **338.** **339.**

340.

Simplify Expressions Using the Order of Operations

In the following exercises, simplify.

341. **342.** **343.**

344. **345.** **346.**

347. **348.**

2.2 Evaluate, Simplify, and Translate Expressions

Evaluate an Expression

In the following exercises, evaluate the following expressions.

349. **350.** **351.**

352.

Identify Terms, Coefficients and Like Terms

In the following exercises, identify the terms in each expression.

353. **354.**

In the following exercises, identify the coefficient of each term.

355. **356.**

In the following exercises, identify the like terms.

357. **358.**

Simplify Expressions by Combining Like Terms

In the following exercises, simplify the following expressions by combining like terms.

359. **360.** **361.**

362. **363.** **364.**

365. **366.**

Translate English Phrases to Algebraic Expressions

In the following exercises, translate the following phrases into algebraic expressions.

367. the difference of and

368. the sum of and twice

369. the product of and

370. the quotient of and

371. times the sum of and

372. less than the product of and

373. Jack bought a sandwich and a coffee. The cost of the sandwich was more than the cost of the coffee. Call the cost of the coffee Write an expression for the cost of the sandwich.

374. The number of poetry books on Brianna's bookshelf is less than twice the number of novels. Call the number of novels Write an expression for the number of poetry books.

2.3 Solve Equations Using the Subtraction and Addition Properties of Equality

Determine Whether a Number is a Solution of an Equation

In the following exercises, determine whether each number is a solution to the equation.

375.

376.

377.

378.

379.

380.

Model the Subtraction Property of Equality

In the following exercises, write the equation modeled by the envelopes and counters and then solve the equation using the subtraction property of equality.

381.

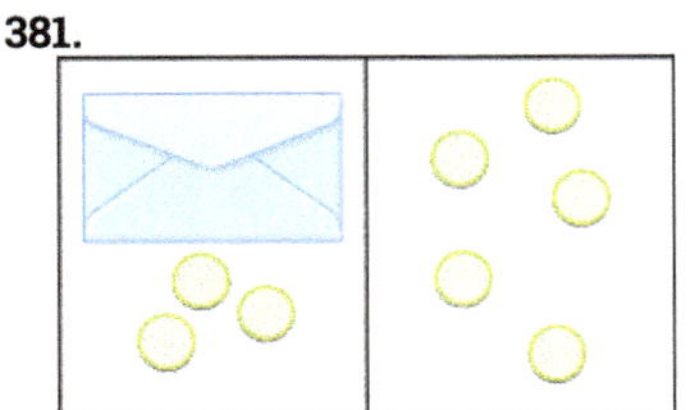

382.

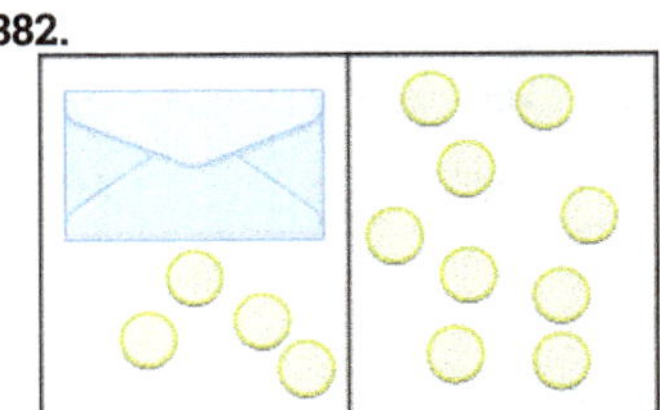

Solve Equations using the Subtraction Property of Equality

In the following exercises, solve each equation using the subtraction property of equality.

383.

384.

385.

386.

Solve Equations using the Addition Property of Equality

In the following exercises, solve each equation using the addition property of equality.

387.

388.

389.

390.

Translate English Sentences to Algebraic Equations

In the following exercises, translate each English sentence into an algebraic equation.

391. The sum of and is equal to

392. The difference of and is equal to

393. The product of and is equal to

394. The quotient of and is equal to

395. Twice the difference of and gives

396. The sum of five times and is

Translate to an Equation and Solve

In the following exercises, translate each English sentence into an algebraic equation and then solve it.

397. Eight more than is equal to

398. less than is

399. The difference of and is

400. The sum of and is

Mixed Practice

In the following exercises, solve each equation.

401.

402.

403.

404.

405.

406.

407.

408.

2.4 Find Multiples and Factors

Identify Multiples of Numbers

In the following exercises, list all the multiples less than for each of the following.

409.

410.

411.

412.

Use Common Divisibility Tests

In the following exercises, using the divisibility tests, determine whether each number is divisible by

413.

414.

415.

416.

Find All the Factors of a Number

In the following exercises, find all the factors of each number.

417.

418.

419.

420.

Identify Prime and Composite Numbers

In the following exercises, identify each number as prime or composite.

421.

422.

423.

424.

2.5 Prime Factorization and the Least Common Multiple

Find the Prime Factorization of a Composite Number

In the following exercises, find the prime factorization of each number.

425. **426.** **427.**

428.

Find the Least Common Multiple of Two Numbers

In the following exercises, find the least common multiple of each pair of numbers.

429. **430.** **431.**

432.

Everyday Math

433. Describe how you have used two topics from The Language of Algebra chapter in your life outside of your math class during the past month.

PRACTICE TEST

In the following exercises, translate from an algebraic equation to English phrases.

434.

435.

In the following exercises, identify each as an expression or equation.

436.

437.

438.

439.

Write in exponential form.

Write in expanded form and then simplify.

In the following exercises, simplify, using the order of operations.

440.

441.

442.

443.

444.

445.

In the following exercises, evaluate each expression.

446.

447.

448.

449.

450. Simplify by combining like terms.

In the following exercises, translate each phrase into an algebraic expression.

451. more than

452. the quotient of and

453. three times the difference of

454. Caroline has fewer earrings on her left ear than on her right ear. Call the number of earrings on her right ear, Write an expression for the number of earrings on her left ear.

In the following exercises, solve each equation.

455.

456.

In the following exercises, translate each English sentence into an algebraic equation and then solve it.

457. less than is

458. the sum of and is

459. List all the multiples of that are less than

460. Find all the factors of

461. Find the prime factorization of

462. Find the LCM (Least Common Multiple) of and

3 INTEGERS

Figure 3.1 The peak of Mount Everest. (credit: Gunther Hagleitner, Flickr)

Chapter Outline

Introduction

At over 29,000 feet, Mount Everest stands as the tallest peak on land. Located along the border of Nepal and China, Mount Everest is also known for its extreme climate. Near the summit, temperatures never rise above freezing. Every year, climbers from around the world brave the extreme conditions in an effort to scale the tremendous height. Only some are successful. Describing the drastic change in elevation the climbers experience and the change in temperatures requires using numbers that extend both above and below zero. In this chapter, we will describe these kinds of numbers and operations using them.

3.1 Introduction to Integers

Learning Objectives

By the end of this section, you will be able to:

- Locate positive and negative numbers on the number line
- Order positive and negative numbers
- Find opposites
- Simplify expressions with absolute value
- Translate word phrases to expressions with integers

BE PREPARED : : 3.1 Before you get started, take this readiness quiz.

Plot on a number line.

If you missed this problem, review Example 1.1.

BE PREPARED : : 3.2 Fill in the appropriate symbol:

If you missed this problem, review Example 2.3.

Locate Positive and Negative Numbers on the Number Line

Do you live in a place that has very cold winters? Have you ever experienced a temperature below zero? If so, you are already familiar with negative numbers. A **negative number** is a number that is less than Very cold temperatures are measured in degrees below zero and can be described by negative numbers. For example, (read as "negative one degree Fahrenheit") is below A minus sign is shown before a number to indicate that it is negative. Figure 3.2 shows which is below

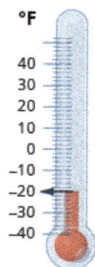

°F
40
30
20
10
0
–10
–20
–30
–40

Figure 3.2 Temperatures below zero are described by negative numbers.

Temperatures are not the only negative numbers. A bank overdraft is another example of a negative number. If a person writes a check for more than he has in his account, his balance will be negative.

Elevations can also be represented by negative numbers. The elevation at sea level is Elevations above sea level are positive and elevations below sea level are negative. The elevation of the Dead Sea, which borders Israel and Jordan, is about below sea level, so the elevation of the Dead Sea can be represented as See Figure 3.3.

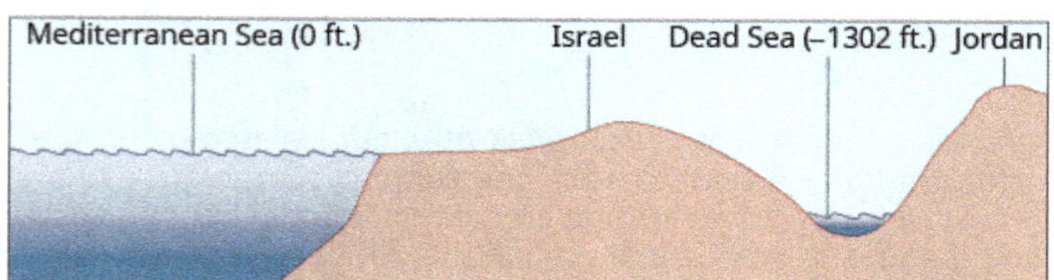

Figure 3.3 The surface of the Mediterranean Sea has an elevation of The diagram shows that nearby mountains have higher (positive) elevations whereas the Dead Sea has a lower (negative) elevation.

Depths below the ocean surface are also described by negative numbers. A submarine, for example, might descend to a depth of Its position would then be as labeled in Figure 3.4.

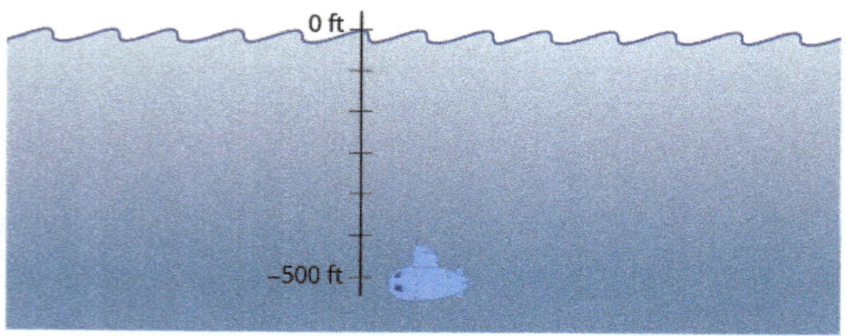

Figure 3.4 Depths below sea level are described by negative numbers. A submarine below sea level is at

Both positive and negative numbers can be represented on a number line. Recall that the number line created in Add Whole Numbers started at and showed the counting numbers increasing to the right as shown in Figure 3.5. The counting numbers on the number line are all positive. We could write a plus sign, before a positive number such as or but it is customary to omit the plus sign and write only the number. If there is no sign, the number is assumed to be positive.

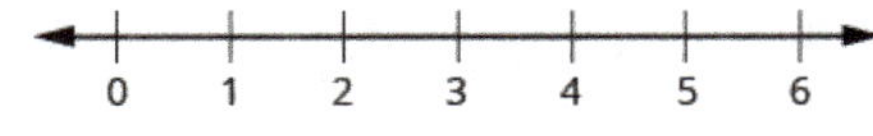

Figure 3.5

Now we need to extend the number line to include negative numbers. We mark several units to the left of zero, keeping the intervals the same width as those on the positive side. We label the marks with negative numbers, starting with at the first mark to the left of at the next mark, and so on. See Figure 3.6.

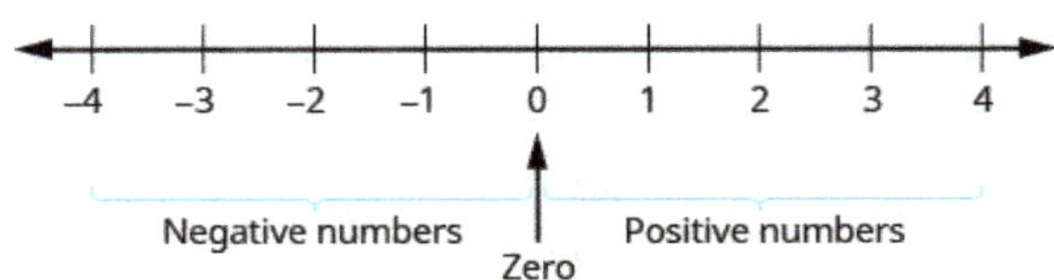

Figure 3.6 On a number line, positive numbers are to the right of zero. Negative numbers are to the left of zero. What about zero? Zero is neither positive nor negative.

The arrows at either end of the line indicate that the number line extends forever in each direction. There is no greatest positive number and there is no smallest negative number.

MANIPULATIVE MATHEMATICS

Doing the Manipulative Mathematics activity "Number Line-part 2" will help you develop a better understanding of integers.

EXAMPLE 3.1

Plot the numbers on a number line:

Solution

Draw a number line. Mark in the center and label several units to the left and right.

To plot start at and count three units to the right. Place a point as shown in Figure 3.7.

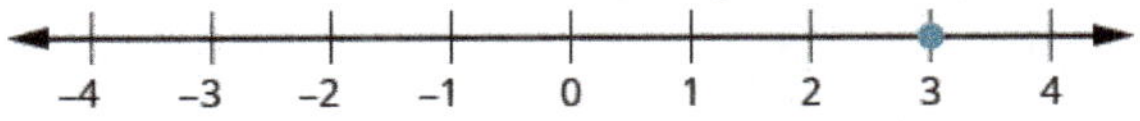

Figure 3.7

To plot start at and count three units to the left. Place a point as shown in Figure 3.8.

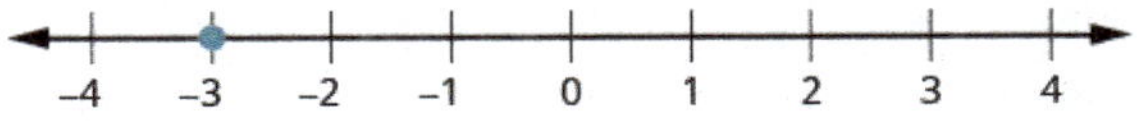

Figure 3.8

To plot start at and count two units to the left. Place a point as shown in Figure 3.9.

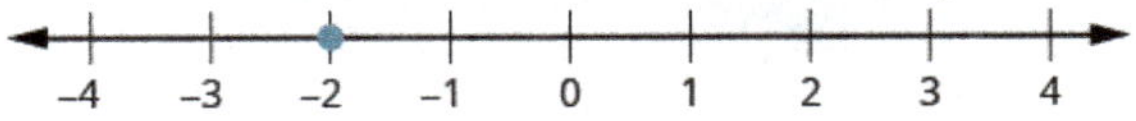

Figure 3.9

TRY IT : : 3.1 Plot the numbers on a number line.

TRY IT : : 3.2 Plot the numbers on a number line.

Order Positive and Negative Numbers

We can use the number line to compare and order positive and negative numbers. Going from left to right, numbers increase in value. Going from right to left, numbers decrease in value. See Figure 3.10.

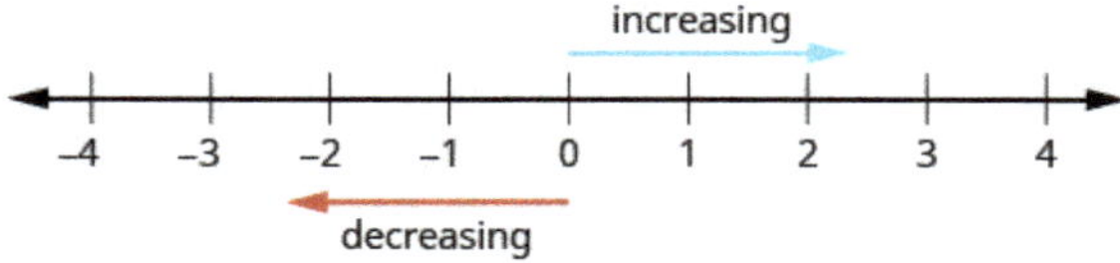

Figure 3.10

Just as we did with positive numbers, we can use inequality symbols to show the ordering of positive and negative numbers. Remember that we use the notation (read *is less than*) when is to the left of on the number line. We write (read *is greater than*) when is to the right of on the number line. This is shown for the numbers and in Figure 3.11.

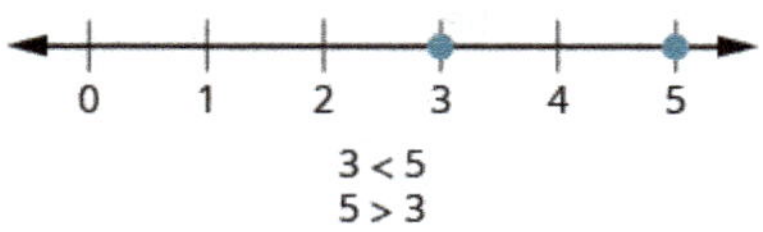

Figure 3.11 The number is to the left of on the number line. So is less than and is greater than

The numbers lines to follow show a few more examples.

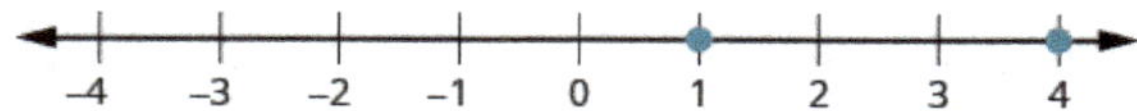

is to the right of on the number line, so

is to the left of on the number line, so

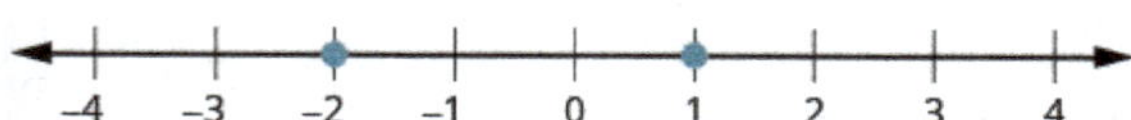

is to the left of on the number line, so

is to the right of on the number line, so

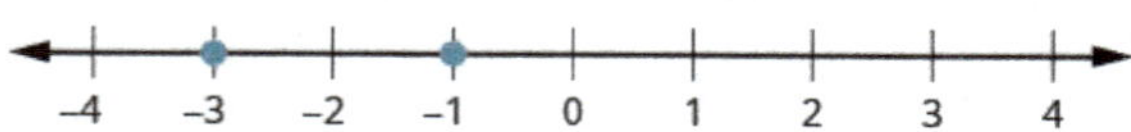

is to the right of on the number line, so

is to the left of on the number line, so

EXAMPLE 3.2

Order each of the following pairs of numbers using or

Solution

Begin by plotting the numbers on a number line as shown in Figure 3.12.

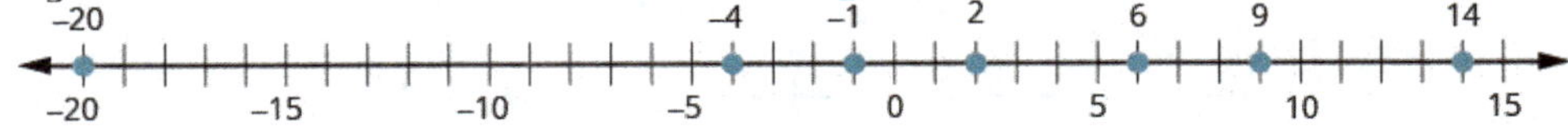

Figure 3.12

Compare 14 and 6.
14 is to the right of 6 on the number line.

Compare −1 and 9.
−1 is to the left of 9 on the number line.

Compare −1 and −4.
−1 is to the right of −4 on the number line.

Compare 2 and −20.
2 is to the right of −20 on the number line.

TRY IT : : 3.3 Order each of the following pairs of numbers using or

TRY IT : : 3.4 Order each of the following pairs of numbers using or

Find Opposites

On the number line, the negative numbers are a mirror image of the positive numbers with zero in the middle. Because the numbers and are the same distance from zero, they are called **opposites**. The opposite of is and the opposite of is as shown in Figure 3.13(a). Similarly, and are opposites as shown in Figure 3.13(b).

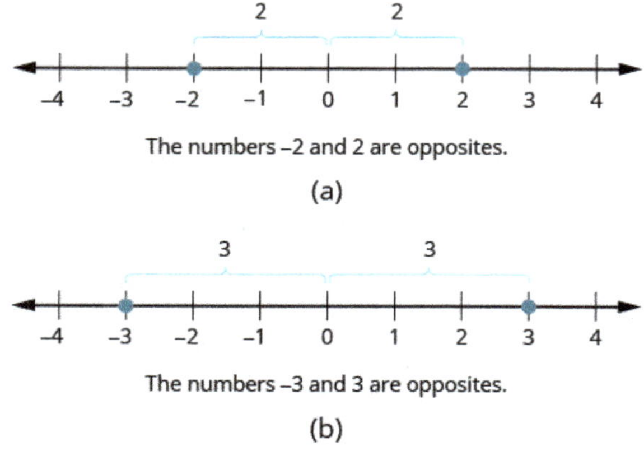

The numbers −2 and 2 are opposites.

(a)

The numbers −3 and 3 are opposites.

(b)

Figure 3.13

Opposite

The opposite of a number is the number that is the same distance from zero on the number line, but on the opposite side of zero.

EXAMPLE 3.3

Find the opposite of each number:

Solution

The number is the same distance from as but on the opposite side of So is the opposite of as shown in Figure 3.14.

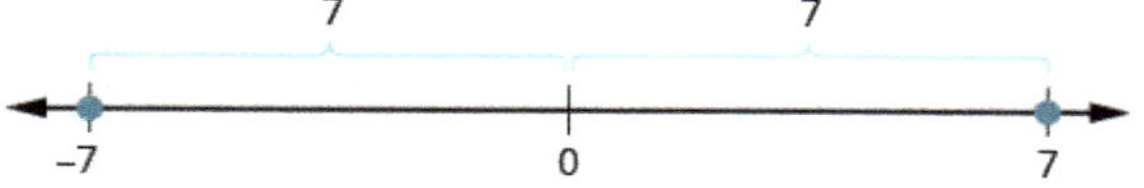

Figure 3.14

The number is the same distance from as , but on the opposite side of So is the opposite of as shown in Figure 3.15.

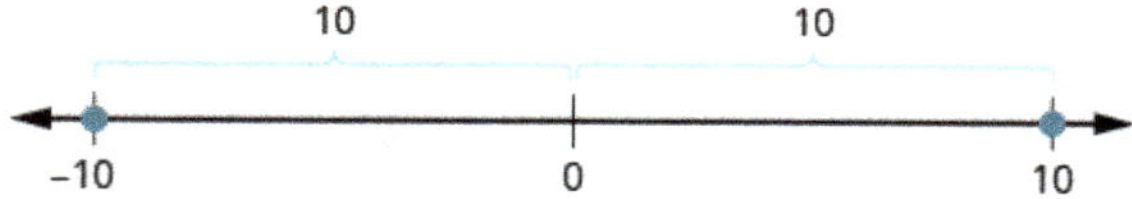

Figure 3.15

TRY IT : : 3.5 Find the opposite of each number:

TRY IT : : 3.6 Find the opposite of each number:

Opposite Notation

Just as the same word in English can have different meanings, the same symbol in algebra can have different meanings.

The specific meaning becomes clear by looking at how it is used. You have seen the symbol in three different ways.

Between two numbers, the symbol indicates the operation of subtraction.
We read as 10 *minus* .

In front of a number, the symbol indicates a negative number.
We read as *negative eight*.

In front of a variable or a number, it indicates the opposite.
We read as *the opposite of* .

Here we have two signs. The sign in the parentheses indicates that the number is negative 2.
The sign outside the parentheses indicates the opposite. We read as *the opposite of*

Opposite Notation

means the opposite of the number

The notation is read *the opposite of*

EXAMPLE 3.4

Simplify:

 **Solution**

The opposite of is

TRY IT : : 3.7 Simplify:

TRY IT : : 3.8 Simplify:

Integers

The set of counting numbers, their opposites, and is the set of integers.

Integers

Integers are counting numbers, their opposites, and zero.

We must be very careful with the signs when evaluating the opposite of a variable.

EXAMPLE 3.5

Evaluate

when when

Solution

To evaluate when , substitute for .

Substitute 8 for *x*.	$-(8)$
Simplify.	

To evaluate when , substitute for .

Substitute −8 for *x*.	$-(-8)$
Simplify.	

TRY IT : : 3.9 Evaluate

TRY IT : : 3.10 Evaluate:

Simplify Expressions with Absolute Value

We saw that numbers such as and are opposites because they are the same distance from on the number line. They are both five units from The distance between and any number on the number line is called the **absolute value** of that number. Because distance is never negative, the absolute value of any number is never negative.

The symbol for absolute value is two vertical lines on either side of a number. So the absolute value of is written as and the absolute value of is written as as shown in Figure 3.16.

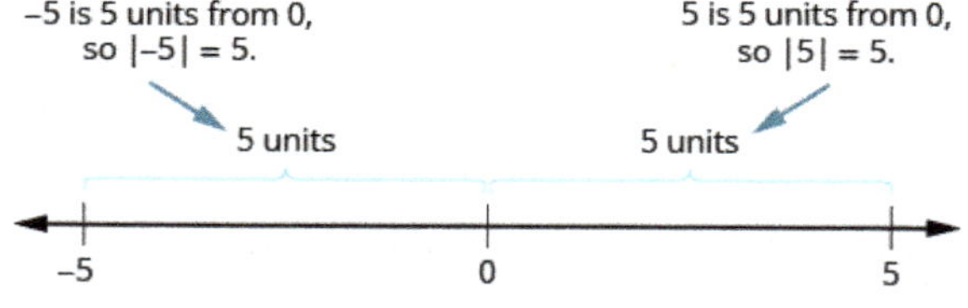
−5 is 5 units from 0,
so |−5| = 5.
5 is 5 units from 0,
so |5| = 5.
5 units
5 units
−5
0
5

Figure 3.16

Absolute Value

The absolute value of a number is its distance from on the number line.

The absolute value of a number is written as

EXAMPLE 3.6

Simplify:

Solution

3 is 3 units from zero.

−44 is 44 units from zero.

0 is already at zero.

TRY IT : : 3.11 Simplify:

TRY IT : : 3.12 Simplify:

We treat absolute value bars just like we treat parentheses in the order of operations. We simplify the expression inside first.

EXAMPLE 3.7

Evaluate:

Solution

To find when	
Substitute −35 for *x*.	$\|-35\|$
Take the absolute value.	

To find when	
Substitute -20 for y.	$\|-(-20)\|$
Simplify.	
Take the absolute value.	

To find when	
Substitute 12 for u.	$-\|12\|$
Take the absolute value.	

To find when	
Substitute -14 for p.	$-\|-14\|$
Take the absolute value.	

Notice that the result is negative only when there is a negative sign outside the absolute value symbol.

TRY IT : : 3.13

Evaluate:

TRY IT : : 3.14

EXAMPLE 3.8

Fill in for each of the following:

Solution

To compare two expressions, simplify each one first. Then compare.

Simplify.
Order.

Simplify.
Order.

Simplify.
Order.

Simplify.
Order.

TRY IT : : 3.15

Fill in

TRY IT : : 3.16

Fill in for each of the following:

Absolute value bars act like grouping symbols. First simplify inside the absolute value bars as much as possible. Then take the absolute value of the resulting number, and continue with any operations outside the absolute value symbols.

EXAMPLE 3.9

Simplify:

Solution

For each expression, follow the order of operations. Begin inside the absolute value symbols just as with parentheses.

	$\lvert 9-3 \rvert$
Simplify inside the absolute value sign.	$\lvert 6 \rvert$
Take the absolute value.	6

	$4\lvert -2 \rvert$
Take the absolute value.	$4 \cdot 2$
Multiply.	8

TRY IT : : 3.17 Simplify:

TRY IT : : 3.18 Simplify:

EXAMPLE 3.10

Simplify:

Solution

For each expression, follow the order of operations. Begin inside the absolute value symbols just as with parentheses.

	$\lvert 8+7 \rvert - \lvert 5+6 \rvert$
Simplify inside each absolute value sign.	$\lvert 15 \rvert - \lvert 11 \rvert$
Subtract.	4

TRY IT : : 3.19 Simplify:

TRY IT : : 3.20 Simplify:

EXAMPLE 3.11

Simplify:

Solution

We use the order of operations. Remember to simplify grouping symbols first, so parentheses inside absolute value symbols would be first.

Simplify in the parentheses first.
Multiply .
Subtract inside the absolute value sign.
Take the absolute value.
Subtract.

TRY IT : : 3.21 Simplify:

TRY IT : : 3.22 Simplify:

Translate Word Phrases into Expressions with Integers

Now we can translate word phrases into expressions with integers. Look for words that indicate a negative sign. For example, the word *negative* in "negative twenty" indicates So does the word *opposite* in "the opposite of

EXAMPLE 3.12

Translate each phrase into an expression with integers:

the opposite of positive fourteen the opposite of negative sixteen

two minus negative seven

Solution

the opposite of fourteen the opposite of −11 negative sixteen two minus negative seven

TRY IT : : 3.23

Translate each phrase into an expression with integers:

the opposite of positive nine the opposite of negative twenty

eleven minus negative four

TRY IT : : 3.24

Translate each phrase into an expression with integers:

the opposite of negative nineteen the opposite of twenty-two negative nine

negative eight minus negative five

As we saw at the start of this section, negative numbers are needed to describe many real-world situations. We'll look at some more applications of negative numbers in the next example.

EXAMPLE 3.13

Translate into an expression with integers:

The temperature is below zero. The football team had a gain of

The elevation of the Dead Sea is below sea level.

A checking account is overdrawn by

Solution

Look for key phrases in each sentence. Then look for words that indicate negative signs. Don't forget to include units of measurement described in the sentence.

	The temperature is 12 degrees Fahrenheit below zero.
Below zero tells us that 12 is a negative number.	

	The football team had a gain of 3 yards.
A *gain* tells us that 3 is a positive number.	yards

	The elevation of the Dead Sea is 1,302 feet below sea level.
Below sea level tells us that 1,302 is a negative number.	feet

	A checking account is overdrawn by $40.
Overdrawn tells us that 40 is a negative number.	

TRY IT : : 3.25 Translate into an expression with integers:

The football team had a gain of

TRY IT : : 3.26 Translate into an expression with integers:

The scuba diver was below the surface of the water.

MEDIA : : ACCESS ADDITIONAL ONLINE RESOURCES

- Introduction to Integers (http://openstaxcollege.org/l/24introinteger)
- Simplifying the Opposites of Negative Integers (http://openstaxcollege.org/l/24neginteger)
- Comparing Absolute Value of Integers (http://openstaxcollege.org/l/24abvalue)
- Comparing Integers Using Inequalities (http://openstaxcollege.org/l/24usinginequal)

3.1 EXERCISES

Practice Makes Perfect

Locate Positive and Negative Numbers on the Number Line

For the following exercises, draw a number line and locate and label the given points on that number line.

1.

2.

3.

4.

Order Positive and Negative Numbers on the Number Line

In the following exercises, order each of the following pairs of numbers, using or

5.

6.

7.

8.

Find Opposites

In the following exercises, find the opposite of each number.

9.

10.

11.

12.

In the following exercises, simplify.

13.

14.

15.

16.

In the following exercises, evaluate.

17.

18.

19.

20.

Simplify Expressions with Absolute Value

In the following exercises, simplify each absolute value expression.

21.

22.

23.

24.

In the following exercises, evaluate each absolute value expression.

25.

26.

27.

28.

In the following exercises, fill in to compare each expression.

29.

30.

31.

32.

In the following exercises, simplify each expression.

33.

34.

35.

36.

37.

38.

39.

40.

41.

42.

Translate Word Phrases into Expressions with Integers

Translate each phrase into an expression with integers. Do not simplify.

43.
the opposite of
the opposite of
negative three
minus negative

44.
the opposite of
the opposite of
negative nine
minus negative

45.
the opposite of
the opposite of
negative twelve
minus negative

46.

the opposite of

the opposite of

negative sixty

minus

47. a temperature of below zero

48. a temperature of below zero

49. an elevation of below sea level

50. an elevation of below sea level

51. a football play loss of

52. a football play gain of

53. a stock gain of

54. a stock loss of

55. a golf score one above par

56. a golf score of below par

Everyday Math

57. Elevation The highest elevation in the United States is Mount McKinley, Alaska, at above sea level. The lowest elevation is Death Valley, California, at below sea level. Use integers to write the elevation of:

Mount McKinley Death Valley

58. Extreme temperatures The highest recorded temperature on Earth is recorded in the Sahara Desert in 1922. The lowest recorded temperature is below recorded in Antarctica in 1983. Use integers to write the:

highest recorded temperature

lowest recorded temperature

59. State budgets In June, 2011, the state of Pennsylvania estimated it would have a budget surplus of That same month, Texas estimated it would have a budget deficit of Use integers to write the budget:

surplus deficit

60. College enrollments Across the United States, community college enrollment grew by students from to In California, community college enrollment declined by students from to Use integers to write the change in enrollment:

growth decline

Writing Exercises

61. Give an example of a negative number from your life experience.

62. What are the three uses of the "–" sign in algebra? Explain how they differ.

Self Check

After completing the exercises, use this checklist to evaluate your mastery of the objectives of this section.

I can...	Confidently	With some help	No-I don't get it!
locate positive and negative numbers on the number line.			
order positive and negative numbers.			
find opposites.			
simplify expressions with absolute value.			
translate word phrases to expressions with integers.			

If most of your checks were:

...confidently. Congratulations! You have achieved the objectives in this section. Reflect on the study skills you used so that you can continue to use them. What did you do to become confident of your ability to do these things? Be specific.

...with some help. This must be addressed quickly because topics you do not master become potholes in your road to success. In math, every topic builds upon previous work. It is important to make sure you have a strong foundation before you move on. Whom can you ask for help? Your fellow classmates and instructor are good resources. Is there a place on campus where math tutors are available? Can your study skills be improved?

...no—I don't get it! This is a warning sign and you must not ignore it. You should get help right away or you will quickly be overwhelmed. See your instructor as soon as you can to discuss your situation. Together you can come up with a plan to get you the help you need.

3.2 Add Integers

Learning Objectives

By the end of this section, you will be able to:

- Model addition of integers
- Simplify expressions with integers
- Evaluate variable expressions with integers
- Translate word phrases to algebraic expressions
- Add integers in applications

BE PREPARED : : 3.3 Before you get started, take this readiness quiz.

Evaluate when

If you missed this problem, review Example 2.13.

Simplify:

If you missed this problem, review Example 2.8.

Translate *the sum of* *and negative* into an algebraic expression.

If you missed this problem, review Table 2.7

Model Addition of Integers

Now that we have located positive and negative numbers on the number line, it is time to discuss arithmetic operations with integers.

Most students are comfortable with the addition and subtraction facts for positive numbers. But doing addition or subtraction with both positive and negative numbers may be more difficult. This difficulty relates to the way the brain learns.

The brain learns best by working with objects in the real world and then generalizing to abstract concepts. Toddlers learn quickly that if they have two cookies and their older brother steals one, they have only one left. This is a concrete example of Children learn their basic addition and subtraction facts from experiences in their everyday lives. Eventually, they know the number facts without relying on cookies.

Addition and subtraction of negative numbers have fewer real world examples that are meaningful to us. Math teachers have several different approaches, such as number lines, banking, temperatures, and so on, to make these concepts real.

We will model addition and subtraction of negatives with two color counters. We let a blue counter represent a positive and a red counter will represent a negative.

positive negative

If we have one positive and one negative counter, the value of the pair is zero. They form a neutral pair. The value of this neutral pair is zero as summarized in Figure 3.17.

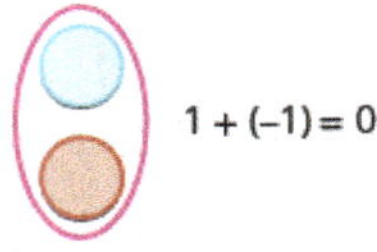

Figure 3.17 A blue counter represents A red counter represents Together they add to zero.

MANIPULATIVE MATHEMATICS

Doing the Manipulative Mathematics activity "Addition of signed Numbers" will help you develop a better understanding of adding integers.

We will model four addition facts using the numbers

EXAMPLE 3.14

Model:

Solution

Interpret the expression.	means the sum of and .
Model the first number. Start with 5 positives.	5
Model the second number. Add 3 positives.	5 3
Count the total number of counters.	8 positives
The sum of 5 and 3 is 8.	

> **TRY IT : : 3.27** Model the expression.

> **TRY IT : : 3.28** Model the expression.

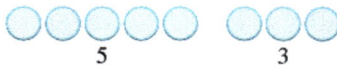

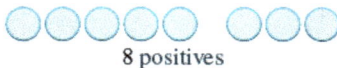

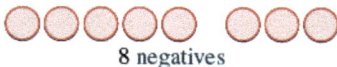

EXAMPLE 3.15

Model:

Solution

Interpret the expression.	means the sum of and .
Model the first number. Start with 5 negatives.	−5
Model the second number. Add 3 negatives.	−5 −3
Count the total number of counters.	8 negatives
The sum of −5 and −3 is −8.	

> **TRY IT : : 3.29** Model the expression.

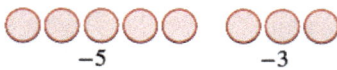

TRY IT : : 3.30 Model the expression.

Example 3.14 and Example 3.15 are very similar. The first example adds positives and positives—both positives. The second example adds negatives and negatives—both negatives. In each case, we got a result of positives or negatives. When the signs are the same, the counters are all the same color.

Now let's see what happens when the signs are different.

EXAMPLE 3.16

Model:

Solution

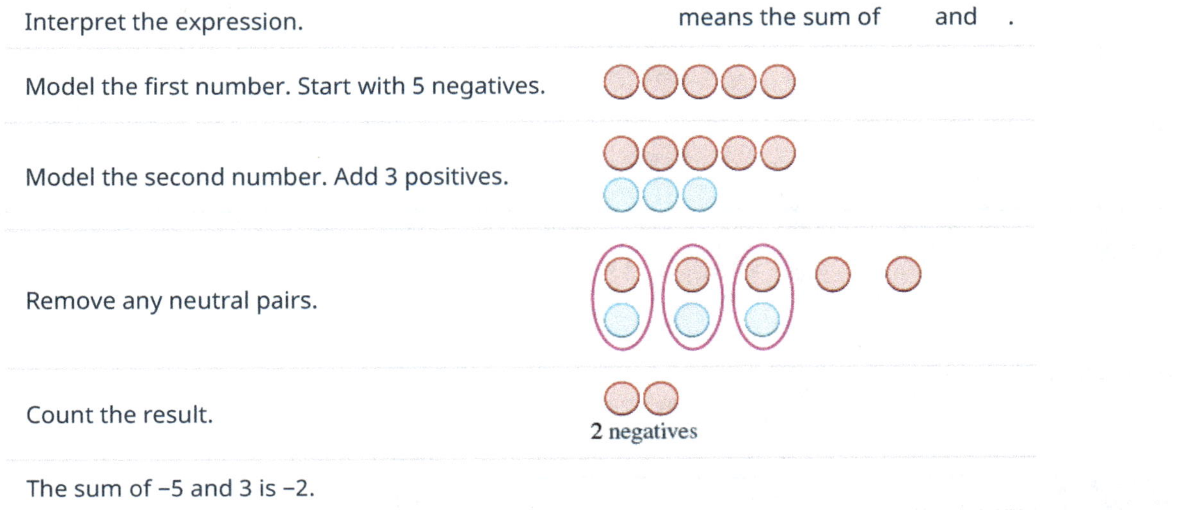

Interpret the expression.	means the sum of and .
Model the first number. Start with 5 negatives.	
Model the second number. Add 3 positives.	
Remove any neutral pairs.	
Count the result.	2 negatives
The sum of −5 and 3 is −2.	

Notice that there were more negatives than positives, so the result is negative.

TRY IT : : 3.31 Model the expression, and then simplify:

TRY IT : : 3.32 Model the expression, and then simplify:

EXAMPLE 3.17

Model:

Solution

Interpret the expression.	means the sum of and .
Model the first number. Start with 5 positives.	
Model the second number. Add 3 negatives.	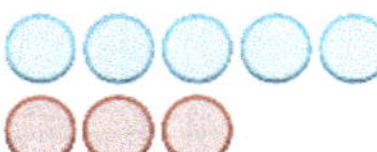
Remove any neutral pairs.	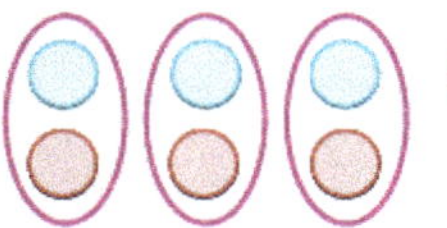 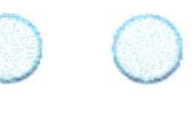
Count the result.	 2 positives
The sum of 5 and −3 is 2.	

> TRY IT : : 3.33 Model the expression, and then simplify:

> TRY IT : : 3.34 Model the expression:

EXAMPLE 3.18 MODELING ADDITION OF POSITIVE AND NEGATIVE INTEGERS

Model each addition.

$4 + 2$ $\quad -3 + 6$ $\quad 4 + (-5)$ $\quad -2 + (-3)$

Solution

Start with 4 positives.	
Add two positives.	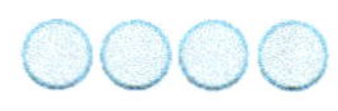
How many do you have?	.

Start with 3 negatives.	
Add 6 positives.	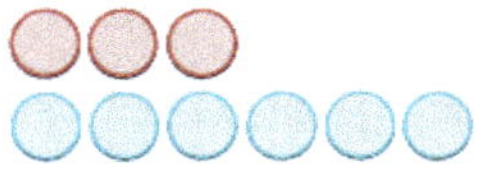
Remove neutral pairs.	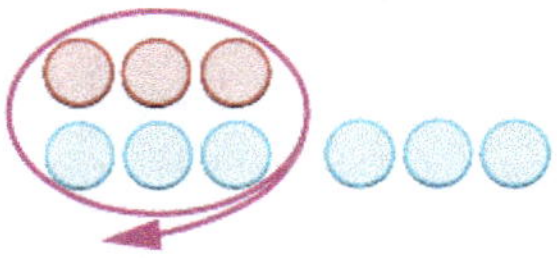
How many are left?	
	.

Start with 4 positives.	
Add 5 negatives.	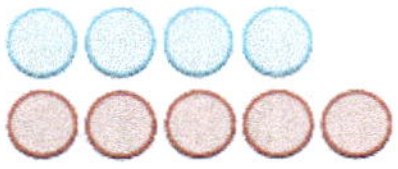
Remove neutral pairs.	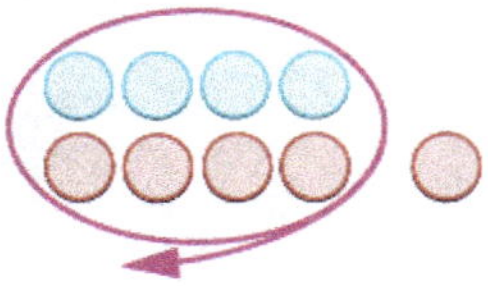
How many are left?	
	.

Start with 2 negatives.	
Add 3 negatives.	
How many do you have?	.

TRY IT : : 3.35 Model each addition.

3 + 4 −1 + 4 4 + (−6) −2 + (−2)

TRY IT : : 3.36

5 + 1 −3 + 7 2 + (−8) −3 + (−4)

Simplify Expressions with Integers

Now that you have modeled adding small positive and negative integers, you can visualize the model in your mind to simplify expressions with any integers.

For example, if you want to add you don't have to count out blue counters and red counters.

Picture blue counters with red counters lined up underneath. Since there would be more negative counters than positive counters, the sum would be negative. Because there are more negative counters.

Let's try another one. We'll add Imagine red counters and more red counters, so we have red counters all together. This means the sum is

Look again at the results of Example 3.14 - Example 3.17.

both positive, sum positive	both negative, sum negative
When the signs are the same, the counters would be all the same color, so add them.	
different signs, more negatives	different signs, more positives
Sum negative	sum positive
When the signs are different, some counters would make neutral pairs; subtract to see how many are left.	

Table 3.1 Addition of Positive and Negative Integers

EXAMPLE 3.19

Simplify:

Solution

Since the signs are different, we subtract from The answer will be negative because there are more negatives than positives.

The signs are different so we subtract from The answer will be positive because there are more positives than negatives

TRY IT : : 3.37 Simplify each expression:

TRY IT : : 3.38 Simplify each expression:

EXAMPLE 3.20

Simplify:

Solution

Since the signs are the same, we add. The answer will be negative because there are only negatives.

TRY IT : : 3.39 Simplify the expression:

TRY IT : : 3.40 Simplify the expression:

The techniques we have used up to now extend to more complicated expressions. Remember to follow the order of operations.

EXAMPLE 3.21

Simplify:

Solution

Simplify inside the parentheses.
Multiply.
Add left to right.

TRY IT : : 3.41 Simplify the expression:

TRY IT : : 3.42 Simplify the expression:

Evaluate Variable Expressions with Integers

Remember that to evaluate an expression means to substitute a number for the variable in the expression. Now we can use negative numbers as well as positive numbers when evaluating expressions.

EXAMPLE 3.22

Evaluate

Solution

Evaluate when

	$-2+7$
Substitute −2 for *x*.	$-2+7$
Simplify.	5

Evaluate when

	$x+7$
Substitute −11 for *x*.	$-11+7$
Simplify.	-4

TRY IT : : 3.43 Evaluate each expression for the given values:

TRY IT : : 3.44 Evaluate each expression for the given values: when

EXAMPLE 3.23

When evaluate

Solution

Evaluate when

	$n+1$
Substitute −5 for *n*.	$-5+1$
Simplify.	-4

Evaluate when

	$-n+1$
Substitute −5 for *n*.	$-(-5)+1$
Simplify.	$5+1$
Add.	6

TRY IT : : 3.45 When evaluate

TRY IT : : 3.46

Next we'll evaluate an expression with two variables.

EXAMPLE 3.24

Evaluate when and

Solution

	$3a+b$
Substitute 12 for *a* and −30 for *b*.	$3(12)+(-30)$
Multiply.	$36+(-30)$
Add.	6

TRY IT : : 3.47 Evaluate the expression:

TRY IT : : 3.48 Evaluate the expression:

EXAMPLE 3.25

Evaluate when and

Solution

This expression has two variables. Substitute for and for

Substitute −18 for *x* and 24 for *y*.

Add inside the parentheses.

Simplify

> TRY IT : : 3.49 Evaluate:

when and

> TRY IT : : 3.50 Evaluate:

when and

Translate Word Phrases to Algebraic Expressions

All our earlier work translating word phrases to algebra also applies to expressions that include both positive and negative numbers. Remember that the phrase *the sum* indicates addition.

EXAMPLE 3.26

Translate and simplify: the sum of and

Solution

The sum of −9 and 5 indicates addition.	the sum of and
Translate.	
Simplify.	

> TRY IT : : 3.51 Translate and simplify the expression:
> the sum of and

> TRY IT : : 3.52 Translate and simplify the expression:
> the sum of and

EXAMPLE 3.27

Translate and simplify: the sum of and increased by

Solution

The phrase *increased by* indicates addition.

The sum of and , increased by

Translate.	
Simplify.	
Add.	

> **TRY IT : : 3.53** Translate and simplify:
> the sum of and increased by

> **TRY IT : : 3.54** Translate and simplify:
> the sum of and increased by

Add Integers in Applications

Recall that we were introduced to some situations in everyday life that use positive and negative numbers, such as temperatures, banking, and sports. For example, a debt of could be represented as Let's practice translating and solving a few applications.

Solving applications is easy if we have a plan. First, we determine what we are looking for. Then we write a phrase that gives the information to find it. We translate the phrase into math notation and then simplify to get the answer. Finally, we write a sentence to answer the question.

EXAMPLE 3.28

The temperature in Buffalo, NY, one morning started at below zero Fahrenheit. By noon, it had warmed up What was the temperature at noon?

Solution

We are asked to find the temperature at noon.

Write a phrase for the temperature.	The temperature warmed up 12 degrees from 7 degrees below zero.
Translate to math notation.	$-7 + 12$
Simplify.	5
Write a sentence to answer the question.	The temperature at noon was 5 degrees Fahrenheit.

> **TRY IT : : 3.55**
> The temperature in Chicago at 5 A.M. was below zero Celsius. Six hours later, it had warmed up What is the temperature at 11 A.M.?

> **TRY IT : : 3.56**
> A scuba diver was swimming below the surface and then dove down another What is her new depth?

EXAMPLE 3.29

A football team took possession of the football on their In the next three plays, they lost gained and then lost On what yard line was the ball at the end of those three plays?

Solution

We are asked to find the yard line the ball was on at the end of three plays.

Write a word phrase for the position of the ball.	Start at 42, then lose 6, gain 4, lose 8.
Translate to math notation.	$42 - 6 + 4 - 8$
Simplify.	32
Write a sentence to answer the question.	At the end of the three plays, the ball is on the 32-yard line.

TRY IT : : 3.57

The Bears took possession of the football on their In the next three plays, they lost gained then lost On what yard line was the ball at the end of those three plays?

TRY IT : : 3.58

The Chargers began with the football on their They gained lost and then gained on the next three plays. Where was the ball at the end of these plays?

MEDIA : : ACCESS ADDITIONAL ONLINE RESOURCES

- Adding Integers with Same Sign Using Color Counters (http://openstaxcollege.org/l/24samesigncount)
- Adding Integers with Different Signs Using Counters (http://openstaxcollege.org/l/24diffsigncount)
- Ex1: Adding Integers (http://openstaxcollege.org/l/24Ex1Add)
- Ex2: Adding Integers (http://openstaxcollege.org/l/24Ex2Add)

3.2 EXERCISES

Practice Makes Perfect

Model Addition of Integers

In the following exercises, model the expression to simplify.

63.

64.

65.

66.

67.

68.

69.

70.

Simplify Expressions with Integers

In the following exercises, simplify each expression.

71.

72.

73.

74.

75.

76.

77.

78.

79.

80.

81.

82.

83.

84.

85.

86.

Evaluate Variable Expressions with Integers

In the following exercises, evaluate each expression.

87. when

88. when

89. when

90. when

91. When evaluate:

92. When evaluate:

93. When evaluate:

94. When evaluate:

95. when, ,

96. when, ,

97. when, ,

98. when, ,

99. when, ,

100. when, ,

101. when, ,

102. when, ,

Translate Word Phrases to Algebraic Expressions

In the following exercises, translate each phrase into an algebraic expression and then simplify.

103. The sum of and

104. The sum of and

105. more than

106. more than

107. added to

108. added to

109. more than the sum of and

110. more than the sum of and

111. the sum of and increased by

112. the sum of and increased by

Add Integers in Applications

In the following exercises, solve.

113. Temperature The temperature in St. Paul, Minnesota was at sunrise. By noon the temperature had risen What was the temperature at noon?

114. Temperature The temperature in Chicago was at 6 am. By afternoon the temperature had risen What was the afternoon temperature?

115. Credit Cards Lupe owes on her credit card. Then she charges more. What is the new balance?

116. Credit Cards Frank owes on his credit card. Then he charges more. What is the new balance?

117. Weight Loss Angie lost the first week of her diet. Over the next three weeks, she lost gained and then lost What was the change in her weight over the four weeks?

118. Weight Loss April lost the first week of her diet. Over the next three weeks, she lost gained and then lost What was the change in her weight over the four weeks?

119. Football The Rams took possession of the football on their own In the next three plays, they lost gained then lost On what yard line was the ball at the end of those three plays?

120. Football The Cowboys began with the ball on their own They gained lost and then gained on the next three plays. Where was the ball at the end of these plays?

121. Calories Lisbeth walked from her house to get a frozen yogurt, and then she walked home. By walking for a total of she burned The frozen yogurt she ate was What was her total calorie gain or loss?

122. Calories Ozzie rode his bike for burning Then he had a iced blended mocha. Represent the change in calories as an integer.

Everyday Math

123. Stock Market The week of September 15, 2008, was one of the most volatile weeks ever for the U.S. stock market. The change in the Dow Jones Industrial Average each day was:

What was the overall change for the week?

124. Stock Market During the week of June 22, 2009, the change in the Dow Jones Industrial Average each day was:

What was the overall change for the week?

Writing Exercises

125. Explain why the sum of and is negative, but the sum of and and is positive.

126. Give an example from your life experience of adding two negative numbers.

Self Check

After completing the exercises, use this checklist to evaluate your mastery of the objectives of this section.

I can...	Confidently	With some help	No-I don't get it!
model addition of integers.			
simplify expressions with integers.			
evaluate variable expressions with integers.			
translate word phrases to algebraic expressions.			
add integers in applications.			

After reviewing this checklist, what will you do to become confident for all objectives?

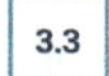

3.3 Subtract Integers

Learning Objectives

By the end of this section, you will be able to:

- Model subtraction of integers
- Simplify expressions with integers
- Evaluate variable expressions with integers
- Translate words phrases to algebraic expressions
- Subtract integers in applications

BE PREPARED : : 3.4 Before you get started, take this readiness quiz.
Simplify:
If you missed this problem, review Example 2.8.

BE PREPARED : : 3.5 Translate *the difference of* *and* into an algebraic expression.
If you missed this problem, review Example 1.36.

BE PREPARED : : 3.6 Add:
If you missed this problem, review Example 3.20.

Model Subtraction of Integers

Remember the story in the last section about the toddler and the cookies? Children learn how to subtract numbers through their everyday experiences. Real-life experiences serve as models for subtracting positive numbers, and in some cases, such as temperature, for adding negative as well as positive numbers. But it is difficult to relate subtracting negative numbers to common life experiences. Most people do not have an intuitive understanding of subtraction when negative numbers are involved. Math teachers use several different models to explain subtracting negative numbers.

We will continue to use counters to model subtraction. Remember, the blue counters represent positive numbers and the red counters represent negative numbers.

Perhaps when you were younger, you read as *five take away three*. When we use counters, we can think of subtraction the same way.

MANIPULATIVE MATHEMATICS

Doing the Manipulative Mathematics activity "Subtraction of Signed Numbers" will help you develop a better understanding of subtracting integers.

We will model four subtraction facts using the numbers and

EXAMPLE 3.30

Model:

Solution

Interpret the expression.	means take away .
Model the first number. Start with 5 positives.	5
Take away the second number. So take away 3 positives.	2 positives
Find the counters that are left.	
	. The difference between and is .

TRY IT : : 3.59 Model the expression:

TRY IT : : 3.60 Model the expression:

EXAMPLE 3.31

Model:

Solution

Interpret the expression.	means take away .
Model the first number. Start with 5 negatives.	−5
Take away the second number. So take away 3 negatives.	2 negatives
Find the number of counters that are left.	
	. The difference between and is .

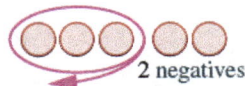

TRY IT : : 3.61 Model the expression:

TRY IT : : 3.62 Model the expression:

Notice that Example 3.30 and Example 3.31 are very much alike.

- First, we subtracted positives from positives to get positives.
- Then we subtracted negatives from negatives to get negatives.

Each example used counters of only one color, and the "take away" model of subtraction was easy to apply.

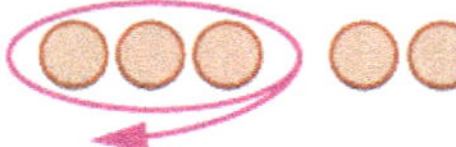

Now let's see what happens when we subtract one positive and one negative number. We will need to use both positive and negative counters and sometimes some neutral pairs, too. Adding a neutral pair does not change the value.

EXAMPLE 3.32

Model:

Solution

Interpret the expression.	means take away .
Model the first number. Start with 5 negatives.	−5
Take away the second number. So we need to take away 3 positives.	
But there are no positives to take away. Add neutral pairs until you have 3 positives.	
Now take away 3 positives.	
Count the number of counters that are left.	8 negatives
	The difference of and is .

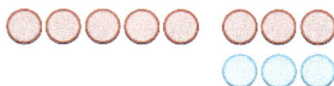

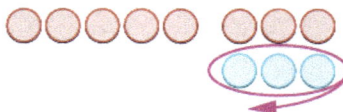

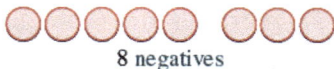

TRY IT : : 3.63 Model the expression:

TRY IT : : 3.64 Model the expression:

EXAMPLE 3.33

Model:

Solution

Interpret the expression.	means take away .
Model the first number. Start with 5 positives.	
Take away the second number, so take away 3 negatives.	
But there are no negatives to take away. Add neutral pairs until you have 3 negatives.	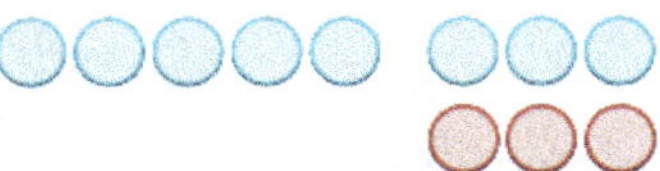
Then take away 3 negatives.	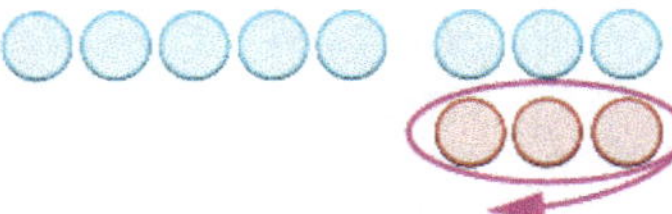
Count the number of counters that are left.	
	The difference of and is .

TRY IT : : 3.65 Model the expression:

TRY IT : : 3.66 Model the expression:

EXAMPLE 3.34

Model each subtraction.

8 − 2 −5 − 4 6 − (−6) −8 − (−3)

Solution

	This means take away .
Start with 8 positives.	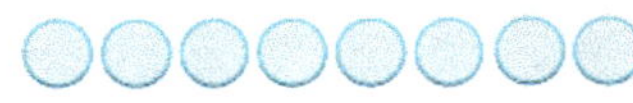
Take away 2 positives.	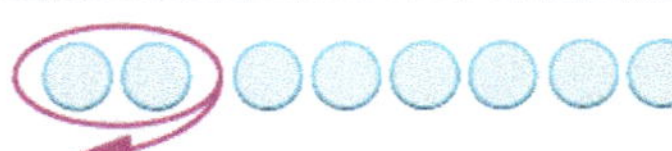
How many are left?	

	This means take away .
Start with 5 negatives.	
You need to take away 4 positives. Add 4 neutral pairs to get 4 positives.	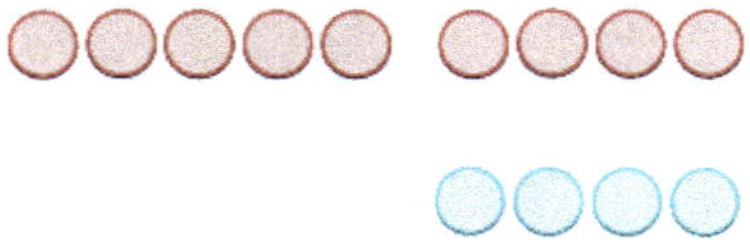
Take away 4 positives.	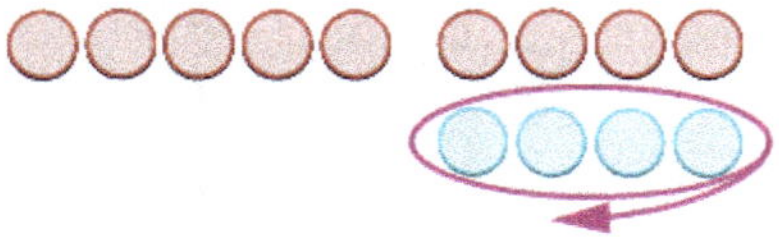
How many are left?	

	This means take away .
Start with 6 positives.	
Add 6 neutrals to get 6 negatives to take away.	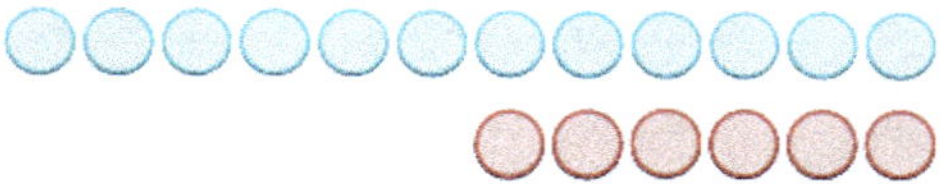
Remove 6 negatives.	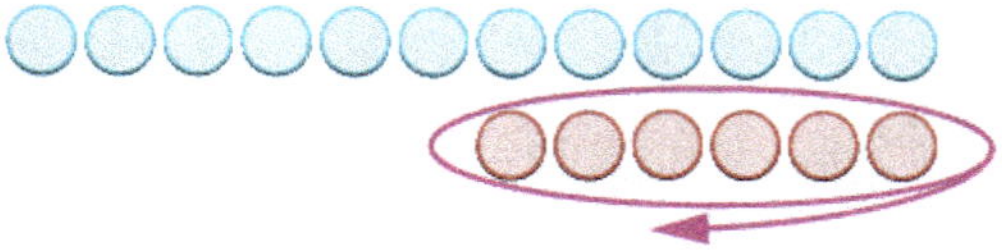
How many are left?	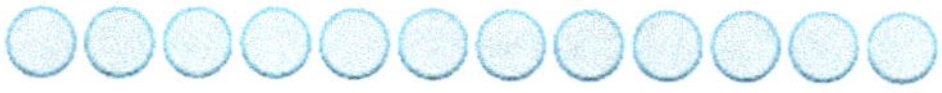

This means take away .	
Start with 8 negatives.	
Take away 3 negatives.	
How many are left?	

> **TRY IT : : 3.67** Model each subtraction.
>
> 7 - (-8) -7 - (-2) 4 - 1 -6 - 8

> **TRY IT : : 3.68** Model each subtraction.
>
> 4 - (-6) -8 - (-1) 7 - 3 -4 - 2

EXAMPLE 3.35

Model each subtraction expression:

Solution

We start with 2 positives.	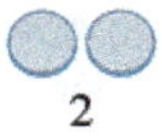 2
We need to take away 8 positives, but we have only 2.	
Add neutral pairs until there are 8 positives to take away.	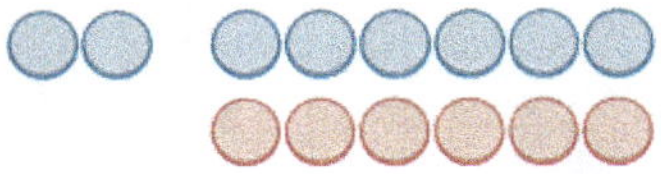
Then take away eight positives.	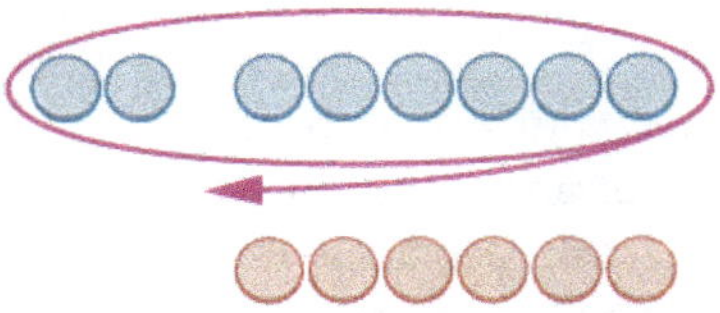
Find the number of counters that are left. There are 6 negatives.	 6 negatives

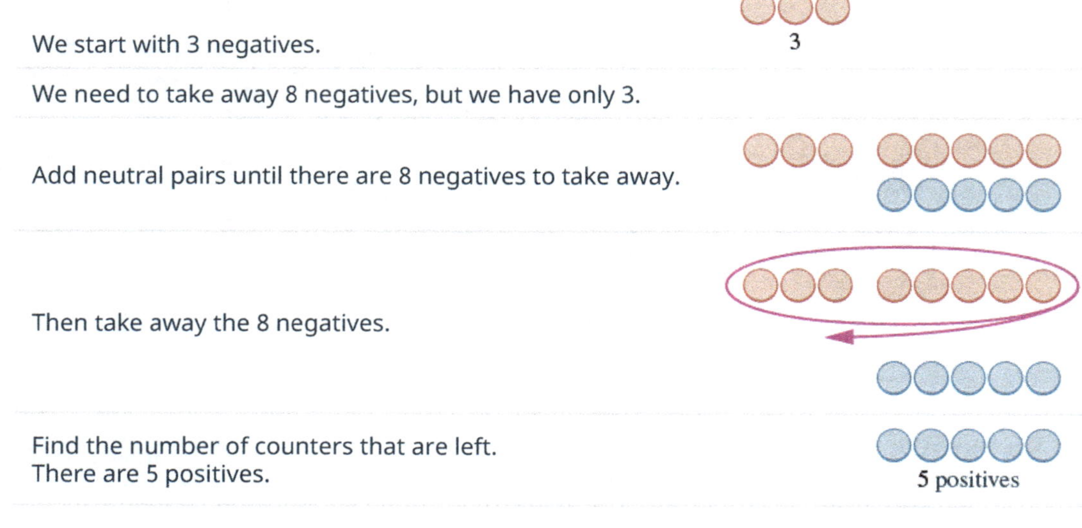

We start with 3 negatives.
3
We need to take away 8 negatives, but we have only 3.
Add neutral pairs until there are 8 negatives to take away.
Then take away the 8 negatives.
Find the number of counters that are left.
There are 5 positives.
5 positives

> **TRY IT : :** 3.69 Model each subtraction expression.

> **TRY IT : :** 3.70 Model each subtraction expression.

Simplify Expressions with Integers

Do you see a pattern? Are you ready to subtract integers without counters? Let's do two more subtractions. We'll think about how we would model these with counters, but we won't actually use the counters.

- Subtract

  Think: We start with negative counters.
  We have to subtract positives, but there are no positives to take away.
  So we add neutral pairs to get the positives. Now we take away the positives.
  So what's left? We have the original negatives plus more negatives from the neutral pair. The result is negatives.

  Notice, that to subtract we added negatives.

- Subtract

  Think: We start with positives.
  We have to subtract negatives, but there are no negatives to take away.
  So we add neutral pairs to the positives. Now we take away the negatives.
  What's left? We have the original positives plus more positives from the neutral pairs. The result is positives.

  Notice that to subtract we added

While we may not always use the counters, especially when we work with large numbers, practicing with them first gave us a concrete way to apply the concept, so that we can visualize and remember how to do the subtraction without the counters.

Have you noticed that subtraction of signed numbers can be done by adding the opposite? You will often see the idea, the Subtraction Property, written as follows:

Subtraction Property

Look at these two examples.

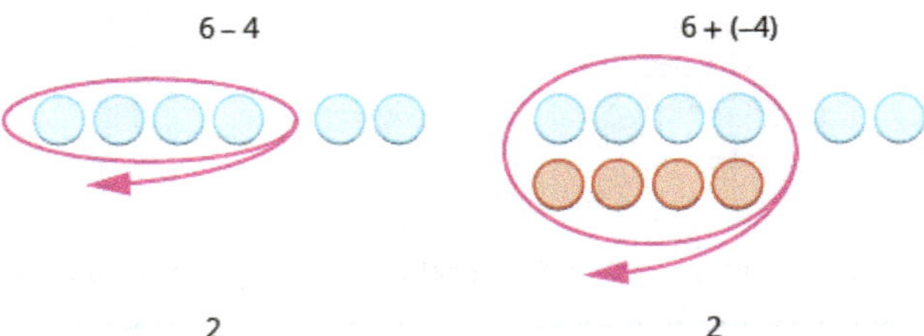

We see that gives the same answer as

Of course, when we have a subtraction problem that has only positive numbers, like the first example, we just do the subtraction. We already knew how to subtract long ago. But knowing that gives the same answer as helps when we are subtracting negative numbers.

EXAMPLE 3.36

Simplify:

Solution

	and
Subtract to simplify.	
Add to simplify.	
Subtracting 8 from 13 is the same as adding −8 to 13.	

	and
Subtract to simplify.	
Add to simplify.	
Subtracting 9 from −17 is the same as adding −9 to −17.	

TRY IT : : 3.71 Simplify each expression:

TRY IT : : 3.72 Simplify each expression:

Now look what happens when we subtract a negative.

8 – (–5)

8 + 5

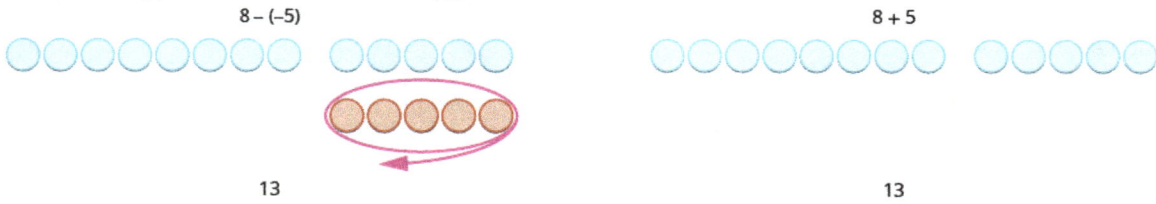

13

13

We see that gives the same result as Subtracting a negative number is like adding a positive.

EXAMPLE 3.37

Simplify:

Solution

	and
Subtract to simplify.	
Add to simplify.	
Subtracting −15 from 9 is the same as adding 15 to 9.	

	and
Subtract to simplify.	
Add to simplify.	
Subtracting −4 from −7 is the same as adding 4 to −7	

TRY IT : : 3.73 Simplify each expression:

TRY IT : : 3.74 Simplify each expression:

Look again at the results of Example 3.30 - Example 3.33.

2 positives	2 negatives
When there would be enough counters of the color to take away, subtract.	
5 negatives, want to subtract 3 positives	5 positives, want to subtract 3 negatives
need neutral pairs	need neutral pairs
When there would not be enough of the counters to take away, add neutral pairs.	

Table 3.4 Subtraction of Integers

EXAMPLE 3.38

Simplify:

Solution

We are taking 58 negatives away from 74 negatives.	
Subtract.	

TRY IT : : 3.75 Simplify the expression:

TRY IT : : 3.76 Simplify the expression:

EXAMPLE 3.39

Simplify:

Solution

We use the order of operations to simplify this expression, performing operations inside the parentheses first. Then we subtract from left to right.

Simplify inside the parentheses first.	$7 - (-4 - 3) - 9$
Subtract from left to right.	$7 - (-7) - 9$
Subtract.	$14 - 9$
	5

TRY IT : : 3.77 Simplify the expression:

TRY IT : : 3.78 Simplify the expression:

EXAMPLE 3.40

Simplify:

Solution

We use the order of operations to simplify this expression. First we multiply, and then subtract from left to right.

Multiply first.	$3 \cdot 7 - 4 \cdot 7 - 5 \cdot 8$
Subtract from left to right.	$21 - 28 - 40$
Subtract.	$-7 - 40$
	-47

TRY IT : : 3.79 Simplify the expression:

TRY IT : : 3.80 Simplify the expression:

Evaluate Variable Expressions with Integers

Now we'll practice evaluating expressions that involve subtracting negative numbers as well as positive numbers.

EXAMPLE 3.41

Evaluate

Solution

To evaluate when , substitute for in the expression.

	$x - 4$
Substitute 3 for *x*.	$3 - 4$
Subtract.	-1

To evaluate when substitute for in the expression.

	$x - 4$
Substitute −6 for *x*.	$-6 - 4$
Subtract.	-10

TRY IT : : 3.81 Evaluate each expression:

TRY IT : : 3.82 Evaluate each expression:

EXAMPLE 3.42

Evaluate

Solution

To evaluate substitute for in the expression.

	$20 - z$
Substitute 12 for *z*.	$20 - 12$
Subtract.	8

To evaluate

	$20 - z$
Substitute −12 for *z*.	$20 - (-12)$
Subtract.	32

TRY IT : : 3.83 Evaluate each expression:

TRY IT : : 3.84 Evaluate each expression:

Translate Word Phrases to Algebraic Expressions

When we first introduced the operation symbols, we saw that the expression may be read in several ways as shown below.

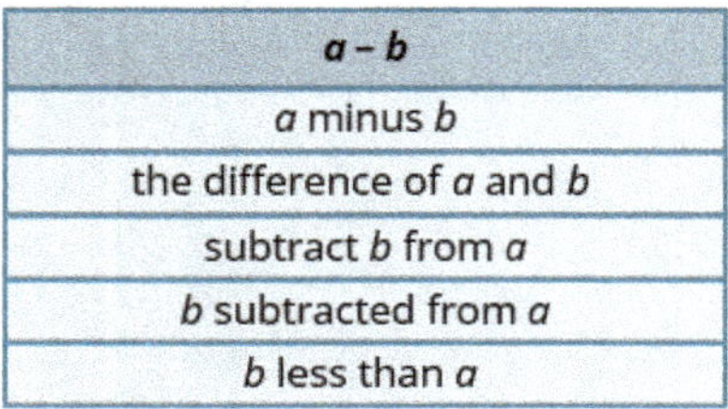

$a - b$
a minus b
the difference of a and b
subtract b from a
b subtracted from a
b less than a

Figure 3.18

Be careful to get and in the right order!

EXAMPLE 3.43

Translate and then simplify:

the difference of and subtract from

Solution

A *difference* means subtraction. Subtract the numbers in the order they are given.

	the difference of 13 and -21
Translate.	$13 - (-21)$
Simplify.	34

Subtract means to take away from

	subtract 24 from -19
Translate.	$-19 - 24$
Simplify.	-43

TRY IT : : 3.85 Translate and simplify:

the difference of and subtract from

TRY IT : : 3.86 Translate and simplify:

the difference of and subtract from

Subtract Integers in Applications

It's hard to find something if we don't know what we're looking for or what to call it. So when we solve an application problem, we first need to determine what we are asked to find. Then we can write a phrase that gives the information to find it. We'll translate the phrase into an expression and then simplify the expression to get the answer. Finally, we summarize the answer in a sentence to make sure it makes sense.

HOW TO : : SOLVE APPLICATION PROBLEMS.

Step 1. Identify what you are asked to find.
Step 2. Write a phrase that gives the information to find it.
Step 3. Translate the phrase to an expression.
Step 4. Simplify the expression.
Step 5. Answer the question with a complete sentence.

EXAMPLE 3.44

In the morning, the temperature in Urbana, Illinois was degrees Fahrenheit. By mid-afternoon, the temperature had dropped to degrees Fahrenheit. What was the difference between the morning and afternoon temperatures?

Solution

Step 1. Identify what we are asked to find.	the difference between the morning and afternoon temperatures
Step 2. Write a phrase that gives the information to find it.	the difference of and
Step 3. Translate the phrase to an expression. The word *difference* indicates subtraction.	
Step 4. Simplify the expression.	
Step 5. Write a complete sentence that answers the question.	The difference in temperature was degrees Fahrenheit.

TRY IT : : 3.87

In the morning, the temperature in Anchorage, Alaska was By mid-afternoon the temperature had dropped to below zero. What was the difference between the morning and afternoon temperatures?

TRY IT : : 3.88

The temperature in Denver was degrees Fahrenheit at lunchtime. By sunset the temperature had dropped to What was the difference between the lunchtime and sunset temperatures?

Geography provides another application of negative numbers with the elevations of places below sea level.

EXAMPLE 3.45

Dinesh hiked from Mt. Whitney, the highest point in California, to Death Valley, the lowest point. The elevation of Mt. Whitney is feet above sea level and the elevation of Death Valley is feet below sea level. What is the difference in elevation between Mt. Whitney and Death Valley?

Solution

Step 1. What are we asked to find?	The difference in elevation between Mt. Whitney and Death Valley
Step 2. Write a phrase.	elevation of Mt. Whitney−elevation of Death Valley
Step 3. Translate.	
Step 4. Simplify.	
Step 5. Write a complete sentence that answers the question.	The difference in elevation is feet.

TRY IT : : 3.89

One day, John hiked to the summit of Haleakala volcano in Hawaii. The next day, while scuba diving, he dove to a cave below sea level. What is the difference between the elevation of the summit of Haleakala and the depth of the cave?

TRY IT : : 3.90

The submarine Nautilus is at below the surface of the water and the submarine Explorer is below the surface of the water. What is the difference in the position of the Nautilus and the Explorer?

Managing your money can involve both positive and negative numbers. You might have overdraft protection on your checking account. This means the bank lets you write checks for more money than you have in your account (as long as they know they can get it back from you!)

EXAMPLE 3.46

Leslie has in her checking account and she writes a check for

What is the balance after she writes the check?

She writes a second check for What is the new balance after this check?

Leslie's friend told her that she had lost a check for that Leslie had given her with her birthday card. What is the balance in Leslie's checking account now?

Solution

What are we asked to find?	The balance of the account
Write a phrase.	minus
Translate	$\$25 - \$8$
Simplify.	$\$17$
Write a sentence answer.	The balance is .

What are we asked to find?	The new balance
Write a phrase.	minus
Translate	$17 − $20
Simplify.	−$3
Write a sentence answer.	She is overdrawn by .

What are we asked to find?	The new balance
Write a phrase.	more than
Translate	−$3 + $10
Simplify.	$7
Write a sentence answer.	The balance is now .

TRY IT : : 3.91

Araceli has in her checking account and writes a check for

What is the balance after she writes the check?

She writes a second check for What is the new balance?

The check for that she sent a charity was never cashed. What is the balance in Araceli's checking account now?

TRY IT : : 3.92

Genevieve's bank account was overdrawn and the balance is

She deposits a check for that she earned babysitting. What is the new balance?

She deposits another check for Is she out of debt yet? What is her new balance?

LINKS TO LITERACY

The Links to Literacy activity "Elevator Magic" will provide you with another view of the topics covered in this section.

MEDIA : : ACCESS ADDITIONAL ONLINE RESOURCES

- Adding and Subtracting Integers (http://openstaxcollege.org/l/24AddSubtrInteg)
- Subtracting Integers with Color Counters (http://openstaxcollege.org/l/24Subtrinteger)
- Subtracting Integers Basics (http://openstaxcollege.org/l/24Subtractbasic)
- Subtracting Integers (http://openstaxcollege.org/l/24introintegerr)
- Integer Application (http://openstaxcollege.org/l/24integerappp)

3.3 EXERCISES

Practice Makes Perfect

Model Subtraction of Integers

In the following exercises, model each expression and simplify.

127. 128. 129.

130. 131. 132.

133. 134.

Simplify Expressions with Integers

In the following exercises, simplify each expression.

135. 136. 137.

138. 139. 140.

141. 142.

In the following exercises, simplify each expression.

143. 144. 145.

146. 147. 148.

149. 150. 151.

152. 153. 154.

155. 156. 157.

158. 159. 160.

161. 162. 163.

164. 165. 166.

167. 168. 169.

170. 171. 172.

173. 174. 175.

176.

177.

178.

Evaluate Variable Expressions with Integers

In the following exercises, evaluate each expression for the given values.

179.

180.

181.

182.

183.

184.

185.

186.

Translate Word Phrases to Algebraic Expressions

In the following exercises, translate each phrase into an algebraic expression and then simplify.

187.

The difference of and

Subtract from

188.

The difference of and

Subtract from

189.

The difference of and

Subtract from

190.

The difference of and

Subtract from

191.

less than

minus

192.

less than

minus

193.

less than

subtracted from

194.

less than

subtracted from

Subtract Integers in Applications

In the following exercises, solve the following applications.

195. Temperature One morning, the temperature in Urbana, Illinois, was By evening, the temperature had dropped What was the temperature that evening?

196. Temperature On Thursday, the temperature in Spincich Lake, Michigan, was By Friday, the temperature had dropped What was the temperature on Friday?

197. Temperature On January 15, the high temperature in Anaheim, California, was That same day, the high temperature in Embarrass, Minnesota was What was the difference between the temperature in Anaheim and the temperature in Embarrass?

198. Temperature On January 21, the high temperature in Palm Springs, California, was and the high temperature in Whitefield, New Hampshire was What was the difference between the temperature in Palm Springs and the temperature in Whitefield?

199. Football At the first down, the Warriors football team had the ball on their On the next three downs, they gained lost and lost What was the yard line at the end of the third down?

200. Football At the first down, the Barons football team had the ball on their On the next three downs, they lost gained and lost What was the yard line at the end of the third down?

201. Checking Account John has in his checking account. He writes a check for What is the new balance in his checking account?

202. Checking Account Ellie has in her checking account. She writes a check for What is the new balance in her checking account?

203. Checking Account Gina has in her checking account. She writes a check for What is the new balance in her checking account?

204. Checking Account Frank has in his checking account. He writes a check for What is the new balance in his checking account?

205. Checking Account Bill has a balance of in his checking account. He deposits to the account. What is the new balance?

206. Checking Account Patty has a balance of in her checking account. She deposits to the account. What is the new balance?

Everyday Math

207. Camping Rene is on an Alpine hike. The temperature is Rene's sleeping bag is rated "comfortable to How much can the temperature change before it is too cold for Rene's sleeping bag?

208. Scuba Diving Shelly's scuba watch is guaranteed to be watertight to She is diving at on the face of an underwater canyon. By how many feet can she change her depth before her watch is no longer guaranteed?

Writing Exercises

209. Explain why the difference of and is

210. Why is the result of subtracting the same as the result of adding

Self Check

After completing the exercises, use this checklist to evaluate your mastery of the objectives of this section.

I can...	Confidently	With some help	No-I don't get it!
model subtraction of integers.			
simplify expressions with integers.			
evaluate variable expressions with integers.			
translate word phrases to algebraic expressions.			
subtract integers in applications.			

What does this checklist tell you about your mastery of this section? What steps will you take to improve?

3.4 Multiply and Divide Integers

Learning Objectives

By the end of this section, you will be able to:

- Multiply integers
- Divide integers
- Simplify expressions with integers
- Evaluate variable expressions with integers
- Translate word phrases to algebraic expressions

BE PREPARED : : 3.7

Before you get started, take this readiness quiz.

Translate the quotient of and into an algebraic expression.

If you missed this problem, review Example 1.67.

BE PREPARED : : 3.8

Add:

If you missed this problem, review Example 3.21.

BE PREPARED : : 3.9

If you missed this problem, review Example 3.23.

Multiply Integers

Since multiplication is mathematical shorthand for repeated addition, our counter model can easily be applied to show multiplication of integers. Let's look at this concrete model to see what patterns we notice. We will use the same examples that we used for addition and subtraction.

We remember that means add times. Here, we are using the model shown in Figure 3.19 just to help us discover the pattern.

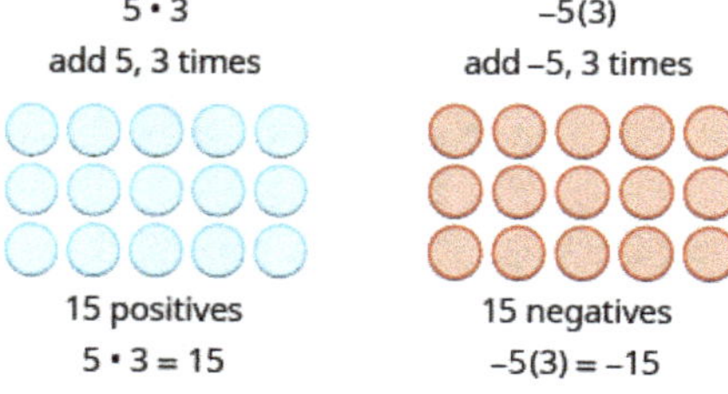

Figure 3.19

Now consider what it means to multiply by It means subtract times. Looking at subtraction as *taking away*, it means to take away times. But there is nothing to take away, so we start by adding neutral pairs as shown in Figure 3.20.

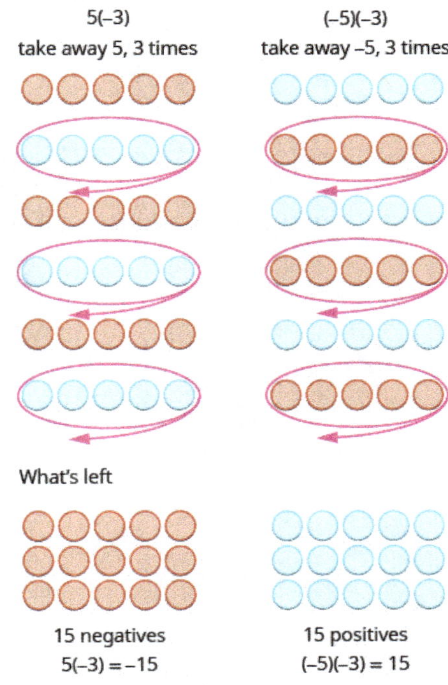

Figure 3.20

In both cases, we started with neutral pairs. In the case on the left, we took away times and the result was To multiply we took away times and the result was So we found that

Notice that for multiplication of two signed numbers, when the signs are the same, the product is positive, and when the signs are different, the product is negative.

Multiplication of Signed Numbers

The sign of the product of two numbers depends on their signs.

Same signs	Product
•Two positives	Positive
•Two negatives	Positive

Different signs	Product
•Positive • negative	Negative
•Negative • positive	Negative

EXAMPLE 3.47

Multiply each of the following:

Solution

Multiply, noting that the signs are different and so the product is negative.

Multiply, noting that the signs are the same and so the product is positive.

Multiply, noting that the signs are different and so the product is negative.

The signs are the same, so the product is positive.

TRY IT : : 3.93 Multiply:

TRY IT : : 3.94 Multiply:

1.
2.
3.
4.

When we multiply a number by the result is the same number. What happens when we multiply a number by Let's multiply a positive number and then a negative number by to see what we get.

Each time we multiply a number by we get its opposite.

Multiplication by

Multiplying a number by gives its opposite.

EXAMPLE 3.48

Multiply each of the following:

Solution

The signs are different, so the product will be negative.
Notice that −7 is the opposite of 7.

The signs are the same, so the product will be positive.
Notice that 11 is the opposite of −11.

TRY IT : : 3.95 Multiply.

TRY IT : : 3.96 Multiply.

Divide Integers

Division is the inverse operation of multiplication. So, because In words, this expression says that can be divided into groups of each because adding five three times gives If we look at some examples of multiplying integers, we might figure out the rules for dividing integers.

Division of signed numbers follows the same rules as multiplication. When the signs are the same, the quotient is positive, and when the signs are different, the quotient is negative.

Division of Signed Numbers

The sign of the quotient of two numbers depends on their signs.

Same signs	Quotient
•Two positives •Two negatives	Positive Positive

Different signs	Quotient
•Positive & negative •Negative & positive	Negative Negative

Remember, you can always check the answer to a division problem by multiplying.

EXAMPLE 3.49

Divide each of the following:

Solution

Divide, noting that the signs are different and so the quotient is negative.

Divide, noting that the signs are the same and so the quotient is positive.

TRY IT : : 3.97 Divide:

TRY IT : : 3.98 Divide:

Just as we saw with multiplication, when we divide a number by the result is the same number. What happens when we divide a number by Let's divide a positive number and then a negative number by to see what we get.

When we divide a number by, we get its opposite.

Division by

Dividing a number by gives its opposite.

EXAMPLE 3.50

Divide each of the following:

Solution

The dividend, 16, is being divided by –1.
Dividing a number by –1 gives its opposite.
Notice that the signs were different, so the result was negative.

The dividend, –20, is being divided by –1.
Dividing a number by –1 gives its opposite.

Notice that the signs were the same, so the quotient was positive.

TRY IT : : 3.99 Divide:

TRY IT : : 3.100 Divide:

Simplify Expressions with Integers

Now we'll simplify expressions that use all four operations–addition, subtraction, multiplication, and division–with integers. Remember to follow the order of operations.

EXAMPLE 3.51

Solution

We use the order of operations. Multiply first and then add and subtract from left to right.

Multiply first.
Add.
Subtract.

TRY IT : : 3.101 Simplify:

TRY IT : : 3.102 Simplify:

EXAMPLE 3.52

Simplify:

Solution

The exponent tells how many times to multiply the base.

The exponent is and the base is We raise to the fourth power.

Write in expanded form.
Multiply.
Multiply.
Multiply.

The exponent is and the base is We raise to the fourth power and then take the opposite.

Write in expanded form.
Multiply.
Multiply.
Multiply.

TRY IT : : 3.103 Simplify:

TRY IT : : 3.104 Simplify:

EXAMPLE 3.53

Solution

According to the order of operations, we simplify inside parentheses first. Then we will multiply and finally we will subtract.

Subtract the parentheses first.
Multiply.
Subtract.

TRY IT : : 3.105 Simplify:

TRY IT : : 3.106 Simplify:

EXAMPLE 3.54

Simplify:

Solution

We simplify the exponent first, then multiply and divide.

Simplify the exponent.
Multiply.
Divide.

TRY IT : : 3.107 Simplify:

TRY IT : : 3.108 Simplify:

EXAMPLE 3.55

Solution

First we will multiply and divide from left to right. Then we will add.

Divide.
Multiply.
Add.

> **TRY IT : : 3.109** Simplify:

> **TRY IT : : 3.110** Simplify:

Evaluate Variable Expressions with Integers

Now we can evaluate expressions that include multiplication and division with integers. Remember that to evaluate an expression, substitute the numbers in place of the variables, and then simplify.

EXAMPLE 3.56

Solution

	$2x^2 - 3x + 8$
Substitute -4 for x.	$2(-4)^2 - 3(-4) + 8$
Simplify exponents.	$2(16) - 3(-4) + 8$
Multiply.	$32 - (-12) + 8$
Subtract.	$44 + 8$
Add.	52

Keep in mind that when we substitute for we use parentheses to show the multiplication. Without parentheses, it would look like

> **TRY IT : : 3.111** Evaluate:

> **TRY IT : : 3.112** Evaluate:

EXAMPLE 3.57

Solution

	$3x + 4y - 6$
Substitute and .	$3(-1) + 4(2) - 6$
Multiply.	$-3 + 8 - 6$
Simplify.	-1

TRY IT : : 3.113 Evaluate: 

TRY IT : : 3.114 Evaluate:

Translate Word Phrases to Algebraic Expressions

Once again, all our prior work translating words to algebra transfers to phrases that include both multiplying and dividing integers. Remember that the key word for multiplication is *product* and for division is *quotient*.

EXAMPLE 3.58

Translate to an algebraic expression and simplify if possible: the product of and

Solution

The word *product* tells us to multiply.

	the product of and
Translate.	
Simplify.	

TRY IT : : 3.115 Translate to an algebraic expression and simplify if possible:

TRY IT : : 3.116 Translate to an algebraic expression and simplify if possible:

EXAMPLE 3.59

Translate to an algebraic expression and simplify if possible: the quotient of and

Solution

The word *quotient* tells us to divide.

	the quotient of −56 and −7
Translate.	
Simplify.	

> **TRY IT : :** 3.117 Translate to an algebraic expression and simplify if possible:

> **TRY IT : :** 3.118 Translate to an algebraic expression and simplify if possible:

MEDIA : : ACCESS ADDITIONAL ONLINE RESOURCES

- Multiplying Integers Using Color Counters (http://openstaxcollege.org/l/24Multiplyinteg)
- Multiplying Integers Using Color Counters With Neutral Pairs (http://openstaxcollege.org/l/24Multiplyneutr)
- Multiplying Integers Basics (http://openstaxcollege.org/l/24Multiplybasic)
- Dividing Integers Basics (http://openstaxcollege.org/l/24Dividebasic)
- Ex. Dividing Integers (http://openstaxcollege.org/l/24Divideinteger)
- Multiplying and Dividing Signed Numbers (http://openstaxcollege.org/l/24Multidivisign)

3.4 EXERCISES

Practice Makes Perfect

Multiply Integers

In the following exercises, multiply each pair of integers.

211. 212. 213.

214. 215. 216.

217. 218. 219.

220. 221. 222.

Divide Integers

In the following exercises, divide.

223. 224. 225.

226. 227. 228.

229. 230. 231.

232.

Simplify Expressions with Integers

In the following exercises, simplify each expression.

233. 234. 235.

236. 237. 238.

239. 240. 241.

242. 243. 244.

245. 246. 247.

248. 249. 250.

251. 252. 253.

254. 255. 256.

257. 258.

Evaluate Variable Expressions with Integers

In the following exercises, evaluate each expression.

259.

260.

261.

262.

263.

264. when

265. when

266. when

267. when and

268. when and

269. when and

270. when and

Translate Word Phrases to Algebraic Expressions

In the following exercises, translate to an algebraic expression and simplify if possible.

271. The product of and 15

272. The product of and

273. The quotient of and

274. The quotient of and

275. The quotient of and the sum of and

276. The quotient of and the sum of and

277. The product of and the difference of

278. The product of and the difference of

Everyday Math

279. Stock market Javier owns shares of stock in one company. On Tuesday, the stock price dropped per share. What was the total effect on Javier's portfolio?

280. Weight loss In the first week of a diet program, eight women lost an average of each. What was the total weight change for the eight women?

Writing Exercises

281. In your own words, state the rules for multiplying two integers.

282. In your own words, state the rules for dividing two integers.

283. Why is

284. Why is

Self Check

After completing the exercises, use this checklist to evaluate your mastery of the objectives of this section.

I can...	Confidently	With some help	No-I don't get it!
multiply integers.			
divide integers.			
simplify expressions with integers.			
evaluate variable expressions with integers.			
translate word phrases to algebraic expressions.			

On a scale of 1–10, how would you rate your mastery of this section in light of your responses on the checklist? How can you improve this?

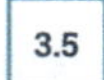

3.5 Solve Equations Using Integers; The Division Property of Equality

Learning Objectives

By the end of this section, you will be able to:

- Determine whether an integer is a solution of an equation
- Solve equations with integers using the Addition and Subtraction Properties of Equality
- Model the Division Property of Equality
- Solve equations using the Division Property of Equality
- Translate to an equation and solve

BE PREPARED : : 3.10 Before you get started, take this readiness quiz.

If you missed this problem, review Example 3.22.

BE PREPARED : : 3.11

If you missed this problem, review Example 2.33.

BE PREPARED : : 3.12 Translate into an algebraic expression *less than*

If you missed this problem, review Table 1.3.

Determine Whether a Number is a Solution of an Equation

In Solve Equations with the Subtraction and Addition Properties of Equality, we saw that a solution of an equation is a value of a variable that makes a true statement when substituted into that equation. In that section, we found solutions that were whole numbers. Now that we've worked with integers, we'll find integer solutions to equations.

The steps we take to determine whether a number is a solution to an equation are the same whether the solution is a whole number or an integer.

HOW TO : : HOW TO DETERMINE WHETHER A NUMBER IS A SOLUTION TO AN EQUATION.

Step 1. Substitute the number for the variable in the equation.

Step 2. Simplify the expressions on both sides of the equation.

Step 3. Determine whether the resulting equation is true.

- If it is true, the number is a solution.
- If it is not true, the number is not a solution.

EXAMPLE 3.60

Determine whether each of the following is a solution of

Solution

Substitute 4 for x in the equation to determine if it is true.

	$2x - 5 = -13$
Substitute 4 for x.	$2(4) - 5 \stackrel{?}{=} -13$
Multiply.	$8 - 5 \stackrel{?}{=} -13$
Subtract.	$3 \neq -13$

Since does not result in a true equation, is not a solution to the equation.

Substitute −4 for x in the equation to determine if it is true.	$2x - 5 = -13$
Substitute −4 for x.	$2(-4) - 5 \stackrel{?}{=} -13$
Multiply.	$-8 - 5 \stackrel{?}{=} -13$
Subtract.	$-13 = -13$ ✓

Since results in a true equation, is a solution to the equation.

Substitute −9 for x in the equation to determine if it is true.

	$2x - 5 = -13$
Substitute −9 for x.	$2(-9) - 5 \stackrel{?}{=} -13$
Multiply.	$-18 - 5 \stackrel{?}{=} -13$
Subtract.	$-23 \neq -13$

Since does not result in a true equation, is not a solution to the equation.

TRY IT : : 3.119 Determine whether each of the following is a solution of

TRY IT : : 3.120 Determine whether each of the following is a solution of

Solve Equations with Integers Using the Addition and Subtraction Properties of Equality

In Solve Equations with the Subtraction and Addition Properties of Equality, we solved equations similar to the two shown here using the Subtraction and Addition Properties of Equality. Now we can use them again with integers.

$$x + 4 = 12 \qquad y - 5 = 9$$

$$x + 4 - 4 = 12 - 4 \qquad y - 5 + 5 = 9 + 5$$

$$x = 8 \qquad y = 14$$

When you add or subtract the same quantity from both sides of an equation, you still have equality.

Properties of Equalities

Subtraction Property of Equality	Addition Property of Equality

EXAMPLE 3.61

Solve:

Solution

	$y + 9 = 5$
Subtract 9 from each side to undo the addition.	$y + 9 - 9 = 5 - 9$
Simplify.	$y = -4$

Check the result by substituting into the original equation.

Substitute −4 for y

Since makes a true statement, we found the solution to this equation.

TRY IT : : 3.121 Solve:

TRY IT : : 3.122 Solve:

EXAMPLE 3.62

Solve:

Solution

	$a - 6 = -8$
Add 6 to each side to undo the subtraction.	$a - 6 + 6 = -8 + 6$
Simplify.	$a = -2$
Check the result by substituting into the original equation:	$a - 6 = -8$
Substitute for	$-2 - 6 \stackrel{?}{=} -8$
	$-8 = -8 \checkmark$

The solution to is

Since makes a true statement, we found the solution to this equation.

TRY IT : : 3.123 Solve:

TRY IT : : 3.124 Solve:

Model the Division Property of Equality

All of the equations we have solved so far have been of the form or We were able to isolate the variable by adding or subtracting the constant term. Now we'll see how to solve equations that involve division.

We will model an equation with envelopes and counters in Figure 3.21.

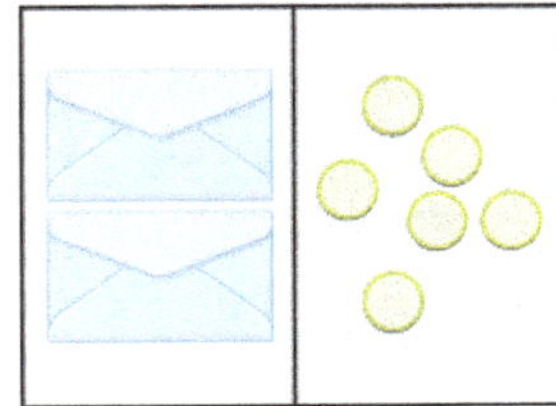

Figure 3.21

Here, there are two identical envelopes that contain the same number of counters. Remember, the left side of the workspace must equal the right side, but the counters on the left side are "hidden" in the envelopes. So how many counters are in each envelope?

To determine the number, separate the counters on the right side into groups of the same size. So counters divided into groups means there must be counters in each group (since

What equation models the situation shown in Figure 3.22? There are two envelopes, and each contains counters. Together, the two envelopes must contain a total of counters. So the equation that models the situation is

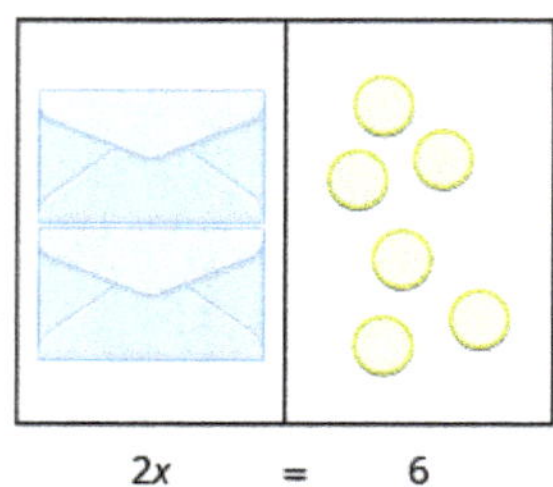

Figure 3.22

We can divide both sides of the equation by as we did with the envelopes and counters.

$$\frac{2x}{2} = \frac{6}{2}$$
$$x = 3$$

We found that each envelope contains Does this check? We know so it works. Three counters in each of two envelopes does equal six.

Figure 3.23 shows another example.

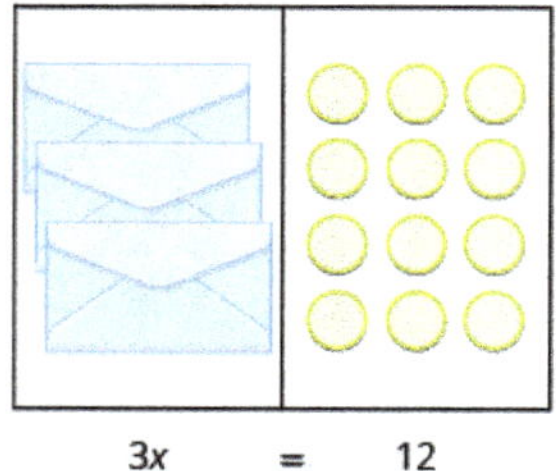

Figure 3.23

Now we have identical envelopes and How many counters are in each envelope? We have to separate the into Since there must be in each envelope. See Figure 3.24.

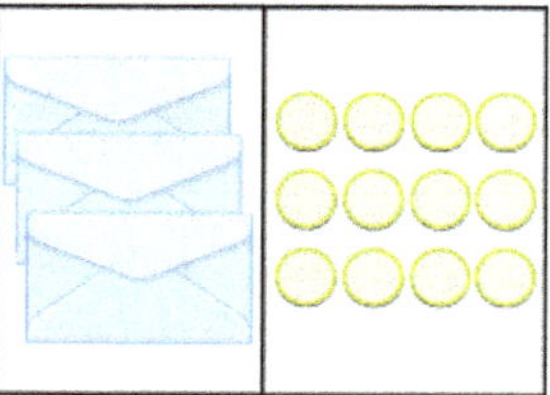

Figure 3.24

The equation that models the situation is We can divide both sides of the equation by

$$\frac{3x}{3} = \frac{12}{3}$$
$$x = 4$$

Does this check? It does because

MANIPULATIVE MATHEMATICS

Doing the Manipulative Mathematics activity “Division Property of Equality” will help you develop a better understanding of how to solve equations using the Division Property of Equality.

EXAMPLE 3.63

Write an equation modeled by the envelopes and counters, and then solve it.

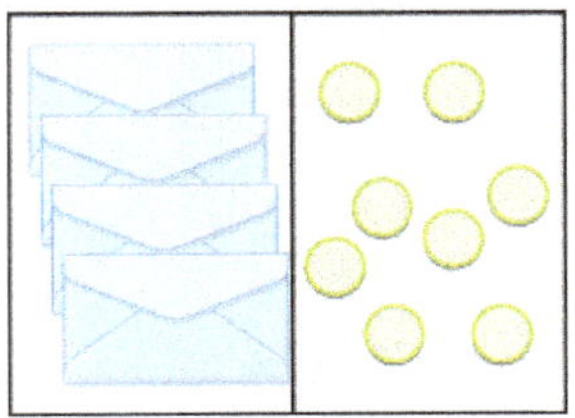

 **Solution**

There are or unknown values, on the left that match the on the right. Let's call the unknown quantity in the envelopes

Write the equation.	$4x = 8$
Divide both sides by 4.	$\frac{4x}{4} = \frac{8}{4}$
Simplify.	$x = 2$

There are in each envelope.

> **TRY IT : :** 3.125 Write the equation modeled by the envelopes and counters. Then solve it.

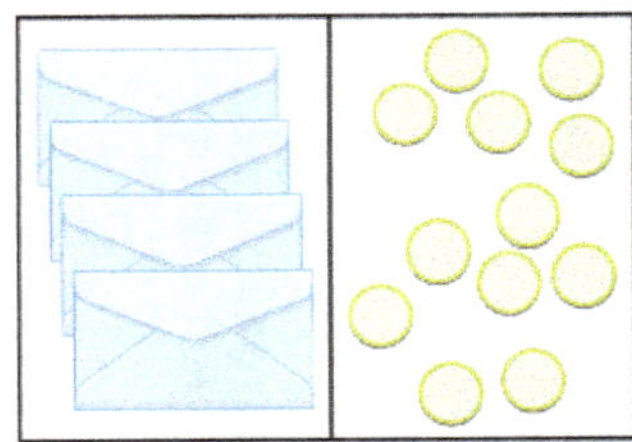

> **TRY IT : :** 3.126 Write the equation modeled by the envelopes and counters. Then solve it.

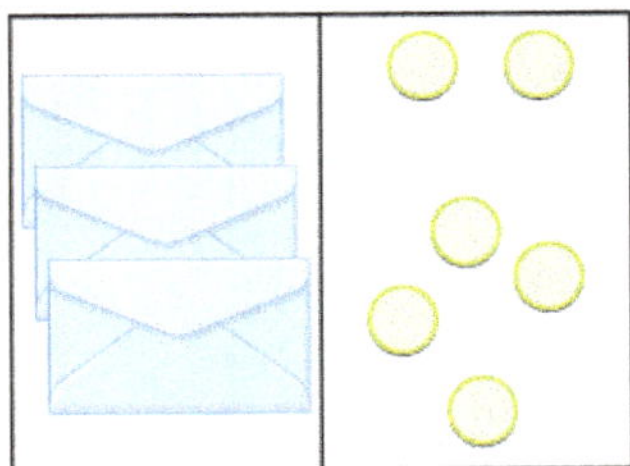

Solve Equations Using the Division Property of Equality

The previous examples lead to the Division Property of Equality. When you divide both sides of an equation by any nonzero number, you still have equality.

Division Property of Equality

EXAMPLE 3.64

Solution

To isolate we need to undo multiplication.

	$7x = -49$
Divide each side by 7.	$\frac{7x}{7} = \frac{-49}{7}$
Simplify.	$x = -7$

Check the solution.

Substitute −7 for x.	

Therefore, is the solution to the equation.

TRY IT : : 3.127 Solve:

TRY IT : : 3.128 Solve:

EXAMPLE 3.65

Solve:

Solution

To isolate we need to undo the multiplication.

	$-3y = 63$
Divide each side by −3.	$\frac{-3y}{-3} = \frac{63}{-3}$
Simplify	$y = -21$

Check the solution.

Substitute −21 for y.	

Since this is a true statement, is the solution to the equation.

 **TRY IT : :** 3.129 Solve:

 **TRY IT : :** 3.130 Solve:

Translate to an Equation and Solve

In the past several examples, we were given an equation containing a variable. In the next few examples, we'll have to first translate word sentences into equations with variables and then we will solve the equations.

EXAMPLE 3.66

Translate and solve: five more than is equal to

Solution

	five more than is equal to
Translate	
Subtract from both sides.	
Simplify.	

Check the answer by substituting it into the original equation.

 **TRY IT : :** 3.131 Translate and solve:

Seven more than is equal to .

 **TRY IT : :** 3.132 Translate and solve:

EXAMPLE 3.67

Translate and solve: the difference of and is

Solution

	the difference of and is
Translate.	
Add to each side.	
Simplify.	

Check the answer by substituting it into the original equation.

> **TRY IT : : 3.133** Translate and solve:
> The difference of and is .

> **TRY IT : : 3.134** Translate and solve:
> The difference of and is .

EXAMPLE 3.68

Translate and solve: the number is the product of and

Solution

	the number of is the product of and
Translate.	
Divide by .	—— ——
Simplify.	

Check the answer by substituting it into the original equation.

> **TRY IT : : 3.135** Translate and solve:
> The number is the product of and .

> **TRY IT : : 3.136** Translate and solve:
> The number is the product of and .

MEDIA : : ACCESS ADDITIONAL ONLINE RESOURCES

- One-Step Equations With Adding Or Subtracting (http://openstaxcollege.org/l/24onestepaddsub)
- One-Step Equations With Multiplying Or Dividing (http://openstaxcollege.org/l/24onestepmuldiv)

3.5 EXERCISES

Practice Makes Perfect

Determine Whether a Number is a Solution of an Equation

In the following exercises, determine whether each number is a solution of the given equation.

285.

286.

287.

288.

Solve Equations Using the Addition and Subtraction Properties of Equality

In the following exercises, solve for the unknown.

289.

290.

291.

292.

293.

294.

295.

296.

297.

298.

299.

300.

Model the Division Property of Equality

In the following exercises, write the equation modeled by the envelopes and counters and then solve it.

301.

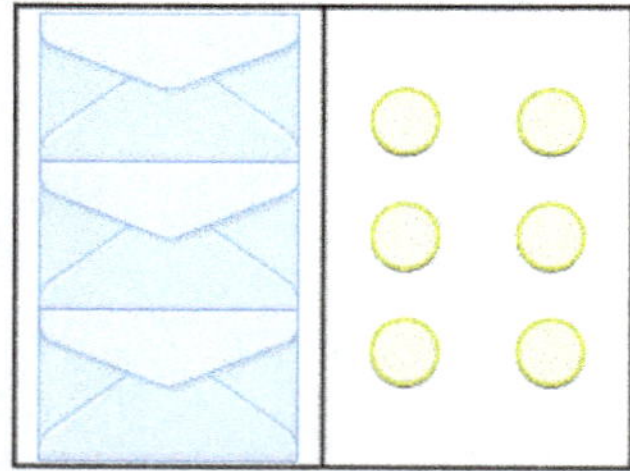

302.

303.

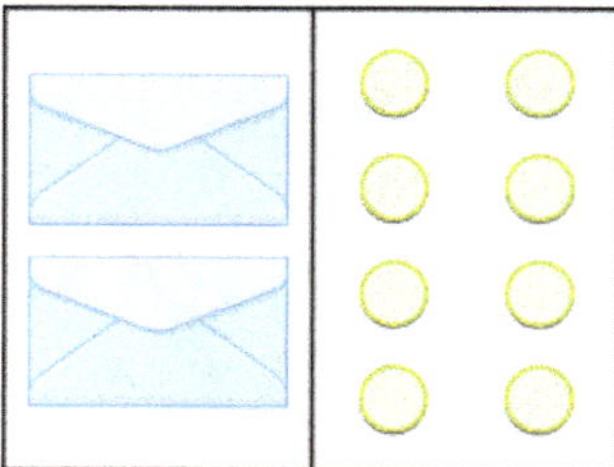

304.

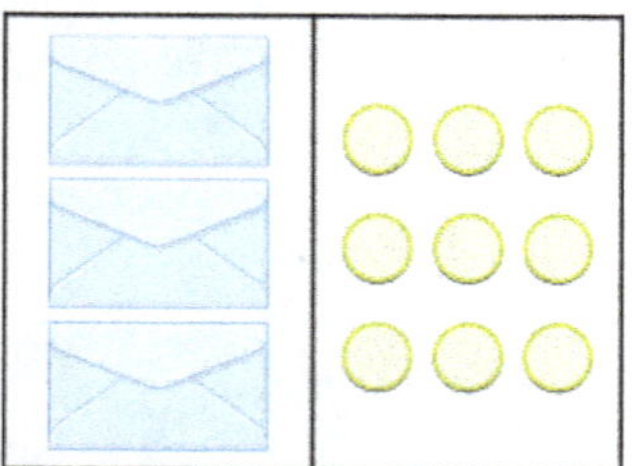

Solve Equations Using the Division Property of Equality

In the following exercises, solve each equation using the division property of equality and check the solution.

305.

306.

307.

308.

309.

310.

311.

312.

313.

314.

315.

316.

Translate to an Equation and Solve

In the following exercises, translate and solve.

317. Four more than is equal to 1.

318. Nine more than is equal to 5.

319. The sum of eight and is .

320. The sum of two and is .

321. The difference of and three is .

322. The difference of and is .

323. The number −42 is the product of −7 and .

324. The number −54 is the product of −9 and .

325. The product of -15 and is 75.

326. The product of −18 and is 36.

327. −6 plus is equal to 4.

328. −2 plus is equal to 1.

329. Nine less than is −4.

330. Thirteen less than is .

Mixed Practice

In the following exercises, solve.

331.

332.

333.

334.

335.

336.

337.

338.

339.

340.

341.

342.

343.

344.

345.

346.

Everyday Math

347. Cookie packaging A package of has equal rows of cookies. Find the number of cookies in each row, by solving the equation

348. Kindergarten class Connie's kindergarten class has She wants them to get into equal groups. Find the number of children in each group, by solving the equation

Writing Exercises

349. Is modeling the Division Property of Equality with envelopes and counters helpful to understanding how to solve the equation Explain why or why not.

350. Suppose you are using envelopes and counters to model solving the equations and Explain how you would solve each equation.

351. Frida started to solve the equation by adding to both sides. Explain why Frida's method will not solve the equation.

352. Raoul started to solve the equation by subtracting from both sides. Explain why Raoul's method will not solve the equation.

Self Check

After completing the exercises, use this checklist to evaluate your mastery of the objectives of this section.

I can...	Confidently	With some help	No-I don't get it!
determine whether an integer is a solution of an equation.			
solve equations with integers using the addition and subtraction properties of equality.			
model division property of equality.			
solve equations using the division property of equality.			
translate to an equation and solve.			

Overall, after looking at the checklist, do you think you are well-prepared for the next Chapter? Why or why not?

CHAPTER 3 REVIEW

KEY TERMS

absolute value The absolute value of a number is its distance from 0 on the number line.

integers Integers are counting numbers, their opposites, and zero ... –3, –2, –1, 0, 1, 2, 3 ...

negative number A negative number is less than zero.

opposites The opposite of a number is the number that is the same distance from zero on the number line, but on the opposite side of zero.

KEY CONCEPTS

3.1 Introduction to Integers

- Opposite Notation
    - means the opposite of the number
    - The notation is read *the opposite of*
- Absolute Value Notation
    - The absolute value of a number is written as .
    - for all numbers.

3.2 Add Integers

- **Addition of Positive and Negative Integers**

both positive, sum positive	both negative, sum negative
When the signs are the same, the counters would be all the same color, so add them.	
different signs, more negatives	different signs, more positives
Sum negative	sum positive
When the signs are different, some counters would make neutral pairs; subtract to see how many are left.	

3.3 Subtract Integers

- **Subtraction of Integers**

2 positives	2 negatives
When there would be enough counters of the color to take away, subtract.	
5 negatives, want to subtract 3 positives	5 positives, want to subtract 3 negatives
need neutral pairs	need neutral pairs
When there would not be enough of the counters to take away, add neutral pairs.	

Table 3.13

- **Subtraction Property**
    -
    -
- **Solve Application Problems**
    - Step 1. Identify what you are asked to find.
    - Step 2. Write a phrase that gives the information to find it.
    - Step 3. Translate the phrase to an expression.
    - Step 4. Simplify the expression.
    - Step 5. Answer the question with a complete sentence.

3.4 Multiply and Divide Integers

- **Multiplication of Signed Numbers**
    - To determine the sign of the product of two signed numbers:

Same Signs	Product
Two positives Two negatives	Positive Positive

Different Signs	Product
Positive • negative Negative • positive	Negative Negative

- **Division of Signed Numbers**
    - To determine the sign of the quotient of two signed numbers:

Same Signs	Quotient
Two positives Two negatives	Positive Positive

Different Signs	Quotient
Positive • negative Negative • Positive	Negative Negative

- **Multiplication by**
  - Multiplying a number by gives its opposite:
- **Division by**
  - Dividing a number by gives its opposite:

3.5 Solve Equations Using Integers; The Division Property of Equality

- **How to determine whether a number is a solution to an equation.**
  - Step 1. Substitute the number for the variable in the equation.
  - Step 2. Simplify the expressions on both sides of the equation.
  - Step 3. Determine whether the resulting equation is true.
    - If it is true, the number is a solution.
    - If it is not true, the number is not a solution.
- **Properties of Equalities**

Subtraction Property of Equality	Addition Property of Equality

- **Division Property of Equality**
  - For any numbers and

    If , then — —.

REVIEW EXERCISES

3.1 Introduction to Integers

Locate Positive and Negative Numbers on the Number Line

In the following exercises, locate and label the integer on the number line.

353.

354.

355.

356.

357.

358.

Order Positive and Negative Numbers

In the following exercises, order each of the following pairs of numbers, using or

359.

360.

361.

362.

363.

364.

Find Opposites

In the following exercises, find the opposite of each number.

365.

366.

367.

368.

In the following exercises, simplify.

369.

370.

In the following exercises, evaluate.

371.

372.

Simplify Absolute Values

In the following exercises, simplify.

373.

374.

375.

376.

377.

378.

In the following exercises, evaluate.

379.

380.

381.

382.

In the following exercises, fill in for each of the following pairs of numbers.

383.

384.

385.

386.

In the following exercises, simplify.

387.

388.

389.

390.

391.

392.

393.

394.

Translate Phrases to Expressions with Integers

In the following exercises, translate each of the following phrases into expressions with positive or negative numbers.

395. the opposite of

396. the opposite of

397. negative

398. minus negative

399. a temperature of below zero

400. an elevation of below sea level

3.2 Add Integers

Model Addition of Integers

In the following exercises, model the following to find the sum.

401.

402.

403.

404.

Simplify Expressions with Integers

In the following exercises, simplify each expression.

405.

406.

407.

408.

409.

410.

411.

412.

Evaluate Variable Expressions with Integers

In the following exercises, evaluate each expression.

413.

414.

415.

416.

Translate Word Phrases to Algebraic Expressions

In the following exercises, translate each phrase into an algebraic expression and then simplify.

417.

418.

419.

420.

Add Integers in Applications

In the following exercises, solve.

421. Temperature On Monday, the high temperature in Denver was Tuesday's high temperature was more. What was the high temperature on Tuesday?

422. Credit Frida owed on her credit card. Then she charged more. What was her new balance?

3.3 Subtract Integers

Model Subtraction of Integers

In the following exercises, model the following.

423.

424.

425.

426.

Simplify Expressions with Integers

In the following exercises, simplify each expression.

427.

428.

429.

430.

431.

432.

433.

434.

Evaluate Variable Expressions with Integers

In the following exercises, evaluate each expression.

435.

436.

437.

438.

Translate Phrases to Algebraic Expressions

In the following exercises, translate each phrase into an algebraic expression and then simplify.

439. the difference of

440. subtract from

Subtract Integers in Applications

In the following exercises, solve the given applications.

441. Temperature One morning the temperature in Bangor, Maine was By afternoon, it had dropped What was the afternoon temperature?

442. Temperature On January 4, the high temperature in Laredo, Texas was and the high in Houlton, Maine was What was the difference in temperature of Laredo and Houlton?

3.4 Multiply and Divide Integers

Multiply Integers

In the following exercises, multiply.

443.

444.

445.

446.

Divide Integers

In the following exercises, divide.

447.

448.

449.

450.

451.

452.

Simplify Expressions with Integers

In the following exercises, simplify each expression.

453.

454.

455.

456.

457.

458.

459.

460.

Evaluate Variable Expressions with Integers

In the following exercises, evaluate each expression.

461.

462.

463.

464.

Translate Word Phrases to Algebraic Expressions

In the following exercises, translate to an algebraic expression and simplify if possible.

465. the product of and

466. the quotient of and the sum of and

3.5 Solve Equations using Integers; The Division Property of Equality

Determine Whether a Number is a Solution of an Equation

In the following exercises, determine whether each number is a solution of the given equation.

467.

468.

Using the Addition and Subtraction Properties of Equality

In the following exercises, solve.

469.

470.

471.

472.

Model the Division Property of Equality

In the following exercises, write the equation modeled by the envelopes and counters. Then solve it.

473.

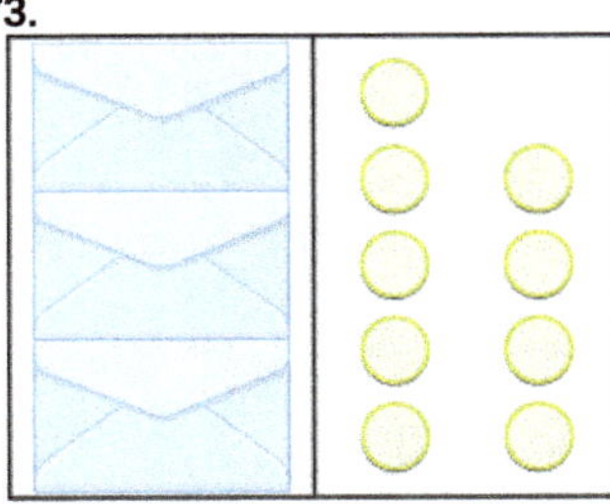

474.

Solve Equations Using the Division Property of Equality

In the following exercises, solve each equation using the division property of equality and check the solution.

475.

476.

477.

478.

Translate to an Equation and Solve.

In the following exercises, translate and solve.

479.

480.

481. Four more than is

482.

Everyday Math

483. Describe how you have used two topics from this chapter in your life outside of your math class during the past month.

PRACTICE TEST

484. Locate and label
and on a number line.

In the following exercises, compare the numbers, using

485.

486.

In the following exercises, find the opposite of each number.

487.

In the following exercises, simplify.

488.

489.

490.

491.

492.

493.

494.

495.

496.

497.

498.

499.

500.

In the following exercises, evaluate.

501.

502.

503.

504.

In the following exercises, translate each phrase into an algebraic expression and then simplify, if possible.

505. the difference of −7 and −4

506. the quotient of and the sum of and

In the following exercises, solve.

507. Early one morning, the temperature in Syracuse was By noon, it had risen What was the temperature at noon?

508. Collette owed on her credit card. Then she charged What was her new balance?

In the following exercises, solve.

509.

510.

511.

In the following exercises, translate and solve.

512.

513.

4 FRACTIONS

Figure 4.1 Bakers combine ingredients to make delicious breads and pastries. (credit: Agustín Ruiz, Flickr)

Chapter Outline

Introduction

Often in life, whole amounts are not exactly what we need. A baker must use a little more than a cup of milk or part of a teaspoon of sugar. Similarly a carpenter might need less than a foot of wood and a painter might use part of a gallon of paint. In this chapter, we will learn about numbers that describe parts of a whole. These numbers, called fractions, are very useful both in algebra and in everyday life. You will discover that you are already familiar with many examples of fractions!

4.1 Visualize Fractions

Learning Objectives

By the end of this section, you will be able to:

- Understand the meaning of fractions
- Model improper fractions and mixed numbers
- Convert between improper fractions and mixed numbers
- Model equivalent fractions
- Find equivalent fractions
- Locate fractions and mixed numbers on the number line
- Order fractions and mixed numbers

BE PREPARED : : 4.1 Before you get started, take this readiness quiz.

Simplify:

If you missed this problem, review Example 2.8.

BE PREPARED : : 4.2 Fill in the blank with or

If you missed this problem, review **Example 3.2**.

Understand the Meaning of Fractions

Andy and Bobby love pizza. On Monday night, they share a pizza equally. How much of the pizza does each one get? Are you thinking that each boy gets half of the pizza? That's right. There is one whole pizza, evenly divided into two parts, so each boy gets one of the two equal parts.

In math, we write — to mean one out of two parts.

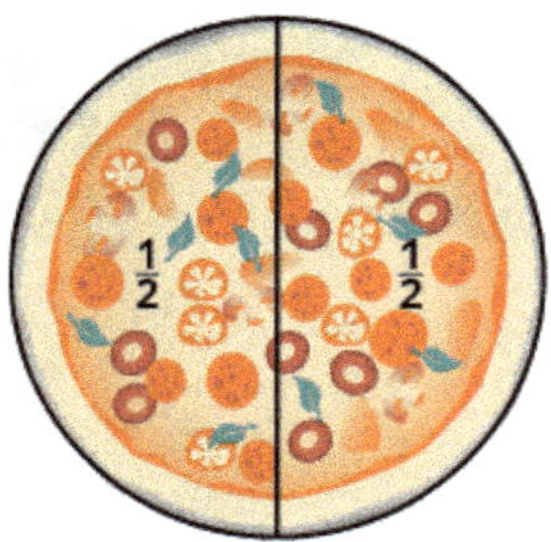

On Tuesday, Andy and Bobby share a pizza with their parents, Fred and Christy, with each person getting an equal amount of the whole pizza. How much of the pizza does each person get? There is one whole pizza, divided evenly into four equal parts. Each person has one of the four equal parts, so each has — of the pizza.

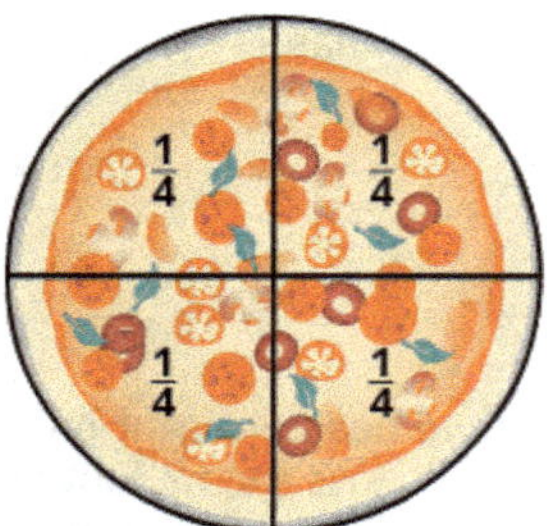

On Wednesday, the family invites some friends over for a pizza dinner. There are a total of people. If they share the pizza equally, each person would get — of the pizza.

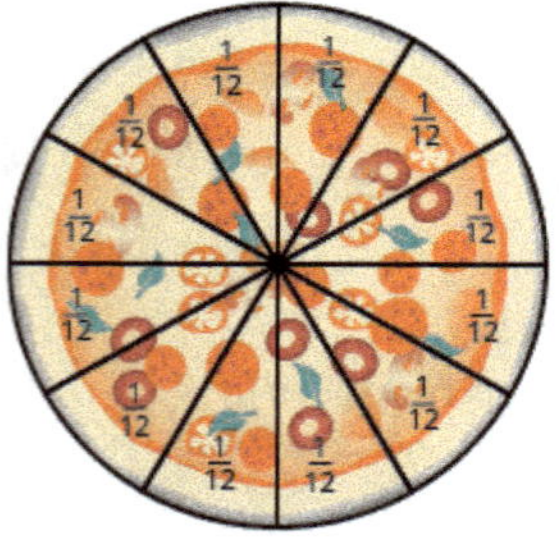

Fractions

A fraction is written — where and are integers and In a fraction, is called the numerator and is called the denominator.

A fraction is a way to represent parts of a whole. The denominator represents the number of equal parts the whole has been divided into, and the numerator represents how many parts are included. The denominator, cannot equal zero because division by zero is undefined.

In Figure 4.2, the circle has been divided into three parts of equal size. Each part represents — of the circle. This type of model is called a fraction circle. Other shapes, such as rectangles, can also be used to model fractions.

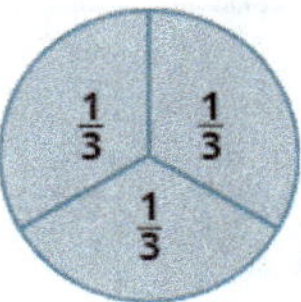

Figure 4.2

MANIPULATIVE MATHEMATICS

Doing the Manipulative Mathematics activity Model Fractions will help you develop a better understanding of fractions, their numerators and denominators.

What does the fraction — represent? The fraction — means two of three equal parts.

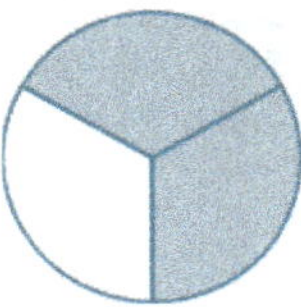

EXAMPLE 4.1

Name the fraction of the shape that is shaded in each of the figures.

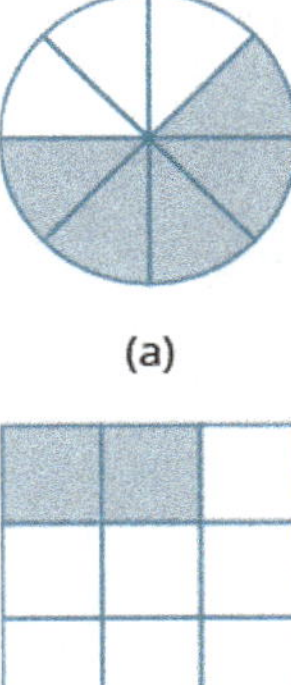

(a)

(b)

Solution

We need to ask two questions. First, how many equal parts are there? This will be the denominator. Second, of these equal parts, how many are shaded? This will be the numerator.

Five out of eight parts are shaded. Therefore, the fraction of the circle that is shaded is —

Two out of nine parts are shaded. Therefore, the fraction of the square that is shaded is —

TRY IT : : 4.1 Name the fraction of the shape that is shaded in each figure:

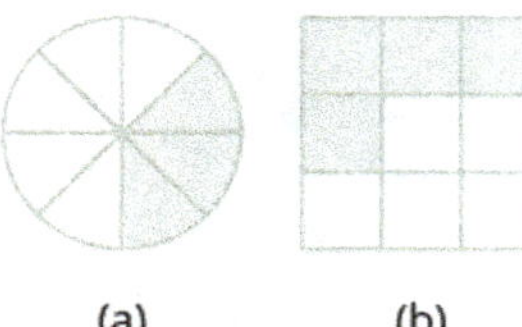

(a) (b)

TRY IT : : 4.2 Name the fraction of the shape that is shaded in each figure:

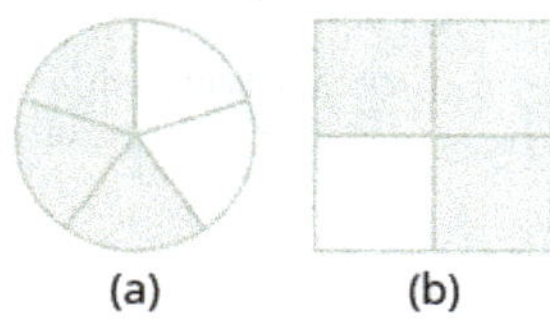

(a) (b)

EXAMPLE 4.2

Shade — of the circle.

Solution

The denominator is so we divide the circle into four equal parts .

The numerator is so we shade three of the four parts .

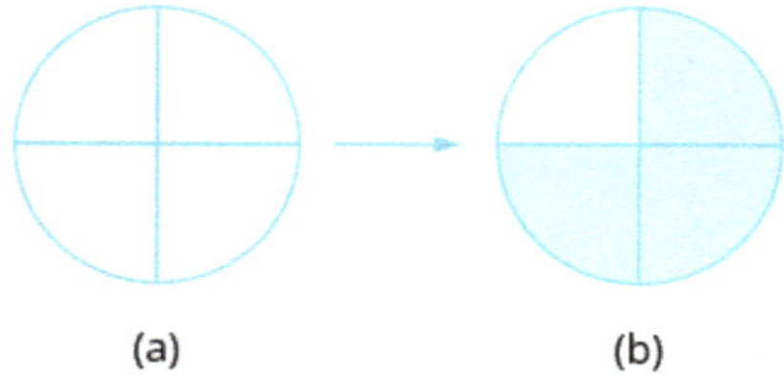

(a) (b)

— of the circle is shaded.

TRY IT : : 4.3 Shade — of the circle.

TRY IT : : 4.4 Shade — of the rectangle.

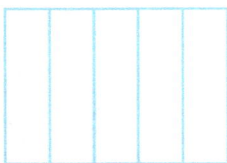

In Example 4.1 and Example 4.2, we used circles and rectangles to model fractions. Fractions can also be modeled as manipulatives called fraction tiles, as shown in Figure 4.3. Here, the whole is modeled as one long, undivided rectangular tile. Beneath it are tiles of equal length divided into different numbers of equally sized parts.

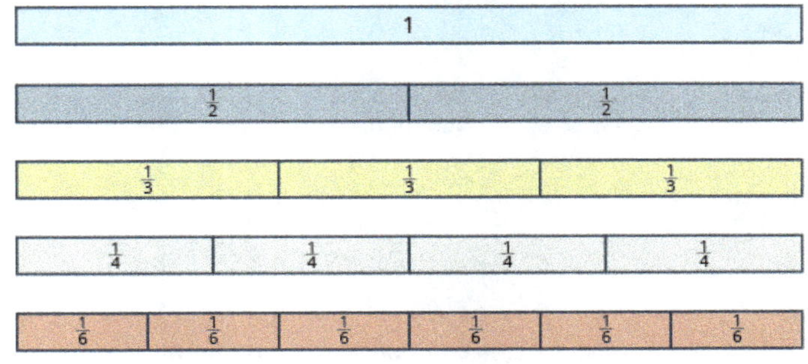

Figure 4.3

We'll be using fraction tiles to discover some basic facts about fractions. Refer to Figure 4.3 to answer the following questions:

How many — tiles does it take to make one whole tile?	It takes two halves to make a whole, so two out of two is —
How many — tiles does it take to make one whole tile?	It takes three thirds, so three out of three is —
How many — tiles does it take to make one whole tile?	It takes four fourths, so four out of four is —
How many — tiles does it take to make one whole tile?	It takes six sixths, so six out of six is —
What if the whole were divided into equal parts? (We have not shown fraction tiles to represent this, but try to visualize it in your mind.) How many — tiles does it take to make one whole tile?	It takes twenty-fourths, so —

It takes twenty-fourths, so —

This leads us to the *Property of One*.

Property of One

Any number, except zero, divided by itself is one.

—

MANIPULATIVE MATHEMATICS

Doing the Manipulative Mathematics activity "Fractions Equivalent to One" will help you develop a better understanding of fractions that are equivalent to one

EXAMPLE 4.3

Use fraction circles to make wholes using the following pieces:

fourths fifths sixths

Solution

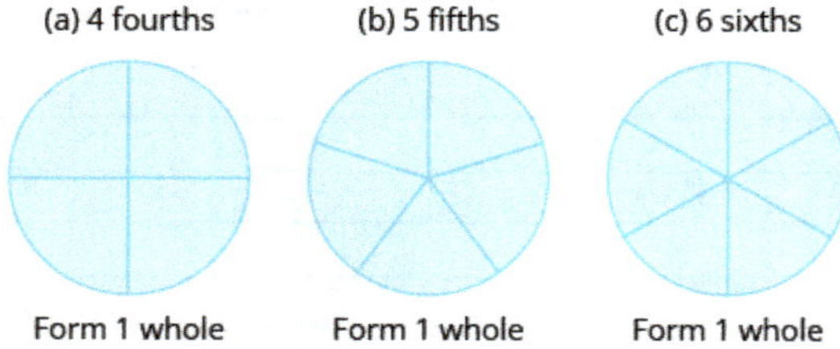

> **TRY IT : : 4.5** Use fraction circles to make wholes with the following pieces: thirds.

> **TRY IT : : 4.6** Use fraction circles to make wholes with the following pieces: eighths.

What if we have more fraction pieces than we need for whole? We'll look at this in the next example.

EXAMPLE 4.4

Use fraction circles to make wholes using the following pieces:

halves fifths thirds

Solution

halves make whole with half left over.

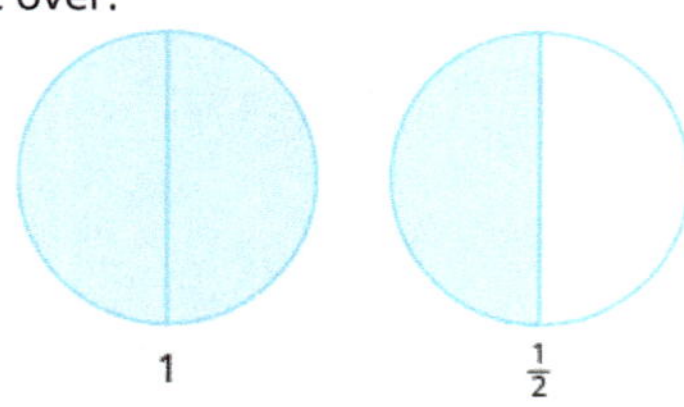

fifths make whole with fifths left over.

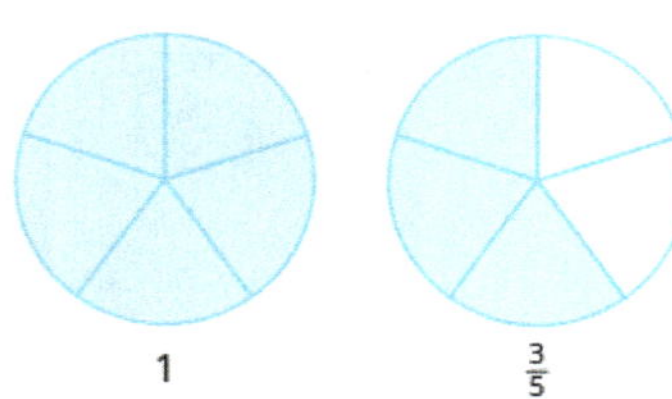

thirds make wholes with third left over.

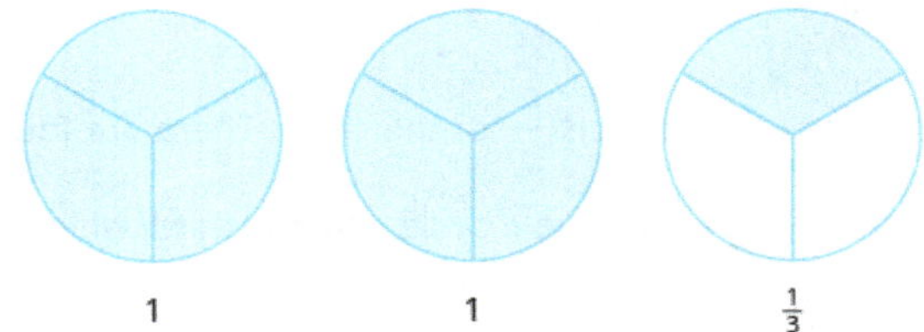

> TRY IT : : 4.7 Use fraction circles to make wholes with the following pieces: thirds.

> TRY IT : : 4.8 Use fraction circles to make wholes with the following pieces: halves.

Model Improper Fractions and Mixed Numbers

In Example 4.4 (b), you had eight equal fifth pieces. You used five of them to make one whole, and you had three fifths left over. Let us use fraction notation to show what happened. You had eight pieces, each of them one fifth, — so altogether you had eight fifths, which we can write as — The fraction — is one whole, plus three fifths, — or — which is read as *one and three-fifths*.

The number — is called a mixed number. A mixed number consists of a whole number and a fraction.

Mixed Numbers

A **mixed number** consists of a whole number and a fraction — where It is written as follows.

—

Fractions such as — — — and — are called improper fractions. In an improper fraction, the numerator is greater than or equal to the denominator, so its value is greater than or equal to one. When a fraction has a numerator that is smaller than the denominator, it is called a proper fraction, and its value is less than one. Fractions such as — — and — are proper fractions.

Proper and Improper Fractions

The fraction — is a **proper fraction** if and an **improper fraction** if

MANIPULATIVE MATHEMATICS

Doing the Manipulative Mathematics activity "Model Improper Fractions" and "Mixed Numbers" will help you develop a better understanding of how to convert between improper fractions and mixed numbers.

EXAMPLE 4.5

Name the improper fraction modeled. Then write the improper fraction as a mixed number.

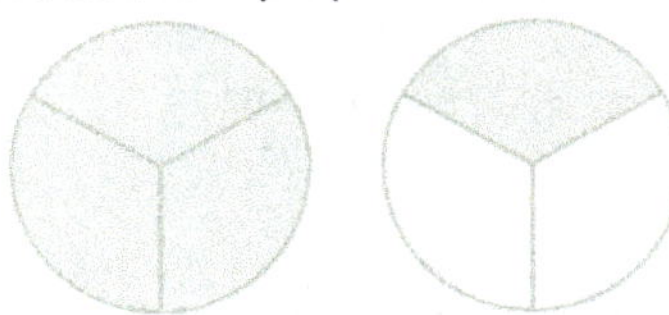

Solution

Each circle is divided into three pieces, so each piece is — of the circle. There are four pieces shaded, so there are four thirds or — The figure shows that we also have one whole circle and one third, which is — So, — —

TRY IT : : 4.9 Name the improper fraction. Then write it as a mixed number.

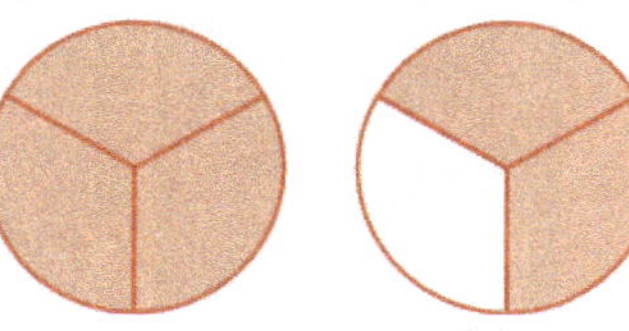

TRY IT : : 4.10 Name the improper fraction. Then write it as a mixed number.

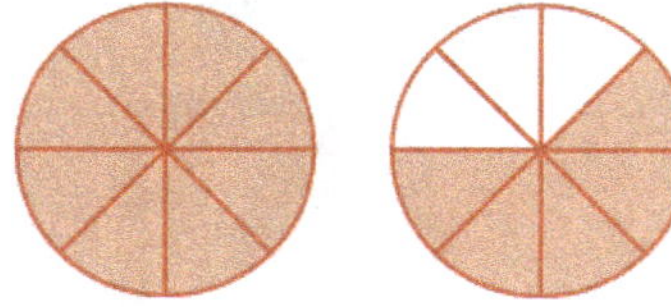

EXAMPLE 4.6

Draw a figure to model —

Solution

The denominator of the improper fraction is Draw a circle divided into eight pieces and shade all of them. This takes care of eight eighths, but we have eighths. We must shade three of the eight parts of another circle.

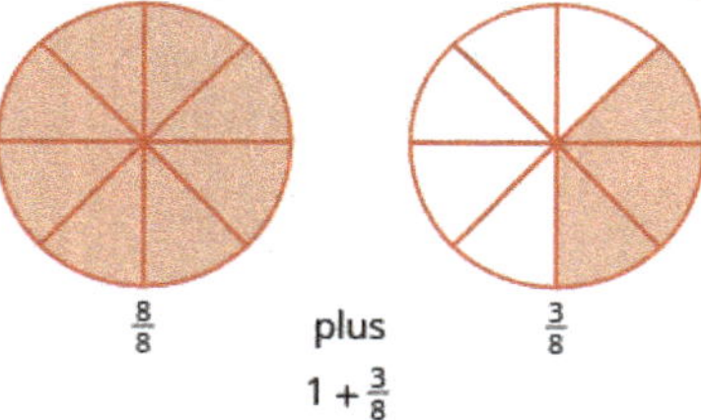

So, — —

TRY IT : : 4.11 Draw a figure to model —

TRY IT : : 4.12 Draw a figure to model —

EXAMPLE 4.7

Use a model to rewrite the improper fraction — as a mixed number.

Solution

We start with sixths — We know that six sixths makes one whole.

—

That leaves us with five more sixths, which is —

So, — —

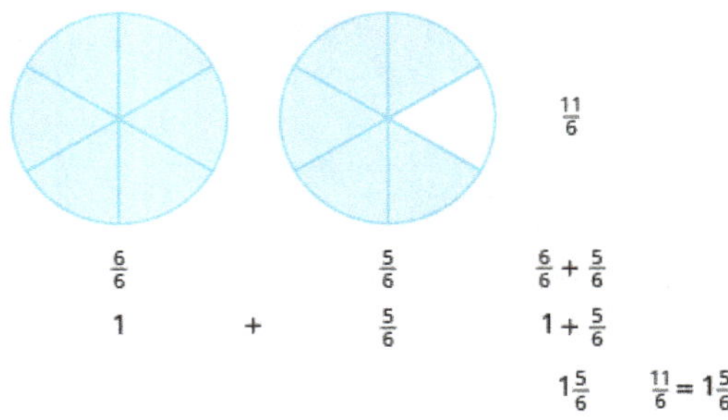

TRY IT : : 4.13 Use a model to rewrite the improper fraction as a mixed number: —

TRY IT : : 4.14 Use a model to rewrite the improper fraction as a mixed number: —

EXAMPLE 4.8

Use a model to rewrite the mixed number — as an improper fraction.

Solution

The mixed number — means one whole plus four fifths. The denominator is so the whole is — Together five fifths and four fifths equals nine fifths.

So, — —

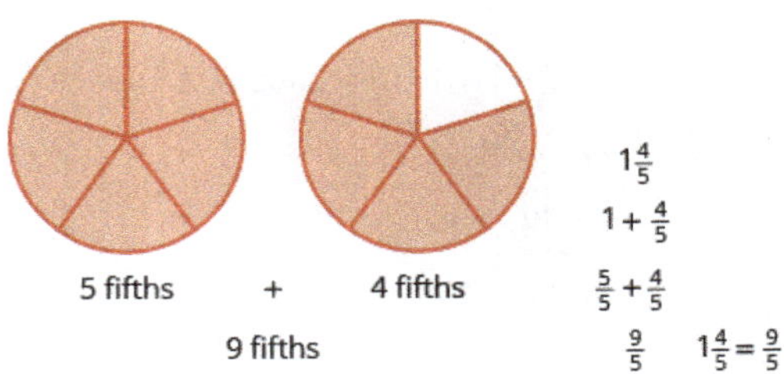

TRY IT : : 4.15 Use a model to rewrite the mixed number as an improper fraction: —

TRY IT : : 4.16 Use a model to rewrite the mixed number as an improper fraction: —

Convert between Improper Fractions and Mixed Numbers

In Example 4.7, we converted the improper fraction — to the mixed number — using fraction circles. We did this by grouping six sixths together to make a whole; then we looked to see how many of the pieces were left. We saw that — made one whole group of six sixths plus five more sixths, showing that — —

The division expression — (which can also be written as)) tells us to find how many groups of are in To convert an improper fraction to a mixed number without fraction circles, we divide.

EXAMPLE 4.9

Convert — to a mixed number.

Solution

	—
Divide the denominator into the numerator.	Remember — means .
	quotient: 1; divisor: 6; $6\overline{)11}$; 6; remainder: 5
Identify the quotient, remainder and divisor.	
Write the mixed number as ———.	—
So, — —	

> **TRY IT : : 4.17** Convert the improper fraction to a mixed number: —

> **TRY IT : : 4.18** Convert the improper fraction to a mixed number: —

HOW TO : : CONVERT AN IMPROPER FRACTION TO A MIXED NUMBER.

Step 1. Divide the denominator into the numerator.
Step 2. Identify the quotient, remainder, and divisor.
Step 3. Write the mixed number as quotient ———.

EXAMPLE 4.10

Convert the improper fraction — to a mixed number.

Solution

	—
Divide the denominator into the numerator.	Remember, — means $\overline{)\quad}$.
Identify the quotient, remainder, and divisor.	divisor → $8\overline{)33}$ with quotient 4, 32, remainder 1
Write the mixed number as quotient ————.	—
	So, — —

> **TRY IT : : 4.19** Convert the improper fraction to a mixed number: —

> **TRY IT : : 4.20** Convert the improper fraction to a mixed number: —

In Example 4.8, we changed — to an improper fraction by first seeing that the whole is a set of five fifths. So we had five fifths and four more fifths.

— — —

Where did the nine come from? There are nine fifths—one whole (five fifths) plus four fifths. Let us use this idea to see how to convert a mixed number to an improper fraction.

EXAMPLE 4.11

Convert the mixed number — to an improper fraction.

Solution

	—
Multiply the whole number by the denominator.	
The whole number is 4 and the denominator is 3.	$\frac{4 \cdot 3 + \square}{\square}$
Simplify.	$\frac{12 + \square}{\square}$
Add the numerator to the product.	
The numerator of the mixed number is 2.	$\frac{12 + 2}{\square}$
Simplify.	$\frac{14}{\square}$
Write the final sum over the original denominator.	
The denominator is 3.	—

TRY IT : : 4.21 Convert the mixed number to an improper fraction: —

TRY IT : : 4.22 Convert the mixed number to an improper fraction: —

HOW TO : : CONVERT A MIXED NUMBER TO AN IMPROPER FRACTION.

Step 1. Multiply the whole number by the denominator.
Step 2. Add the numerator to the product found in Step 1.
Step 3. Write the final sum over the original denominator.

EXAMPLE 4.12

Convert the mixed number — to an improper fraction.

Solution

—

Multiply the whole number by the denominator.	
The whole number is 10 and the denominator is 7.	$\frac{10 \cdot 7 + \square}{\square}$
Simplify.	$\frac{70 + \square}{\square}$
Add the numerator to the product.	
The numerator of the mixed number is 2.	$\frac{70 + 2}{\square}$
Simplify.	$\frac{72}{\square}$
Write the final sum over the original denominator.	
The denominator is 7.	—

TRY IT : : 4.23 Convert the mixed number to an improper fraction: —

TRY IT : : 4.24 Convert the mixed number to an improper fraction: —

Model Equivalent Fractions

Let's think about Andy and Bobby and their favorite food again. If Andy eats — of a pizza and Bobby eats — of the pizza, have they eaten the same amount of pizza? In other words, does — — We can use fraction tiles to find out whether Andy and Bobby have eaten *equivalent* parts of the pizza.

Equivalent Fractions

Equivalent fractions are fractions that have the same value.

Fraction tiles serve as a useful model of equivalent fractions. You may want to use fraction tiles to do the following activity. Or you might make a copy of Figure 4.3 and extend it to include eighths, tenths, and twelfths.

Start with a — tile. How many fourths equal one-half? How many of the — tiles exactly cover the — tile?

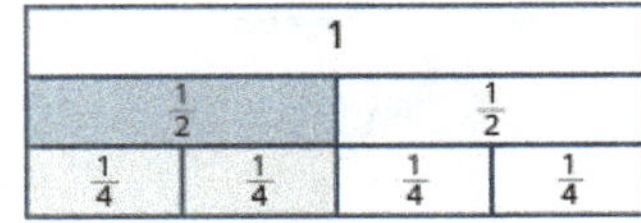

Since two — tiles cover the — tile, we see that — is the same as — or — —

How many of the — tiles cover the — tile?

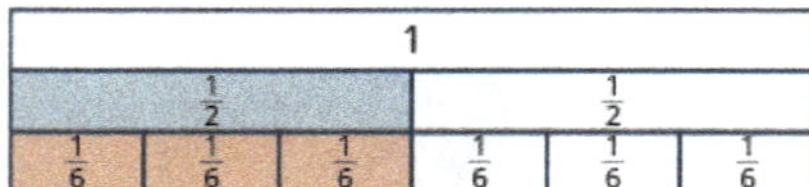

Since three — tiles cover the — tile, we see that — is the same as —

So, — — The fractions are equivalent fractions.

MANIPULATIVE MATHEMATICS

Doing the activity "Equivalent Fractions" will help you develop a better understanding of what it means when two fractions are equivalent.

EXAMPLE 4.13

Use fraction tiles to find equivalent fractions. Show your result with a figure.

How many eighths equal one-half? How many tenths equal one-half?

How many twelfths equal one-half?

Solution

It takes four — tiles to exactly cover the — tile, so — —

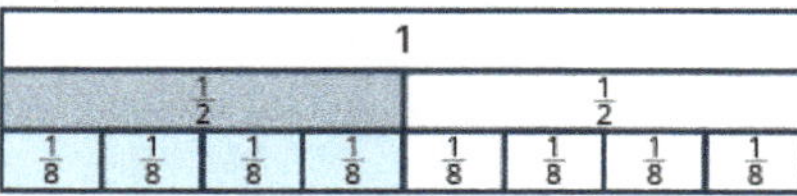

It takes five — tiles to exactly cover the — tile, so — —

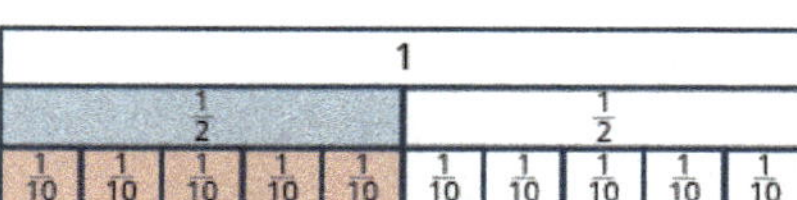

It takes six — tiles to exactly cover the — tile, so — —

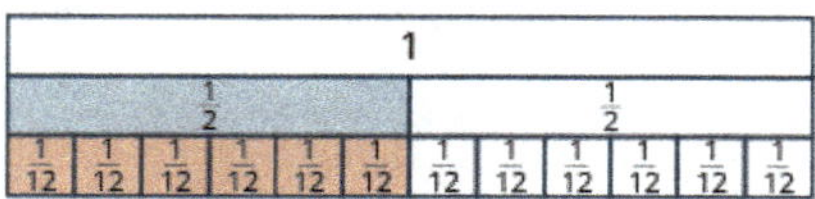

Suppose you had tiles marked — How many of them would it take to equal — Are you thinking ten tiles? If you are, you're right, because — —

We have shown that — — — — — — and — are all equivalent fractions.

TRY IT : : 4.25 Use fraction tiles to find equivalent fractions: How many eighths equal one-fourth?

TRY IT : : 4.26 Use fraction tiles to find equivalent fractions: How many twelfths equal one-fourth?

Find Equivalent Fractions

We used fraction tiles to show that there are many fractions equivalent to — For example, — — and — are all equivalent to — When we lined up the fraction tiles, it took four of the — tiles to make the same length as a — tile. This showed that — — See Example 4.13.

We can show this with pizzas, too. Figure 4.4(a) shows a single pizza, cut into two equal pieces with — shaded. Figure 4.4(b) shows a second pizza of the same size, cut into eight pieces with — shaded.

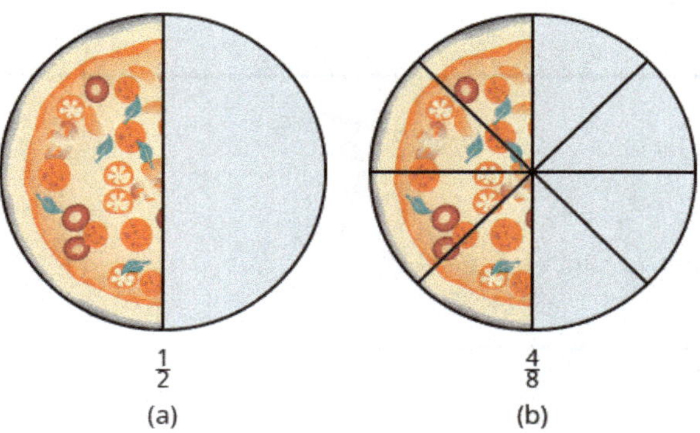

Figure 4.4

This is another way to show that — is equivalent to —

How can we use mathematics to change — into — How could you take a pizza that is cut into two pieces and cut it into eight pieces? You could cut each of the two larger pieces into four smaller pieces! The whole pizza would then be cut into eight pieces instead of just two. Mathematically, what we've described could be written as:

$$\frac{1 \cdot 4}{2 \cdot 4} = \frac{4}{8}$$

These models lead to the Equivalent Fractions Property, which states that if we multiply the numerator and denominator of a fraction by the same number, the value of the fraction does not change.

Equivalent Fractions Property

If and are numbers where and then

— ——

When working with fractions, it is often necessary to express the same fraction in different forms. To find equivalent forms of a fraction, we can use the Equivalent Fractions Property. For example, consider the fraction one-half.

$$\frac{1 \cdot 3}{2 \cdot 3} = \frac{3}{6} \quad \text{so} \quad \frac{1}{2} = \frac{3}{6}$$

$$\frac{1 \cdot 2}{2 \cdot 2} = \frac{2}{4} \quad \text{so} \quad \frac{1}{2} = \frac{2}{4}$$

$$\frac{1 \cdot 10}{2 \cdot 10} = \frac{10}{20} \quad \text{so} \quad \frac{1}{2} = \frac{10}{20}$$

So, we say that — — — and —— are equivalent fractions.

EXAMPLE 4.14

Find three fractions equivalent to —

Solution

To find a fraction equivalent to — we multiply the numerator and denominator by the same number (but not zero). Let us multiply them by and

$$\frac{2\cdot 2}{5\cdot 2}=\frac{4}{10}\qquad\frac{2\cdot 3}{5\cdot 3}=\frac{6}{15}\qquad\frac{2\cdot 5}{5\cdot 5}=\frac{10}{25}$$

So, — — and — are equivalent to —

TRY IT : : 4.27 Find three fractions equivalent to —

TRY IT : : 4.28 Find three fractions equivalent to —

EXAMPLE 4.15

Find a fraction with a denominator of that is equivalent to —

Solution

To find equivalent fractions, we multiply the numerator and denominator by the same number. In this case, we need to multiply the denominator by a number that will result in

Since we can multiply by to get we can find the equivalent fraction by multiplying both the numerator and denominator by

$$\frac{2}{7}=\frac{2\cdot 3}{7\cdot 3}=\frac{6}{21}$$

TRY IT : : 4.29 Find a fraction with a denominator of that is equivalent to —

TRY IT : : 4.30 Find a fraction with a denominator of that is equivalent to —

Locate Fractions and Mixed Numbers on the Number Line

Now we are ready to plot fractions on a number line. This will help us visualize fractions and understand their values.

MANIPULATIVE MATHEMATICS

Doing the Manipulative Mathematics activity "Number Line Part " will help you develop a better understanding of the location of fractions on the number line.

Let us locate — — — — — and — on the number line.

We will start with the whole numbers and because they are the easiest to plot.

The proper fractions listed are — and — We know proper fractions have values less than one, so — and — are located between the whole numbers and The denominators are both so we need to divide the segment of the number line between and into five equal parts. We can do this by drawing four equally spaced marks on the number line, which we can then label as — — — and —

Now plot points at — and —

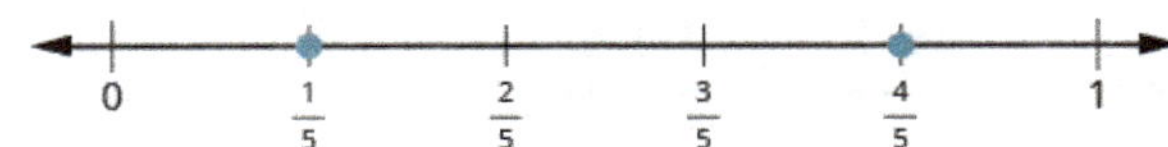

The only mixed number to plot is — Between what two whole numbers is — Remember that a mixed number is a whole number plus a proper fraction, so — Since it is greater than but not a whole unit greater, — is between and We need to divide the portion of the number line between and into three equal pieces (thirds) and plot — at the first mark.

Finally, look at the improper fractions — — and — Locating these points will be easier if you change each of them to a mixed number.

— — — — — —

Here is the number line with all the points plotted.

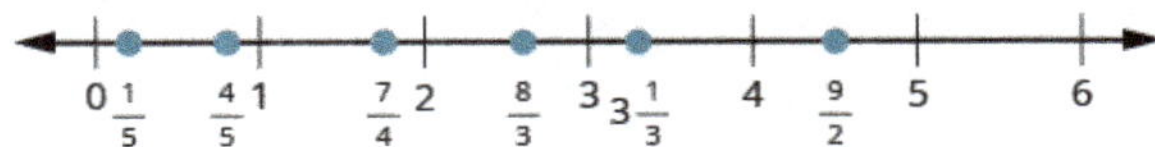

EXAMPLE 4.16

Locate and label the following on a number line: — — — — and —

Solution

Start by locating the proper fraction — It is between and To do this, divide the distance between and into four equal parts. Then plot —

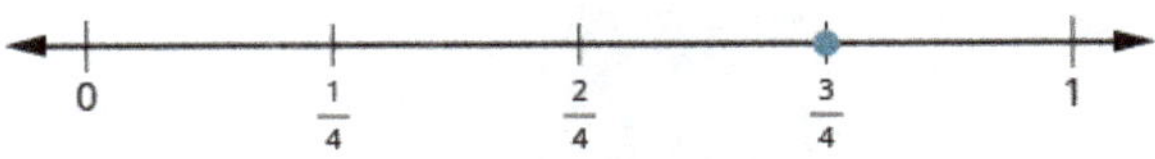

Next, locate the mixed number — It is between and on the number line. Divide the number line between and into five equal parts, and then plot — one-fifth of the way between and .

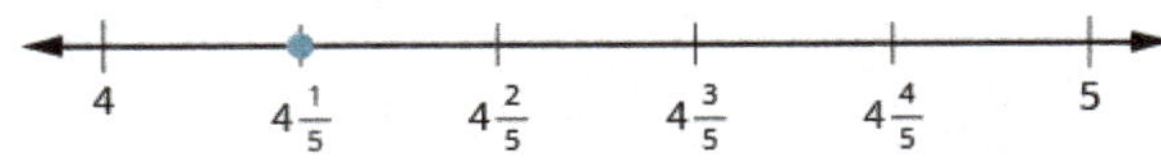

Now locate the improper fractions — and —.

It is easier to plot them if we convert them to mixed numbers first.

— — — —

Divide the distance between and into thirds.

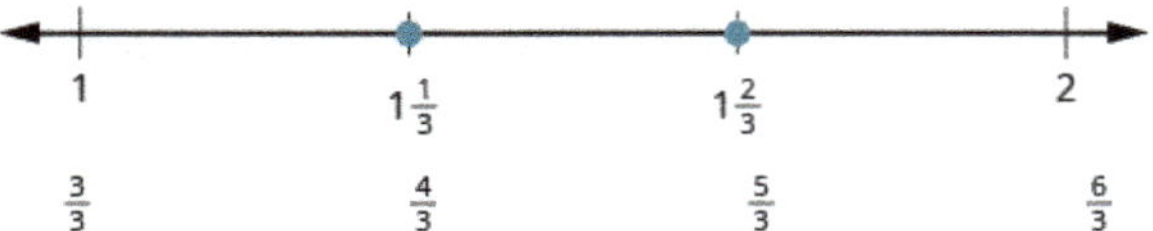

Next let us plot — We write it as a mixed number, — —. Plot it between and

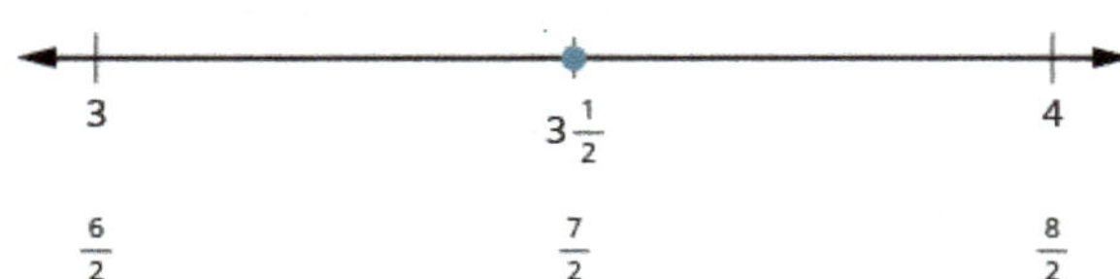

The number line shows all the numbers located on the number line.

> **TRY IT : :** 4.31 Locate and label the following on a number line: — — — — —

> **TRY IT : :** 4.32 Locate and label the following on a number line: — — — — —

In Introduction to Integers, we defined the opposite of a number. It is the number that is the same distance from zero on the number line but on the opposite side of zero. We saw, for example, that the opposite of is and the opposite of is

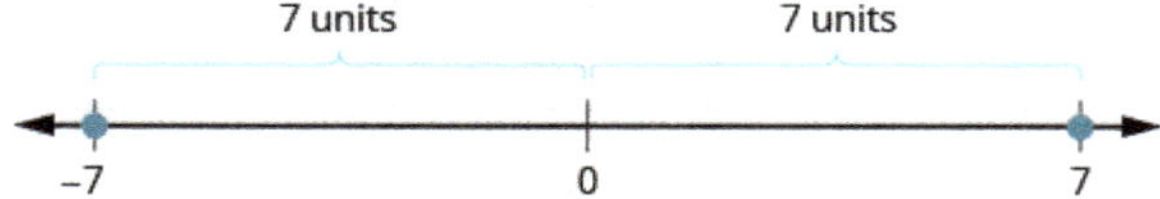

Fractions have opposites, too. The opposite of — is — It is the same distance from on the number line, but on the opposite side of

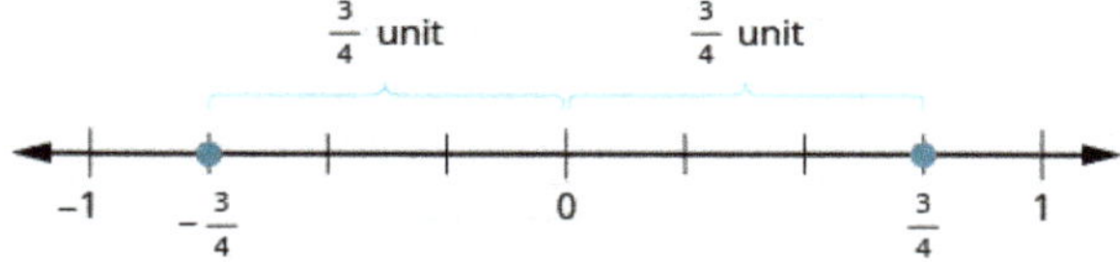

Thinking of negative fractions as the opposite of positive fractions will help us locate them on the number line. To locate — on the number line, first think of where — is located. It is an improper fraction, so we first convert it to the mixed number — and see that it will be between and on the number line. So its opposite, — will be between and on the number line.

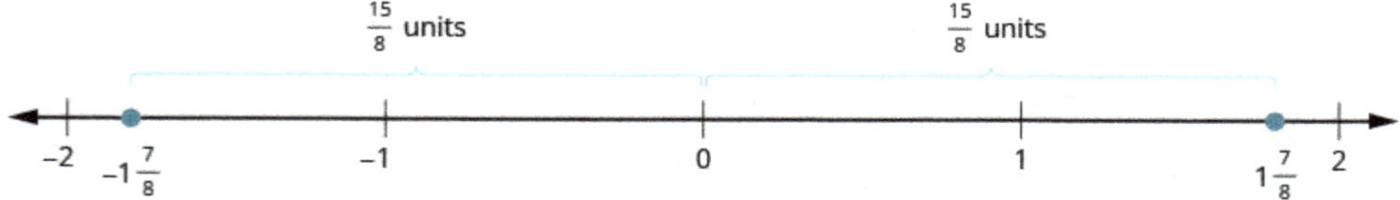

EXAMPLE 4.17

Locate and label the following on the number line: — — — — — and —

Solution

Draw a number line. Mark in the middle and then mark several units to the left and right.

To locate — divide the interval between and into four equal parts. Each part represents one-quarter of the distance. So plot — at the first mark.

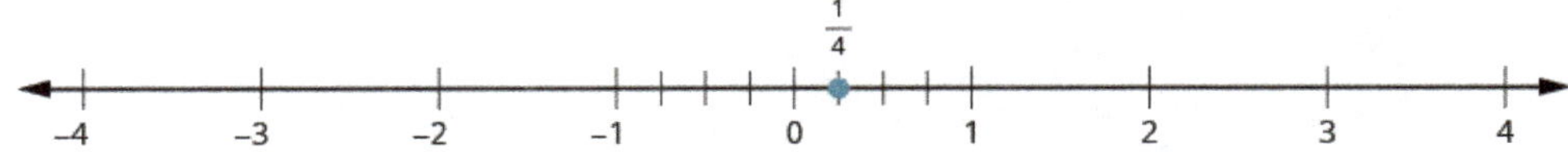

To locate — divide the interval between and into four equal parts. Plot — at the first mark to the left of

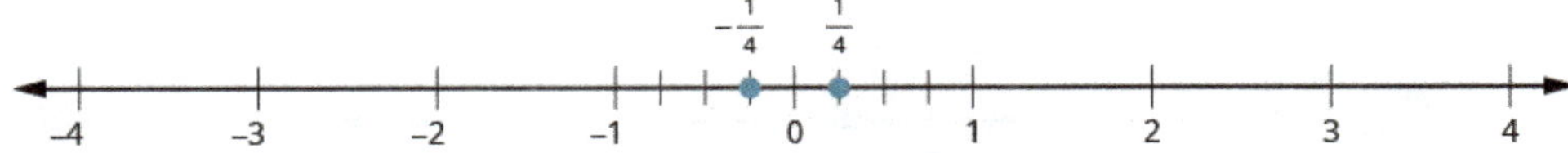

Since — is between and divide the interval between and into three equal parts. Plot — at the first mark to the right of Then since — is the opposite of — it is between and Divide the interval between and into three equal parts. Plot — at the first mark to the left of

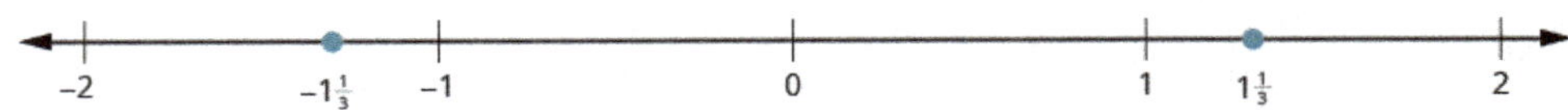

To locate — and — it may be helpful to rewrite them as the mixed numbers — and —

Since — is between and divide the interval between and into two equal parts. Plot — at the mark. Then since — is between and divide the interval between and into two equal parts. Plot — at the mark.

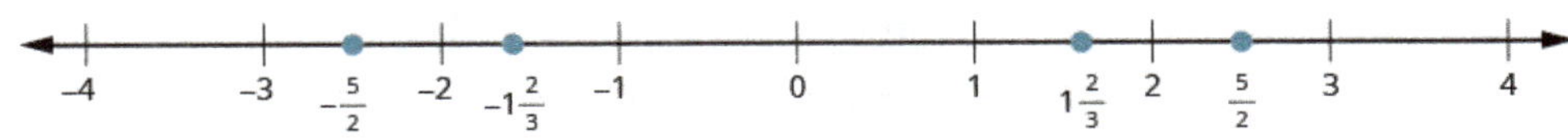

TRY IT : : 4.33 Locate and label each of the given fractions on a number line:

— — — — — —

TRY IT : : 4.34 Locate and label each of the given fractions on a number line:

— — — — — —

Order Fractions and Mixed Numbers

We can use the inequality symbols to order fractions. Remember that means that is to the right of on the number line. As we move from left to right on a number line, the values increase.

EXAMPLE 4.18

Order each of the following pairs of numbers, using or

— — — — ——

Solution

—

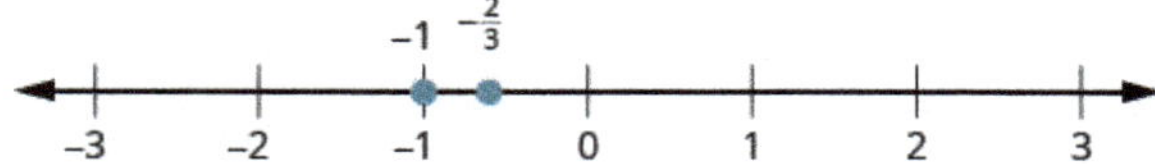

—

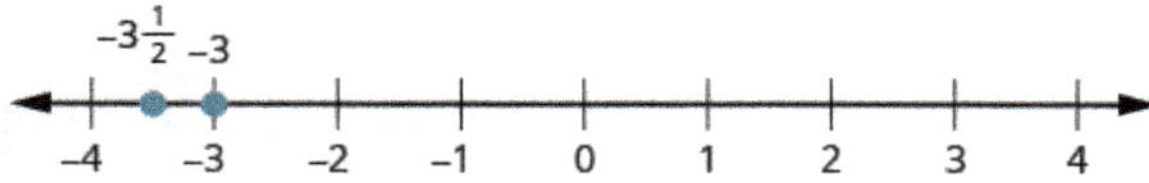

— —

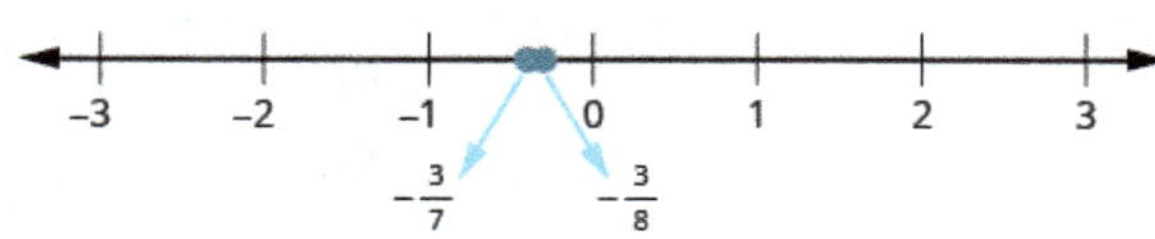

——

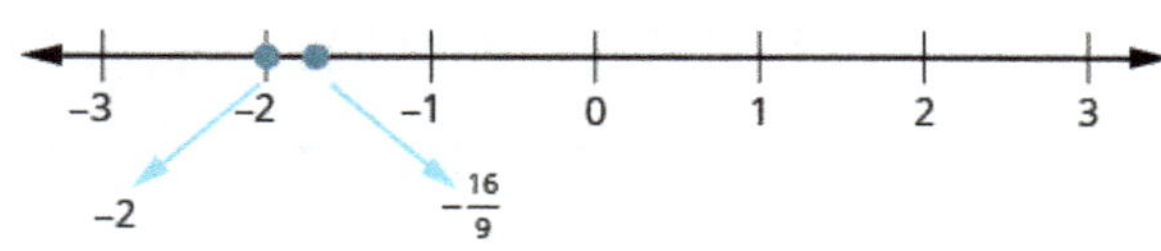

TRY IT : : 4.35 Order each of the following pairs of numbers, using or

— — — — —

TRY IT : : 4.36 Order each of the following pairs of numbers, using or

— — — — —

MEDIA : : ACCESS ADDITIONAL ONLINE RESOURCES

- Introduction to Fractions (http://www.openstax.org/l/24Introtofract)
- Identify Fractions Using Pattern Blocks (http://www.openstax.org/l/24FractPattBloc)

4.1 EXERCISES

Practice Makes Perfect

In the following exercises, name the fraction of each figure that is shaded.

1.

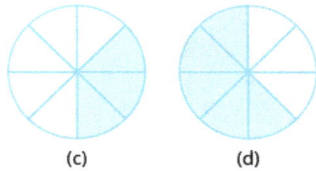

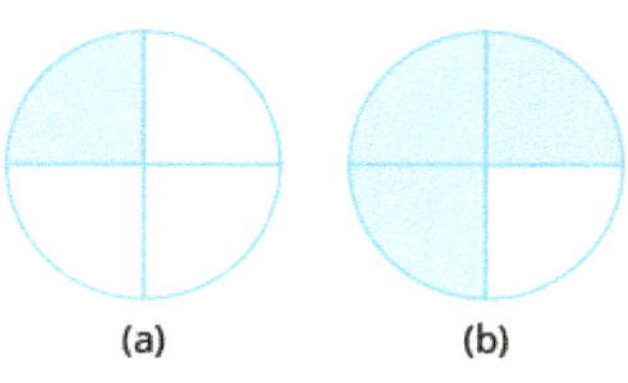

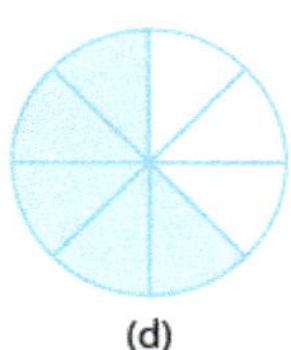

2.

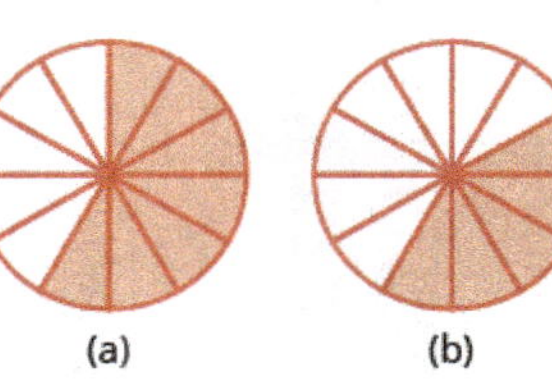

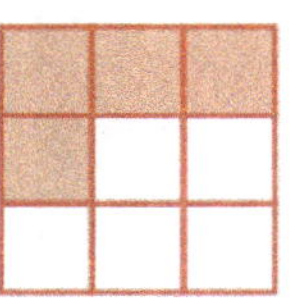

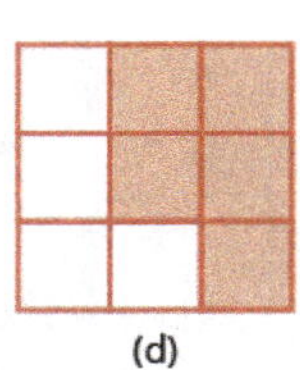

In the following exercises, shade parts of circles or squares to model the following fractions.

3. —

4. —

5. —

6. —

7. —

8. —

9. —

10. —

In the following exercises, use fraction circles to make wholes using the following pieces.

11. thirds

12. eighths

13. sixths

14. thirds

15. fifths

16. fourths

In the following exercises, name the improper fractions. Then write each improper fraction as a mixed number.

17.

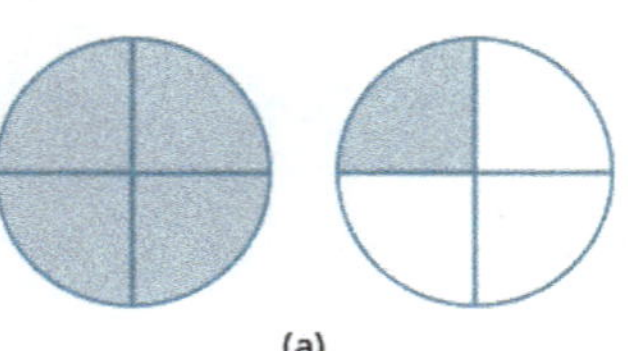

(a)

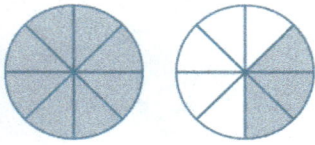

(b)

(c)

18.

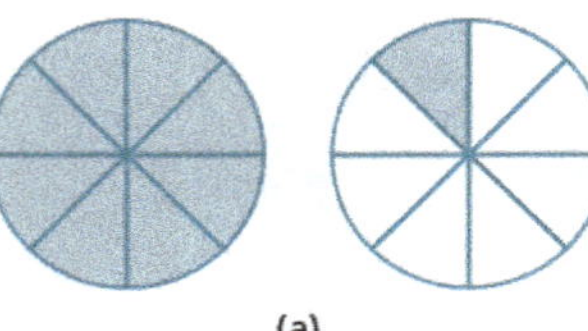

(a)

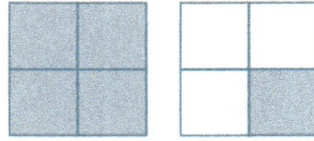

(b)

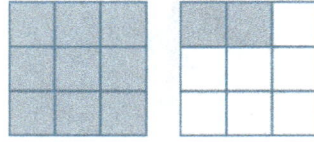

(c)

19.

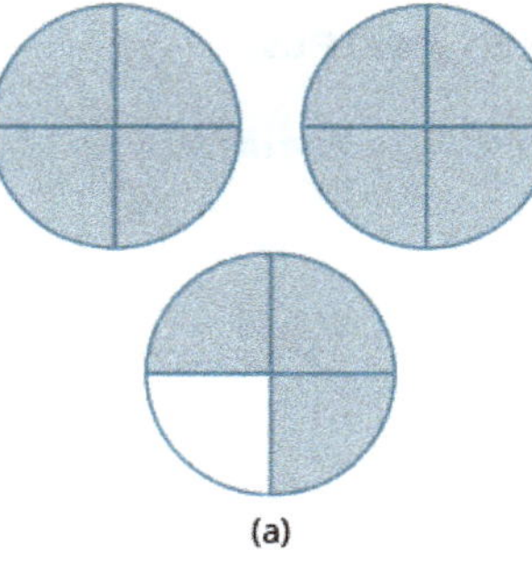

(a)

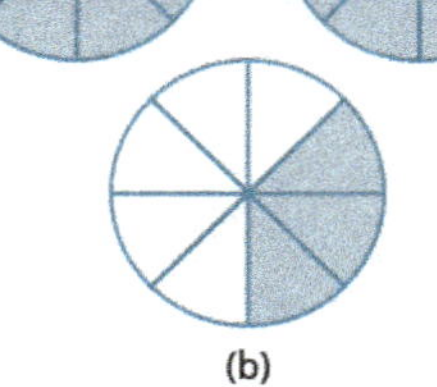

(b)

In the following exercises, draw fraction circles to model the given fraction.

20. —

21. —

22. —

23. —

24. —

25. —

26. —

27. —

In the following exercises, rewrite the improper fraction as a mixed number.

28. —

29. —

30. —

31. —

32. —

33. —

34. —

35. —

In the following exercises, rewrite the mixed number as an improper fraction.

36. —

37. —

38. —

39. —

40. —

41. —

42. —

43. —

In the following exercises, use fraction tiles or draw a figure to find equivalent fractions.

44. How many sixths equal one-third?

45. How many twelfths equal one-third?

46. How many eighths equal three-fourths?

47. How many twelfths equal three-fourths?

48. How many fourths equal three-halves?

49. How many sixths equal three-halves?

In the following exercises, find three fractions equivalent to the given fraction. Show your work, using figures or algebra.

50. —

51. —

52. —

53. —

54. —

55. —

In the following exercises, plot the numbers on a number line.

56. — — —

57. — — —

58. — — —

59. — — —

60. — —

61. — —

62. — — — — — —

63. — — — — — —

In the following exercises, order each of the following pairs of numbers, using or

64. —

65. —

66. —

67. —

68. — —

69. — —

70. —

71. —

Everyday Math

72. Music Measures A choreographed dance is broken into counts. A — count has one step in a count, a — count has two steps in a count and a — count has three steps in a count. How many steps would be in a — count? What type of count has four steps in it?

73. Music Measures Fractions are used often in music. In — time, there are four quarter notes in one measure.

How many measures would eight quarter notes make?

The song "Happy Birthday to You" has quarter notes. How many measures are there in "Happy Birthday to You?"

74. Baking Nina is making five pans of fudge to serve after a music recital. For each pan, she needs — cup of walnuts.

How many cups of walnuts does she need for five pans of fudge?

Do you think it is easier to measure this amount when you use an improper fraction or a mixed number? Why?

Writing Exercises

75. Give an example from your life experience (outside of school) where it was important to understand fractions.

76. Explain how you locate the improper fraction —

on a number line on which only the whole numbers from through are marked.

Self Check

After completing the exercises, use this checklist to evaluate your mastery of the objectives of this section.

I can...	Confidently	With some help	No-I don't get it!
understand the meaning of fractions.			
model improper fractions and mixed numbers.			
convert between improper fractions and mixed numbers.			
model equivalent fractions.			
find equivalent fractions.			
locate fractions and mixed numbers on the number line.			
order fractions and mixed numbers.			

If most of your checks were:

...confidently. Congratulations! You have achieved the objectives in this section. Reflect on the study skills you used so that you can continue to use them. What did you do to become confident of your ability to do these things? Be specific.

...with some help. This must be addressed quickly because topics you do not master become potholes in your road to success. In math, every topic builds upon previous work. It is important to make sure you have a strong foundation before you move on. Whom can you ask for help? Your fellow classmates and instructor are good resources. Is there a place on campus where math tutors are available? Can your study skills be improved?

...no—I don't get it! This is a warning sign and you must not ignore it. You should get help right away or you will quickly be overwhelmed. See your instructor as soon as you can to discuss your situation. Together you can come up with a plan to get you the help you need.

4.2 Multiply and Divide Fractions

Learning Objectives

By the end of this section, you will be able to:

- Simplify fractions
- Multiply fractions
- Find reciprocals
- Divide fractions

BE PREPARED : : 4.3 Before you get started, take this readiness quiz.

Find the prime factorization of
If you missed this problem, review Example 2.48.

BE PREPARED : : 4.4 Draw a model of the fraction —

If you missed this problem, review Example 4.2.

BE PREPARED : : 4.5 Find two fractions equivalent to —

If you missed this problem, review Example 4.14.

Simplify Fractions

In working with equivalent fractions, you saw that there are many ways to write fractions that have the same value, or represent the same part of the whole. How do you know which one to use? Often, we'll use the fraction that is in *simplified* form.

A fraction is considered simplified if there are no common factors, other than in the numerator and denominator. If a fraction does have common factors in the numerator and denominator, we can reduce the fraction to its simplified form by removing the common factors.

Simplified Fraction

A fraction is considered simplified if there are no common factors in the numerator and denominator.

For example,

- — is simplified because there are no common factors of and
- —— is not simplified because is a common factor of and

The process of simplifying a fraction is often called *reducing the fraction*. In the previous section, we used the Equivalent Fractions Property to find equivalent fractions. We can also use the Equivalent Fractions Property in reverse to simplify fractions. We rewrite the property to show both forms together.

Equivalent Fractions Property

If are numbers where then

Notice that is a common factor in the numerator and denominator. Anytime we have a common factor in the numerator and denominator, it can be removed.

HOW TO : : SIMPLIFY A FRACTION.

Step 1. Rewrite the numerator and denominator to show the common factors. If needed, factor the numerator and denominator into prime numbers.

Step 2. Simplify, using the equivalent fractions property, by removing common factors.

Step 3. Multiply any remaining factors.

EXAMPLE 4.19

Simplify: —

Solution

To simplify the fraction, we look for any common factors in the numerator and the denominator.

Notice that 5 is a factor of both 10 and 15.	—
Factor the numerator and denominator.	$\frac{2 \cdot 5}{3 \cdot 5}$
Remove the common factors.	$\frac{2 \cdot \cancel{5}}{3 \cdot \cancel{5}}$
Simplify.	—

TRY IT : : 4.37 Simplify: —.

TRY IT : : 4.38 Simplify: —.

To simplify a negative fraction, we use the same process as in Example 4.19. Remember to keep the negative sign.

EXAMPLE 4.20

Simplify: —

Solution

We notice that 18 and 24 both have factors of 6.	—
Rewrite the numerator and denominator showing the common factor.	$-\frac{3 \cdot 6}{4 \cdot 6}$
Remove common factors.	$-\frac{3 \cdot \cancel{6}}{4 \cdot \cancel{6}}$
Simplify.	—

TRY IT : : 4.39 Simplify:

TRY IT : : 4.40 Simplify:

After simplifying a fraction, it is always important to check the result to make sure that the numerator and denominator do not have any more factors in common. Remember, the definition of a simplified fraction: *a fraction is considered simplified if there are no common factors in the numerator and denominator*.

When we simplify an improper fraction, there is no need to change it to a mixed number.

EXAMPLE 4.21

Simplify:

Solution

Rewrite the numerator and denominator, showing the common factors, 8.	$\frac{7 \cdot 8}{4 \cdot 8}$
Remove common factors.	$\frac{7 \cdot \cancel{8}}{4 \cdot \cancel{8}}$
Simplify.	

TRY IT : : 4.41 Simplify:

TRY IT : : 4.42 Simplify:

HOW TO : : SIMPLIFY A FRACTION.

Step 1. Rewrite the numerator and denominator to show the common factors. If needed, factor the numerator and denominator into prime numbers.

Step 2. Simplify, using the equivalent fractions property, by removing common factors.

Step 3. Multiply any remaining factors

Sometimes it may not be easy to find common factors of the numerator and denominator. A good idea, then, is to factor the numerator and the denominator into prime numbers. (You may want to use the factor tree method to identify the prime factors.) Then divide out the common factors using the Equivalent Fractions Property.

EXAMPLE 4.22

Simplify:

Solution

Use factor trees to factor the numerator and denominator.

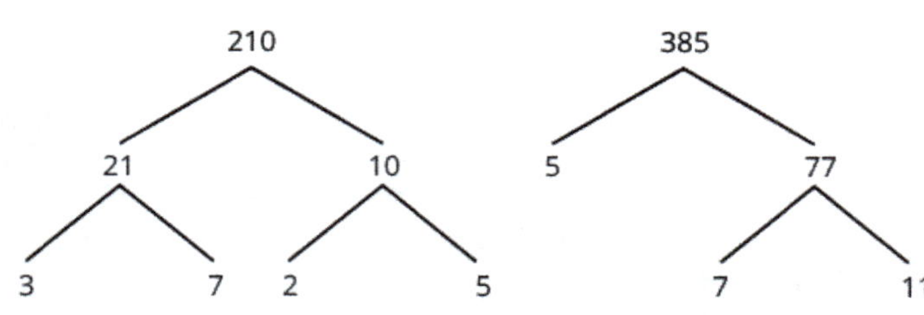

Rewrite the numerator and denominator as the product of the primes.

Remove the common factors. $\frac{2 \cdot 3 \cdot \cancel{5} \cdot \cancel{7}}{\cancel{5} \cdot \cancel{7} \cdot 11}$

Simplify.

Multiply any remaining factors.

TRY IT : : 4.43 Simplify:

TRY IT : : 4.44 Simplify:

We can also simplify fractions containing variables. If a variable is a common factor in the numerator and denominator, we remove it just as we do with an integer factor.

EXAMPLE 4.23

Simplify:

Solution

Rewrite numerator and denominator showing common factors.

Remove common factors.

Simplify.

TRY IT : : 4.45 Simplify:

TRY IT : : 4.46 Simplify: —

Multiply Fractions

A model may help you understand multiplication of fractions. We will use fraction tiles to model — — To multiply — and — think — of —

Start with fraction tiles for three-fourths. To find one-half of three-fourths, we need to divide them into two equal groups. Since we cannot divide the three — tiles evenly into two parts, we exchange them for smaller tiles.

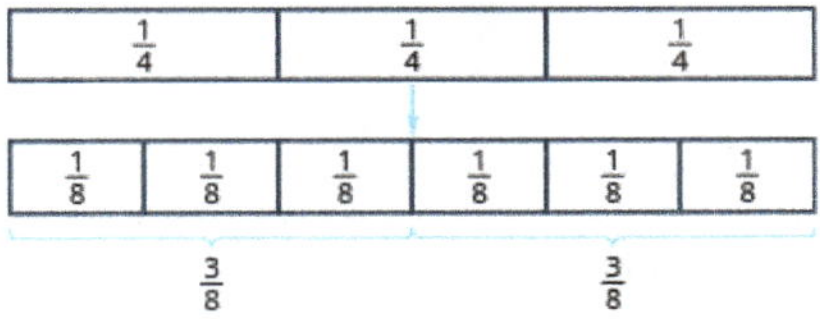

We see — is equivalent to — Taking half of the six — tiles gives us three — tiles, which is —

Therefore,

— — —

MANIPULATIVE MATHEMATICS

Doing the Manipulative Mathematics activity "Model Fraction Multiplication" will help you develop a better understanding of how to multiply fractions.

EXAMPLE 4.24

Use a diagram to model — —

Solution

First shade in — of the rectangle.

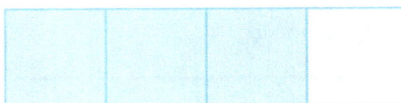

We will take — of this — so we heavily shade — of the shaded region.

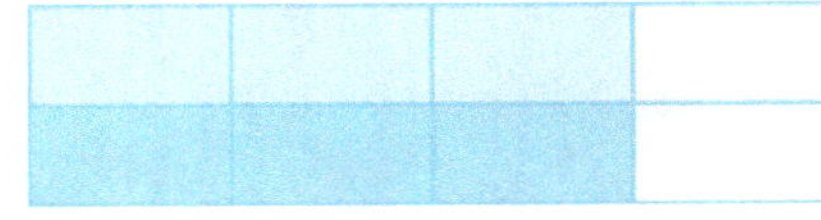

Notice that out of the pieces are heavily shaded. This means that — of the rectangle is heavily shaded.

Therefore, — of — is — or — — —

TRY IT : : 4.47 Use a diagram to model: — —

> **TRY IT : : 4.48** Use a diagram to model: — —

Look at the result we got from the model in Example 4.24. We found that — — — Do you notice that we could have gotten the same answer by multiplying the numerators and multiplying the denominators?

	— —
Multiply the numerators, and multiply the denominators.	— —
Simplify.	—

This leads to the definition of fraction multiplication. To multiply fractions, we multiply the numerators and multiply the denominators. Then we write the fraction in simplified form.

Fraction Multiplication

If are numbers where then

— — —

EXAMPLE 4.25

Multiply, and write the answer in simplified form: — —

Solution

	— —
Multiply the numerators; multiply the denominators.	—
Simplify.	—

There are no common factors, so the fraction is simplified.

> **TRY IT : : 4.49** Multiply, and write the answer in simplified form: — —

> **TRY IT : : 4.50** Multiply, and write the answer in simplified form: — —

When multiplying fractions, the properties of positive and negative numbers still apply. It is a good idea to determine the sign of the product as the first step. In Example 4.26 we will multiply two negatives, so the product will be positive.

EXAMPLE 4.26

Multiply, and write the answer in simplified form: — —

Solution

The signs are the same, so the product is positive. Multiply the numerators, multiply the denominators.	
Simplify.	
Look for common factors in the numerator and denominator. Rewrite showing common factors.	$\frac{5 \cdot \cancel{2}}{12 \cdot \cancel{2}}$
Remove common factors.	

Another way to find this product involves removing common factors earlier.

Determine the sign of the product. Multiply.	
Show common factors and then remove them.	$\frac{5 \cdot \cancel{2}}{4 \cdot \cancel{2} \cdot 3}$
Multiply remaining factors.	

We get the same result.

TRY IT : : 4.51 Multiply, and write the answer in simplified form:

TRY IT : : 4.52 Multiply, and write the answer in simplified form:

EXAMPLE 4.27

Multiply, and write the answer in simplified form:

Solution

	— —
Determine the sign of the product; multiply.	— —
Are there any common factors in the numerator and the denominator? We know that 7 is a factor of 14 and 21, and 5 is a factor of 20 and 15.	
Rewrite showing common factors.	$-\frac{2 \cdot \cancel{7} \cdot 4 \cdot \cancel{5}}{3 \cdot \cancel{5} \cdot 3 \cdot \cancel{7}}$
Remove the common factors.	—
Multiply the remaining factors.	—

TRY IT : : 4.53 Multiply, and write the answer in simplified form: — —

TRY IT : : 4.54 Multiply, and write the answer in simplified form: — —

When multiplying a fraction by an integer, it may be helpful to write the integer as a fraction. Any integer, can be written as — So, — for example.

EXAMPLE 4.28

Multiply, and write the answer in simplified form:

—

—

Solution

	—
Write 56 as a fraction.	— —
Determine the sign of the product; multiply.	—
Simplify.	

Write −20x as a fraction.	
Determine the sign of the product; multiply.	
Show common factors and then remove them.	$-\frac{12 \cdot 4 \cdot 5x}{5 \cdot 1}$
Multiply remaining factors; simplify.	−48x

TRY IT : : 4.55 Multiply, and write the answer in simplified form:

TRY IT : : 4.56 Multiply, and write the answer in simplified form:

Find Reciprocals

The fractions and are related to each other in a special way. So are and Do you see how? Besides looking like upside-down versions of one another, if we were to multiply these pairs of fractions, the product would be

Such pairs of numbers are called reciprocals.

Reciprocal

The **reciprocal** of the fraction is where and

A number and its reciprocal have a product of

To find the reciprocal of a fraction, we invert the fraction. This means that we place the numerator in the denominator and the denominator in the numerator.

To get a positive result when multiplying two numbers, the numbers must have the same sign. So reciprocals must have the same sign.

$$\frac{a}{b} \cdot \frac{b}{a} = 1 \text{ positive}$$

$$3 \cdot \frac{1}{3} = 1 \quad \text{and} \quad -3 \cdot \left(-\frac{1}{3}\right) = 1$$

both positive both negative

To find the reciprocal, keep the same sign and invert the fraction. The number zero does not have a reciprocal. Why? A number and its reciprocal multiply to Is there any number so that No. So, the number does not have a reciprocal.

EXAMPLE 4.29

Find the reciprocal of each number. Then check that the product of each number and its reciprocal is

Solution

To find the reciprocals, we keep the sign and invert the fractions.

Find the reciprocal of .	The reciprocal of is .
Check:	
Multiply the number and its reciprocal.	
Multiply numerators and denominators.	
Simplify.	

Find the reciprocal of .	
Simplify.	
Check:	

Find the reciprocal of .	
Check:	

Find the reciprocal of .	
Write as a fraction.	—
Write the reciprocal of —.	—
Check:	—

TRY IT : : 4.57 Find the reciprocal:

— — —

TRY IT : : 4.58 Find the reciprocal:

— — —

In a previous chapter, we worked with opposites and absolute values. Table 4.1 compares opposites, absolute values, and reciprocals.

Opposite	Absolute Value	Reciprocal
has opposite sign	is never negative	has same sign, fraction inverts

Table 4.1

EXAMPLE 4.30

Fill in the chart for each fraction in the left column:

Number	Opposite	Absolute Value	Reciprocal
—			
—			
—			

Solution

To find the opposite, change the sign. To find the absolute value, leave the positive numbers the same, but take the opposite of the negative numbers. To find the reciprocal, keep the sign the same and invert the fraction.

Number	Opposite	Absolute Value	Reciprocal
—	—	—	—
—	—	—	
—	—	—	—
			—

TRY IT : : 4.59 Fill in the chart for each number given:

Number	Opposite	Absolute Value	Reciprocal
—			
—			
—			

TRY IT : : 4.60 Fill in the chart for each number given:

Number	Opposite	Absolute Value	Reciprocal
—			
—			
—			

Divide Fractions

Why is We previously modeled this with counters. How many groups of counters can be made from a group of counters?

There are groups of counters. In other words, there are four in So,

What about dividing fractions? Suppose we want to find the quotient: — — We need to figure out how many — there are in — We can use fraction tiles to model this division. We start by lining up the half and sixth fraction tiles as shown in Figure 4.5. Notice, there are three — tiles in — so — —

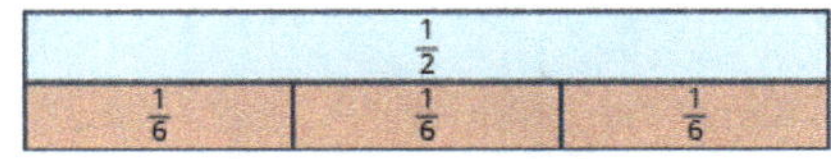

Figure 4.5

MANIPULATIVE MATHEMATICS

Doing the Manipulative Mathematics activity "Model Fraction Division" will help you develop a better understanding of dividing fractions.

EXAMPLE 4.31

Model: — —

Solution

We want to determine how many — are in — Start with one — tile. Line up — tiles underneath the — tile.

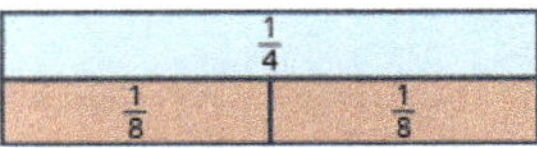

There are two — in —

So, — —

TRY IT : : 4.61 Model: — —

TRY IT : : 4.62 Model: — —

EXAMPLE 4.32

Model: —

Solution

We are trying to determine how many — there are in We can model this as shown.

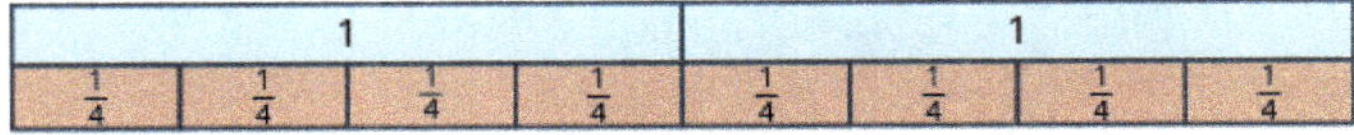

Because there are eight — in —

TRY IT : : 4.63 Model: —

TRY IT : : 4.64 Model: —

Let's use money to model — in another way. We often read — as a 'quarter', and we know that a quarter is one-fourth of a dollar as shown in Figure 4.6. So we can think of — as, "How many quarters are there in two dollars?" One dollar is quarters, so dollars would be quarters. So again, —

Figure 4.6 The U.S. coin called a quarter is worth one-fourth of a dollar.

Using fraction tiles, we showed that — — Notice that — — also. How are — and — related? They are reciprocals. This leads us to the procedure for fraction division.

Fraction Division

If are numbers where then

— — — —

To divide fractions, multiply the first fraction by the reciprocal of the second.

We need to say to be sure we don't divide by zero.

EXAMPLE 4.33

Divide, and write the answer in simplified form: — —

Solution

	— —
Multiply the first fraction by the reciprocal of the second.	— —
Multiply. The product is negative.	—

TRY IT : : 4.65 Divide, and write the answer in simplified form: — —

TRY IT : : 4.66 Divide, and write the answer in simplified form: — —

EXAMPLE 4.34

Divide, and write the answer in simplified form: — —

Solution

Multiply the first fraction by the reciprocal of the second.	
Multiply.	

> TRY IT : : 4.67 Divide, and write the answer in simplified form:

> TRY IT : : 4.68 Divide, and write the answer in simplified form:

EXAMPLE 4.35

Divide, and write the answer in simplified form:

Solution

Multiply the first fraction by the reciprocal of the second.	
Multiply. Remember to determine the sign first.	
Rewrite to show common factors.	
Remove common factors and simplify.	

> TRY IT : : 4.69 Divide, and write the answer in simplified form:

> TRY IT : : 4.70 Divide, and write the answer in simplified form:

EXAMPLE 4.36

Divide, and write the answer in simplified form:

Solution

Multiply the first fraction by the reciprocal of the second.	
Multiply.	
Rewrite showing common factors.	$\frac{\cancel{7}\cdot\cancel{9}\cdot 3}{\cancel{9}\cdot 2\cdot\cancel{7}\cdot 2}$
Remove common factors.	
Simplify.	

> **TRY IT : :** 4.71 Divide, and write the answer in simplified form:

> **TRY IT : :** 4.72 Divide, and write the answer in simplified form:

MEDIA : : ACCESS ADDITIONAL ONLINE RESOURCES

- Simplifying Fractions (http://www.openstax.org/l/24SimplifyFrac)
- Multiplying Fractions (Positive Only) (http://www.openstax.org/l/24MultiplyFrac)
- Multiplying Signed Fractions (http://www.openstax.org/l/24MultSigned)
- Dividing Fractions (Positive Only) (http://www.openstax.org/l/24DivideFrac)
- Dividing Signed Fractions (http://www.openstax.org/l/24DivideSign)

4.2 EXERCISES

Practice Makes Perfect

Simplify Fractions

In the following exercises, simplify each fraction. Do not convert any improper fractions to mixed numbers.

77.
78.
79.
80.
81.
82.
83.
84.
85.
86.
87.
88.
89.
90.
91.
92.
93.
94.

Multiply Fractions

In the following exercises, use a diagram to model.

95.
96.
97.
98.

In the following exercises, multiply, and write the answer in simplified form.

99.
100.
101.
102.
103.
104.
105.
106.
107.
108.
109.
110.
111.
112.
113.
114.
115.
116.
117.
118.
119.

120. —

121. —

122. —

123. —

124. —

125. —

126. —

Find Reciprocals

In the following exercises, find the reciprocal.

127. —

128. —

129. —

130. —

131. —

132.

133.

134.

135.

136. Fill in the chart.

	Opposite	Absolute Value	Reciprocal
—			
—			
—			

137. Fill in the chart.

	Opposite	Absolute Value	Reciprocal
—			
—			
—			

Divide Fractions

In the following exercises, model each fraction division.

138. — —

139. — —

140. —

141. —

In the following exercises, divide, and write the answer in simplified form.

142. — —

143. — —

144. — —

145. — —

146. — —

147. — —

148. — —

149. — —

150. — —

151. — —

152. — —

153. — —

154. — —

155. — —

156. — —

157. — —

158. — —

159. — —

160. —

161. —

162. —

163. —

164. —

165. —

166. — — —

167. — — —

Everyday Math

168. Baking A recipe for chocolate chip cookies calls for — cup brown sugar. Imelda wants to double the recipe.

How much brown sugar will Imelda need? Show your calculation. Write your result as an improper fraction and as a mixed number.

Measuring cups usually come in sets of — — — — cup. Draw a diagram to show two different ways that Imelda could measure the brown sugar needed to double the recipe.

169. Baking Nina is making pans of fudge to serve after a music recital. For each pan, she needs — cup of condensed milk.

How much condensed milk will Nina need? Show your calculation. Write your result as an improper fraction and as a mixed number.

Measuring cups usually come in sets of — — — — cup. Draw a diagram to show two different ways that Nina could measure the condensed milk she needs.

170. Portions Don purchased a bulk package of candy that weighs pounds. He wants to sell the candy in little bags that hold — pound. How many little bags of candy can he fill from the bulk package?

171. Portions Kristen has — yards of ribbon. She wants to cut it into equal parts to make hair ribbons for her daughter's dolls. How long will each doll's hair ribbon be?

Writing Exercises

172. Explain how you find the reciprocal of a fraction.

173. Explain how you find the reciprocal of a negative fraction.

174. Rafael wanted to order half a medium pizza at a restaurant. The waiter told him that a medium pizza could be cut into or slices. Would he prefer out of slices or out of slices? Rafael replied that since he wasn't very hungry, he would prefer out of slices. Explain what is wrong with Rafael's reasoning.

175. Give an example from everyday life that demonstrates how — — —

Self Check

After completing the exercises, use this checklist to evaluate your mastery of the objectives of this section.

I can...	Confidently	With some help	No-I don't get it!
simplify fractions.			
multiply fractions.			
find reciprocals.			
divide fractions.			

After reviewing this checklist, what will you do to become confident for all objectives?

4.3 Multiply and Divide Mixed Numbers and Complex Fractions

Learning Objectives

By the end of this section, you will be able to:

- Multiply and divide mixed numbers
- Translate phrases to expressions with fractions
- Simplify complex fractions
- Simplify expressions written with a fraction bar

BE PREPARED : : 4.6

Before you get started, take this readiness quiz.

Divide and reduce, if possible:

If you missed this problem, review Example 3.21.

BE PREPARED : : 4.7

Multiply and write the answer in simplified form: — —.

If you missed this problem, review Example 4.25.

BE PREPARED : : 4.8

Convert — into an improper fraction.

If you missed this problem, review Example 4.11.

Multiply and Divide Mixed Numbers

In the previous section, you learned how to multiply and divide fractions. All of the examples there used either proper or improper fractions. What happens when you are asked to multiply or divide mixed numbers? Remember that we can convert a mixed number to an improper fraction. And you learned how to do that in Visualize Fractions.

EXAMPLE 4.37

Multiply: — —

Solution

	— —
Convert — to an improper fraction.	— —
Multiply.	—
Look for common factors.	—
Remove common factors.	—
Simplify.	—

Notice that we left the answer as an improper fraction, and did not convert it to a mixed number. In algebra, it is preferable to write answers as improper fractions instead of mixed numbers. This avoids any possible confusion between and

TRY IT : : 4.73 Multiply, and write your answer in simplified form:

TRY IT : : 4.74 Multiply, and write your answer in simplified form:

HOW TO : : MULTIPLY OR DIVIDE MIXED NUMBERS.

Step 1. Convert the mixed numbers to improper fractions.

Step 2. Follow the rules for fraction multiplication or division.

Step 3. Simplify if possible.

EXAMPLE 4.38

Multiply, and write your answer in simplified form:

Solution

Convert mixed numbers to improper fractions.	
Multiply.	
Look for common factors.	
Remove common factors.	
Simplify.	

TRY IT : : 4.75 Multiply, and write your answer in simplified form.

TRY IT : : 4.76 Multiply, and write your answer in simplified form.

EXAMPLE 4.39

Divide, and write your answer in simplified form:

Solution

Convert mixed numbers to improper fractions.	
Multiply the first fraction by the reciprocal of the second.	
Multiply.	
Look for common factors.	
Remove common factors.	
Simplify.	

TRY IT : : 4.77 Divide, and write your answer in simplified form:

TRY IT : : 4.78 Divide, and write your answer in simplified form:

EXAMPLE 4.40

Divide:

Solution

Convert mixed numbers to improper fractions.	
Multiply the first fraction by the reciprocal of the second.	
Multiply.	
Look for common factors.	
Remove common factors.	
Simplify.	

TRY IT : : 4.79 Divide, and write your answer in simplified form: — —

TRY IT : : 4.80 Divide, and write your answer in simplified form: — —

Translate Phrases to Expressions with Fractions

The words *quotient* and *ratio* are often used to describe fractions. In Subtract Whole Numbers, we defined quotient as the result of division. The quotient of is the result you get from dividing or — Let's practice translating some phrases into algebraic expressions using these terms.

EXAMPLE 4.41

Translate the phrase into an algebraic expression: "the quotient of and

Solution

The keyword is *quotient*; it tells us that the operation is division. Look for the words *of* and *and* to find the numbers to divide.

This tells us that we need to divide by —

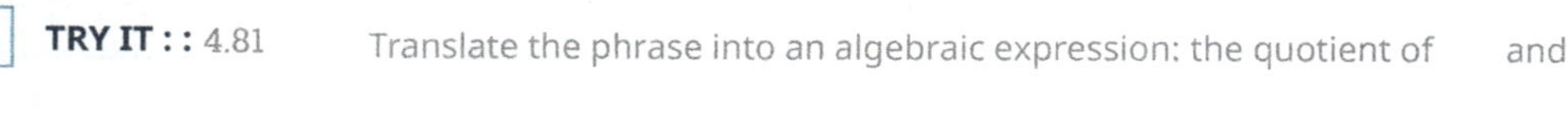

TRY IT : : 4.81 Translate the phrase into an algebraic expression: the quotient of and

TRY IT : : 4.82 Translate the phrase into an algebraic expression: the quotient of and

EXAMPLE 4.42

Translate the phrase into an algebraic expression: the quotient of the difference of and and

Solution

We are looking for the *quotient* of the *difference* of and , and This means we want to divide the difference of and by

TRY IT : : 4.83

Translate the phrase into an algebraic expression: the quotient of the difference of and and

TRY IT : : 4.84

Translate the phrase into an algebraic expression: the quotient of the sum of and and

Simplify Complex Fractions

Our work with fractions so far has included proper fractions, improper fractions, and mixed numbers. Another kind of fraction is called **complex fraction**, which is a fraction in which the numerator or the denominator contains a fraction.

Some examples of complex fractions are:

To simplify a complex fraction, remember that the fraction bar means division. So the complex fraction can be written as

EXAMPLE 4.43

Simplify:

Solution

Rewrite as division.	
Multiply the first fraction by the reciprocal of the second.	
Multiply.	
Look for common factors.	
Remove common factors and simplify.	

TRY IT : : 4.85 Simplify:

TRY IT : : 4.86 Simplify:

HOW TO : : SIMPLIFY A COMPLEX FRACTION.

Step 1. Rewrite the complex fraction as a division problem.
Step 2. Follow the rules for dividing fractions.
Step 3. Simplify if possible.

EXAMPLE 4.44

Simplify:

Solution

Rewrite as division.	
Multiply the first fraction by the reciprocal of the second.	
Multiply; the product will be negative.	
Look for common factors.	
Remove common factors and simplify.	

TRY IT : : 4.87 Simplify:

TRY IT : : 4.88 Simplify:

EXAMPLE 4.45

Simplify:

Solution

Rewrite as division.	
Multiply the first fraction by the reciprocal of the second.	
Multiply.	
Look for common factors.	
Remove common factors and simplify.	

> **TRY IT : : 4.89** Simplify:

> **TRY IT : : 4.90** Simplify:

EXAMPLE 4.46

Simplify:

Solution

Rewrite as division.	
Change the mixed number to an improper fraction.	
Multiply the first fraction by the reciprocal of the second.	
Multiply.	
Look for common factors.	
Remove common factors and simplify.	

> **TRY IT : : 4.91** Simplify:

> **TRY IT : : 4.92** Simplify:

Simplify Expressions with a Fraction Bar

Where does the negative sign go in a fraction? Usually, the negative sign is placed in front of the fraction, but you will sometimes see a fraction with a negative numerator or denominator. Remember that fractions represent division. The fraction could be the result of dividing a negative by a positive, or of dividing a positive by a negative.

When the numerator and denominator have different signs, the quotient is negative.

$$\frac{-1}{3} = -\frac{1}{3} \qquad \frac{\text{negative}}{\text{positive}} = \text{negative} \qquad \frac{1}{-3} = -\frac{1}{3} \qquad \frac{\text{positive}}{\text{negative}} = \text{negative}$$

If *both* the numerator and denominator are negative, then the fraction itself is positive because we are dividing a negative by a negative.

Placement of Negative Sign in a Fraction

For any positive numbers

EXAMPLE 4.47

Which of the following fractions are equivalent to

Solution

The quotient of a positive and a negative is a negative, so is negative. Of the fractions listed, are also negative.

TRY IT : : 4.93 Which of the following fractions are equivalent to

TRY IT : : 4.94 Which of the following fractions are equivalent to

Fraction bars act as grouping symbols. The expressions above and below the fraction bar should be treated as if they were in parentheses. For example, means The order of operations tells us to simplify the numerator and the denominator first—as if there were parentheses—before we divide.

We'll add fraction bars to our set of grouping symbols from Use the Language of Algebra to have a more complete set here.

Grouping Symbols

Parentheses	( )
Brackets	[ ]
Braces	{ }
Absolute value	||
Fraction Bar	$\frac{\square}{\square}$

HOW TO : : SIMPLIFY AN EXPRESSION WITH A FRACTION BAR.

Step 1. Simplify the numerator.
Step 2. Simplify the denominator.
Step 3. Simplify the fraction.

EXAMPLE 4.48

Simplify:

Solution

Simplify the expression in the numerator.	
Simplify the expression in the denominator.	
Simplify the fraction.	6

TRY IT : : 4.95 Simplify:

TRY IT : : 4.96 Simplify:

EXAMPLE 4.49

Simplify:

Solution

Use the order of operations. Multiply in the numerator and use the exponent in the denominator.	
Simplify the numerator and the denominator.	
Simplify the fraction.	

TRY IT : : 4.97 Simplify:

TRY IT : : 4.98 Simplify:

EXAMPLE 4.50

Simplify:

Solution

Use the order of operations (parentheses first, then exponents).	
Simplify the numerator and denominator.	
Simplify the fraction.	

TRY IT : : 4.99 Simplify:

TRY IT : : 4.100 Simplify:

EXAMPLE 4.51

Simplify:

Solution

Multiply.	
Simplify.	
Divide.	

TRY IT : : 4.101 Simplify:

TRY IT : : 4.102 Simplify:

MEDIA : : ACCESS ADDITIONAL ONLINE RESOURCES

- Division Involving Mixed Numbers (http://www.openstax.org/l/24DivisionMixed)
- Evaluate a Complex Fraction (http://www.openstax.org/l/24ComplexFrac)

4.3 EXERCISES

Practice Makes Perfect

Multiply and Divide Mixed Numbers

In the following exercises, multiply and write the answer in simplified form.

176.

177.

178.

179.

180.

181.

182.

183.

In the following exercises, divide, and write your answer in simplified form.

184.

185.

186.

187.

188.

189.

190.

191.

Translate Phrases to Expressions with Fractions

In the following exercises, translate each English phrase into an algebraic expression.

192. the quotient of and

193. the quotient of and

194. the quotient of and

195. the quotient of and

196. the quotient of and the sum of and

197. the quotient of and the difference of and

Simplify Complex Fractions

In the following exercises, simplify the complex fraction.

198.

199.

200.

201.

202.

203.

204.

205.

206.

207.

208.

209.

210.

211.

212.

213.

Simplify Expressions with a Fraction Bar

In the following exercises, identify the equivalent fractions.

214. Which of the following fractions are equivalent to

215. Which of the following fractions are equivalent to

216. Which of the following fractions are equivalent to

217. Which of the following fractions are equivalent to

In the following exercises, simplify.

218.

219.

220.

221.

222.

223.

224.

225.

226.

227.

228.

229.

230.

231.

232.

233.

234.

235.

236.

237.

238.

239.

240.

241.

242.

243.

244.

245.

246.

247.

Everyday Math

248. Baking A recipe for chocolate chip cookies calls for — cups of flour. Graciela wants to double the recipe.

1. How much flour will Graciela need? Show your calculation. Write your result as an improper fraction and as a mixed number.
2. Measuring cups usually come in sets with cups for — — — — cup. Draw a diagram to show two different ways that Graciela could measure out the flour needed to double the recipe.

249. Baking A booth at the county fair sells fudge by the pound. Their award winning "Chocolate Overdose" fudge contains — cups of chocolate chips per pound.

How many cups of chocolate chips are in a half-pound of the fudge?

The owners of the booth make the fudge in -pound batches. How many chocolate chips do they need to make a -pound batch? Write your results as improper fractions and as a mixed numbers.

Writing Exercises

250. Explain how to find the reciprocal of a mixed number.

251. Explain how to multiply mixed numbers.

252. Randy thinks that — — is — Explain what is wrong with Randy's thinking.

253. Explain why — — and — are equivalent.

Self Check

After completing the exercises, use this checklist to evaluate your mastery of the objectives of this section.

I can...	Confidently	With some help	No-I don't get it!
multiply and divide mixed numbers.			
translate phrases to expressions with fractions.			
simplify complex fractions.			
simplify expressions written with a fraction bar.			

What does this checklist tell you about your mastery of this section? What steps will you take to improve?

4.4 Add and Subtract Fractions with Common Denominators

Learning Objectives

By the end of this section, you will be able to:

- Model fraction addition
- Add fractions with a common denominator
- Model fraction subtraction
- Subtract fractions with a common denominator

BE PREPARED : : 4.9 Before you get started, take this readiness quiz.

Simplify:

If you missed this problem, review Example 2.22.

BE PREPARED : : 4.10 Draw a model of the fraction —

If you missed this problem, review Example 4.2.

BE PREPARED : : 4.11 Simplify: ——

If you missed this problem, review Example 4.48.

Model Fraction Addition

How many quarters are pictured? One quarter plus quarters equals quarters.

Remember, quarters are really fractions of a dollar. Quarters are another way to say fourths. So the picture of the coins shows that

— — —

Let's use fraction circles to model the same example, — —

Start with one — piece.
$\frac{1}{4}$
Add two more — pieces.
+
$+\frac{2}{4}$
The result is — .
$\frac{3}{4}$

So again, we see that

— — —

MANIPULATIVE MATHEMATICS

Doing the Manipulative Mathematics activity "Model Fraction Addition" will help you develop a better understanding of adding fractions

EXAMPLE 4.52

Use a model to find the sum — —

Solution

Start with three — pieces.	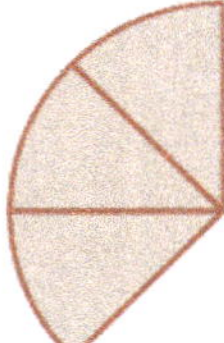	$\frac{3}{8}$
Add two — pieces.	+ 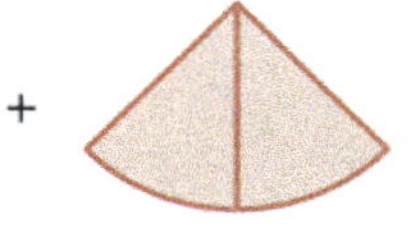	$+\frac{2}{8}$
How many — pieces are there?	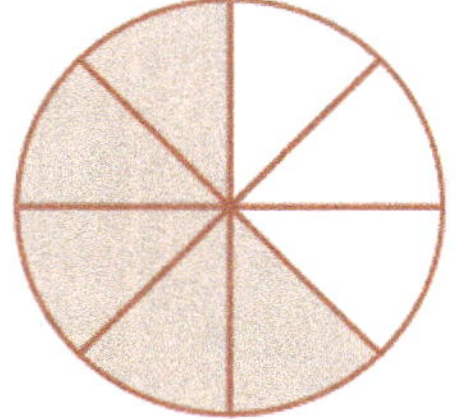	$\frac{5}{8}$

There are five — pieces, or five-eighths. The model shows that — — —

> **TRY IT : : 4.103** Use a model to find each sum. Show a diagram to illustrate your model.
>
> — —

> **TRY IT : : 4.104** Use a model to find each sum. Show a diagram to illustrate your model.
>
> — —

Add Fractions with a Common Denominator

Example 4.52 shows that to add the same-size pieces—meaning that the fractions have the same denominator—we just add the number of pieces.

Fraction Addition

If are numbers where then

— — ——

To add fractions with a common denominators, add the numerators and place the sum over the common denominator.

EXAMPLE 4.53

Find the sum: — —

Solution

	— —
Add the numerators and place the sum over the common denominator.	——
Simplify.	—

TRY IT : : 4.105 Find each sum: — —

TRY IT : : 4.106 Find each sum: — —

EXAMPLE 4.54

Find the sum: — —

Solution

	— —
Add the numerators and place the sum over the common denominator.	——

Note that we cannot simplify this fraction any more. Since are not like terms, we cannot combine them.

TRY IT : : 4.107 Find the sum: — —

TRY IT : : 4.108 Find the sum: — —

EXAMPLE 4.55

Find the sum: — —

Solution

We will begin by rewriting the first fraction with the negative sign in the numerator.

— ——

Rewrite the first fraction with the negative in the numerator.	
Add the numerators and place the sum over the common denominator.	
Simplify the numerator.	
Rewrite with negative sign in front of the fraction.	

> **TRY IT : : 4.109** Find the sum:

> **TRY IT : : 4.110** Find the sum:

EXAMPLE 4.56

Find the sum:

Solution

Add the numerators and place the sum over the common denominator.	
Combine like terms.	

> **TRY IT : : 4.111** Find the sum:

> **TRY IT : : 4.112** Find the sum:

EXAMPLE 4.57

Find the sum:

Solution

	— —
Add the numerators and place the sum over the common denominator.	———
Add.	—
Simplify the fraction.	–

TRY IT : : 4.113 Find each sum: — —

TRY IT : : 4.114 Find each sum: — —

Model Fraction Subtraction

Subtracting two fractions with common denominators is much like adding fractions. Think of a pizza that was cut into slices. Suppose five pieces are eaten for dinner. This means that, after dinner, there are seven pieces (or — of the pizza) left in the box. If Leonardo eats of these remaining pieces (or — of the pizza), how much is left? There would be pieces left (or — of the pizza).

— — —

Let's use fraction circles to model the same example, — —

Start with seven — pieces. Take away two — pieces. How many twelfths are left?

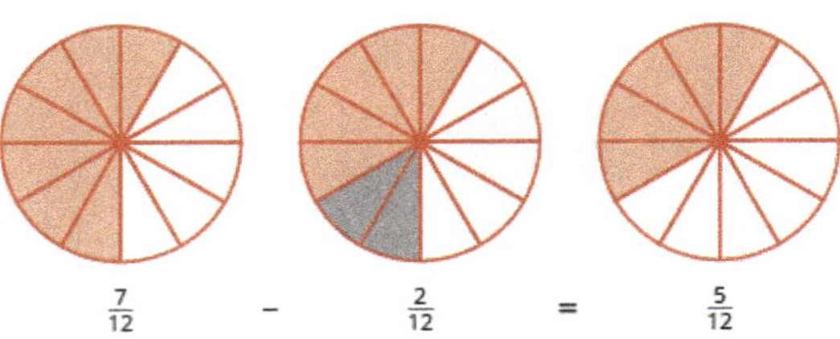

Again, we have five twelfths, —

MANIPULATIVE MATHEMATICS

Doing the Manipulative Mathematics activity "Model Fraction Subtraction" will help you develop a better understanding of subtracting fractions.

EXAMPLE 4.58

Use fraction circles to find the difference: — —

Solution

Start with four — pieces. Take away one — piece. Count how many fifths are left. There are three — pieces left.

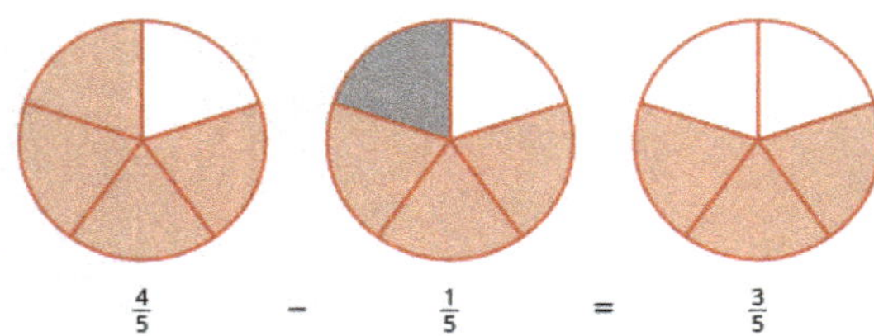

> **TRY IT : : 4.115** Use a model to find each difference. Show a diagram to illustrate your model.
>
> — —

> **TRY IT : : 4.116** Use a model to find each difference. Show a diagram to illustrate your model.
>
> — —

Subtract Fractions with a Common Denominator

We subtract fractions with a common denominator in much the same way as we add fractions with a common denominator.

Fraction Subtraction

If are numbers where then

— — ——

To subtract fractions with a common denominators, we subtract the numerators and place the difference over the common denominator.

EXAMPLE 4.59

Find the difference: — —

Solution

	— —
Subtract the numerators and place the difference over the common denominator.	——
Simplify the numerator.	—
Simplify the fraction by removing common factors.	—

> **TRY IT : : 4.117** Find the difference: — —

> **TRY IT : :** 4.118 Find the difference: — —

EXAMPLE 4.60

Find the difference: — —

Solution

	— —
Subtract the numerators and place the difference over the common denominator.	——

The fraction is simplified because we cannot combine the terms in the numerator.

> **TRY IT : :** 4.119 Find the difference: — —

> **TRY IT : :** 4.120 Find the difference: — —

EXAMPLE 4.61

Find the difference: — —

Solution

Remember, the fraction — can be written as ——

	— —
Subtract the numerators.	———
Simplify.	——
Rewrite with the negative sign in front of the fraction.	—

> **TRY IT : :** 4.121 Find the difference: — —

> **TRY IT : :** 4.122 Find the difference: — —

Now lets do an example that involves both addition and subtraction.

EXAMPLE 4.62

Simplify: — — —

Solution

Combine the numerators over the common denominator.	
Simplify the numerator, working left to right.	
Subtract the terms in the numerator.	
Rewrite with the negative sign in front of the fraction.	

TRY IT : : 4.123 Simplify:

TRY IT : : 4.124 Simplify:

MEDIA : : ACCESS ADDITIONAL ONLINE RESOURCES

- Adding Fractions With Pattern Blocks (http://www.openstax.org/l/24AddFraction)
- Adding Fractions With Like Denominators (http://www.openstax.org/l/24AddLikeDenom)
- Subtracting Fractions With Like Denominators (http://www.openstax.org/l/24SubtrLikeDeno)

4.4 EXERCISES

Practice Makes Perfect

Model Fraction Addition

In the following exercises, use a model to add the fractions. Show a diagram to illustrate your model.

254. — —

255. — —

256. — —

257. — —

Add Fractions with a Common Denominator

In the following exercises, find each sum.

258. — —

259. — —

260. — —

261. — —

262. — —

263. — —

264. — —

265. — —

266. — —

267. — —

268. — —

269. — —

270. — —

271. — —

272. — —

273. — —

274. — —

275. — —

276. — — —

277. — — —

Model Fraction Subtraction

In the following exercises, use a model to subtract the fractions. Show a diagram to illustrate your model.

278. — —

279. — —

Subtract Fractions with a Common Denominator

In the following exercises, find the difference.

280. — —

281. — —

282. — —

283. — —

284. — —

285. — —

286. — —

287. — —

288. — —

289. — —

290. — —

291. — —

292. — —

293. — —

294. — —

295. — —

296. — —

297. — —

298. — —

299. — —

300. — —

301. — —

302. — —

303. — —

Mixed Practice

In the following exercises, perform the indicated operation and write your answers in simplified form.

304. — —

305. — —

306. — —

307. — —

308. — —

309. — —

310. — —

311. — —

Everyday Math

312. Trail Mix Jacob is mixing together nuts and raisins to make trail mix. He has — of a pound of nuts and — of a pound of raisins. How much trail mix can he make?

313. Baking Janet needs — of a cup of flour for a recipe she is making. She only has — of a cup of flour and will ask to borrow the rest from her next-door neighbor. How much flour does she have to borrow?

Writing Exercises

314. Greg dropped his case of drill bits and three of the bits fell out. The case has slots for the drill bits, and the slots are arranged in order from smallest to largest. Greg needs to put the bits that fell out back in the case in the empty slots. Where do the three bits go? Explain how you know.

Bits in case: —, —, __, __, —, —, __, —, —, —.

Bits that fell out: —, —, —.

315. After a party, Lupe has — of a cheese pizza, — of a pepperoni pizza, and — of a veggie pizza left. Will all the slices fit into pizza box? Explain your reasoning.

Self Check

After completing the exercises, use this checklist to evaluate your mastery of the objectives of this section.

I can...	Confidently	With some help	No-I don't get it!
model fraction addition.			
add fractions with a common denominator.			
model fraction subtraction.			
subtract fractions with a common denominator.			
find the least common denominator (LCD).			
convert fractions to equivalent fractions with the LCD.			

On a scale of 1–10, how would you rate your mastery of this section in light of your responses on the checklist? How can you improve this?

4.5 Add and Subtract Fractions with Different Denominators

Learning Objectives

By the end of this section, you will be able to:

- Find the least common denominator (LCD)
- Convert fractions to equivalent fractions with the LCD
- Add and subtract fractions with different denominators
- Identify and use fraction operations
- Use the order of operations to simplify complex fractions
- Evaluate variable expressions with fractions

BE PREPARED : : 4.12 Before you get started, take this readiness quiz.

Find two fractions equivalent to —

If you missed this problem, review Example 4.14.

BE PREPARED : : 4.13 Simplify: ————

If you missed this problem, review Example 4.48.

Find the Least Common Denominator

In the previous section, we explained how to add and subtract fractions with a common denominator. But how can we add and subtract fractions with unlike denominators?

Let's think about coins again. Can you add one quarter and one dime? You could say there are two coins, but that's not very useful. To find the total value of one quarter plus one dime, you change them to the same kind of unit—cents. One quarter equals cents and one dime equals cents, so the sum is cents. See Figure 4.7.

Figure 4.7 Together, a quarter and a dime are worth cents, or —— of a dollar.

Similarly, when we add fractions with different denominators we have to convert them to equivalent fractions with a common denominator. With the coins, when we convert to cents, the denominator is Since there are cents in one dollar, cents is —— and cents is —— So we add —— —— to get —— which is cents.

You have practiced adding and subtracting fractions with common denominators. Now let's see what you need to do with fractions that have different denominators.

First, we will use fraction tiles to model finding the common denominator of — and —

We'll start with one — tile and — tile. We want to find a common fraction tile that we can use to match *both* — and — exactly.

If we try the — pieces, of them exactly match the — piece, but they do not exactly match the — piece.

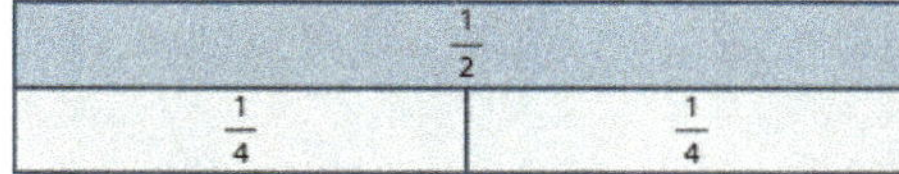

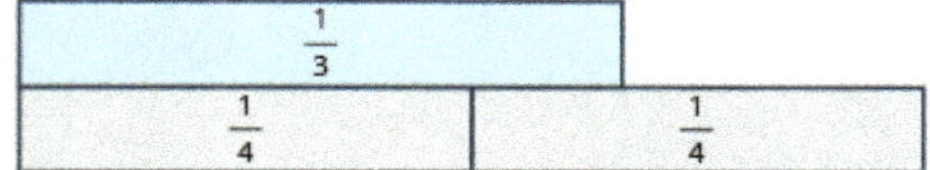

If we try the — pieces, they do not exactly cover the — piece or the — piece.

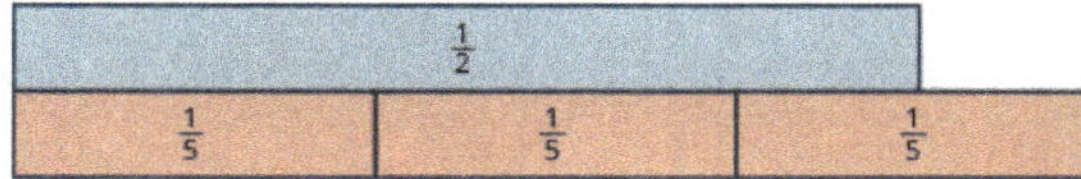

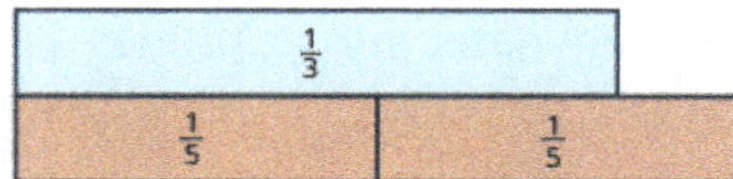

If we try the — pieces, we see that exactly of them cover the — piece, and exactly of them cover the — piece.

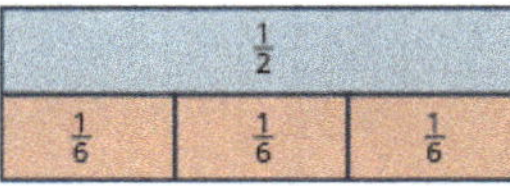

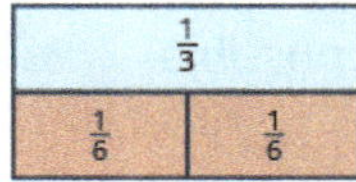

If we were to try the — pieces, they would also work.

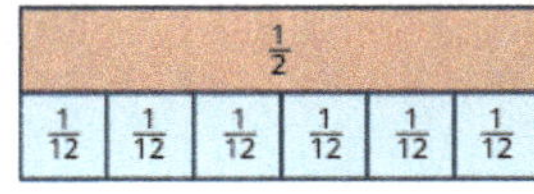

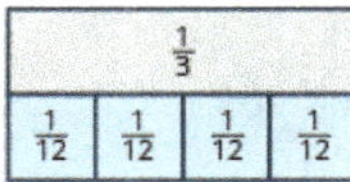

Even smaller tiles, such as — and — would also exactly cover the — piece and the — piece.

The denominator of the largest piece that covers both fractions is the **least common denominator (LCD)** of the two fractions. So, the least common denominator of — and — is

Notice that all of the tiles that cover — and — have something in common: Their denominators are common multiples of and the denominators of — and — The least common multiple (LCM) of the denominators is and so we say that is the least common denominator (LCD) of the fractions — and —

MANIPULATIVE MATHEMATICS

Doing the Manipulative Mathematics activity "Finding the Least Common Denominator" will help you develop a better understanding of the LCD.

Least Common Denominator

The **least common denominator (LCD)** of two fractions is the least common multiple (LCM) of their denominators.

To find the LCD of two fractions, we will find the LCM of their denominators. We follow the procedure we used earlier to find the LCM of two numbers. We only use the denominators of the fractions, not the numerators, when finding the LCD.

EXAMPLE 4.63

Find the LCD for the fractions — and —

Solution

Factor each denominator into its primes.	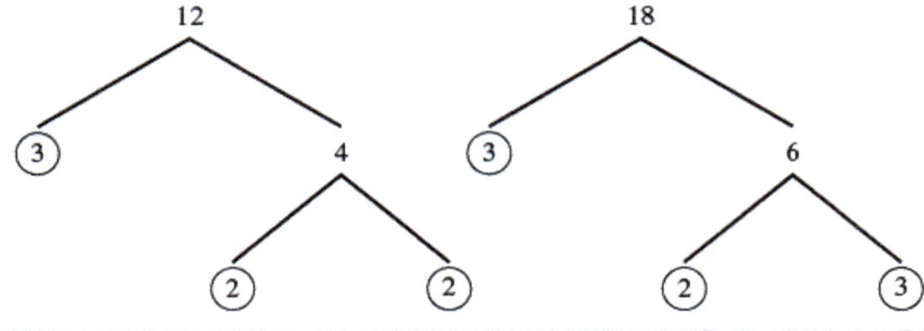
List the primes of 12 and the primes of 18 lining them up in columns when possible.	$12 = 2 \cdot 2 \cdot 3$ $18 = 2 \cdot \quad 3 \cdot 3$
Bring down the columns.	$12 = 2 \cdot 2 \cdot 3$ $18 = 2 \cdot \quad 3 \cdot 3$ $\text{LCM} = 2 \cdot 2 \cdot 3 \cdot 3$
Multiply the factors. The product is the LCM.	
The LCM of 12 and 18 is 36, so the LCD of —— and —— is 36.	LCD of —— and —— is 36.

TRY IT : : 4.125 Find the least common denominator for the fractions: —— and ——

TRY IT : : 4.126 Find the least common denominator for the fractions: —— and ——

To find the LCD of two fractions, find the LCM of their denominators. Notice how the steps shown below are similar to the steps we took to find the LCM.

HOW TO : : FIND THE LEAST COMMON DENOMINATOR (LCD) OF TWO FRACTIONS.

Step 1. Factor each denominator into its primes.
Step 2. List the primes, matching primes in columns when possible.
Step 3. Bring down the columns.
Step 4. Multiply the factors. The product is the LCM of the denominators.
Step 5. The LCM of the denominators is the LCD of the fractions.

EXAMPLE 4.64

Find the least common denominator for the fractions —— and ——

Solution

To find the LCD, we find the LCM of the denominators.

Find the LCM of and

$$\begin{aligned} 15 &= \qquad\quad 3 \cdot 5 \\ 24 &= 2 \cdot 2 \cdot 2 \cdot 3 \\ \hline \text{LCD} &= 2 \cdot 2 \cdot 2 \cdot 3 \cdot 5 \\ \text{LCD} &= 120 \end{aligned}$$

The LCM of and is So, the LCD of — and — is

TRY IT : : 4.127 Find the least common denominator for the fractions: — and —

TRY IT : : 4.128 Find the least common denominator for the fractions: — and —

Convert Fractions to Equivalent Fractions with the LCD

Earlier, we used fraction tiles to see that the LCD of — — is We saw that three — pieces exactly covered — and two — pieces exactly covered — so

— — — —

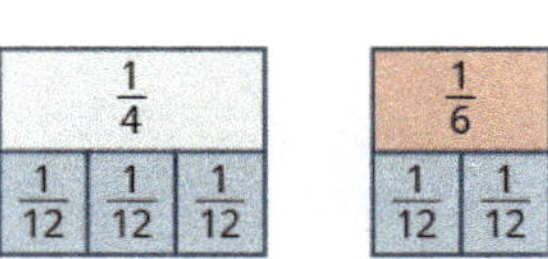

We say that — — are equivalent fractions and also that — — are equivalent fractions.

We can use the Equivalent Fractions Property to algebraically change a fraction to an equivalent one. Remember, two fractions are equivalent if they have the same value. The Equivalent Fractions Property is repeated below for reference.

Equivalent Fractions Property

If are whole numbers where

— — — —

To add or subtract fractions with different denominators, we will first have to convert each fraction to an equivalent fraction with the LCD. Let's see how to change — — to equivalent fractions with denominator without using models.

EXAMPLE 4.65

Convert — — to equivalent fractions with denominator their LCD.

Solution

Find the LCD.	The LCD of — and — is 12.
Find the number to multiply 4 to get 12.	$4 \cdot 3 = 12$
Find the number to multiply 6 to get 12.	$6 \cdot 2 = 12$
Use the Equivalent Fractions Property to convert each fraction to an equivalent fraction with the LCD, multiplying both the numerator and denominator of each fraction by the same number.	$\frac{1}{4}$ $\frac{1}{6}$ $\frac{1 \cdot 3}{4 \cdot 3}$ $\frac{1 \cdot 2}{6 \cdot 2}$
Simplify the numerators and denominators.	$\frac{3}{12}$ $\frac{2}{12}$

We do not reduce the resulting fractions. If we did, we would get back to our original fractions and lose the common denominator.

TRY IT : : 4.129 Change to equivalent fractions with the LCD:

— and —

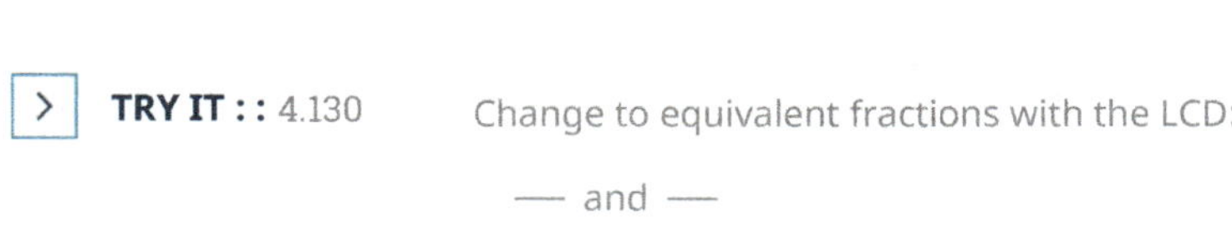

TRY IT : : 4.130 Change to equivalent fractions with the LCD:

— and —

HOW TO : : CONVERT TWO FRACTIONS TO EQUIVALENT FRACTIONS WITH THEIR LCD AS THE COMMON DENOMINATOR.

Step 1. Find the LCD.

Step 2. For each fraction, determine the number needed to multiply the denominator to get the LCD.

Step 3. Use the Equivalent Fractions Property to multiply both the numerator and denominator by the number you found in Step 2.

Step 4. Simplify the numerator and denominator.

EXAMPLE 4.66

Convert — and — to equivalent fractions with denominator their LCD.

Solution

	The LCD is 120. We will start at Step 2.
Find the number that must multiply 15 to get 120.	$15 \cdot 8 = 120$
Find the number that must multiply 24 to get 120.	$24 \cdot 5 = 120$
Use the Equivalent Fractions Property.	$\frac{8 \cdot 8}{15 \cdot 8} \quad \frac{11 \cdot 5}{24 \cdot 5}$
Simplify the numerators and denominators.	$\frac{64}{120} \quad \frac{55}{120}$

TRY IT : : 4.131 Change to equivalent fractions with the LCD:

— and — LCD

TRY IT : : 4.132 Change to equivalent fractions with the LCD:

— and — LCD

Add and Subtract Fractions with Different Denominators

Once we have converted two fractions to equivalent forms with common denominators, we can add or subtract them by adding or subtracting the numerators.

HOW TO : : ADD OR SUBTRACT FRACTIONS WITH DIFFERENT DENOMINATORS.

Step 1. Find the LCD.
Step 2. Convert each fraction to an equivalent form with the LCD as the denominator.
Step 3. Add or subtract the fractions.
Step 4. Write the result in simplified form.

EXAMPLE 4.67

Add: — —

Solution

— —

Find the LCD of 2, 3.

$$\begin{array}{r} 2 = 2 \\ 3 = \quad 3 \\ \hline \text{LCD} = 2 \cdot 3 \\ \text{LCD} = 6 \end{array}$$

Change into equivalent fractions with the LCD 6.	$\frac{1 \cdot 3}{2 \cdot 3} + \frac{1 \cdot 2}{3 \cdot 2}$
Simplify the numerators and denominators.	— —
Add.	—

Remember, always check to see if the answer can be simplified. Since and have no common factors, the fraction — cannot be reduced.

TRY IT : : 4.133 Add: — —

TRY IT : : 4.134 Add: — —

EXAMPLE 4.68

Subtract: — —

Solution

— —

Find the LCD of 2 and 4.

$$\begin{array}{r} 2 = 2 \\ 4 = 2 \cdot 2 \\ \hline \text{LCD} = 2 \cdot 2 \\ \text{LCD} = 4 \end{array}$$

Rewrite as equivalent fractions using the LCD 4.	$\frac{1 \cdot 2}{2 \cdot 2} - \left(-\frac{1}{4}\right)$
Simplify the first fraction.	— —
Subtract.	———
Simplify.	—

One of the fractions already had the least common denominator, so we only had to convert the other fraction.

> **TRY IT : :** 4.135 Simplify: — —

> **TRY IT : :** 4.136 Simplify: — —

EXAMPLE 4.69

Add: — —

Solution

— —

Find the LCD of 12 and 18.

$$\begin{array}{l} 12 = 2 \cdot 2 \cdot 3 \\ 18 = 2 \cdot \quad 3 \cdot 3 \\ \hline \text{LCD} = 2 \cdot 2 \cdot 3 \cdot 3 \\ \text{LCD} = 36 \end{array}$$

Rewrite as equivalent fractions with the LCD. $\frac{7 \cdot 3}{12 \cdot 3} + \frac{5 \cdot 2}{18 \cdot 2}$

Simplify the numerators and denominators. — —

Add. —

Because is a prime number, it has no factors in common with The answer is simplified.

> **TRY IT : :** 4.137 Add: — —

> **TRY IT : :** 4.138 Add: — —

When we use the Equivalent Fractions Property, there is a quick way to find the number you need to multiply by to get the LCD. Write the factors of the denominators and the LCD just as you did to find the LCD. The "missing" factors of each denominator are the numbers you need.

missing factors

$$\begin{array}{l} 12 = 2 \cdot 2 \cdot 3 \\ 18 = 2 \cdot \quad 3 \cdot 3 \\ \hline \text{LCD} = 2 \cdot 2 \cdot 3 \cdot 3 \\ \text{LCD} = 36 \end{array}$$

The LCD, has factors of and factors of

Twelve has two factors of but only one of —so it is 'missing' one We multiplied the numerator and denominator of — by to get an equivalent fraction with denominator

Eighteen is missing one factor of —so you multiply the numerator and denominator — by to get an equivalent fraction with denominator We will apply this method as we subtract the fractions in the next example.

EXAMPLE 4.70

Subtract: — —

Solution

— —

Find the LCD.

$$\begin{array}{l} 15 = \qquad\qquad 3 \cdot 5 \\ \underline{24 = 2 \cdot 2 \cdot 2 \cdot 3} \\ \text{LCD} = 2 \cdot 2 \cdot 2 \cdot 3 \cdot 5 \\ \text{LCD} = 120 \end{array}$$

15 is 'missing' three factors of 2
24 is 'missing' a factor of 5

Rewrite as equivalent fractions with the LCD.	$\frac{7 \cdot 8}{15 \cdot 8} - \frac{19 \cdot 5}{24 \cdot 5}$
Simplify each numerator and denominator.	— —
Subtract.	—
Rewrite showing the common factor of 3.	—
Remove the common factor to simplify.	—

TRY IT : : 4.139 Subtract: — —

TRY IT : : 4.140 Subtract: — —

EXAMPLE 4.71

Add: — —

Solution

Find the LCD.

$$\begin{array}{l} 30 = 2 \cdot 3 \cdot 5 \\ 42 = 2 \cdot 3 \cdot \quad 7 \\ \hline \text{LCD} = 2 \cdot 3 \cdot 5 \cdot 7 \\ \text{LCD} = 210 \end{array}$$

Rewrite as equivalent fractions with the LCD. $-\frac{11 \cdot 7}{30 \cdot 7} + \frac{23 \cdot 5}{42 \cdot 5}$

Simplify each numerator and denominator.

Add.

Rewrite showing the common factor of 2.

Remove the common factor to simplify.

TRY IT : : 4.141 Add:

TRY IT : : 4.142 Add:

In the next example, one of the fractions has a variable in its numerator. We follow the same steps as when both numerators are numbers.

EXAMPLE 4.72

Add:

Solution

The fractions have different denominators.

Find the LCD. $5 = \quad\quad 5$ $8 = 2 \cdot 2 \cdot 2$ $\text{LCD} = 2 \cdot 2 \cdot 2 \cdot 5$ $\text{LCD} = 40$	
Rewrite as equivalent fractions with the LCD.	$\frac{3 \cdot 8}{5 \cdot 8} + \frac{x \cdot 5}{8 \cdot 5}$
Simplify the numerators and denominators.	
Add.	

We cannot add and since they are not like terms, so we cannot simplify the expression any further.

TRY IT : : 4.143 Add:

TRY IT : : 4.144 Add:

Identify and Use Fraction Operations

By now in this chapter, you have practiced multiplying, dividing, adding, and subtracting fractions. The following table summarizes these four fraction operations. Remember: You need a common denominator to add or subtract fractions, but not to multiply or divide fractions

Summary of Fraction Operations

Fraction multiplication: Multiply the numerators and multiply the denominators.

Fraction division: Multiply the first fraction by the reciprocal of the second.

Fraction addition: Add the numerators and place the sum over the common denominator. If the fractions have different denominators, first convert them to equivalent forms with the LCD.

Fraction subtraction: Subtract the numerators and place the difference over the common denominator. If the fractions have different denominators, first convert them to equivalent forms with the LCD.

EXAMPLE 4.73

Simplify:

Solution

First we ask ourselves, "What is the operation?"

The operation is addition.

Do the fractions have a common denominator? No.

Find the LCD. $4 = 2 \cdot 2$ $6 = 2 \cdot \quad 3$ $\text{LCD} = 2 \cdot 2 \cdot 3$ $\text{LCD} = 12$	
Rewrite each fraction as an equivalent fraction with the LCD.	$-\frac{1 \cdot 3}{4 \cdot 3} + \frac{1 \cdot 2}{6 \cdot 2}$
Simplify the numerators and denominators.	
Add the numerators and place the sum over the common denominator.	
Check to see if the answer can be simplified. It cannot.	

The operation is division. We do not need a common denominator.

To divide fractions, multiply the first fraction by the reciprocal of the second.	
Multiply.	
Simplify.	

> **TRY IT : : 4.145** Simplify each expression:

> **TRY IT : : 4.146** Simplify each expression:

EXAMPLE 4.74

Simplify:

Solution

The operation is subtraction. The fractions do not have a common denominator.

Rewrite each fraction as an equivalent fraction with the LCD, 30.

Subtract the numerators and place the difference over the common denominator.

The operation is multiplication; no need for a common denominator.

To multiply fractions, multiply the numerators and multiply the denominators.

Rewrite, showing common factors.

Remove common factors to simplify.

> **TRY IT : :** 4.147 Simplify:

> **TRY IT : :** 4.148 Simplify:

Use the Order of Operations to Simplify Complex Fractions

In Multiply and Divide Mixed Numbers and Complex Fractions, we saw that a complex fraction is a fraction in which the numerator or denominator contains a fraction. We simplified complex fractions by rewriting them as division problems. For example,

Now we will look at complex fractions in which the numerator or denominator can be simplified. To follow the order of operations, we simplify the numerator and denominator separately first. Then we divide the numerator by the denominator.

HOW TO : : SIMPLIFY COMPLEX FRACTIONS.

Step 1. Simplify the numerator.

Step 2. Simplify the denominator.

Step 3. Divide the numerator by the denominator.

Step 4. Simplify if possible.

EXAMPLE 4.75

Simplify:

Solution

Simplify the numerator.	
Simplify the term with the exponent in the denominator.	
Add the terms in the denominator.	
Divide the numerator by the denominator.	
Rewrite as multiplication by the reciprocal.	
Multiply.	

TRY IT : : 4.149

Simplify: .

TRY IT : : 4.150

Simplify: .

EXAMPLE 4.76

Simplify:

Solution

Rewrite numerator with the LCD of 6 and denominator with LCD of 12.	
Add in the numerator. Subtract in the denominator.	
Divide the numerator by the denominator.	
Rewrite as multiplication by the reciprocal.	
Rewrite, showing common factors.	
Simplify.	2

> TRY IT : : 4.151 Simplify: .

> TRY IT : : 4.152 Simplify: .

Evaluate Variable Expressions with Fractions

We have evaluated expressions before, but now we can also evaluate expressions with fractions. Remember, to evaluate an expression, we substitute the value of the variable into the expression and then simplify.

EXAMPLE 4.77

Evaluate when

Solution

To evaluate when substitute for in the expression.

Substitute $-\frac{1}{3}$ for x.	$-\frac{1}{3}+\frac{1}{3}$
Simplify.	

To evaluate — when — we substitute — for in the expression.

Substitute $-\frac{3}{4}$ for x.	$-\frac{3}{4}+\frac{1}{3}$
Rewrite as equivalent fractions with the LCD, 12.	
Simplify the numerators and denominators.	
Add.	

TRY IT : : 4.153 Evaluate: — when

TRY IT : : 4.154 Evaluate: — when

EXAMPLE 4.78

Evaluate — when —

Solution

We substitute — for in the expression.

Substitute $-\frac{2}{3}$ for y.	$-\frac{2}{3}-\frac{5}{6}$
Rewrite as equivalent fractions with the LCD, 6.	
Subtract.	
Simplify.	

TRY IT : : 4.155 Evaluate: — when —

TRY IT : : 4.156 Evaluate: — when —

EXAMPLE 4.79

Evaluate when — and —

Solution

Substitute the values into the expression. In the exponent applies only to

	$2x^2y$
Substitute $\frac{1}{4}$ for x and $-\frac{2}{3}$ for y.	$2\left(\frac{1}{4}\right)^2\left(-\frac{2}{3}\right)$
Simplify exponents first.	$2\left(\frac{1}{16}\right)\left(-\frac{2}{3}\right)$
Multiply. The product will be negative.	$-\frac{2}{1}\cdot\frac{1}{16}\cdot\frac{2}{3}$
Simplify.	$-\frac{4}{48}$
Remove the common factors.	$-\frac{1\cdot\cancel{4}}{\cancel{4}\cdot 12}$
Simplify.	$-\frac{1}{12}$

TRY IT : : 4.157 Evaluate. when — and —

TRY IT : : 4.158 Evaluate. when — and —

EXAMPLE 4.80

Evaluate —— when and

Solution

We substitute the values into the expression and simplify.

	——
Substitute −4 for p, −2 for q and 8 for r.	$\frac{-4+(-2)}{8}$
Add in the numerator first.	—
Simplify.	—

TRY IT : : 4.159 Evaluate: —— when

TRY IT : : 4.160 Evaluate: —— when

4.5 EXERCISES

Practice Makes Perfect

Find the Least Common Denominator (LCD)

In the following exercises, find the least common denominator (LCD) for each set of fractions.

316. — —

317. — —

318. — —

319. — —

320. — —

321. — —

322. — —

323. — —

324. — — —

325. — — —

Convert Fractions to Equivalent Fractions with the LCD

In the following exercises, convert to equivalent fractions using the LCD.

326. — —

327. — —

328. — —

329. — —

330. — —

331. — —

332. — — —

333. — — —

Add and Subtract Fractions with Different Denominators

In the following exercises, add or subtract. Write the result in simplified form.

334. — —

335. — —

336. — —

337. — —

338. — —

339. — —

340. — —

341. — —

342. — —

343. — —

344. — —

345. — —

346. — —

347. — —

348. — —

349. — —

350. — —

351. — —

352. — —

353. — —

354. — —

355. — —

356. — —

357. — —

358.

359.

360.

361.

362.

363.

364.

365.

366.

367.

368.

369.

Identify and Use Fraction Operations

In the following exercises, perform the indicated operations. Write your answers in simplified form.

370.

371.

372.

373.

374.

375.

376.

377.

378.

379.

380.

381.

382.

383.

384.

385.

386.

387.

388.

389.

Use the Order of Operations to Simplify Complex Fractions

In the following exercises, simplify.

390.

391.

392.

393.

394.

395.

396.

397.

398.

399.

400.

401.

Mixed Practice

In the following exercises, simplify.

402.

403.

404.

405.

406.

407.

408.

409.

410.

411.

412.

413.

414.

415.

In the following exercises, evaluate the given expression. Express your answers in simplified form, using improper fractions if necessary.

416. when

417. when

418. when

419. when

420. when

421. when

422. when

423. when

424. when

425.

426.

427.

428.

429.

430.

431.

Everyday Math

432. Decorating Laronda is making covers for the throw pillows on her sofa. For each pillow cover, she needs —— yard of print fabric and — yard of solid fabric. What is the total amount of fabric Laronda needs for each pillow cover?

433. Baking Vanessa is baking chocolate chip cookies and oatmeal cookies. She needs — cups of sugar for the chocolate chip cookies, and — cups for the oatmeal cookies How much sugar does she need altogether?

Writing Exercises

434. Explain why it is necessary to have a common denominator to add or subtract fractions.

435. Explain how to find the LCD of two fractions.

Self Check

After completing the exercises, use this checklist to evaluate your mastery of the objectives of this section.

I can...	Confidently	With some help	No-I don't get it!
add and subtract fractions with different denominators.			
identify and use fraction operations.			
use the order of operations to simplify complex fractions.			
evaluate variable expressions with fractions.			

After looking at the checklist, do you think you are well prepared for the next section? Why or why not?

4.6 Add and Subtract Mixed Numbers

Learning Objectives

By the end of this section, you will be able to:

- Model addition of mixed numbers with a common denominator
- Add mixed numbers with a common denominator
- Model subtraction of mixed numbers
- Subtract mixed numbers with a common denominator
- Add and subtract mixed numbers with different denominators

BE PREPARED : : 4.14 Before you get started, take this readiness quiz.

Draw figure to model —

If you missed this problem, review Example 4.6.

BE PREPARED : : 4.15 Change — to a mixed number.

If you missed this problem, review Example 4.9.

BE PREPARED : : 4.16 Change — to an improper fraction.

If you missed this problem, review Example 4.11.

Model Addition of Mixed Numbers with a Common Denominator

So far, we've added and subtracted proper and improper fractions, but not mixed numbers. Let's begin by thinking about addition of mixed numbers using money.

If Ron has dollar and quarter, he has — dollars.

If Don has dollars and quarter, he has — dollars.

What if Ron and Don put their money together? They would have dollars and quarters. They add the dollars and add the quarters. This makes — dollars. Because two quarters is half a dollar, they would have and a half dollars, or — dollars.

—

—

— —

When you added the dollars and then added the quarters, you were adding the whole numbers and then adding the fractions.

— —

We can use fraction circles to model this same example:

		— —
Start with —.	one whole and one — pieces	$1\frac{1}{4}$
Add — more.	two wholes and one — pieces	$+\ 2\frac{1}{4}$
The sum is:	three wholes and two —'s	$3\frac{2}{4} = 3\frac{1}{2}$

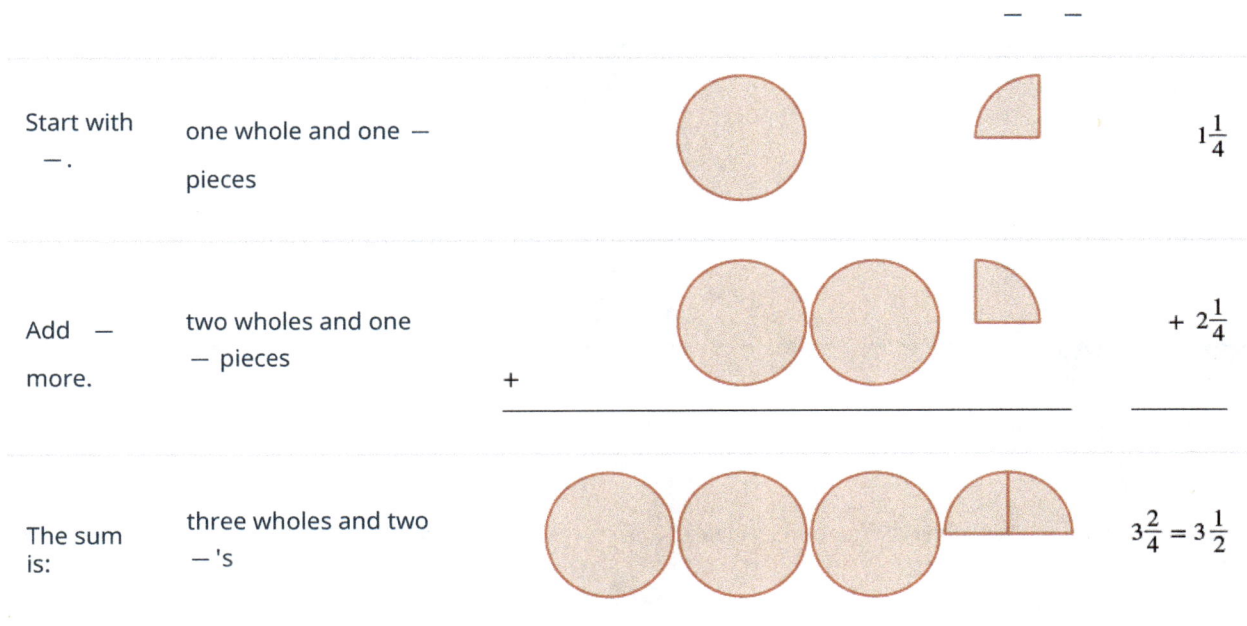

MANIPULATIVE MATHEMATICS

Doing the Manipulative Mathematics activity "Model Mixed Number Addition/Subtraction" will help you develop a better understanding of adding and subtracting mixed numbers.

EXAMPLE 4.81

Model — — and give the sum.

Solution

We will use fraction circles, whole circles for the whole numbers and — pieces for the fractions.

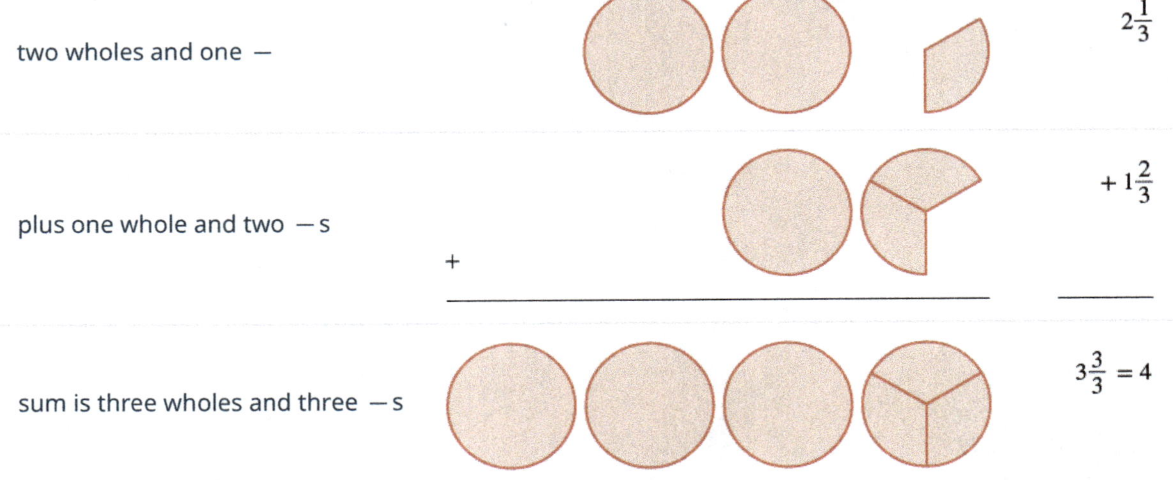

two wholes and one —
$2\frac{1}{3}$
plus one whole and two —s
+
$+\ 1\frac{2}{3}$
sum is three wholes and three —s
$3\frac{3}{3} = 4$

This is the same as wholes. So, — —

TRY IT : : 4.161 Use a model to add the following. Draw a picture to illustrate your model.

— —

TRY IT : : 4.162 Use a model to add the following. Draw a picture to illustrate your model.

— —

EXAMPLE 4.82

Model — — and give the sum as a mixed number.

Solution

We will use fraction circles, whole circles for the whole numbers and — pieces for the fractions.

one whole and three —	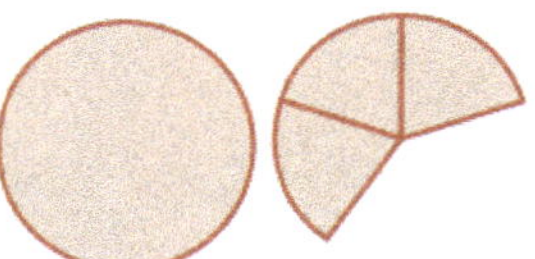	$1\frac{3}{5}$
plus two wholes and three — .	+ 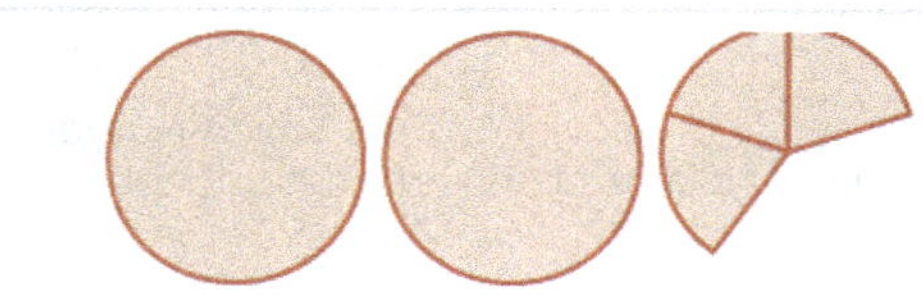	$+2\frac{3}{5}$
sum is three wholes and six —	  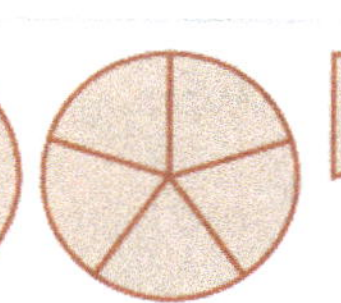	$3\frac{6}{5}=4\frac{1}{5}$

Adding the whole circles and fifth pieces, we got a sum of — We can see that — is equivalent to — so we add that to the to get —

TRY IT : : 4.163 Model, and give the sum as a mixed number. Draw a picture to illustrate your model.

— —

TRY IT : : 4.164 Model, and give the sum as a mixed number. Draw a picture to illustrate your model.

— —

Add Mixed Numbers

Modeling with fraction circles helps illustrate the process for adding mixed numbers: We add the whole numbers and add the fractions, and then we simplify the result, if possible.

HOW TO : : ADD MIXED NUMBERS WITH A COMMON DENOMINATOR.

Step 1. Add the whole numbers.

Step 2. Add the fractions.

Step 3. Simplify, if possible.

EXAMPLE 4.83

Add: — —

Solution

	— —
Add the whole numbers.	$\begin{array}{r} 3\frac{4}{9} \\ +2\frac{2}{9} \\ \hline 5 \end{array}$
Add the fractions.	$\begin{array}{r} 3\frac{4}{9} \\ +2\frac{2}{9} \\ \hline 5\frac{6}{9} \end{array}$
Simplify the fraction.	$\begin{array}{r} 3\frac{4}{9} \\ +2\frac{2}{9} \\ \hline 5\frac{6}{9} = 5\frac{2}{3} \end{array}$

TRY IT : : 4.165 Find the sum: — —

TRY IT : : 4.166 Find the sum: — —

In Example 4.83, the sum of the fractions was a proper fraction. Now we will work through an example where the sum is an improper fraction.

EXAMPLE 4.84

Find the sum: — —

Solution

Add the whole numbers and then add the fractions.	
Rewrite — as an improper fraction.	
Add.	
Simplify.	

> **TRY IT : : 4.167** Find the sum:

> **TRY IT : : 4.168** Find the sum:

An alternate method for adding mixed numbers is to convert the mixed numbers to improper fractions and then add the improper fractions. This method is usually written horizontally.

EXAMPLE 4.85

Add by converting the mixed numbers to improper fractions:

Solution

Convert to improper fractions.	
Add the fractions.	
Simplify the numerator.	
Rewrite as a mixed number.	
Simplify the fraction.	

Since the problem was given in mixed number form, we will write the sum as a mixed number.

TRY IT : : 4.169 Find the sum by converting the mixed numbers to improper fractions:

TRY IT : : 4.170 Find the sum by converting the mixed numbers to improper fractions:

Table 4.2 compares the two methods of addition, using the expression as an example. Which way do you prefer?

Mixed Numbers	Improper Fractions

Table 4.2

Model Subtraction of Mixed Numbers

Let's think of pizzas again to model subtraction of mixed numbers with a common denominator. Suppose you just baked a whole pizza and want to give your brother half of the pizza. What do you have to do to the pizza to give him half? You have to cut it into at least two pieces. Then you can give him half.

We will use fraction circles (pizzas!) to help us visualize the process.

Start with one whole.

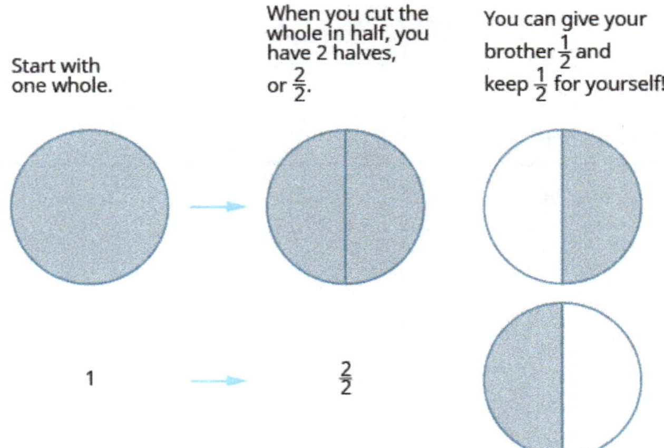
Start with one whole.
When you cut the whole in half, you have 2 halves, or $\frac{2}{2}$.
You can give your brother $\frac{1}{2}$ and keep $\frac{1}{2}$ for yourself!
1
$\frac{2}{2}$

Algebraically, you would write:

$$\begin{array}{r} 1 \\ -\frac{1}{2} \\ \hline \end{array} \rightarrow \begin{array}{r} \frac{2}{2} \\ -\frac{1}{2} \\ \hline \end{array} \rightarrow \begin{array}{r} \frac{2}{2} \\ -\frac{1}{2} \\ \hline \frac{1}{2} \end{array}$$

EXAMPLE 4.86

Use a model to subtract: −

Solution

	Model	Math Notation
Rewrite vertically. Start with one whole.		$\begin{array}{r} 1 \\ -\frac{1}{3} \\ \hline \end{array}$
Since $\frac{1}{3}$ has denominator 3, cut the whole into 3 pieces. The 1 whole becomes $\frac{3}{3}$.		$\begin{array}{r} \frac{3}{3} \\ -\frac{1}{3} \\ \hline \end{array}$
Take away $\frac{1}{3}$. There are $\frac{2}{3}$ left.		$\begin{array}{r} \frac{3}{3} \\ -\frac{1}{3} \\ \hline \frac{2}{3} \end{array}$

TRY IT : : 4.171 Use a model to subtract: −

TRY IT : : 4.172 Use a model to subtract: −

What if we start with more than one whole? Let's find out.

EXAMPLE 4.87

Use a model to subtract: −

Solution

	Model	Math Notation
Rewrite vertically. Start with two wholes.		$\begin{array}{r} 2 \\ -\frac{3}{4} \\ \hline \end{array}$
Since $\frac{3}{4}$ has denominator 4, cut one of the wholes into 4 pieces. You have 1 whole and $\frac{4}{4}$.		$\begin{array}{r} 1\frac{4}{4} \\ -\frac{3}{4} \\ \hline \end{array}$
Take away $\frac{3}{4}$. There is $1\frac{1}{4}$ left.		$\begin{array}{r} 1\frac{4}{4} \\ -\frac{3}{4} \\ \hline 1\frac{1}{4} \end{array}$

TRY IT : : 4.173 Use a model to subtract: —

TRY IT : : 4.174 Use a model to subtract: —

In the next example, we'll subtract more than one whole.

EXAMPLE 4.88

Use a model to subtract: —

Solution

	Model	Math Notation
Rewrite vertically. Start with two wholes.		$\begin{array}{r} 2 \\ -1\frac{2}{5} \\ \hline \end{array}$
Since $\frac{2}{5}$ has denominator 5, cut one of the wholes into 5 pieces. You have 1 whole and $\frac{5}{5}$.		$\begin{array}{r} 1\frac{5}{5} \\ -1\frac{2}{5} \\ \hline \end{array}$
Take away $1\frac{2}{5}$. There is $\frac{3}{5}$ left.		$\begin{array}{r} 1\frac{5}{5} \\ -1\frac{2}{5} \\ \hline \frac{3}{5} \end{array}$

TRY IT : : 4.175 Use a model to subtract: —

TRY IT : : 4.176 Use a model to subtract: —

What if you start with a mixed number and need to subtract a fraction? Think about this situation: You need to put three quarters in a parking meter, but you have only a bill and one quarter. What could you do? You could change the dollar bill into quarters. The value of quarters is the same as one dollar bill, but the quarters are more useful for the parking meter. Now, instead of having a bill and one quarter, you have quarters and can put quarters in the meter.

This models what happens when we subtract a fraction from a mixed number. We subtracted three quarters from one dollar and one quarter.

We can also model this using fraction circles, much like we did for addition of mixed numbers.

EXAMPLE 4.89

Use a model to subtract: — —

Solution

Rewrite vertically. Start with one whole and one fourth.	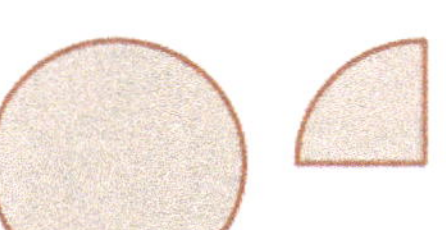	$\begin{array}{r} 1\frac{1}{4} \\ -\frac{3}{4} \\ \hline \end{array}$
Since the fractions have denominator 4, cut the whole into 4 pieces. You now have — and — which is —.	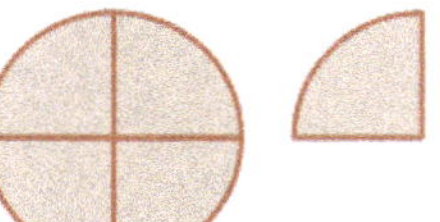	$\begin{array}{r} \frac{5}{4} \\ -\frac{3}{4} \\ \hline \end{array}$
Take away —. There is — left.		$\begin{array}{r} \frac{5}{4} \\ -\frac{3}{4} \\ \hline \frac{2}{4} = \frac{1}{2} \end{array}$

TRY IT : : 4.177 Use a model to subtract. Draw a picture to illustrate your model.

— —

TRY IT : : 4.178 Use a model to subtract. Draw a picture to illustrate your model.

— —

Subtract Mixed Numbers with a Common Denominator

Now we will subtract mixed numbers without using a model. But it may help to picture the model in your mind as you read the steps.

HOW TO : :

Subtract mixed numbers with common denominators.

Step 1. Rewrite the problem in vertical form.

Step 2. Compare the two fractions.

- If the top fraction is larger than the bottom fraction, go to Step 3.
- If not, in the top mixed number, take one whole and add it to the fraction part, making a mixed number with an improper fraction.

Step 3. Subtract the fractions.

Step 4. Subtract the whole numbers.

Step 5. Simplify, if possible.

EXAMPLE 4.90

Find the difference: — —

Solution

	$5\frac{3}{5} - 2\frac{4}{5}$
Rewrite the problem in vertical form.	$\begin{array}{r} 5\frac{3}{5} \\ -\,2\frac{4}{5} \\ \hline \end{array}$
Since — is less than —, take 1 from the 5 and add it to the — — — —	$\begin{array}{r} 5\frac{3}{5} \\ -\,2\frac{4}{5} \\ \hline \end{array} \longrightarrow \begin{array}{r} 4\frac{8}{5} \\ -\,2\frac{4}{5} \\ \hline \end{array}$
Subtract the fractions.	$\begin{array}{r} 4\frac{8}{5} \\ -\,2\frac{4}{5} \\ \hline \frac{4}{5} \end{array}$
Subtract the whole parts. The result is in simplest form.	$\begin{array}{r} 4\frac{8}{5} \\ -\,2\frac{4}{5} \\ \hline 2\frac{4}{5} \end{array}$

Since the problem was given with mixed numbers, we leave the result as mixed numbers.

TRY IT : : 4.179 Find the difference: — —

TRY IT : : 4.180 Find the difference: — —

Just as we did with addition, we could subtract mixed numbers by converting them first to improper fractions. We should write the answer in the form it was given, so if we are given mixed numbers to subtract we will write the answer as a mixed number.

HOW TO : : SUBTRACT MIXED NUMBERS WITH COMMON DENOMINATORS AS IMPROPER FRACTIONS.

Step 1. Rewrite the mixed numbers as improper fractions.

Step 2. Subtract the numerators.

Step 3. Write the answer as a mixed number, simplifying the fraction part, if possible.

EXAMPLE 4.91

Find the difference by converting to improper fractions:

— —

Solution

	— —
Rewrite as improper fractions.	— —
Subtract the numerators.	—
Rewrite as a mixed number.	—

TRY IT : : 4.181 Find the difference by converting the mixed numbers to improper fractions:

— —

TRY IT : : 4.182 Find the difference by converting the mixed numbers to improper fractions:

— —

Add and Subtract Mixed Numbers with Different Denominators

To add or subtract mixed numbers with different denominators, we first convert the fractions to equivalent fractions with the LCD. Then we can follow all the steps we used above for adding or subtracting fractions with like denominators.

EXAMPLE 4.92

Add: — —

Solution

Since the denominators are different, we rewrite the fractions as equivalent fractions with the LCD, Then we will add and simplify.

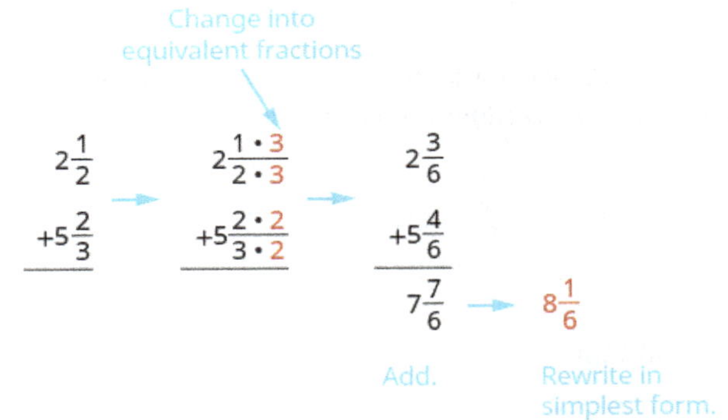
Change into
equivalent fractions
Add.
Rewrite in
simplest form.

We write the answer as a mixed number because we were given mixed numbers in the problem.

TRY IT : : 4.183 Add: — —

TRY IT : : 4.184 Add: — —

EXAMPLE 4.93

Subtract: — —

Solution

Since the denominators of the fractions are different, we will rewrite them as equivalent fractions with the LCD Once in that form, we will subtract. But we will need to borrow first.

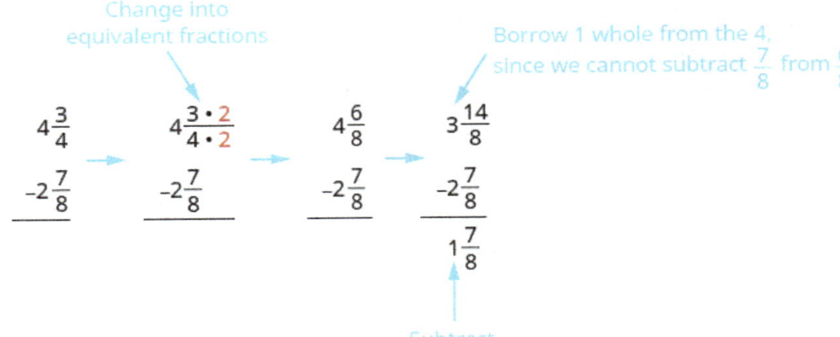
Change into
equivalent fractions
Borrow 1 whole from the 4,
since we cannot subtract $\frac{7}{8}$ from $\frac{6}{8}$.
Subtract.

We were given mixed numbers, so we leave the answer as a mixed number.

TRY IT : : 4.185 Find the difference: — —

TRY IT : : 4.186 Find the difference: — —

EXAMPLE 4.94

Subtract: — —

Solution

We can see the answer will be negative since we are subtracting from Generally, when we know the answer will be negative it is easier to subtract with improper fractions rather than mixed numbers.

Change to equivalent fractions with the LCD.	
Rewrite as improper fractions.	
Subtract.	
Rewrite as a mixed number.	

TRY IT : : 4.187 Subtract:

TRY IT : : 4.188 Subtract:

MEDIA : : ACCESS ADDITIONAL ONLINE RESOURCES

- Adding Mixed Numbers (http://www.openstax.org/l/24AddMixed)
- Subtracting Mixed Numbers (http://www.openstax.org/l/24SubtractMixed)

4.6 EXERCISES

Practice Makes Perfect

Model Addition of Mixed Numbers

In the following exercises, use a model to find the sum. Draw a picture to illustrate your model.

436. — —

437. — —

438. — —

439. — —

Add Mixed Numbers with a Common Denominator

In the following exercises, add.

440. — —

441. — —

442. — —

443. — —

444. — —

445. — —

446. — —

447. — —

Model Subtraction of Mixed Numbers

In the following exercises, use a model to find the difference. Draw a picture to illustrate your model.

448. — —

449. — —

Subtract Mixed Numbers with a Common Denominator

In the following exercises, find the difference.

450. — —

451. — —

452. — —

453. — —

454. — —

455. — —

456. — —

457. — —

Add and Subtract Mixed Numbers with Different Denominators

In the following exercises, write the sum or difference as a mixed number in simplified form.

458. — —

459. — —

460. — —

461. — —

462. — —

463. — —

464. — —

465. — —

Mixed Practice

In the following exercises, perform the indicated operation and write the result as a mixed number in simplified form.

466. — —

467. — —

468. — —

469. — —

470. — —

471. — —

472. — —

473. — —

474. — —

475. — —

476. —

477. —

478. — —

479. — —

480. — —

481. — —

482. — —

483. — —

484. — —

485. — —

486. — —

487. — —

488. — —

489. — —

Everyday Math

490. Sewing Renata is sewing matching shirts for her husband and son. According to the patterns she will use, she needs — yards of fabric for her husband's shirt and — yards of fabric for her son's shirt. How much fabric does she need to make both shirts?

491. Sewing Pauline has — yards of fabric to make a jacket. The jacket uses — yards. How much fabric will she have left after making the jacket?

492. Printing Nishant is printing invitations on his computer. The paper is — inches wide, and he sets the print area to have a —-inch border on each side. How wide is the print area on the sheet of paper?

493. Framing a picture Tessa bought a picture frame for her son's graduation picture. The picture is inches wide. The picture frame is — inches wide on each side. How wide will the framed picture be?

Writing Exercises

494. Draw a diagram and use it to explain how to add — —

495. Edgar will have to pay in tolls to drive to the city.

Explain how he can make change from a bill before he leaves so that he has the exact amount he needs.

How is Edgar's situation similar to how you subtract —

496. Add — — twice, first by leaving them as mixed numbers and then by rewriting as improper fractions. Which method do you prefer, and why?

497. Subtract — — twice, first by leaving them as mixed numbers and then by rewriting as improper fractions. Which method do you prefer, and why?

Self Check

After completing the exercises, use this checklist to evaluate your mastery of the objectives of this section.

I can...	Confidently	With some help	No-I don't get it!
model addition of mixed numbers common with a denominator.			
add mixed numbers with a common denominator.			
model subtraction of mixed numbers.			
subtract mixed numbers with a common denominator.			
add and subtract mixed numbers with different denominators.			

After reviewing this checklist, what will you do to become confident for all objectives?

4.7 Solve Equations with Fractions

Learning Objectives

By the end of this section, you will be able to:

- Determine whether a fraction is a solution of an equation
- Solve equations with fractions using the Addition, Subtraction, and Division Properties of Equality
- Solve equations using the Multiplication Property of Equality
- Translate sentences to equations and solve

BE PREPARED : : 4.17

Before you get started, take this readiness quiz. If you miss a problem, go back to the section listed and review the material.

Evaluate

If you missed this problem, review Example 3.23.

BE PREPARED : : 4.18 Solve:

If you missed this problem, review Example 3.61.

BE PREPARED : : 4.19 Solve:

If you missed this problem, review Example 4.28.

Determine Whether a Fraction is a Solution of an Equation

As we saw in Solve Equations with the Subtraction and Addition Properties of Equality and Solve Equations Using Integers; The Division Property of Equality, a solution of an equation is a value that makes a true statement when substituted for the variable in the equation. In those sections, we found whole number and integer solutions to equations. Now that we have worked with fractions, we are ready to find fraction solutions to equations.

The steps we take to determine whether a number is a solution to an equation are the same whether the solution is a whole number, an integer, or a fraction.

HOW TO : : DETERMINE WHETHER A NUMBER IS A SOLUTION TO AN EQUATION.

Step 1. Substitute the number for the variable in the equation.

Step 2. Simplify the expressions on both sides of the equation.

Step 3. Determine whether the resulting equation is true. If it is true, the number is a solution. If it is not true, the number is not a solution.

EXAMPLE 4.95

Determine whether each of the following is a solution of

Solution

	$x - \frac{3}{10} = \frac{1}{2}$
Substitute 1 for *x*.	$1 - \frac{3}{10} \stackrel{?}{=} \frac{1}{2}$
Change to fractions with a LCD of 10.	$\frac{10}{10} - \frac{3}{10} \stackrel{?}{=} \frac{5}{10}$
Subtract.	$\frac{7}{10} \neq \frac{5}{10}$

Since does not result in a true equation, is not a solution to the equation.

	$x - \frac{3}{10} = \frac{1}{2}$
Substitute $\frac{4}{5}$ for *x*.	$\frac{4}{5} - \frac{3}{10} \stackrel{?}{=} \frac{1}{2}$
	$\frac{8}{10} - \frac{3}{10} \stackrel{?}{=} \frac{5}{10}$
Subtract.	$\frac{5}{10} = \frac{5}{10}$ ✓

Since — results in a true equation, — is a solution to the equation — —

	$x - \frac{3}{10} = \frac{1}{2}$
Substitute $-\frac{4}{5}$ for *x*.	$-\frac{4}{5} - \frac{3}{10} \stackrel{?}{=} \frac{1}{2}$
	$-\frac{8}{10} - \frac{3}{10} \stackrel{?}{=} \frac{5}{10}$
Subtract.	$\frac{11}{10} \neq \frac{5}{10}$

Since — does not result in a true equation, — is not a solution to the equation.

TRY IT : : 4.189 Determine whether each number is a solution of the given equation.

— —

— —

TRY IT : : 4.190 Determine whether each number is a solution of the given equation.

Solve Equations with Fractions using the Addition, Subtraction, and Division Properties of Equality

In Solve Equations with the Subtraction and Addition Properties of Equality and Solve Equations Using Integers; The Division Property of Equality, we solved equations using the Addition, Subtraction, and Division Properties of Equality. We will use these same properties to solve equations with fractions.

Addition, Subtraction, and Division Properties of Equality

For any numbers

	Addition Property of Equality
	Subtraction Property of Equality
	Division Property of Equality

Table 4.3

In other words, when you add or subtract the same quantity from both sides of an equation, or divide both sides by the same quantity, you still have equality.

EXAMPLE 4.96

Solve:

Solution

	$y+\frac{9}{16}=\frac{5}{16}$
Subtract from each side to undo the addition.	$y+\frac{9}{16}-\frac{9}{16}=\frac{5}{16}-\frac{9}{16}$
Simplify on each side of the equation.	$y+0=-\frac{4}{16}$
Simplify the fraction.	$y=-\frac{1}{4}$
Check:	$y+\frac{9}{16}=\frac{5}{16}$
Substitute .	$-\frac{1}{4}+\frac{9}{16}\stackrel{?}{=}\frac{5}{16}$
Rewrite as fractions with the LCD.	$-\frac{4}{16}+\frac{9}{16}\stackrel{?}{=}\frac{5}{16}$
Add.	$\frac{5}{16}=\frac{5}{16}$ ✓

Since — makes —— —— a true statement, we know we have found the solution to this equation.

TRY IT : : 4.191 Solve: —— ——

TRY IT : : 4.192 Solve: —— ——

We used the Subtraction Property of Equality in Example 4.96. Now we'll use the Addition Property of Equality.

EXAMPLE 4.97

Solve: — —

Solution

	$a-\frac{5}{9}=-\frac{8}{9}$
Add — from each side to undo the addition.	$a-\frac{5}{9}+\frac{5}{9}=-\frac{8}{9}+\frac{5}{9}$
Simplify on each side of the equation.	$a+0=-\frac{3}{9}$
Simplify the fraction.	$a=-\frac{1}{3}$
Check:	$a-\frac{5}{9}=-\frac{8}{9}$
Substitute —.	$-\frac{1}{3}-\frac{5}{9}\stackrel{?}{=}-\frac{8}{9}$
Change to common denominator.	$-\frac{3}{9}-\frac{5}{9}\stackrel{?}{=}-\frac{8}{9}$
Subtract.	$-\frac{8}{9}=-\frac{8}{9}$ ✓

Since — makes the equation true, we know that — is the solution to the equation.

TRY IT : : 4.193 Solve: — —

TRY IT : : 4.194 Solve: — —

The next example may not seem to have a fraction, but let's see what happens when we solve it.

EXAMPLE 4.98

Solve:

Solution

Divide both sides by 10 to undo the multiplication.	— —
Simplify.	—
Check:	
Substitute — into the original equation.	—
Simplify.	/ 7
Multiply.	

The solution to the equation was the fraction — We leave it as an improper fraction.

TRY IT : : 4.195 Solve:

TRY IT : : 4.196 Solve:

Solve Equations with Fractions Using the Multiplication Property of Equality

Consider the equation — We want to know what number divided by gives So to "undo" the division, we will need to multiply by The *Multiplication Property of Equality* will allow us to do this. This property says that if we start with two equal quantities and multiply both by the same number, the results are equal.

The Multiplication Property of Equality

For any numbers and

If you multiply both sides of an equation by the same quantity, you still have equality.

Let's use the Multiplication Property of Equality to solve the equation —

EXAMPLE 4.99

Solve: —

Solution

	$\frac{x}{7} = -9$
Use the Multiplication Property of Equality to multiply both sides by . This will isolate the variable.	$7 \cdot \frac{x}{7} = 7(-9)$
Multiply.	$\frac{7x}{7} = -63$
Simplify.	$x = -63$
Check. Substitute −63 for *x* for in the original equation.	$\frac{-63}{7} \stackrel{?}{=} -9$
The equation is true.	$-9 = -9$ ✓

TRY IT : : 4.197 Solve: —

TRY IT : : 4.198 Solve: —

EXAMPLE 4.100

Solve: ——

Solution

Here, is divided by We must multiply by to isolate

	$\frac{p}{-8} = -40$
Multiply both sides by	$-8\left(\frac{p}{-8}\right) = -8(-40)$
Multiply.	$\frac{-8p}{-8} = 320$
Simplify.	$p = 320$
Check:	
Substitute .	$\frac{320}{-8} \stackrel{?}{=} -40$
The equation is true.	$-40 = -40$ ✓

TRY IT : : 4.199 Solve: ——

TRY IT : : 4.200 Solve: ——

Solve Equations with a Coefficient of

Look at the equation Does it look as if is already isolated? But there is a negative sign in front of so it is not isolated.

There are three different ways to isolate the variable in this type of equation. We will show all three ways in Example 4.101.

EXAMPLE 4.101

Solve:

Solution

One way to solve the equation is to rewrite as and then use the Division Property of Equality to isolate

	$-y = 15$
Rewrite as .	$-1y = 15$
Divide both sides by −1.	$\frac{-1y}{-1} = \frac{15}{-1}$
Simplify each side.	$y = -15$

Another way to solve this equation is to multiply both sides of the equation by

	$y = 15$
Multiply both sides by −1.	$-1(-y) = -1(15)$
Simplify each side.	$y = -15$

The third way to solve the equation is to read as "the opposite of What number has as its opposite? The opposite of is So

For all three methods, we isolated is isolated and solved the equation.

Check:

	$y = 15$
Substitute .	$-(-15) \stackrel{?}{=} (15)$
Simplify. The equation is true.	$15 = 15$ ✓

 **TRY IT : :** 4.201 Solve:

 **TRY IT : :** 4.202 Solve:

Solve Equations with a Fraction Coefficient

When we have an equation with a fraction coefficient we can use the Multiplication Property of Equality to make the coefficient equal to

For example, in the equation:

—

The coefficient of is — To solve for we need its coefficient to be Since the product of a number and its reciprocal is our strategy here will be to isolate by multiplying by the reciprocal of — We will do this in Example 4.102.

EXAMPLE 4.102

Solve: —

Solution

	$\frac{3}{4}x = 24$
Multiply both sides by the reciprocal of the coefficient.	$\frac{4}{3} \cdot \frac{3}{4}x = \frac{4}{3} \cdot 24$
Simplify.	$1x = \frac{4}{3} \cdot \frac{24}{1}$
Multiply.	$x = 32$
Check:	$\frac{3}{4}x = 24$
Substitute .	$\frac{3}{4} \cdot 32 \stackrel{?}{=} 24$
Rewrite as a fraction.	$\frac{3}{4} \cdot \frac{32}{1} \stackrel{?}{=} 24$
Multiply. The equation is true.	$24 = 24$ ✓

Notice that in the equation — we could have divided both sides by — to get by itself. Dividing is the same as multiplying by the reciprocal, so we would get the same result. But most people agree that multiplying by the reciprocal is easier.

TRY IT : : 4.203 Solve: —

TRY IT : : 4.204 Solve: —

EXAMPLE 4.103

Solve: —

Solution

The coefficient is a negative fraction. Remember that a number and its reciprocal have the same sign, so the reciprocal of the coefficient must also be negative.

	$\frac{3}{8}w = 72$
Multiply both sides by the reciprocal of —.	$-\frac{8}{3}\left(-\frac{3}{8}w\right) = \left(-\frac{8}{3}\right)72$
Simplify; reciprocals multiply to one.	$1w = -\frac{8}{3} \cdot \frac{72}{1}$
Multiply.	$w = -192$
Check:	$-\frac{3}{8}w = 72$
Let .	$-\frac{3}{8}(-192) \stackrel{?}{=} 72$
Multiply. It checks.	$72 = 72$ ✓

> **TRY IT : :** 4.205 Solve: —

> **TRY IT : :** 4.206 Solve: —

Translate Sentences to Equations and Solve

Now we have covered all four properties of equality—subtraction, addition, division, and multiplication. We'll list them all together here for easy reference.

Subtraction Property of Equality: For any real numbers and if then	**Addition Property of Equality:** For any real numbers and if then
Division Property of Equality: For any numbers and where if then — —	**Multiplication Property of Equality:** For any real numbers and if then

When you add, subtract, multiply or divide the same quantity from both sides of an equation, you still have equality.

In the next few examples, we'll translate sentences into equations and then solve the equations. It might be helpful to review the translation table in Evaluate, Simplify, and Translate Expressions.

EXAMPLE 4.104

Translate and solve: divided by is

Solution

Translate.	n divided by 6 is 24 $\frac{n}{6} = -24$
Multiply both sides by .	$6 \cdot \frac{n}{6} = 6(-24)$
Simplify.	$n = -144$
Check: Is divided by equal to ?	
Translate.	$\frac{-144}{6} \overset{?}{=} -24$
Simplify. It checks.	$-24 = -24$ ✓

> **TRY IT : : 4.207** Translate and solve: divided by is equal to

> **TRY IT : : 4.208** Translate and solve: divided by is equal to

EXAMPLE 4.105

Translate and solve: The quotient of and is

Solution

Translate.	The quotient of q and −5 is 70 $\frac{q}{-5} = 70$
Multiply both sides by .	$-5\left(\frac{q}{-5}\right) = -5(70)$
Simplify.	$q = -350$
Check: Is the quotient of and equal to ?	
Translate.	$\frac{-350}{-5} \overset{?}{=} 70$
Simplify. It checks.	$70 = 70$ ✓

> **TRY IT : : 4.209** Translate and solve: The quotient of and is

> **TRY IT : : 4.210** Translate and solve: The quotient of and is

EXAMPLE 4.106

Translate and solve: Two-thirds of is

Solution

Translate.	Two-thirds of f is 18 $\frac{2}{3}f = 18$
Multiply both sides by —.	$\frac{3}{2} \cdot \frac{2}{3}f = \frac{3}{2} \cdot 18$
Simplify.	$f = 27$
Check:	Is two-thirds of equal to ?
Translate.	$\frac{2}{3}(27) \stackrel{?}{=} 18$
Simplify. It checks.	$18 = 18$ ✓

TRY IT : : 4.211 Translate and solve: Two-fifths of is

TRY IT : : 4.212 Translate and solve: Three-fourths of is

EXAMPLE 4.107

Translate and solve: The quotient of and — is —

Solution

	The quotient of and — is —.
Translate.	— — —
Multiply both sides by — to isolate .	— — — — —
Simplify.	—
Remove common factors and multiply.	—
Check:	
Is the quotient of — and — equal to —?	— — — —
Rewrite as division.	— — —
Multiply the first fraction by the reciprocal of the second.	— — —
Simplify.	— —

Our solution checks.

TRY IT : : 4.213 Translate and solve. The quotient of and — is —

TRY IT : : 4.214 Translate and solve The quotient of and — is —

EXAMPLE 4.108

Translate and solve: The sum of three-eighths and is three and one-half.

Solution

Translate.	The sum of three-eighths and x is three and one-half $\frac{3}{8}+x = 3\frac{1}{2}$
Use the Subtraction Property of Equality to subtract — from both sides.	$\frac{3}{8}+x-\frac{3}{8}=3\frac{1}{2}-\frac{3}{8}$
Combine like terms on the left side.	$x=3\frac{1}{2}-\frac{3}{8}$
Convert mixed number to improper fraction.	$x=\frac{7}{2}-\frac{3}{8}$
Convert to equivalent fractions with LCD of 8.	$x=\frac{28}{8}-\frac{3}{8}$
Subtract.	$x=\frac{25}{8}$
Write as a mixed number.	$x=3\frac{1}{8}$

We write the answer as a mixed number because the original problem used a mixed number.

Check:

Is the sum of three-eighths and — equal to three and one-half?

	— — —
Add.	— —
Simplify.	— —

The solution checks.

TRY IT : : 4.215 Translate and solve: The sum of five-eighths and is one-fourth.

TRY IT : : 4.216 Translate and solve: The difference of one-and-three-fourths and is five-sixths.

MEDIA : : ACCESS ADDITIONAL ONLINE RESOURCES

- Solve One Step Equations With Fractions (http://www.openstax.org/l/24SolveOneStep)
- Solve One Step Equations With Fractions by Adding or Subtracting (http://www.openstax.org/l/24OneStepAdd)
- Solve One Step Equations With Fraction by Multiplying (http://www.openstax.org/l/24OneStepMulti)

4.7 EXERCISES

Practice Makes Perfect

Determine Whether a Fraction is a Solution of an Equation

In the following exercises, determine whether each number is a solution of the given equation.

498.

499.

500.

501.

Solve Equations with Fractions using the Addition, Subtraction, and Division Properties of Equality

In the following exercises, solve.

502.

503.

504.

505.

506.

507.

508.

509.

510.

511.

512.

513.

514.

515.

516.

517.

Solve Equations with Fractions Using the Multiplication Property of Equality

In the following exercises, solve.

518.

519.

520.

521.

522.

523.

524.

525.

526.

527.

528.

529.

530.

531.

532.

533. —

534. —

535. —

536. —

537. —

Mixed Practice

In the following exercises, solve.

538.

539.

540. —

541. —

542. — —

543. — —

544. — —

545. — —

546. — —

547. — —

548. —

549. —

Translate Sentences to Equations and Solve

In the following exercises, translate to an algebraic equation and solve.

550. divided by eight is

551. divided by six is

552. divided by is

553. divided by is

554. The quotient of and is

555. The quotient of and is

556. The quotient of and twelve is

557. The quotient of and nine is

558. Three-fourths of is

559. Two-fifths of is

560. Seven-tenths of is

561. Four-ninths of is

562. divided by equals negative

563. The quotient of and is

564. Three-fourths of is

565. The quotient of and — is —

566. The sum of five-sixths and is —

567. The sum of three-fourths and is —

568. The difference of and one-fourth is —

569. The difference of and one-third is —

Everyday Math

570. Shopping Teresa bought a pair of shoes on sale for The sale price was — of the regular price. Find the regular price of the shoes by solving the equation —

571. Playhouse The table in a child's playhouse is — of an adult-size table. The playhouse table is inches high. Find the height of an adult-size table by solving the equation —

Writing Exercises

572. Example 4.100 describes three methods to solve the equation Which method do you prefer? Why?

573. Richard thinks the solution to the equation — is Explain why Richard is wrong.

Self Check

After completing the exercises, use this checklist to evaluate your mastery of the objectives of this section.

I can...	Confidently	With some help	No-I don't get it!
determine whether a fraction is a solution of an equation.			
solve equations with fractions using the addition, subtraction, and division properties of equality.			
solve equations using the multiplication property of equality.			
translate sentences to equations and solve.			

Overall, after looking at the checklist, do you think you are well-prepared for the next Chapter? Why or why not?

CHAPTER 4 REVIEW

KEY TERMS

complex fraction A complex fraction is a fraction in which the numerator or the denominator contains a fraction.

equivalent fractions Equivalent fractions are two or more fractions that have the same value.

fraction A fraction is written —. in a fraction, is the numerator and is the denominator. A fraction represents parts of a whole. The denominator is the number of equal parts the whole has been divided into, and the numerator indicates how many parts are included.

least common denominator (LCD) The least common denominator (LCD) of two fractions is the least common multiple (LCM) of their denominators.

mixed number A mixed number consists of a whole number and a fraction — where . It is written as —, where .

proper and improper fractions The fraction — is *proper* if and *improper* if .

reciprocal The reciprocal of the fraction — is — where and .

simplified fraction A fraction is considered simplified if there are no common factors in the numerator and denominator.

KEY CONCEPTS

4.1 Visualize Fractions

- **Property of One**
    - Any number, except zero, divided by itself is one.
      — , where .
- **Mixed Numbers**
    - A **mixed number** consists of a whole number and a fraction — where .
    - It is written as follows: —
- **Proper and Improper Fractions**
    - The fraction is a proper fraction if and an improper fraction if .
- **Convert an improper fraction to a mixed number.**
  Step 1. Divide the denominator into the numerator.
  Step 2. Identify the quotient, remainder, and divisor.
  Step 3. Write the mixed number as quotient ————.
- **Convert a mixed number to an improper fraction.**
  Step 1. Multiply the whole number by the denominator.
  Step 2. Add the numerator to the product found in Step 1.
  Step 3. Write the final sum over the original denominator.
- **Equivalent Fractions Property**
    - If and are numbers where , , then — ——.

4.2 Multiply and Divide Fractions

- **Equivalent Fractions Property**

- If are numbers where , , then — —— and —— —.

- **Simplify a fraction.**

  Step 1. Rewrite the numerator and denominator to show the common factors. If needed, factor the numerator and denominator into prime numbers.

  Step 2. Simplify, using the equivalent fractions property, by removing common factors.

  Step 3. Multiply any remaining factors.

- **Fraction Multiplication**
  - If and are numbers where and , then — — —.

- **Reciprocal**
  - A number and its reciprocal have a product of . — —
  -

Opposite	Absolute Value	Reciprocal
has opposite sign	is never negative	has same sign, fraction inverts

Table 4.4

- **Fraction Division**
  - If and are numbers where , and , then

    — — — —

  - To divide fractions, multiply the first fraction by the reciprocal of the second.

4.3 Multiply and Divide Mixed Numbers and Complex Fractions

- **Multiply or divide mixed numbers.**

  Step 1. Convert the mixed numbers to improper fractions.

  Step 2. Follow the rules for fraction multiplication or division.

  Step 3. Simplify if possible.

- **Simplify a complex fraction.**

  Step 1. Rewrite the complex fraction as a division problem.

  Step 2. Follow the rules for dividing fractions.

  Step 3. Simplify if possible.

- **Placement of negative sign in a fraction.**
  - For any positive numbers and , — — —.

- **Simplify an expression with a fraction bar.**

  Step 1. Simplify the numerator.

  Step 2. Simplify the denominator.

  Step 3. Simplify the fraction.

4.4 Add and Subtract Fractions with Common Denominators

- **Fraction Addition**
  - If and are numbers where , then — — ——.
  - To add fractions, add the numerators and place the sum over the common denominator.

- **Fraction Subtraction**

- If and are numbers where , then — — ——.
- To subtract fractions, subtract the numerators and place the difference over the common denominator.

4.5 Add and Subtract Fractions with Different Denominators

- **Find the least common denominator (LCD) of two fractions.**
  Step 1. Factor each denominator into its primes.
  Step 2. List the primes, matching primes in columns when possible.
  Step 3. Bring down the columns.
  Step 4. Multiply the factors. The product is the LCM of the denominators.
  Step 5. The LCM of the denominators is the LCD of the fractions.
- **Equivalent Fractions Property**
  - If , and are whole numbers where , then
    — —— and —— —
- **Convert two fractions to equivalent fractions with their LCD as the common denominator.**
  Step 1. Find the LCD.
  Step 2. For each fraction, determine the number needed to multiply the denominator to get the LCD.
  Step 3. Use the Equivalent Fractions Property to multiply the numerator and denominator by the number from Step 2.
  Step 4. Simplify the numerator and denominator.
- **Add or subtract fractions with different denominators.**
  Step 1. Find the LCD.
  Step 2. Convert each fraction to an equivalent form with the LCD as the denominator.
  Step 3. Add or subtract the fractions.
  Step 4. Write the result in simplified form.
- **Summary of Fraction Operations**
  - **Fraction multiplication:** Multiply the numerators and multiply the denominators.
    — — ——
  - **Fraction division:** Multiply the first fraction by the reciprocal of the second.
    — — — —
  - **Fraction addition:** Add the numerators and place the sum over the common denominator. If the fractions have different denominators, first convert them to equivalent forms with the LCD.
    — — ——
  - **Fraction subtraction:** Subtract the numerators and place the difference over the common denominator. If the fractions have different denominators, first convert them to equivalent forms with the LCD.
    — — ——
- **Simplify complex fractions.**
  Step 1. Simplify the numerator.
  Step 2. Simplify the denominator.
  Step 3. Divide the numerator by the denominator.
  Step 4. Simplify if possible.

4.6 Add and Subtract Mixed Numbers

- **Add mixed numbers with a common denominator.**
  Step 1. Add the whole numbers.
  Step 2. Add the fractions.

Step 3. Simplify, if possible.

- **Subtract mixed numbers with common denominators.**

  Step 1. Rewrite the problem in vertical form.

  Step 2. Compare the two fractions.
  If the top fraction is larger than the bottom fraction, go to Step 3.
  If not, in the top mixed number, take one whole and add it to the fraction part, making a mixed number with an improper fraction.

  Step 3. Subtract the fractions.

  Step 4. Subtract the whole numbers.

  Step 5. Simplify, if possible.

- **Subtract mixed numbers with common denominators as improper fractions.**

  Step 1. Rewrite the mixed numbers as improper fractions.

  Step 2. Subtract the numerators.

  Step 3. Write the answer as a mixed number, simplifying the fraction part, if possible.

4.7 Solve Equations with Fractions

- **Determine whether a number is a solution to an equation.**

  Step 1. Substitute the number for the variable in the equation.

  Step 2. Simplify the expressions on both sides of the equation.

  Step 3. Determine whether the resulting equation is true. If it is true, the number is a solution. If it is not true, the number is not a solution.

- **Addition, Subtraction, and Division Properties of Equality**
  - For any numbers a, b, and c,
    if , then . Addition Property of Equality
  - if , then . Subtraction Property of Equality
  - if , then — —, . Division Property of Equality

- **The Multiplication Property of Equality**
  - For any numbers and , then .
  - If you multiply both sides of an equation by the same quantity, you still have equality.

REVIEW EXERCISES

4.1 Visualize Fractions

In the following exercises, name the fraction of each figure that is shaded.

574.

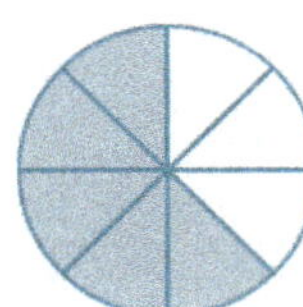

575.

In the following exercises, name the improper fractions. Then write each improper fraction as a mixed number.

576.

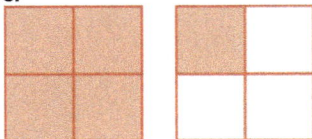

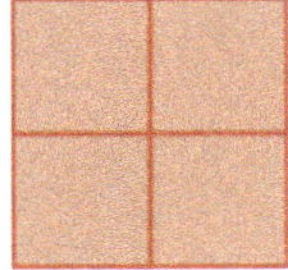

577.

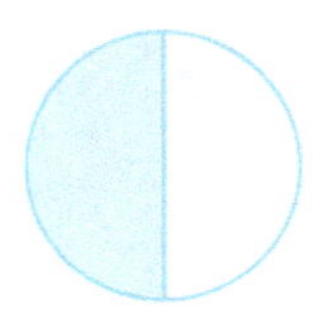

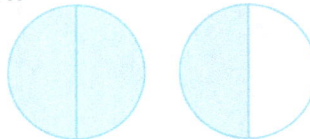

In the following exercises, convert the improper fraction to a mixed number.

578. —

579. —

In the following exercises, convert the mixed number to an improper fraction.

580. —

581. —

582. Find three fractions equivalent to — Show your work, using figures or algebra.

583. Find three fractions equivalent to — Show your work, using figures or algebra.

In the following exercises, locate the numbers on a number line.

584. — — —

585. — — — — — —

In the following exercises, order each pair of numbers, using or

586. —

587. —

4.2 Multiply and Divide Fractions

In the following exercises, simplify.

588. —

589. —

590. —

591. —

In the following exercises, multiply.

592. — —

593. — —

594. — —

595. —

596. —

597. — —

In the following exercises, find the reciprocal.

598. —

599. —

600.

601. —

602. Fill in the chart.

	Opposite	Absolute Value	Reciprocal
—			
—			
—			

In the following exercises, divide.

603. — —

604. — —

605. —

606. —

607. — —

4.3 Multiply and Divide Mixed Numbers and Complex Fractions

In the following exercises, perform the indicated operation.

608. — —

609. — —

610. —

611. — —

In the following exercises, translate the English phrase into an algebraic expression.

612. the quotient of and

613. the quotient of and the difference of and

In the following exercises, simplify the complex fraction

614. —

615. —

616. —

617. —

In the following exercises, simplify.

618. —

619. —

620. —

4.4 Add and Subtract Fractions with Common Denominators

In the following exercises, add.

621. — —

622. — —

623. — —

624.

625.

In the following exercises, subtract.

626.

627.

628.

629.

630.

631.

4.5 Add and Subtract Fractions with Different Denominators

In the following exercises, find the least common denominator.

632.

633.

634.

635.

In the following exercises, change to equivalent fractions using the given LCD.

636.

637.

638.

639.

In the following exercises, perform the indicated operations and simplify.

640.

641.

642.

643.

644.

645.

646.

647.

648.

649.

650.

In the following exercises, evaluate.

651. when

652. when

4.6 Add and Subtract Mixed Numbers

In the following exercises, perform the indicated operation.

653.

654.

655.

656.

657.

658.

659.

660.

4.7 Solve Equations with Fractions

In the following exercises, determine whether the each number is a solution of the given equation.

661.

662.

In the following exercises, solve the equation.

663.

664.

665.

666.

667.

In the following exercises, translate and solve.

668. The sum of two-thirds and is

669. The difference of and one-tenth is

670. The quotient of and is

671. Three-eighths of is

PRACTICE TEST

Convert the improper fraction to a mixed number.

672.

Convert the mixed number to an improper fraction.

673.

Locate the numbers on a number line.

674.

In the following exercises, simplify.

675.

676.

677.

678.

679.

680.

681.

682.

683.

684.

685.

686.

687.

688.

689.

690.

691.

692.

693.

694.

695.

696.

697.

Evaluate.

698. when

In the following exercises, solve the equation.

699.

700.

701.

702. —

703. —

704. Translate and solve: The quotient of and is

Solve for

Figure 5.1 The price of a gallon of gasoline is written as a decimal number. (credit: Mark Turnauckus, Flickr)

Chapter Outline

Introduction

Gasoline price changes all the time. They might go down for a period of time, but then they usually rise again. One thing that stays the same is that the price is not usually a whole number. Instead, it is shown using a decimal point to describe the cost in dollars and cents. We use decimal numbers all the time, especially when dealing with money. In this chapter, we will explore decimal numbers and how to perform operations using them.

5.1 Decimals

Learning Objectives

By the end of this section, you will be able to:

- Name decimals
- Write decimals
- Convert decimals to fractions or mixed numbers
- Locate decimals on the number line
- Order decimals
- Round decimals

BE PREPARED : : 5.1 Before you get started, take this readiness quiz.

Name the number in words.

If you missed this problem, review Example 1.4.

BE PREPARED : : 5.2 Round to the nearest ten.

If you missed this problem, review Example 1.9.

BE PREPARED : : 5.3

Locate — on a number line.

If you missed this problem, review **Example 4.16**.

Name Decimals

You probably already know quite a bit about decimals based on your experience with money. Suppose you buy a sandwich and a bottle of water for lunch. If the sandwich costs , the bottle of water costs , and the total sales tax is , what is the total cost of your lunch?

\$3.45	Sandwich
\$1.25	Water
+ \$0.33	Tax
\$5.03	Total

The total is Suppose you pay with a bill and pennies. Should you wait for change? No, and pennies is the same as

Because each penny is worth — of a dollar. We write the value of one penny as since

—

Writing a number with a decimal is known as decimal notation. It is a way of showing parts of a whole when the whole is a power of ten. In other words, decimals are another way of writing fractions whose denominators are powers of ten. Just as the counting numbers are based on powers of ten, decimals are based on powers of ten. Table 5.1 shows the counting numbers.

Counting number	Name
	One
	Ten
	One hundred
	One thousand
	Ten thousand

Table 5.1

How are decimals related to fractions? Table 5.2 shows the relation.

Decimal	Fraction	Name
	—	One tenth
	—	One hundredth
	—	One thousandth
	—	One ten-thousandth

Table 5.2

When we name a whole number, the name corresponds to the place value based on the powers of ten. In Whole Numbers, we learned to read as *ten thousand*. Likewise, the names of the decimal places correspond to their fraction values. Notice how the place value names in Figure 5.2 relate to the names of the fractions from Table 5.2.

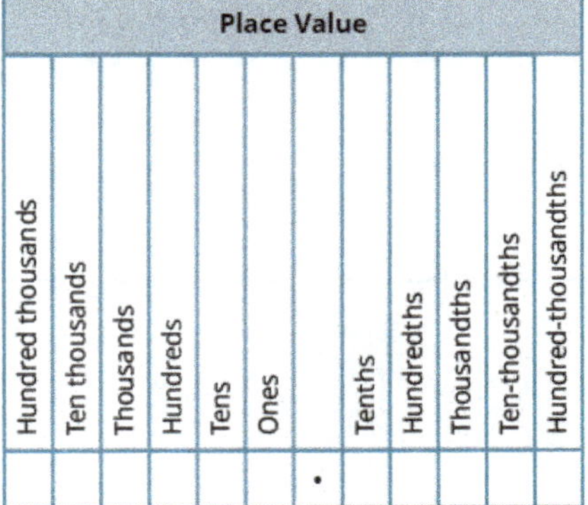

Figure 5.2 This chart illustrates place values to the left and right of the decimal point.

Notice two important facts shown in Figure 5.2.

- The "th" at the end of the name means the number is a fraction. "One thousand" is a number larger than one, but "one thousandth" is a number smaller than one.
- The tenths place is the first place to the right of the decimal, but the tens place is two places to the left of the decimal.

Remember that lunch? We read as *five dollars and three cents*. Naming decimals (those that don't represent money) is done in a similar way. We read the number as *five and three hundredths*.

We sometimes need to translate a number written in decimal notation into words. As shown in Figure 5.3, we write the amount on a check in both words and numbers.

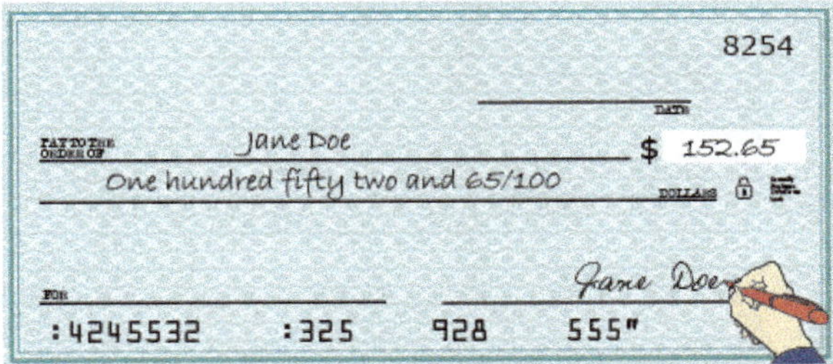

Figure 5.3 When we write a check, we write the amount as a decimal number as well as in words. The bank looks at the check to make sure both numbers match. This helps prevent errors.

Let's try naming a decimal, such as 15.68.	
We start by naming the number to the left of the decimal.	fifteen_____
We use the word "and" to indicate the decimal point.	fifteen and____
Then we name the number to the right of the decimal point as if it were a whole number.	fifteen and sixty-eight____
Last, name the decimal place of the last digit.	fifteen and sixty-eight hundredths

The number is read *fifteen and sixty-eight hundredths*.

HOW TO : : NAME A DECIMAL NUMBER.

- Name the number to the left of the decimal point.
- Write "and" for the decimal point.
- Name the "number" part to the right of the decimal point as if it were a whole number.
- Name the decimal place of the last digit.

EXAMPLE 5.1

Name each decimal:

Solution

	4.3
Name the number to the left of the decimal point.	four____
Write "and" for the decimal point.	four and____
Name the number to the right of the decimal point as if it were a whole number.	four and three____
Name the decimal place of the last digit.	four and three tenths

	2.45
Name the number to the left of the decimal point.	two____
Write "and" for the decimal point.	two and____
Name the number to the right of the decimal point as if it were a whole number.	two and forty-five____
Name the decimal place of the last digit.	two and forty-five hundredths

	0.009
Name the number to the left of the decimal point.	Zero is the number to the left of the decimal; it is not included in the name.
Name the number to the right of the decimal point as if it were a whole number.	nine____
Name the decimal place of the last digit.	nine thousandths

Name the number to the left of the decimal point.	negative fifteen
Write "and" for the decimal point.	negative fifteen and____
Name the number to the right of the decimal point as if it were a whole number.	negative fifteen and five hundred seventy-one____
Name the decimal place of the last digit.	negative fifteen and five hundred seventy-one thousandths

> **TRY IT : :** 5.1 Name each decimal:

> **TRY IT : :** 5.2 Name each decimal:

Write Decimals

Now we will translate the name of a decimal number into decimal notation. We will reverse the procedure we just used.

Let's start by writing the number six and seventeen hundredths:

	six and seventeen hundredths
The word *and* tells us to place a decimal point.	___.___
The word before *and* is the whole number; write it to the left of the decimal point.	6.____
The decimal part is seventeen hundredths. Mark two places to the right of the decimal point for hundredths.	6._ _
Write the numerals for seventeen in the places marked.	6.17

EXAMPLE 5.2

Write fourteen and thirty-seven hundredths as a decimal.

Solution

	fourteen and thirty-seven hundredths
Place a decimal point under the word 'and'.	______. ________
Translate the words before 'and' into the whole number and place it to the left of the decimal point.	14. ________
Mark two places to the right of the decimal point for "hundredths".	14.__ __
Translate the words after "and" and write the number to the right of the decimal point.	14.37
	Fourteen and thirty-seven hundredths is written 14.37.

TRY IT : : 5.3 Write as a decimal: thirteen and sixty-eight hundredths.

TRY IT : : 5.4 Write as a decimal: five and eight hundred ninety-four thousandths.

HOW TO : : WRITE A DECIMAL NUMBER FROM ITS NAME.

Step 1. Look for the word "and"—it locates the decimal point.

Step 2. Mark the number of decimal places needed to the right of the decimal point by noting the place value indicated by the last word.

- Place a decimal point under the word "and." Translate the words before "and" into the whole number and place it to the left of the decimal point.
- If there is no "and," write a "0" with a decimal point to its right.

Step 3. Translate the words after "and" into the number to the right of the decimal point. Write the number in the spaces—putting the final digit in the last place.

Step 4. Fill in zeros for place holders as needed.

The second bullet in Step 2 is needed for decimals that have no whole number part, like 'nine thousandths'. We recognize them by the words that indicate the place value after the decimal – such as 'tenths' or 'hundredths.' Since there is no whole number, there is no 'and.' We start by placing a zero to the left of the decimal and continue by filling in the numbers to the right, as we did above.

EXAMPLE 5.3

Write twenty-four thousandths as a decimal.

Solution

	twenty-four thousandths
Look for the word "and".	There is no "and" so start with 0 0.
To the right of the decimal point, put three decimal places for thousandths.	0. _ _ _ tenths hundredths thousandths
Write the number 24 with the 4 in the thousandths place.	0. _ 2 4 tenths hundredths thousandths
Put zeros as placeholders in the remaining decimal places.	0.024
	So, twenty-four thousandths is written 0.024

TRY IT : : 5.5 Write as a decimal: fifty-eight thousandths.

TRY IT : : 5.6 Write as a decimal: sixty-seven thousandths.

Before we move on to our next objective, think about money again. We know that is the same as The way we write depends on the context. In the same way, integers can be written as decimals with as many zeros as needed to the right of the decimal.

Convert Decimals to Fractions or Mixed Numbers

We often need to rewrite decimals as fractions or mixed numbers. Let's go back to our lunch order to see how we can convert decimal numbers to fractions. We know that means dollars and cents. Since there are cents in one dollar, cents means —— of a dollar, so ——

We convert decimals to fractions by identifying the place value of the farthest right digit. In the decimal the is in the hundredths place, so is the denominator of the fraction equivalent to

——

For our lunch, we can write the decimal as a mixed number.

——

Notice that when the number to the left of the decimal is zero, we get a proper fraction. When the number to the left of the decimal is not zero, we get a mixed number.

HOW TO : : CONVERT A DECIMAL NUMBER TO A FRACTION OR MIXED NUMBER.

Step 1. Look at the number to the left of the decimal.
- If it is zero, the decimal converts to a proper fraction.
- If it is not zero, the decimal converts to a mixed number.
  - Write the whole number.

Step 2. Determine the place value of the final digit.

Step 3. Write the fraction.
- numerator—the 'numbers' to the right of the decimal point
- denominator—the place value corresponding to the final digit

Step 4. Simplify the fraction, if possible.

EXAMPLE 5.4

Write each of the following decimal numbers as a fraction or a mixed number:

Solution

	4.09
There is a 4 to the left of the decimal point. Write "4" as the whole number part of the mixed number.	$4\frac{\square}{\square}$
Determine the place value of the final digit.	4. 0 (tenths) 9 (hundredths)
Write the fraction. Write 9 in the numerator as it is the number to the right of the decimal point.	$4\frac{9}{\square}$
Write 100 in the denominator as the place value of the final digit, 9, is hundredth.	$4\frac{9}{100}$
The fraction is in simplest form.	So, $4.09 = 4\frac{9}{100}$

Did you notice that the number of zeros in the denominator is the same as the number of decimal places?

	3.7
There is a 3 to the left of the decimal point. Write "3" as the whole number part of the mixed number.	$3\frac{\square}{\square}$
Determine the place value of the final digit.	3. 7 tenths
Write the fraction. Write 7 in the numerator as it is the number to the right of the decimal point.	$3\frac{7}{\square}$
Write 10 in the denominator as the place value of the final digit, 7, is tenths.	$3\frac{7}{10}$
The fraction is in simplest form.	So, $3.7 = 3\frac{7}{10}$

	−0.286
There is a 0 to the left of the decimal point. Write a negative sign before the fraction.	$-\frac{\square}{\square}$
Determine the place value of the final digit and write it in the denominator.	−0. 2 8 6 tenths hundredths thousandths
Write the fraction. Write 286 in the numerator as it is the number to the right of the decimal point. Write 1,000 in the denominator as the place value of the final digit, 6, is thousandths.	$-\frac{286}{1000}$
We remove a common factor of 2 to simplify the fraction.	$-\frac{143}{500}$

> **TRY IT : :** 5.7 Write as a fraction or mixed number. Simplify the answer if possible.

> **TRY IT : :** 5.8 Write as a fraction or mixed number. Simplify the answer if possible.

Locate Decimals on the Number Line

Since decimals are forms of fractions, locating decimals on the number line is similar to locating fractions on the number line.

EXAMPLE 5.5

Locate on a number line.

Solution

The decimal is equivalent to — so is located between and On a number line, divide the interval between and into equal parts and place marks to separate the parts.

Label the marks We write as and as so that the numbers are consistently in tenths. Finally, mark on the number line.

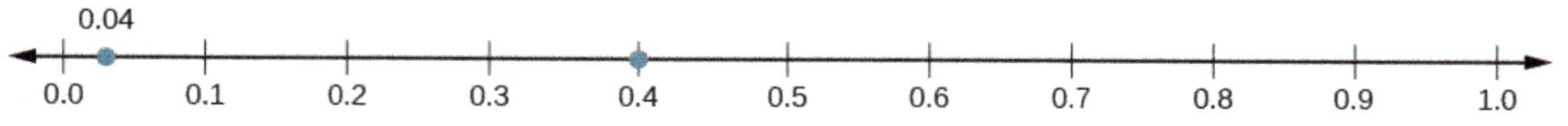

> **TRY IT : : 5.9** Locate on a number line.

> **TRY IT : : 5.10** Locate on a number line.

EXAMPLE 5.6

Locate on a number line.

Solution

The decimal is equivalent to —— so it is located between and On a number line, mark off and label the multiples of in the interval between and (, , etc.) and mark between and a little closer to .

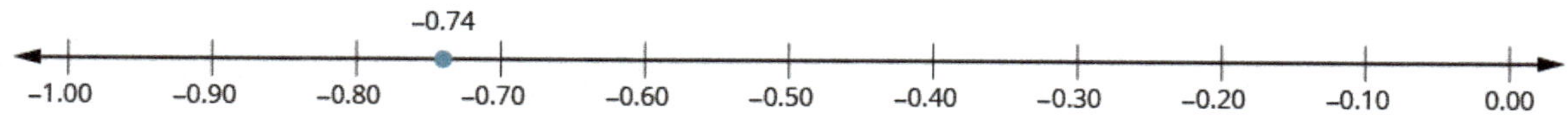

> **TRY IT : : 5.11** Locate on a number line.

> **TRY IT : : 5.12** Locate on a number line.

Order Decimals

Which is larger, or

If you think of this as money, you know that (forty cents) is greater than (four cents). So,

In previous chapters, we used the number line to order numbers.

Where are and located on the number line?

0.04

0.0 0.1 0.2 0.3 0.4 0.5 0.6 0.7 0.8 0.9 1.0

We see that is to the right of So we know

How does compare to This doesn't translate into money to make the comparison easy. But if we convert and to fractions, we can tell which is larger.

Convert to fractions.		
We need a common denominator to compare them.	$\frac{31 \cdot 10}{100 \cdot 10}$	

Because we know that Therefore,

Notice what we did in converting to a fraction—we started with the fraction and ended with the equivalent fraction Converting back to a decimal gives So is equivalent to Writing zeros at the end of a decimal does not change its value.

If two decimals have the same value, they are said to be equivalent decimals.

We say and are equivalent decimals.

Equivalent Decimals

Two decimals are **equivalent decimals** if they convert to equivalent fractions.

Remember, writing zeros at the end of a decimal does not change its value.

HOW TO : : ORDER DECIMALS.

Step 1. Check to see if both numbers have the same number of decimal places. If not, write zeros at the end of the one with fewer digits to make them match.

Step 2. Compare the numbers to the right of the decimal point as if they were whole numbers.

Step 3. Order the numbers using the appropriate inequality sign.

EXAMPLE 5.7

Order the following decimals using

Solution

Check to see if both numbers have the same number of decimal places. They do not, so write one zero at the right of 0.6.

Compare the numbers to the right of the decimal point as if they were whole numbers.

Order the numbers using the appropriate inequality sign.

Check to see if both numbers have the same number of decimal places. They do not, so write one zero at the right of 0.83.

Compare the numbers to the right of the decimal point as if they were whole numbers.

Order the numbers using the appropriate inequality sign.

> **TRY IT : :** 5.13 Order each of the following pairs of numbers, using

> **TRY IT : :** 5.14 Order each of the following pairs of numbers, using

When we order negative decimals, it is important to remember how to order negative integers. Recall that larger numbers are to the right on the number line. For example, because lies to the right of on the number line, we know that Similarly, smaller numbers lie to the left on the number line. For example, because lies to the left of on the number line, we know that

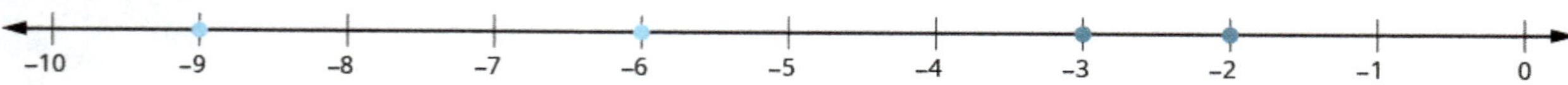

If we zoomed in on the interval between and we would see in the same way that

EXAMPLE 5.8

Use to order.

Solution

Write the numbers one under the other, lining up the decimal points.

They have the same number of digits.

Since tenth is greater than tenths.

TRY IT : : 5.15 Order each of the following pairs of numbers, using

TRY IT : : 5.16 Order each of the following pairs of numbers, using

Round Decimals

In the United States, gasoline prices are usually written with the decimal part as thousandths of a dollar. For example, a gas station might post the price of unleaded gas at per gallon. But if you were to buy exactly one gallon of gas at this price, you would pay , because the final price would be rounded to the nearest cent. In Whole Numbers, we saw that we round numbers to get an approximate value when the exact value is not needed. Suppose we wanted to round to the nearest dollar. Is it closer to or to What if we wanted to round to the nearest ten cents; is it closer to or to The number lines in Figure 5.4 can help us answer those questions.

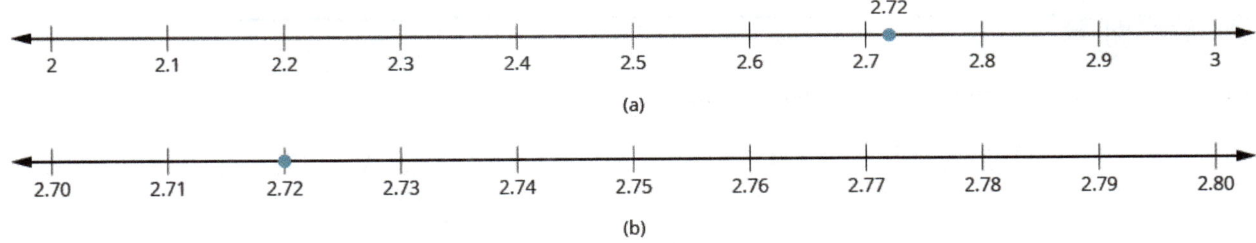

Figure 5.4 We see that is closer to than to So, rounded to the nearest whole number is We see that is closer to than So we say that rounded to the nearest tenth is

Can we round decimals without number lines? Yes! We use a method based on the one we used to round whole numbers.

HOW TO : : ROUND A DECIMAL.

Step 1. Locate the given place value and mark it with an arrow.

Step 2. Underline the digit to the right of the given place value.

Step 3. Is this digit greater than or equal to

- Yes - add to the digit in the given place value.
- No - do not change the digit in the given place value

Step 4. Rewrite the number, removing all digits to the right of the given place value.

EXAMPLE 5.9

Round to the nearest hundredth.

Solution

	18.379
Locate the hundredths place and mark it with an arrow.	hundredths place ↓ 18.379
Underline the digit to the right of the 7.	hundredths place ↓ 18.379
Because 9 is greater than or equal to 5, add 1 to the 7.	18.379 delete add 1
Rewrite the number, deleting all digits to the right of the hundredths place.	18.38
	18.38 is 18.379 rounded to the nearest hundredth.

TRY IT : : 5.17 Round to the nearest hundredth:

TRY IT : : 5.18 Round to the nearest hundredth:

EXAMPLE 5.10

Round to the nearest tenth whole number.

Solution

Round 18.379 to the nearest tenth.

	18.379
Locate the tenths place and mark it with an arrow.	tenths place ↓ 18.379
Underline the digit to the right of the tenths digit.	tenths place ↓ 18.3<u>7</u>9
Because 7 is greater than or equal to 5, add 1 to the 3.	18.379 add 1, delete
Rewrite the number, deleting all digits to the right of the tenths place.	18.4
	So, 18.379 rounded to the nearest tenth is 18.4.

Round 18.379 to the nearest whole number.

	18.379
Locate the ones place and mark it with an arrow.	ones place ↓ 18.379
Underline the digit to the right of the ones place.	ones place ↓ 18.<u>3</u>79
Since 3 is not greater than or equal to 5, do not add 1 to the 8.	18.379 do not add 1, delete
Rewrite the number, deleting all digits to the right of the ones place.	18
	So 18.379 rounded to the nearest whole number is 18.

TRY IT : : 5.19 Round to the nearest hundredth tenth whole number.

TRY IT : : 5.20 Round to the nearest thousandth hundredth tenth.

MEDIA : : ACCESS ADDITIONAL ONLINE RESOURCES

- Introduction to Decimal Notation (http://www.openstax.org/l/24decmlnotat)
- Write a Number in Decimal Notation from Words (http://www.openstax.org/l/24word2dcmlnot)
- Identify Decimals on the Number Line (http://www.openstax.org/l/24decmlnumline)
- Rounding Decimals (http://www.openstax.org/l/24rounddecml)
- Writing a Decimal as a Simplified Fraction (http://www.openstax.org/l/24decmlsimpfrac)

5.1 EXERCISES

Practice Makes Perfect

Name Decimals

In the following exercises, name each decimal.

1.

2.

3.

4.

5.

6.

7.

8.

9.

10.

11.

12.

Write Decimals

In the following exercises, translate the name into a decimal number.

13. Eight and three hundredths

14. Nine and seven hundredths

15. Twenty-nine and eighty-one hundredths

16. Sixty-one and seventy-four hundredths

17. Seven tenths

18. Six tenths

19. One thousandth

20. Nine thousandths

21. Twenty-nine thousandths

22. Thirty-five thousandths

23. Negative eleven and nine ten-thousandths

24. Negative fifty-nine and two ten-thousandths

25. Thirteen and three hundred ninety-five ten thousandths

26. Thirty and two hundred seventy-nine thousandths

Convert Decimals to Fractions or Mixed Numbers

In the following exercises, convert each decimal to a fraction or mixed number.

27.

28.

29.

30.

31.

32.

33.

34.

35.

36.

37.

38.

39.

40.

41.

42.

43.

44.

45.

46.

47.

48.

49.

50.

Locate Decimals on the Number Line

In the following exercises, locate each number on a number line.

51.

52.

53.

54. **55.** **56.**

57. **58.**

Order Decimals

In the following exercises, order each of the following pairs of numbers, using

59. **60.** **61.**

62. **63.** **64.**

65. **66.** **67.**

68. **69.** **70.**

Round Decimals

In the following exercises, round each number to the nearest tenth.

71. **72.** **73.**

74.

In the following exercises, round each number to the nearest hundredth.

75. **76.** **77.**

78. **79.** **80.**

81. **82.**

In the following exercises, round each number to the nearest hundredth tenth whole number.

83. **84.** **85.**

86.

Everyday Math

87. Salary Increase Danny got a raise and now makes a year. Round this number to the nearest:

dollar

thousand dollars

ten thousand dollars.

88. New Car Purchase Selena's new car cost Round this number to the nearest:

dollar

thousand dollars

ten thousand dollars.

89. Sales Tax Hyo Jin lives in San Diego. She bought a refrigerator for and when the clerk calculated the sales tax it came out to exactly Round the sales tax to the nearest

penny dollar.

90. Sales Tax Jennifer bought a dining room set for her home in Cincinnati. She calculated the sales tax to be exactly Round the sales tax to the nearest penny dollar.

Writing Exercises

91. How does your knowledge of money help you learn about decimals?

92. Explain how you write “three and nine hundredths” as a decimal.

93. Jim ran a race in Tim ran the same race in Who had the faster time, Jim or Tim? How do you know?

94. Gerry saw a sign advertising postcards marked for sale at What is wrong with the advertised price?

Self Check

After completing the exercises, use this checklist to evaluate your mastery of the objectives of this section.

I can...	Confidently	With some help	No-I don't get it!
name decimals.			
write decimals.			
convert decimals to fractions or mixed numbers.			
locate decimals on the number line.			
order decimals.			
round decimals.			

If most of your checks were:

...confidently. Congratulations! You have achieved the objectives in this section. Reflect on the study skills you used so that you can continue to use them. What did you do to become confident of your ability to do these things? Be specific.

...with some help. This must be addressed quickly because topics you do not master become potholes in your road to success. In math, every topic builds upon previous work. It is important to make sure you have a strong foundation before you move on. Whom can you ask for help? Your fellow classmates and instructor are good resources. Is there a place on campus where math tutors are available? Can your study skills be improved?

...no—I don't get it! This is a warning sign and you must not ignore it. You should get help right away or you will quickly be overwhelmed. See your instructor as soon as you can to discuss your situation. Together you can come up with a plan to get you the help you need.

5.2 Decimal Operations

Learning Objectives

By the end of this section, you will be able to:

- Add and subtract decimals
- Multiply decimals
- Divide decimals
- Use decimals in money applications

BE PREPARED : : 5.4 Before you get started, take this readiness quiz.

Simplify ——

If you missed this problem, review Example 4.19.

BE PREPARED : : 5.5 Multiply —— ——

If you missed this problem, review Example 4.25.

BE PREPARED : : 5.6 Divide

If you missed this problem, review Example 3.49.

Add and Subtract Decimals

Let's take one more look at the lunch order from the start of Decimals, this time noticing how the numbers were added together.

\$3.45	Sandwich
\$1.25	Water
+ \$0.33	Tax
\$5.03	Total

All three items (sandwich, water, tax) were priced in dollars and cents, so we lined up the dollars under the dollars and the cents under the cents, with the decimal points lined up between them. Then we just added each column, as if we were adding whole numbers. By lining up decimals this way, we can add or subtract the corresponding place values just as we did with whole numbers.

HOW TO : : ADD OR SUBTRACT DECIMALS.

Step 1. Write the numbers vertically so the decimal points line up.

Step 2. Use zeros as place holders, as needed.

Step 3. Add or subtract the numbers as if they were whole numbers. Then place the decimal in the answer under the decimal points in the given numbers.

EXAMPLE 5.11

Add:

Solution

Write the numbers vertically so the decimal points line up.	
Place holders are not needed since both numbers have the same number of decimal places.	
Add the numbers as if they were whole numbers. Then place the decimal in the answer under the decimal points in the given numbers.	

TRY IT : : 5.21 Add:

TRY IT : : 5.22 Add:

EXAMPLE 5.12

Add:

Solution

Write the numbers vertically so the decimal points line up.	$\begin{array}{r} 23.5 \\ +\,41.38 \\ \hline \end{array}$
Place 0 as a place holder after the 5 in 23.5, so that both numbers have two decimal places.	$\begin{array}{r} 23.50 \\ +\,41.38 \\ \hline \end{array}$
Add the numbers as if they were whole numbers. Then place the decimal in the answer under the decimal points in the given numbers.	$\begin{array}{r} 23.50 \\ +\,41.38 \\ \hline 64.88 \end{array}$

TRY IT : : 5.23 Add:

TRY IT : : 5.24 Add:

How much change would you get if you handed the cashier a bill for a purchase? We will show the steps to calculate this in the next example.

EXAMPLE 5.13

Subtract:

Solution

Write the numbers vertically so the decimal points line up. Remember 20 is a whole number, so place the decimal point after the 0.	$\begin{array}{r} 20. \\ -\ 14.65 \\ \hline \end{array}$
Place two zeros after the decimal point in 20, as place holders so that both numbers have two decimal places.	$\begin{array}{r} 20.00 \\ -\ 14.65 \\ \hline \end{array}$
Subtract the numbers as if they were whole numbers. Then place the decimal in the answer under the decimal points in the given numbers.	$\begin{array}{r} \overset{1}{\cancel{2}}\overset{\overset{9}{\cancel{10}}}{\cancel{0}}.\overset{\overset{9}{\cancel{10}}}{\cancel{0}}\overset{10}{\cancel{0}} \\ -\ 14.65 \\ \hline 5.35 \end{array}$

> TRY IT : : 5.25 Subtract:

> TRY IT : : 5.26 Subtract:

EXAMPLE 5.14

Subtract:

Solution

If we subtract from the answer will be negative since To subtract easily, we can subtract from Then we will place the negative sign in the result.

Write the numbers vertically so the decimal points line up.	$\begin{array}{r} 7.4 \\ -\ 2.51 \\ \hline \end{array}$
Place zero after the 4 in 7.4 as a place holder, so that both numbers have two decimal places.	$\begin{array}{r} 7.40 \\ -\ 2.51 \\ \hline \end{array}$
Subtract and place the decimal in the answer.	$\begin{array}{r} 7.40 \\ -\ 2.51 \\ \hline 4.89 \end{array}$
Remember that we are really subtracting so the answer is negative.	

> TRY IT : : 5.27 Subtract:

> TRY IT : : 5.28 Subtract:

Multiply Decimals

Multiplying decimals is very much like multiplying whole numbers—we just have to determine where to place the decimal point. The procedure for multiplying decimals will make sense if we first review multiplying fractions.

Do you remember how to multiply fractions? To multiply fractions, you multiply the numerators and then multiply the denominators.

So let's see what we would get as the product of decimals by converting them to fractions first. We will do two examples side-by-side in Table 5.3. Look for a pattern.

	A	B
Convert to fractions.		
Multiply.		
Convert back to decimals.		

Table 5.3

There is a pattern that we can use. In A, we multiplied two numbers that each had one decimal place, and the product had two decimal places. In B, we multiplied a number with one decimal place by a number with two decimal places, and the product had three decimal places.

How many decimal places would you expect for the product of If you said "five", you recognized the pattern. When we multiply two numbers with decimals, we count all the decimal places in the factors—in this case two plus three—to get the number of decimal places in the product—in this case five.

$$\underbrace{(0.01)}_{\text{2 places}}\ \underbrace{(0.004)}_{\text{3 places}} = \underbrace{0.00004}_{\text{5 places}}$$

$$\left(\frac{1}{100}\right)\left(\frac{4}{1000}\right) = \frac{4}{100{,}000}$$

Once we know how to determine the number of digits after the decimal point, we can multiply decimal numbers without converting them to fractions first. The number of decimal places in the product is the sum of the number of decimal places in the factors.

The rules for multiplying positive and negative numbers apply to decimals, too, of course.

Multiplying Two Numbers

When multiplying two numbers,

- if their signs are the same, the product is positive.
- if their signs are different, the product is negative.

When you multiply signed decimals, first determine the sign of the product and then multiply as if the numbers were both positive. Finally, write the product with the appropriate sign.

HOW TO : : MULTIPLY DECIMAL NUMBERS.

Step 1. Determine the sign of the product.

Step 2. Write the numbers in vertical format, lining up the numbers on the right.

Step 3. Multiply the numbers as if they were whole numbers, temporarily ignoring the decimal points.

Step 4. Place the decimal point. The number of decimal places in the product is the sum of the number of decimal places in the factors. If needed, use zeros as placeholders.

Step 5. Write the product with the appropriate sign.

EXAMPLE 5.15

Multiply:

Solution

Determine the sign of the product. The signs are the same.	The product will be positive.
Write the numbers in vertical format, lining up the numbers on the right.	$\begin{array}{r} 4.075 \\ \times\ 3.9 \\ \hline \end{array}$
Multiply the numbers as if they were whole numbers, temporarily ignoring the decimal points.	$\begin{array}{r} 4.075 \\ \times\ 3.9 \\ \hline 36675 \\ 12225 \\ \hline 158925 \end{array}$
Place the decimal point. Add the number of decimal places in the factors Place the decimal point 4 places from the right.	$\begin{array}{rl} 4.075 & \text{3 places} \\ \times\ 3.9 & \text{1 place} \\ \hline 36675 & \\ 12225 & \\ \hline 158925 & \text{4 places} \end{array}$
The product is positive.	

> TRY IT : : 5.29 Multiply:

> TRY IT : : 5.30 Multiply:

EXAMPLE 5.16

Multiply:

Solution

The signs are different.	The product will be negative.
Write in vertical format, lining up the numbers on the right.	
Multiply.	
Place the decimal point 3 places from the right. (−8.2)(5.19) 1 place 2 places	
The product is negative.	

> **TRY IT : : 5.31** Multiply:

> **TRY IT : : 5.32** Multiply:

In the next example, we'll need to add several placeholder zeros to properly place the decimal point.

EXAMPLE 5.17

Multiply:

Solution

The product is positive.	
Write in vertical format, lining up the numbers on the right.	$\begin{array}{r} 0.045 \\ \times\, 0.03 \\ \hline \end{array}$
Multiply.	$\begin{array}{r} 0.045 \\ \times\, 0.03 \\ \hline 135 \end{array}$
The decimal point must be 5 places from the right. (0.03)(0.045) 2 places 3 places Add zeros as needed to get the 5 places.	$\begin{array}{r} 0.045 \\ \times\, 0.03 \\ \hline 0.00135 \end{array}$
The product is positive.	

TRY IT : : 5.33 Multiply:

TRY IT : : 5.34 Multiply:

Multiply by Powers of

In many fields, especially in the sciences, it is common to multiply decimals by powers of Let's see what happens when we multiply by some powers of

1.9436(10)	1.9436(100)	1.9436(1000)
$\begin{array}{r}1.9436\\ \times\quad 10\\ \hline 19.4360\end{array}$	$\begin{array}{r}1.9436\\ \times\quad 100\\ \hline 194.3600\end{array}$	$\begin{array}{r}1.9436\\ \times\quad 1000\\ \hline 1943.6000\end{array}$

Look at the results without the final zeros. Do you notice a pattern?

The number of places that the decimal point moved is the same as the number of zeros in the power of ten. Table 5.4 summarizes the results.

Multiply by	Number of zeros	Number of places decimal point moves
		place to the right
		places to the right
		places to the right
		places to the right

Table 5.4

We can use this pattern as a shortcut to multiply by powers of ten instead of multiplying using the vertical format. We can count the zeros in the power of and then move the decimal point that same of places to the right.

So, for example, to multiply by move the decimal point places to the right.

$$45.86 \times 100 = 4586.$$

Sometimes when we need to move the decimal point, there are not enough decimal places. In that case, we use zeros as placeholders. For example, let's multiply by We need to move the decimal point places to the right. Since there is only one digit to the right of the decimal point, we must write a in the hundredths place.

$$2.4 \times 100 = 240.$$

HOW TO : : MULTIPLY A DECIMAL BY A POWER OF 10.

Step 1. Move the decimal point to the right the same number of places as the number of zeros in the power of

Step 2. Write zeros at the end of the number as placeholders if needed.

EXAMPLE 5.18

Multiply by factors of

Solution

By looking at the number of zeros in the multiple of ten, we see the number of places we need to move the decimal to the right.

There is 1 zero in 10, so move the decimal point 1 place to the right.	5.63
There are 2 zeros in 100, so move the decimal point 2 places to the right.	 5.63
There are 3 zeros in 1000, so move the decimal point 3 places to the right.	5.63
A zero must be added at the end.	

> **TRY IT : : 5.35** Multiply by factors of

> **TRY IT : : 5.36** Multiply by factors of

Divide Decimals

Just as with multiplication, division of decimals is very much like dividing whole numbers. We just have to figure out where the decimal point must be placed.

To understand decimal division, let's consider the multiplication problem

Remember, a multiplication problem can be rephrased as a division problem. So we can write

We can think of this as "If we divide 8 tenths into four groups, how many are in each group?" Figure 5.5 shows that there are four groups of two-tenths in eight-tenths. So

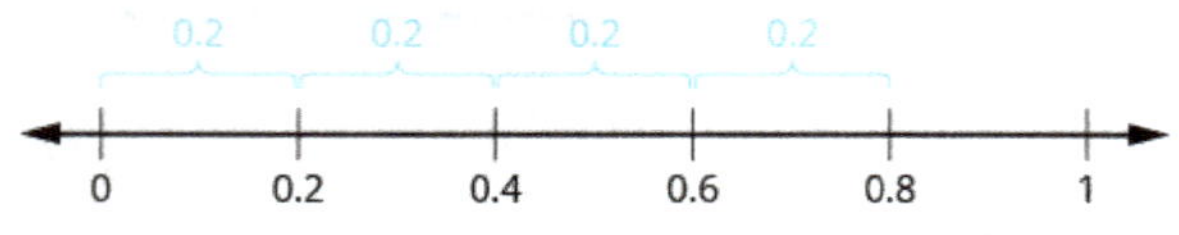

Figure 5.5

Using long division notation, we would write

$$4\overline{)0.8}\quad\text{quotient } 0.2$$

Notice that the decimal point in the quotient is directly above the decimal point in the dividend.

To divide a decimal by a whole number, we place the decimal point in the quotient above the decimal point in the dividend and then divide as usual. Sometimes we need to use extra zeros at the end of the dividend to keep dividing until there is no remainder.

HOW TO : : DIVIDE A DECIMAL BY A WHOLE NUMBER.

Step 1. Write as long division, placing the decimal point in the quotient above the decimal point in the dividend.

Step 2. Divide as usual.

EXAMPLE 5.19

Divide:

Solution

Write as long division, placing the decimal point in the quotient above the decimal point in the dividend.	$3\overline{)0.12}$
Divide as usual. Since 3 does not go into 0 or 1 we use zeros as placeholders.	quotient 0.04; $3\overline{)0.12}$; $\underline{12}$; 0

TRY IT : : 5.37 Divide:

TRY IT : : 5.38 Divide:

In everyday life, we divide whole numbers into decimals—money—to find the price of one item. For example, suppose a case of water bottles cost To find the price per water bottle, we would divide by and round the answer to the nearest cent (hundredth).

EXAMPLE 5.20

Divide:

Solution

Place the decimal point in the quotient above the decimal point in the dividend.	$24\overline{)3.99}$
Divide as usual. When do we stop? Since this division involves money, we round it to the nearest cent (hundredth). To do this, we must carry the division to the thousandths place.	$\begin{array}{r} 0.166 \\ 24\overline{)3.990} \\ \underline{24} \\ 159 \\ \underline{144} \\ 150 \\ \underline{144} \\ 6 \end{array}$
Round to the nearest cent.	

This means the price per bottle is cents.

TRY IT : : 5.39 Divide:

TRY IT : : 5.40 Divide:

Divide a Decimal by Another Decimal

So far, we have divided a decimal by a whole number. What happens when we divide a decimal by another decimal? Let's look at the same multiplication problem we looked at earlier, but in a different way.

Remember, again, that a multiplication problem can be rephrased as a division problem. This time we ask, "How many times does go into Because we can say that goes into four times. This means that divided by is

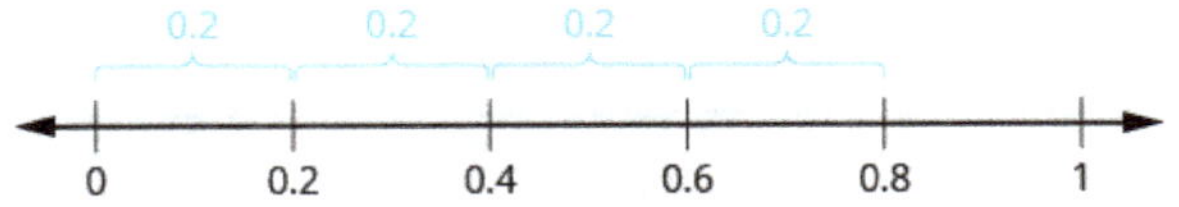

We would get the same answer, if we divide by both whole numbers. Why is this so? Let's think about the division problem as a fraction.

We multiplied the numerator and denominator by and ended up just dividing by To divide decimals, we multiply both the numerator and denominator by the same power of to make the denominator a whole number. Because of the Equivalent Fractions Property, we haven't changed the value of the fraction. The effect is to move the decimal points in the numerator and denominator the same number of places to the right.

We use the rules for dividing positive and negative numbers with decimals, too. When dividing signed decimals, first determine the sign of the quotient and then divide as if the numbers were both positive. Finally, write the quotient with the appropriate sign.

It may help to review the vocabulary for division:

a	$\div$	b	$\frac{a \text{ dividend}}{b \text{ divisor}}$	b	$\overline{)\,a}$
dividend		divisor		divisor	dividend

HOW TO : : DIVIDE DECIMAL NUMBERS.

Step 1. Determine the sign of the quotient.

Step 2. Make the divisor a whole number by moving the decimal point all the way to the right. Move the decimal point in the dividend the same number of places to the right, writing zeros as needed.

Step 3. Divide. Place the decimal point in the quotient above the decimal point in the dividend.

Step 4. Write the quotient with the appropriate sign.

EXAMPLE 5.21

Divide:

Solution

Determine the sign of the quotient.	The quotient will be negative.
Make the divisor the whole number by 'moving' the decimal point all the way to the right. 'Move' the decimal point in the dividend the same number of places to the right.	$3.4\overline{)2.89}$
Divide. Place the decimal point in the quotient above the decimal point in the dividend. Add zeros as needed until the remainder is zero.	$\begin{array}{r} 0.85 \\ 34.\overline{)28.90} \\ \underline{272} \\ 170 \\ \underline{170} \\ 0 \end{array}$
Write the quotient with the appropriate sign.	

TRY IT : : 5.41 Divide:

TRY IT : : 5.42 Divide:

EXAMPLE 5.22

Divide:

Solution

The signs are the same.	The quotient is positive.
Make the divisor a whole number by 'moving' the decimal point all the way to the right. 'Move' the decimal point in the dividend the same number of places.	$0.06\overline{)25.65}$
Divide. Place the decimal point in the quotient above the decimal point in the dividend.	$\begin{array}{r} 427.5 \\ 006.\overline{)2565.0} \\ \underline{-24} \\ 16 \\ \underline{-12} \\ 45 \\ \underline{-42} \\ 30 \\ \underline{30} \end{array}$
Write the quotient with the appropriate sign.	

> TRY IT : : 5.43 Divide:

> TRY IT : : 5.44 Divide:

Now we will divide a whole number by a decimal number.

EXAMPLE 5.23

Divide:

Solution

The signs are the same.	The quotient is positive.
Make the divisor a whole number by 'moving' the decimal point all the way to the right. Move the decimal point in the dividend the same number of places, adding zeros as needed.	$0.05\overline{)4.00}$
Divide. Place the decimal point in the quotient above the decimal point in the dividend.	$\begin{array}{r} 80. \\ 5\overline{)400.} \\ \underline{40} \\ 00 \\ \underline{00} \end{array}$
Write the quotient with the appropriate sign.	

We can relate this example to money. How many nickels are there in four dollars? Because there are nickels in

> TRY IT : : 5.45 Divide:

TRY IT : : 5.46 Divide:

Use Decimals in Money Applications

We often apply decimals in real life, and most of the applications involving money. The Strategy for Applications we used in The Language of Algebra gives us a plan to follow to help find the answer. Take a moment to review that strategy now.

Strategy for Applications

1. Identify what you are asked to find.
2. Write a phrase that gives the information to find it.
3. Translate the phrase to an expression.
4. Simplify the expression.
5. Answer the question with a complete sentence.

EXAMPLE 5.24

Paul received for his birthday. He spent on a video game. How much of Paul's birthday money was left?

Solution

What are you asked to find?	How much did Paul have left?
Write a phrase.	\$50 less \$31.64
Translate.	
Simplify.	18.36
Write a sentence.	Paul has \$18.36 left.

TRY IT : : 5.47

Nicole earned for babysitting her cousins, then went to the bookstore and spent on books and coffee. How much of her babysitting money was left?

TRY IT : : 5.48

Amber bought a pair of shoes for and a purse for The sales tax was How much did Amber spend?

EXAMPLE 5.25

Jessie put gallons of gas in her car. One gallon of gas costs How much does Jessie owe for the gas? (Round the answer to the nearest cent.)

Solution

What are you asked to find?	How much did Jessie owe for all the gas?
Write a phrase.	8 times the cost of one gallon of gas
Translate.	
Simplify.	$28.232
Round to the nearest cent.	$28.23
Write a sentence.	Jessie owes $28.23 for her gas purchase.

TRY IT : : 5.49

Hector put gallons of gas into his car. One gallon of gas costs How much did Hector owe for the gas? Round to the nearest cent.

TRY IT : : 5.50

Christopher bought pizzas for the team. Each pizza cost How much did all the pizzas cost?

EXAMPLE 5.26

Four friends went out for dinner. They shared a large pizza and a pitcher of soda. The total cost of their dinner was If they divide the cost equally, how much should each friend pay?

Solution

What are you asked to find?	How much should each friend pay?
Write a phrase.	$31.76 divided equally among the four friends.
Translate to an expression.	
Simplify.	$7.94
Write a sentence.	Each friend should pay $7.94 for his share of the dinner.

TRY IT : : 5.51

Six friends went out for dinner. The total cost of their dinner was If they divide the bill equally, how much should each friend pay?

TRY IT : : 5.52

Chad worked hours last week and his paycheck was How much does he earn per hour?

Be careful to follow the order of operations in the next example. Remember to multiply before you add.

EXAMPLE 5.27

Marla buys bananas that cost each and oranges that cost each. How much is the total cost of the fruit?

Solution

What are you asked to find?	How much is the total cost of the fruit?
Write a phrase.	6 times the cost of each banana plus 4 times the cost of each orange
Translate to an expression.	
Simplify.	
Add.	$3.28
Write a sentence.	Marla's total cost for the fruit is $3.28.

TRY IT : : 5.53

Suzanne buys cans of beans that cost each and cans of corn that cost each. How much is the total cost of these groceries?

TRY IT : : 5.54

Lydia bought movie tickets for the family. She bought two adult tickets for each and four children's tickets for each. How much did the tickets cost Lydia in all?

LINKS TO LITERACY

The *Links to Literacy* activity "Alexander Who Used to be Rich Last Sunday" will provide you with another view of the topics covered in this section.

MEDIA : : ACCESS ADDITIONAL ONLINE RESOURCES

- Adding and Subtracting Decimals (http://www.openstax.org/l/24addsubdecmls)
- Multiplying Decimals (http://www.openstax.org/l/24multdecmls)
- Multiplying by Powers of Ten (http://www.openstax.org/l/24multpowten)
- Dividing Decimals (http://www.openstax.org/l/24divddecmls)
- Dividing by Powers of Ten (http://www.openstax.org/l/24divddecmlss)

5.2 EXERCISES

Practice Makes Perfect

Add and Subtract Decimals

In the following exercises, add or subtract.

95.

96.

97.

98.

99.

100.

101.

102.

103.

104.

105.

106.

107.

108.

109.

110.

111.

112.

113.

114.

115.

116.

117.

118.

119.

120.

121.

122.

Multiply Decimals

In the following exercises, multiply.

123.

124.

125.

126.

127.

128.

129.

130.

131.

132.

133.

134.

135.

136.

137.

138.

139.

140.

141.

142.

Divide Decimals

In the following exercises, divide.

143.

144.

145.

146.

147.

148.

149.

150.

151.

152.

153.

154.

155.

156.

157.

158.

159.

160.

161.

162.

Mixed Practice

In the following exercises, simplify.

163.

164.

165.

166.

167.

168.

169.

170.

171.

172.

173.

174.

175.

176.

Use Decimals in Money Applications

In the following exercises, use the strategy for applications to solve.

177. Spending money Brenda got from the ATM. She spent on a pair of earrings. How much money did she have left?

178. Spending money Marissa found in her pocket. She spent on a smoothie. How much of the did she have left?

179. Shopping Adam bought a t-shirt for and a book for The sales tax was How much did Adam spend?

180. Restaurant Roberto's restaurant bill was for the entrée and for the drink. He left a tip. How much did Roberto spend?

181. Coupon Emily bought a box of cereal that cost She had a coupon for off, and the store doubled the coupon. How much did she pay for the box of cereal?

182. Coupon Diana bought a can of coffee that cost She had a coupon for off, and the store doubled the coupon. How much did she pay for the can of coffee?

183. Diet Leo took part in a diet program. He weighed pounds at the start of the program. During the first week, he lost pounds. During the second week, he had lost pounds. The third week, he gained pounds. The fourth week, he lost pounds. What did Leo weigh at the end of the fourth week?

184. Snowpack On April the snowpack at the ski resort was meters deep, but the next few days were very warm. By April the snow depth was meters less. On April it snowed and added meters of snow. What was the total depth of the snow?

185. Coffee Noriko bought coffees for herself and her co-workers. Each coffee was How much did she pay for all the coffees?

186. Subway Fare Arianna spends per day on subway fare. Last week she rode the subway days. How much did she spend for the subway fares?

187. Income Mayra earns per hour. Last week she worked hours. How much did she earn?

188. Income Peter earns per hour. Last week he worked hours. How much did he earn?

189. Hourly Wage Alan got his first paycheck from his new job. He worked hours and earned How much does he earn per hour?

190. Hourly Wage Maria got her first paycheck from her new job. She worked hours and earned How much does she earn per hour?

191. Restaurant Jeannette and her friends love to order mud pie at their favorite restaurant. They always share just one piece of pie among themselves. With tax and tip, the total cost is How much does each girl pay if the total number sharing the mud pie is

192. Pizza Alex and his friends go out for pizza and video games once a week. They share the cost of a pizza equally. How much does each person pay if the total number sharing the pizza is

193. Fast Food At their favorite fast food restaurant, the Carlson family orders burgers that cost each and orders of fries at each. What is the total cost of the order?

194. Home Goods Chelsea needs towels to take with her to college. She buys bath towels that cost each and washcloths that cost each. What is the total cost for the bath towels and washcloths?

195. Zoo The Lewis and Chousmith families are planning to go to the zoo together. Adult tickets cost and children's tickets cost What will the total cost be for adults and children?

196. Ice Skating Jasmine wants to have her birthday party at the local ice skating rink. It will cost per child and per adult. What will the total cost be for children and adults?

Everyday Math

197. Paycheck Annie has two jobs. She gets paid per hour for tutoring at City College and per hour at a coffee shop. Last week she tutored for hours and worked at the coffee shop for hours.

How much did she earn?

If she had worked all hours as a tutor instead of working both jobs, how much more would she have earned?

198. Paycheck Jake has two jobs. He gets paid per hour at the college cafeteria and at the art gallery. Last week he worked hours at the cafeteria and hours at the art gallery.

How much did he earn?

If he had worked all hours at the art gallery instead of working both jobs, how much more would he have earned?

Writing Exercises

199. At the 2010 winter Olympics, two skiers took the silver and bronze medals in the Men's Super-G ski event. Miller's time was minute seconds and Weibrecht's time was minute seconds. Find the difference in their times and then write the name of that decimal.

200. Find the quotient of and explain in words all the steps taken.

Self Check

After completing the exercises, use this checklist to evaluate your mastery of the objectives of this section.

I can...	Confidently	With some help	No-I don't get it!
add and subtract decimals.			
multiply decimals.			
divide decimals.			
use decimals in money applications.			

After reviewing this checklist, what will you do to become confident for all objectives?

Decimals and Fractions

Learning Objectives

By the end of this section, you will be able to:

- Convert fractions to decimals
- Order decimals and fractions
- Simplify expressions using the order of operations
- Find the circumference and area of circles

BE PREPARED : : 5.7 Before you get started, take this readiness quiz.
Divide:
If you missed this problem, review Example 5.19.

BE PREPARED : : 5.8 Order using or
If you missed this problem, review Example 5.7.

BE PREPARED : : 5.9 Order using or
If you missed this problem, review Example 5.8.

Convert Fractions to Decimals

In Decimals, we learned to convert decimals to fractions. Now we will do the reverse—convert fractions to decimals. Remember that the fraction bar indicates division. So — can be written or $\overline{)\,}$ This means that we can convert a fraction to a decimal by treating it as a division problem.

Convert a Fraction to a Decimal

To convert a fraction to a decimal, divide the numerator of the fraction by the denominator of the fraction.

EXAMPLE 5.28

Write the fraction — as a decimal.

Solution

A fraction bar means division, so we can write the fraction — using division.	$4\overline{)3}$
Divide.	$\begin{array}{r} 0.75 \\ 4\overline{)3.00} \\ \underline{28} \\ 20 \\ \underline{20} \\ 0 \end{array}$
	So the fraction — is equal to

> **TRY IT : :** 5.55 Write the fraction as a decimal: —

> **TRY IT : :** 5.56 Write the fraction as a decimal: —

EXAMPLE 5.29

Write the fraction — as a decimal.

Solution

The value of this fraction is negative. After dividing, the value of the decimal will be negative. We do the division ignoring the sign, and then write the negative sign in the answer.	—
Divide by	$\begin{array}{r} 3.5 \\ 2\overline{)7.0} \\ \underline{6} \\ 10 \\ \underline{10} \\ 0 \end{array}$
	So, —

> **TRY IT : :** 5.57 Write the fraction as a decimal: —

> **TRY IT : :** 5.58 Write the fraction as a decimal: —

Repeating Decimals

So far, in all the examples converting fractions to decimals the division resulted in a remainder of zero. This is not always the case. Let's see what happens when we convert the fraction — to a decimal. First, notice that — is an improper fraction. Its value is greater than The equivalent decimal will also be greater than

We divide by

$$\begin{array}{r} 1.333... \\ 3\overline{)4.000} \\ \underline{3} \\ 10 \\ \underline{9} \\ 10 \\ \underline{9} \\ 10 \\ \underline{9} \\ 1 \end{array}$$

No matter how many more zeros we write, there will always be a remainder of and the threes in the quotient will go on forever. The number is called a repeating decimal. Remember that the "..." means that the pattern repeats.

Repeating Decimal

A **repeating decimal** is a decimal in which the last digit or group of digits repeats endlessly.

How do you know how many 'repeats' to write? Instead of writing we use a shorthand notation by placing a line over the digits that repeat. The repeating decimal is written The line above the tells you that the repeats endlessly. So

For other decimals, two or more digits might repeat. Table 5.5 shows some more examples of repeating decimals.

	is the repeating digit
	is the repeating digit
	is the repeating block
	is the repeating block

Table 5.5

EXAMPLE 5.30

Write — as a decimal.

Solution

Divide by

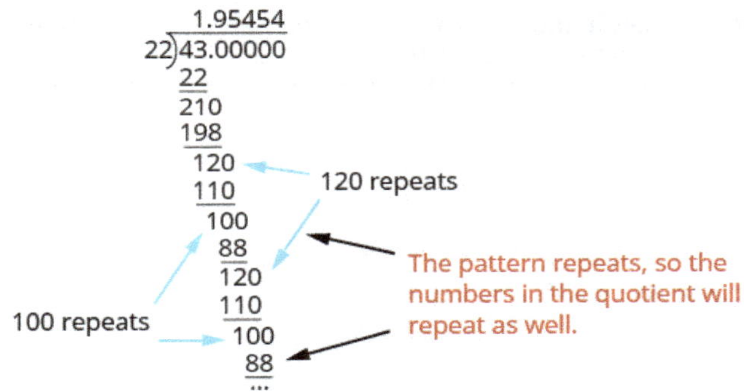

1.95454
22)43.00000
22
210
198
120
110
100
88
120
110
100
88
...
120 repeats
100 repeats
The pattern repeats, so the numbers in the quotient will repeat as well.

Notice that the differences of and repeat, so there is a repeat in the digits of the quotient; will repeat endlessly. The first decimal place in the quotient, is not part of the pattern. So,

—

> **TRY IT : : 5.59** Write as a decimal: —

> **TRY IT : : 5.60** Write as a decimal: —

It is useful to convert between fractions and decimals when we need to add or subtract numbers in different forms. To add a fraction and a decimal, for example, we would need to either convert the fraction to a decimal or the decimal to a fraction.

EXAMPLE 5.31

Simplify: —

Solution

—

Change — to a decimal.	$\begin{array}{r} 0.875 \\ 8\overline{)7.000} \\ \underline{64} \\ 60 \\ \underline{56} \\ 40 \\ \underline{40} \\ 0 \end{array}$
Add.	

TRY IT : : 5.61 Simplify: —

TRY IT : : 5.62 Simplify: —

Order Decimals and Fractions

In Decimals, we compared two decimals and determined which was larger. To compare a decimal to a fraction, we will first convert the fraction to a decimal and then compare the decimals.

EXAMPLE 5.32

Order — using or

Solution

—

Convert — to a decimal.	
Compare to	
Rewrite with the original fraction.	—

TRY IT : : 5.63 Order each of the following pairs of numbers, using or

—

TRY IT : : 5.64 Order each of the following pairs of numbers, using or

—

When ordering negative numbers, remember that larger numbers are to the right on the number line and any positive number is greater than any negative number.

EXAMPLE 5.33

Order — using or

Solution

	—
Convert — to a decimal.	
Compare to .	
Rewrite the inequality with the original fraction.	—

TRY IT : : 5.65 Order each of the following pairs of numbers, using or

—

TRY IT : : 5.66 Order each of the following pairs of numbers, using or

—

EXAMPLE 5.34

Write the numbers — — in order from smallest to largest.

Solution

	— —
Convert the fractions to decimals.	
Write the smallest decimal number first.	
Write the next larger decimal number in the middle place.	
Write the last decimal number (the larger) in the third place.	
Rewrite the list with the original fractions.	— —

TRY IT : : 5.67 Write each set of numbers in order from smallest to largest: — —

TRY IT : : 5.68 Write each set of numbers in order from smallest to largest: — —

Simplify Expressions Using the Order of Operations

The order of operations introduced in Use the Language of Algebra also applies to decimals. Do you remember what the phrase "Please excuse my dear Aunt Sally" stands for?

EXAMPLE 5.35

Simplify the expressions:

—

Solution

Simplify inside parentheses.	
Multiply.	

	—
Simplify inside parentheses.	—
Write as a fraction.	— —
Multiply.	—
Simplify.	

TRY IT : : 5.69 Simplify: —

TRY IT : : 5.70 Simplify: —

EXAMPLE 5.36

Simplify each expression:

—

Solution

Simplify exponents.	
Divide.	
Multiply.	
Add.	
Subtract.	

	—
Simplify exponents.	—
Multiply.	—
Convert — to a decimal.	
Add.	

TRY IT : : 5.71 Simplify:

TRY IT : : 5.72 Simplify: —

Find the Circumference and Area of Circles

The properties of circles have been studied for over years. All circles have exactly the same shape, but their sizes are affected by the length of the **radius**, a line segment from the center to any point on the circle. A line segment that passes through a circle's center connecting two points on the circle is called a **diameter**. The diameter is twice as long as the radius. See Figure 5.6.

The size of a circle can be measured in two ways. The distance around a circle is called its **circumference**.

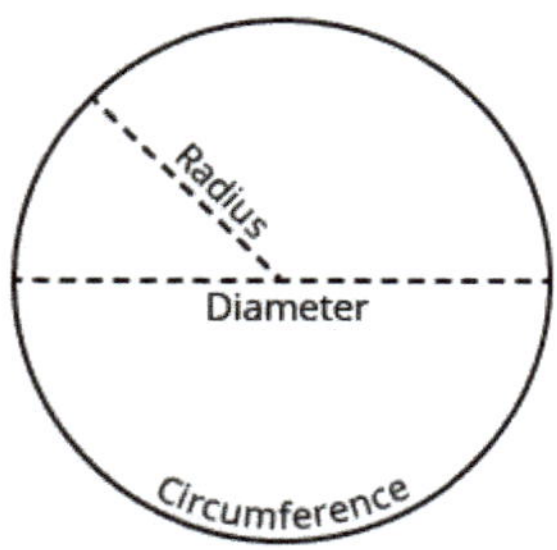

Figure 5.6

Archimedes discovered that for circles of all different sizes, dividing the circumference by the diameter always gives the same number. The value of this number is pi, symbolized by Greek letter (pronounced pie). However, the exact value of cannot be calculated since the decimal never ends or repeats (we will learn more about numbers like this in The Properties of Real Numbers.)

MANIPULATIVE MATHEMATICS

Doing the Manipulative Mathematics activity Pi Lab will help you develop a better understanding of pi.

If we want the exact circumference or area of a circle, we leave the symbol in the answer. We can get an approximate answer by substituting as the value of We use the symbol to show that the result is approximate, not exact.

Properties of Circles

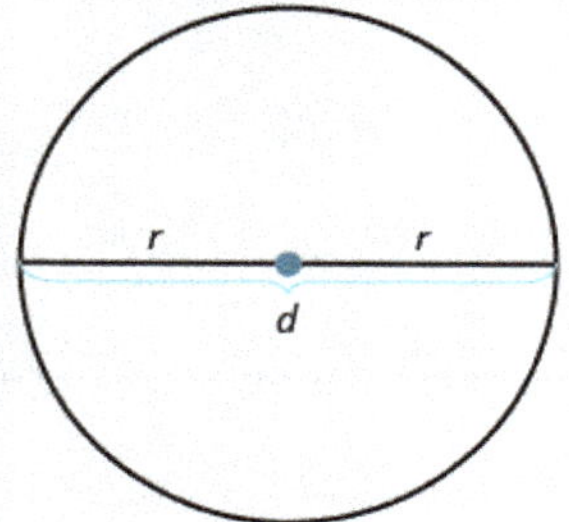

Since the diameter is twice the radius, another way to find the circumference is to use the formula

Suppose we want to find the exact area of a circle of radius inches. To calculate the area, we would evaluate the formula for the area when inches and leave the answer in terms of

We write after the So the exact value of the area is square inches.

To approximate the area, we would substitute

Remember to use square units, such as square inches, when you calculate the area.

EXAMPLE 5.37

A circle has radius centimeters. Approximate its circumference and area.

Solution

Find the circumference when
Write the formula for circumference.
Substitute 3.14 for and 10 for , .
Multiply.

Find the area when
Write the formula for area.
Substitute 3.14 for and 10 for .
Multiply.

TRY IT : : 5.73 A circle has radius inches. Approximate its circumference and area.

TRY IT : : 5.74 A circle has radius feet. Approximate its circumference and area.

EXAMPLE 5.38

A circle has radius centimeters. Approximate its circumference and area.

Solution

Find the circumference when
Write the formula for circumference.
Substitute 3.14 for and 42.5 for
Multiply.

Find the area when .

Write the formula for area.

Substitute 3.14 for and 42.5 for .

Multiply.

> **TRY IT : :** 5.75 A circle has radius centimeters. Approximate its circumference and area.

> **TRY IT : :** 5.76 A circle has radius meters. Approximate its circumference and area.

Approximate with a Fraction

Convert the fraction — to a decimal. If you use your calculator, the decimal number will fill up the display and show But if we round that number to two decimal places, we get the decimal approximation of When we have a circle with radius given as a fraction, we can substitute — for instead of And, since — is also an approximation of we will use the symbol to show we have an approximate value.

EXAMPLE 5.39

A circle has radius — meter. Approximate its circumference and area.

Solution

Find the circumference when —

Write the formula for circumference.

Substitute — for and — for . — —

Multiply. —

Find the area when —

Write the formula for area.

Substitute — for and — for . — —

Multiply. —

TRY IT : : 5.77 A circle has radius —— meters. Approximate its circumference and area.

TRY IT : : 5.78 A circle has radius —— inches. Approximate its circumference and area.

MEDIA : : ACCESS ADDITIONAL ONLINE RESOURCES

- Converting a Fraction to a Decimal - Part 2 (http://www.openstax.org/l/24convfrac2dec)
- Convert a Fraction to a Decimal (repeating) (http://www.openstax.org/l/24convfr2decrep)
- Compare Fractions and Decimals using Inequality Symbols (http://www.openstax.org/l/24compfrcdec)
- Determine the Area of a Circle (http://www.openstax.org/l/24circlearea)
- Determine the Circumference of a Circle (http://www.openstax.org/l/24circlecirc)

5.3 EXERCISES

Practice Makes Perfect

Convert Fractions to Decimals

In the following exercises, convert each fraction to a decimal.

201. —
202. —
203. —
204. —
205. —
206. —
207. —
208. —
209. —
210. —
211. —
212. —
213. —
214. —
215. —
216. —

In the following exercises, simplify the expression.

217. —
218. —
219. —
220. —
221. —
222. —

Order Decimals and Fractions

In the following exercises, order each pair of numbers, using or

223. —
224. —
225. —
226. —
227. —
228. —
229. —
230. —
231. —
232. —
233. —
234. —

In the following exercises, write each set of numbers in order from least to greatest.

235. — —
236. — —
237. — —
238. — —
239. — —
240. — —
241. — —
242. — —

Simplify Expressions Using the Order of Operations

In the following exercises, simplify.

243.

244.

245.

246.

247. —

248. —

249. —

250. —

251.

252.

253.

254.

255. —

256. —

257. — —

258. — —

Mixed Practice

In the following exercises, simplify. Give the answer as a decimal.

259. —

260. —

261. —

262. —

263. — —

264. — —

265. —

266. —

Find the Circumference and Area of Circles

In the following exercises, approximate the circumference and area of each circle. If measurements are given in fractions, leave answers in fraction form.

267.

268.

269.

270.

271.

272.

273.

274.

275. —

276. —

277. —

278. —

279. —

280. —

Everyday Math

281. Kelly wants to buy a pair of boots that are on sale for — of the original price. The original price of the boots is What is the sale price of the shoes?

282. An architect is planning to put a circular mosaic in the entry of a new building. The mosaic will be in the shape of a circle with radius of feet. How many square feet of tile will be needed for the mosaic? (Round your answer up to the next whole number.)

Writing Exercises

283. Is it easier for you to convert a decimal to a fraction or a fraction to a decimal? Explain.

284. Describe a situation in your life in which you might need to find the area or circumference of a circle.

Self Check

After completing the exercises, use this checklist to evaluate your mastery of the objectives of this section.

I can…	Confidently	With some help	No-I don't get it!
convert fractions to decimals.			
order decimals and fractions.			
simplify expressions using the order of operations.			
find the circumference and area of circles.			

What does this checklist tell you about your mastery of this section? What steps will you take to improve?

Solve Equations with Decimals

Learning Objectives

By the end of this section, you will be able to:

- Determine whether a decimal is a solution of an equation
- Solve equations with decimals
- Translate to an equation and solve

BE PREPARED : : 5.10 Before you get started, take this readiness quiz.

Evaluate — —

If you missed this problem, review Example 4.77.

BE PREPARED : : 5.11 Evaluate when

If you missed this problem, review Example 3.41.

BE PREPARED : : 5.12 Solve —

If you missed this problem, review Example 4.99.

Determine Whether a Decimal is a Solution of an Equation

Solving equations with decimals is important in our everyday lives because money is usually written with decimals. When applications involve money, such as shopping for yourself, making your family's budget, or planning for the future of your business, you'll be solving equations with decimals.

Now that we've worked with decimals, we are ready to find solutions to equations involving decimals. The steps we take to determine whether a number is a solution to an equation are the same whether the solution is a whole number, an integer, a fraction, or a decimal. We'll list these steps here again for easy reference.

HOW TO : : DETERMINE WHETHER A NUMBER IS A SOLUTION TO AN EQUATION.

Step 1. Substitute the number for the variable in the equation.

Step 2. Simplify the expressions on both sides of the equation.

Step 3. Determine whether the resulting equation is true.

- If so, the number is a solution.
- If not, the number is not a solution.

EXAMPLE 5.40

Determine whether each of the following is a solution of

Solution

	$x - 0.7 = 1.5$
Substitute 1 for x.	$1 - 0.7 \stackrel{?}{=} 1.5$
Subtract.	$0.3 \neq 1.5$

Since does not result in a true equation, is not a solution to the equation.

	$x - 0.7 = 1.5$
Substitute −0.8 for x.	$-0.8 - 0.7 \stackrel{?}{=} 1.5$
Subtract.	$-1.5 \neq 1.5$

Since does not result in a true equation, is not a solution to the equation.

	$x - 0.7 = 1.5$
Substitute 2.2 for x.	$2.2 - 0.7 \stackrel{?}{=} 1.5$
Subtract.	$1.5 = 1.5$ ✓

Since results in a true equation, is a solution to the equation.

TRY IT : : 5.79 Determine whether each value is a solution of the given equation.

TRY IT : : 5.80 Determine whether each value is a solution of the given equation.

Solve Equations with Decimals

In previous chapters, we solved equations using the Properties of Equality. We will use these same properties to solve equations with decimals.

Properties of Equality

Subtraction Property of Equality For any numbers If then	**Addition Property of Equality** For any numbers If then
The Division Property of Equality For any numbers If then — —	**The Multiplication Property of Equality** For any numbers If then

When you add, subtract, multiply or divide the same quantity from both sides of an equation, you still have equality.

EXAMPLE 5.41

Solve:

Solution

We will use the Subtraction Property of Equality to isolate the variable.

	$y + 2.3 = -4.7$
Subtract 2.3 from each side, to undo the addition.	$y + 2.3 - 2.3 = -4.7 - 2.3$
Simplify.	$y = -7$
Check:	$y + 2.3 = -4.7$
Substitute $y = -7$.	$-7 + 2.3 \stackrel{?}{=} -4.7$
Simplify.	$-4.7 = -4.7$ ✓

Since makes a true statement, we know we have found a solution to this equation.

TRY IT : : 5.81 Solve:

TRY IT : : 5.82 Solve:

EXAMPLE 5.42

Solve:

Solution

We will use the Addition Property of Equality.

	$a - 4.75 = -1.39$
Add 4.75 to each side, to undo the subtraction.	$a - 4.75 + 4.75 = -1.39 + 4.75$
Simplify.	$a = 3.36$
Check:	$a - 4.75 = -1.39$
Substitute $a = 3.36$.	$3.36 - 4.75 \stackrel{?}{=} -1.39$
	$-1.39 = -1.39$ ✓

Since the result is a true statement, is a solution to the equation.

TRY IT : : 5.83 Solve:

TRY IT : : 5.84 Solve:

EXAMPLE 5.43

Solve:

Solution

We will use the Division Property of Equality.

Use the Properties of Equality to find a value for

	$-4.8 = 0.8n$
We must divide both sides by 0.8 to isolate *n*.	$\frac{-4.8}{0.8} = \frac{0.8n}{0.8}$
Simplify.	$-6 = n$
Check:	$-4.8 = 0.8n$
Substitute $n = -6$.	$-4.8 \stackrel{?}{=} 0.8(-6)$
	$-4.8 = -4.8$ ✓

Since makes a true statement, we know we have a solution.

TRY IT : : 5.85 Solve:

TRY IT : : 5.86 Solve:

EXAMPLE 5.44

Solve:

Solution

We will use the Multiplication Property of Equality.

	$\frac{p}{-1.8} = -6.5$
Here, *p* is divided by −1.8. We must multiply by −1.8 to isolate *p*	$-1.8\left(\frac{p}{-1.8}\right) = -1.8(-6.5)$
Multiply.	$p = 11.7$
Check:	$\frac{p}{-1.8} = -6.5$
Substitute $p = 11.7$.	$\frac{11.7}{-1.8} \stackrel{?}{=} -6.5$
	$-6.5 = -6.5$ ✓

A solution to is

TRY IT : : 5.87 Solve:

TRY IT : : 5.88 Solve:

Translate to an Equation and Solve

Now that we have solved equations with decimals, we are ready to translate word sentences to equations and solve. Remember to look for words and phrases that indicate the operations to use.

EXAMPLE 5.45

Translate and solve: The difference of and is

Solution

Translate.	The difference of n and 4.3 is 2.1. $n - 4.3 = 2.1$
Add to both sides of the equation.	$n - 4.3 + 4.3 = 2.1 + 4.3$
Simplify.	$n = 6.4$
Check: Is the difference of and 4.3 equal to 2.1?	
Let : Is the difference of 6.4 and 4.3 equal to 2.1?	
Translate.	$6.4 - 4.3 \stackrel{?}{=} 2.1$
Simplify.	$2.1 = 2.1$ ✓

TRY IT : : 5.89 Translate and solve: The difference of and is

TRY IT : : 5.90 Translate and solve: The difference of and is

EXAMPLE 5.46

Translate and solve: The product of and is

Solution

Translate.	The product of 3.1 and x is 5.27. $-3.1x = 5.27$
Divide both sides by .	$\frac{-3.1x}{-3.1} = \frac{5.27}{-3.1}$
Simplify.	$x = -1.7$
Check: Is the product of −3.1 and equal to ?	
Let : Is the product of and equal to ?	
Translate.	$-3.1(-1.7) \stackrel{?}{=} 5.27$
Simplify.	$5.27 = 5.27$ ✓

> **TRY IT : :** 5.91 Translate and solve: The product of and is

> **TRY IT : :** 5.92 Translate and solve: The product of and is

EXAMPLE 5.47

Translate and solve: The quotient of and is

Solution

Translate.		The quotient of p and −2.4 is 6.5. $\frac{p}{-2.4} = 6.5$
Multiply both sides by .		$-2.4\left(\frac{p}{-2.4}\right) = -2.4(6.5)$
Simplify.		$p = -15.6$
Check:	Is the quotient of and equal to ?	
Let	Is the quotient of and equal to ?	
Translate.	$\frac{-15.6}{-2.4} \stackrel{?}{=} 6.5$	
Simplify.	$6.5 = 6.5$ ✓	

> **TRY IT : :** 5.93 Translate and solve: The quotient of and is

> **TRY IT : :** 5.94 Translate and solve: The quotient of and is

EXAMPLE 5.48

Translate and solve: The sum of and is

Solution

Translate.		The sum of n and 2.9 is 1.7. $n + 2.9 = 1.7$
Subtract from each side.		$n + 2.9 - 2.9 = 1.7 - 2.9$
Simplify.		$n = -1.2$
Check:	Is the sum and equal to ?	
Let	Is the sum and equal to ?	
Translate.	$-1.2 + 2.9 \stackrel{?}{=} 1.7$	
Simplify.	$1.7 = 1.7$ ✓	

TRY IT : : 5.95 Translate and solve: The sum of and is

TRY IT : : 5.96 Translate and solve: The sum of and is

MEDIA : : ACCESS ADDITIONAL ONLINE RESOURCES

- Solving One Step Equations Involving Decimals (http://openstaxcollege.org/l/24eqwithdec)
- Solve a One Step Equation With Decimals by Adding and Subtracting (http://openstaxcollege.org/l/24eqnwdecplsmin)
- Solve a One Step Equation With Decimals by Multiplying (http://openstaxcollege.org/l/24eqnwdecmult)
- Solve a One Step Equation With Decimals by Dividing (http://openstaxcollege.org/l/24eqnwdecdiv)

5.4 EXERCISES

Practice Makes Perfect

Determine Whether a Decimal is a Solution of an Equation

In the following exercises, determine whether each number is a solution of the given equation.

285.

286.

287. —

288.

Solve Equations with Decimals

In the following exercises, solve the equation.

289.

290.

291.

292.

293.

294.

295.

296.

297.

298.

299.

300.

301.

302.

303.

304.

305.

306.

307.

308.

309.

310.

311.

312.

313.

314.

315.

316.

317. —

318. —

319. —

320. —

321. —

322. —

323. —

324. —

Mixed Practice

In the following exercises, solve the equation. Then check your solution.

325.

326. – –

327.

328. – —

329.

330.

331. – —

332. —

333. – —

334. —

335. – –

336.

337. —

338.

339.

340.

341.

342. – —

343. —

344.

Translate to an Equation and Solve

In the following exercises, translate and solve.

345. The difference of and is

346. The difference and is

347. The product of and is

348. The product of and is

349. The quotient of and is

350. The quotient of and is

351. The sum of and is

352. The sum of and is

Everyday Math

353. Shawn bought a pair of shoes on sale for . Solve the equation to find the original price of the shoes,

354. Mary bought a new refrigerator. The total price including sales tax was Find the retail price, of the refrigerator before tax by solving the equation

Writing Exercises

355. Think about solving the equation but do not actually solve it. Do you think the solution should be greater than or less than Explain your reasoning. Then solve the equation to see if your thinking was correct.

356. Think about solving the equation but do not actually solve it. Do you think the solution should be greater than or less than Explain your reasoning. Then solve the equation to see if your thinking was correct.

Self Check

After completing the exercises, use this checklist to evaluate your mastery of the objectives of this section.

I can...	Confidently	With some help	No-I don't get it!
determine whether a decimal is a solution of an equation.			
solve equations with decimals.			
translate to an equation and solve.			

On a scale of 1-10, how would you rate your mastery of this section in light of your responses on the checklist? How can you improve this?

5.5 Averages and Probability

Learning Objectives

By the end of this section, you will be able to:

- Calculate the mean of a set of numbers
- Find the median of a set of numbers
- Find the mode of a set of numbers
- Apply the basic definition of probability

BE PREPARED : : 5.13 Before you get started, take this readiness quiz.

Simplify:

If you missed this problem, review Example 4.48.

BE PREPARED : : 5.14 Simplify:

If you missed this problem, review Example 2.8.

BE PREPARED : : 5.15 Convert — to a decimal.

If you missed this problem, review Example 5.28.

One application of decimals that arises often is finding the *average* of a set of numbers. What do you think of when you hear the word *average*? Is it your grade point average, the average rent for an apartment in your city, the batting average of a player on your favorite baseball team? The average is a typical value in a set of numerical data. Calculating an average sometimes involves working with decimal numbers. In this section, we will look at three different ways to calculate an average.

Calculate the Mean of a Set of Numbers

The mean is often called the arithmetic average. It is computed by dividing the sum of the values by the number of values. Students want to know the mean of their test scores. Climatologists report that the mean temperature has, or has not, changed. City planners are interested in the mean household size.

Suppose Ethan's first three test scores were To find the mean score, he would add them and divide by

His mean test score is points.

The Mean

The **mean** of a set of numbers is the arithmetic average of the numbers.

HOW TO : : CALCULATE THE MEAN OF A SET OF NUMBERS.

Step 1. Write the formula for the mean

Step 2. Find the sum of all the values in the set. Write the sum in the numerator.

Step 3. Count the number, of values in the set. Write this number in the denominator.

Step 4. Simplify the fraction.

Step 5. Check to see that the mean is reasonable. It should be greater than the least number and less than the greatest number in the set.

EXAMPLE 5.49

Find the mean of the numbers

 **Solution**

Write the formula for the mean:	
Write the sum of the numbers in the numerator.	
Count how many numbers are in the set. There are 5 numbers in the set, so .	
Add the numbers in the numerator.	
Then divide.	
Check to see that the mean is 'typical': 10 is neither less than 6 nor greater than 15.	The mean is 10.

TRY IT : : 5.97 Find the mean of the numbers:

TRY IT : : 5.98 Find the mean of the numbers:

EXAMPLE 5.50

The ages of the members of a family who got together for a birthday celebration were years. Find the mean age.

Solution

Write the formula for the mean:	
Write the sum of the numbers in the numerator.	
Count how many numbers are in the set. Call this and write it in the denominator.	
Simplify the fraction.	

Is 'typical'? Yes, it is neither less than nor greater than The mean age is years.

TRY IT : : 5.99

The ages of the four students in Ben's carpool are Find the mean age of the students.

TRY IT : : 5.100

Yen counted the number of emails she received last week. The numbers were Find the mean number of emails.

Did you notice that in the last example, while all the numbers were whole numbers, the mean was a number with one decimal place? It is customary to report the mean to one more decimal place than the original numbers. In the next example, all the numbers represent money, and it will make sense to report the mean in dollars and cents.

EXAMPLE 5.51

For the past four months, Daisy's cell phone bills were Find the mean cost of Daisy's cell phone bills.

Solution

Write the formula for the mean.	
Count how many numbers are in the set. Call this and write it in the denominator.	
Write the sum of all the numbers in the numerator.	
Simplify the fraction.	

Does seem 'typical' of this set of numbers? Yes, it is neither less than nor greater than The mean cost of her cell phone bill was

TRY IT : : 5.101

Last week Ray recorded how much he spent for lunch each workday. He spent
Find the mean of how much he spent each day.

TRY IT : : 5.102

Lisa has kept the receipts from the past four trips to the gas station. The receipts show the following amounts:
Find the mean.

Find the Median of a Set of Numbers

When Ann, Bianca, Dora, Eve, and Francine sing together on stage, they line up in order of their heights. Their heights, in inches, are shown in Table 5.6.

Ann	Bianca	Dora	Eve	Francine

Table 5.6

Dora is in the middle of the group. Her height, is the *median* of the girls' heights. Half of the heights are less than or equal to Dora's height, and half are greater than or equal. The median is the middle value.

Median

The **median** of a set of data values is the middle value.

- Half the data values are less than or equal to the median.
- Half the data values are greater than or equal to the median.

What if Carmen, the pianist, joins the singing group on stage? Carmen is inches tall, so she fits in the height order between Bianca and Dora. Now the data set looks like this:

There is no single middle value. The heights of the six girls can be divided into two equal parts.

Statisticians have agreed that in cases like this the median is the mean of the two values closest to the middle. So the median is the mean of ——— The median height is inches.

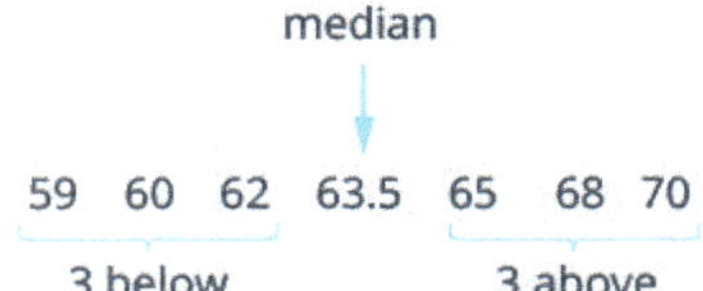

Notice that when the number of girls was the median was the third height, but when the number of girls was the median was the mean of the third and fourth heights. In general, when the number of values is odd, the median will be the one value in the middle, but when the number is even, the median is the mean of the two middle values.

HOW TO : : FIND THE MEDIAN OF A SET OF NUMBERS.

Step 1. List the numbers from smallest to largest.

Step 2. Count how many numbers are in the set. Call this

Step 3. Is odd or even?

- If is an odd number, the median is the middle value.
- If is an even number, the median is the mean of the two middle values.

EXAMPLE 5.52

Find the median of

Solution

List the numbers in order from smallest to largest.	9, 11, 12, 13, 15, 18, 19
Count how many numbers are in the set. Call this .	
Is odd or even?	odd
The median is the middle value.	median 9, 11, 12, **13,** 15, 18, 19 3 below 3 above
The middle is the number in the 4th position.	So the median of the data is 13.

TRY IT : : 5.103 Find the median of the data set:

TRY IT : : 5.104 Find the median of the data set:

EXAMPLE 5.53

Kristen received the following scores on her weekly math quizzes:

Find her median score.

Solution

Find the median of 83, 79, 85, 86, 92, 100, 76, 90, 88, and 64.	
List the numbers in order from smallest to largest.	64, 76, 79, 83, 85, 86, 88, 90, 92, 100
Count the number of data values in the set. Call this	
Is odd or even?	even
The median is the mean of the two middle values, the 5th and 6th numbers.	64, 76, 79, 83, **85**, **86**, 88, 90, 92, 100 5 numbers 5 numbers
Find the mean of 85 and 86.	
	Kristen's median score is 85.5.

TRY IT : : 5.105 Find the median of the data set:

TRY IT : : 5.106 Find the median of the data set:

Identify the Mode of a Set of Numbers

The *average* is one number in a set of numbers that is somehow typical of the whole set of numbers. The mean and median are both often called the average. Yes, it can be confusing when the word average refers to two different numbers, the mean and the median! In fact, there is a third number that is also an average. This average is the **mode**. The mode of a set of numbers is the number that occurs the most. The **frequency**, is the number of times a number occurs. So the mode of a set of numbers is the number with the highest frequency.

Mode

The **mode** of a set of numbers is the number with the highest frequency.

Suppose Jolene kept track of the number of miles she ran since the start of the month, as shown in Figure 5.7.

Sunday	Monday	Tuesday	Wednesday	Thursday	Friday	Saturday
				1 2 mi New Year's day	2	3 15 mi
4 8 mi	5	6 3 mi	7 8 mi	8	9 5 mi	10 8 mi
11	12	13	14	15	16	17

Figure 5.7

If we list the numbers in order it is easier to identify the one with the highest frequency.

Jolene ran miles three times, and every other distance is listed only once. So the mode of the data is miles.

HOW TO : : IDENTIFY THE MODE OF A SET OF NUMBERS.

Step 1. List the data values in numerical order.

Step 2. Count the number of times each value appears.

Step 3. The mode is the value with the highest frequency.

EXAMPLE 5.54

The ages of students in a college math class are listed below. Identify the mode.

Solution

The ages are already listed in order. We will make a table of frequencies to help identify the age with the highest frequency.

Age	18	19	20	21	22	23	24	25	29	30	40	44
Frequency	4	3	7	2	5	1	2	1	1	1	1	1

Now look for the highest frequency. The highest frequency is which corresponds to the age So the mode of the ages in this class is years.

TRY IT : : 5.107

The number of sick days employees used last year: Identify the mode.

TRY IT : : 5.108

The number of handbags owned by women in a book club: Identify the mode.

EXAMPLE 5.55

The data lists the heights (in inches) of students in a statistics class. Identify the mode.

Solution

List each number with its frequency.

Number	56	60	61	62	63	64	65	66	67	70	74
Frequency	1	2	2	1	5	4	3	5	6	1	1

Now look for the highest frequency. The highest frequency is which corresponds to the height inches. So the mode of this set of heights is inches.

TRY IT : : 5.109

The ages of the students in a statistics class are listed here: , , , , , , , , , , , , , , , , , , , . What is the mode?

TRY IT : : 5.110

Students listed the number of members in their household as follows: , , , , , , , , , , , , , , , , , , . What is the mode?

Some data sets do not have a mode because no value appears more than any other. And some data sets have more than one mode. In a given set, if two or more data values have the same highest frequency, we say they are all modes.

Use the Basic Definition of Probability

The probability of an event tells us how likely that event is to occur. We usually write probabilities as fractions or decimals. For example, picture a fruit bowl that contains five pieces of fruit - three bananas and two apples.

If you want to choose one piece of fruit to eat for a snack and don't care what it is, there is a — probability you will choose a banana, because there are three bananas out of the total of five pieces of fruit. The probability of an event is the number of favorable outcomes divided by the total number of outcomes.

$$\text{Probability of an event} = \frac{\text{number of favorable outcomes}}{\text{total number of outcomes}}$$

$$\text{Probability of choosing a banana} = \frac{3}{5}$$

There are 3 bananas.
There are 5 pieces of fruit.

Probability

The **probability** of an event is the number of favorable outcomes divided by the total number of outcomes possible.

Converting the fraction — to a decimal, we would say there is a probability of choosing a banana.

—

This basic definition of probability assumes that all the outcomes are equally likely to occur. If you study probabilities in a later math class, you'll learn about several other ways to calculate probabilities.

EXAMPLE 5.56

The ski club is holding a raffle to raise money. They sold tickets. All of the tickets are placed in a jar. One ticket will be pulled out of the jar at random, and the winner will receive a prize. Cherie bought one raffle ticket.

Find the probability she will win the prize.

Convert the fraction to a decimal.

Solution

What are you asked to find?	The probability Cherie wins the prize.
What is the number of favorable outcomes?	1, because Cherie has 1 ticket.
Use the definition of probability.	
Substitute into the numerator and denominator.	

Convert the fraction to a decimal.	
Write the probability as a fraction.	
Convert the fraction to a decimal.	

TRY IT : : 5.111

Ignaly is attending a fashion show where the guests are seated at tables of ten. One guest from each table will be selected at random to receive a door prize. Find the probability Ignaly will win the door prize for her table. Convert the fraction to a decimal.

TRY IT : : 5.112

Hoang is among people available to sit on a jury. One person will be chosen at random from the Find the probability Hoang will be chosen. Convert the fraction to a decimal.

EXAMPLE 5.57

Three women and five men interviewed for a job. One of the candidates will be offered the job.

Find the probability the job is offered to a woman.

Convert the fraction to a decimal.

Solution

What are you asked to find?	The probability the job is offered to a woman.
What is the number of favorable outcomes?	3, because there are three women.
What are the total number of outcomes?	8, because 8 people interviewed.
Use the definition of probability.	
Substitute into the numerator and denominator.	
Convert the fraction to a decimal.	
Write the probability as a fraction.	
Convert the fraction to a decimal.	

TRY IT : : 5.113

A bowl of Halloween candy contains chocolate candies and lemon candies. Tanya will choose one piece of candy at random. Find the probability Tanya will choose a chocolate candy. Convert the fraction to a decimal.

TRY IT : : 5.114

Dan has pairs of black socks and pairs of blue socks. He will choose one pair at random to wear tomorrow. Find the probability Dan will choose a pair of black socks Convert the fraction to a decimal.

MEDIA : : ACCESS ADDITIONAL ONLINE RESOURCES

- Mean, Median, and Mode (http://www.openstax.org/l/24meanmedmode)
- Find the Mean of a Data Set (http://www.openstax.org/l/24meandataset)
- Find the Median of a Data Set (http://www.openstax.org/l/24meddataset)
- Find the Mode of a Data Set (http://www.openstax.org/l/24modedataset)

5.5 EXERCISES

Practice Makes Perfect

Calculate the Mean of a Set of Numbers

In the following exercises, find the mean.

357. , , , ,

358. , , , , ,

359. , , , , ,

360. , , , , and

361. , , ,

362. , , ,

363. , , , , ,

364. , , , , , ,

365. Four girls leaving a mall were asked how much money they had just spent. The amounts were , , , and . Find the mean amount of money spent.

366. Juan bought shirts to wear to his new job. The costs of the shirts were , , , , and . Find the mean cost.

367. The number of minutes it took Jim to ride his bike to school for each of the past six days was , , , , , and . Find the mean number of minutes.

368. Norris bought six books for his classes this semester. The costs of the books were , , , , , and . Find the mean cost.

369. The top eight hitters in a softball league have batting averages of , , , , , , , and . Find the mean of the batting averages. Round your answer to the nearest thousandth.

370. The monthly snowfall at a ski resort over a six-month period was and inches. Find the mean snowfall.

Find the Median of a Set of Numbers

In the following exercises, find the median.

371. , , , ,

372. , , , ,

373. , , , , , , , ,

374. , , , , , , ,

375. , , , , , , ,

376. , , , , , , ,

377. , , , , ,

378. , , , , , , ,

379. Last week Ray recorded how much he spent for lunch each workday. He spent , , , , and . Find the median.

380. Michaela is in charge of 6 two-year olds at a daycare center. Their ages, in months, are , , , , , and . Find the median age.

381. Brian is teaching a swim class for three-year olds. Their ages, in months, are Find the median age.

382. Sal recorded the amount he spent for gas each week for the past weeks. The amounts were and Find the median amount.

Identify the Mode of a Set of Numbers

In the following exercises, identify the mode.

383. , , , , , , , , , ,

384. , , , , , , , , , ,

385. , , , , , , , , , , , , , , , , , , ,

386. , , , , , , , , , , , , , , , , , , ,

387. The number of children per house on one block: , , , , , , , , , , , ,

388. The number of movies watched each month last year: , , , , , , , , , , ,

389. The number of units being taken by students in one class: , , , , , , , , , , , .

390. The number of hours of sleep per night for the past two weeks: , , , , , , , , , , , , .

Use the Basic Definition of Probability

In the following exercises, express the probability as both a fraction and a decimal. (Round to three decimal places, if necessary.)

391. Josue is in a book club with members. One member is chosen at random each month to select the next month's book. Find the probability that Josue will be chosen next month.

392. Jessica is one of eight kindergarten teachers at Mandela Elementary School. One of the kindergarten teachers will be selected at random to attend a summer workshop. Find the probability that Jessica will be selected.

393. There are people who work in Dane's department. Next week, one person will be selected at random to bring in doughnuts. Find the probability that Dane will be selected. Round your answer to the nearest thousandth.

394. Monica has two strawberry yogurts and six banana yogurts in her refrigerator. She will choose one yogurt at random to take to work. Find the probability Monica will choose a strawberry yogurt.

395. Michel has four rock CDs and six country CDs in his car. He will pick one CD to play on his way to work. Find the probability Michel will pick a rock CD.

396. Noah is planning his summer camping trip. He can't decide among six campgrounds at the beach and twelve campgrounds in the mountains, so he will choose one campground at random. Find the probability that Noah will choose a campground at the beach.

397. Donovan is considering transferring to a He is considering out-of state colleges and colleges in his state. He will choose one college at random to visit during spring break. Find the probability that Donovan will choose an out-of-state college.

398. There are number combinations possible in the Mega Millions lottery. One winning jackpot ticket will be chosen at random. Brent chooses his favorite number combination and buys one ticket. Find the probability Brent will win the jackpot. Round the decimal to the first digit that is not zero, then write the name of the decimal.

Everyday Math

399. Joaquin gets paid every Friday. His paychecks for the past Fridays were Find the mean, median, and mode.

400. The cash register receipts each day last week at a coffee shop were Find the mean, median, and mode.

Writing Exercises

401. Explain in your own words the difference between the mean, median, and mode of a set of numbers.

402. Make an example of probability that relates to your life. Write your answer as a fraction and explain what the numerator and denominator represent.

Self Check

After completing the exercises, use this checklist to evaluate your mastery of the objectives of this section.

I can...	Confidently	With some help	No-I don't get it!
calculate the mean of a set of numbers.			
find the median of a set of numbers.			
find the mode of a set of numbers.			
use the basic definition of probability.			

After looking at the checklist, do you think you are well prepared for the next section? Why or why not?

5.6 Ratios and Rate

Learning Objectives

By the end of this section, you will be able to:

- Write a ratio as a fraction
- Write a rate as a fraction
- Find unit rates
- Find unit price
- Translate phrases to expressions with fractions

BE PREPARED : : 5.16

Before you get started, take this readiness quiz.

Simplify: —

If you missed this problem, review Example 4.19.

BE PREPARED : : 5.17

Divide:

If you missed this problem, review Example 5.19.

BE PREPARED : : 5.18

Simplify: —

If you missed this problem, review Example 4.43.

Write a Ratio as a Fraction

When you apply for a mortgage, the loan officer will compare your total debt to your total income to decide if you qualify for the loan. This comparison is called the debt-to-income ratio. A **ratio** compares two quantities that are measured with the same unit. If we compare and , the ratio is written as —

Ratios

A **ratio** compares two numbers or two quantities that are measured with the same unit. The ratio of to is written —

In this section, we will use the fraction notation. When a ratio is written in fraction form, the fraction should be simplified. If it is an improper fraction, we do not change it to a mixed number. Because a ratio compares two quantities, we would leave a ratio as — instead of simplifying it to so that we can see the two parts of the ratio.

EXAMPLE 5.58

Write each ratio as a fraction:

Solution

Write as a fraction with the first number in the numerator and the second in the denominator.	—
Simplify the fraction.	–

Write as a fraction with the first number in the numerator and the second in the denominator.	—
Simplify.	–

We leave the ratio in as an improper fraction.

TRY IT : : 5.115 Write each ratio as a fraction:

TRY IT : : 5.116 Write each ratio as a fraction:

Ratios Involving Decimals

We will often work with ratios of decimals, especially when we have ratios involving money. In these cases, we can eliminate the decimals by using the Equivalent Fractions Property to convert the ratio to a fraction with whole numbers in the numerator and denominator.

For example, consider the ratio We can write it as a fraction with decimals and then multiply the numerator and denominator by to eliminate the decimals.

$$\frac{0.8}{0.05}$$

$$\frac{(0.8)100}{(0.05)100}$$

$$\frac{80}{5}$$

Do you see a shortcut to find the equivalent fraction? Notice that — and — The least common denominator of — and — is By multiplying the numerator and denominator of — by we 'moved' the decimal two places to the right to get the equivalent fraction with no decimals. Now that we understand the math behind the process, we can find the fraction with no decimals like this:

$$\frac{0.80}{0.05}$$

"Move" the decimal 2 places.	—
Simplify.	—

You do not have to write out every step when you multiply the numerator and denominator by powers of ten. As long as you move both decimal places the same number of places, the ratio will remain the same.

EXAMPLE 5.59

Write each ratio as a fraction of whole numbers:

Solution

Write as a fraction.	——
Rewrite as an equivalent fraction without decimals, by moving both decimal points 1 place to the right.	——
Simplify.	–

So is equivalent to –

The numerator has one decimal place and the denominator has To clear both decimals we need to move the decimal places to the right.

Write as a fraction.	——
Move both decimals right two places.	——
Simplify.	–

So is equivalent to –

TRY IT : : 5.117 Write each ratio as a fraction:

TRY IT : : 5.118 Write each ratio as a fraction:

Some ratios compare two mixed numbers. Remember that to divide mixed numbers, you first rewrite them as improper fractions.

EXAMPLE 5.60

Write the ratio of — — as a fraction.

Solution

	— —
Write as a fraction.	—
Convert the numerator and denominator to improper fractions.	—
Rewrite as a division of fractions.	— —
Invert the divisor and multiply.	— —
Simplify.	—

TRY IT : : 5.119 Write each ratio as a fraction: — —

TRY IT : : 5.120 Write each ratio as a fraction: — —

Applications of Ratios

One real-world application of ratios that affects many people involves measuring cholesterol in blood. The ratio of total cholesterol to HDL cholesterol is one way doctors assess a person's overall health. A ratio of less than to is considered good.

EXAMPLE 5.61

Hector's total cholesterol is mg/dl and his HDL cholesterol is mg/dl. Find the ratio of his total cholesterol to his HDL cholesterol. Assuming that a ratio less than to is considered good, what would you suggest to Hector?

Solution

First, write the words that express the ratio. We want to know the ratio of Hector's total cholesterol to his HDL cholesterol.

Write as a fraction.	
Substitute the values.	
Simplify.	

Is Hector's cholesterol ratio ok? If we divide by we obtain approximately so Hector's cholesterol ratio is high! Hector should either lower his total cholesterol or raise his HDL cholesterol.

TRY IT : : 5.121 Find the patient's ratio of total cholesterol to HDL cholesterol using the given information.

Total cholesterol is mg/dL and HDL cholesterol is mg/dL.

TRY IT : : 5.122 Find the patient's ratio of total cholesterol to HDL cholesterol using the given information.

Total cholesterol is mg/dL and HDL cholesterol is mg/dL.

Ratios of Two Measurements in Different Units

To find the ratio of two measurements, we must make sure the quantities have been measured with the same unit. If the measurements are not in the same units, we must first convert them to the same units.

We know that to simplify a fraction, we divide out common factors. Similarly in a ratio of measurements, we divide out the common unit.

EXAMPLE 5.62

The Americans with Disabilities Act (ADA) Guidelines for wheel chair ramps require a maximum vertical rise of inch for every foot of horizontal run. What is the ratio of the rise to the run?

Solution

In a ratio, the measurements must be in the same units. We can change feet to inches, or inches to feet. It is usually easier to convert to the smaller unit, since this avoids introducing more fractions into the problem.

Write the words that express the ratio.

	Ratio of the rise to the run
Write the ratio as a fraction.	
Substitute in the given values.	
Convert 1 foot to inches.	
Simplify, dividing out common factors and units.	

So the ratio of rise to run is to This means that the ramp should rise inch for every inches of horizontal run to comply with the guidelines.

TRY IT : : 5.123 Find the ratio of the first length to the second length: inches to foot.

> **TRY IT : :** 5.124 Find the ratio of the first length to the second length: foot to inches.

Write a Rate as a Fraction

Frequently we want to compare two different types of measurements, such as miles to gallons. To make this comparison, we use a **rate**. Examples of rates are miles in hours, words in minutes, and dollars per ounces.

Rate

A **rate** compares two quantities of different units. A rate is usually written as a fraction.

When writing a fraction as a rate, we put the first given amount with its units in the numerator and the second amount with its units in the denominator. When rates are simplified, the units remain in the numerator and denominator.

EXAMPLE 5.63

Bob drove his car miles in hours. Write this rate as a fraction.

Solution

Write as a fraction, with 525 miles in the numerator and 9 hours in the denominator.	

So miles in hours is equivalent to

> **TRY IT : :** 5.125 Write the rate as a fraction: miles in hours.

> **TRY IT : :** 5.126 Write the rate as a fraction: miles in hours.

Find Unit Rates

In the last example, we calculated that Bob was driving at a rate of This tells us that every three hours, Bob will travel miles. This is correct, but not very useful. We usually want the rate to reflect the number of miles in one hour. A rate that has a denominator of unit is referred to as a **unit rate**.

Unit Rate

A **unit rate** is a rate with denominator of unit.

Unit rates are very common in our lives. For example, when we say that we are driving at a speed of miles per hour we mean that we travel miles in hour. We would write this rate as miles/hour (read miles per hour). The common abbreviation for this is mph. Note that when no number is written before a unit, it is assumed to be

So miles/hour really means

Two rates we often use when driving can be written in different forms, as shown:

Example	Rate	Write	Abbreviate	Read
miles in hour	———	miles/hour	mph	
miles to gallon	———	miles/gallon	mpg	

Another example of unit rate that you may already know about is hourly pay rate. It is usually expressed as the amount of money earned for one hour of work. For example, if you are paid for each hour you work, you could write that your hourly (unit) pay rate is (read per hour.)

To convert a rate to a unit rate, we divide the numerator by the denominator. This gives us a denominator of

EXAMPLE 5.64

Anita was paid last week for working What is Anita's hourly pay rate?

Solution

Start with a rate of dollars to hours. Then divide.	
Write as a rate.	———
Divide the numerator by the denominator.	———
Rewrite as a rate.	

Anita's hourly pay rate is per hour.

TRY IT : : 5.127 Find the unit rate: for hours.

TRY IT : : 5.128 Find the unit rate: for hours.

EXAMPLE 5.65

Sven drives his car miles, using gallons of gasoline. How many miles per gallon does his car get?

Solution

Start with a rate of miles to gallons. Then divide.

Write as a rate.	———
Divide 455 by 14 to get the unit rate.	———

Sven's car gets miles/gallon, or mpg.

TRY IT : : 5.129 Find the unit rate: miles to gallons of gas.

TRY IT : : 5.130 Find the unit rate: miles to gallons of gas.

Find Unit Price

Sometimes we buy common household items 'in bulk', where several items are packaged together and sold for one price. To compare the prices of different sized packages, we need to find the unit price. To find the unit price, divide the total price by the number of items. A **unit price** is a unit rate for one item.

Unit price

A **unit price** is a unit rate that gives the price of one item.

EXAMPLE 5.66

The grocery store charges for a case of bottles of water. What is the unit price?

Solution

What are we asked to find? We are asked to find the unit price, which is the price per bottle.

Write as a rate.	
Divide to find the unit price.	
Round the result to the nearest penny.	

The unit price is approximately per bottle. Each bottle costs about

TRY IT : : 5.131 Find the unit price. Round your answer to the nearest cent if necessary.

of juice boxes for

TRY IT : : 5.132 Find the unit price. Round your answer to the nearest cent if necessary.

of bottles of ice tea for

Unit prices are very useful if you comparison shop. The *better buy* is the item with the lower unit price. Most grocery stores list the unit price of each item on the shelves.

EXAMPLE 5.67

Paul is shopping for laundry detergent. At the grocery store, the liquid detergent is priced at for loads of laundry and the same brand of powder detergent is priced at for loads.

Which is the better buy, the liquid or the powder detergent?

Solution

To compare the prices, we first find the unit price for each type of detergent.

	Liquid	Powder
Write as a rate.		
Find the unit price.		
Round to the nearest cent.		

Now we compare the unit prices. The unit price of the liquid detergent is about per load and the unit price of the powder detergent is about per load. The powder is the better buy.

TRY IT : : 5.133

Find each unit price and then determine the better buy. Round to the nearest cent if necessary.

Brand A Storage Bags, for count, or Brand B Storage Bags, for count

TRY IT : : 5.134

Find each unit price and then determine the better buy. Round to the nearest cent if necessary.

Brand C Chicken Noodle Soup, for ounces, or Brand D Chicken Noodle Soup, for ounces

Notice in Example 5.67 that we rounded the unit price to the nearest cent. Sometimes we may need to carry the division to one more place to see the difference between the unit prices.

Translate Phrases to Expressions with Fractions

Have you noticed that the examples in this section used the comparison words *ratio of, to, per, in, for, on*, and *from*? When you translate phrases that include these words, you should think either ratio or rate. If the units measure the same quantity (length, time, etc.), you have a ratio. If the units are different, you have a rate. In both cases, you write a fraction.

EXAMPLE 5.68

Translate the word phrase into an algebraic expression:

miles per hours students to teachers dollars for hours

Solution

Write as a rate.	

Write as a rate.	

Write as a rate.

TRY IT : : 5.135 Translate the word phrase into an algebraic expression.

miles per hours parents to students dollars for minutes

TRY IT : : 5.136 Translate the word phrase into an algebraic expression.

miles per hours students to buses dollars for hours

MEDIA : : ACCESS ADDITIONAL ONLINE RESOURCES

- Ratios (http://www.openstax.org/l/24ratios)
- Write Ratios as a Simplified Fractions Involving Decimals and Fractions (http://www.openstax.org/l/24ratiosimpfrac)
- Write a Ratio as a Simplified Fraction (http://www.openstax.org/l/24ratiosimp)
- Rates and Unit Rates (http://www.openstax.org/l/24rates)
- Unit Rate for Cell Phone Plan (http://www.openstax.org/l/24unitrate)

5.6 EXERCISES

Practice Makes Perfect

Write a Ratio as a Fraction

In the following exercises, write each ratio as a fraction.

403. to

404. to

405. to

406. to

407. to

408. to

409. to

410. to

411. to

412. to

413. — to —

414. — to —

415. — to —

416. — to —

417. to

418. to

419. to

420. to

421. ounces to ounces

422. ounces to ounces

423. feet to feet

424. feet to feet

425. milligrams to milligrams

426. milligrams to milligrams

427. total cholesterol of to HDL cholesterol of

428. total cholesterol of to HDL cholesterol of

429. inches to foot

430. inches to foot

Write a Rate as a Fraction

In the following exercises, write each rate as a fraction.

431. calories per ounces

432. calories per ounces

433. pounds per square inches

434. pounds per square inches

435. miles in hours

436. miles in hours

437. for hours

438. for hours

Find Unit Rates

In the following exercises, find the unit rate. Round to two decimal places, if necessary.

439. calories per ounces

440. calories per ounces

441. pounds per square inches

442. pounds per square inches

443. miles in hours

444. miles in hours

445. for hours

446. for hours

447. miles on gallons of gas

448. miles on gallons of gas

449. pounds in weeks

450. pounds in weeks

451. beats in minute

452. beats in minute

453. The bindery at a printing plant assembles magazines in hours. How many magazines are assembled in one hour?

454. The pressroom at a printing plant prints sections in hours. How many sections are printed per hour?

Find Unit Price

In the following exercises, find the unit price. Round to the nearest cent.

455. Soap bars at for

456. Soap bars at for

457. Women's sports socks at pairs for

458. Men's dress socks at pairs for

459. Snack packs of cookies at for

460. Granola bars at for

461. CD-RW discs at for

462. CDs at for

463. The grocery store has a special on macaroni and cheese. The price is for boxes. How much does each box cost?

464. The pet store has a special on cat food. The price is for cans. How much does each can cost?

In the following exercises, find each unit price and then identify the better buy. Round to three decimal places.

465. Mouthwash, size for or size for

466. Toothpaste, ounce size for or size for

467. Breakfast cereal, ounces for or ounces for

468. Breakfast Cereal, ounces for or ounces for

469. Ketchup, regular bottle for or squeeze bottle for

470. Mayonnaise regular bottle for or squeeze bottle for

471. Cheese for lb. block or for — lb. block

472. Candy for a lb. bag or for — lb. of loose candy

Translate Phrases to Expressions with Fractions

In the following exercises, translate the English phrase into an algebraic expression.

473. miles per hours

474. feet per seconds

475. for lbs.

476. beats in minutes

477. calories in ounces

478. minutes for dollars

479. the ratio of and

480. the ratio of and

Everyday Math

481. One elementary school in Ohio has students and teachers. Write the student-to-teacher ratio as a unit rate.

482. The average American produces about pounds of paper trash per year How many pounds of paper trash does the average American produce each day? (Round to the nearest tenth of a pound.)

483. A popular fast food burger weighs ounces and contains calories, grams of fat, grams of carbohydrates, and grams of protein. Find the unit rate of calories per ounce grams of fat per ounce grams of carbohydrates per ounce grams of protein per ounce. Round to two decimal places.

484. A chocolate mocha coffee with whipped cream contains calories, grams of fat, grams of carbohydrates, and grams of protein. Find the unit rate of calories per ounce grams of fat per ounce grams of carbohydrates per ounce grams of protein per ounce.

Writing Exercises

485. Would you prefer the ratio of your income to your friend's income to be or Explain your reasoning.

486. The parking lot at the airport charges for every minutes. How much does it cost to park for hour? Explain how you got your answer to part . Was your reasoning based on the unit cost or did you use another method?

487. Kathryn ate a cup of frozen yogurt and then went for a swim. The frozen yogurt had calories. Swimming burns calories per hour. For how many minutes should Kathryn swim to burn off the calories in the frozen yogurt? Explain your reasoning.

488. Mollie had a cappuccino at her neighborhood coffee shop. The cappuccino had calories. If Mollie walks for one hour, she burns calories. For how many minutes must Mollie walk to burn off the calories in the cappuccino? Explain your reasoning.

Self Check

After completing the exercises, use this checklist to evaluate your mastery of the objectives of this section.

I can…	Confidently	With some help	No-I don't get it!
write a ratio as a fraction.			
write a rate as a fraction.			
find unit rates.			
find unit price.			
translate phrases to expressions with fractions.			

After reviewing this checklist, what will you do to become confident for all objectives?

5.7 Simplify and Use Square Roots

Learning Objectives

By the end of this section, you will be able to:

- Simplify expressions with square roots
- Estimate square roots
- Approximate square roots
- Simplify variable expressions with square roots
- Use square roots in applications

BE PREPARED : : 5.19 Before you get started, take this readiness quiz.

Simplify:

If you missed this problem, review **Example 3.52**.

BE PREPARED : : 5.20 Round to the nearest hundredth.

If you missed this problem, review **Example 5.9**.

BE PREPARED : : 5.21 Evaluate for

If you missed this problem, review **Example 2.14**.

Simplify Expressions with Square Roots

To start this section, we need to review some important vocabulary and notation.

Remember that when a number is multiplied by itself, we can write this as which we read aloud as

For example, is read as

We call the *square* of because Similarly, is the square of because

Square of a Number

If then is the square of

Modeling Squares

Do you know why we use the word *square*? If we construct a square with three tiles on each side, the total number of tiles would be nine.

This is why we say that the square of three is nine.

The number is called a perfect square because it is the square of a whole number.

MANIPULATIVE MATHEMATICS

Doing the Manipulative Mathematics activity Square Numbers will help you develop a better understanding of perfect square numbers

The chart shows the squares of the counting numbers through You can refer to it to help you identify the perfect squares.

Number	n	1	2	3	4	5	6	7	8	9	10	11	12	13	14	15
Square	n^2	1	4	9	16	25	36	49	64	81	100	121	144	169	196	225

Perfect Squares

A **perfect square** is the square of a whole number.

What happens when you square a negative number?

When we multiply two negative numbers, the product is always positive. So, the square of a negative number is always positive.

The chart shows the squares of the negative integers from to

Number	n	–1	–2	–3	–4	–5	–6	–7	–8	–9	–10	–11	–12	–13	–14	–15
Square	n^2	1	4	9	16	25	36	49	64	81	100	121	144	169	196	225

Did you notice that these squares are the same as the squares of the positive numbers?

Square Roots

Sometimes we will need to look at the relationship between numbers and their squares in reverse. Because we say is the square of We can also say that is a square root of

Square Root of a Number

A number whose square is is called a square root of

If then is a **square root** of

Notice also, so is also a square root of Therefore, both and are square roots of

So, every positive number has two square roots: one positive and one negative.

What if we only want the positive square root of a positive number? The *radical sign,* $\sqrt{\ }$ stands for the positive square root. The positive square root is also called the **principal square root**.

Square Root Notation

$\sqrt{\ }$ is read as "the square root of

$\sqrt{\ }$

radical sign → $\sqrt{m}$ ← radicand

We can also use the radical sign for the square root of zero. Because $\sqrt{\ }$ Notice that zero has only one square root.

The chart shows the square roots of the first perfect square numbers.

$\sqrt{1}$	$\sqrt{4}$	$\sqrt{9}$	$\sqrt{16}$	$\sqrt{25}$	$\sqrt{36}$	$\sqrt{49}$	$\sqrt{64}$	$\sqrt{81}$	$\sqrt{100}$	$\sqrt{121}$	$\sqrt{144}$	$\sqrt{169}$	$\sqrt{196}$	$\sqrt{225}$
1	2	3	4	5	6	7	8	9	10	11	12	13	14	15

EXAMPLE 5.69

Simplify: $\sqrt{}$ $\sqrt{}$

Solution

$\sqrt{}$
Since

$\sqrt{}$
Since

> **TRY IT : :** 5.137 Simplify: $\sqrt{}$ $\sqrt{}$

> **TRY IT : :** 5.138 Simplify: $\sqrt{}$ $\sqrt{}$

Every positive number has two square roots and the radical sign indicates the positive one. We write $\sqrt{}$ If we want to find the negative square root of a number, we place a negative in front of the radical sign. For example, $\sqrt{}$

EXAMPLE 5.70

Simplify. $\sqrt{}$ $\sqrt{}$

Solution

$\sqrt{}$
The negative is in front of the radical sign.

$\sqrt{}$
The negative is in front of the radical sign.

> **TRY IT : :** 5.139 Simplify: $\sqrt{}$ $\sqrt{}$

> **TRY IT : :** 5.140 Simplify: $\sqrt{}$ $\sqrt{}$

Square Root of a Negative Number

Can we simplify $\sqrt{\ }$ Is there a number whose square is

None of the numbers that we have dealt with so far have a square that is Why? Any positive number squared is positive, and any negative number squared is also positive. In the next chapter we will see that all the numbers we work with are called the real numbers. So we say there is no real number equal to $\sqrt{\ }$ If we are asked to find the square root of any negative number, we say that the solution is not a real number.

EXAMPLE 5.71

Simplify: $\sqrt{\ }$ $\sqrt{\ }$

Solution

There is no real number whose square is Therefore, $\sqrt{\ }$ is not a real number.

The negative is in front of the radical sign, so we find the opposite of the square root of

	$\sqrt{\ }$
The negative is in front of the radical.	

TRY IT : : 5.141 Simplify: $\sqrt{\ }$ $\sqrt{\ }$

TRY IT : : 5.142 Simplify: $\sqrt{\ }$ $\sqrt{\ }$

Square Roots and the Order of Operations

When using the order of operations to simplify an expression that has square roots, we treat the radical sign as a grouping symbol. We simplify any expressions under the radical sign before performing other operations.

EXAMPLE 5.72

Simplify: $\sqrt{\ }$ $\sqrt{\ }$ $\sqrt{\ }$

Solution

Use the order of operations.	
	$\sqrt{\ }$ $\sqrt{\ }$
Simplify each radical.	
Add.	

Use the order of operations.	
	$\sqrt{}$
Add under the radical sign.	$\sqrt{}$
Simplify.	

> **TRY IT : : 5.143** Simplify: $\sqrt{}$ $\sqrt{}$ $\sqrt{}$

> **TRY IT : : 5.144** Simplify: $\sqrt{}$ $\sqrt{}$ $\sqrt{}$

Notice the different answers in parts and of Example 5.72. It is important to follow the order of operations correctly. In , we took each square root first and then added them. In , we added under the radical sign first and then found the square root.

Estimate Square Roots

So far we have only worked with square roots of perfect squares. The square roots of other numbers are not whole numbers.

Number	Square root
4	$\sqrt{4} = 2$
5	$\sqrt{5}$
6	$\sqrt{6}$
7	$\sqrt{7}$
8	$\sqrt{8}$
9	$\sqrt{9} = 3$

We might conclude that the square roots of numbers between and will be between and and they will not be whole numbers. Based on the pattern in the table above, we could say that $\sqrt{}$ is between and Using inequality symbols, we write

$$\sqrt{}$$

EXAMPLE 5.73

Estimate $\sqrt{}$ between two consecutive whole numbers.

Solution

Think of the perfect squares closest to Make a small table of these perfect squares and their squares roots.

Number	Square root
36	6
49	7
64	8
81	9

60 $\sqrt{60}$

$\sqrt{}$	$\sqrt{}$

> TRY IT : : 5.145 Estimate $\sqrt{\ }$ between two consecutive whole numbers.

> TRY IT : : 5.146 Estimate $\sqrt{\ }$ between two consecutive whole numbers.

Approximate Square Roots with a Calculator

There are mathematical methods to approximate square roots, but it is much more convenient to use a calculator to find square roots. Find the $\sqrt{\ }$ or $\sqrt{\ }$ key on your calculator. You will to use this key to approximate square roots. When you use your calculator to find the square root of a number that is not a perfect square, the answer that you see is not the exact number. It is an approximation, to the number of digits shown on your calculator's display. The symbol for an approximation is and it is read *approximately*.

Suppose your calculator has a display. Using it to find the square root of will give This is the approximate square root of When we report the answer, we should use the "approximately equal to" sign instead of an equal sign.

$$\sqrt{\ }$$

You will seldom use this many digits for applications in algebra. So, if you wanted to round $\sqrt{\ }$ to two decimal places, you would write

$$\sqrt{\ }$$

How do we know these values are approximations and not the exact values? Look at what happens when we square them.

The squares are close, but not exactly equal, to

EXAMPLE 5.74

Round $\sqrt{\ }$ to two decimal places using a calculator.

Solution

$\sqrt{\ }$
Use the calculator square root key.
Round to two decimal places.
$\sqrt{\ }$

> TRY IT : : 5.147 Round $\sqrt{\ }$ to two decimal places.

> TRY IT : : 5.148 Round $\sqrt{\ }$ to two decimal places.

Simplify Variable Expressions with Square Roots

Expressions with square root that we have looked at so far have not had any variables. What happens when we have to find a square root of a variable expression?

Consider $\sqrt{\ }$ where Can you think of an expression whose square is

$$\sqrt{\ }$$

When we use a variable in a square root expression, for our work, we will assume that the variable represents a nonnegative number. In every example and exercise that follows, each variable in a square root expression is greater than or equal to zero.

EXAMPLE 5.75

Simplify: $\sqrt{}$

Solution

Think about what we would have to square to get . Algebraically,

$\sqrt{}$

Since

TRY IT : : 5.149 Simplify: $\sqrt{}$

TRY IT : : 5.150 Simplify: $\sqrt{}$

EXAMPLE 5.76

Simplify: $\sqrt{}$

Solution

$\sqrt{}$

TRY IT : : 5.151 Simplify: $\sqrt{}$

TRY IT : : 5.152 Simplify: $\sqrt{}$

EXAMPLE 5.77

Simplify: $\sqrt{}$

Solution

$\sqrt{}$

TRY IT : : 5.153 Simplify: $\sqrt{}$

TRY IT : : 5.154 Simplify: $\sqrt{}$

EXAMPLE 5.78

Simplify: $\sqrt{}$

Solution

$\sqrt{}$

TRY IT : : 5.155 Simplify: $\sqrt{}$

TRY IT : : 5.156 Simplify: $\sqrt{}$

Use Square Roots in Applications

As you progress through your college courses, you'll encounter several applications of square roots. Once again, if we use our strategy for applications, it will give us a plan for finding the answer!

HOW TO : : USE A STRATEGY FOR APPLICATIONS WITH SQUARE ROOTS.

Step 1. Identify what you are asked to find.
Step 2. Write a phrase that gives the information to find it.
Step 3. Translate the phrase to an expression.
Step 4. Simplify the expression.
Step 5. Write a complete sentence that answers the question.

Square Roots and Area

We have solved applications with area before. If we were given the length of the sides of a square, we could find its area by squaring the length of its sides. Now we can find the length of the sides of a square if we are given the area, by finding the square root of the area.

If the area of the square is square units, the length of a side is $\sqrt{}$ units. See Table 5.7.

Area (square units)	Length of side (units)
	$\sqrt{}$
	$\sqrt{}$
	$\sqrt{}$

Table 5.7

EXAMPLE 5.79

Mike and Lychelle want to make a square patio. They have enough concrete for an area of square feet. To the nearest tenth of a foot, how long can a side of their square patio be?

Solution

We know the area of the square is square feet and want to find the length of the side. If the area of the square is square units, the length of a side is $\sqrt{}$ units.

What are you asked to find?	The length of each side of a square patio
Write a phrase.	The length of a side
Translate to an expression.	$\sqrt{}$
Evaluate $\sqrt{}$ when .	$\sqrt{}$
Use your calculator.	
Round to one decimal place.	
Write a sentence.	Each side of the patio should be feet.

> **TRY IT : : 5.157**
>
> Katie wants to plant a square lawn in her front yard. She has enough sod to cover an area of square feet. To the nearest tenth of a foot, how long can a side of her square lawn be?

> **TRY IT : : 5.158**
>
> Sergio wants to make a square mosaic as an inlay for a table he is building. He has enough tile to cover an area of square centimeters. How long can a side of his mosaic be?

Square Roots and Gravity

Another application of square roots involves gravity. On Earth, if an object is dropped from a height of feet, the time in seconds it will take to reach the ground is found by evaluating the expression $\frac{\sqrt{}}{}$ For example, if an object is dropped from a height of feet, we can find the time it takes to reach the ground by evaluating $\frac{\sqrt{}}{}$

	$\frac{\sqrt{}}{}$
Take the square root of 64.	$-$
Simplify the fraction.	

It would take seconds for an object dropped from a height of feet to reach the ground.

EXAMPLE 5.80

Christy dropped her sunglasses from a bridge feet above a river. How many seconds does it take for the sunglasses to reach the river?

Solution

What are you asked to find?	The number of seconds it takes for the sunglasses to reach the river
Write a phrase.	The time it will take to reach the river
Translate to an expression.	√
Evaluate √ when .	√
Find the square root of 400.	—
Simplify.	
Write a sentence.	It will take 5 seconds for the sunglasses to reach the river.

TRY IT : : 5.159

A helicopter drops a rescue package from a height of feet. How many seconds does it take for the package to reach the ground?

TRY IT : : 5.160

A window washer drops a squeegee from a platform feet above the sidewalk. How many seconds does it take for the squeegee to reach the sidewalk?

Square Roots and Accident Investigations

Police officers investigating car accidents measure the length of the skid marks on the pavement. Then they use square roots to determine the speed, in miles per hour, a car was going before applying the brakes. According to some formulas, if the length of the skid marks is feet, then the speed of the car can be found by evaluating √

EXAMPLE 5.81

After a car accident, the skid marks for one car measured feet. To the nearest tenth, what was the speed of the car (in mph) before the brakes were applied?

Solution

What are you asked to find?	The speed of the car before the brakes were applied
Write a phrase.	The speed of the car
Translate to an expression.	√
Evaluate √ when	√
Multiply.	√
Use your calculator.	
Round to tenths.	
Write a sentence.	The speed of the car was approximately 67.5 miles per hour.

TRY IT : : 5.161

An accident investigator measured the skid marks of a car and found their length was feet. To the nearest tenth, what was the speed of the car before the brakes were applied?

TRY IT : : 5.162

The skid marks of a vehicle involved in an accident were feet long. To the nearest tenth, how fast had the vehicle been going before the brakes were applied?

LINKS TO LITERACY

The *Links to Literacy* activity "Sea Squares" will provide you with another view of the topics covered in this section.

MEDIA : : ACCESS ADDITIONAL ONLINE RESOURCES

- Introduction to Square Roots (http://www.openstax.org/l/24introsqroots)
- Estimating Square Roots with a Calculator (http://www.openstax.org/l/24estsqrtcalc)

5.7 EXERCISES

Practice Makes Perfect

Simplify Expressions with Square Roots

In the following exercises, simplify.

489. $\sqrt{\ }$

490. $\sqrt{\ }$

491. $\sqrt{\ }$

492. $\sqrt{\ }$

493. $\sqrt{\ }$

494. $\sqrt{\ }$

495. $\sqrt{\ }$

496. $\sqrt{\ }$

497. $\sqrt{\ }$

498. $\sqrt{\ }$

499. $\sqrt{\ }$

500. $\sqrt{\ }$

501. $\sqrt{\ }$

502. $\sqrt{\ }$

503. $\sqrt{\ }$ $\sqrt{\ }$

504. $\sqrt{\ }$ $\sqrt{\ }$

Estimate Square Roots

In the following exercises, estimate each square root between two consecutive whole numbers.

505. $\sqrt{\ }$

506. $\sqrt{\ }$

507. $\sqrt{\ }$

508. $\sqrt{\ }$

Approximate Square Roots with a Calculator

In the following exercises, use a calculator to approximate each square root and round to two decimal places.

509. $\sqrt{\ }$

510. $\sqrt{\ }$

511. $\sqrt{\ }$

512. $\sqrt{\ }$

Simplify Variable Expressions with Square Roots

In the following exercises, simplify. (Assume all variables are greater than or equal to zero.)

513. $\sqrt{\ }$

514. $\sqrt{\ }$

515. $\sqrt{\ }$

516. $\sqrt{\ }$

517. $\sqrt{\ }$

518. $\sqrt{\ }$

519. $\sqrt{\ }$

520. $\sqrt{\ }$

Use Square Roots in Applications

In the following exercises, solve. Round to one decimal place.

521. Landscaping Reed wants to have a square garden plot in his backyard. He has enough compost to cover an area of square feet. How long can a side of his garden be?

522. Landscaping Vince wants to make a square patio in his yard. He has enough concrete to pave an area of square feet. How long can a side of his patio be?

523. Gravity An airplane dropped a flare from a height of feet above a lake. How many seconds did it take for the flare to reach the water?

524. Gravity A hang glider dropped his cell phone from a height of feet. How many seconds did it take for the cell phone to reach the ground?

525. Gravity A construction worker dropped a hammer while building the Grand Canyon skywalk, feet above the Colorado River. How many seconds did it take for the hammer to reach the river?

526. Accident investigation The skid marks from a car involved in an accident measured feet. What was the speed of the car before the brakes were applied?

527. Accident investigation The skid marks from a car involved in an accident measured feet. What was the speed of the car before the brakes were applied?

528. Accident investigation An accident investigator measured the skid marks of one of the vehicles involved in an accident. The length of the skid marks was feet. What was the speed of the vehicle before the brakes were applied?

529. Accident investigation An accident investigator measured the skid marks of one of the vehicles involved in an accident. The length of the skid marks was feet. What was the speed of the vehicle before the brakes were applied?

Everyday Math

530. Decorating Denise wants to install a square accent of designer tiles in her new shower. She can afford to buy square centimeters of the designer tiles. How long can a side of the accent be?

531. Decorating Morris wants to have a square mosaic inlaid in his new patio. His budget allows for tiles. Each tile is square with an area of one square inch. How long can a side of the mosaic be?

Writing Exercises

532. Why is there no real number equal to $\sqrt{}$

533. What is the difference between and $\sqrt{}$

Self Check

After completing the exercises, use this checklist to evaluate your mastery of the objectives of this section.

I can...	Confidently	With some help	No-I don't get it!
simplify expressions with square roots.			
estimate square roots.			
approximate square roots.			
simplify variable expressions with square roots.			
use square roots in applications.			

Overall, after looking at the checklist, do you think you are well-prepared for the next Chapter? Why or why not?

CHAPTER 5 REVIEW

KEY TERMS

circumference of a circle The distance around a circle is called its **circumference**.

diameter of a circle A diameter of a circle is a line segment that passes through a circle's center connecting two points on the circle.

equivalent decimals Two decimals are equivalent decimals if they convert to equivalent fractions.

mean The mean of a set of numbers is the arithmetic average of the numbers. The formula is

median The median of a set of data values is the middle value.

- Half the data values are less than or equal to the median.
- Half the data values are greater than or equal to the median.

mode The mode of a set of numbers is the number with the highest frequency.

radius of a circle A radius of a circle is a line segment from the center to any point on the circle.

rate A rate compares two quantities of different units. A rate is usually written as a fraction.

ratio A ratio compares two numbers or two quantities that are measured with the same unit. The ratio of to is written to , —, or .

repeating decimal A repeating decimal is a decimal in which the last digit or group of digits repeats endlessly.

unit price A **unit price** is a unit rate that gives the price of one item.

unit rate A unit rate is a rate with denominator of 1 unit.

KEY CONCEPTS

5.1 Decimals

- **Name a decimal number.**

  Step 1. Name the number to the left of the decimal point.

  Step 2. Write "and" for the decimal point.

  Step 3. Name the "number" part to the right of the decimal point as if it were a whole number.

  Step 4. Name the decimal place of the last digit.

- **Write a decimal number from its name.**

  Step 1. Look for the word "and"—it locates the decimal point.
  Place a decimal point under the word "and." Translate the words before "and" into the whole number and place it to the left of the decimal point.
  If there is no "and," write a "0" with a decimal point to its right.

  Step 2. Mark the number of decimal places needed to the right of the decimal point by noting the place value indicated by the last word.

  Step 3. Translate the words after "and" into the number to the right of the decimal point. Write the number in the spaces—putting the final digit in the last place.

  Step 4. Fill in zeros for place holders as needed.

- **Convert a decimal number to a fraction or mixed number.**

  Step 1. Look at the number to the left of the decimal.
  If it is zero, the decimal converts to a proper fraction.
  If it is not zero, the decimal converts to a mixed number.
  Write the whole number.

  Step 2. Determine the place value of the final digit.

  Step 3. Write the fraction. numerator—the 'numbers' to the right of the decimal point denominator—the place value corresponding to the final digit

  Step 4. Simplify the fraction, if possible.

- **Order decimals.**

Step 1. Check to see if both numbers have the same number of decimal places. If not, write zeros at the end of the one with fewer digits to make them match.

Step 2. Compare the numbers to the right of the decimal point as if they were whole numbers.

Step 3. Order the numbers using the appropriate inequality sign.

- **Round a decimal.**

Step 1. Locate the given place value and mark it with an arrow.

Step 2. Underline the digit to the right of the given place value.

Step 3. Is this digit greater than or equal to 5?
Yes - add 1 to the digit in the given place value.
No - do not change the digit in the given place value

Step 4. Rewrite the number, removing all digits to the right of the given place value.

5.2 Decimal Operations

- **Add or subtract decimals.**

Step 1. Write the numbers vertically so the decimal points line up.

Step 2. Use zeros as place holders, as needed.

Step 3. Add or subtract the numbers as if they were whole numbers. Then place the decimal in the answer under the decimal points in the given numbers.

- **Multiply decimal numbers.**

Step 1. Determine the sign of the product.

Step 2. Write the numbers in vertical format, lining up the numbers on the right.

Step 3. Multiply the numbers as if they were whole numbers, temporarily ignoring the decimal points.

Step 4. Place the decimal point. The number of decimal places in the product is the sum of the number of decimal places in the factors. If needed, use zeros as placeholders.

Step 5. Write the product with the appropriate sign.

- **Multiply a decimal by a power of 10.**

Step 1. Move the decimal point to the right the same number of places as the number of zeros in the power of 10.

Step 2. Write zeros at the end of the number as placeholders if needed.

- **Divide a decimal by a whole number.**

Step 1. Write as long division, placing the decimal point in the quotient above the decimal point in the dividend.

Step 2. Divide as usual.

- **Divide decimal numbers.**

Step 1. Determine the sign of the quotient.

Step 2. Make the divisor a whole number by moving the decimal point all the way to the right. Move the decimal point in the dividend the same number of places to the right, writing zeros as needed.

Step 3. Divide. Place the decimal point in the quotient above the decimal point in the dividend.

Step 4. Write the quotient with the appropriate sign.

- **Strategy for Applications**

Step 1. Identify what you are asked to find.

Step 2. Write a phrase that gives the information to find it.

Step 3. Translate the phrase to an expression.

Step 4. Simplify the expression.

Step 5. Answer the question with a complete sentence.

5.3 Decimals and Fractions

- **Convert a Fraction to a Decimal** To convert a fraction to a decimal, divide the numerator of the fraction by the denominator of the fraction.
- **Properties of Circles**

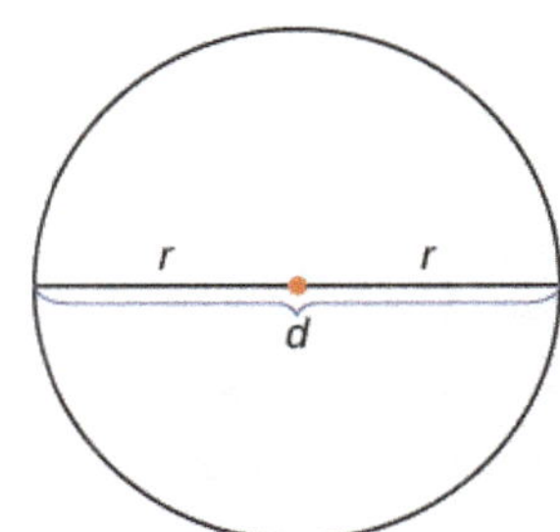

is the length of the radius
is the length of the diameter
The circumference is .
The area is .

5.4 Solve Equations with Decimals

- **Determine whether a number is a solution to an equation.**
  - Substitute the number for the variable in the equation.
  - Simplify the expressions on both sides of the equation.
  - Determine whether the resulting equation is true.
    If so, the number is a solution.
    If not, the number is not a solution.
- **Properties of Equality**

Subtraction Property of Equality	**Addition Property of Equality**
For any numbers , , and ,	For any numbers , , and ,
Division of Property of Equality	**Multiplication Property of Equality**
For any numbers , , and , — —	For any numbers , , and ,

Table 5.8

5.5 Averages and Probability

- **Calculate the mean of a set of numbers.**

  Step 1. Write the formula for the mean

  Step 2. Find the sum of all the values in the set. Write the sum in the numerator.

  Step 3. Count the number, n, of values in the set. Write this number in the denominator.

  Step 4. Simplify the fraction.

  Step 5. Check to see that the mean is reasonable. It should be greater than the least number and less than the greatest number in the set.
- **Find the median of a set of numbers.**

  Step 1. List the numbers from least to greatest.

  Step 2. Count how many numbers are in the set. Call this .

Step 3. Is odd or even?
If is an odd number, the median is the middle value.
If is an even number, the median is the mean of the two middle values

- **Identify the mode of a set of numbers.**
  Step 1. List the data values in numerical order.
  Step 2. Count the number of times each value appears.
  Step 3. The mode is the value with the highest frequency.

5.7 Simplify and Use Square Roots

- **Square Root Notation** $\sqrt{\ }$ is read 'the square root of '
  If , then $\sqrt{\ }$, for .

radical sign → $\sqrt{m}$ ← radicand

- **Use a strategy for applications with square roots.**
  - Identify what you are asked to find.
  - Write a phrase that gives the information to find it.
  - Translate the phrase to an expression.
  - Simplify the expression.
  - Write a complete sentence that answers the question.

REVIEW EXERCISES

5.1 Decimals

Name Decimals

In the following exercises, name each decimal.

534.

535.

536.

537.

538.

539.

Write Decimals

In the following exercises, write as a decimal.

540. three tenths

541. nine hundredths

542. twenty-seven hundredths

543. ten and thirty-five thousandths

544. negative twenty and three tenths

545. negative five hundredths

Convert Decimals to Fractions or Mixed Numbers

In the following exercises, convert each decimal to a fraction. Simplify the answer if possible.

546.

547.

548.

549.

Locate Decimals on the Number Line

550.

Order Decimals

In the following exercises, order each of the following pairs of numbers, using or

551.

552.

553.

554.

Round Decimals

In the following exercises, round each number to the nearest: hundredth tenth whole number.

555.

556.

557.

5.2 Decimal Operations

Add and Subtract Decimals

In the following exercises, add or subtract.

558.

559.

560.

561.

562.

563.

Multiply Decimals

In the following exercises, multiply.

564.

565.

566.

567.

Divide Decimals

In the following exercises, divide.

568.

569.

570.

571.

572.

573.

Use Decimals in Money Applications

In the following exercises, use the strategy for applications to solve.

574. Miranda got from her ATM. She spent on lunch and on a book. How much money did she have left? Round to the nearest cent if necessary.

575. Jessie put gallons of gas in her car. One gallon of gas costs How much did Jessie owe for all the gas?

576. A pack of water bottles cost How much did each bottle cost?

577. Alice bought a roll of paper towels that cost She had a coupon for off, and the store doubled the coupon. How much did Alice pay for the paper towels?

5.3 Decimals and Fractions

Convert Fractions to Decimals

In the following exercises, convert each fraction to a decimal.

578. —

579. —

580. —

581. —

582. —

583. —

Order Decimals and Fractions

In the following exercises, order each pair of numbers, using or

584. —

585. —

586. —

587. —

588. —

589. —

In the following exercises, write each set of numbers in order from least to greatest.

590. — —

591. — —

Simplify Expressions Using the Order of Operations

In the following exercises, simplify

592.

593. —

594.

595. —

596.

597. — —

Find the Circumference and Area of Circles

In the following exercises, approximate the circumference and area of each circle.

598.

599.

600. —

601.

5.4 Solve Equations with Decimals

Determine Whether a Decimal is a Solution of an Equation

In the following exercises, determine whether the each number is a solution of the given equation.

602.

603.

604. —

605.

Solve Equations with Decimals

In the following exercises, solve.

606.

607.

608.

609.

610.

611.

612.

613.

614.

615. —

616. —

617. —

Translate to an Equation and Solve

In the following exercises, translate and solve.

618. The difference of and is

619. The product of and is

620. The quotient of and is

621. The sum of and is

5.5 Averages and Probability

Find the Mean of a Set of Numbers

In the following exercises, find the mean of the numbers.

622.

623. , , , and

624. Each workday last week, Yoshie kept track of the number of minutes she had to wait for the bus. She waited minutes. Find the mean.

625. In the last three months, Raul's water bills were Find the mean.

Find the Median of a Set of Numbers

In the following exercises, find the median.

626. , , , ,

627. , , , , , , ,

628. The ages of the eight men in Jerry's model train club are Find the median age.

629. The number of clients at Miranda's beauty salon each weekday last week were Find the median number of clients.

Find the Mode of a Set of Numbers

In the following exercises, identify the mode of the numbers.

630. , , , , , , ,

631. The number of siblings of a group of students: , , , , , , , , , , ,

Use the Basic Definition of Probability

In the following exercises, solve. (Round decimals to three places.)

632. The Sustainability Club sells tickets to a raffle, and Albert buys one ticket. One ticket will be selected at random to win the grand prize. Find the probability Albert will win the grand prize. Express your answer as a fraction and as a decimal.

633. Luc has to read novels and short stories for his literature class. The professor will choose one reading at random for the final exam. Find the probability that the professor will choose a novel for the final exam. Express your answer as a fraction and as a decimal.

5.6 Ratios and Rate

Write a Ratio as a Fraction

In the following exercises, write each ratio as a fraction. Simplify the answer if possible.

634. to

635. to

636. to

637. to

638. — —

639. — —

640. ounces to ounces

641. inches to feet

Write a Rate as a Fraction

In the following exercises, write each rate as a fraction. Simplify the answer if possible.

642. calories per ounces

643. pounds per square inches

644. miles in hours

645. for hours

Find Unit Rates

In the following exercises, find the unit rate.

646. calories per ounces

647. pounds per square inches

648. miles in hours

649. for hours

Find Unit Price

In the following exercises, find the unit price.

650. t-shirts: for

651. Highlighters: for

652. An office supply store sells a box of pens for The box contains pens. How much does each pen cost?

653. Anna bought a pack of kitchen towels for How much did each towel cost? Round to the nearest cent if necessary.

In the following exercises, find each unit price and then determine the better buy.

654. Shampoo: ounces for or ounces for

655. Vitamins: tablets for or for

Translate Phrases to Expressions with Fractions

In the following exercises, translate the English phrase into an algebraic expression.

656. miles per

657. adults to children

658. the ratio of and the difference of and

659. the ratio of and the sum of and

5.7 Simplify and Use Square Roots

Simplify Expressions with Square Roots

In the following exercises, simplify.

660. $\sqrt{}$

661. $\sqrt{}$

662. $\sqrt{}$

663. $\sqrt{}$

664. $\sqrt{}$

665. $\sqrt{}$

666. $\sqrt{}$ $\sqrt{}$

667. $\sqrt{}$

Estimate Square Roots

In the following exercises, estimate each square root between two consecutive whole numbers.

668. $\sqrt{}$

669. $\sqrt{}$

Approximate Square Roots

In the following exercises, approximate each square root and round to two decimal places.

670. $\sqrt{}$

671. $\sqrt{}$

Simplify Variable Expressions with Square Roots

In the following exercises, simplify. (Assume all variables are greater than or equal to zero.)

672. $\sqrt{}$

673. $\sqrt{}$

674. $\sqrt{}$

675. $\sqrt{}$

676. $\sqrt{}$

677. $\sqrt{}$

678. $\sqrt{}$

679. $\sqrt{}$

Use Square Roots in Applications

In the following exercises, solve. Round to one decimal place.

680. **Art** Diego has square inch tiles. He wants to use them to make a square mosaic. How long can each side of the mosaic be?

681. **Landscaping** Janet wants to plant a square flower garden in her yard. She has enough topsoil to cover an area of square feet. How long can a side of the flower garden be?

682. **Gravity** A hiker dropped a granola bar from a lookout spot feet above a valley. How long did it take the granola bar to reach the valley floor?

683. **Accident investigation** The skid marks of a car involved in an accident were feet. How fast had the car been going before applying the brakes?

PRACTICE TEST

684. Write six and thirty-four thousandths as a decimal.

685. Write as a fraction.

686. Write — as a decimal.

687. Round to the nearest tenth hundredth whole number

688. Write the numbers — — in order from smallest to largest.

In the following exercises, simplify each expression.

689.

690.

691.

692.

693.

694.

695.

696.

697.

698.

699. —

700. —

In the following exercises, solve.

701.

702. —

703.

704.

705. Three friends went out to dinner and agreed to split the bill evenly. The bill was How much should each person pay?

706. A circle has radius Find the circumference and area.

707. The ages, in months, of children in a preschool class are: , , , , , , , , ,

Find the mean median mode

708. Of the nurses in Doreen's department, are women and are men. One of the nurses will be assigned at random to work an extra shift next week. Find the probability a woman nurse will be assigned the extra shift. Convert the fraction to a decimal.

709. Find each unit price and then the better buy.
Laundry detergent: ounces for or ounces for

In the following exercises, simplify.

710. $\sqrt{}$

711. $\sqrt{}$

712. Estimate $\sqrt{}$ to between two whole numbers.

713. Yanet wants a square patio in her backyard. She has square feet of tile. How long can a side of the patio be?

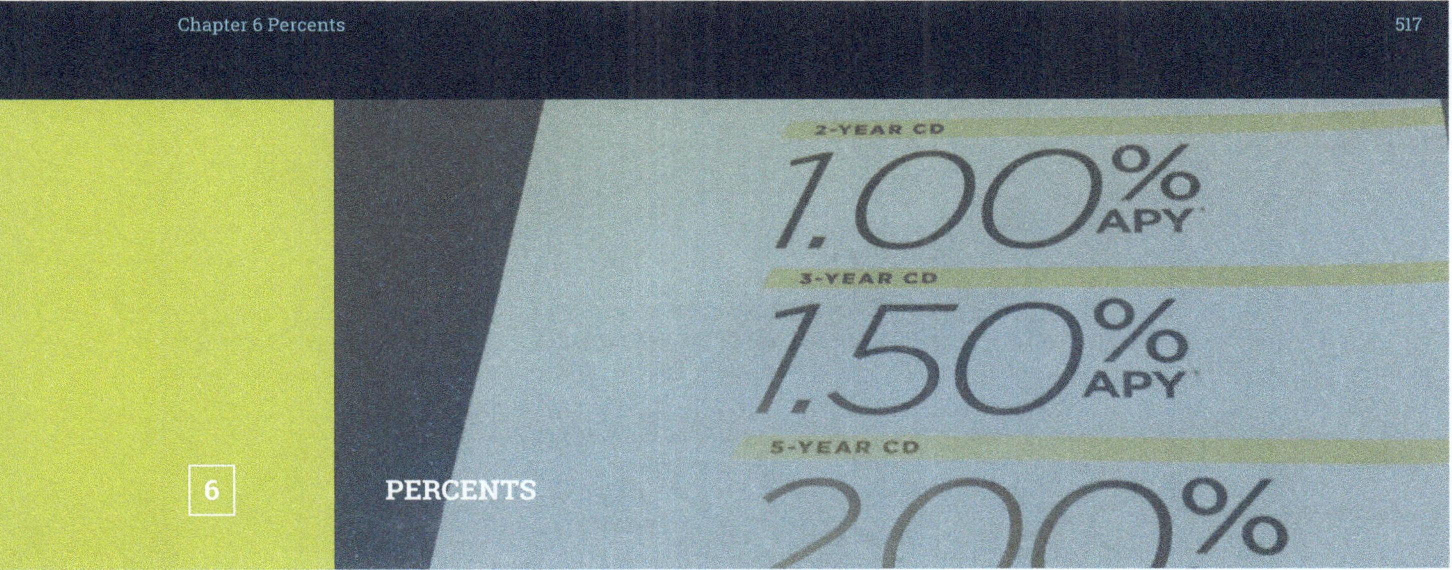

6 PERCENTS

Figure 6.1 Banks provide money for savings and charge money for loans. The interest on savings and loans is usually given as a percent. (credit: Mike Mozart, Flickr)

Chapter Outline

Introduction

When you deposit money in a savings account at a bank, it earns additional money. Figuring out how your money will grow involves understanding and applying concepts of percents. In this chapter, we will find out what percents are and how we can use them to solve problems.

6.1 Understand Percent

Learning Objectives

By the end of this section, you will be able to:

- Use the definition of percent
- Convert percents to fractions and decimals
- Convert decimals and fractions to percents

BE PREPARED : : 6.1 Before you get started, take this readiness quiz.

Translate "the ratio of to into an algebraic expression.

If you missed this problem, review **Table 2.7**.

BE PREPARED : : 6.2 Write — as a decimal.

If you missed this problem, review **Example 5.28**.

BE PREPARED : : 6.3 Write as a fraction.
If you missed this problem, review Example 5.4.

Use the Definition of Percent

How many cents are in one dollar? There are cents in a dollar. How many years are in a century? There are years in a century. Does this give you a clue about what the word "percent" means? It is really two words, "per cent," and means per one hundred. A **percent** is a ratio whose denominator is We use the percent symbol to show percent.

Percent

A percent is a ratio whose denominator is

According to data from the American Association of Community Colleges about of community college students are female. This means out of every community college students are female, as Figure 6.2 shows. Out of the squares on the grid, are shaded, which we write as the ratio ——

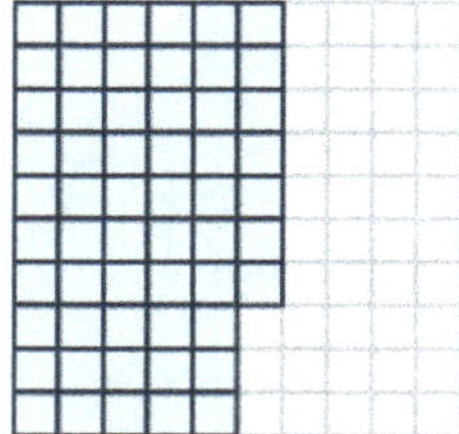

Figure 6.2 Among every community college students, are female.

Similarly, means a ratio of —— means a ratio of —— and means a ratio of —— In words, "one hundred percent" means the total is —— and since —— we see that means whole.

EXAMPLE 6.1

According to the Public Policy Institute of California of parents of public school children would like their youngest child to earn a graduate degree. Write this percent as a ratio.

Solution

The amount we want to convert is 44%.	
Write the percent as a ratio. Remember that *percent* means per 100.	——

TRY IT : : 6.1 Write the percent as a ratio.
According to a survey, of college students have a smartphone.

TRY IT : : 6.2 Write the percent as a ratio.
A study found that of U.S. teens send text messages regularly.

EXAMPLE 6.2

In according to a U.S. Department of Education report, out of every first-time freshmen college students at public institutions took at least one remedial course. Write this as a ratio and then as a percent.

Solution

The amount we want to convert is out of .	out of
Write as a ratio.	——
Convert the 21 per 100 to percent.	

> **TRY IT : : 6.3**
>
> Write as a ratio and then as a percent: The American Association of Community Colleges reported that out of full-time community college students balance their studies with full-time or part time employment.

> **TRY IT : : 6.4**
>
> Write as a ratio and then as a percent: In response to a student survey, out of Santa Ana College students expressed a goal of earning an Associate's degree or transferring to a four-year college.

Convert Percents to Fractions and Decimals

Since percents are ratios, they can easily be expressed as fractions. Remember that percent means per so the denominator of the fraction is

HOW TO : : CONVERT A PERCENT TO A FRACTION.

Step 1. Write the percent as a ratio with the denominator

Step 2. Simplify the fraction if possible.

EXAMPLE 6.3

Convert each percent to a fraction:

Solution

Write as a ratio with denominator 100.	——
Simplify.	—

Write as a ratio with denominator 100.	
Simplify.	

> **TRY IT : : 6.5** Convert each percent to a fraction:

> **TRY IT : : 6.6** Convert each percent to a fraction:

The previous example shows that a percent can be greater than We saw that means or These are improper fractions, and their values are greater than one.

EXAMPLE 6.4

Convert each percent to a fraction:

Solution

Write as a ratio with denominator 100.	
Clear the decimal by multiplying numerator and denominator by 10.	
Multiply.	
Rewrite showing common factors.	
Simplify.	

	—
Write as a ratio with denominator 100.	—
Write the numerator as an improper fraction.	—
Rewrite as fraction division, replacing 100 with —.	— —
Multiply by the reciprocal.	— —
Simplify.	—

> **TRY IT : : 6.7** Convert each percent to a fraction:
>
> —

> **TRY IT : : 6.8** Convert each percent to a fraction:
>
> —

In Decimals, we learned how to convert fractions to decimals. To convert a percent to a decimal, we first convert it to a fraction and then change the fraction to a decimal.

HOW TO : : CONVERT A PERCENT TO A DECIMAL.

Step 1. Write the percent as a ratio with the denominator

Step 2. Convert the fraction to a decimal by dividing the numerator by the denominator.

EXAMPLE 6.5

Convert each percent to a decimal:

Solution

Because we want to change to a decimal, we will leave the fractions with denominator instead of removing common factors.

Write as a ratio with denominator 100.	
Change the fraction to a decimal by dividing the numerator by the denominator.	

Write as a ratio with denominator 100.	
Change the fraction to a decimal by dividing the numerator by the denominator.	

> **TRY IT : : 6.9** Convert each percent to a decimal:

> **TRY IT : : 6.10** Convert each percent to a decimal:

EXAMPLE 6.6

Convert each percent to a decimal:

Solution

Write as a ratio with denominator 100.	
Change the fraction to a decimal by dividing the numerator by the denominator.	

Write as a ratio with denominator 100.	
Change the fraction to a decimal by dividing the numerator by the denominator.	

TRY IT : : 6.11 Convert each percent to a decimal:

TRY IT : : 6.12 Convert each percent to a decimal:

Let's summarize the results from the previous examples in Table 6.1, and look for a pattern we could use to quickly convert a percent number to a decimal number.

Percent	Decimal

Table 6.1

Do you see the pattern?

To convert a percent number to a decimal number, we move the decimal point two places to the left and remove the sign. (Sometimes the decimal point does not appear in the percent number, but just like we can think of the integer as we can think of as) Notice that we may need to add zeros in front of the number when moving the decimal to the left.

Figure 6.3 uses the percents in Table 6.1 and shows visually how to convert them to decimals by moving the decimal point two places to the left.

Percent	Decimal
006.%	0.06
078.%	0.78
135.%	1.35
012.5%	0.125

Figure 6.3

EXAMPLE 6.7

Among a group of business leaders, believe that poor math and science education in the U.S. will lead to higher unemployment rates.

Convert the percent to: a fraction a decimal

Solution

Write as a ratio with denominator 100.

Change the fraction to a decimal by dividing the numerator by the denominator.

TRY IT : : 6.13 Convert the percent to: a fraction and a decimal

Twitter's share of web traffic jumped when one celebrity tweeted live on air.

TRY IT : : 6.14

Convert the percent to: a fraction and a decimal

The U.S. Census estimated that in of the population of Boston age or older have a bachelor's or higher degrees.

EXAMPLE 6.8

There are four suits of cards in a deck of cards—hearts, diamonds, clubs, and spades. The probability of randomly choosing a heart from a shuffled deck of cards is Convert the percent to:

a fraction a decimal

Figure 6.4 (credit: Riles32807, Wikimedia Commons)

Solution

Write as a ratio with denominator 100.	
Simplify.	

—

Change the fraction to a decimal by dividing the numerator by the denominator.

TRY IT : : 6.15 Convert the percent to: a fraction, and a decimal

The probability that it will rain Monday is

TRY IT : : 6.16 Convert the percent to: a fraction, and a decimal

The probability of getting heads three times when tossing a coin three times is

Convert Decimals and Fractions to Percents

To convert a decimal to a percent, remember that percent means per hundred. If we change the decimal to a fraction whose denominator is it is easy to change that fraction to a percent.

HOW TO : : CONVERT A DECIMAL TO A PERCENT.

Step 1. Write the decimal as a fraction.

Step 2. If the denominator of the fraction is not rewrite it as an equivalent fraction with denominator

Step 3. Write this ratio as a percent.

EXAMPLE 6.9

Convert each decimal to a percent:

Solution

Write as a fraction. The denominator is 100.	—
Write this ratio as a percent.	

The denominator is 100.	—
Write this ratio as a percent.	

TRY IT : : 6.17 Convert each decimal to a percent:

TRY IT : : 6.18 Convert each decimal to a percent:

To convert a mixed number to a percent, we first write it as an improper fraction.

EXAMPLE 6.10

Convert each decimal to a percent:

Solution

Write as a fraction.	
Write as an improper fraction. The denominator is 100.	
Write this ratio as a percent.	

Notice that since the result is more than

Write as a fraction. The denominator is 1,000.	
Divide the numerator and denominator by 10, so that the denominator is 100.	
Write this ratio as a percent.	

TRY IT : : 6.19 Convert each decimal to a percent:

TRY IT : : 6.20 Convert each decimal to a percent:

Let's summarize the results from the previous examples in Table 6.2 so we can look for a pattern.

Decimal	Percent

Table 6.2

Do you see the pattern? To convert a decimal to a percent, we move the decimal point two places to the right and then add the percent sign.

Figure 6.5 uses the decimal numbers in Table 6.2 and shows visually to convert them to percents by moving the decimal point two places to the right and then writing the sign.

Percent	Decimal
006.%	0.06
078.%	0.78
135.%	1.35
012.5%	0.125

Figure 6.5

In Decimals, we learned how to convert fractions to decimals. Now we also know how to change decimals to percents. So to convert a fraction to a percent, we first change it to a decimal and then convert that decimal to a percent.

HOW TO : : CONVERT A FRACTION TO A PERCENT.

Step 1. Convert the fraction to a decimal.

Step 2. Convert the decimal to a percent.

EXAMPLE 6.11

Convert each fraction or mixed number to a percent: — — —

Solution

To convert a fraction to a decimal, divide the numerator by the denominator.

Change to a decimal.	—
Write as a percent by moving the decimal two places.	0.75

Change to a decimal.	—
Write as a percent by moving the decimal two places.	1.375

Write as an improper fraction.	—
Change to a decimal.	—
Write as a percent.	2.20

Notice that we needed to add zeros at the end of the number when moving the decimal two places to the right.

> TRY IT : : 6.21 Convert each fraction or mixed number to a percent: — — —

> TRY IT : : 6.22 Convert each fraction or mixed number to a percent: — — —

Sometimes when changing a fraction to a decimal, the division continues for many decimal places and we will round off the quotient. The number of decimal places we round to will depend on the situation. If the decimal involves money, we round to the hundredths place. For most other cases in this book we will round the number to the nearest thousandth, so the percent will be rounded to the nearest tenth.

EXAMPLE 6.12

Convert — to a percent.

Solution

To change a fraction to a decimal, we divide the numerator by the denominator.

—

Change to a decimal—rounding to the nearest thousandth.	
Write as a percent.	

> TRY IT : : 6.23 Convert the fraction to a percent: —

> TRY IT : : 6.24 Convert the fraction to a percent: —

When we first looked at fractions and decimals, we saw that fractions converted to a repeating decimal. When we converted the fraction — to a decimal, we wrote the answer as We will use this same notation, as well as fraction notation, when we convert fractions to percents in the next example.

EXAMPLE 6.13

An article in a medical journal claimed that approximately — of American adults are obese. Convert the fraction — to a percent.

Solution

	—
Change to a decimal.	$\begin{array}{r} 0.33\ldots \\ 3\overline{)1.00} \\ \underline{9} \\ 10 \\ \underline{9} \\ 1 \end{array}$
Write as a repeating decimal.	
Write as a percent.	—

We could also write the percent as .

TRY IT : : 6.25

Convert the fraction to a percent:

According to the U.S. Census Bureau, about — of United States housing units have just bedroom.

TRY IT : : 6.26

Convert the fraction to a percent:

According to the U.S. Census Bureau, about — of Colorado residents speak a language other than English at home.

6.1 EXERCISES

Practice Makes Perfect

Use the Definition of Percents

In the following exercises, write each percent as a ratio.

1. In the unemployment rate for those with only a high school degree was

2. In among the unemployed, were long-term unemployed.

3. The unemployment rate for those with Bachelor's degrees was in

4. The unemployment rate in Michigan in was

In the following exercises, write as

a ratio and a percent

5. out of nursing candidates received their degree at a community college.

6. out of firefighters and law enforcement officers were educated at a community college.

7. out of first-time freshmen students attend a community college.

8. out of full-time community college faculty have a master's degree.

Convert Percents to Fractions and Decimals

In the following exercises, convert each percent to a fraction and simplify all fractions.

9.

10.

11.

12.

13.

14.

15.

16.

17.

18.

19.

20.

21. —

22. —

23. —

24. —

In the following exercises, convert each percent to a decimal.

25.

26.

27.

28.

29.

30.

31.

32.

33.

34.

35.

36.

37.

38.

39.

40.

In the following exercises, convert each percent to

a simplified fraction and a decimal

41. In of home sales had owner financing. (*Source:* Bloomberg Businessweek, 5/23–29/2011)

42. In of the United States population was of Asian descent. (*Source:* www.census.gov)

43. According to government data, in the number of cell phones in India was of the population.

44. According to the U.S. Census Bureau, among Americans age or older who had doctorate degrees in are women.

45. A couple plans to have two children. The probability they will have two girls is

46. Javier will choose one digit at random from through The probability he will choose is

47. According to the local weather report, the probability of thunderstorms in New York City on July is

48. A club sells tickets to a raffle. Osbaldo bought one ticket. The probability he will win the raffle is

Convert Decimals and Fractions to Percents

In the following exercises, convert each decimal to a percent.

49.

50.

51.

52.

53.

54.

55.

56.

57.

58.

59.

60.

61.

62.

63.

64.

In the following exercises, convert each fraction to a percent.

65. —

66. —

67. —

68. —

69. —

70. —

71. —

72. —

73. —

74. —

75. —

76. —

77. —

78. —

79. —

80. —

In the following exercises, convert each fraction to a percent.

81. — of washing machines needed repair.

82. — of dishwashers needed repair.

In the following exercises, convert each fraction to a percent.

83. According to the National Center for Health Statistics, in — of American adults were obese.

84. The U.S. Census Bureau estimated that in of Americans lived in the same house as they did year before.

In the following exercises, complete the table.

85.

Fraction	Decimal	Percent
—		
—		

Table 6.3

86.

Fraction	Decimal	Percent
—		
—		

Table 6.4

Everyday Math

87. Sales tax Felipa says she has an easy way to estimate the sales tax when she makes a purchase. The sales tax in her city is She knows this is a little less than

Convert to a fraction.

Use your answer from to estimate the sales tax Felipa would pay on a dress.

88. Savings Ryan has of each paycheck automatically deposited in his savings account.

Write as a fraction.

Use your answer from to find the amount that goes to savings from Ryan's paycheck.

Amelio is shopping for textbooks online. He found three sellers that are offering a book he needs for the same price, including shipping. To decide which seller to buy from he is comparing their customer satisfaction ratings. The ratings are given in the chart.

Use the chart to answer the following questions

Seller	Rating

89. Write seller rating as a fraction and a decimal.

90. Write seller rating as a percent and a decimal.

91. Write seller rating as a percent and a decimal.

92. Which seller should Amelio buy from and why?

Writing Exercises

93. Convert to fractions. Do you notice a pattern? Explain what the pattern is.

94. Convert — — — — — — — — and — to percents. Do you notice a pattern? Explain what the pattern is.

95. When the Szetos sold their home, the selling price was of what they had paid for the house ago. Explain what means in this context.

96. According to cnn.com, cell phone use in was of what it had been in Explain what means in this context.

Self Check

After completing the exercises, use this checklist to evaluate your mastery of the objectives of this section.

I can...	Confidently	With some help	No-I don't get it!
use the definition of percent.			
convert percents to fractions and decimals.			
convert decimals and fractions to percents.			

If most of your checks were:

...confidently. Congratulations! You have achieved the objectives in this section. Reflect on the study skills you used so that you can continue to use them. What did you do to become confident of your ability to do these things? Be specific.

...with some help. This must be addressed quickly because topics you do not master become potholes in your road to success. In math, every topic builds upon previous work. It is important to make sure you have a strong foundation before you move on. Whom can you ask for help? Your fellow classmates and instructor are good resources. Is there a place on campus where math tutors are available? Can your study skills be improved?

...no—I don't get it! This is a warning sign and you must not ignore it. You should get help right away or you will quickly be overwhelmed. See your instructor as soon as you can to discuss your situation. Together you can come up with a plan to get you the help you need.

6.2 Solve General Applications of Percent

Learning Objectives

By the end of this section, you will be able to:

- Translate and solve basic percent equations
- Solve applications of percent
- Find percent increase and percent decrease

BE PREPARED : : 6.4 Before you get started, take this readiness quiz.

Translate and solve: — of is

If you missed this problem, review Example 4.105.

BE PREPARED : : 6.5 Simplify:

If you missed this problem, review Example 5.15.

BE PREPARED : : 6.6 Solve:

If you missed this problem, review Example 5.43.

Translate and Solve Basic Percent Equations

We will solve percent equations by using the methods we used to solve equations with fractions or decimals. In the past, you may have solved percent problems by setting them up as proportions. That was the best method available when you did not have the tools of algebra. Now as a prealgebra student, you can translate word sentences into algebraic equations, and then solve the equations.

We'll look at a common application of percent—tips to a server at a restaurant—to see how to set up a basic percent application.

When Aolani and her friends ate dinner at a restaurant, the bill came to They wanted to leave a tip. What amount would the tip be?

To solve this, we want to find what *amount* is of The is called the *base*. The amount of the tip would be or See Figure 6.6. To find the amount of the tip, we multiplied the percent by the base.

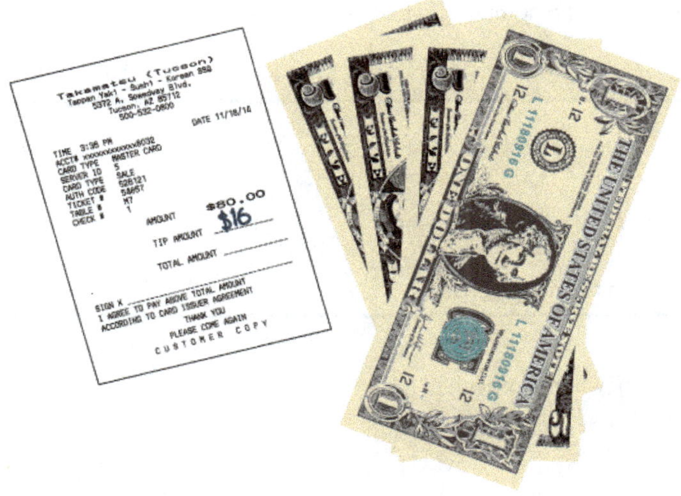

Figure 6.6 A tip for an restaurant bill comes out to

In the next examples, we will find the amount. We must be sure to change the given percent to a decimal when we translate the words into an equation.

EXAMPLE 6.14

What number is of

Solution

Translate into algebra. Let the number.	What number is 35% of 90? $n = 0.35 \cdot 90$
Multiply.	$n = 31.5$
	is of

TRY IT : : 6.27 What number is of

TRY IT : : 6.28 What number is of

EXAMPLE 6.15

of is what number?

Solution

Translate into algebra. Let the number.	125% of 28 is what number? $1.25 \cdot 28 = a$
Multiply.	$35 = a$
	of is .

Remember that a percent over is a number greater than We found that of is which is greater than

> **TRY IT : : 6.29** of is what number?

> **TRY IT : : 6.30** of is what number?

In the next examples, we are asked to find the base.

EXAMPLE 6.16

Translate and solve: is of what number?

Solution

Translate. Let the number.	36 is 75% of what number? $36 = 0.75 \cdot b$
Divide both sides by 0.75.	$\frac{36}{0.75} = \frac{0.75b}{0.75}$
Simplify.	$48 = b$ 36 is 75% of 48.

> **TRY IT : : 6.31** is of what number?

> **TRY IT : : 6.32** is of what number?

EXAMPLE 6.17

of what number is

Solution

Translate. Let the number.	6.5% of what number is $1.17? $0.065 \cdot b = 1.17$
Divide both sides by 0.065.	$\frac{0.065n}{0.065} = \frac{1.17}{0.065}$
Simplify.	$n = 18$ 6.5% of $18 is $1.17.

> **TRY IT : : 6.33** of what number is

> **TRY IT : : 6.34** of what number is

In the next examples, we will solve for the percent.

EXAMPLE 6.18

What percent of is

Solution

Translate into algebra. Let the percent.	What percent of 36 is 9? $p \cdot 36 = 9$
Divide by 36.	$\frac{36p}{36} = \frac{9}{36}$
Simplify.	$p = \frac{1}{4}$
Convert to decimal form.	$p = 0.25$
Convert to percent.	$p = 25\%$ 25% of 36 is 9.

> **TRY IT : : 6.35** What percent of is

> **TRY IT : : 6.36** What percent of is

EXAMPLE 6.19

is what percent of

Solution

Translate into algebra. Let the percent.	144 is what percent of 96? $144 = p \cdot 96$
Divide by 96.	$\frac{144}{96} = \frac{96p}{96}$
Simplify.	$1.5 = p$
Convert to percent.	$150\% = p$ 144 is 150% of 96.

> **TRY IT : : 6.37** is what percent of

> **TRY IT : : 6.38** is what percent of

Solve Applications of Percent

Many applications of percent occur in our daily lives, such as tips, sales tax, discount, and interest. To solve these applications we'll translate to a basic percent equation, just like those we solved in the previous examples in this section. Once you translate the sentence into a percent equation, you know how to solve it.

We will update the strategy we used in our earlier applications to include equations now. Notice that we will translate a sentence into an equation.

HOW TO : : SOLVE AN APPLICATION

Step 1. Identify what you are asked to find and choose a variable to represent it.
Step 2. Write a sentence that gives the information to find it.
Step 3. Translate the sentence into an equation.
Step 4. Solve the equation using good algebra techniques.
Step 5. Check the answer in the problem and make sure it makes sense.
Step 6. Write a complete sentence that answers the question.

Now that we have the strategy to refer to, and have practiced solving basic percent equations, we are ready to solve percent applications. Be sure to ask yourself if your final answer makes sense—since many of the applications we'll solve involve everyday situations, you can rely on your own experience.

EXAMPLE 6.20

Dezohn and his girlfriend enjoyed a dinner at a restaurant, and the bill was They want to leave an tip. If the tip will be of the total bill, how much should the tip be?

Solution

What are you asked to find?	the amount of the tip
Choose a variable to represent it.	Let amount of tip.
Write a sentence that give the information to find it.	The tip is 18% of the total bill.
Translate the sentence into an equation.	The tip is 18% of \$68.50. $t = 0.18 \cdot \$68.50$
Multiply.	$t = 12.33$
Check. Is this answer reasonable?	
If we approximate the bill to \$70 and the percent to 20%, we would have a tip of \$14. So a tip of \$12.33 seems reasonable.	
Write a complete sentence that answers the question.	The couple should leave a tip of \$12.33.

TRY IT : : 6.39

Cierra and her sister enjoyed a special dinner in a restaurant, and the bill was If she wants to leave of the total bill as her tip, how much should she leave?

TRY IT : : 6.40

Kimngoc had lunch at her favorite restaurant. She wants to leave of the total bill as her tip. If her bill was how much will she leave for the tip?

EXAMPLE 6.21

The label on Masao's breakfast cereal said that one serving of cereal provides milligrams (mg) of potassium, which is of the recommended daily amount. What is the total recommended daily amount of potassium?

Nutrition Facts
Serving Size: 1 cup (47g)
Servings Per Container: About 7

Amount Per Serving	Cereal	With Milk
Calories	180	230
Calories from Fat	10	20
	% Daily Value*	
Total Fat 1g	2%	2%
Saturated Fat 0g	0%	0%
Trans Fat 0g		
Polyunsaturated Fat 0.5g		
Monounsaturated Fat 0.5g		
Cholesterol 0mg	0%	0%
Sodium 190mg	8%	11%
Potassium 85mg	2%	8%
Total Carbohydrate 40g	13%	15%
Dietary Fiber 1g	4%	4%
Sugars 8g		
Protein 3g		

Solution

What are you asked to find?	the total amount of potassium recommended
Choose a variable to represent it.	Let total amount of potassium.
Write a sentence that gives the information to find it.	85 mg is 2% of the total amount.
Translate the sentence into an equation.	$\underbrace{85\text{ mg}}_{85}\ \underbrace{\text{is}}_{=}\ \underbrace{2\%}_{0.02}\ \underbrace{\text{of}}_{\cdot}\ \underbrace{a}_{a}?$
Divide both sides by 0.02.	$\frac{85}{0.02} = \frac{0.02a}{0.02}$
Simplify.	$4{,}250 = a$
Check: Is this answer reasonable?	
Yes. 2% is a small percent and 85 is a small part of 4,250.	
Write a complete sentence that answers the question.	The amount of potassium that is recommended is 4250 mg.

TRY IT : : 6.41

One serving of wheat square cereal has grams of fiber, which is of the recommended daily amount. What is the total recommended daily amount of fiber?

TRY IT : : 6.42

One serving of rice cereal has mg of sodium, which is of the recommended daily amount. What is the total recommended daily amount of sodium?

EXAMPLE 6.22

Mitzi received some gourmet brownies as a gift. The wrapper said each brownie was calories, and had calories of fat. What percent of the total calories in each brownie comes from fat?

Solution

What are you asked to find?	the percent of the total calories from fat
Choose a variable to represent it.	Let percent from fat.
Write a sentence that gives the information to find it.	What percent of 480 is 240?
Translate the sentence into an equation.	$\underbrace{\text{What percent}}_{p}\ \underbrace{\text{of}}_{\cdot}\ \underbrace{480}_{480}\ \underbrace{\text{is}}_{=}\ \underbrace{240?}_{240}$
Divide both sides by 480.	$\frac{p \cdot 480}{480} = \frac{240}{480}$
Simplify.	$p = 0.5$
Convert to percent form.	$p = 50\%$
Check. Is this answer reasonable?	
Yes. 240 is half of 480, so 50% makes sense.	
Write a complete sentence that answers the question.	Of the total calories in each brownie, 50% is fat.

TRY IT : : 6.43

Veronica is planning to make muffins from a mix. The package says each muffin will be calories and calories will be from fat. What percent of the total calories is from fat? (Round to the nearest whole percent.)

TRY IT : : 6.44

The brownie mix Ricardo plans to use says that each brownie will be calories, and calories are from fat. What percent of the total calories are from fat?

Find Percent Increase and Percent Decrease

People in the media often talk about how much an amount has increased or decreased over a certain period of time. They usually express this increase or decrease as a percent.

To find the **percent increase**, first we find the amount of increase, which is the difference between the new amount and the original amount. Then we find what percent the amount of increase is of the original amount.

HOW TO : : FIND PERCENT INCREASE.

Step 1. Find the amount of increase.

-

Step 2. Find the percent increase as a percent of the original amount.

EXAMPLE 6.23

In the California governor proposed raising community college fees from per unit to per unit. Find the percent increase. (Round to the nearest tenth of a percent.)

Solution

What are you asked to find?	the percent increase
Choose a variable to represent it.	Let percent.
Find the amount of increase.	$\underbrace{36}_{\text{new amount}} - \underbrace{26}_{\text{original amount}} = \underbrace{10}_{\text{increase}}$
Find the percent increase.	The increase is what percent of the original amount?
Translate to an equation.	$\underbrace{10}_{10}\ \underbrace{\text{is}}_{=}\ \underbrace{\text{what percent}}_{p}\ \underbrace{\text{of}}_{\cdot}\ \underbrace{26?}_{26}$
Divide both sides by 26.	$\frac{10}{26} = \frac{26p}{26}$
Round to the nearest thousandth.	$0.384 = p$
Convert to percent form.	$38.4\% = p$
Write a complete sentence.	The new fees represent a 38.4% increase over the old fees.

TRY IT : : 6.45

In the IRS increased the deductible mileage cost to cents from cents. Find the percent increase. (Round to the nearest tenth of a percent.)

TRY IT : : 6.46

In the standard bus fare in Chicago was In the standard bus fare was Find the percent increase. (Round to the nearest tenth of a percent.)

Finding the **percent decrease** is very similar to finding the percent increase, but now the amount of decrease is the difference between the original amount and the final amount. Then we find what percent the amount of decrease is of the original amount.

HOW TO : : FIND PERCENT DECREASE.

Step 1. Find the amount of decrease.

Step 2. Find the percent decrease as a percent of the original amount.

EXAMPLE 6.24

The average price of a gallon of gas in one city in June was The average price in that city in July was Find the percent decrease.

Solution

What are you asked to find?	the percent decrease
Choose a variable to represent it.	Let percent.
Find the amount of decrease.	$\underbrace{3.71}_{\text{original amount}} - \underbrace{3.64}_{\text{new amount}} = \underbrace{0.07}_{\text{increase}}$
Find the percent of decrease.	The decrease is what percent of the original amount?
Translate to an equation.	$\underbrace{0.07}_{0.07}\ \underbrace{\text{is}}_{=}\ \underbrace{\text{what percent}}_{p}\ \underbrace{\text{of}}_{\cdot}\ \underbrace{3.71}_{3.71}$
Divide both sides by 3.71.	$\frac{0.07}{3.71} = \frac{3.71p}{3.71}$
Round to the nearest thousandth.	$0.019 = p$
Convert to percent form.	$1.9\% = p$
Write a complete sentence.	The price of gas decreased 1.9%.

TRY IT : : 6.47

The population of one city was about in The population of the city is projected to be about in Find the percent decrease. (Round to the nearest tenth of a percent.)

TRY IT : : 6.48

Last year Sheila's salary was Because of furlough days, this year her salary was Find the percent decrease. (Round to the nearest tenth of a percent.)

MEDIA : : ACCESS ADDITIONAL ONLINE RESOURCES

- Percent Increase and Percent Decrease Visualization (http://www.openstax.org/l/24percentincdec)

6.2 EXERCISES

Practice Makes Perfect

Translate and Solve Basic Percent Equations

In the following exercises, translate and solve.

97. What number is of

98. What number is of

99. What number is of

100. What number is of

101. of is what number?

102. of is what number?

103. of is what number?

104. of is what number?

105. is of what number?

106. is of what number?

107. is of what number?

108. is of what number?

109. of what number is

110. of what number is

111. of what number is

112. of what number is

113. What percent of is

114. What percent of is

115. What percent of is

116. What percent of is

117. is what percent of

118. is what percent of

119. is what percent of

120. is what percent of

Solve Applications of Percents

In the following exercises, solve the applications of percents.

121. Geneva treated her parents to dinner at their favorite restaurant. The bill was She wants to leave of the total bill as a tip. How much should the tip be?

122. When Hiro and his co-workers had lunch at a restaurant the bill was They want to leave of the total bill as a tip. How much should the tip be?

123. Trong has of each paycheck automatically deposited to his savings account. His last paycheck was How much money was deposited to Trong's savings account?

124. Cherise deposits of each paycheck into her retirement account. Her last paycheck was How much did Cherise deposit into her retirement account?

125. One serving of oatmeal has grams of fiber, which is of the recommended daily amount. What is the total recommended daily amount of fiber?

126. One serving of trail mix has grams of carbohydrates, which is of the recommended daily amount. What is the total recommended daily amount of carbohydrates?

127. A bacon cheeseburger at a popular fast food restaurant contains milligrams (mg) of sodium, which is of the recommended daily amount. What is the total recommended daily amount of sodium?

128. A grilled chicken salad at a popular fast food restaurant contains milligrams (mg) of sodium, which is of the recommended daily amount. What is the total recommended daily amount of sodium?

129. The nutrition fact sheet at a fast food restaurant says the fish sandwich has calories, and calories are from fat. What percent of the total calories is from fat?

130. The nutrition fact sheet at a fast food restaurant says a small portion of chicken nuggets has calories, and calories are from fat. What percent of the total calories is from fat?

131. Emma gets paid per month. She pays a month for rent. What percent of her monthly pay goes to rent?

132. Dimple gets paid per month. She pays a month for rent. What percent of her monthly pay goes to rent?

Find Percent Increase and Percent Decrease

In the following exercises, find the percent increase or percent decrease.

133. Tamanika got a raise in her hourly pay, from to Find the percent increase.

134. Ayodele got a raise in her hourly pay, from to Find the percent increase.

135. Annual student fees at the University of California rose from about in to about in Find the percent increase.

136. The price of a share of one stock rose from to Find the percent increase.

137. According to Time magazine annual global seafood consumption rose from pounds per person in to pounds per person today. Find the percent increase. (Round to the nearest tenth of a percent.)

138. In one month, the median home price in the Northeast rose from to Find the percent increase. (Round to the nearest tenth of a percent.)

139. A grocery store reduced the price of a loaf of bread from to Find the percent decrease.

140. The price of a share of one stock fell from to Find the percent decrease.

141. Hernando's salary was last year. This year his salary was cut to Find the percent decrease.

142. From to the population of Detroit fell from about to about Find the percent decrease. (Round to the nearest tenth of a percent.)

143. In one month, the median home price in the West fell from to Find the percent decrease. (Round to the nearest tenth of a percent.)

144. Sales of video games and consoles fell from million to million in one year. Find the percent decrease. (Round to the nearest tenth of a percent.)

Everyday Math

145. Tipping At the campus coffee cart, a medium coffee costs MaryAnne brings with her when she buys a cup of coffee and leaves the change as a tip. What percent tip does she leave?

146. Late Fees Alison was late paying her credit card bill of She was charged a late fee. What was the amount of the late fee?

Writing Exercises

147. Without solving the problem is of what number", think about what the solution might be. Should it be a number that is greater than or less than Explain your reasoning.

148. Without solving the problem "What is of think about what the solution might be. Should it be a number that is greater than or less than Explain your reasoning.

149. After returning from vacation, Alex said he should have packed fewer shorts and more shirts. Explain what Alex meant.

150. Because of road construction in one city, commuters were advised to plan their Monday morning commute to take of their usual commuting time. Explain what this means.

Self Check

After completing the exercises, use this checklist to evaluate your mastery of the objectives of this section.

I can...	Confidently	With some help	No-I don't get it!
translate and solve basic percent equations.			
solve applications of percent.			
find percent increase and percent decrease.			

After reviewing this checklist, what will you do to become confident for all objectives?

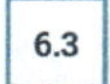

6.3 Solve Sales Tax, Commission, and Discount Applications

Learning Objectives

By the end of this section, you will be able to:

- Solve sales tax applications
- Solve commission applications
- Solve discount applications
- Solve mark-up applications

BE PREPARED : : 6.7 Before you get started, take this readiness quiz.

Solve through multiplication.

If you missed this problem, review Example 5.17.

BE PREPARED : : 6.8 Solve through division.

If you missed this problem, review Example 5.22.

Solve Sales Tax Applications

Sales tax and commissions are applications of percent in our everyday lives. To solve these applications, we will follow the same strategy we used in the section on decimal operations. We show it again here for easy reference.

HOW TO : :

Solve an application

Step 1. Identify what you are asked to find and choose a variable to represent it.

Step 2. Write a sentence that gives the information to find it.

Step 3. Translate the sentence into an equation.

Step 4. Solve the equation using good algebra techniques.

Step 5. Check the answer in the problem and make sure it makes sense.

Step 6. Write a complete sentence that answers the question.

Remember that whatever the application, once we write the sentence with the given information (Step 2), we can translate it to a percent equation and then solve it.

Do you pay a tax when you shop in your city or state? In many parts of the United States, sales tax is added to the purchase price of an item. See Figure 6.7. The sales tax is determined by computing a percent of the purchase price.

To find the sales tax multiply the purchase price by the sales tax rate. Remember to convert the sales tax rate from a percent to a decimal number. Once the sales tax is calculated, it is added to the purchase price. The result is the total cost—this is what the customer pays.

BOULEVARD
ONE MISSION STREET
SAN FRANCISCO, CA 94105
(415) 543-6084
DINING ROOM

5213 KEN

Tbl 32/1	Chk 5265	Gst 2
	Apr14'08 08:05PM	

1 D SOUP	13.75
1 D LOBSTER LINGUI	18.75
1 D LAMB	32.00
1 1/2 GL SAUV BLANC	4.25
1 D BANANAS FOSTER	9.50
SUBTOTAL	78.25
Tax (8%)	6.26
Total	84.51

BOULEVARD COOKBOOKS
ARE NOW AVAILABLE
PLEASE ASK YOUR SERVER
THANK YOU FOR DINING WITH US

Figure 6.7 The sales tax is calculated as a percent of the purchase price.

Sales Tax

The sales tax is a percent of the purchase price.

EXAMPLE 6.25

Cathy bought a bicycle in Washington, where the sales tax rate was of the purchase price. What was the sales tax and the total cost of a bicycle if the purchase price of the bicycle was

Solution

Identify what you are asked to find.	What is the sales tax?
Choose a variable to represent it.	Let sales tax.
Write a sentence that gives the information to find it.	The sales tax is 6.5% of the purchase price.
Translate into an equation. (Remember to change the percent to a decimal).	The sales tax is 6.5% of the \$392 purchase price. $t = 0.065 \cdot 392$
Simplify.	$t = 25.48$
Check: Is this answer reasonable?	
Yes, because the sales tax amount is less than 10% of the purchase price.	
Write a complete sentence that answers the question.	The sales tax is \$25.48.

Identify what you are asked to find.	What is the total cost of the bicycle?
Choose a variable to represent it.	Let total cost of bicycle.
Write a sentence that gives the information to find it.	The total cost is the purchase price plus the sales tax.
Translate into an equation.	The total cost is \$392 plus \$25.48 $c = 392 + 25.48$
Simplify.	$c = 417.48$
Check: Is this answer reasonable?	
Yes, because the total cost is a little more than the purchase price.	
Write a complete sentence that answers the question.	The total cost of the bicycle is \$417.48.

 **TRY IT : :** 6.49

Find the sales tax and the total cost: Alexandra bought a television set for in Boston, where the sales tax rate was of the purchase price.

 **TRY IT : :** 6.50

Find the sales tax and the total cost: Kim bought a winter coat for in St. Louis, where the sales tax rate was of the purchase price.

EXAMPLE 6.26

Evelyn bought a new smartphone for plus tax. She was surprised when she got the receipt and saw that the tax was What was the sales tax rate for this purchase?

Solution

Identify what you are asked to find.	What is the sales tax rate?
Choose a variable to represent it.	Let sales tax.
Write a sentence that gives the information to find it.	What percent of the price is the sales tax?
Translate into an equation.	What percent of the \$499 price is the \$42.42 tax? $r \cdot 499 = 42.42$
Divide.	$\frac{499r}{499} = \frac{42.42}{499}$
Simplify.	$r = 0.085$
Check. Is this answer reasonable?	
Yes, because 8.5% is close to 10%. 10% of \$500 is \$50, which is close to \$42.42.	
Write a complete sentence that answers the question.	The sales tax rate is 8.5%.

TRY IT : : 6.51

Diego bought a new car for He was surprised that the dealer then added What was the sales tax rate for this purchase?

TRY IT : : 6.52 What is the sales tax rate if a purchase will have of sales tax added to it?

Solve Commission Applications

Sales people often receive a **commission**, or percent of total sales, for their sales. Their income may be just the commission they earn, or it may be their commission added to their hourly wages or salary. The commission they earn is calculated as a certain percent of the price of each item they sell. That percent is called the **rate of commission**.

Commission

A commission is a percentage of total sales as determined by the rate of commission.

To find the commission on a sale, multiply the rate of commission by the total sales. Just as we did for computing sales tax, remember to first convert the rate of commission from a percent to a decimal.

EXAMPLE 6.27

Helene is a realtor. She receives commission when she sells a house. How much commission will she receive for selling a house that costs

Solution

Identify what you are asked to find.	What is the commission?
Choose a variable to represent it.	Let the commission.
Write a sentence that gives the information to find it.	The commission is 3% of the price.
Translate into an equation.	The commission is 3% of the \$260,000 price. $c = 0.03 \cdot 260{,}000$
Simplify.	$c = 7800$
Check. Is this answer reasonable?	
Yes. 1% of \$260,000 is \$2,600, and \$7,800 is three times \$2,600.	
Write a complete sentence that answers the question.	Helene will receive a commission of \$7,800.

TRY IT : : 6.53

Bob is a travel agent. He receives commission when he books a cruise for a customer. How much commission will he receive for booking a cruise?

TRY IT : : 6.54

Fernando receives commission when he makes a computer sale. How much commission will he receive for selling a computer for

EXAMPLE 6.28

Rikki earned commission when she sold a stove. What rate of commission did she get?

Solution

Identify what you are asked to find.	What is the rate of commission?
Choose a variable to represent it.	Let the rate of commission.
Write a sentence that gives the information to find it.	The commission is what percent of the sale?
Translate into an equation.	The \$87 commission is what percent of the \$1450 sale? $87 = 0.03 \cdot 1450$
Divide.	$\frac{87}{1450} = \frac{1450r}{1450}$
Simplify.	$0.06 = r$
Change to percent form.	$r = 6\%$
Check if this answer is reasonable.	
Yes. A 10% commission would have been \$145. The 6% commission, \$87, is a little more than half of that.	
Write a complete sentence that answers the question.	The commission was 6% of the price of the stove.

> **TRY IT : :** 6.55
>
> Homer received commission when he sold a car for What rate of commission did he get?

> **TRY IT : :** 6.56
>
> Bernice earned commission when she sold an living room set. What rate of commission did she get?

Solve Discount Applications

Applications of discount are very common in retail settings Figure 6.8. When you buy an item on sale, the **original price** of the item has been reduced by some dollar amount. The **discount rate**, usually given as a percent, is used to determine the amount of the discount. To determine the **amount of discount**, we multiply the discount rate by the original price. We summarize the discount model in the box below.

Figure 6.8 Applications of discounts are common in everyday life. (credit: Charleston's TheDigitel, Flickr)

Discount

An amount of discount is a percent off the original price.

The sale price should always be less than the original price. In some cases, the amount of discount is a fixed dollar amount. Then we just find the sale price by subtracting the amount of discount from the original price.

EXAMPLE 6.29

Jason bought a pair of sunglasses that were on sale for off. The original price of the sunglasses was What was the sale price of the sunglasses?

Solution

Identify what you are asked to find.	What is the sale price?
Choose a variable to represent it.	Let the sale price.
Write a sentence that gives the information to find it.	The sale price is the original price minus the discount.
Translate into an equation.	$\underbrace{\text{The sale price}}_{s}\ \underbrace{\text{is}}_{=}\ \underbrace{\text{the original \$39 price}}_{39}\ \underbrace{\text{minus}}_{-}\ \underbrace{\text{the \$10 discount.}}_{10}$
Simplify.	$s = 29$
Check if this answer is reasonable.	
Yes. The sale price, \$29, is less than the original price, \$39.	
Write a complete sentence that answers the question.	The sale price of the sunglasses was \$29.

TRY IT : : 6.57

Marta bought a dishwasher that was on sale for off. The original price of the dishwasher was What was the sale price of the dishwasher?

TRY IT : : 6.58

Orlando bought a pair of shoes that was on sale for off. The original price of the shoes was What was the sale price of the shoes?

In Example 6.29, the amount of discount was a set amount, In Example 6.30 the discount is given as a percent of the original price.

EXAMPLE 6.30

Elise bought a dress that was discounted off of the original price of What was the amount of discount and the sale price of the dress?

Solution

Before beginning, you may find it helpful to organize the information in a list.
Original price = \$140
Discount rate = 35%

Amount of discount = ?

Identify what you are asked to find.	What is the amount of discount?
Choose a variable to represent it.	Let the amount of discount.
Write a sentence that gives the information to find it.	The discount is 35% of the original price.
Translate into an equation.	The discount is 35% of the \$140 original price. $s = 0.35 \cdot 140$
Simplify.	$d = 49$
Check if this answer is reasonable.	
Yes. A \$49 discount is reasonable for a \$140 dress.	
Write a complete sentence that answers the question.	The amount of discount was \$49.

Original price = \$140
Amount of discount = \$49
Sale price = ?

Identify what you are asked to find.	What is the sale price of the dress?
Choose a variable to represent it.	Let the sale price.
Write a sentence that gives the information to find it.	The sale price is the original price minus the discount.
Translate into an equation.	The sale price is the \$140 minus the \$49 discount. $s = 140 - 49$
Simplify.	$s = 91$
Check if this answer is reasonable.	
Yes. The sale price, \$91, is less than the original price, \$140.	
Write a complete sentence that answers the question.	The sale price of the dress was \$91.

TRY IT : : 6.59

Find the amount of discount and the sale price: Sergio bought a belt that was discounted from an original price of

TRY IT : : 6.60

Find the amount of discount and the sale price: Oscar bought a barbecue grill that was discounted from an original price of

There may be times when you buy something on sale and want to know the discount rate. The next example will show this case.

EXAMPLE 6.31

Jeannette bought a swimsuit at a sale price of The original price of the swimsuit was Find the amount of discount and discount rate.

Solution

Before beginning, you may find it helpful to organize the information in a list.
Original price = $31
Amount of discount = ?
Sale price = $13.95

Identify what you are asked to find.	What is the amount of discount?
Choose a variable to represent it.	Let the amount of discount.
Write a sentence that gives the information to find it.	The discount is the original price minus the sale price.
Translate into an equation.	The discount is the $31 original price minus the $13.95 sale price. $s = 31 - 13.95$
Simplify.	$d = 17.05$
Check if this answer is reasonable.	
Yes. The $17.05 discount is less than the original price.	
Write a complete sentence that answers the question.	The amount of discount was $17.05.

Before beginning, you may find it helpful to organize the information in a list.
Original price = $31
Amount of discount = $17.05
Discount rate = ?

Identify what you are asked to find.	What is the discount rate?
Choose a variable to represent it.	Let the discount rate.
Write a sentence that gives the information to find it.	The discount is what percent of the original price?
Translate into an equation.	The discount of $17.05 is what percent of the $31 original price. $s = r \cdot 31$
Divide.	$\frac{17.05}{31} = \frac{r(31)}{31}$
Simplify.	$0.55 = r$
Check if this answer is reasonable.	
The rate of discount was a little more than 50% and the amount of discount is a little more than half of $31.	
Write a complete sentence that answers the question.	The rate of discount was 55%.

TRY IT : : 6.61

Find the amount of discount and the discount rate: Lena bought a kitchen table at the sale price of The original price of the table was

TRY IT : : 6.62

Find the amount of discount and the discount rate: Nick bought a multi-room air conditioner at a sale price of The original price of the air conditioner was

Solve Mark-up Applications

Applications of mark-up are very common in retail settings. The price a retailer pays for an item is called the **wholesale price**. The retailer then adds a **mark-up** to the wholesale price to get the **list price**, the price he sells the item for. The mark-up is usually calculated as a percent of the wholesale price. The percent is called the **mark-up rate**. To determine the amount of mark-up, multiply the mark-up rate by the wholesale price. We summarize the mark-up model in the box below.

Mark-up

The mark-up is the amount added to the wholesale price.

The list price should always be more than the wholesale price.

EXAMPLE 6.32

Adam's art gallery bought a photograph at the wholesale price of Adam marked the price up Find the amount of mark-up and the list price of the photograph.

Solution

Identify what you are asked to find.	What is the amount of mark-up?
Choose a variable to represent it.	Let the amount of each mark-up.
Write a sentence that gives the information to find it.	The mark-up is 40% of the wholesale price.
Translate into an equation.	The mark-up is 40% of the \$250 wholesale price. $\underbrace{\text{The mark-up}}_{m}\ \underbrace{\text{is}}_{=}\ \underbrace{40\%}_{0.40}\ \underbrace{\text{of}}_{\cdot}\ \underbrace{\text{the \$250 wholesale price.}}_{250}$
Simplify.	$m = 100$
Check if this answer is reasonable.	
Yes. The markup rate is less than 50% and \$100 is less than half of \$250.	
Write a complete sentence that answers the question.	The mark-up on the photograph was \$100.

Identify what you are asked to find.	What is the list price?
Choose a variable to represent it.	Let the list price.
Write a sentence that gives the information to find it.	The list price is the wholesale price plus the mark-up.
Translate into an equation.	The list price is the \$250 wholesale price plus the \$100 mark-up. $p = 250 + 100$
Simplify.	$p = 350$
Check if this answer is reasonable.	
Yes. The list price, \$350, is more than the wholesale price, \$250.	
Write a complete sentence that answers the question.	The list price of the photograph was \$350.

TRY IT : : 6.63

Jim's music store bought a guitar at wholesale price Jim marked the price up Find the amount of mark-up and the list price.

TRY IT : : 6.64

The Auto Resale Store bought Pablo's Toyota for They marked the price up Find the amount of mark-up and the list price.

6.3 EXERCISES

Practice Makes Perfect

Solve Sales Tax Applications

In the following exercises, find the sales tax and the total cost.

151. The cost of a pair of boots was The sales tax rate is of the purchase price.

152. The cost of a refrigerator was The sales tax rate is of the purchase price.

153. The cost of a microwave oven was The sales tax rate is of the purchase price.

154. The cost of a tablet computer is The sales tax rate is of the purchase price.

155. The cost of a file cabinet is The sales tax rate is of the purchase price.

156. The cost of a luggage set The sales tax rate is of the purchase price.

157. The cost of a dresser The sales tax rate is of the purchase price.

158. The cost of a sofa is The sales tax rate is of the purchase price.

In the following exercises, find the sales tax rate.

159. Shawna bought a mixer for The sales tax on the purchase was

160. Orphia bought a coffee table for The sales tax on the purchase was

161. Bopha bought a bedroom set for The sales tax on the purchase was

162. Ruth bought a washer and dryer set for The sales tax on the purchase was

Solve Commission Applications

In the following exercises, find the commission.

163. Christopher sold his dinette set for through an online site, which charged him of the selling price as commission. How much was the commission?

164. Michele rented a booth at a craft fair, which charged her commission on her sales. One day her total sales were How much was the commission?

165. Farrah works in a jewelry store and receives commission when she makes a sale. How much commission will she receive for selling a ring?

166. Jamal works at a car dealership and receives commission when he sells a car. How much commission will he receive for selling a car?

167. Hector receives commission when he sells an insurance policy. How much commission will he receive for selling a policy for

168. Denise receives commission when she books a tour at the travel agency. How much commission will she receive for booking a tour with total cost

In the following exercises, find the rate of commission.

169. Dontay is a realtor and earned commission on the sale of a house. What is his rate of commission?

170. Nevaeh is a cruise specialist and earned commission after booking a cruise that cost What is her rate of commission?

171. As a waitress, Emily earned in tips on sales of last Saturday night. What was her rate of commission?

172. Alejandra earned commission on weekly sales of as a salesperson at the computer store. What is her rate of commission?

173. Maureen earned commission when she sold a car. What was the rate of commission?

174. Lucas earned commission when he brought a job to his office. What was the rate of commission?

Solve Discount Applications

In the following exercises, find the sale price.

175. Perla bought a cellphone that was on sale for off. The original price of the cellphone was

176. Sophie saw a dress she liked on sale for off. The original price of the dress was

177. Rick wants to buy a tool set with original price Next week the tool set will be on sale for off.

178. Angelo's store is having a sale on TV sets. One set, with an original price of is selling for off.

In the following exercises, find the amount of discount and the sale price.

179. Janelle bought a beach chair on sale at off. The original price was

180. Errol bought a skateboard helmet on sale at off. The original price was

181. Kathy wants to buy a camera that lists for The camera is on sale with a discount.

182. Colleen bought a suit that was discounted from an original price of

183. Erys bought a treadmill on sale at off. The original price was

184. Jay bought a guitar on sale at off. The original price was

In the following exercises, find the amount of discount and the discount rate. (Round to the nearest tenth of a percent if needed.)

185. Larry and Donna bought a sofa at the sale price of The original price of the sofa was

186. Hiroshi bought a lawnmower at the sale price of The original price of the lawnmower is

187. Patty bought a baby stroller on sale for The original price of the stroller was

188. Bill found a book he wanted on sale for The original price of the book was

189. Nikki bought a patio set on sale for The original price was

190. Stella bought a dinette set on sale for The original price was

Solve Mark-up Applications

In the following exercises, find the amount of the mark-up and the list price.

191. Daria bought a bracelet at wholesale cost to sell in her handicraft store. She marked the price up

192. Regina bought a handmade quilt at wholesale cost to sell in her quilt store. She marked the price up

193. Tom paid a pound for tomatoes to sell at his produce store. He added a mark-up.

194. Flora paid her supplier a stem for roses to sell at her flower shop. She added an mark-up.

195. Alan bought a used bicycle for After re-conditioning it, he added mark-up and then advertised it for sale.

196. Michael bought a classic car for He restored it, then added mark-up before advertising it for sale.

Everyday Math

197. Coupons Yvonne can use two coupons for the same purchase at her favorite department store. One coupon gives her off and the other gives her off. She wants to buy a bedspread that sells for

Calculate the discount price if Yvonne uses the coupon first and then takes off.

Calculate the discount price if Yvonne uses the off coupon first and then uses the coupon.

In which order should Yvonne use the coupons?

198. Cash Back Jason can buy a bag of dog food for at two different stores. One store offers cash back on the purchase plus off his next purchase. The other store offers cash back.

Calculate the total savings from the first store, including the savings on the next purchase.

Calculate the total savings from the second store.

Which store should Jason buy the dog food from? Why?

Writing Exercises

199. Priam bought a jacket that was on sale for off. The original price of the jacket was While the sales clerk figured the price by calculating the amount of discount and then subtracting that amount from Priam found the price faster by calculating of

Explain why Priam was correct.

Will Priam's method work for any original price?

200. Roxy bought a scarf on sale for off. The original price of the scarf was Roxy claimed that the price she paid for the scarf was the same as the amount she saved. Was Roxy correct? Explain.

Self Check

After completing the exercises, use this checklist to evaluate your mastery of the objectives of this section.

I can...	Confidently	With some help	No-I don't get it!
solve sales tax applications.			
solve commission applications.			
solve discount applications.			
solve mark-up applications.			

What does this checklist tell you about your mastery of this section? What steps will you take to improve?

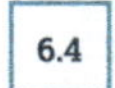

Solve Simple Interest Applications

Learning Objectives

By the end of this section, you will be able to:

- Use the simple interest formula
- Solve simple interest applications

BE PREPARED : : 6.9 Before you get started, take this readiness quiz.

Solve

If you missed this problem, review Example 5.43.

BE PREPARED : : 6.10 Solve

If you missed this problem, review Example 5.44.

Use the Simple Interest Formula

Do you know that banks pay you to let them keep your money? The money you put in the bank is called the **principal**, and the bank pays you **interest**, The interest is computed as a certain percent of the principal; called the **rate of interest**, The rate of interest is usually expressed as a percent per year, and is calculated by using the decimal equivalent of the percent. The variable for time, represents the number of years the money is left in the account.

Simple Interest

If an amount of money, the principal, is invested for a period of years at an annual interest rate the amount of interest, earned is

where

Interest earned according to this formula is called **simple interest**.

The formula we use to calculate simple interest is To use the simple interest formula we substitute in the values for variables that are given, and then solve for the unknown variable. It may be helpful to organize the information by listing all four variables and filling in the given information.

EXAMPLE 6.33

Find the simple interest earned after years on at an interest rate of

Solution

Organize the given information in a list.

We will use the simple interest formula to find the interest.

Write the formula.

Substitute the given information. Remember to write the percent in decimal form.

Simplify.

Check your answer. Is $90 a reasonable interest earned on $500 in 3 years?

In 3 years the money earned 18%. If we rounded to 20%, the interest would have been 500(0.20) or $100. Yes, $90 is reasonable.

Write a complete sentence that answers the question. The simple interest is $90.

TRY IT : : 6.65 Find the simple interest earned after years on at an interest rate of

TRY IT : : 6.66 Find the simple interest earned after years on at an interest rate of

In the next example, we will use the simple interest formula to find the principal.

EXAMPLE 6.34

Find the principal invested if interest was earned in years at an interest rate of

Solution

Organize the given information in a list.

We will use the simple interest formula to find the principal.

Write the formula.

Substitute the given information.

Divide.

Simplify.

Check your answer. Is it reasonable that $2,225 would earn $178 in 2 years?

Write a complete sentence that answers the question. The principal is $2,225.

TRY IT : : 6.67 Find the principal invested if interest was earned in years at an interest rate of

TRY IT : : 6.68

Find the principal invested if interest was earned in years at an interest rate of

Now we will solve for the rate of interest.

EXAMPLE 6.35

Find the rate if a principal of earned interest in years.

Solution

Organize the given information.

We will use the simple interest formula to find the rate.

Write the formula.	
Substitute the given information.	
Multiply.	
Divide.	
Simplify.	
Write as a percent.	
Check your answer. Is 11.5% a reasonable rate if $3,772 was earned in 4 years?	
Write a complete sentence that answers the question.	The rate was 11.5%.

TRY IT : : 6.69 Find the rate if a principal of earned interest in years.

TRY IT : : 6.70 Find the rate if a principal of earned interest in years.

Solve Simple Interest Applications

Applications with simple interest usually involve either investing money or borrowing money. To solve these applications, we continue to use the same strategy for applications that we have used earlier in this chapter. The only difference is that in place of translating to get an equation, we can use the simple interest formula.

We will start by solving a simple interest application to find the interest.

EXAMPLE 6.36

Nathaly deposited in her bank account where it will earn interest. How much interest will Nathaly earn in years?

Solution

We are asked to find the Interest,

Organize the given information in a list.

Write the formula.	
Substitute the given information.	
Simplify.	
Check your answer. Is $2,500 a reasonable interest on $12,500 over 5 years?	
At 4% interest per year, in 5 years the interest would be 20% of the principal. Is 20% of $12,500 equal to $2,500? Yes.	
Write a complete sentence that answers the question.	The interest is $2,500.

TRY IT : : 6.71

Areli invested a principal of in her bank account with interest rate How much interest did she earn in years?

TRY IT : : 6.72

Susana invested a principal of in her bank account with interest rate How much interest did she earn in years?

There may be times when you know the amount of interest earned on a given principal over a certain length of time, but you don't know the rate. For instance, this might happen when family members lend or borrow money among themselves instead of dealing with a bank. In the next example, we'll show how to solve for the rate.

EXAMPLE 6.37

Loren lent his brother to help him buy a car. In his brother paid him back the plus in interest. What was the rate of interest?

Solution

We are asked to find the rate of interest,

Organize the given information.

Write the formula.	
Substitute the given information.	
Multiply.	
Divide.	
Simplify.	
Change to percent form.	
Check your answer. Is 5.5% a reasonable interest rate to pay your brother?	
Write a complete sentence that answers the question.	The rate of interest was 5.5%.

TRY IT : : 6.73

Jim lent his sister to help her buy a house. In years, she paid him the plus interest. What was the rate of interest?

TRY IT : : 6.74

Hang borrowed from her parents to pay her tuition. In years, she paid them interest in addition to the she borrowed. What was the rate of interest?

There may be times when you take a loan for a large purchase and the amount of the principal is not clear. This might happen, for instance, in making a car purchase when the dealer adds the cost of a warranty to the price of the car. In the next example, we will solve a simple interest application for the principal.

EXAMPLE 6.38

Eduardo noticed that his new car loan papers stated that with an interest rate of he would pay in interest over years. How much did he borrow to pay for his car?

Solution

We are asked to find the principal,

Organize the given information.

Write the formula.	
Substitute the given information.	
Multiply.	
Divide.	
Simplify.	
Check your answer. Is $17,590 a reasonable amount to borrow to buy a car?	
Write a complete sentence that answers the question.	The amount borrowed was $17,590.

TRY IT : : 6.75

Sean's new car loan statement said he would pay in interest from an interest rate of over years. How much did he borrow to buy his new car?

TRY IT : : 6.76

In years, Gloria's bank account earned interest at How much had she deposited in the account?

In the simple interest formula, the rate of interest is given as an annual rate, the rate for one year. So the units of time must be in years. If the time is given in months, we convert it to years.

EXAMPLE 6.39

Caroline got as graduation gifts and invested it in a certificate of deposit that earned interest. How much interest did this investment earn?

Solution

We are asked to find the interest,

Organize the given information.

Write the formula.	
Substitute the given information, converting 10 months to — of a year.	—
Multiply.	
Check your answer. Is $15.75 a reasonable amount of interest?	
If Caroline had invested the $900 for a full year at 2% interest, the amount of interest would have been $18. Yes, $15.75 is reasonable.	
Write a complete sentence that answers the question.	The interest earned was $15.75.

TRY IT : : 6.77

Adriana invested for months in an account that paid interest. How much interest did she earn?

TRY IT : : 6.78

Milton invested for months in an account that paid interest How much interest did he earn?

6.4 EXERCISES

Practice Makes Perfect

Use the Simple Interest Formula

In the following exercises, use the simple interest formula to fill in the missing information.

201.

Interest	Principal	Rate	Time (years)

Table 6.5

202.

Interest	Principal	Rate	Time (years)

Table 6.6

203.

Interest	Principal	Rate	Time (years)

Table 6.7

204.

Interest	Principal	Rate	Time (years)

Table 6.8

205.

Interest	Principal	Rate	Time (years)

Table 6.9

206.

Interest	Principal	Rate	Time (years)

Table 6.10

In the following exercises, solve the problem using the simple interest formula.

207. Find the simple interest earned after years on at an interest rate of

208. Find the simple interest earned after years on at an interest rate of

209. Find the simple interest earned after years on at an interest rate of

210. Find the simple interest earned after years on at an interest rate of

211. Find the simple interest earned after years on at an interest rate of

212. Find the simple interest earned after years on at an interest rate of

213. Find the principal invested if interest was earned in years at an interest rate of

214. Find the principal invested if interest was earned in years at an interest rate of

215. Find the principal invested if interest was earned in years at an interest rate of

216. Find the principal invested if interest was earned in years at an interest rate of

217. Find the principal invested if interest was earned in years at an interest rate of

218. Find the principal invested if interest was earned in years at an interest rate of

219. Find the rate if a principal of earned interest in years.

220. Find the rate if a principal of earned interest in years.

221. Find the rate if a principal of earned interest in years.

222. Find the rate if a principal of earned interest in years.

Solve Simple Interest Applications

In the following exercises, solve the problem using the simple interest formula.

223. Casey deposited in a bank account with interest rate How much interest was earned in years?

224. Terrence deposited in a bank account with interest rate How much interest was earned in years?

225. Robin deposited in a bank account with interest rate How much interest was earned in years?

226. Carleen deposited in a bank account with interest rate How much interest was earned in years?

227. Hilaria borrowed from her grandfather to pay for college. Five years later, she paid him back the plus interest. What was the rate of interest?

228. Kenneth lent his niece to buy a computer. Two years later, she paid him back the plus interest. What was the rate of interest?

229. Lebron lent his daughter to help her buy a condominium. When she sold the condominium four years later, she paid him the plus interest. What was the rate of interest?

230. Pablo borrowed to start a business. Three years later, he repaid the plus interest. What was the rate of interest?

231. In years, a bank account that paid earned interest. What was the principal of the account?

232. In years, a bond that paid earned interest. What was the principal of the bond?

233. Joshua's computer loan statement said he would pay in interest for a year loan at How much did Joshua borrow to buy the computer?

234. Margaret's car loan statement said she would pay in interest for a year loan at How much did Margaret borrow to buy the car?

235. Caitlin invested in an certificate of deposit paying interest. How much interest did she earn form this investment?

236. Diego invested in a certificate of deposit paying interest. How much interest did he earn form this investment?

237. Airin borrowed from her parents for the down payment on a car and promised to pay them back in months at a rate of interest. How much interest did she owe her parents?

238. Yuta borrowed from his brother to pay for his textbooks and promised to pay him back in months at a rate of interest. How much interest did Yuta owe his brother?

Everyday Math

239. Interest on savings Find the interest rate your local bank pays on savings accounts.

What is the interest rate?

Calculate the amount of interest you would earn on a principal of for years.

240. Interest on a loan Find the interest rate your local bank charges for a car loan.

What is the interest rate?

Calculate the amount of interest you would pay on a loan of for years.

Writing Exercises

241. Why do banks pay interest on money deposited in savings accounts?

242. Why do banks charge interest for lending money?

Self Check

After completing the exercises, use this checklist to evaluate your mastery of the objectives of this section.

I can...	Confidently	With some help	No-I don't get it!
use the simple interest formula.			
solve simple interest applications.			

On a scale of 1–10, how would you rate your mastery of this section in light of your responses on the checklist? How can you improve this?

6.5 Solve Proportions and their Applications

Learning Objectives

By the end of this section, you will be able to:

- Use the definition of proportion
- Solve proportions
- Solve applications using proportions
- Write percent equations as proportions
- Translate and solve percent proportions

BE PREPARED : : 6.11 Before you get started, take this readiness quiz.

Simplify: —

If you missed this problem, review Example 4.44.

BE PREPARED : : 6.12 Solve: —

If you missed this problem, review Example 4.99.

BE PREPARED : : 6.13 Write as a rate: Sale rode his bike miles in hours.
If you missed this problem, review Example 5.63.

Use the Definition of Proportion

In the section on Ratios and Rates we saw some ways they are used in our daily lives. When two ratios or rates are equal, the equation relating them is called a **proportion**.

Proportion

A proportion is an equation of the form — — where

The proportion states two ratios or rates are equal. The proportion is read is to as is to

The equation — — is a proportion because the two fractions are equal. The proportion — — is read is to as is to

If we compare quantities with units, we have to be sure we are comparing them in the right order. For example, in the proportion we compare the number of students to the number of teachers. We put students in the numerators and teachers in the denominators.

EXAMPLE 6.40

Write each sentence as a proportion:

is to as is to hits in at bats is the same as hits in at-bats.

for ounces is equivalent to for ounces.

Solution

	3 is to 7 as 15 is to 35.
Write as a proportion.	

	5 hits in 8 at-bats is the same as 30 hits in 48 at-bats.
Write each fraction to compare hits to at-bats.	
Write as a proportion.	

	\$1.50 for 6 ounces is equivalent to \$2.25 for 9 ounces.
Write each fraction to compare dollars to ounces.	
Write as a proportion.	

TRY IT : : 6.79

Write each sentence as a proportion:

is to as is to hits in at-bats is the same as hits in at-bats.

for ounces is equivalent to for ounces.

TRY IT : : 6.80

Write each sentence as a proportion:

is to as is to adults for children is the same as adults for children.

for ounces is equivalent to for ounces.

Look at the proportions — — and — — From our work with equivalent fractions we know these equations are true. But how do we know if an equation is a proportion with equivalent fractions if it contains fractions with larger numbers?

To determine if a proportion is true, we find the **cross products** of each proportion. To find the cross products, we multiply each denominator with the opposite numerator (diagonally across the equal sign). The results are called a cross products because of the cross formed. The cross products of a proportion are equal.

$8 \cdot 1 = 8 \quad 2 \cdot 4 = 8 \qquad 9 \cdot 2 = 18 \quad 3 \cdot 6 = 18$

$$\frac{1}{2} = \frac{4}{8} \qquad \frac{2}{3} = \frac{6}{9}$$

Cross Products of a Proportion

For any proportion of the form — — where its cross products are equal.

$$a \cdot d = b \cdot c$$

$$\frac{a}{b} = \frac{c}{d}$$

Cross products can be used to test whether a proportion is true. To test whether an equation makes a proportion, we find the cross products. If they are the equal, we have a proportion.

EXAMPLE 6.41

Determine whether each equation is a proportion:

Solution

To determine if the equation is a proportion, we find the cross products. If they are equal, the equation is a proportion.

	$\frac{4}{9} = \frac{12}{28}$
Find the cross products.	$\frac{4}{9} = \frac{12}{28}$

Since the cross products are not equal, the equation is not a proportion.

	$\frac{17.5}{37.5} = \frac{7}{15}$
Find the cross products.	$\frac{17.5}{37.5} = \frac{7}{15}$

Since the cross products are equal, the equation is a proportion.

> **TRY IT : :** 6.81 Determine whether each equation is a proportion:

> **TRY IT : :** 6.82 Determine whether each equation is a proportion:

Solve Proportions

To solve a proportion containing a variable, we remember that the proportion is an equation. All of the techniques we have used so far to solve equations still apply. In the next example, we will solve a proportion by multiplying by the Least Common Denominator (LCD) using the Multiplication Property of Equality.

EXAMPLE 6.42

Solve: — —

Solution

	$\frac{x}{63} = \frac{4}{7}$
To isolate , multiply both sides by the LCD, 63.	$63\left(\frac{x}{63}\right) = 63\left(\frac{4}{7}\right)$
Simplify.	$x = \frac{9 \cdot \cancel{7} \cdot 4}{\cancel{7}}$
Divide the common factors.	$x = 36$
Check: To check our answer, we substitute into the original proportion.	
	$\frac{x}{63} = \frac{4}{7}$
Substitute $x = 36$	$\frac{36}{63} \stackrel{?}{=} \frac{4}{7}$
Show common factors.	$\frac{4 \cdot 9}{7 \cdot 9} \stackrel{?}{=} \frac{4}{7}$
Simplify.	$\frac{4}{7} = \frac{4}{7}$ ✓

TRY IT : : 6.83 Solve the proportion: — —

TRY IT : : 6.84 Solve the proportion: — —

When the variable is in a denominator, we'll use the fact that the cross products of a proportion are equal to solve the proportions.

We can find the cross products of the proportion and then set them equal. Then we solve the resulting equation using our familiar techniques.

EXAMPLE 6.43

Solve: — —

Solution

Notice that the variable is in the denominator, so we will solve by finding the cross products and setting them equal.

	$\frac{144}{a} = \frac{9}{4}$
Find the cross products and set them equal.	$4 \cdot 144 = a \cdot 9$
Simplify.	$576 = 9a$
Divide both sides by 9.	$\frac{576}{9} = \frac{9a}{9}$
Simplify.	$64 = a$
Check your answer.	
	$\frac{144}{a} = \frac{9}{4}$
Substitute $a = 64$	$\frac{144}{64} \stackrel{?}{=} \frac{9}{4}$
Show common factors..	$\frac{9 \cdot 16}{4 \cdot 16} \stackrel{?}{=} \frac{9}{4}$
Simplify.	$\frac{9}{4} = \frac{9}{4}$ ✓

Another method to solve this would be to multiply both sides by the LCD, Try it and verify that you get the same solution.

TRY IT : : 6.85 Solve the proportion: — —

TRY IT : : 6.86 Solve the proportion: — —

EXAMPLE 6.44

Solve: — —

Solution

Find the cross products and set them equal.	$\frac{52}{91} = \frac{-4}{y}$
	$y \cdot 52 = 91(-4)$
Simplify.	$52y = -364$
Divide both sides by 52.	$\frac{52y}{52} = \frac{-364}{52}$
Simplify.	$y = -7$
Check:	
	$\frac{52}{91} = \frac{-4}{y}$
Substitute $y = -7$	$\frac{52}{91} \stackrel{?}{=} \frac{-4}{-7}$
Show common factors.	
Simplify.	$\frac{4}{7} = \frac{4}{7}$ ✓

TRY IT : : 6.87 Solve the proportion: — —

TRY IT : : 6.88 Solve the proportion: — —

Solve Applications Using Proportions

The strategy for solving applications that we have used earlier in this chapter, also works for proportions, since proportions are equations. When we set up the proportion, we must make sure the units are correct—the units in the numerators match and the units in the denominators match.

EXAMPLE 6.45

When pediatricians prescribe acetaminophen to children, they prescribe milliliters (ml) of acetaminophen for every pounds of the child's weight. If Zoe weighs pounds, how many milliliters of acetaminophen will her doctor prescribe?

Solution

Identify what you are asked to find.	How many ml of acetaminophen the doctor will prescribe
Choose a variable to represent it.	Let ml of acetaminophen.
Write a sentence that gives the information to find it.	If 5 ml is prescribed for every 25 pounds, how much will be prescribed for 80 pounds?
Translate into a proportion.	$\frac{\text{ml}}{\text{pounds}} = \frac{\text{ml}}{\text{pounds}}$
Substitute given values—be careful of the units.	$\frac{5}{25} = \frac{a}{80}$
Multiply both sides by 80.	$80 \cdot \frac{5}{25} = 80 \cdot \frac{a}{80}$
Multiply and show common factors.	$\frac{16 \cdot 5 \cdot 5}{5 \cdot 5} = \frac{80a}{80}$
Simplify.	$16 = a$
Check if the answer is reasonable.	
Yes. Since 80 is about 3 times 25, the medicine should be about 3 times 5.	
Write a complete sentence.	The pediatrician would prescribe 16 ml of acetaminophen to Zoe.

You could also solve this proportion by setting the cross products equal.

TRY IT : : 6.89

Pediatricians prescribe milliliters (ml) of acetaminophen for every pounds of a child's weight. How many milliliters of acetaminophen will the doctor prescribe for Emilia, who weighs pounds?

TRY IT : : 6.90

For every kilogram (kg) of a child's weight, pediatricians prescribe milligrams (mg) of a fever reducer. If Isabella weighs kg, how many milligrams of the fever reducer will the pediatrician prescribe?

EXAMPLE 6.46

One brand of microwave popcorn has calories per serving. A whole bag of this popcorn has servings. How many calories are in a whole bag of this microwave popcorn?

Solution

Identify what you are asked to find.	How many calories are in a whole bag of microwave popcorn?
Choose a variable to represent it.	Let number of calories.
Write a sentence that gives the information to find it.	If there are 120 calories per serving, how many calories are in a whole bag with 3.5 servings?
Translate into a proportion.	$\frac{\text{calories}}{\text{serving}} = \frac{\text{calories}}{\text{serving}}$
Substitute given values.	$\frac{120}{1} = \frac{c}{3.5}$
Multiply both sides by 3.5.	$(3.5)\left(\frac{120}{1}\right) = (3.5)\left(\frac{c}{3.5}\right)$
Multiply.	$420 = c$
Check if the answer is reasonable.	
Yes. Since 3.5 is between 3 and 4, the total calories should be between 360 (3·120) and 480 (4·120).	
Write a complete sentence.	The whole bag of microwave popcorn has 420 calories.

TRY IT : : 6.91

Marissa loves the Caramel Macchiato at the coffee shop. The oz. medium size has calories. How many calories will she get if she drinks the large oz. size?

TRY IT : : 6.92

Yaneli loves Starburst candies, but wants to keep her snacks to calories. If the candies have calories for pieces, how many pieces can she have in her snack?

EXAMPLE 6.47

Josiah went to Mexico for spring break and changed dollars into Mexican pesos. At that time, the exchange rate had U.S. is equal to Mexican pesos. How many Mexican pesos did he get for his trip?

Solution

Identify what you are asked to find.	How many Mexican pesos did Josiah get?
Choose a variable to represent it.	Let number of pesos.
Write a sentence that gives the information to find it.	If \$1 U.S. is equal to 12.54 Mexican pesos, then \$325 is how many pesos?
Translate into a proportion.	$\frac{\$}{\text{pesos}} = \frac{\$}{\text{pesos}}$
Substitute given values.	$\frac{1}{12.54} = \frac{325}{p}$
The variable is in the denominator, so find the cross products and set them equal.	$p \cdot 1 = 12.54(325)$
Simplify.	$c = 4{,}075.5$
Check if the answer is reasonable.	
Yes, \$100 would be \$1,254 pesos. \$325 is a little more than 3 times this amount.	
Write a complete sentence.	Josiah has 4075.5 pesos for his spring break trip.

TRY IT : : 6.93

Yurianna is going to Europe and wants to change dollars into Euros. At the current exchange rate, US is equal to Euro. How many Euros will she have for her trip?

TRY IT : : 6.94

Corey and Nicole are traveling to Japan and need to exchange into Japanese yen. If each dollar is yen, how many yen will they get?

Write Percent Equations As Proportions

Previously, we solved percent equations by applying the properties of equality we have used to solve equations throughout this text. Some people prefer to solve percent equations by using the proportion method. The proportion method for solving percent problems involves a percent proportion. A **percent proportion** is an equation where a percent is equal to an equivalent ratio.

For example, —— and we can simplify —— — Since the equation —— — shows a percent equal to an equivalent ratio, we call it a **percent proportion**. Using the vocabulary we used earlier:

Percent Proportion

The amount is to the base as the percent is to

If we restate the problem in the words of a proportion, it may be easier to set up the proportion:

We could also say:

First we will practice translating into a percent proportion. Later, we'll solve the proportion.

EXAMPLE 6.48

Translate to a proportion. What number is of

Solution

If you look for the word "of", it may help you identify the base.

Identify the parts of the percent proportion.	What number (amount) is 75% (percent) of 90? (base)
Restate as a proportion.	What number out of 90 is the same as 75 out of 100?
Set up the proportion. Let .	

TRY IT : : 6.95 Translate to a proportion: What number is of

TRY IT : : 6.96 Translate to a proportion: What number is of

EXAMPLE 6.49

Translate to a proportion. is of what number?

Solution

Identify the parts of the percent proportion.	19 (amount) is 25% (percent) of what number ? (base)
Restate as a proportion.	19 out of what number is the same as 25 out of 100?
Set up the proportion. Let .	

TRY IT : : 6.97 Translate to a proportion: is of what number?

TRY IT : : 6.98 Translate to a proportion: is of what number?

EXAMPLE 6.50

Translate to a proportion. What percent of is

Solution

Identify the parts of the percent proportion.	What percent (percent) of 27 (base) is 9? (amount)
Restate as a proportion.	9 out of 27 is the same as what number out of 100?
Set up the proportion. Let .	— —

TRY IT : : 6.99 Translate to a proportion: What percent of is

TRY IT : : 6.100 Translate to a proportion: What percent of is

Translate and Solve Percent Proportions

Now that we have written percent equations as proportions, we are ready to solve the equations.

EXAMPLE 6.51

Translate and solve using proportions: What number is of

Solution

Identify the parts of the percent proportion.	What number (amount) is 45% (percent) of 80? (base)
Restate as a proportion.	What number out of 80 is the same as 45 out of 100?
Set up the proportion. Let number.	$\frac{n}{80} = \frac{45}{100}$
Find the cross products and set them equal.	$100 \cdot n = 80 \cdot 45$
Simplify.	$100n = 3{,}600$
Divide both sides by 100.	$\frac{100n}{100} = \frac{3{,}600}{100}$
Simplify.	$n = 36$
Check if the answer is reasonable.	
Yes. 45 is a little less than half of 100 and 36 is a little less than half 80.	
Write a complete sentence that answers the question.	36 is 45% of 80.

TRY IT : : 6.101 Translate and solve using proportions: What number is of

TRY IT : : 6.102 Translate and solve using proportions: What number is of

In the next example, the percent is more than which is more than one whole. So the unknown number will be more than the base.

EXAMPLE 6.52

Translate and solve using proportions: of is what number?

Solution

Identify the parts of the percent proportion.	125% (percent) is 25 (base) of what number? (amount)
Restate as a proportion.	What number out of 25 is the same as 125 out of 100?
Set up the proportion. Let number.	$\frac{n}{25} = \frac{125}{100}$
Find the cross products and set them equal.	$100 \cdot n = 25 \cdot 125$
Simplify.	$100n = 3{,}125$
Divide both sides by 100.	$\frac{100n}{100} = \frac{3{,}125}{100}$
Simplify.	$n = 31.25$
Check if the answer is reasonable.	
Yes. 125 is more than 100 and 31.25 is more than 25.	
Write a complete sentence that answers the question.	125% of 25 is 31.25.

TRY IT : : 6.103 Translate and solve using proportions: of is what number?

TRY IT : : 6.104 Translate and solve using proportions: of is what number?

Percents with decimals and money are also used in proportions.

EXAMPLE 6.53

Translate and solve: of what number is

Solution

Identify the parts of the percent proportion.	6.5% (percent) of what number (base) is \$1.56? (amount)
Restate as a proportion.	\$1.56 out of what number is the same as 6.5 out of 100?
Set up the proportion. Let number.	$\frac{1.56}{n}=\frac{6.5}{100}$
Find the cross products and set them equal.	$100(1.56)=n\cdot 6.5$
Simplify.	$156=6.5n$
Divide both sides by 6.5 to isolate the variable.	$\frac{156}{6.5}=\frac{6.5n}{6.5}$
Simplify.	$24=n$
Check if the answer is reasonable.	
Yes. 6.5% is a small amount and \$1.56 is much less than \$24.	
Write a complete sentence that answers the question.	6.5% of \$24 is \$1.56.

TRY IT : : 6.105 Translate and solve using proportions: of what number is

TRY IT : : 6.106 Translate and solve using proportions: of what number is

EXAMPLE 6.54

Translate and solve using proportions: What percent of is

Solution

Identify the parts of the percent proportion.	What percent (percent) of 72 (base) is 9? (amount)
Restate as a proportion.	9 out of 72 is the same as what number out of 100?
Set up the proportion. Let number.	$\frac{9}{72}=\frac{n}{100}$
Find the cross products and set them equal.	$72\cdot n=100\cdot 9$
Simplify.	$72n=900$
Divide both sides by 72.	$\frac{72n}{72}=\frac{900}{72}$
Simplify.	$n=12.5$
Check if the answer is reasonable.	
Yes. 9 is — of 72 and — is 12.5%.	
Write a complete sentence that answers the question.	12.5% of 72 is 9.

TRY IT : : 6.107 Translate and solve using proportions: What percent of is

TRY IT : : 6.108 Translate and solve using proportions: What percent of is

6.5 EXERCISES

Practice Makes Perfect

Use the Definition of Proportion

In the following exercises, write each sentence as a proportion.

243. is to as is to

244. is to as is to

245. is to as is to

246. is to as is to

247. wins in games is the same as wins in games.

248. wins in games is the same as wins in games.

249. campers to counselor is the same as campers to counselors.

250. campers to counselor is the same as campers to counselors.

251. for ounces is the same as for ounces.

252. for ounces is the same as for ounces.

253. for pounds is the same as for pounds.

254. for pounds is the same as for pounds.

In the following exercises, determine whether each equation is a proportion.

255. — ——

256. — ——

257. — —

258. – —

259. — ——

260. — ——

261. —— ———

262. —— ——

Solve Proportions

In the following exercises, solve each proportion.

263. — –

264. — —

265. — –

266. — –

267. – ——

268. – ——

269. —— ——

270. —— ——

271. — ——

272. — ——

273. — –

274. — –

275. —— ——

276. —— ——

277. – —

278. – –

Solve Applications Using Proportions

In the following exercises, solve the proportion problem.

279. Pediatricians prescribe milliliters (ml) of acetaminophen for every pounds of a child's weight. How many milliliters of acetaminophen will the doctor prescribe for Jocelyn, who weighs pounds?

280. Brianna, who weighs kg, just received her shots and needs a pain killer. The pain killer is prescribed for children at milligrams (mg) for every kilogram (kg) of the child's weight. How many milligrams will the doctor prescribe?

281. At the gym, Carol takes her pulse for sec and counts beats. How many beats per minute is this? Has Carol met her target heart rate of beats per minute?

282. Kevin wants to keep his heart rate at beats per minute while training. During his workout he counts beats in seconds. How many beats per minute is this? Has Kevin met his target heart rate?

283. A new energy drink advertises calories for ounces. How many calories are in ounces of the drink?

284. One ounce can of soda has calories. If Josiah drinks the big ounce size from the local mini-mart, how many calories does he get?

285. Karen eats — cup of oatmeal that counts for points on her weight loss program. Her husband, Joe, can have points of oatmeal for breakfast. How much oatmeal can he have?

286. An oatmeal cookie recipe calls for — cup of butter to make dozen cookies. Hilda needs to make dozen cookies for the bake sale. How many cups of butter will she need?

287. Janice is traveling to Canada and will change US dollars into Canadian dollars. At the current exchange rate, US is equal to Canadian. How many Canadian dollars will she get for her trip?

288. Todd is traveling to Mexico and needs to exchange into Mexican pesos. If each dollar is worth pesos, how many pesos will he get for his trip?

289. Steve changed into Euros. How many Euros did he receive per US dollar?

290. Martha changed US into Australian dollars. How many Australian dollars did she receive per US dollar?

291. At the laundromat, Lucy changed into quarters. How many quarters did she get?

292. When she arrived at a casino, Gerty changed into nickels. How many nickels did she get?

293. Jesse's car gets miles per gallon of gas. If Las Vegas is miles away, how many gallons of gas are needed to get there and then home? If gas is per gallon, what is the total cost of the gas for the trip?

294. Danny wants to drive to Phoenix to see his grandfather. Phoenix is miles from Danny's home and his car gets miles per gallon. How many gallons of gas will Danny need to get to and from Phoenix? If gas is per gallon, what is the total cost for the gas to drive to see his grandfather?

295. Hugh leaves early one morning to drive from his home in Chicago to go to Mount Rushmore, miles away. After hours, he has gone miles. At that rate, how long will the whole drive take?

296. Kelly leaves her home in Seattle to drive to Spokane, a distance of miles. After hours, she has gone miles. At that rate, how long will the whole drive take?

297. Phil wants to fertilize his lawn. Each bag of fertilizer covers about square feet of lawn. Phil's lawn is approximately square feet. How many bags of fertilizer will he have to buy?

298. April wants to paint the exterior of her house. One gallon of paint covers about square feet, and the exterior of the house measures approximately square feet. How many gallons of paint will she have to buy?

Write Percent Equations as Proportions

In the following exercises, translate to a proportion.

299. What number is of

300. What number is of

301. What number is of

302. What number is of

303. is of what number?

304. is of what number?

305. is of what number?

306. is of what number?

307. What percent of is

308. What percent of is

309. What percent of is

310. What percent of is

Translate and Solve Percent Proportions

In the following exercises, translate and solve using proportions.

311. What number is of

312. What number is of

313. of is what number?

314. of is what number?

315. of is what number?

316. of is what number?

317. What is of

318. What is of

319. of what number is

320. of what number is

321. is of what number?

322. is of what number?

323. What percent of is

324. What percent of is

325. What percent of is

326. What percent of is

Everyday Math

327. Mixing a concentrate Sam bought a large bottle of concentrated cleaning solution at the warehouse store. He must mix the concentrate with water to make a solution for washing his windows. The directions tell him to mix ounces of concentrate with ounces of water. If he puts ounces of concentrate in a bucket, how many ounces of water should he add? How many ounces of the solution will he have altogether?

328. Mixing a concentrate Travis is going to wash his car. The directions on the bottle of car wash concentrate say to mix ounces of concentrate with ounces of water. If Travis puts ounces of concentrate in a bucket, how much water must he mix with the concentrate?

Writing Exercises

329. To solve “what number is of do you prefer to use an equation like you did in the section on Decimal Operations or a proportion like you did in this section? Explain your reason.

330. To solve “what percent of is do you prefer to use an equation like you did in the section on Decimal Operations or a proportion like you did in this section? Explain your reason.

Self Check

After completing the exercises, use this checklist to evaluate your mastery of the objectives of this section.

I can...	Confidently	With some help	No-I don't get it!
use definition of proportion.			
solve proportions.			
solve applications using proportions.			
write percent equations as proportions.			
translate and solve percent proportions.			

Overall, after looking at the checklist, do you think you are well-prepared for the next Chapter? Why or why not?

CHAPTER 6 REVIEW

KEY TERMS

commission A commission is a percentage of total sales as determined by the rate of commission.

discount An amount of discount is a percent off the original price, determined by the discount rate.

mark-up The mark-up is the amount added to the wholesale price, determined by the mark-up rate.

percent A perfecnt is a ratio whose denominator is .

percent decrease The percent decrease is the percent the amount of decrease is of the original amount.

percent increase The percent increase is the percent the amount of increase is of the original amount.

proportion A proportion is an equation of the form — —, where , .The proportion states two ratios or rates are equal. The proportion is read " is to , as is to ".

sales tax The sales tax is a percent of the purchase price.

simple interest If an amount of money, , the principal, is invested for a period of years at an annual interest rate , the amount of interest, , earned is . Interest earned according to this formula is called simple interest.

KEY CONCEPTS

6.1 Understand Percent

- **Convert a percent to a fraction.**
  Step 1. Write the percent as a ratio with the denominator 100.
  Step 2. Simplify the fraction if possible.
- **Convert a percent to a decimal.**
  Step 1. Write the percent as a ratio with the denominator 100.
  Step 2. Convert the fraction to a decimal by dividing the numerator by the denominator.
- **Convert a decimal to a percent.**
  Step 1. Write the decimal as a fraction.
  Step 2. If the denominator of the fraction is not 100, rewrite it as an equivalent fraction with denominator 100.
  Step 3. Write this ratio as a percent.
- **Convert a fraction to a percent.**
  Step 1. Convert the fraction to a decimal.
  Step 2. Convert the decimal to a percent.

6.2 Solve General Applications of Percent

- **Solve an application.**
  Step 1. Identify what you are asked to find and choose a variable to represent it.
  Step 2. Write a sentence that gives the information to find it.
  Step 3. Translate the sentence into an equation.
  Step 4. Solve the equation using good algebra techniques.
  Step 5. Write a complete sentence that answers the question.
  Step 6. Check the answer in the problem and make sure it makes sense.
- **Find percent increase.**
  Step 1. Find the amount of increase:

  Step 2. Find the percent increase as a percent of the original amount.
- **Find percent decrease.**

Step 1. Find the amount of decrease.

Step 2. Find the percent decrease as a percent of the original amount.

6.3 Solve Sales Tax, Commission, and Discount Applications

- **Sales Tax** The sales tax is a percent of the purchase price.
  -
  -
- **Commission** A commission is a percentage of total sales as determined by the rate of commission.
  -
- **Discount** An amount of discount is a percent off the original price, determined by the discount rate.
  -
  -
- **Mark-up** The mark-up is the amount added to the wholesale price, determined by the mark-up rate.
  -
  -

6.4 Solve Simple Interest Applications

- **Simple interest**
  - If an amount of money, , the principal, is invested for a period of years at an annual interest rate , the amount of interest, , earned is
  - Interest earned according to this formula is called **simple interest**.

6.5 Solve Proportions and their Applications

- **Proportion**
  - A proportion is an equation of the form — —, where , .The proportion states two ratios or rates are equal. The proportion is read " is to , as is to ".
- **Cross Products of a Proportion**
  - For any proportion of the form — —, where , its cross products are equal: .
- **Percent Proportion**
  - The amount is to the base as the percent is to 100.

REVIEW EXERCISES

6.1 Understand Percent

In the following exercises, write each percent as a ratio.

331. admission rate for the university

332. rate of college students with student loans

In the following exercises, write as a ratio and then as a percent.

333. out of architects are women.

334. out of every nurses are men.

In the following exercises, convert each percent to a fraction.

335.

336.

337.

338. —

In the following exercises, convert each percent to a decimal.

339.

340.

341.

342. 4.9%

In the following exercises, convert each percent to a simplified fraction and a decimal.

343. In of the United States population was age or over. (Source: www.census.gov)

344. In of the United States population was under years old. (Source: www.census.gov)

345. When a die is tossed, the probability it will land with an even number of dots on the top side is

346. A couple plans to have three children. The probability they will all be girls is

In the following exercises, convert each decimal to a percent.

347.

348.

349.

350.

351.

352.

In the following exercises, convert each fraction to a percent.

353. —

354. —

355. —

356. —

357. According to the Centers for Disease Control, — of adults do not take a vitamin or supplement.

358. According to the Centers for Disease Control, among adults who do take a vitamin or supplement, — take a multivitamin.

In the following exercises, translate and solve.

359. What number is of

360. of is what number?

361. is of what number?

362. is of what number?

363. of what number is

364. of what number is

365. What percent of is

366. What percent of is

6.2 Solve General Applications of Percents

In the following exercises, solve.

367. When Aurelio and his family ate dinner at a restaurant, the bill was Aurelio wants to leave of the total bill as a tip. How much should the tip be?

368. One granola bar has grams of fiber, which is of the recommended daily amount. What is the total recommended daily amount of fiber?

369. The nutrition label on a package of granola bars says that each granola bar has calories, and calories are from fat. What percent of the total calories is from fat?

370. Elsa gets paid per month. Her car payment is What percent of her monthly pay goes to her car payment?

In the following exercises, solve.

371. Jorge got a raise in his hourly pay, from to Find the percent increase.

372. Last year Bernard bought a new car for This year the car is worth Find the percent decrease.

6.3 Solve Sales Tax, Commission, and Discount Applications

In the following exercises, find the sales tax the total cost.

373. The cost of a lawn mower was The sales tax rate is of the purchase price.

374. The cost of a water heater is The sales tax rate is of the purchase price.

In the following exercises, find the sales tax rate.

375. Andy bought a piano for The sales tax on the purchase was

376. Nahomi bought a purse for The sales tax on the purchase was

In the following exercises, find the commission.

377. Ginny is a realtor. She receives commission when she sells a house. How much commission will she receive for selling a house for

378. Jackson receives commission when he sells a dinette set. How much commission will he receive for selling a dinette set for

In the following exercises, find the rate of commission.

379. Ruben received commission when he sold a painting at the art gallery where he works. What was the rate of commission?

380. Tori received for selling a membership at her gym. What was her rate of commission?

In the following exercises, find the sale price.

381. Aya bought a pair of shoes that was on sale for off. The original price of the shoes was

382. Takwanna saw a cookware set she liked on sale for off. The original price of the cookware was

In the following exercises, find the amount of discount and the sale price.

383. Nga bought a microwave for her office. The microwave was discounted from an original price of

384. Jarrett bought a tie that was discounted from an original price of

In the following exercises, find the amount of discount the discount rate. (Round to the nearest tenth of a percent if needed.)

385. Hilda bought a bedspread on sale for The original price of the bedspread was

386. Tyler bought a phone on sale for The original price of the phone was

In the following exercises, find

the amount of the mark-up the list price

387. Manny paid a pound for apples. He added mark-up before selling them at his produce stand. What price did he charge for the apples?

388. It cost Noelle for the materials she used to make a purse. She added a mark-up before selling it at her friend's store. What price did she ask for the purse?

6.4 Solve Simple Interest Applications

In the following exercises, solve the simple interest problem.

389. Find the simple interest earned after years on invested at an interest rate of

390. Find the simple interest earned after years on invested at an interest rate of

391. Find the principal invested if interest was earned in years at an interest rate of

392. Find the interest rate if interest was earned from a principal of invested for years.

393. Kazuo deposited in a bank account with interest rate How much interest was earned in years?

394. Brent invested in a friend's business. In years the friend paid him the plus interest. What was the rate of interest?

395. Fresia lent her son for college expenses. Three years later he repaid her the plus interest. What was the rate of interest?

396. In years, a bond that paid earned interest. What was the principal of the bond?

6.5 Solve Proportions and their Applications

In the following exercises, write each sentence as a proportion.

397. is to as is to

398. miles to gallons is the same as miles to gallons.

399. teacher to students is the same as teachers to students.

400. for ounces is the same as for ounces.

In the following exercises, determine whether each equation is a proportion.

401.

402.

403.

404.

In the following exercises, solve each proportion.

405.

406.

407.

408.

In the following exercises, solve the proportion problem.

409. The children's dosage of acetaminophen is milliliters (ml) for every pounds of a child's weight. How many milliliters of acetaminophen will be prescribed for a pound child?

410. After a workout, Dennis takes his pulse for sec and counts beats. How many beats per minute is this?

411. An ounce serving of ice cream has calories. If Lavonne eats ounces of ice cream, how many calories does she get?

412. Alma is going to Europe and wants to exchange into Euros. If each dollar is Euros, how many Euros will Alma get?

413. Zack wants to drive from Omaha to Denver, a distance of miles. If his car gets miles to the gallon, how many gallons of gas will Zack need to get to Denver?

414. Teresa is planning a party for people. Each gallon of punch will serve people. How many gallons of punch will she need?

In the following exercises, translate to a proportion.

415. What number is of

416. is of what number?

417. What percent of is

418. What percent of is

In the following exercises, translate and solve using proportions.

419. What number is of

420. of what number is

421. is of what number?

422. What percent of is

PRACTICE TEST

In the following exercises, convert each percent to a decimal a simplified fraction.

423.

424.

425.

In the following exercises, convert each fraction to a percent. (Round to decimal places if needed.)

426. —

427. —

428. —

In the following exercises, solve the percent problem.

429. is what percent of

430. What number is of

431. of what number is

432. Yuki's monthly paycheck is She pays for rent. What percent of her paycheck goes to rent?

433. The total number of vehicles on one freeway dropped from to Find the percent decrease (round to the nearest tenth of a percent).

434. Kyle bought a bicycle in Denver where the sales tax was of the purchase price. The purchase price of the bicycle was What was the total cost?

435. Mara received commission when she sold a suit. What was her rate of commission?

436. Kiyoshi bought a television set on sale for The original price was Find:

the amount of discount

the discount rate (round to the nearest tenth of a percent)

437. Oxana bought a dresser at a garage sale for She refinished it, then added a markup before advertising it for sale. What price did she ask for the dresser?

438. Find the simple interest earned after years on invested at an interest rate of

439. Brenda borrowed from her brother. Two years later, she repaid the plus interest. What was the rate of interest?

440. Write as a proportion: gallons to miles is the same as gallons to miles.

441. Solve for a: — —

442. Vin read pages of a book in minutes. At that rate, how long will it take him to read pages?

Figure 7.1 Quiltmakers know that by rearranging the same basic blocks the resulting quilts can look very different. What happens when we rearrange the numbers in an expression? Does the resulting value change? We will answer these questions in this chapter as we will learn about the properties of numbers. (credit: Hans, Public Domain)

Chapter Outline

Introduction

A quilt is formed by sewing many different pieces of fabric together. The pieces can vary in color, size, and shape. The combinations of different kinds of pieces provide for an endless possibility of patterns. Much like the pieces of fabric, mathematicians distinguish among different types of numbers. The kinds of numbers in an expression provide for an endless possibility of outcomes. We have already described counting numbers, whole numbers, and integers. In this chapter, we will learn about other types of numbers and their properties.

7.1 Rational and Irrational Numbers

Learning Objectives

By the end of this section, you will be able to:

- Identify rational numbers and irrational numbers
- Classify different types of real numbers

BE PREPARED : : 7.1 Before you get started, take this readiness quiz.

Write as an improper fraction.

If you missed this problem, review Example 5.4.

BE PREPARED : : 7.2 Write —— as a decimal.

If you missed this problem, review Example 5.30.

☑ **BE PREPARED : :** 7.3 Simplify: √
If you missed this problem, review Example 5.69.

Identify Rational Numbers and Irrational Numbers

Congratulations! You have completed the first six chapters of this book! It's time to take stock of what you have done so far in this course and think about what is ahead. You have learned how to add, subtract, multiply, and divide whole numbers, fractions, integers, and decimals. You have become familiar with the language and symbols of algebra, and have simplified and evaluated algebraic expressions. You have solved many different types of applications. You have established a good solid foundation that you need so you can be successful in algebra.

In this chapter, we'll make sure your skills are firmly set. We'll take another look at the kinds of numbers we have worked with in all previous chapters. We'll work with properties of numbers that will help you improve your number sense. And we'll practice using them in ways that we'll use when we solve equations and complete other procedures in algebra.

We have already described numbers as counting numbers, whole numbers, and integers. Do you remember what the difference is among these types of numbers?

counting numbers	
whole numbers	
integers	

Rational Numbers

What type of numbers would you get if you started with all the integers and then included all the fractions? The numbers you would have form the set of rational numbers. A **rational number** is a number that can be written as a ratio of two integers.

Rational Numbers

A rational number is a number that can be written in the form — where and are integers and

All fractions, both positive and negative, are rational numbers. A few examples are

— — — —

Each numerator and each denominator is an integer.

We need to look at all the numbers we have used so far and verify that they are rational. The definition of rational numbers tells us that all fractions are rational. We will now look at the counting numbers, whole numbers, integers, and decimals to make sure they are rational.

Are integers rational numbers? To decide if an integer is a rational number, we try to write it as a ratio of two integers. An easy way to do this is to write it as a fraction with denominator one.

— — —

Since any integer can be written as the ratio of two integers, all integers are rational numbers. Remember that all the counting numbers and all the whole numbers are also integers, and so they, too, are rational.

What about decimals? Are they rational? Let's look at a few to see if we can write each of them as the ratio of two integers. We've already seen that integers are rational numbers. The integer could be written as the decimal So, clearly, some decimals are rational.

Think about the decimal Can we write it as a ratio of two integers? Because means — we can write it as an improper fraction, — So is the ratio of the integers and It is a rational number.

In general, any decimal that ends after a number of digits (such as or is a rational number. We can use the reciprocal (or multiplicative inverse) of the place value of the last digit as the denominator when writing the decimal as a fraction.

EXAMPLE 7.1

Write each as the ratio of two integers:

Solution

Write the integer as a fraction with denominator 1.	

Write the decimal as a mixed number.	
Then convert it to an improper fraction.	

Convert the mixed number to an improper fraction.	

TRY IT : : 7.1 Write each as the ratio of two integers:

TRY IT : : 7.2 Write each as the ratio of two integers:

Let's look at the decimal form of the numbers we know are rational. We have seen that every integer is a rational number, since for any integer, We can also change any integer to a decimal by adding a decimal point and a zero.

We have also seen that every fraction is a rational number. Look at the decimal form of the fractions we just considered.

What do these examples tell you? Every rational number can be written both as a ratio of integers and as a decimal that either stops or repeats. The table below shows the numbers we looked at expressed as a ratio of integers and as a decimal.

Rational Numbers		
	Fractions	**Integers**
Number	— — — —	
Ratio of Integer	— — — —	— — — — — —
Decimal number		

Irrational Numbers

Are there any decimals that do not stop or repeat? Yes. The number (the Greek letter pi, pronounced 'pie'), which is very important in describing circles, has a decimal form that does not stop or repeat.

Similarly, the decimal representations of square roots of whole numbers that are not perfect squares never stop and never repeat. For example,

A decimal that does not stop and does not repeat cannot be written as the ratio of integers. We call this kind of number an **irrational number**.

Irrational Number

An irrational number is a number that cannot be written as the ratio of two integers. Its decimal form does not stop and does not repeat.

Let's summarize a method we can use to determine whether a number is rational or irrational.

If the decimal form of a number

- stops or repeats, the number is rational.
- does not stop and does not repeat, the number is irrational.

EXAMPLE 7.2

Identify each of the following as rational or irrational:

Solution

The bar above the indicates that it repeats. Therefore, is a repeating decimal, and is therefore a rational number.

This decimal stops after the , so it is a rational number.

The ellipsis means that this number does not stop. There is no repeating pattern of digits. Since the number doesn't stop and doesn't repeat, it is irrational.

TRY IT : : 7.3 Identify each of the following as rational or irrational:

TRY IT : : 7.4 Identify each of the following as rational or irrational:

Let's think about square roots now. Square roots of perfect squares are always whole numbers, so they are rational. But the decimal forms of square roots of numbers that are not perfect squares never stop and never repeat, so these square roots are irrational.

EXAMPLE 7.3

Identify each of the following as rational or irrational:

$\sqrt{\ }$ $\sqrt{\ }$

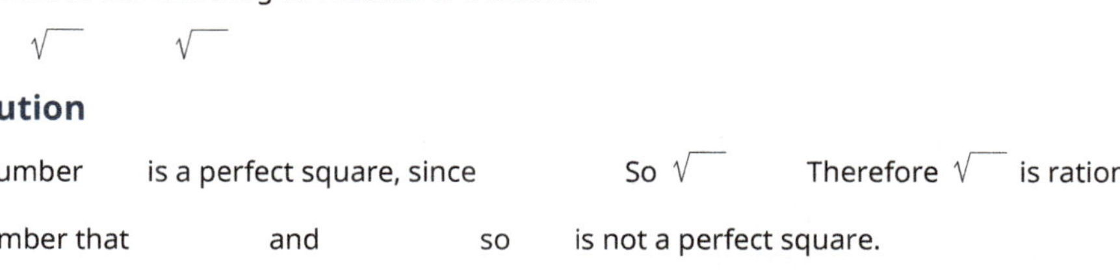

Solution

The number is a perfect square, since So $\sqrt{\ }$ Therefore $\sqrt{\ }$ is rational.

Remember that and so is not a perfect square.

This means $\sqrt{\ }$ is irrational.

TRY IT : : 7.5 Identify each of the following as rational or irrational:

$\sqrt{\ }$ $\sqrt{\ }$

TRY IT : : 7.6 Identify each of the following as rational or irrational:

$\sqrt{\ }$ $\sqrt{\ }$

Classify Real Numbers

We have seen that all counting numbers are whole numbers, all whole numbers are integers, and all integers are rational numbers. Irrational numbers are a separate category of their own. When we put together the rational numbers and the irrational numbers, we get the set of **real numbers**.

Figure 7.2 illustrates how the number sets are related.

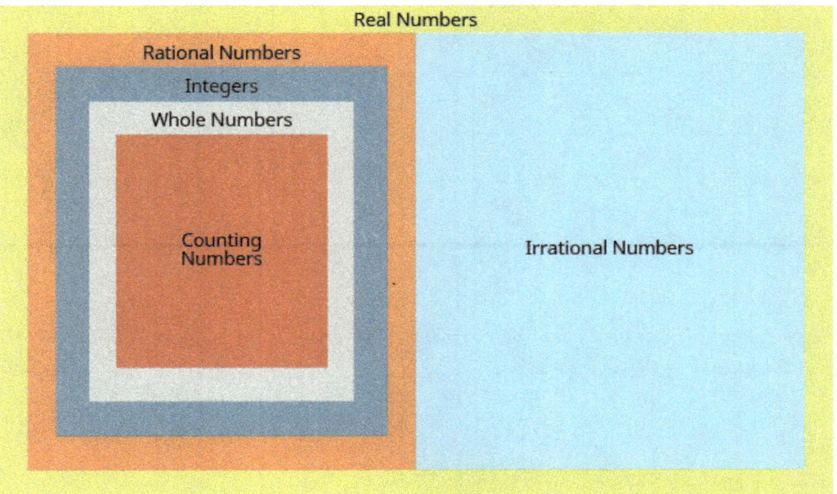

Figure 7.2 This diagram illustrates the relationships between the different types of real numbers.

Real Numbers

Real numbers are numbers that are either rational or irrational.

Does the term "real numbers" seem strange to you? Are there any numbers that are not "real", and, if so, what could they be? For centuries, the only numbers people knew about were what we now call the real numbers. Then mathematicians discovered the set of *imaginary numbers*. You won't encounter imaginary numbers in this course, but you will later on in your studies of algebra.

EXAMPLE 7.4

Determine whether each of the numbers in the following list is a whole number, integer, rational number, irrational number, and real number.

— √ √

Solution

The whole numbers are The number is the only whole number given.

The integers are the whole numbers, their opposites, and From the given numbers, and are integers. Also, notice that is the square of so √ So the integers are √

Since all integers are rational, the numbers √ are also rational. Rational numbers also include fractions and decimals that terminate or repeat, so — are rational.

The number is not a perfect square, so √ is irrational.

All of the numbers listed are real.

We'll summarize the results in a table.

Number	Whole	Integer	Rational	Irrational	Real
—					
√					
√					

> **TRY IT : : 7.7**
>
> Determine whether each number is a whole number, integer, rational number, irrational number, and real number: √ — √

> **TRY IT : : 7.8**
>
> Determine whether each number is a whole number, integer, rational number, irrational number, and real number: √ — √

MEDIA : : ACCESS ADDITIONAL ONLINE RESOURCES

- Sets of Real Numbers (http://www.openstax.org/l/24RealNumber)
- Real Numbers (http://www.openstax.org/l/24RealNumbers)

7.1 EXERCISES

Practice Makes Perfect

Rational Numbers

In the following exercises, write as the ratio of two integers.

1.

2.

3.

4.

In the following exercises, determine which of the given numbers are rational and which are irrational.

5. , ,

6. , ,

7. , ,

8. ,

In the following exercises, identify whether each number is rational or irrational.

9. √ √

10. √ √

11. √ √

12. √ √

Classifying Real Numbers

In the following exercises, determine whether each number is whole, integer, rational, irrational, and real.

13. , , —, √ ,

14. , −, √ , , —,

15. √ , , −, , , −

Everyday Math

16. Field trip All the graders at Lincoln Elementary School will go on a field trip to the science museum. Counting all the children, teachers, and chaperones, there will be people. Each bus holds people.

How many buses will be needed?

Why must the answer be a whole number?

Why shouldn't you round the answer the usual way?

17. Child care Serena wants to open a licensed child care center. Her state requires that there be no more than children for each teacher. She would like her child care center to serve children.

How many teachers will be needed?

Why must the answer be a whole number?

Why shouldn't you round the answer the usual way?

Writing Exercises

18. In your own words, explain the difference between a rational number and an irrational number.

19. Explain how the sets of numbers (counting, whole, integer, rational, irrationals, reals) are related to each other.

Self Check

After completing the exercises, use this checklist to evaluate your mastery of the objectives of this section.

I can...	Confidently	With some help	No-I don't get it!
identify rational and irrational numbers.			
classify different types of real numbers.			

If most of your checks were:

...confidently. Congratulations! You have achieved the objectives in this section. Reflect on the study skills you used so that you can continue to use them. What did you do to become confident of your ability to do these things? Be specific.

...with some help. This must be addressed quickly because topics you do not master become potholes in your road to success. In math, every topic builds upon previous work. It is important to make sure you have a strong foundation before you move on. Whom can you ask for help? Your fellow classmates and instructor are good resources. Is there a place on campus where math tutors are available? Can your study skills be improved?

...no—I don't get it! This is a warning sign and you must not ignore it. You should get help right away or you will quickly be overwhelmed. See your instructor as soon as you can to discuss your situation. Together you can come up with a plan to get you the help you need.

7.2 Commutative and Associative Properties

Learning Objectives

By the end of this section, you will be able to:

- Use the commutative and associative properties
- Evaluate expressions using the commutative and associative properties
- Simplify expressions using the commutative and associative properties

BE PREPARED : : 7.4 Before you get started, take this readiness quiz.

Simplify:

If you missed this problem, review Example 2.22.

BE PREPARED : : 7.5 Multiply: —

If you missed this problem, review Example 4.28.

BE PREPARED : : 7.6 Find the opposite of

If you missed this problem, review Example 3.3.

In the next few sections, we will take a look at the properties of real numbers. Many of these properties will describe things you already know, but it will help to give names to the properties and define them formally. This way we'll be able to refer to them and use them as we solve equations in the next chapter.

Use the Commutative and Associative Properties

Think about adding two numbers, such as and

The results are the same.

Notice, the order in which we add does not matter. The same is true when multiplying and

Again, the results are the same! The order in which we multiply does not matter.

These examples illustrate the commutative properties of addition and multiplication.

Commutative Properties

Commutative Property of Addition: if and are real numbers, then

Commutative Property of Multiplication: if and are real numbers, then

The commutative properties have to do with order. If you change the order of the numbers when adding or multiplying, the result is the same.

EXAMPLE 7.5

Use the commutative properties to rewrite the following expressions:

Solution

Use the commutative property of addition to change the order.

Use the commutative property of multiplication to change the order.

TRY IT : : 7.9 Use the commutative properties to rewrite the following:

TRY IT : : 7.10 Use the commutative properties to rewrite the following:

What about subtraction? Does order matter when we subtract numbers? Does give the same result as

Since changing the order of the subtraction did not give the same result, we can say that subtraction is not commutative. Let's see what happens when we divide two numbers. Is division commutative?

Since changing the order of the division did not give the same result, division is not commutative.

Addition and multiplication are commutative. Subtraction and division are not commutative.

Suppose you were asked to simplify this expression.

How would you do it and what would your answer be?

Some people would think and then Others might start with and then

Both ways give the same result, as shown in Figure 7.3. (Remember that parentheses are grouping symbols that indicate which operations should be done first.)

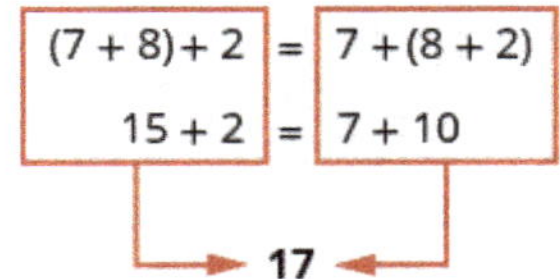

Figure 7.3

When adding three numbers, changing the grouping of the numbers does not change the result. This is known as the Associative Property of Addition.

The same principle holds true for multiplication as well. Suppose we want to find the value of the following expression:

—

Changing the grouping of the numbers gives the same result, as shown in Figure 7.4.

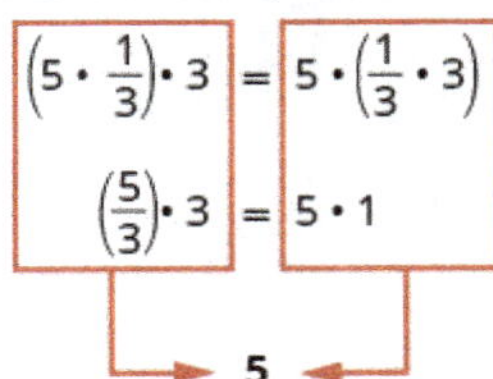

Figure 7.4

When multiplying three numbers, changing the grouping of the numbers does not change the result. This is known as the Associative Property of Multiplication.

If we multiply three numbers, changing the grouping does not affect the product.

You probably know this, but the terminology may be new to you. These examples illustrate the *Associative Properties*.

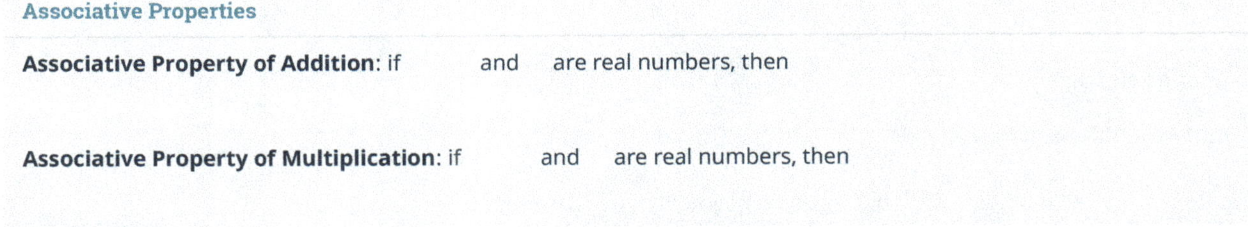

Associative Properties

Associative Property of Addition: if and are real numbers, then

Associative Property of Multiplication: if and are real numbers, then

EXAMPLE 7.6

Use the associative properties to rewrite the following:

—

Solution

Change the grouping.

Notice that is so the addition will be easier if we group as shown on the right.

Change the grouping.

Notice that — is The multiplication will be easier if we group as shown on the right.

> **TRY IT : : 7.11** Use the associative properties to rewrite the following:

> **TRY IT : : 7.12** Use the associative properties to rewrite the following:

Besides using the associative properties to make calculations easier, we will often use it to simplify expressions with variables.

EXAMPLE 7.7

Use the Associative Property of Multiplication to simplify:

Solution

Change the grouping.	
Multiply in the parentheses.	

Notice that we can multiply but we could not multiply without having a value for

> **TRY IT : : 7.13** Use the Associative Property of Multiplication to simplify the given expression:

> **TRY IT : : 7.14** Use the Associative Property of Multiplication to simplify the given expression:

Evaluate Expressions using the Commutative and Associative Properties

The commutative and associative properties can make it easier to evaluate some algebraic expressions. Since order does not matter when adding or multiplying three or more terms, we can rearrange and re-group terms to make our work easier, as the next several examples illustrate.

EXAMPLE 7.8

Evaluate each expression when —

Solution

	$x + 0.37 + (-x)$
Substitute — for .	$\frac{7}{8} + 0.37 + \left(-\frac{7}{8}\right)$
Convert fractions to decimals.	$0.875 + 0.37 + (-0.875)$
Add left to right.	$1.245 - 0.875$
Subtract.	0.37

	$x + (-x) + 0.37$
Substitute — for x.	$\frac{7}{8} + \left(-\frac{7}{8}\right) + 0.37$
Add opposites first.	0.37

What was the difference between part and part ? Only the order changed. By the Commutative Property of Addition, But wasn't part much easier?

> **TRY IT : : 7.15** Evaluate each expression when —

> **TRY IT : : 7.16** Evaluate each expression when —

Let's do one more, this time with multiplication.

EXAMPLE 7.9

Evaluate each expression when

— — — —

Solution

	$\frac{4}{3}\left(\frac{3}{4}n\right)$
Substitute 17 for n.	$\frac{4}{3}\left(\frac{3}{4} \cdot 17\right)$
Multiply in the parentheses first.	$\frac{4}{3}\left(\frac{51}{4}\right)$
Multiply again.	17

	$\left(\frac{4}{3} \cdot \frac{3}{4}\right)n$
Substitute 17 for n.	$\left(\frac{4}{3} \cdot \frac{3}{4}\right) \cdot 17$
Multiply. The product of reciprocals is 1.	$(1) \cdot 17$
Multiply again.	17

What was the difference between part and part here? Only the grouping changed. By the Associative Property of Multiplication, — — — — By carefully choosing how to group the factors, we can make the work easier.

> **TRY IT : :** 7.17 Evaluate each expression when — — — —

> **TRY IT : :** 7.18 Evaluate each expression when — — — —

Simplify Expressions Using the Commutative and Associative Properties

When we have to simplify algebraic expressions, we can often make the work easier by applying the Commutative or Associative Property first instead of automatically following the order of operations. Notice that in Example 7.8 part was easier to simplify than part because the opposites were next to each other and their sum is Likewise, part in Example 7.9 was easier, with the reciprocals grouped together, because their product is In the next few examples, we'll use our number sense to look for ways to apply these properties to make our work easier.

EXAMPLE 7.10

Simplify:

Solution

Notice the first and third terms are opposites, so we can use the commutative property of addition to reorder the terms.

Re-order the terms.	
Add left to right.	
Add.	

> **TRY IT : :** 7.19 Simplify:

> **TRY IT : :** 7.20 Simplify:

Now we will see how recognizing reciprocals is helpful. Before multiplying left to right, look for reciprocals—their product is

EXAMPLE 7.11

Simplify: — — —

Solution

Notice the first and third terms are reciprocals, so we can use the Commutative Property of Multiplication to reorder the factors.

	— — —
Re-order the terms.	— — —
Multiply left to right.	—
Multiply.	—

TRY IT : : 7.21 Simplify: — — —

TRY IT : : 7.22 Simplify: — — —

In expressions where we need to add or subtract three or more fractions, combine those with a common denominator first.

EXAMPLE 7.12

Simplify: — — —

Solution

Notice that the second and third terms have a common denominator, so this work will be easier if we change the grouping.

	— — —
Group the terms with a common denominator.	— — —
Add in the parentheses first.	— —
Simplify the fraction.	—
Add.	—
Convert to an improper fraction.	—

TRY IT : : 7.23 Simplify: — — —

> TRY IT : : 7.24 Simplify: — —— ——

When adding and subtracting three or more terms involving decimals, look for terms that combine to give whole numbers.

EXAMPLE 7.13

Simplify:

Solution

Notice that the sum of the second and third coefficients is a whole number.

Change the grouping.
Add in the parentheses first.
Add.

Many people have good number sense when they deal with money. Think about adding cents and cent. Do you see how this applies to adding

> TRY IT : : 7.25 Simplify:

> TRY IT : : 7.26 Simplify:

No matter what you are doing, it is always a good idea to think ahead. When simplifying an expression, think about what your steps will be. The next example will show you how using the Associative Property of Multiplication can make your work easier if you plan ahead.

EXAMPLE 7.14

Simplify the expression:

Solution

Notice that multiplying is easier than multiplying because it gives a whole number. (Think about having quarters—that makes

Regroup.
Multiply in the brackets first.
Multiply.

> TRY IT : : 7.27 Simplify:

> TRY IT : : 7.28 Simplify:

When simplifying expressions that contain variables, we can use the commutative and associative properties to re-order or regroup terms, as shown in the next pair of examples.

EXAMPLE 7.15

Simplify:

Solution

Use the associative property of multiplication to re-group.
Multiply in the parentheses.

TRY IT : : 7.29 Simplify:

TRY IT : : 7.30 Simplify:

In The Language of Algebra, we learned to combine like terms by rearranging an expression so the like terms were together. We simplified the expression by rewriting it as and then simplified it to We were using the Commutative Property of Addition.

EXAMPLE 7.16

Simplify:

Solution

Use the Commutative Property of Addition to re-order so that like terms are together.

Re-order terms.
Combine like terms.

TRY IT : : 7.31 Simplify:

TRY IT : : 7.32 Simplify:

LINKS TO LITERACY

The *Links to Literacy* activity, "Each Orange Had 8 Slices" will provide you with another view of the topics covered in this section.

7.2 EXERCISES

Practice Makes Perfect

Use the Commutative and Associative Properties

In the following exercises, use the commutative properties to rewrite the given expression.

20. **21.** **22.**

23. **24.** **25.**

26. **27.** **28.**

29. **30.** **31.**

In the following exercises, use the associative properties to rewrite the given expression.

32. **33.** **34.**

35. **36.** **37.**

38. — **39.** — **40.**

41. **42.** **43.**

Evaluate Expressions using the Commutative and Associative Properties

In the following exercises, evaluate each expression for the given value.

44. If — evaluate:

45. If — evaluate:

46. If — evaluate:

47. If — evaluate:

48. If evaluate:

— — — —

49. If evaluate:

— —

— —

50. If evaluate:

— —

— —

51. If evaluate:

— —

— —

Simplify Expressions Using the Commutative and Associative Properties

In the following exercises, simplify.

52. **53.** **54.** — — —

55. — — —

56. — — —

57. — — —

58. — — —

59. — — —

60. —

61. —

62. — — —

63. — — —

64. — — —

65. — — —

66.

67.

68.

69.

70.

71.

72.

73.

74.

75.

76. —

77. —

78.

79.

80.

81.

82. — — — —

83. — — — —

84.

85.

Everyday Math

86. Stamps Allie and Loren need to buy stamps. Allie needs four stamps and nine stamps. Loren needs eight stamps and three stamps.

How much will Allie's stamps cost?

How much will Loren's stamps cost?

What is the total cost of the girls' stamps?

How many stamps do the girls need altogether? How much will they cost?

How many stamps do the girls need altogether? How much will they cost?

87. Counting Cash Grant is totaling up the cash from a fundraising dinner. In one envelope, he has twenty-three bills, eighteen bills, and thirty-four bills. In another envelope, he has fourteen bills, nine bills, and twenty-seven bills.

How much money is in the first envelope?

How much money is in the second envelope?

What is the total value of all the cash?

What is the value of all the bills?

What is the value of all bills?

What is the value of all bills?

Writing Exercises

88. In your own words, state the Commutative Property of Addition and explain why it is useful.

89. In your own words, state the Associative Property of Multiplication and explain why it is useful.

Self Check

After completing the exercises, use this checklist to evaluate your mastery of the objectives of this section.

I can...	Confidently	With some help	No-I don't get it!
use the Commutative and Associative Properties.			
evaluate expressions using the Commutative and Associative Properties.			
simplify expressions using the Commutative and Associative Properties.			

After reviewing this checklist, what will you do to become confident for all objectives?

7.3 Distributive Property

Learning Objectives

By the end of this section, you will be able to:

- Simplify expressions using the distributive property
- Evaluate expressions using the distributive property

BE PREPARED : : 7.7 Before you get started, take this readiness quiz.

Multiply:

If you missed this problem, review Example 5.15

BE PREPARED : : 7.8 Simplify:

If you missed this problem, review Example 3.51

BE PREPARED : : 7.9 Combine like terms:

If you missed this problem, review Example 2.22.

Simplify Expressions Using the Distributive Property

Suppose three friends are going to the movies. They each need that is, dollars and quarter. How much money do they need all together? You can think about the dollars separately from the quarters.

$3 \times 9 = 27$

$3 \times \$0.25 = \$0.75$

They need times so and times quarter, so cents. In total, they need

If you think about doing the math in this way, you are using the Distributive Property.

Distributive Property

If are real numbers, then

Back to our friends at the movies, we could show the math steps we take to find the total amount of money they need like this:

In algebra, we use the Distributive Property to remove parentheses as we simplify expressions. For example, if we are asked to simplify the expression the order of operations says to work in the parentheses first. But we cannot add and since they are not like terms. So we use the Distributive Property, as shown in Example 7.19.

EXAMPLE 7.17

Simplify:

Solution

Distribute.	
Multiply.	

> **TRY IT : :** 7.33 Simplify:

> **TRY IT : :** 7.34 Simplify:

Some students find it helpful to draw in arrows to remind them how to use the Distributive Property. Then the first step in Example 7.17 would look like this:

$$3(x+4)$$

$$3 \cdot x + 3 \cdot 4$$

EXAMPLE 7.18

Simplify:

Solution

	$6(5y+1)$
Distribute.	$6 \cdot 5y + 6 \cdot 1$
Multiply.	$30y + 6$

> **TRY IT : :** 7.35 Simplify:

> **TRY IT : :** 7.36 Simplify:

The distributive property can be used to simplify expressions that look slightly different from Here are two other forms.

Distributive Property

If are real numbers, then

Other forms

EXAMPLE 7.19

Simplify:

Solution

	$2(x-3)$
Distribute.	$2 \cdot x - 2 \cdot 3$
Multiply.	$2x - 6$

TRY IT : : 7.37 Simplify:

TRY IT : : 7.38 Simplify:

Do you remember how to multiply a fraction by a whole number? We'll need to do that in the next two examples.

EXAMPLE 7.20

Simplify: —

Solution

	$\frac{3}{4}(n+12)$
Distribute.	$\frac{3}{4} \cdot n + \frac{3}{4} \cdot 12$
Simplify.	$\frac{3}{4}n + 9$

TRY IT : : 7.39 Simplify: —

TRY IT : : 7.40 Simplify: —

EXAMPLE 7.21

Simplify: — —

Solution

	$8\left(\frac{3}{8}x+\frac{1}{4}\right)$
Distribute.	$8\cdot\frac{3}{8}x+8\cdot\frac{1}{4}$
Multiply.	$3x+2$

TRY IT : : 7.41 Simplify: — —

TRY IT : : 7.42 Simplify: — —

Using the Distributive Property as shown in the next example will be very useful when we solve money applications later.

EXAMPLE 7.22

Simplify:

Solution

	$100(0.3+0.25q)$
Distribute.	$100(0.3)+100(0.25q)$
Multiply.	$30+25q$

TRY IT : : 7.43 Simplify:

TRY IT : : 7.44 Simplify:

In the next example we'll multiply by a variable. We'll need to do this in a later chapter.

EXAMPLE 7.23

Simplify:

Solution

	$m(n-4)$
Distribute.	$m\cdot n-m\cdot 4$
Multiply.	$mn-4m$

Notice that we wrote We can do this because of the Commutative Property of Multiplication. When a term is the product of a number and a variable, we write the number first.

TRY IT : : 7.45 Simplify:

TRY IT : : 7.46 Simplify:

The next example will use the 'backwards' form of the Distributive Property,

EXAMPLE 7.24

Simplify:

Solution

	$(x+8)p$
Distribute.	$px + 8p$

TRY IT : : 7.47 Simplify:

TRY IT : : 7.48 Simplify:

When you distribute a negative number, you need to be extra careful to get the signs correct.

EXAMPLE 7.25

Simplify:

Solution

	$-2(4y+1)$
Distribute.	$-2 \cdot 4y + (-2) \cdot 1$
Simplify.	$-8y - 2$

TRY IT : : 7.49 Simplify:

TRY IT : : 7.50 Simplify:

EXAMPLE 7.26

Simplify:

Solution

	$-11(4-3a)$
Distribute.	$-11 \cdot 4 - (-11) \cdot 3a$
Multiply.	$-44 + (-33a)$
Simplify.	$-44 + 33a$

You could also write the result as Do you know why?

TRY IT : : 7.51 Simplify:

TRY IT : : 7.52 Simplify:

In the next example, we will show how to use the Distributive Property to find the opposite of an expression. Remember,

EXAMPLE 7.27

Simplify:

Solution

	$-(y+5)$
Multiplying by −1 results in the opposite.	$-1(y+5)$
Distribute.	$-1\cdot y+(-1)\cdot 5$
Simplify.	$-y+(-5)$
Simplify.	$-y-5$

TRY IT : : 7.53 Simplify:

TRY IT : : 7.54 Simplify:

Sometimes we need to use the Distributive Property as part of the order of operations. Start by looking at the parentheses. If the expression inside the parentheses cannot be simplified, the next step would be multiply using the distributive property, which removes the parentheses. The next two examples will illustrate this.

EXAMPLE 7.28

Simplify:

Solution

	$8-2(x+3)$
Distribute.	$8-2\cdot x-2\cdot 3$
Multiply.	$8-2x-6$
Combine like terms.	$-2x+2$

TRY IT : : 7.55 Simplify:

TRY IT : : 7.56 Simplify:

EXAMPLE 7.29

Simplify:

Solution

	$4(x-8)-(x+3)$
Distribute.	$4x-32-x-3$
Combine like terms.	$3x-35$

TRY IT : : 7.57 Simplify:

TRY IT : : 7.58 Simplify:

Evaluate Expressions Using the Distributive Property

Some students need to be convinced that the Distributive Property always works.

In the examples below, we will practice evaluating some of the expressions from previous examples; in part , we will evaluate the form with parentheses, and in part we will evaluate the form we got after distributing. If we evaluate both expressions correctly, this will show that they are indeed equal.

EXAMPLE 7.30

When evaluate:

Solution

Substitute 10 for y.	$6(5\cdot10+1)$
Simplify in the parentheses.	
Multiply.	

	$6\cdot5y+6\cdot1$
Substitute 10 for y.	$6\cdot5\cdot10+6\cdot1$
Simplify.	$300+6$
Add.	306

Notice, the answers are the same. When

Try it yourself for a different value of

TRY IT : : 7.59 Evaluate when

TRY IT : : 7.60 Evaluate when

EXAMPLE 7.31

When evaluate

Solution

Substitute 3 for *y*.	$-2(4 \cdot 3 + 1)$
Simplify in the parentheses.	
Multiply.	

Substitute 3 for *y*.	$-2 \cdot 4 \cdot 3 + (-2) \cdot 1$
Multiply.	
Subtract.	
The answers are the same. When	

TRY IT : : 7.61 Evaluate when

TRY IT : : 7.62 Evaluate when

EXAMPLE 7.32

When evaluate and to show that

Solution

Substitute 35 for *y*.	$-(35 + 5)$
Add in the parentheses.	
Simplify.	

Substitute 35 for *y*.	$-35 - 5$
Simplify.	
The answers are the same when demonstrating that	

TRY IT : : 7.63 Evaluate when to show that

TRY IT : : 7.64 Evaluate when to show that

MEDIA : : ACCESS ADDITIONAL ONLINE RESOURCES

- Model Distribution (http://www.openstax.org/l/24ModelDist)
- The Distributive Property (http://www.openstax.org/l/24DistProp)

7.3 EXERCISES

Practice Makes Perfect

Simplify Expressions Using the Distributive Property

In the following exercises, simplify using the distributive property.

90.

91.

92.

93.

94.

95.

96.

97.

98. —

99. —

100. —

101. —

102. — —

103. — —

104. — —

105. — —

106.

107.

108.

109.

110.

111.

112.

113.

114.

115.

116.

117.

118.

119.

120.

121.

122.

123.

124.

125.

126.

127.

128.

129.

130.

131.

132.

133.

134.

135.

136.

137.

138.

139.

140.

141.

Evaluate Expressions Using the Distributive Property

In the following exercises, evaluate both expressions for the given value.

142. If evaluate

143. If evaluate

144. If — evaluate

—

—

145. If — evaluate

—

—

146. If — evaluate

147. If — evaluate

148. If evaluate

149. If evaluate

150. If evaluate

151. If evaluate

152. If evaluate

153. If evaluate

Everyday Math

154. Buying by the case Joe can buy his favorite ice tea at a convenience store for per bottle. At the grocery store, he can buy a case of bottles for

Use the distributive property to find the cost of bottles bought individually at the convenience store. (Hint: notice that is)

Is it a bargain to buy the iced tea at the grocery store by the case?

155. Multi-pack purchase Adele's shampoo sells for per bottle at the drug store. At the warehouse store, the same shampoo is sold as a for

Show how you can use the distributive property to find the cost of bottles bought individually at the drug store.

How much would Adele save by buying the at the warehouse store?

Writing Exercises

156. Simplify — using the distributive property and explain each step.

157. Explain how you can multiply without paper or a calculator by thinking of as and then using the distributive property.

Self Check

After completing the exercises, use this checklist to evaluate your mastery of the objectives of this section.

I can...	Confidently	With some help	No-I don't get it!
simplify expressions using the Distributive Property.			
evaluate expressions using the Distributive Property.			

What does this checklist tell you about your mastery of this section? What steps will you take to improve?

7.4 Properties of Identity, Inverses, and Zero

Learning Objectives

By the end of this section, you will be able to:

- Recognize the identity properties of addition and multiplication
- Use the inverse properties of addition and multiplication
- Use the properties of zero
- Simplify expressions using the properties of identities, inverses, and zero

BE PREPARED : : 7.10 Before you get started, take this readiness quiz.

Find the opposite of

If you missed this problem, review **Example 3.3**.

BE PREPARED : : 7.11 Find the reciprocal of —

If you missed this problem, review **Example 4.29**.

BE PREPARED : : 7.12 Multiply: — —

If you missed this problem, review **Example 4.27**.

Recognize the Identity Properties of Addition and Multiplication

What happens when we add zero to any number? Adding zero doesn't change the value. For this reason, we call the **additive identity**.

For example,

What happens when you multiply any number by one? Multiplying by one doesn't change the value. So we call the **multiplicative identity**.

For example,

—

—

Identity Properties

The **identity property of addition**: for any real number

The **identity property of multiplication**: for any real number

EXAMPLE 7.33

Identify whether each equation demonstrates the identity property of addition or multiplication.

Solution

We are adding 0.	We are using the identity property of addition.

We are multiplying by 1.	We are using the identity property of multiplication.

TRY IT : : 7.65

Identify whether each equation demonstrates the identity property of addition or multiplication:

TRY IT : : 7.66

Identify whether each equation demonstrates the identity property of addition or multiplication:

Use the Inverse Properties of Addition and Multiplication

What number added to 5 gives the additive identity, 0?
We know 5 + (−5) = 0
What number added to −6 gives the additive identity, 0?
We know −6 + 6 = 0

Notice that in each case, the missing number was the opposite of the number.

We call the **additive inverse** of The opposite of a number is its additive inverse. A number and its opposite add to which is the additive identity.

What number multiplied by — gives the multiplicative identity, In other words, two-thirds times what results in

—	We know $\frac{2}{3} \cdot \frac{3}{2} = 1$

What number multiplied by gives the multiplicative identity, In other words two times what results in

We know $2 \cdot \frac{1}{2} = 1$

Notice that in each case, the missing number was the reciprocal of the number.

We call — the **multiplicative inverse** of The reciprocal of a number is its multiplicative inverse. A number and its reciprocal multiply to which is the multiplicative identity.

We'll formally state the Inverse Properties here:

Inverse Properties

Inverse Property of Addition for any real number

Inverse Property of Multiplication for any real number

—

—

EXAMPLE 7.34

Find the additive inverse of each expression: — .

Solution

To find the additive inverse, we find the opposite.

The additive inverse of is its opposite, The additive inverse of — is its opposite, —

The additive inverse of is its opposite,

TRY IT : : 7.67 Find the additive inverse: — .

TRY IT : : 7.68 Find the additive inverse: — .

EXAMPLE 7.35

Find the multiplicative inverse: — .

Solution

To find the multiplicative inverse, we find the reciprocal.

The multiplicative inverse of is its reciprocal, —

The multiplicative inverse of — is its reciprocal,

To find the multiplicative inverse of we first convert to a fraction, — Then we find the reciprocal, —

TRY IT : : 7.69 Find the multiplicative inverse: — .

TRY IT : : 7.70 Find the multiplicative inverse: — .

Use the Properties of Zero

We have already learned that zero is the additive identity, since it can be added to any number without changing the number's identity. But zero also has some special properties when it comes to multiplication and division.

Multiplication by Zero

What happens when you multiply a number by Multiplying by makes the product equal zero. The product of any real number and is

Multiplication by Zero

For any real number

EXAMPLE 7.36

Simplify: — .

Solution

The product of any real number and 0 is 0.

—

The product of any real number and 0 is 0.

The product of any real number and 0 is 0.

TRY IT : : 7.71 Simplify: —

TRY IT : : 7.72 Simplify: —

Dividing with Zero

What about dividing with Think about a real example: if there are no cookies in the cookie jar and three people want to share them, how many cookies would each person get? There are cookies to share, so each person gets cookies.

Remember that we can always check division with the related multiplication fact. So, we know that

Division of Zero

For any real number except — and

Zero divided by any real number except zero is zero.

EXAMPLE 7.37

Simplify: — —.

Solution

Zero divided by any real number, except 0, is zero.

—

Zero divided by any real number, except 0, is zero.

—

Zero divided by any real number, except 0, is zero.

TRY IT : : 7.73 Simplify: — —.

TRY IT : : 7.74 Simplify: — .

Now let's think about dividing a number *by* zero. What is the result of dividing by Think about the related multiplication fact. Is there a number that multiplied by gives

Since any real number multiplied by equals there is no real number that can be multiplied by to obtain We can conclude that there is no answer to and so we say that division by zero is undefined.

Division by Zero

For any real number — and are undefined.

Division *by* zero is undefined.

EXAMPLE 7.38

Simplify: —— — .

Solution

Division by zero is undefined.	undefined

	——
Division by zero is undefined.	undefined

	—
Division by zero is undefined.	undefined

TRY IT : : 7.75 Simplify: —— — .

TRY IT : : 7.76 Simplify: —— ——

We summarize the properties of zero.

Properties of Zero

Multiplication by Zero: For any real number

Division by Zero: For any real number

— Zero divided by any real number, except itself, is zero.

— is undefined. Division by zero is undefined.

Simplify Expressions using the Properties of Identities, Inverses, and Zero

We will now practice using the properties of identities, inverses, and zero to simplify expressions.

EXAMPLE 7.39

Simplify:

Solution

Notice the additive inverses, and .
Add.

TRY IT : : 7.77 Simplify:

TRY IT : : 7.78 Simplify:

EXAMPLE 7.40

Simplify:

Solution

Regroup, using the associative property.
Multiply.
Simplify; 1 is the multiplicative identity.

TRY IT : : 7.79 Simplify:

TRY IT : : 7.80 Simplify:

EXAMPLE 7.41

Simplify: ——, where .

Solution

Zero divided by any real number except itself is zero.

TRY IT : : 7.81 Simplify: ——, where .

TRY IT : : 7.82 Simplify: ——, where .

EXAMPLE 7.42

Simplify: ——

Solution

Division by zero is undefined.	undefined

TRY IT : : 7.83 Simplify: ——

TRY IT : : 7.84 Simplify: ——

EXAMPLE 7.43

Simplify: — —

Solution

We cannot combine the terms in parentheses, so we multiply the two fractions first.

Multiply; the product of reciprocals is 1.
Simplify by recognizing the multiplicative identity.

TRY IT : : 7.85 Simplify: — —

TRY IT : : 7.86 Simplify: — —

All the properties of real numbers we have used in this chapter are summarized in Table 7.1.

Property	Of Addition	Of Multiplication
Commutative Property		
If *a* and *b* are real numbers then…		
Associative Property		
If *a*, *b*, and *c* are real numbers then…		
Identity Property	is the additive identity	is the multiplicative identity
For any real number *a*,		
Inverse Property	is the additive inverse of	is the multiplicative inverse of
For any real number *a*,		
Distributive Property	If are real numbers, then	
Properties of Zero		
For any real number *a*,		
For any real number	— — is undefined	

Table 7.1 Properties of Real Numbers

▶ **MEDIA : :** ACCESS ADDITIONAL ONLINE RESOURCES

- Multiplying and Dividing Involving Zero (http://www.openstax.org/l/24MultDivZero)

7.4 EXERCISES

Practice Makes Perfect

Recognize the Identity Properties of Addition and Multiplication

In the following exercises, identify whether each example is using the identity property of addition or multiplication.

158.

159. — —

160.

161.

Use the Inverse Properties of Addition and Multiplication

In the following exercises, find the multiplicative inverse.

162.

163.

164.

165.

166. —

167. —

168. —

169. —

170.

171.

172.

173.

Use the Properties of Zero

In the following exercises, simplify using the properties of zero.

174.

175. —

176. —

177.

178. —

179. —

180. —

181. —

182. —

183.

184.

185. — —

Simplify Expressions using the Properties of Identities, Inverses, and Zero

In the following exercises, simplify using the properties of identities, inverses, and zero.

186.

187.

188.

189.

190.

191.

192.

193.

194. ———, where

195. ———, where

196. ————, where

197. ————, where

198. — , where —

199. — , where —

200. ————, where

201. ———, where **202.** ———, where **203.** ———, where

204. — — , where — —

205. — — , where — —

206. — —

207. — —

208. —

209. —

Everyday Math

210. Insurance copayment Carrie had to have fillings done. Each filling cost Her dental insurance required her to pay of the cost. Calculate Carrie's cost

by finding her copay for each filling, then finding her total cost for fillings, and

by multiplying

Which of the Properties of Real Numbers did you use for part (b)?

211. Cooking time Helen bought a turkey for her family's Thanksgiving dinner and wants to know what time to put the turkey in the oven. She wants to allow minutes per pound cooking time.

Calculate the length of time needed to roast the turkey by multiplying to find the number of minutes and then multiplying the product by — to convert minutes into hours.

Multiply —

Which of the Properties of Real Numbers allows you to multiply — instead of —

Writing Exercises

212. In your own words, describe the difference between the additive inverse and the multiplicative inverse of a number.

213. How can the use of the properties of real numbers make it easier to simplify expressions?

Self Check

After completing the exercises, use this checklist to evaluate your mastery of the objectives of this section.

I can...	Confidently	With some help	No-I don't get it!
recognize the Identity Properties of Addition and Multiplication.			
use the Inverse Properties of Addition and Multiplication.			
use the Properties of Zero.			
simplify expressions using the properties of identities, inverses, and zero.			

On a scale of 1-10, how would you rate your mastery of this section in light of your responses on the checklist? How can you improve this?

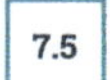

Systems of Measurement

Learning Objectives

By the end of this section, you will be able to:

- Make unit conversions in the U.S. system
- Use mixed units of measurement in the U.S. system
- Make unit conversions in the metric system
- Use mixed units of measurement in the metric system
- Convert between the U.S. and the metric systems of measurement
- Convert between Fahrenheit and Celsius temperatures

BE PREPARED : : 7.13 Before you get started, take this readiness quiz.

Multiply:

If you missed this problem, review Example 5.18.

BE PREPARED : : 7.14

Simplify: —

If you missed this problem, review Example 4.20.

BE PREPARED : : 7.15

Multiply: — —

If you missed this problem, review Example 4.27.

In this section we will see how to convert among different types of units, such as feet to miles or kilograms to pounds. The basic idea in all of the unit conversions will be to use a form of the multiplicative identity, to change the units but not the value of a quantity.

Make Unit Conversions in the U.S. System

There are two systems of measurement commonly used around the world. Most countries use the metric system. The United States uses a different system of measurement, usually called the U.S. system. We will look at the U.S. system first.

The U.S. system of measurement uses units of inch, foot, yard, and mile to measure length and pound and ton to measure weight. For capacity, the units used are cup, pint, quart and gallons. Both the U.S. system and the metric system measure time in seconds, minutes, or hours.

The equivalencies among the basic units of the U.S. system of measurement are listed in Table 7.2. The table also shows, in parentheses, the common abbreviations for each measurement.

U.S. System Units	
Length	**Volume**
foot (ft) = inches (in) yard (yd) = feet (ft) mile (mi) = feet (ft)	teaspoons (t) = tablespoon (T) Tablespoons (T) = cup (C) cup (C) = fluid ounces (fl oz) pint (pt) = cups (C) quart (qt) = pints (pt) gallon (gal) = quarts (qt)
Weight	**Time**
pound (lb) = ounces (oz) ton = pounds (lb)	minute (min) = seconds (s) hour (h) = minutes (min) day = hours (h) week (wk) = days year (yr) = days

Table 7.2

In many real-life applications, we need to convert between units of measurement. We will use the identity property of multiplication to do these conversions. We'll restate the Identity Property of Multiplication here for easy reference.

To use the identity property of multiplication, we write in a form that will help us convert the units. For example, suppose we want to convert inches to feet. We know that foot is equal to inches, so we can write as the fraction When we multiply by this fraction, we do not change the value but just change the units.

But also equals How do we decide whether to multiply by or We choose the fraction that will make the units we want to convert *from* divide out. For example, suppose we wanted to convert inches to feet. If we choose the fraction that has inches in the denominator, we can eliminate the inches.

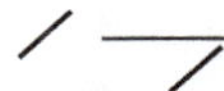

On the other hand, if we wanted to convert feet to inches, we would choose the fraction that has feet in the denominator.

We treat the unit words like factors and 'divide out' common units like we do common factors.

HOW TO : : MAKE UNIT CONVERSIONS.

Step 1. Multiply the measurement to be converted by write as a fraction relating the units given and the units needed.

Step 2. Multiply.

Step 3. Simplify the fraction, performing the indicated operations and removing the common units.

EXAMPLE 7.44

Mary Anne is inches tall. What is her height in feet?

Solution

Convert 66 inches into feet.	
Multiply the measurement to be converted by 1.	inches
Write 1 as a fraction relating the units given and the units needed.	
Multiply.	
Simplify the fraction.	

Notice that the when we simplified the fraction, we first divided out the inches.

Mary Anne is feet tall.

> TRY IT : : 7.87 Lexie is inches tall. Convert her height to feet.

> TRY IT : : 7.88 Rene bought a hose that is yards long. Convert the length to feet.

When we use the Identity Property of Multiplication to convert units, we need to make sure the units we want to change from will divide out. Usually this means we want the conversion fraction to have those units in the denominator.

EXAMPLE 7.45

Ndula, an elephant at the San Diego Safari Park, weighs almost tons. Convert her weight to pounds.

Figure 7.5 (credit: Guldo Da Rozze, Flickr)

Solution

We will convert tons into pounds, using the equivalencies in Table 7.2. We will use the Identity Property of Multiplication, writing as the fraction

Multiply the measurement to be converted by 1.	
Write 1 as a fraction relating tons and pounds.	
Simplify.	
Multiply.	
	Ndula weighs almost 6,400 pounds.

> **TRY IT : : 7.89** Arnold's SUV weighs about tons. Convert the weight to pounds.

> **TRY IT : : 7.90** A cruise ship weighs tons. Convert the weight to pounds.

Sometimes to convert from one unit to another, we may need to use several other units in between, so we will need to multiply several fractions.

EXAMPLE 7.46

Juliet is going with her family to their summer home. She will be away for weeks. Convert the time to minutes.

Solution

To convert weeks into minutes, we will convert weeks to days, days to hours, and then hours to minutes. To do this, we will multiply by conversion factors of

Write 1 as .	$\frac{9\text{ wk}}{1} \cdot \frac{7\text{ days}}{1\text{ wk}} \cdot \frac{24\text{ hr}}{1\text{ day}} \cdot \frac{60\text{ min}}{1\text{ hr}}$
Cancel common units.	$\frac{9\cancel{\text{ wk}}}{1} \cdot \frac{7\cancel{\text{ days}}}{1\cancel{\text{ wk}}} \cdot \frac{24\cancel{\text{ hr}}}{1\cancel{\text{ day}}} \cdot \frac{60\text{ min}}{1\cancel{\text{ hr}}}$
Multiply.	
	Juliet will be away for 90,720 minutes.

> **TRY IT : : 7.91**
> The distance between Earth and the moon is about miles. Convert this length to yards.

> **TRY IT : : 7.92** A team of astronauts spends weeks in space. Convert the time to minutes.

EXAMPLE 7.47

How many fluid ounces are in gallon of milk?

Figure 7.6 (credit: www.bluewaikiki.com, Flickr)

Solution

Use conversion factors to get the right units: convert gallons to quarts, quarts to pints, pints to cups, and cups to fluid ounces.

	1 gallon
Multiply the measurement to be converted by 1.	
Simplify.	
Multiply.	
Simplify.	128 fluid ounces
	There are 128 fluid ounces in a gallon.

TRY IT : : 7.93 How many cups are in gallon?

TRY IT : : 7.94 How many teaspoons are in cup?

Use Mixed Units of Measurement in the U.S. System

Performing arithmetic operations on measurements with mixed units of measures requires care. Be sure to add or subtract like units.

EXAMPLE 7.48

Charlie bought three steaks for a barbecue. Their weights were ounces, pound ounces, and pound ounces. How many total pounds of steak did he buy?

Figure 7.7 (credit: Helen Penjam, Flickr)

Solution

We will add the weights of the steaks to find the total weight of the steaks.

Add the ounces. Then add the pounds.	14 ounces 1 pound 2 ounces + 1 pound 6 ounces 2 pounds 22 ounces
Convert 22 ounces to pounds and ounces.	
Add the pounds.	2 pounds + 1 pound, 6 ounces 3 pounds, 6 ounces
	Charlie bought 3 pounds 6 ounces of steak.

TRY IT : : 7.95

Laura gave birth to triplets weighing pounds ounces, pounds ounces, and pounds ounces. What was the total birth weight of the three babies?

TRY IT : : 7.96

Seymour cut two pieces of crown molding for his family room that were feet inches and feet inches. What was the total length of the molding?

EXAMPLE 7.49

Anthony bought four planks of wood that were each feet inches long. If the four planks are placed end-to-end, what is the total length of the wood?

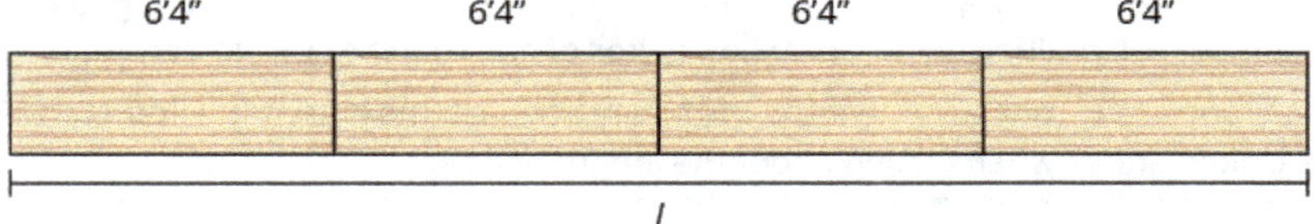

Solution

We will multiply the length of one plank by to find the total length.

Multiply the inches and then the feet.	6 feet 4 inches × 4 24 feet 16 inches
Convert 16 inches to feet.	24 feet + 1 foot 4 inches
Add the feet.	25 feet 4 inches
	Anthony bought 25 feet 4 inches of wood.

TRY IT : : 7.97

Henri wants to triple his spaghetti sauce recipe, which calls for pound ounces of ground turkey. How many pounds of ground turkey will he need?

 **TRY IT : :** 7.98

Joellen wants to double a solution of gallons quarts. How many gallons of solution will she have in all?

Make Unit Conversions in the Metric System

In the metric system, units are related by powers of The root words of their names reflect this relation. For example, the basic unit for measuring length is a meter. One kilometer is meters; the prefix *kilo-* means thousand. One centimeter is —— of a meter, because the prefix *centi-* means one one-hundredth (just like one cent is —— of one dollar).

The equivalencies of measurements in the metric system are shown in Table 7.3. The common abbreviations for each measurement are given in parentheses.

Metric Measurements		
Length	**Mass**	**Volume/Capacity**
kilometer (km) = m	kilogram (kg) = g	kiloliter (kL) = L
hectometer (hm) = m	hectogram (hg) = g	hectoliter (hL) = L
dekameter (dam) = m	dekagram (dag) = g	dekaliter (daL) = L
meter (m) = m	gram (g) = g	liter (L) = L
decimeter (dm) = m	decigram (dg) = g	deciliter (dL) = L
centimeter (cm) = m	centigram (cg) = g	centiliter (cL) = L
millimeter (mm) = m	milligram (mg) = g	milliliter (mL) = L
meter = centimeters	gram = centigrams	liter = centiliters
meter = millimeters	gram = milligrams	liter = milliliters

Table 7.3

To make conversions in the metric system, we will use the same technique we did in the U.S. system. Using the identity property of multiplication, we will multiply by a conversion factor of one to get to the correct units.

Have you ever run a or race? The lengths of those races are measured in kilometers. The metric system is commonly used in the United States when talking about the length of a race.

EXAMPLE 7.50

Nick ran a race. How many meters did he run?

Figure 7.8 (credit: William Warby, Flickr)

Solution

We will convert kilometers to meters using the Identity Property of Multiplication and the equivalencies in Table 7.3.

	10 kilometers
Multiply the measurement to be converted by 1.	$10 \text{ km} \cdot 1$
Write 1 as a fraction relating kilometers and meters.	$10 \text{ km} \cdot \frac{1000 \text{ m}}{1 \text{ km}}$
Simplify.	$\frac{10 \cancel{\text{km}} \cdot 1000 \text{ m}}{1 \cancel{\text{km}}}$
Multiply.	10,000 m
	Nick ran 10,000 meters.

TRY IT : : 7.99 Sandy completed her first race. How many meters did she run?

TRY IT : : 7.100 Herman bought a rug meters in length. How many centimeters is the length?

EXAMPLE 7.51

Eleanor's newborn baby weighed grams. How many kilograms did the baby weigh?

Solution

We will convert grams to kilograms.

	3200 grams
Multiply the measurement to be converted by 1.	$3200 \text{ g} \cdot 1$
Write 1 as a fraction relating kilograms and grams.	$3200 \text{ g} \cdot \frac{1 \text{ kg}}{1000 \text{ g}}$
Simplify.	$3200 \cancel{\text{g}} \cdot \frac{1 \text{ kg}}{1000 \cancel{\text{g}}}$
Multiply.	$\frac{3200 \text{ kilograms}}{1000}$
Divide.	3.2 kilograms
	The baby weighed kilograms.

TRY IT : : 7.101 Kari's newborn baby weighed grams. How many kilograms did the baby weigh?

TRY IT : : 7.102

Anderson received a package that was marked grams. How many kilograms did this package weigh?

Since the metric system is based on multiples of ten, conversions involve multiplying by multiples of ten. In Decimal Operations, we learned how to simplify these calculations by just moving the decimal.

To multiply by we move the decimal to the right places, respectively. To multiply by we move the decimal to the left places respectively.

We can apply this pattern when we make measurement conversions in the metric system.

In Example 7.51, we changed grams to kilograms by multiplying by —— This is the same as moving the decimal places to the left.

$$3200 \cdot \frac{1}{1000} \qquad 3200.$$

$$3.2 \qquad 3.2$$

EXAMPLE 7.52

Convert:

liters to kiloliters liters to milliliters.

Solution

We will convert liters to kiloliters. In Table 7.3, we see that

	350 L
Multiply by 1, writing 1 as a fraction relating liters to kiloliters.	$350\text{ L} \cdot \frac{1\text{ kL}}{1000\text{ L}}$
Simplify.	$350\text{ L} \cdot \frac{1\text{ kL}}{1000\text{ L}}$
Move the decimal 3 units to the left.	$350\text{ L} \cdot \frac{1\text{ kL}}{1000\text{ L}}$
	0.35 kL

We will convert liters to milliliters. In Table 7.3, we see that

	4.1 L
Multiply by 1, writing 1 as a fraction relating milliliters to liters.	$4.1\text{ L} \cdot \frac{1000\text{ mL}}{1\text{ L}}$
Simplify.	$4.1\text{ L} \cdot \frac{1000\text{ mL}}{1\text{ L}}$
Move the decimal 3 units to the left.	4.100 mL
	4100 mL

TRY IT : : 7.103 Convert: L to kL L to mL.

TRY IT : : 7.104 Convert: hL to L L to cL.

Use Mixed Units of Measurement in the Metric System

Performing arithmetic operations on measurements with mixed units of measures in the metric system requires the same care we used in the U.S. system. But it may be easier because of the relation of the units to the powers of We still must make sure to add or subtract like units.

EXAMPLE 7.53

Ryland is meters tall. His younger brother is centimeters tall. How much taller is Ryland than his younger brother?

Solution

We will subtract the lengths in meters. Convert centimeters to meters by moving the decimal places to the left; cm is the same as m.

Now that both measurements are in meters, subtract to find out how much taller Ryland is than his brother.

Ryland is meters taller than his brother.

TRY IT : : 7.105

Mariella is meters tall. Her daughter is centimeters tall. How much taller is Mariella than her daughter? Write the answer in centimeters.

TRY IT : : 7.106

The fence around Hank's yard is meters high. Hank is centimeters tall. How much shorter than the fence is Hank? Write the answer in meters.

EXAMPLE 7.54

Dena's recipe for lentil soup calls for milliliters of olive oil. Dena wants to triple the recipe. How many liters of olive oil will she need?

Solution

We will find the amount of olive oil in milliliters then convert to liters.

	Triple 150 mL
Translate to algebra.	
Multiply.	
Convert to liters.	
Simplify.	
	Dena needs 0.45 liter of olive oil.

TRY IT : : 7.107

A recipe for Alfredo sauce calls for milliliters of milk. Renata is making pasta with Alfredo sauce for a big party and needs to multiply the recipe amounts by How many liters of milk will she need?

TRY IT : : 7.108

To make one pan of baklava, Dorothea needs grams of filo pastry. If Dorothea plans to make pans of baklava, how many kilograms of filo pastry will she need?

Convert Between U.S. and Metric Systems of Measurement

Many measurements in the United States are made in metric units. A drink may come in bottles, calcium may come in capsules, and we may run a race. To work easily in both systems, we need to be able to convert between the two systems.

Table 7.4 shows some of the most common conversions.

Conversion Factors Between U.S. and Metric Systems					
Length		**Weight**		**Volume**	
in =	cm	lb =	kg	qt =	L
ft =	m	oz =	g	fl oz =	mL
yd =	m				
mi =	km				
m =	ft	kg =	lb	L =	qt

Table 7.4

We make conversions between the systems just as we do within the systems—by multiplying by unit conversion factors.

EXAMPLE 7.55

Lee's water bottle holds mL of water. How many fluid ounces are in the bottle? Round to the nearest tenth of an ounce.

Solution

	500 mL
Multiply by a unit conversion factor relating mL and ounces.	
Simplify.	
Divide.	
	The water bottle holds 16.7 fluid ounces.

TRY IT : : 7.109 How many quarts of soda are in a bottle?

TRY IT : : 7.110 How many liters are in quarts of milk?

The conversion factors in Table 7.4 are not exact, but the approximations they give are close enough for everyday purposes. In Example 7.55, we rounded the number of fluid ounces to the nearest tenth.

EXAMPLE 7.56

Soleil lives in Minnesota but often travels in Canada for work. While driving on a Canadian highway, she passes a sign that says the next rest stop is in kilometers. How many miles until the next rest stop? Round your answer to the nearest mile.

Solution

	100 kilometers
Multiply by a unit conversion factor relating kilometers and miles.	
Simplify.	
Divide.	62 mi
	It is about 62 miles to the next rest stop.

TRY IT : : 7.111

The height of Mount Kilimanjaro is meters. Convert the height to feet. Round to the nearest foot.

TRY IT : : 7.112

The flight distance from New York City to London is kilometers. Convert the distance to miles. Round to the nearest mile.

Convert Between Fahrenheit and Celsius Temperatures

Have you ever been in a foreign country and heard the weather forecast? If the forecast is for What does that mean?

The U.S. and metric systems use different scales to measure temperature. The U.S. system uses degrees Fahrenheit, written The metric system uses degrees Celsius, written Figure 7.9 shows the relationship between the two systems.

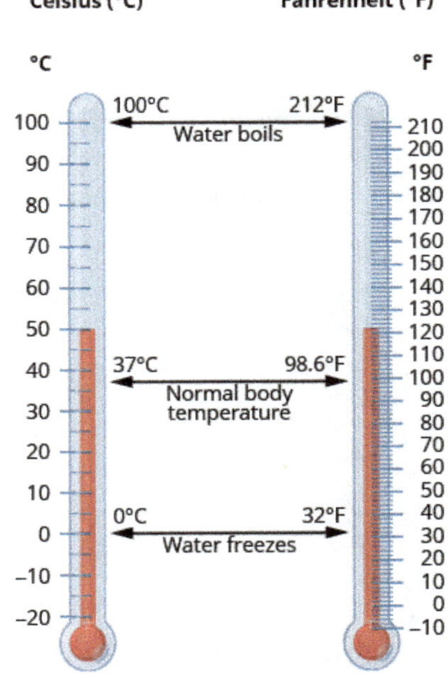

Figure 7.9 A temperature of is equivalent to

If we know the temperature in one system, we can use a formula to convert it to the other system.

Temperature Conversion

To convert from Fahrenheit temperature, to Celsius temperature, use the formula

—

To convert from Celsius temperature, to Fahrenheit temperature, use the formula

—

EXAMPLE 7.57

Convert into degrees Celsius.

Solution

We will substitute into the formula to find

Use the formula for converting °F to °C	—
Substitute 50 for F.	$C = \frac{5}{9}(50 - 32)$
Simplify in parentheses.	—
Multiply.	
	A temperature of 50°F is equivalent to 10°C.

TRY IT : : 7.113 Convert the Fahrenheit temperatures to degrees Celsius:

TRY IT : : 7.114 Convert the Fahrenheit temperatures to degrees Celsius:

EXAMPLE 7.58

The weather forecast for Paris predicts a high of Convert the temperature into degrees Fahrenheit.

Solution

We will substitute into the formula to find

Use the formula for converting °F to °C	—
Substitute 20 for C.	$F = \frac{9}{5}(20) + 32$
Multiply.	
Add.	
	So 20°C is equivalent to 68°F.

TRY IT : : 7.115 Convert the Celsius temperatures to degrees Fahrenheit:

The temperature in Helsinki, Finland was

TRY IT : : 7.116 Convert the Celsius temperatures to degrees Fahrenheit:

The temperature in Sydney, Australia was

MEDIA : : ACCESS ADDITIONAL ONLINE RESOURCES

- American Unit Conversion (http://www.openstax.org/l/24USConversion)
- Time Conversions (http://www.openstax.org/l/24TimeConversio)
- Metric Unit Conversions (http://www.openstax.org/l/24MetricConvers)
- American and Metric Conversions (http://www.openstax.org/l/24UStoMetric)
- Convert from Celsius to Fahrenheit (http://www.openstax.org/l/24CtoFdegrees)
- Convert from Fahrenheit to Celsius (http://www.openstax.org/l/24FtoCdegrees)

7.5 EXERCISES

Practice Makes Perfect

Make Unit Conversions in the U.S. System

In the following exercises, convert the units.

214. A park bench is feet long. Convert the length to inches.

215. A floor tile is feet wide. Convert the width to inches.

216. A ribbon is inches long. Convert the length to feet.

217. Carson is inches tall. Convert his height to feet.

218. Jon is feet inches tall. Convert his height to inches.

219. Faye is feet inches tall. Convert her height to inches.

220. A football field is feet wide. Convert the width to yards.

221. On a baseball diamond, the distance from home plate to first base is yards. Convert the distance to feet.

222. Ulises lives miles from school. Convert the distance to feet.

223. Denver, Colorado, is feet above sea level. Convert the height to miles.

224. A killer whale weighs tons. Convert the weight to pounds.

225. Blue whales can weigh as much as tons. Convert the weight to pounds.

226. An empty bus weighs pounds. Convert the weight to tons.

227. At take-off, an airplane weighs pounds. Convert the weight to tons.

228. The voyage of the *Mayflower* took months and days. Convert the time to days (30 days = 1 month).

229. Lynn's cruise lasted days and hours. Convert the time to hours.

230. Rocco waited — hours for his appointment. Convert the time to seconds.

231. Misty's surgery lasted — hours. Convert the time to seconds.

232. How many teaspoons are in a pint?

233. How many tablespoons are in a gallon?

234. JJ's cat, Posy, weighs pounds. Convert her weight to ounces.

235. April's dog, Beans, weighs pounds. Convert his weight to ounces.

236. Baby Preston weighed pounds ounces at birth. Convert his weight to ounces.

237. Baby Audrey weighed pounds ounces at birth. Convert her weight to ounces.

238. Crista will serve cups of juice at her son's party. Convert the volume to gallons.

239. Lance needs cups of water for the runners in a race. Convert the volume to gallons.

Use Mixed Units of Measurement in the U.S. System

In the following exercises, solve and write your answer in mixed units.

240. Eli caught three fish. The weights of the fish were pounds ounces, pound ounces, and pounds ounces. What was the total weight of the three fish?

241. Judy bought pound ounces of almonds, pounds ounces of walnuts, and ounces of cashews. What was the total weight of the nuts?

242. One day Anya kept track of the number of minutes she spent driving. She recorded trips of

How much time (in hours and minutes) did Anya spend driving?

243. Last year Eric went on business trips. The number of days of each was How much time (in weeks and days) did Eric spend on business trips last year?

244. Renee attached a extension cord to her computer's power cord. What was the total length of the cords?

245. Fawzi's SUV is feet inches tall. If he puts a box on top of his SUV, what is the total height of the SUV and the box?

246. Leilani wants to make placemats. For each placemat she needs inches of fabric. How many yards of fabric will she need for the placemats?

247. Mireille needs to cut inches of ribbon for each of the girls in her dance class. How many yards of ribbon will she need altogether?

Make Unit Conversions in the Metric System

In the following exercises, convert the units.

248. Ghalib ran kilometers. Convert the length to meters.

249. Kitaka hiked kilometers. Convert the length to meters.

250. Estrella is meters tall. Convert her height to centimeters.

251. The width of the wading pool is meters. Convert the width to centimeters.

252. Mount Whitney is meters tall. Convert the height to kilometers.

253. The depth of the Mariana Trench is meters. Convert the depth to kilometers.

254. June's multivitamin contains milligrams of calcium. Convert this to grams.

255. A typical ruby-throated hummingbird weights grams. Convert this to milligrams.

256. One stick of butter contains grams of fat. Convert this to milligrams.

257. One serving of gourmet ice cream has grams of fat. Convert this to milligrams.

258. The maximum mass of an airmail letter is kilograms. Convert this to grams.

259. Dimitri's daughter weighed kilograms at birth. Convert this to grams.

260. A bottle of wine contained milliliters. Convert this to liters.

261. A bottle of medicine contained milliliters. Convert this to liters.

Use Mixed Units of Measurement in the Metric System

In the following exercises, solve and write your answer in mixed units.

262. Matthias is meters tall. His son is centimeters tall. How much taller, in centimeters, is Matthias than his son?

263. Stavros is meters tall. His sister is centimeters tall. How much taller, in centimeters, is Stavros than his sister?

264. A typical dove weighs grams. A typical duck weighs kilograms. What is the difference, in grams, of the weights of a duck and a dove?

265. Concetta had a bag of flour. She used grams of flour to make biscotti. How many kilograms of flour are left in the bag?

266. Harry mailed packages that weighed grams each. What was the total weight of the packages in kilograms?

267. One glass of orange juice provides milligrams of potassium. Linda drinks one glass of orange juice every morning. How many grams of potassium does Linda get from her orange juice in days?

268. Jonas drinks milliliters of water times a day. How many liters of water does Jonas drink in a day?

269. One serving of whole grain sandwich bread provides grams of protein. How many milligrams of protein are provided by servings of whole grain sandwich bread?

Convert Between U.S. and Metric Systems

In the following exercises, make the unit conversions. Round to the nearest tenth.

270. Bill is inches tall. Convert his height to centimeters.

271. Frankie is inches tall. Convert his height to centimeters.

272. Marcus passed a football yards. Convert the pass length to meters.

273. Connie bought yards of fabric to make drapes. Convert the fabric length to meters.

274. Each American throws out an average of pounds of garbage per year. Convert this weight to kilograms (2.20 pounds = 1 kilogram).

275. An average American will throw away pounds of trash over his or her lifetime. Convert this weight to kilograms (2.20 pounds = 1 kilogram).

276. A run is kilometers long. Convert this length to miles.

277. Kathryn is meters tall. Convert her height to feet.

278. Dawn's suitcase weighed kilograms. Convert the weight to pounds.

279. Jackson's backpack weighs kilograms. Convert the weight to pounds.

280. Ozzie put gallons of gas in his truck. Convert the volume to liters.

281. Bernard bought gallons of paint. Convert the volume to liters.

Convert between Fahrenheit and Celsius

In the following exercises, convert the Fahrenheit temperature to degrees Celsius. Round to the nearest tenth.

282.

283.

284.

285.

286.

287.

288.

289.

In the following exercises, convert the Celsius temperatures to degrees Fahrenheit. Round to the nearest tenth.

290.

291.

292.

293.

294.

295.

296.

297.

Everyday Math

298. Nutrition Julian drinks one can of soda every day. Each can of soda contains grams of sugar. How many kilograms of sugar does Julian get from soda in year?

299. Reflectors The reflectors in each lane-marking stripe on a highway are spaced yards apart. How many reflectors are needed for a one-mile-long stretch of highway?

Writing Exercises

300. Some people think that to Fahrenheit is the ideal temperature range.

What is your ideal temperature range? Why do you think so?

Convert your ideal temperatures from Fahrenheit to Celsius.

301. Did you grow up using the U.S. customary or the metric system of measurement? Describe two examples in your life when you had to convert between systems of measurement. Which system do you think is easier to use? Explain.

Self Check

After completing the exercises, use this checklist to evaluate your mastery of the objectives of this section.

I can…	Confidently	With some help	No-I don't get it!
make unit conversions in the U.S. system.			
use mixed units of measurement in the U.S. system.			
use mixed units of measurement in the metric system.			
convert between the U.S. and the metric systems of measurement.			
convert between Fahrenheit and Celsius temperatures.			

Overall, after looking at the checklist, do you think you are well-prepared for the next chapter? Why or why not?

CHAPTER 7 REVIEW

KEY TERMS

Additive Identity The **additive identity** is 0. When zero is added to any number, it does not change the value.

Additive Inverse The opposite of a number is its **additive inverse**. The **additive inverse** of a is .

Irrational number An **irrational number** is a number that cannot be written as the ratio of two integers. Its decimal form does not stop and does not repeat.

Multiplicative Identity The **multiplicative identity** is 1. When one multiplies any number, it does not change the value.

Multiplicative Inverse The reciprocal of a number is its **multiplicative inverse**. The **multiplicative inverse** of *a* is —.

Rational number A **rational number** is a number that can be written in the form —, where *p* and *q* are integers and . Its decimal form stops or repeats.

Real number a **real number** is a number that is either rational or irrational.

KEY CONCEPTS

7.1 Rational and Irrational Numbers

- **Real numbers**
    - 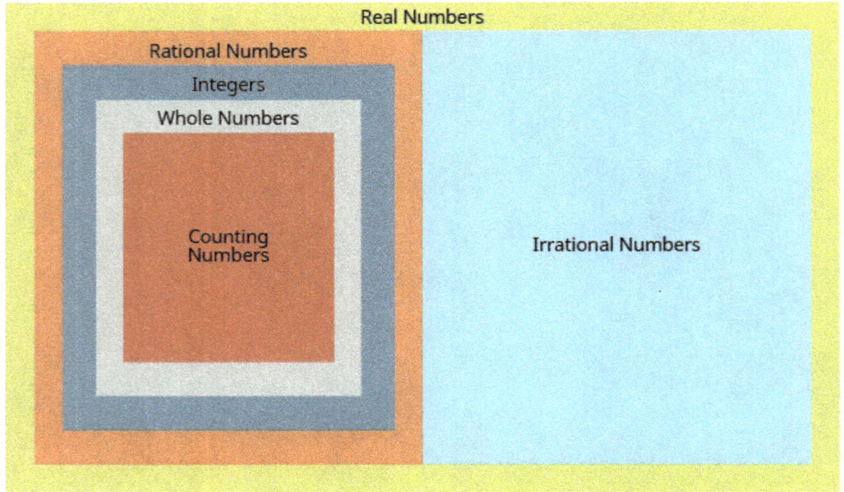

7.2 Commutative and Associative Properties

- **Commutative Properties**
    - **Commutative Property of Addition:**
        - If are real numbers, then
    - **Commutative Property of Multiplication:**
        - If are real numbers, then
- **Associative Properties**
    - **Associative Property of Addition:**
        - If are real numbers then
    - **Associative Property of Multiplication:**
        - If are real numbers then

7.3 Distributive Property

- **Distributive Property:**
  - If are real numbers then
    -
    -
    -

7.4 Properties of Identity, Inverses, and Zero

- **Identity Properties**
  - **Identity Property of Addition:** For any real number a: **0** is the **additive identity**
  - **Identity Property of Multiplication:** For any real number a: **1** is the **multiplicative identity**
- **Inverse Properties**
  - **Inverse Property of Addition:** For any real number a: is the **additive inverse** of a
  - **Inverse Property of Multiplication:** For any real number a: — — is the **multiplicative inverse** of a
- **Properties of Zero**
  - **Multiplication by Zero:** For any real number a,
  - **Division of Zero:** For any real number a, —
  - **Division by Zero:** For any real number a, — is undefined and is undefined. Division by zero is undefined.

REVIEW EXERCISES

7.1 Rational and Irrational Numbers

In the following exercises, write as the ratio of two integers.

302.

303.

304.

305.

In the following exercises, determine which of the numbers is rational.

306.

307.

In the following exercises, identify whether each given number is rational or irrational.

308. $\sqrt{}$ $\sqrt{}$

309. $\sqrt{}$ $\sqrt{}$

In the following exercises, list the whole numbers, integers, rational numbers, irrational numbers, real numbers for each set of numbers.

310. — $\sqrt{}$

311. — $\sqrt{}$ —

7.2 Commutative and Associative Properties

In the following exercises, use the commutative property to rewrite the given expression.

312.

313.

314.

315.

In the following exercises, use the associative property to rewrite the given expression.

316.

317.

318.

319. –

In the following exercises, evaluate each expression for the given value.

320. If —— evaluate:

321. If – evaluate:

322. If evaluate:

– –

– –

323. If evaluate:

– –

– –

In the following exercises, simplify using the commutative and associative properties.

324.

325. – —— –

326. —— —— ——

327. –

328. —— – –

329.

330.

331. –

332.

333.

7.3 Distributive Property

In the following exercises, simplify using the distributive property.

334.

335.

336.

337.

338. –

339.

340.

341.

In the following exercises, evaluate using the distributive property.

342. If evaluate to show that

343. If — evaluate — and — to show that — —

344. If evaluate and to show that

345. If evaluate and to show that

7.4 Properties of Identities, Inverses, and Zero

In the following exercises, identify whether each example is using the identity property of addition or multiplication.

346.

347.

348.

349.

In the following exercises, find the additive inverse.

350.

351.

352. —

353. —

In the following exercises, find the multiplicative inverse.

354. —

355.

356. —

357. —

In the following exercises, simplify.

358.

359. —

360. —

361. —

362.

363.

364. — —

365. —

366. — —

367.

7.5 Systems of Measurement

In the following exercises, convert between U.S. units. Round to the nearest tenth.

368. A floral arbor is feet tall. Convert the height to inches.

369. A picture frame is inches wide. Convert the width to feet.

370. Kelly is feet inches tall. Convert her height to inches.

371. A playground is feet wide. Convert the width to yards.

372. The height of Mount Shasta is feet. Convert the height to miles.

373. Shamu weighs tons. Convert the weight to pounds.

374. The play lasted — hours. Convert the time to minutes.

375. How many tablespoons are in a quart?

376. Naomi's baby weighed pounds ounces at birth. Convert the weight to ounces.

377. Trinh needs cups of paint for her class art project. Convert the volume to gallons.

In the following exercises, solve, and state your answer in mixed units.

378. John caught lobsters. The weights of the lobsters were pound ounces, pound ounces, pounds ounces, and pounds ounces. What was the total weight of the lobsters?

379. Every day last week, Pedro recorded the amount of time he spent reading. He read for minutes. How much time, in hours and minutes, did Pedro spend reading?

380. Fouad is feet inches tall. If he stands on a rung of a ladder feet inches high, how high off the ground is the top of Fouad's head?

381. Dalila wants to make pillow covers. Each cover takes inches of fabric. How many yards and inches of fabric does she need for pillow covers?

In the following exercises, convert between metric units.

382. Donna is meters tall. Convert her height to centimeters.

383. Mount Everest is meters tall. Convert the height to kilometers.

384. One cup of yogurt contains milligrams of calcium. Convert this to grams.

385. One cup of yogurt contains grams of protein. Convert this to milligrams.

386. Sergio weighed kilograms at birth. Convert this to grams.

387. A bottle of water contained milliliters. Convert this to liters.

In the following exercises, solve.

388. Minh is meters tall. His daughter is centimeters tall. How much taller, in meters, is Minh than his daughter?

389. Selma had a bottle of water. If she drank milliliters, how much water, in milliliters, was left in the bottle?

390. One serving of cranberry juice contains grams of sugar. How many kilograms of sugar are in servings of cranberry juice?

391. One ounce of tofu provides grams of protein. How many milligrams of protein are provided by ounces of tofu?

In the following exercises, convert between U.S. and metric units. Round to the nearest tenth.

392. Majid is inches tall. Convert his height to centimeters.

393. A college basketball court is feet long. Convert this length to meters.

394. Caroline walked kilometers. Convert this length to miles.

395. Lucas weighs kilograms. Convert his weight to pounds.

396. Steve's car holds liters of gas. Convert this to gallons.

397. A box of books weighs pounds. Convert this weight to kilograms.

In the following exercises, convert the Fahrenheit temperatures to degrees Celsius. Round to the nearest tenth.

398.

399.

400.

401.

In the following exercises, convert the Celsius temperatures to degrees Fahrenheit. Round to the nearest tenth.

402.

403.

404.

405.

PRACTICE TEST

406. For the numbers list the rational numbers and irrational numbers.

407. Is √ rational or irrational?

408. From the numbers — — √ which are integers rational irrational real numbers?

409. Rewrite using the commutative property:

410. Rewrite the expression using the associative property:

411. Rewrite the expression using the associative property:

412. Evaluate —— when

413. For the number — find the additive inverse multiplicative inverse.

In the following exercises, simplify the given expression.

414. — —

415.

416.

417. — — —

418. —

419.

420.

421.

422.

423. —

424.

425.

426.

427. —

428. —

429. —

In the following exercises, solve using the appropriate unit conversions.

430. Azize walked — miles. Convert this distance to feet.

431. One cup of milk contains milligrams of calcium. Convert this to grams.

432. Larry had phone customer phone calls yesterday. The calls lasted minutes. How much time, in hours and minutes, did Larry spend on the phone?

433. Janice ran kilometers. Convert this distance to miles. Round to the nearest hundredth of a mile.

434. Yolie is inches tall. Convert her height to centimeters. Round to the nearest centimeter.

435. Use the formula — to convert to degrees

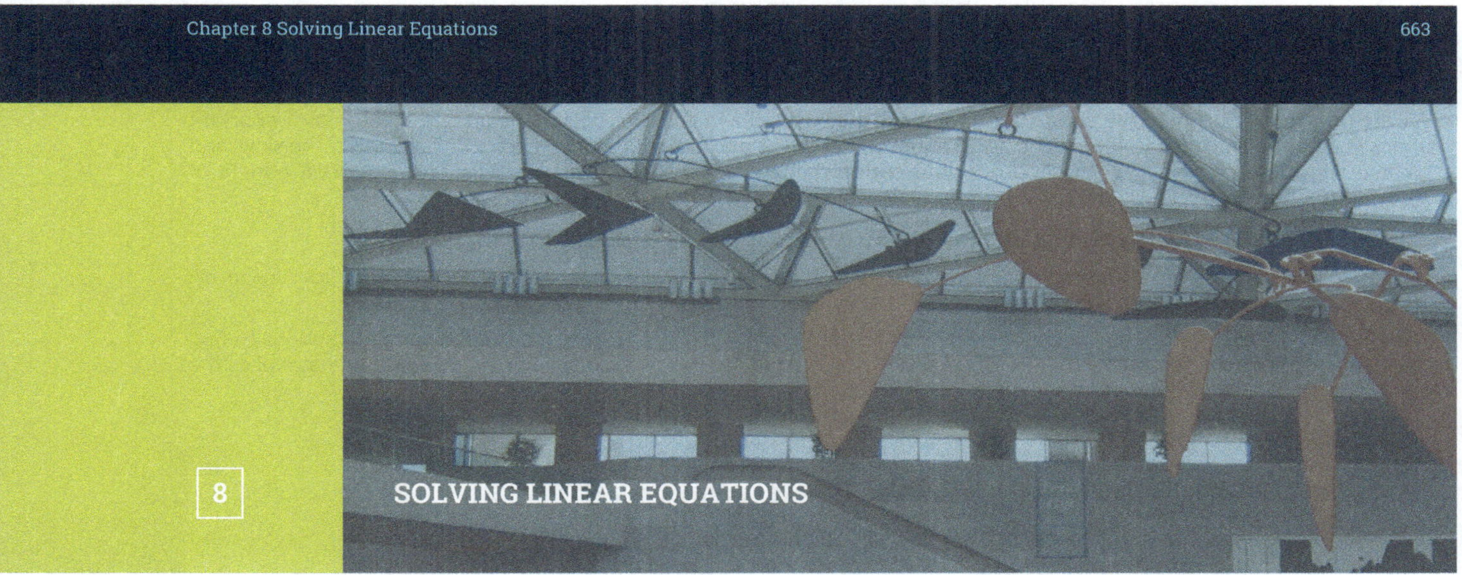

Figure 8.1 A Calder mobile is balanced and has several elements on each side. (credit: paurian, Flickr)

Chapter Outline

Introduction

Teetering high above the floor, this amazing mobile remains aloft thanks to its carefully balanced mass. Any shift in either direction could cause the mobile to become lopsided, or even crash downward. In this chapter, we will solve equations by keeping quantities on both sides of an equal sign in perfect balance.

8.1 Solve Equations Using the Subtraction and Addition Properties of Equality

Learning Objectives

By the end of this section, you will be able to:

- Solve equations using the Subtraction and Addition Properties of Equality
- Solve equations that need to be simplified
- Translate an equation and solve
- Translate and solve applications

BE PREPARED : : 8.1 Before you get started, take this readiness quiz.

Solve:

If you missed this problem, review Example 2.33.

BE PREPARED : : 8.2 Translate into algebra 'five less than

If you missed this problem, review Example 2.24.

BE PREPARED : : 8.3 Is a solution to

If you missed this problem, review Example 2.28.

We are now ready to "get to the good stuff." You have the basics down and are ready to begin one of the most important topics in algebra: solving equations. The applications are limitless and extend to all careers and fields. Also, the skills and techniques you learn here will help improve your critical thinking and problem-solving skills. This is a great benefit of studying mathematics and will be useful in your life in ways you may not see right now.

Solve Equations Using the Subtraction and Addition Properties of Equality

We began our work solving equations in previous chapters. It has been a while since we have seen an equation, so we will review some of the key concepts before we go any further.

We said that solving an equation is like discovering the answer to a puzzle. The purpose in solving an equation is to find the value or values of the variable that make each side of the equation the same. Any value of the variable that makes the equation true is called a solution to the equation. It is the answer to the puzzle.

Solution of an Equation

A **solution of an equation** is a value of a variable that makes a true statement when substituted into the equation.

In the earlier sections, we listed the steps to determine if a value is a solution. We restate them here.

HOW TO : : DETERMINE WHETHER A NUMBER IS A SOLUTION TO AN EQUATION.

Step 1. Substitute the number for the variable in the equation.

Step 2. Simplify the expressions on both sides of the equation.

Step 3. Determine whether the resulting equation is true.

- If it is true, the number is a solution.
- If it is not true, the number is not a solution.

EXAMPLE 8.1

Determine whether — is a solution for

Solution

	$4y + 3 = 8y$
Substitute $\frac{3}{4}$ for y.	$4\left(\frac{3}{4}\right) + 3 \stackrel{?}{=} 8\left(\frac{3}{4}\right)$
Multiply.	$3 + 3 \stackrel{?}{=} 6$
Add.	$6 = 6 \checkmark$

Since — results in a true equation, — is a solution to the equation

TRY IT : : 8.1 Is — a solution for

TRY IT : : 8.2 Is — a solution for

We introduced the Subtraction and Addition Properties of Equality in Solving Equations Using the Subtraction and Addition Properties of Equality. In that section, we modeled how these properties work and then applied them to solving equations with whole numbers. We used these properties again each time we introduced a new system of numbers. Let's review those properties here.

Subtraction and Addition Properties of Equality

Subtraction Property of Equality

For all real numbers and if then

Addition Property of Equality

For all real numbers and if then

When you add or subtract the same quantity from both sides of an equation, you still have equality.

We introduced the Subtraction Property of Equality earlier by modeling equations with envelopes and counters. Figure 8.2 models the equation

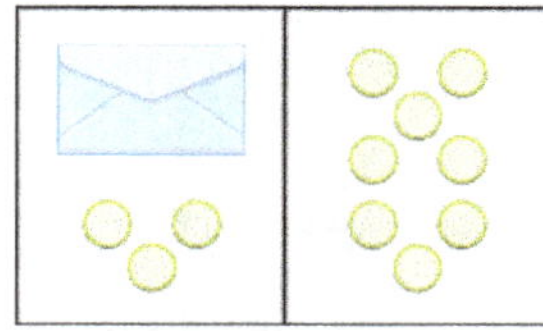

Figure 8.2

The goal is to isolate the variable on one side of the equation. So we 'took away' from both sides of the equation and found the solution

Some people picture a balance scale, as in Figure 8.3, when they solve equations.

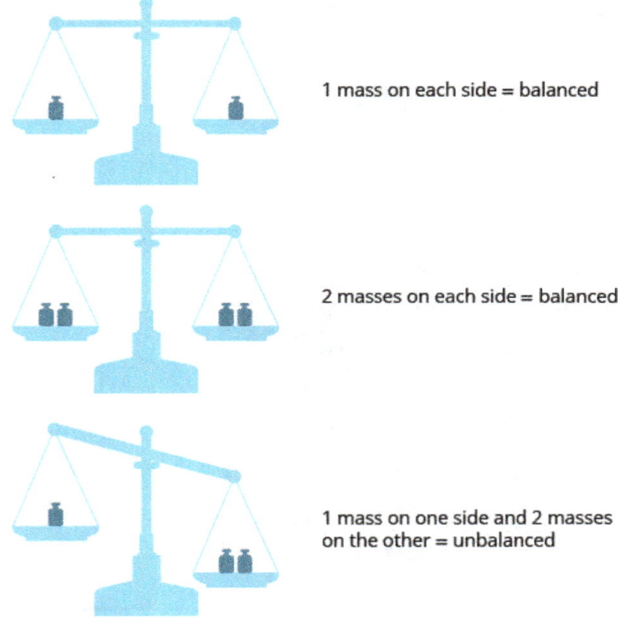
1 mass on each side = balanced
2 masses on each side = balanced
1 mass on one side and 2 masses on the other = unbalanced

Figure 8.3

The quantities on both sides of the equal sign in an equation are equal, or balanced. Just as with the balance scale, whatever you do to one side of the equation you must also do to the other to keep it balanced.

Let's review how to use Subtraction and Addition Properties of Equality to solve equations. We need to isolate the variable on one side of the equation. And we check our solutions by substituting the value into the equation to make sure we have a true statement.

EXAMPLE 8.2

Solve:

Solution

To isolate we undo the addition of by using the Subtraction Property of Equality.

	$x - 11 = -3$
Subtract 11 from each side to "undo" the addition.	$x + 11 - 11 = -3 - 11$
Simplify.	$x = -14$
Check:	$x - 11 = -3$
Substitute .	$-14 + 11 \stackrel{?}{=} -3$
	$-3 = -3$ ✓

Since makes a true statement, we know that it is a solution to the equation.

TRY IT : : 8.3 Solve:

TRY IT : : 8.4 Solve:

In the original equation in the previous example, was added to the , so we subtracted to 'undo' the addition. In the next example, we will need to 'undo' subtraction by using the Addition Property of Equality.

EXAMPLE 8.3

Solve:

Solution

	$m + 4 = -5$
Subtract 4 from each side to "undo" the addition.	$m + 4 - 4 = -5 - 4$
Simplify.	$m = -9$
Check:	$m + 4 = -5$
Substitute .	$-9 + 4 \stackrel{?}{=} -5$
	$-5 = -5$ ✓
	The solution to is .

TRY IT : : 8.5 Solve:

TRY IT : : 8.6 Solve:

Now let's review solving equations with fractions.

EXAMPLE 8.4

Solve: — —

Solution

	$n - \frac{3}{8} = \frac{1}{2}$
Use the Addition Property of Equality.	$n - \frac{3}{8} + \frac{3}{8} = \frac{1}{2} + \frac{3}{8}$
Find the LCD to add the fractions on the right.	$n - \frac{3}{8} + \frac{3}{8} = \frac{4}{8} + \frac{3}{8}$
Simplify	$n = \frac{7}{8}$
Check:	$n - \frac{3}{8} = \frac{1}{2}$
Substitute $n = \frac{7}{8}$.	$\frac{7}{8} - \frac{3}{8} \stackrel{?}{=} \frac{1}{2}$
Subtract.	$\frac{4}{8} \stackrel{?}{=} \frac{1}{2}$
Simplify.	$\frac{1}{2} = \frac{1}{2}$ ✓
The solution checks.	

TRY IT : : 8.7 Solve: — —

TRY IT : : 8.8 Solve: — —

In Solve Equations with Decimals, we solved equations that contained decimals. We'll review this next.

EXAMPLE 8.5

Solve

Solution

	$a - 3.7 = 4.3$
Use the Addition Property of Equality.	$a - 3.7 + 3.7 = 4.3 + 3.7$
Add.	$a = 8$
Check:	$a - 3.7 = 4.3$
Substitute .	$8 - 3.7 \stackrel{?}{=} 4.3$
Simplify.	$4.3 = 4.3$ ✓
The solution checks.	

TRY IT : : 8.9 Solve:

TRY IT : : 8.10 Solve:

Solve Equations That Need to Be Simplified

In the examples up to this point, we have been able to isolate the variable with just one operation. Many of the equations we encounter in algebra will take more steps to solve. Usually, we will need to simplify one or both sides of an equation before using the Subtraction or Addition Properties of Equality. You should always simplify as much as possible before trying to isolate the variable.

EXAMPLE 8.6

Solve:

Solution

The left side of the equation has an expression that we should simplify before trying to isolate the variable.

	$3x - 7 - 2x - 4 = 1$
Rearrange the terms, using the Commutative Property of Addition.	$3x - 2x - 7 - 4 = 1$
Combine like terms.	$x - 11 = 1$
Add 11 to both sides to isolate .	$x - 11 + 11 = 1 + 11$
Simplify.	$x = 12$
Check. Substitute into the original equation. $3x - 7 - 2x - 4 = 1$ $3(12) - 7 - 2(12) - 4 = 1$ $36 - 7 - 24 - 4 = 1$ $29 - 24 - 4 = 1$ $5 - 4 = 1$ $1 = 1$ ✓	

The solution checks.

TRY IT : : 8.11 Solve:

TRY IT : : 8.12 Solve:

EXAMPLE 8.7

Solve:

Solution

The left side of the equation has an expression that we should simplify.

	$3(n - 4) - 2n = -3$
Distribute on the left.	$3n - 12 - 2n = -3$
Use the Commutative Property to rearrange terms.	$3n - 2n - 12 = -3$
Combine like terms.	$n - 12 = -3$
Isolate n using the Addition Property of Equality.	$n - 12 + 12 = -3 + 12$
Simplify.	$n = 9$

Check.
Substitute into the original equation.

$$3(n - 4) - 2n = -3$$
$$3(9 - 4) - 2 \cdot 9 = -3$$
$$3(5) - 18 = -3$$
$$15 - 18 = -3$$
$$-3 = -3 \checkmark$$

The solution checks.

> **TRY IT : : 8.13** Solve:

> **TRY IT : : 8.14** Solve:

EXAMPLE 8.8

Solve:

Solution

Both sides of the equation have expressions that we should simplify before we isolate the variable.

	$2(3k - 1) - 5k = -2 - 7$
Distribute on the left, subtract on the right.	$6k - 2 - 5k = -9$
Use the Commutative Property of Addition.	$6k - 5k - 2 = -9$
Combine like terms.	$k - 2 = -9$
Undo subtraction by using the Addition Property of Equality.	$k - 2 + 2 = -9 + 2$
Simplify.	$k = -7$

Check.
Let

$$2(3k - 1) - 5k = -2 - 7$$
$$2(3(-7) - 1) - 5(-7) = -2 - 7$$
$$2(-21 - 1) - 5(-7) = -9$$
$$2(-22) + 35 = -9$$
$$-44 + 35 = -9$$
$$-9 = -9 \checkmark$$

The solution checks.

TRY IT : : 8.15 Solve:

TRY IT : : 8.16 Solve:

Translate an Equation and Solve

In previous chapters, we translated word sentences into equations. The first step is to look for the word (or words) that translate(s) to the equal sign. Table 8.1 reminds us of some of the words that translate to the equal sign.

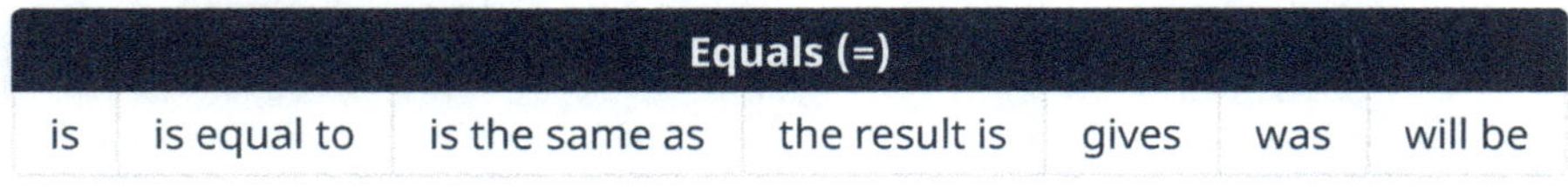

Equals (=)						
is	is equal to	is the same as	the result is	gives	was	will be

Table 8.1

Let's review the steps we used to translate a sentence into an equation.

HOW TO : : TRANSLATE A WORD SENTENCE TO AN ALGEBRAIC EQUATION.

Step 1. Locate the "equals" word(s). Translate to an equal sign.

Step 2. Translate the words to the left of the "equals" word(s) into an algebraic expression.

Step 3. Translate the words to the right of the "equals" word(s) into an algebraic expression.

Now we are ready to try an example.

EXAMPLE 8.9

Translate and solve: five more than is equal to

Solution

Translate.	Five more than x is equal to 26 $x + 5 = 26$
Subtract 5 from both sides.	$x + 5 - 5 = 26 - 5$
Simplify.	$x = 21$
Check: Is five more than ? $21 + 5 \stackrel{?}{=} 26$ $26 = 26 \checkmark$ The solution checks.	

TRY IT : : 8.17 Translate and solve: Eleven more than is equal to

TRY IT : : 8.18 Translate and solve: Twelve less than is equal to

EXAMPLE 8.10

Translate and solve: The difference of and is

Solution

Translate.	The difference of $5p$ and $4p$ $5p - 4p$	is $=$	23 23
Simplify.	$p = 23$		
Check: $5p - 4p = 23$ $5(23) - 4(23) \stackrel{?}{=} 23$ $115 - 22 \stackrel{?}{=} 23$ $23 = 23$ ✓			
The solution checks.			

TRY IT : : 8.19 Translate and solve: The difference of and is

TRY IT : : 8.20 Translate and solve: The difference of and is

Translate and Solve Applications

In most of the application problems we solved earlier, we were able to find the quantity we were looking for by simplifying an algebraic expression. Now we will be using equations to solve application problems. We'll start by restating the problem in just one sentence, assign a variable, and then translate the sentence into an equation to solve. When assigning a variable, choose a letter that reminds you of what you are looking for.

EXAMPLE 8.11

The Robles family has two dogs, Buster and Chandler. Together, they weigh pounds.

Chandler weighs pounds. How much does Buster weigh?

Solution

Read the problem carefully.	
Identify what you are asked to find, and choose a variable to represent it.	How much does Buster weigh? Let Buster's weight
Write a sentence that gives the information to find it.	Buster's weight plus Chandler's weight equals 71 pounds.
We will restate the problem, and then include the given information.	Buster's weight plus 28 equals 71.
Translate the sentence into an equation, using the variable .	$b + 28 = 71$
Solve the equation using good algebraic techniques.	$b + 28 - 28 = 71 - 28$ $b = 43$
Check the answer in the problem and make sure it makes sense.	
Is 43 pounds a reasonable weight for a dog? Yes. Does Buster's weight plus Chandler's weight equal 71 pounds?	
Write a complete sentence that answers the question, "How much does Buster weigh?"	Buster weighs 43 pounds

TRY IT : : 8.21

Translate into an algebraic equation and solve: The Pappas family has two cats, Zeus and Athena. Together, they weigh pounds. Zeus weighs pounds. How much does Athena weigh?

TRY IT : : 8.22

Translate into an algebraic equation and solve: Sam and Henry are roommates. Together, they have books. Sam has books. How many books does Henry have?

HOW TO : : DEVISE A PROBLEM-SOLVING STRATEGY.

Step 1. Read the problem. Make sure you understand all the words and ideas.
Step 2. Identify what you are looking for.
Step 3. Name what you are looking for. Choose a variable to represent that quantity.
Step 4. Translate into an equation. It may be helpful to restate the problem in one sentence with all the important information. Then, translate the English sentence into an algebra equation.
Step 5. Solve the equation using good algebra techniques.
Step 6. Check the answer in the problem and make sure it makes sense.
Step 7. Answer the question with a complete sentence.

EXAMPLE 8.12

Shayla paid for her new car. This was less than the sticker price. What was the sticker price of the car?

Solution

What are you asked to find?	"What was the sticker price of the car?"
Assign a variable.	Let the sticker price of the car.
Write a sentence that gives the information to find it.	$24,575 is $875 less than the sticker price $24,575 is $875 less than
Translate into an equation.	$24{,}575 = s - 875$
Solve.	$24{,}575 + 875 = s - 875 + 875$ $24{,}575 = s$
Check:	
Is $875 less than $25,450 equal to $24,575?	
Write a sentence that answers the question.	The sticker price was $25,450.

TRY IT : : 8.23

Translate into an algebraic equation and solve: Eddie paid for his new car. This was less than the sticker price. What was the sticker price of the car?

TRY IT : : 8.24

Translate into an algebraic equation and solve: The admission price for the movies during the day is This is less than the price at night. How much does the movie cost at night?

The *Links to Literacy* activity, "The 100-pound Problem", will provide you with another view of the topics covered in this section.

MEDIA : : ACCESS ADDITIONAL ONLINE RESOURCES

- Solving One Step Equations By Addition and Subtraction (http://www.openstax.org/l/24Solveonestep)
- Solve One Step Equations By Add and Subtract Whole Numbers (Variable on Left) (http://www.openstax.org/l/24SolveByAdd)
- Solve One Step Equations By Add and Subtract Whole Numbers (Variable on Right) (http://www.openstax.org/l/24AddSubtrWhole)

8.1 EXERCISES

Practice Makes Perfect

Solve Equations Using the Subtraction and Addition Properties of Equality

In the following exercises, determine whether the given value is a solution to the equation.

1. Is — a solution of

2. Is — a solution of

3. Is — a solution of

4. Is — a solution of

In the following exercises, solve each equation.

5.

6.

7. — —

8. — —

9.

10.

11.

12.

13. —

14. —

15.

16.

17.

18.

19. — —

20. — —

21. — —

Solve Equations that Need to be Simplified

In the following exercises, solve each equation.

22.

23.

24.

25.

26.

27.

28.

29.

30.

31.

32.

33.

34.

35.

36.

37.

38.

39.

40.

41.

Translate to an Equation and Solve

In the following exercises, translate to an equation and then solve.

42. Five more than is equal to

43. The sum of and is

44. Ten less than is

45. Three less than is

46. The sum of and is

47. Eight more than is equal to

48. The difference of and is

49. The difference of and is

50. The difference of and − is −

51. The difference of and − is —

52. The sum of and is

53. The sum of and is

Translate and Solve Applications

In the following exercises, translate into an equation and solve.

54. Pilar drove from home to school and then to her aunt's house, a total of miles. The distance from Pilar's house to school is miles. What is the distance from school to her aunt's house?

55. Jeff read a total of pages in his English and Psychology textbooks. He read pages in his English textbook. How many pages did he read in his Psychology textbook?

56. Pablo's father is years older than his mother. Pablo's mother is years old. How old is his father?

57. Eva's daughter is years younger than her son. Eva's son is years old. How old is her daughter?

58. Allie weighs pounds less than her twin sister Lorrie. Allie weighs pounds. How much does Lorrie weigh?

59. For a family birthday dinner, Celeste bought a turkey that weighed pounds less than the one she bought for Thanksgiving. The birthday dinner turkey weighed pounds. How much did the Thanksgiving turkey weigh?

60. The nurse reported that Tricia's daughter had gained pounds since her last checkup and now weighs pounds. How much did Tricia's daughter weigh at her last checkup?

61. Connor's temperature was degrees higher this morning than it had been last night. His temperature this morning was degrees. What was his temperature last night?

62. Melissa's math book cost less than her art book cost. Her math book cost How much did her art book cost?

63. Ron's paycheck this week was less than his paycheck last week. His paycheck this week was How much was Ron's paycheck last week?

Everyday Math

64. Baking Kelsey needs — cup of sugar for the cookie recipe she wants to make. She only has — cup of sugar and will borrow the rest from her neighbor. Let equal the amount of sugar she will borrow. Solve the equation — — to find the amount of sugar she should ask to borrow.

65. Construction Miguel wants to drill a hole for a — screw. The screw should be — inch larger than the hole. Let equal the size of the hole he should drill. Solve the equation — — to see what size the hole should be.

Writing Exercises

66. Is a solution to the equation How do you know?

67. Write a word sentence that translates the equation and then make up an application that uses this equation in its solution.

Self Check

After completing the exercises, use this checklist to evaluate your mastery of the objectives of this section.

I can...	Confidently	With some help	No-I don't get it!
solve equations using the Subtraction and Addition Properties of Equality.			
solve equations that need to be simplified.			
translate an equation and solve.			
translate and solve applications.			

If most of your checks were:

...confidently. Congratulations! You have achieved the objectives in this section. Reflect on the study skills you used so that you can continue to use them. What did you do to become confident of your ability to do these things? Be specific.

...with some help. This must be addressed quickly because topics you do not master become potholes in your road to success. In math, every topic builds upon previous work. It is important to make sure you have a strong foundation before you move on. Whom can you ask for help? Your fellow classmates and instructor are good resources. Is there a place on campus where math tutors are available? Can your study skills be improved?

...no—I don't get it! This is a warning sign and you must not ignore it. You should get help right away or you will quickly be overwhelmed. See your instructor as soon as you can to discuss your situation. Together you can come up with a plan to get you the help you need.

8.2 Solve Equations Using the Division and Multiplication Properties of Equality

Learning Objectives

By the end of this section, you will be able to:

- Solve equations using the Division and Multiplication Properties of Equality
- Solve equations that need to be simplified

BE PREPARED : : 8.4 Before you get started, take this readiness quiz.

Simplify: —

If you missed this problem, review Example 4.28.

BE PREPARED : : 8.5 What is the reciprocal of —

If you missed this problem, review Example 4.29.

BE PREPARED : : 8.6 Evaluate when

If you missed this problem, review Example 3.56.

Solve Equations Using the Division and Multiplication Properties of Equality

We introduced the Multiplication and Division Properties of Equality in Solve Equations Using Integers; The Division Property of Equality and Solve Equations with Fractions. We modeled how these properties worked using envelopes and counters and then applied them to solving equations (See Solve Equations Using Integers; The Division Property of Equality). We restate them again here as we prepare to use these properties again.

Division and Multiplication Properties of Equality

Division Property of Equality: For all real numbers and if then — —

Multiplication Property of Equality: For all real numbers if then

When you divide or multiply both sides of an equation by the same quantity, you still have equality.

Let's review how these properties of equality can be applied in order to solve equations. Remember, the goal is to 'undo' the operation on the variable. In the example below the variable is multiplied by so we will divide both sides by to 'undo' the multiplication.

EXAMPLE 8.13

Solve:

Solution

We use the Division Property of Equality to divide both sides by

	$4x = -28$
Divide both sides by 4 to undo the multiplication.	$\frac{4x}{4} = \frac{-28}{4}$
Simplify.	$x = -7$
Check your answer. Let .	
$4x = -28$	
$4(-7) \stackrel{?}{=} -28$	
$-28 = -28\ \checkmark$	

Since this is a true statement, is a solution to

TRY IT : : 8.25 Solve:

TRY IT : : 8.26 Solve:

In the previous example, to 'undo' multiplication, we divided. How do you think we 'undo' division?

EXAMPLE 8.14

Solve: —

Solution

Here is divided by We can multiply both sides by to isolate

	$\frac{a}{-7} = -42$
Multiply both sides by .	$-7\left(\frac{a}{-7}\right) = -7(-42)$ $\frac{-7a}{-7} = 294$
Simplify.	$a = 294$
Check your answer. Let .	
$\frac{a}{-7} = -42$	
$\frac{294}{-7} \stackrel{?}{=} -42$	
$-42 = -42\ \checkmark$	

TRY IT : : 8.27 Solve: —

TRY IT : : 8.28 Solve: —

EXAMPLE 8.15

Solve:

Solution

Remember is equivalent to

	$-r = 2$
Rewrite as .	$-1r = 2$
Divide both sides by .	$\frac{-1r}{-1} = \frac{2}{-1}$
	$r = -2$
Check.	$-r = 2$
Substitute	$-(-2) \stackrel{?}{=} 2$
Simplify.	$2 = 2 ✓$

In Solve Equations with Fractions, we saw that there are two other ways to solve

We could multiply both sides by

We could take the opposite of both sides.

TRY IT : : 8.29 Solve:

TRY IT : : 8.30 Solve:

EXAMPLE 8.16

Solve: —

Solution

Since the product of a number and its reciprocal is our strategy will be to isolate by multiplying by the reciprocal of —

	$\frac{2}{3}x = 18$
Multiply by the reciprocal of —.	$\frac{3}{2} \cdot \frac{2}{3}x = \frac{3}{2} \cdot 18$
Reciprocals multiply to one.	$1x = \frac{3}{2} \cdot \frac{18}{1}$
Multiply.	$x = 27$
Check your answer. Let	
$\frac{2}{3}x = 18$	
$\frac{2}{3} \cdot 27 \stackrel{?}{=} 18$	
$18 = 18$ ✓	

Notice that we could have divided both sides of the equation — by — to isolate While this would work, multiplying by the reciprocal requires fewer steps.

TRY IT : : 8.31 Solve: —

TRY IT : : 8.32 Solve: —

Solve Equations That Need to be Simplified

Many equations start out more complicated than the ones we've just solved. First, we need to simplify both sides of the equation as much as possible

EXAMPLE 8.17

Solve:

Solution

Start by combining like terms to simplify each side.

	$8x + 9x - 5x = -3 + 15$
Combine like terms.	$12x = 12$
Divide both sides by 12 to isolate x.	$\frac{12x}{12} = \frac{12}{12}$
Simplify.	$x = 1$
Check your answer. Let	
$8x + 9x - 5x = -3 + 15$	
$8 \cdot 1 + 9 \cdot 1 - 5 \cdot 1 \stackrel{?}{=} -3 + 15$	
$8 + 9 - 5 \stackrel{?}{=} -3 + 15$	
$12 = 12$ ✓	

TRY IT : : 8.33 Solve:

TRY IT : : 8.34 Solve:

EXAMPLE 8.18

Solve:

Solution

Simplify each side by combining like terms.

	$11 - 20 = 17y - 8y - 6y$
Simplify each side.	$-9 = 3y$
Divide both sides by 3 to isolate y.	$\frac{-9}{3} = \frac{3y}{3}$
Simplify.	$-3 = y$
Check your answer. Let	
$11 - 20 = 17y - 8y - 6y$	
$11 - 20 \stackrel{?}{=} 17(-3) - 8(-3) - 6(-3)$	
$11 - 20 \stackrel{?}{=} -51 + 24 + 18$	
$-9 = -9$ ✓	

Notice that the variable ended up on the right side of the equal sign when we solved the equation. You may prefer to take one more step to write the solution with the variable on the left side of the equal sign.

TRY IT : : 8.35 Solve:

TRY IT : : 8.36 Solve:

EXAMPLE 8.19

Solve:

Solution

Remember—always simplify each side first.

	$-3(n-2)-6=21$
Distribute.	$-3n+6-6=21$
Simplify.	$-3n=21$
Divide both sides by -3 to isolate n.	$\frac{-3n}{-3}=\frac{21}{-3}$ $n=-7$
Check your answer. Let .	
$-3(n-2)-6=21$	
$-3(-7-2)-6\stackrel{?}{=}21$	
$-3(-9)-6\stackrel{?}{=}21$	
$27-6\stackrel{?}{=}21$	
$21=21$ ✓	

TRY IT : : 8.37 Solve:

TRY IT : : 8.38 Solve:

LINKS TO LITERACY

The *Links to Literacy* activity, "Everybody Wins" will provide you with another view of the topics covered in this section.

MEDIA : : ACCESS ADDITIONAL ONLINE RESOURCES

- Solving One Step Equation by Mult/Div. Integers (Var on Left) (http://www.openstax.org/l/24OneStepMultiL)
- Solving One Step Equation by Mult/Div. Integers (Var on Right) (http://www.openstax.org/l/24OneStepMultiR)
- Solving One Step Equation in the Form: −x = −a (http://www.openstax.org/l/24xa)

8.2 EXERCISES

Practice Makes Perfect

Solve Equations Using the Division and Multiplication Properties of Equality

In the following exercises, solve each equation for the variable using the Division Property of Equality and check the solution.

68.

69.

70.

71.

72.

73.

74.

75.

76.

77.

78.

79.

In the following exercises, solve each equation for the variable using the Multiplication Property of Equality and check the solution.

80. —

81. —

82. —

83. —

84. —

85. —

86. —

87. —

88. —

89. —

90. —

91. —

92. — —

93. — —

Solve Equations That Need to be Simplified

In the following exercises, solve the equation.

94.

95.

96.

97.

98.

99.

100. — —

101. — —

102.

103.

Everyday Math

104. Balloons Ramona bought balloons for a party. She wants to make equal bunches. Find the number of balloons in each bunch, by solving the equation

105. Teaching Connie's kindergarten class has children. She wants them to get into equal groups. Find the number of children in each group, by solving the equation

106. Ticket price Daria paid for children's tickets at the ice skating rink. Find the price of each ticket, by solving the equation

107. Unit price Nishant paid for a pack of juice bottles. Find the price of each bottle, by solving the equation

108. Fuel economy Tania's SUV gets half as many miles per gallon (mpg) as her husband's hybrid car. The SUV gets Find the miles per gallons, of the hybrid car, by solving the equation —

109. Fabric The drill team used yards of fabric to make flags for one-third of the members. Find how much fabric, they would need to make flags for the whole team by solving the equation —

Writing Exercises

110. Frida started to solve the equation by adding to both sides. Explain why Frida's method will result in the correct solution.

111. Emiliano thinks is the solution to the equation — Explain why he is wrong.

Self Check

After completing the exercises, use this checklist to evaluate your mastery of the objectives of this section.

I can...	Confidently	With some help	No-I don't get it!
solve equations using the Division and Multiplication Properties of Equality.			
solve equations that need to be simplified.			

After reviewing this checklist, what will you do to become confident for all objectives?

8.3 Solve Equations with Variables and Constants on Both Sides

Learning Objectives

By the end of this section, you will be able to:

- Solve an equation with constants on both sides
- Solve an equation with variables on both sides
- Solve an equation with variables and constants on both sides
- Solve equations using a general strategy

BE PREPARED : : 8.7

Before you get started, take this readiness quiz.

Simplify:

If you missed this problem, review **Example 2.22**.

BE PREPARED : : 8.8

Solve:

If you missed this problem, review **Example 2.31**.

BE PREPARED : : 8.9

Solve:

If you missed this problem, review **Example 3.65**.

Solve an Equation with Constants on Both Sides

You may have noticed that in all the equations we have solved so far, all the variable terms were on only one side of the equation with the constants on the other side. This does not happen all the time—so now we'll see how to solve equations where the variable terms and/or constant terms are on both sides of the equation.

Our strategy will involve choosing one side of the equation to be the variable side, and the other side of the equation to be the constant side. Then, we will use the Subtraction and Addition Properties of Equality, step by step, to get all the variable terms together on one side of the equation and the constant terms together on the other side.

By doing this, we will transform the equation that started with variables and constants on both sides into the form We already know how to solve equations of this form by using the Division or Multiplication Properties of Equality.

EXAMPLE 8.20

Solve:

Solution

In this equation, the variable is only on the left side. It makes sense to call the left side the variable side. Therefore, the right side will be the constant side. We'll write the labels above the equation to help us remember what goes where.

	variable constant $4x + 6 = -14$
Since the left side is the variable side, the 6 is out of place. We must "undo" adding 6 by subtracting 6, and to keep the equality we must subtract 6 from both sides. Use the Subtraction Property of Equality.	$4x + 6 - 6 = -14 - 6$
Simplify.	$4x = -20$
Now all the s are on the left and the constant on the right.	
Use the Division Property of Equality.	$\frac{4x}{4} = \frac{-20}{4}$
Simplify.	$x = -5$
Check:	$4x + 6 = -14$
Let .	$4(-5) + 6 = -14$
	$-20 + 6 = -14$
	$-14 = -14$ ✓

> TRY IT : : 8.39 Solve:

> TRY IT : : 8.40 Solve:

EXAMPLE 8.21

Solve:

Solution

Notice that the variable is only on the left side of the equation, so this will be the variable side and the right side will be the constant side. Since the left side is the variable side, the is out of place. It is subtracted from the so to 'undo' subtraction, add to both sides.

	variable constant $2y - 7 = 15$
Add 7 to both sides.	$2y - 7 + 7 = 15 + 7$
Simplify.	$2y = 22$
The variables are now on one side and the constants on the other.	
Divide both sides by 2.	$\frac{2y}{2} = \frac{22}{2}$
Simplify.	$y = 11$
Check:	$2y - 7 = 15$
Substitute: .	$2 \cdot 11 - 7 \stackrel{?}{=} 15$
	$22 - 7 \stackrel{?}{=} 15$
	$15 = 15$ ✓

TRY IT : : 8.41 Solve:

TRY IT : : 8.42 Solve:

Solve an Equation with Variables on Both Sides

What if there are variables on both sides of the equation? We will start like we did above—choosing a variable side and a constant side, and then use the Subtraction and Addition Properties of Equality to collect all variables on one side and all constants on the other side. Remember, what you do to the left side of the equation, you must do to the right side too.

EXAMPLE 8.22

Solve:

Solution

Here the variable, is on both sides, but the constants appear only on the right side, so let's make the right side the "constant" side. Then the left side will be the "variable" side.

	variable constant $5x = 4x + 7$
We don't want any variables on the right, so subtract the .	$5x - 4x = 4x - 4x + 7$
Simplify.	$x = 7$
We have all the variables on one side and the constants on the other. We have solved the equation.	
Check:	$5x = 4x + 7$
Substitute 7 for .	$5(7) \stackrel{?}{=} 4(7) + 7$
	$35 \stackrel{?}{=} 28 + 7$
	$35 = 35 \checkmark$

TRY IT : : 8.43 Solve:

TRY IT : : 8.44 Solve:

EXAMPLE 8.23

Solve:

Solution

The only constant, is on the left side of the equation and variable, is on both sides. Let's leave the constant on the left and collect the variables to the right.

	constant variable $5y - 8 = 7y$
Subtract from both sides.	$5y - 5y - 8 = 7y - 5y$
Simplify.	$-8 = 2y$
We have the variables on the right and the constants on the left. Divide both sides by 2.	$\frac{-8}{2} = \frac{2y}{2}$
Simplify.	$-4 = y$
Rewrite with the variable on the left.	$y = -4$
Check: Let .	
$5y - 8 = 7y$	
$5(-4) - 8 \overset{?}{=} 7(-4)$	
$-20 - 8 \overset{?}{=} -28$	
$-28 = -28$ ✓	

> **TRY IT : :** 8.45 Solve:

> **TRY IT : :** 8.46 Solve:

EXAMPLE 8.24

Solve:

Solution

The only constant, is on the right, so let the left side be the variable side.

	variable side constant side $7x = -x + 24$
Remove the from the right side by adding to both sides.	$7x + x = -x + x + 24$
Simplify.	$8x = 24$
All the variables are on the left and the constants are on the right. Divide both sides by 8.	$\frac{8x}{8} = \frac{24}{8}$
Simplify.	$x = 3$
Check: Substitute .	
$7x = -x + 24$ $7(3) \overset{?}{=} -(3) + 24$ $21 = 21$ ✓	

> **TRY IT : :** 8.47 Solve:

TRY IT : : 8.48 Solve:

Solve Equations with Variables and Constants on Both Sides

The next example will be the first to have variables *and* constants on both sides of the equation. As we did before, we'll collect the variable terms to one side and the constants to the other side.

EXAMPLE 8.25

Solve:

Solution

Start by choosing which side will be the variable side and which side will be the constant side. The variable terms are and Since is greater than make the left side the variable side and so the right side will be the constant side.

	$7x + 5 = 6x + 2$
Collect the variable terms to the left side by subtracting from both sides.	$7x - 6x + 5 = 6x - 6x + 2$
Simplify.	$x + 5 = 2$
Now, collect the constants to the right side by subtracting 5 from both sides.	$x + 5 - 5 = 2 - 5$
Simplify.	$x = -3$
The solution is .	
Check: Let .	

$$7x + 5 = 6x + 2$$
$$7(-3) + 5 \stackrel{?}{=} 6(-3) + 2$$
$$-21 + 5 \stackrel{?}{=} -18 + 2$$
$$-16 = -16 \checkmark$$

TRY IT : : 8.49 Solve:

TRY IT : : 8.50 Solve:

We'll summarize the steps we took so you can easily refer to them.

HOW TO : : SOLVE AN EQUATION WITH VARIABLES AND CONSTANTS ON BOTH SIDES.

Step 1. Choose one side to be the variable side and then the other will be the constant side.

Step 2. Collect the variable terms to the variable side, using the Addition or Subtraction Property of Equality.

Step 3. Collect the constants to the other side, using the Addition or Subtraction Property of Equality.

Step 4. Make the coefficient of the variable using the Multiplication or Division Property of Equality.

Step 5. Check the solution by substituting it into the original equation.

It is a good idea to make the variable side the one in which the variable has the larger coefficient. This usually makes the arithmetic easier.

EXAMPLE 8.26

Solve:

Solution

We have on the left and on the right. Since make the left side the "variable" side.

	$6n - 2 = -3n + 7$
We don't want variables on the right side—add to both sides to leave only constants on the right.	$6n + 3n - 2 = -3n + 3n + 7$
Combine like terms.	$9n - 2 = 7$
We don't want any constants on the left side, so add 2 to both sides.	$9n - 2 + 2 = 7 + 2$
Simplify.	$9n = 9$
The variable term is on the left and the constant term is on the right. To get the coefficient of to be one, divide both sides by 9.	$\frac{9n}{9} = \frac{9}{9}$
Simplify.	$n = 1$
Check: Substitute 1 for .	
$6n - 2 = -3n + 7$ $6(1) - 2 \stackrel{?}{=} -3(1) + 7$ $4 = 4 \checkmark$	

TRY IT : : 8.51 Solve:

TRY IT : : 8.52 Solve:

EXAMPLE 8.27

Solve:

Solution

This equation has on the left and on the right. Since make the right side the variable side and the left side the constant side.

	$2a-7=5a+8$
Subtract from both sides to remove the variable term from the left.	$2a-2a-7=5a-2a+8$
Combine like terms.	$-7=3a+8$
Subtract 8 from both sides to remove the constant from the right.	$-7-8=3a+8-8$
Simplify.	$-15=3a$
Divide both sides by 3 to make 1 the coefficient of .	$\frac{-15}{3}=\frac{3a}{3}$
Simplify.	$-5=a$
Check: Let .	

$$\begin{aligned} 2a-7&=5a+8 \\ 2(-5)-7&\stackrel{?}{=}5(-5)+8 \\ -10-7&\stackrel{?}{=}-25+8 \\ -17&=-17\ \checkmark \end{aligned}$$

Note that we could have made the left side the variable side instead of the right side, but it would have led to a negative coefficient on the variable term. While we could work with the negative, there is less chance of error when working with positives. The strategy outlined above helps avoid the negatives!

TRY IT : : 8.53 Solve:

TRY IT : : 8.54 Solve:

To solve an equation with fractions, we still follow the same steps to get the solution.

EXAMPLE 8.28

Solve: — —

Solution

Since — — make the left side the variable side and the right side the constant side.

	$\frac{3}{2}x + 5 = \frac{1}{2}x - 3$
Subtract — from both sides.	$\frac{3}{2}x - \frac{1}{2}x + 5 = \frac{1}{2}x - \frac{1}{2}x - 3$
Combine like terms.	$x + 5 = -3$
Subtract 5 from both sides.	$x + 5 - 5 = -3 - 5$
Simplify.	$x = -8$
Check: Let .	

$$\frac{3}{2}x + 5 = \frac{1}{2}x - 3$$
$$\frac{3}{2}(-8) + 5 \stackrel{?}{=} \frac{1}{2}(-8) - 3$$
$$-12 + 5 \stackrel{?}{=} -4 - 3$$
$$-7 = -7 \checkmark$$

> **TRY IT : :** 8.55 Solve: — —

> **TRY IT : :** 8.56 Solve: — —

We follow the same steps when the equation has decimals, too.

EXAMPLE 8.29

Solve:

Solution

Since make the left side the variable side and the right side the constant side.

	$3.4x + 4 = 1.6x - 5$
Subtract from both sides.	$3.4x - 1.6x + 4 = 1.6x - 1.6x - 5$
Combine like terms.	$1.8x + 4 = -5$
Subtract 4 from both sides.	$1.8x + 4 - 4 = -5 - 4$
Simplify.	$1.8x = -9$
Use the Division Property of Equality.	$\frac{1.8x}{1.8} = \frac{-9}{1.8}$
Simplify.	$x = -5$
Check: Let .	

$$3.4x + 4 = 1.6x - 5$$
$$3.4(-5) + 4 \stackrel{?}{=} 1.6(-5) - 5$$
$$-17 + 4 \stackrel{?}{=} -8 - 5$$
$$-13 = -13 \checkmark$$

> **TRY IT : :** 8.57 Solve:

> **TRY IT : :** 8.58 Solve:

Solve Equations Using a General Strategy

Each of the first few sections of this chapter has dealt with solving one specific form of a linear equation. It's time now to lay out an overall strategy that can be used to solve *any* linear equation. We call this the *general strategy*. Some equations won't require all the steps to solve, but many will. Simplifying each side of the equation as much as possible first makes the rest of the steps easier.

HOW TO : : USE A GENERAL STRATEGY FOR SOLVING LINEAR EQUATIONS.

Step 1. Simplify each side of the equation as much as possible. Use the Distributive Property to remove any parentheses. Combine like terms.

Step 2. Collect all the variable terms to one side of the equation. Use the Addition or Subtraction Property of Equality.

Step 3. Collect all the constant terms to the other side of the equation. Use the Addition or Subtraction Property of Equality.

Step 4. Make the coefficient of the variable term to equal to Use the Multiplication or Division Property of Equality. State the solution to the equation.

Step 5. Check the solution. Substitute the solution into the original equation to make sure the result is a true statement.

EXAMPLE 8.30

Solve:

Solution

	$3(x+2)=18$
Simplify each side of the equation as much as possible. Use the Distributive Property.	$3x+6=18$
Collect all variable terms on one side of the equation—all s are already on the left side.	
Collect constant terms on the other side of the equation. Subtract 6 from each side	$3x+6-6=18-6$
Simplify.	$3x=12$
Make the coefficient of the variable term equal to 1. Divide each side by 3.	$\frac{3x}{3}=\frac{12}{3}$
Simplify.	$x=4$
Check: Let .	

$$3(x+2)=18$$
$$3(4+2)\stackrel{?}{=}18$$
$$3(6)\stackrel{?}{=}18$$
$$18=18\checkmark$$

> **TRY IT : :** 8.59 Solve:

> **TRY IT : :** 8.60 Solve:

EXAMPLE 8.31

Solve:

Solution

	$-(x+5)=7$
Simplify each side of the equation as much as possible by distributing. The only term is on the left side, so all variable terms are on the left side of the equation.	$-x-5=7$
Add 5 to both sides to get all constant terms on the right side of the equation.	$-x-5+5=7+5$
Simplify.	$-x=12$
Make the coefficient of the variable term equal to 1 by multiplying both sides by -1.	$-1(-x)=-1(12)$
Simplify.	$x=-12$
Check: Let .	

$$-(x+5)=7$$
$$-(-12+5)\stackrel{?}{=}7$$
$$-(-7)\stackrel{?}{=}7$$
$$7=7\checkmark$$

> **TRY IT : :** 8.61 Solve:

> **TRY IT : :** 8.62 Solve:

EXAMPLE 8.32

Solve:

Solution

	$4(x-2)+5=-3$
Simplify each side of the equation as much as possible. Distribute.	$4x-8+5=-3$
Combine like terms	$4x-3=-3$
The only is on the left side, so all variable terms are on one side of the equation.	
Add 3 to both sides to get all constant terms on the other side of the equation.	$4x-3+3=-3+3$
Simplify.	$4x=0$
Make the coefficient of the variable term equal to 1 by dividing both sides by 4.	$\frac{4x}{4}=\frac{0}{4}$
Simplify.	$x=0$
Check: Let .	
$4(x-2)+5=-3$ $4(0-2)+5\stackrel{?}{=}-3$ $4(-2)+5\stackrel{?}{=}-3$ $-8+5\stackrel{?}{=}-3$ $-3=-3$ ✓	

TRY IT : : 8.63 Solve:

TRY IT : : 8.64 Solve:

EXAMPLE 8.33

Solve:

Solution

Be careful when distributing the negative.

	$8 - 2(3y + 5) = 0$
Simplify—use the Distributive Property.	$8 - 6y - 10 = 0$
Combine like terms.	$-6y - 2 = 0$
Add 2 to both sides to collect constants on the right.	$-6y - 2 + 2 = 0 + 2$
Simplify.	$-6y = 2$
Divide both sides by −6.	$\frac{-6y}{-6} = \frac{2}{-6}$
Simplify.	$y = -\frac{1}{3}$
Check: Let —.	

$$8 - 2(3y + 5) = 0$$
$$8 - 2\left[3\left(-\frac{1}{3}\right) + 5\right] = 0$$
$$8 - 2(-1 + 5) \stackrel{?}{=} 0$$
$$8 - 2(4) \stackrel{?}{=} 0$$
$$8 - 8 \stackrel{?}{=} 0$$
$$0 = 0 \checkmark$$

TRY IT : : 8.65 Solve:

TRY IT : : 8.66 Solve:

EXAMPLE 8.34

Solve:

Solution

	$3(x-2)-5=4(2x+1)+5$
Distribute.	$3x-6-5=8x+4+5$
Combine like terms.	$3x-11=8x+9$
Subtract to get all the variables on the right since .	$3x-3x-11=8x-3x+9$
Simplify.	$-11=5x+9$
Subtract 9 to get the constants on the left.	$-11-9=5x+9-9$
Simplify.	$-20=5x$
Divide by 5.	$\frac{-20}{5}=\frac{5x}{5}$
Simplify.	$-4=x$
Check: Substitute: .	

$$\begin{aligned} 3(x-2)-5 &= 4(2x+1)+5 \\ 3(-4-2)-5 &\stackrel{?}{=} 4[2(-4)+1]+5 \\ 3(-6)-5 &\stackrel{?}{=} 4(-8+1)+5 \\ -18-5 &\stackrel{?}{=} 4(-7)+5 \\ -23 &\stackrel{?}{=} -28+5 \\ -23 &= -23\checkmark \end{aligned}$$

TRY IT : : 8.67 Solve:

TRY IT : : 8.68 Solve:

EXAMPLE 8.35

Solve: —

Solution

	$\frac{1}{2}(6x - 2) = 5 - x$
Distribute.	$3x - 1 = 5 - x$
Add to get all the variables on the left.	$3x - 1 + x = 5 - x + x$
Simplify.	$4x - 1 = 5$
Add 1 to get constants on the right.	$4x - 1 + 1 = 5 + 1$
Simplify.	$4x = 6$
Divide by 4.	$\frac{4x}{4} = \frac{6}{4}$
Simplify.	$x = \frac{3}{2}$
Check: Let —.	

$$\frac{1}{2}(6x - 2) = 5 - x$$
$$\frac{1}{2}\left(6 \cdot \frac{3}{2} - 2\right) \stackrel{?}{=} 5 - \frac{3}{2}$$
$$\frac{1}{2}(9 - 2) \stackrel{?}{=} \frac{10}{2} - \frac{3}{2}$$
$$\frac{1}{2}(7) \stackrel{?}{=} \frac{7}{2}$$
$$\frac{7}{2} = \frac{7}{2} \checkmark$$

TRY IT : : 8.69 Solve: —

TRY IT : : 8.70 Solve: —

In many applications, we will have to solve equations with decimals. The same general strategy will work for these equations.

EXAMPLE 8.36

Solve:

Solution

	$0.24(100x + 5) = 0.4(30x + 15)$
Distribute.	$24x + 1.2 = 12x + 6$
Subtract to get all the s to the left.	$24x + 1.2 - 12x = 12x + 6 - 12x$
Simplify.	$12x + 1.2 = 6$
Subtract 1.2 to get the constants to the right.	$12x + 1.2 - 1.2 = 6 - 1.2$
Simplify.	$12x = 4.8$
Divide.	$\frac{12x}{12} = \frac{4.8}{12}$
Simplify.	$x = 0.4$
Check: Let .	

$$0.24(100x + 5) = 0.4(30x + 15)$$
$$0.24(100(0.4) + 5) \stackrel{?}{=} 0.4(30(0.4) + 15)$$
$$0.24(40 + 5) \stackrel{?}{=} 0.4(12 + 15)$$
$$0.24(45) \stackrel{?}{=} 0.4(27)$$
$$10.8 = 10.8 \checkmark$$

TRY IT : : 8.71 Solve:

TRY IT : : 8.72 Solve:

MEDIA : : ACCESS ADDITIONAL ONLINE RESOURCES

- Solving Multi-Step Equations (http://www.openstax.org/l/24SolveMultStep)
- Solve an Equation with Variable Terms on Both Sides (http://www.openstax.org/l/24SolveEquatVar)
- Solving Multi-Step Equations (L5.4) (http://www.openstax.org/l/24MultiStepEqu)
- Solve an Equation with Variables and Parentheses on Both Sides (http://www.openstax.org/l/24EquVarParen)

8.3 EXERCISES

Practice Makes Perfect

Solve an Equation with Constants on Both Sides

In the following exercises, solve the equation for the variable.

112.

113.

114.

115.

116.

117.

118.

119.

120.

121.

122.

123.

Solve an Equation with Variables on Both Sides

In the following exercises, solve the equation for the variable.

124.

125.

126.

127.

128.

129.

130.

131.

132. —

133. —

134.

135.

Solve an Equation with Variables and Constants on Both Sides

In the following exercises, solve the equations for the variable.

136.

137.

138.

139.

140.

141.

142.

143.

144.

145.

146.

147.

148. — —

149. — —

150. — —

151. — —

152. — —

153. — —

154. — —

155. — —

156.

157.

158.

159.

160.

161.

Solve an Equation Using the General Strategy

In the following exercises, solve the linear equation using the general strategy.

162.

163.

164.

165.

166.

167.

168.

169.

170.

171.

172. −

173. −

174.

175.

176.

177.

178.

179.

180.

181.

182.

183.

184.

185.

186.

187.

188.

189.

190.

191.

192.

193.

194.

195.

196.

197.

198.

Everyday Math

199. Making a fence Jovani has a fence around the rectangular garden in his backyard. The perimeter of the fence is feet. The length is feet more than the width. Find the width, by solving the equation

200. Concert tickets At a school concert, the total value of tickets sold was Student tickets sold for and adult tickets sold for The number of adult tickets sold was less than times the number of student tickets. Find the number of student tickets sold, by solving the equation

201. Coins Rhonda has in nickels and dimes. The number of dimes is one less than twice the number of nickels. Find the number of nickels, by solving the equation

202. Fencing Micah has feet of fencing to make a rectangular dog pen in his yard. He wants the length to be feet more than the width. Find the length, by solving the equation

Writing Exercises

203. When solving an equation with variables on both sides, why is it usually better to choose the side with the larger coefficient as the variable side?

204. Solve the equation explaining all the steps of your solution.

205. What is the first step you take when solving the equation Explain why this is your first step.

206. Solve the equation — explaining all the steps of your solution as in the examples in this section.

207. Using your own words, list the steps in the General Strategy for Solving Linear Equations.

208. Explain why you should simplify both sides of an equation as much as possible before collecting the variable terms to one side and the constant terms to the other side.

Self Check

After completing the exercises, use this checklist to evaluate your mastery of the objectives of this section.

I can...	Confidently	With some help	No-I don't get it!
solve an equation with constants on both sides.			
solve an equation with variables on both sides.			
solve an equation with variables and constants on both sides.			

What does this checklist tell you about your mastery of this section? What steps will you take to improve?

8.4 Solve Equations with Fraction or Decimal Coefficients

Learning Objectives

By the end of this section, you will be able to:

- Solve equations with fraction coefficients
- Solve equations with decimal coefficients

BE PREPARED : : 8.10 Before you get started, take this readiness quiz.

Multiply: —

If you missed this problem, review Example 4.28

BE PREPARED : : 8.11 Find the LCD of — —

If you missed this problem, review Example 4.63

BE PREPARED : : 8.12 Multiply: by

If you missed this problem, review Example 5.18

Solve Equations with Fraction Coefficients

Let's use the General Strategy for Solving Linear Equations introduced earlier to solve the equation — — —

	$\frac{1}{8}x + \frac{1}{2} = \frac{1}{4}$
To isolate the term, subtract — from both sides.	$\frac{1}{8}x + \frac{1}{2} - \frac{1}{2} = \frac{1}{4} - \frac{1}{2}$
Simplify the left side.	$\frac{1}{8}x = \frac{1}{4} - \frac{1}{2}$
Change the constants to equivalent fractions with the LCD.	$\frac{1}{8}x = \frac{1}{4} - \frac{2}{4}$
Subtract.	$\frac{1}{8}x = -\frac{1}{4}$
Multiply both sides by the reciprocal of —.	$\frac{8}{1} \cdot \frac{1}{8}x = \frac{8}{1}\left(-\frac{1}{4}\right)$
Simplify.	$x = -2$

This method worked fine, but many students don't feel very confident when they see all those fractions. So we are going to show an alternate method to solve equations with fractions. This alternate method eliminates the fractions.

We will apply the Multiplication Property of Equality and multiply both sides of an equation by the least common denominator of *all* the fractions in the equation. The result of this operation will be a new equation, equivalent to the first, but with no fractions. This process is called *clearing the equation of fractions*. Let's solve the same equation again, but this time use the method that clears the fractions.

EXAMPLE 8.37

Solve: — — —

Solution

Find the least common denominator of *all* the fractions in the equation.	$\frac{1}{8}x+\frac{1}{2}=\frac{1}{4}$ LCD $=8$
Multiply both sides of the equation by that LCD, 8. This clears the fractions.	$8\left(\frac{1}{8}x+\frac{1}{2}\right)=8\left(\frac{1}{4}\right)$
Use the Distributive Property.	$8\cdot\frac{1}{8}x+8\cdot\frac{1}{2}=8\cdot\frac{1}{4}$
Simplify — and notice, no more fractions!	$x+4=2$
Solve using the General Strategy for Solving Linear Equations.	$x+4-4=2-4$
Simplify.	$x=-2$
Check: Let $\frac{1}{8}x+\frac{1}{2}=\frac{1}{4}$ $\frac{1}{8}(-2)+\frac{1}{2}\stackrel{?}{=}\frac{1}{4}$ $-\frac{2}{8}+\frac{1}{2}\stackrel{?}{=}\frac{1}{4}$ $-\frac{2}{8}+\frac{4}{8}\stackrel{?}{=}\frac{1}{4}$ $\frac{2}{4}\stackrel{?}{=}\frac{1}{4}$ $\frac{1}{4}=\frac{1}{4}$ ✓	

TRY IT : : 8.73 Solve: — — —

TRY IT : : 8.74 Solve: — — —

Notice in Example 8.37 that once we cleared the equation of fractions, the equation was like those we solved earlier in this chapter. We changed the problem to one we already knew how to solve! We then used the General Strategy for Solving Linear Equations.

HOW TO : : SOLVE EQUATIONS WITH FRACTION COEFFICIENTS BY CLEARING THE FRACTIONS.

Step 1. Find the least common denominator of *all* the fractions in the equation.

Step 2. Multiply both sides of the equation by that LCD. This clears the fractions.

Step 3. Solve using the General Strategy for Solving Linear Equations.

EXAMPLE 8.38

Solve: — — —

Solution

We want to clear the fractions by multiplying both sides of the equation by the LCD of all the fractions in the equation.

Find the least common denominator of *all* the fractions in the equation.	$7 = \frac{1}{2}x + \frac{3}{4}x - \frac{2}{3}x$ LCD = 12
Multiply both sides of the equation by 12.	$12(7) = 12 \cdot \frac{1}{2}x + \frac{3}{4}x - \frac{2}{3}x$
Distribute.	$12(7) = 12 \cdot \frac{1}{2}x + 12 \cdot \frac{3}{4}x - 12 \cdot \frac{2}{3}x$
Simplify — and notice, no more fractions!	$84 = 6x + 9x - 8x$
Combine like terms.	$84 = 7x$
Divide by 7.	$\frac{84}{7} = \frac{7x}{7}$
Simplify.	$12 = x$
Check: Let	

$7 = \frac{1}{2}x + \frac{3}{4}x - \frac{2}{3}x$

$7 \stackrel{?}{=} \frac{1}{2}(12) + \frac{3}{4}(12) - \frac{2}{3}(12)$

$7 \stackrel{?}{=} 6 + 9 - 8$

$7 = 7\checkmark$

TRY IT : : 8.75 Solve: — — —

TRY IT : : 8.76 Solve: — — —

In the next example, we'll have variables and fractions on both sides of the equation.

EXAMPLE 8.39

Solve: — — —

Solution

Find the LCD of all the fractions in the equation.	$x+\frac{1}{3}=\frac{1}{6}x-\frac{1}{2}$, LCD $=6$
Multiply both sides by the LCD.	$6\left(x+\frac{1}{3}\right)=6\left(\frac{1}{6}x-\frac{1}{2}\right)$
Distribute.	$6\cdot x+6\cdot\frac{1}{3}=6\cdot\frac{1}{6}x-6\cdot\frac{1}{2}$
Simplify — no more fractions!	$6x+2=x-3$
Subtract from both sides.	$6x-x+2=x-x-3$
Simplify.	$5x+2=-3$
Subtract 2 from both sides.	$5x+2-2=-3-2$
Simplify.	$5x=-5$
Divide by 5.	$\frac{5x}{5}=\frac{-5}{5}$
Simplify.	$x=-1$
Check: Substitute	

$$x+\frac{1}{3}=\frac{1}{6}x-\frac{1}{2}$$
$$(-1)+\frac{1}{3}\stackrel{?}{=}\frac{1}{6}(-1)-\frac{1}{2}$$
$$(-1)+\frac{1}{3}\stackrel{?}{=}-\frac{1}{6}-\frac{1}{2}$$
$$-\frac{3}{3}+\frac{1}{3}\stackrel{?}{=}-\frac{1}{6}-\frac{3}{6}$$
$$-\frac{2}{3}\stackrel{?}{=}-\frac{4}{6}$$
$$-\frac{2}{3}=-\frac{2}{3}\checkmark$$

> **TRY IT : :** 8.77 Solve: — — —

> **TRY IT : :** 8.78 Solve: — — —

In Example 8.40, we'll start by using the Distributive Property. This step will clear the fractions right away!

EXAMPLE 8.40

Solve: —

Solution

	$1 = \frac{1}{2}(4x + 2)$
Distribute.	$1 = \frac{1}{2} \cdot 4x + \frac{1}{2} \cdot 2$
Simplify. Now there are no fractions to clear!	$1 = 2x + 1$
Subtract 1 from both sides.	$1 - 1 = 2x + 1 - 1$
Simplify.	$0 = 2x$
Divide by 2.	$\frac{0}{2} = \frac{2x}{2}$
Simplify.	$0 = x$
Check: Let	

$1 = \frac{1}{2}(4x + 2)$

$1 \stackrel{?}{=} \frac{1}{2}(4(0) + 2)$

$1 \stackrel{?}{=} \frac{1}{2}(2)$

$1 \stackrel{?}{=} \frac{2}{2}$

$1 = 1$ ✓

TRY IT : : 8.79 Solve: —

TRY IT : : 8.80 Solve: —

Many times, there will still be fractions, even after distributing.

EXAMPLE 8.41

Solve: — —

Solution

	$\frac{1}{2}(y-5)=\frac{1}{4}(y-1)$
Distribute.	$\frac{1}{2}\cdot y-\frac{1}{2}\cdot 5=\frac{1}{4}\cdot y-\frac{1}{4}\cdot 1$
Simplify.	$\frac{1}{2}y-\frac{5}{2}=\frac{1}{4}y-\frac{1}{4}$
Multiply by the LCD, 4.	$4\left(\frac{1}{2}y-\frac{5}{2}\right)=4\left(\frac{1}{4}y-\frac{1}{4}\right)$
Distribute.	$4\cdot\frac{1}{2}y-4\cdot\frac{5}{2}=4\cdot\frac{1}{4}y-4\cdot\frac{1}{4}$
Simplify.	$2y-10=y-1$
Collect the terms to the left.	$2y-10-y=y-1-y$
Simplify.	$y-10=-1$
Collect the constants to the right.	$y-10+10=-1+10$
Simplify.	$y=9$
Check: Substitute for	

$$\frac{1}{2}(y-5)=\frac{1}{4}(y-1)$$
$$\frac{1}{2}(9-5)\stackrel{?}{=}\frac{1}{4}(9-1)$$
$$\frac{1}{2}(4)\stackrel{?}{=}\frac{1}{4}(8)$$
$$2=2\checkmark$$

TRY IT : : 8.81 Solve: — —

TRY IT : : 8.82 Solve: — —

Solve Equations with Decimal Coefficients

Some equations have decimals in them. This kind of equation will occur when we solve problems dealing with money and percent. But decimals are really another way to represent fractions. For example, — and — So, when we have an equation with decimals, we can use the same process we used to clear fractions—multiply both sides of the equation by the least common denominator.

EXAMPLE 8.42

Solve:

Solution

The only decimal in the equation is Since — the LCD is We can multiply both sides by to clear the decimal.

	$0.8x - 5 = 7$
Multiply both sides by the LCD.	$10(0.8x - 5) = 10(7)$
Distribute.	$10(0.8x) - 10(5) = 10(7)$
Multiply, and notice, no more decimals!	$8x - 50 = 70$
Add 50 to get all constants to the right.	$8x - 50 + 50 = 70 + 50$
Simplify.	$8x = 120$
Divide both sides by 8.	$\frac{8x}{8} = \frac{120}{8}$
Simplify.	$x = 15$
Check: Let	

$$0.8(15) - 5 \stackrel{?}{=} 7$$
$$12 - 5 \stackrel{?}{=} 7$$
$$7 = 7 \checkmark$$

TRY IT : : 8.83 Solve:

TRY IT : : 8.84 Solve:

EXAMPLE 8.43

Solve:

Solution

Look at the decimals and think of the equivalent fractions.

Notice, the LCD is

By multiplying by the LCD we will clear the decimals.

	$0.06x + 0.02 = 0.25x - 1.5$
Multiply both sides by 100.	$100(0.06x + 0.02) = 100(0.25x - 1.5)$
Distribute.	$100(0.06x) + 100(0.02) = 100(0.25x) - 100(1.5)$
Multiply, and now no more decimals.	$6x + 2 = 25x - 150$
Collect the variables to the right.	$6x - 6x + 2 = 25x - 6x - 150$
Simplify.	$2 = 19x - 150$
Collect the constants to the left.	$2 + 150 = 19x - 150 + 150$
Simplify.	$152 = 19x$
Divide by 19.	$\frac{152}{19} = \frac{19x}{19}$
Simplify.	$8 = x$
Check: Let	

$$0.06(8) + 0.02 = 0.25(8) - 1.5$$
$$0.48 + 0.02 = 2.00 - 1.5$$
$$0.50 = 0.50 \checkmark$$

TRY IT : : 8.85 Solve:

TRY IT : : 8.86 Solve:

The next example uses an equation that is typical of the ones we will see in the money applications in the next chapter. Notice that we will distribute the decimal first before we clear all decimals in the equation.

EXAMPLE 8.44

Solve:

Solution

	$0.25x + 0.05(x + 3) = 2.85$
Distribute first.	$0.25x + 0.05x + 0.15 = 2.85$
Combine like terms.	$0.30x + 0.15 = 2.85$
To clear decimals, multiply by 100.	$100(0.30x + 0.15) = 100(2.85)$
Distribute.	$30x + 15 = 285$
Subtract 15 from both sides.	$30x + 15 - 15 = 285 - 15$
Simplify.	$30x = 270$
Divide by 30.	$\frac{30x}{30} = \frac{270}{30}$
Simplify.	$x = 9$
Check: Let	

$$0.25x + 0.05(x + 3) = 2.85$$
$$0.25(9) + 0.05(9 + 3) \stackrel{?}{=} 2.85$$
$$2.25 + 0.05(12) \stackrel{?}{=} 2.85$$
$$2.25 + 0.60 \stackrel{?}{=} 2.85$$
$$2.85 = 2.85 \checkmark$$

TRY IT : : 8.87 Solve:

TRY IT : : 8.88 Solve:

MEDIA : : ACCESS ADDITIONAL ONLINE RESOURCES

- Solve an Equation with Fractions with Variable Terms on Both Sides (http://www.openstax.org/l/24FracVariTerm)
- Ex 1: Solve an Equation with Fractions with Variable Terms on Both Sides (http://www.openstax.org/l/24Ex1VariTerms)
- Ex 2: Solve an Equation with Fractions with Variable Terms on Both Sides (http://www.openstax.org/l/24Ex2VariTerms)
- Solving Multiple Step Equations Involving Decimals (http://www.openstax.org/l/24EquawithDec)
- Ex: Solve a Linear Equation With Decimals and Variables on Both Sides (http://www.openstax.org/l/24LinEquaDec)
- Ex: Solve an Equation with Decimals and Parentheses (http://www.openstax.org/l/24DecParens)

8.4 EXERCISES

Practice Makes Perfect

Solve equations with fraction coefficients

In the following exercises, solve the equation by clearing the fractions.

209.

210.

211.

212.

213.

214.

215.

216.

217.

218.

219.

220.

221.

222.

223.

224.

225.

226.

227.

228.

229.

230.

231.

232.

Solve Equations with Decimal Coefficients

In the following exercises, solve the equation by clearing the decimals.

233.

234.

235.

236.

237.

238.

239.

240.

241.

242.

243.

244.

245.

246.

247.

248.

Everyday Math

249. Coins Taylor has in dimes and pennies. The number of pennies is more than the number of dimes. Solve the equation for the number of dimes.

250. Stamps Travis bought worth of stamps and stamps. The number of stamps was less than the number of stamps. Solve the equation for to find the number of stamps Travis bought.

Writing Exercises

251. Explain how to find the least common denominator of — — —

252. If an equation has several fractions, how does multiplying both sides by the LCD make it easier to solve?

253. If an equation has fractions only on one side, why do you have to multiply both sides of the equation by the LCD?

254. In the equation what is the LCD? How do you know?

Self Check

After completing the exercises, use this checklist to evaluate your mastery of the objectives of this section.

I can...	Confidently	With some help	No-I don't get it!
solve equations using a general strategy.			
solve equations with fraction coefficients.			
solve equations with decimal coefficients.			

Overall, after looking at the checklist, do you think you are well-prepared for the next Chapter? Why or why not?

CHAPTER 8 REVIEW

KEY TERMS

solution of an equation A solution of an equation is a value of a variable that makes a true statement when substituted into the equation.

KEY CONCEPTS

8.1 Solve Equations Using the Subtraction and Addition Properties of Equality

- **Determine whether a number is a solution to an equation.**

  Step 1. Substitute the number for the variable in the equation.

  Step 2. Simplify the expressions on both sides of the equation.

  Step 3. Determine whether the resulting equation is true.

  If it is true, the number is a solution.
  If it is not true, the number is not a solution.

- **Subtraction and Addition Properties of Equality**
  - **Subtraction Property of Equality**
    For all real numbers *a*, *b*, and *c*,
    if $a = b$ then .
  - **Addition Property of Equality**
    For all real numbers *a*, *b*, and *c*,
    if $a = b$ then .

- **Translate a word sentence to an algebraic equation.**

  Step 1. Locate the "equals" word(s). Translate to an equal sign.

  Step 2. Translate the words to the left of the "equals" word(s) into an algebraic expression.

  Step 3. Translate the words to the right of the "equals" word(s) into an algebraic expression.

- **Problem-solving strategy**

  Step 1. Read the problem. Make sure you understand all the words and ideas.

  Step 2. Identify what you are looking for.

  Step 3. Name what you are looking for. Choose a variable to represent that quantity.

  Step 4. Translate into an equation. It may be helpful to restate the problem in one sentence with all the important information. Then, translate the English sentence into an algebra equation.

  Step 5. Solve the equation using good algebra techniques.

  Step 6. Check the answer in the problem and make sure it makes sense.

  Step 7. Answer the question with a complete sentence.

8.2 Solve Equations Using the Division and Multiplication Properties of Equality

- **Division and Multiplication Properties of Equality**
  - **Division Property of Equality:** For all real numbers *a*, *b*, *c*, and , if , then — —.
  - **Multiplication Property of Equality:** For all real numbers *a*, *b*, *c*, if , then .

8.3 Solve Equations with Variables and Constants on Both Sides

- **Solve an equation with variables and constants on both sides**

  Step 1. Choose one side to be the variable side and then the other will be the constant side.

  Step 2. Collect the variable terms to the variable side, using the Addition or Subtraction Property of Equality.

  Step 3. Collect the constants to the other side, using the Addition or Subtraction Property of Equality.

  Step 4. Make the coefficient of the variable 1, using the Multiplication or Division Property of Equality.

  Step 5. Check the solution by substituting into the original equation.

- **General strategy for solving linear equations**

  Step 1. Simplify each side of the equation as much as possible. Use the Distributive Property to remove any parentheses. Combine like terms.

  Step 2. Collect all the variable terms to one side of the equation. Use the Addition or Subtraction Property of Equality.

  Step 3. Collect all the constant terms to the other side of the equation. Use the Addition or Subtraction Property of Equality.

  Step 4. Make the coefficient of the variable term to equal to 1. Use the Multiplication or Division Property of Equality. State the solution to the equation.

  Step 5. Check the solution. Substitute the solution into the original equation to make sure the result is a true statement.

8.4 Solve Equations with Fraction or Decimal Coefficients

- **Solve equations with fraction coefficients by clearing the fractions.**

  Step 1. Find the least common denominator of *all* the fractions in the equation.

  Step 2. Multiply both sides of the equation by that LCD. This clears the fractions.

  Step 3. Solve using the General Strategy for Solving Linear Equations.

REVIEW EXERCISES

8.1 Solve Equations using the Subtraction and Addition Properties of Equality

In the following exercises, determine whether the given number is a solution to the equation.

255.

256.

257.

258.

In the following exercises, solve the equation using the Subtraction Property of Equality.

259.

260.

261. — —

262.

In the following exercises, solve the equation using the Addition Property of Equality.

263.

264.

265. — —

266.

In the following exercises, solve the equation.

267.

268.

269. —

270.

271.

272.

273.

274.

In the following exercises, translate each English sentence into an algebraic equation and then solve it.

275. The sum of and is

276. Four less than is

In the following exercises, translate into an algebraic equation and solve.

277. Rochelle's daughter is years old. Her son is years younger. How old is her son?

278. Tan weighs pounds. Minh weighs pounds more than Tan. How much does Minh weigh?

279. Peter paid to go to the movies, which was less than he paid to go to a concert. How much did he pay for the concert?

280. Elissa earned this week, which was more than she earned last week. How much did she earn last week?

8.2 Solve Equations using the Division and Multiplication Properties of Equality

In the following exercises, solve each equation using the Division Property of Equality.

281.

282.

283.

284.

In the following exercises, solve each equation using the Multiplication Property of Equality.

285. —

286. ——

287. —

288. — —

In the following exercises, solve each equation.

289.

290. —

291.

292. — —

293.

294.

8.3 Solve Equations with Variables and Constants on Both Sides

In the following exercises, solve the equations with constants on both sides.

295.

296.

297.

298.

In the following exercises, solve the equations with variables on both sides.

299.

300.

301.

302. —

In the following exercises, solve the equations with constants and variables on both sides.

303.

304.

305.

306. — —

In the following exercises, solve each linear equation using the general strategy.

307.

308.

309.

310.

311.

312. —

313.

314.

315.

316.

8.4 Solve Equations with Fraction or Decimal Coefficients

In the following exercises, solve each equation by clearing the fractions.

317. — — —

318. — —

319. — — — —

320. — —

In the following exercises, solve each equation by clearing the decimals.

321.

322.

323.

324.

PRACTICE TEST

325. Determine whether each number is a solution to the equation.

—

In the following exercises, solve each equation.

326.

327.

328.

329.

330.

331. —

332.

333.

334.

335.

336.

337.

338. —

339.

340.

341. — — —

342.

343. Translate and solve: The difference of twice and is

344. Samuel paid for gas this week, which was less than he paid last week. How much did he pay last week?

Figure 9.1 Note the many individual shapes in this building. (credit: Bert Kaufmann, Flickr)

Chapter Outline

Introduction

We are surrounded by all sorts of geometry. Architects use geometry to design buildings. Artists create vivid images out of colorful geometric shapes. Street signs, automobiles, and product packaging all take advantage of geometric properties. In this chapter, we will begin by considering a formal approach to solving problems and use it to solve a variety of common problems, including making decisions about money. Then we will explore geometry and relate it to everyday situations, using the problem-solving strategy we develop.

9.1 Use a Problem Solving Strategy

Learning Objectives

By the end of this section, you will be able to:

- Approach word problems with a positive attitude
- Use a problem solving strategy for word problems
- Solve number problems

BE PREPARED : : 9.1 Before you get started, take this readiness quiz.

Translate less than twice into an algebraic expression.

If you missed this problem, review Example 2.25.

BE PREPARED : : 9.2 Solve: —

If you missed this problem, review Example 8.16.

BE PREPARED : : 9.3 Solve:
If you missed this problem, review Example 8.20.

Approach Word Problems with a Positive Attitude

The world is full of word problems. How much money do I need to fill the car with gas? How much should I tip the server at a restaurant? How many socks should I pack for vacation? How big a turkey do I need to buy for Thanksgiving dinner, and what time do I need to put it in the oven? If my sister and I buy our mother a present, how much will each of us pay?

Now that we can solve equations, we are ready to apply our new skills to word problems. Do you know anyone who has had negative experiences in the past with word problems? Have you ever had thoughts like the student in Figure 9.2?

Figure 9.2 Negative thoughts about word problems can be barriers to success.

When we feel we have no control, and continue repeating negative thoughts, we set up barriers to success. We need to calm our fears and change our negative feelings.

Start with a fresh slate and begin to think positive thoughts like the student in Figure 9.3. Read the positive thoughts and say them out loud.

Figure 9.3 When it comes to word problems, a positive attitude is a big step toward success.

If we take control and believe we can be successful, we will be able to master word problems.

Think of something that you can do now but couldn't do three years ago. Whether it's driving a car, snowboarding, cooking a gourmet meal, or speaking a new language, you have been able to learn and master a new skill. Word problems are no different. Even if you have struggled with word problems in the past, you have acquired many new math skills that will help you succeed now!

Use a Problem-solving Strategy for Word Problems

In earlier chapters, you translated word phrases into algebraic expressions, using some basic mathematical vocabulary and symbols. Since then you've increased your math vocabulary as you learned about more algebraic procedures, and you've had more practice translating from words into algebra.

You have also translated word sentences into algebraic equations and solved some word problems. The word problems applied math to everyday situations. You had to restate the situation in one sentence, assign a variable, and then write an equation to solve. This method works as long as the situation is familiar to you and the math is not too complicated.

Now we'll develop a strategy you can use to solve any word problem. This strategy will help you become successful with word problems. We'll demonstrate the strategy as we solve the following problem.

EXAMPLE 9.1

Pete bought a shirt on sale for which is one-half the original price. What was the original price of the shirt?

Solution

Step 1. **Read** the problem. Make sure you understand all the words and ideas. You may need to read the problem two or more times. If there are words you don't understand, look them up in a dictionary or on the Internet.

- *In this problem, do you understand what is being discussed? Do you understand every word?*

Step 2. **Identify** what you are looking for. It's hard to find something if you are not sure what it is! Read the problem again and look for words that tell you what you are looking for!

- *In this problem, the words "what was the original price of the shirt" tell you that what you are looking for: the original price of the shirt.*

Step 3. **Name** what you are looking for. Choose a variable to represent that quantity. You can use any letter for the variable, but it may help to choose one that helps you remember what it represents.

- *Let the original price of the shirt*

Step 4. **Translate** into an equation. It may help to first restate the problem in one sentence, with all the important information. Then translate the sentence into an equation.

18	is	one-half	of	the original price .
18	=	$\frac{1}{2}$	·	p

Step 5. **Solve** the equation using good algebra techniques. Even if you know the answer right away, using algebra will better prepare you to solve problems that do not have obvious answers.

Write the equation.	$18 = \frac{1}{2}p$
Multiply both sides by 2.	$2 \cdot 18 = 2 \cdot \frac{1}{2}p$
Simplify.	$36 = p$

Step 6. **Check** the answer in the problem and make sure it makes sense.

- *We found that which means the original price was Does make sense in the problem? Yes, because is one-half of and the shirt was on sale at half the original price.*

Step 7. **Answer** the question with a complete sentence.

- *The problem asked "What was the original price of the shirt?" The answer to the question is: "The original price of the shirt was*

If this were a homework exercise, our work might look like this:

Let $p =$ the original price.

18 is one-half the original price.

$$18 = \frac{1}{2}p$$

$$2 \cdot 18 = 2 \cdot \frac{1}{2}p$$

$$36 = p$$

Check:

Is $36 a reasonable price for a shirt? *Yes.*

Is 18 one-half of 36? *Yes.*

The original price of the shirt was $36.

TRY IT : : 9.1

Joaquin bought a bookcase on sale for which was two-thirds the original price. What was the original price of the bookcase?

TRY IT : : 9.2

Two-fifths of the people in the senior center dining room are men. If there are men, what is the total number of people in the dining room?

We list the steps we took to solve the previous example.

Problem-Solving Strategy

Step 1. **Read** the word problem. Make sure you understand all the words and ideas. You may need to read the problem two or more times. If there are words you don't understand, look them up in a dictionary or on the internet.

Step 2. **Identify** what you are looking for.

Step 3. **Name** what you are looking for. Choose a variable to represent that quantity.

Step 4. **Translate** into an equation. It may be helpful to first restate the problem in one sentence before translating.

Step 5. **Solve** the equation using good algebra techniques.
Step 6. **Check** the answer in the problem. Make sure it makes sense.
Step 7. **Answer** the question with a complete sentence.

Let's use this approach with another example.

EXAMPLE 9.2

Yash brought apples and bananas to a picnic. The number of apples was three more than twice the number of bananas. Yash brought apples to the picnic. How many bananas did he bring?

Solution

Step 1. **Read** the problem.	
Step 2. **Identify** what you are looking for.	How many bananas did he bring?
Step 3. **Name** what you are looking for. Choose a variable to represent the number of bananas.	Let
Step 4. **Translate.** Restate the problem in one sentence with all the important information. Translate into an equation.	The number of apples was 3 more than twice the number of bananas. $11 = 3 + 2b$
Step 5. **Solve** the equation.	$11 = 2b + 3$
Subtract 3 from each side.	$11 - 3 = 2b + 3 - 3$
Simplify.	$8 = 2b$
Divide each side by 2.	$\frac{8}{2} = \frac{2b}{2}$
Simplify.	$4 = b$
Step 6. **Check:** First, is our answer reasonable? Yes, bringing four bananas to a picnic seems reasonable. The problem says the number of apples was three more than twice the number of bananas. If there are four bananas, does that make eleven apples? Twice 4 bananas is 8. Three more than 8 is 11.	
Step 7. **Answer** the question.	Yash brought 4 bananas to the picnic.

TRY IT : : 9.3

Guillermo bought textbooks and notebooks at the bookstore. The number of textbooks was more than the number of notebooks. He bought textbooks. How many notebooks did he buy?

TRY IT : : 9.4

Gerry worked Sudoku puzzles and crossword puzzles this week. The number of Sudoku puzzles he completed is seven more than the number of crossword puzzles. He completed Sudoku puzzles. How many crossword puzzles did he complete?

In Solve Sales Tax, Commission, and Discount Applications, we learned how to translate and solve basic percent equations and used them to solve sales tax and commission applications. In the next example, we will apply our Problem Solving Strategy to more applications of percent.

EXAMPLE 9.3

Nga's car insurance premium increased by which was of the original cost. What was the original cost of the premium?

Solution

Step 1. **Read** the problem. Remember, if there are words you don't understand, look them up.	
Step 2. **Identify** what you are looking for.	the original cost of the premium
Step 3. **Name.** Choose a variable to represent the original cost of premium.	Let
Step 4. **Translate.** Restate as one sentence. Translate into an equation.	\$60 was 8% of the original cost. $60 = 0.08 \cdot c$
Step 5. **Solve** the equation.	$60 = 0.08c$
Divide both sides by 0.08.	$\frac{60}{0.08} = \frac{0.08c}{0.08}$
Simplify.	
Step 6. **Check:** Is our answer reasonable? Yes, a \$750 premium on auto insurance is reasonable. Now let's check our algebra. Is 8% of 750 equal to 60? $750 = c$ $0.08(750) = 60$ $60 = 60$ ✓	
Step 7. **Answer** the question.	The original cost of Nga's premium was \$750.

TRY IT : : 9.5

Pilar's rent increased by The increase was What was the original amount of Pilar's rent?

TRY IT : : 9.6

Steve saves of his paycheck each month. If he saved last month, how much was his paycheck?

Solve Number Problems

Now we will translate and solve **number problems**. In number problems, you are given some clues about one or more numbers, and you use these clues to build an equation. Number problems don't usually arise on an everyday basis, but they provide a good introduction to practicing the Problem Solving Strategy. Remember to look for clue words such as *difference*, *of*, and *and*.

EXAMPLE 9.4

The difference of a number and six is Find the number.

Solution

Step 1. **Read** the problem. Do you understand all the words?	
Step 2. **Identify** what you are looking for.	the number
Step 3. **Name.** Choose a variable to represent the number.	Let
Step 4. **Translate.** Restate as one sentence. Translate into an equation.	The difference of a number and 6 is 13. $n - 6 = 13$
Step 5. **Solve** the equation. Add 6 to both sides. Simplify.	$n - 6 = 13$ $n - 6 + 6 = 13 + 6$ $n = 19$
Step 6. **Check:** The difference of 19 and 6 is 13. It checks.	
Step 7. **Answer** the question.	The number is 19.

 **TRY IT : :** 9.7 The difference of a number and eight is Find the number.

TRY IT : : 9.8 The difference of a number and eleven is Find the number.

EXAMPLE 9.5

The sum of twice a number and seven is Find the number.

Solution

Step 1. **Read** the problem.	
Step 2. **Identify** what you are looking for.	the number
Step 3. **Name.** Choose a variable to represent the number.	Let
Step 4. **Translate.** Restate the problem as one sentence. Translate into an equation.	The sum of twice a number and 7 is 15. $2 \cdot n + 7 = 15$
Step 5. **Solve** the equation.	$2n + 7 = 15$
Subtract 7 from each side and simplify.	$2n = 8$
Divide each side by 2 and simplify.	$n = 4$
Step 6. **Check:** is the sum of twice 4 and 7 equal to 15? $2 \cdot 4 + 7 = 15$ $8 + 7 = 15$ $15 = 15\checkmark$	
Step 7. **Answer** the question.	The number is 4.

> **TRY IT : :** 9.9 The sum of four times a number and two is Find the number.

> **TRY IT : :** 9.10 The sum of three times a number and seven is Find the number.

Some number word problems ask you to find two or more numbers. It may be tempting to name them all with different variables, but so far we have only solved equations with one variable. We will define the numbers in terms of the same variable. Be sure to read the problem carefully to discover how all the numbers relate to each other.

EXAMPLE 9.6

One number is five more than another. The sum of the numbers is twenty-one. Find the numbers.

Solution

Step 1. **Read** the problem.	
Step 2. **Identify** what you are looking for.	You are looking for two numbers.
Step 3. **Name.** Choose a variable to represent the first number. What do you know about the second number? Translate.	Let One number is five more than another.
Step 4. **Translate.** Restate the problem as one sentence with all the important information. Translate into an equation. Substitute the variable expressions.	The sum of the numbers is 21. The sum of the 1st number and the 2nd number is 21. $\underbrace{1^{st}\text{ number}}_{n} + \underbrace{2^{nd}\text{ number}}_{n+5} = 21$
Step 5. **Solve** the equation.	$n + n + 5 = 21$
Combine like terms.	$2n + 5 = 21$
Subtract five from both sides and simplify.	$2n = 16$
Divide by two and simplify.	$n = 8$ 1st number
Find the second number too.	$n + 5$ 2nd number
Substitute $n = 8$	$8 + 5$
	13
Step 6. **Check:**	
Do these numbers check in the problem? Is one number 5 more than the other? Is thirteen, 5 more than 8? Yes. Is the sum of the two numbers 21?	$13 \stackrel{?}{=} 8 + 5$ $13 = 13 ✓$ $8 + 13 \stackrel{?}{=} 21$ $21 = 21 ✓$
Step 7. **Answer** the question.	The numbers are 8 and 13.

> **TRY IT : :** 9.11
>
> One number is six more than another. The sum of the numbers is twenty-four. Find the numbers.

> **TRY IT : :** 9.12
>
> The sum of two numbers is fifty-eight. One number is four more than the other. Find the numbers.

EXAMPLE 9.7

The sum of two numbers is negative fourteen. One number is four less than the other. Find the numbers.

Solution

Step 1. **Read** the problem.	
Step 2. **Identify** what you are looking for.	two numbers
Step 3. **Name.** Choose a variable. What do you know about the second number? Translate.	Let n = 1st number One number is 4 less than the other. n - 4 = 2nd number
Step 4. **Translate.** Write as one sentence. Translate into an equation. Substitute the variable expressions.	The sum of two numbers is negative fourteen. 1st number + 2nd number = −14 $n + n - 4 = -14$
Step 5. **Solve** the equation.	$n + n - 4 = -14$
Combine like terms.	$2n - 4 = -14$
Add 4 to each side and simplify.	$2n = -10$
Divide by 2.	$n = -5$ 1st number
Substitute to find the 2nd number.	$n - 4$ 2nd number
	$-5 - 4$
	-9
Step 6. **Check:**	
Is −9 four less than −5?	$-5 - 4 \stackrel{?}{=} -9$ $-9 = -9$ ✓
Is their sum −14?	$-5 + (-9) \stackrel{?}{=} -14$ $-14 = -14$ ✓
Step 7. **Answer** the question.	The numbers are −5 and −9.

> **TRY IT : :** 9.13
>
> The sum of two numbers is negative twenty-three. One number is less than the other. Find the numbers.

TRY IT : : 9.14

The sum of two numbers is negative eighteen. One number is more than the other. Find the numbers.

EXAMPLE 9.8

One number is ten more than twice another. Their sum is one. Find the numbers.

Solution

Step 1. **Read** the problem.	
Step 2. **Identify** what you are looking for.	two numbers
Step 3. **Name.** Choose a variable. One number is ten more than twice another.	Let x = 1st number $2x + 10$ = 2nd number
Step 4. **Translate.** Restate as one sentence.	Their sum is one.
Translate into an equation	The sum of the two numbers is 1. $x + (2x + 10) = 1$
Step 5. **Solve** the equation.	$x + 2x + 10 = 1$
Combine like terms.	$3x + 10 = 1$
Subtract 10 from each side.	$3x = -9$
Divide each side by 3 to get the first number.	$x = -3$
Substitute to get the second number.	$2x + 10$
	$2(-3) + 10$
	4
Step 6. **Check.**	
	$2(-3) + 10 \stackrel{?}{=} 4$
Is 4 ten more than twice −3?	$-6 + 10 = 4$
	$4 = 4 ✓$
Is their sum 1?	$-3 + 4 \stackrel{?}{=} 1$
	$1 = 1 ✓$
Step 7. **Answer** the question.	The numbers are −3 and 4.

TRY IT : : 9.15 One number is eight more than twice another. Their sum is negative four. Find the numbers.

TRY IT : : 9.16

One number is three more than three times another. Their sum is negative five. Find the numbers.

Consecutive integers are integers that immediately follow each other. Some examples of consecutive integers are:

Notice that each number is one more than the number preceding it. So if we define the first integer as the next consecutive integer is The one after that is one more than so it is or

EXAMPLE 9.9

The sum of two consecutive integers is Find the numbers.

Solution

Step 1. **Read** the problem.	
Step 2. **Identify** what you are looking for.	two consecutive integers
Step 3. **Name.**	Let n = 1st integer n + 1 = next consecutive integer
Step 4. **Translate.** Restate as one sentence. Translate into an equation.	The sum of the integers is 47. $n+n+1 = 47$
Step 5. **Solve** the equation.	$n+n+1=47$
Combine like terms.	$2n+1=47$
Subtract 1 from each side.	$2n=46$
Divide each side by 2.	$n=23$ 1st integer
Substitute to get the second number.	$n+1$ 2nd integer
	$23+1$
	24
Step 6. **Check:**	$23+24\stackrel{?}{=}47$ $47=47\checkmark$
Step 7. **Answer** the question.	The two consecutive integers are 23 and 24.

TRY IT : : 9.17 The sum of two consecutive integers is Find the numbers.

TRY IT : : 9.18 The sum of two consecutive integers is Find the numbers.

EXAMPLE 9.10

Find three consecutive integers whose sum is

Solution

Step 1. **Read** the problem.	
Step 2. **Identify** what you are looking for.	three consecutive integers
Step 3. **Name.**	Let $n = 1^{st}$ integer $n + 1 = 2^{nd}$ consecutive integer $n + 2 = 3^{rd}$ consecutive integer
Step 4. **Translate.** Restate as one sentence. Translate into an equation.	The sum of the three integers is 42. $n \quad + \quad n+1 \quad + \quad n+2 \quad = \quad 42$
Step 5. **Solve** the equation.	$n+n+1+n+2=42$
Combine like terms.	$3n+3=42$
Subtract 3 from each side.	$3n=39$
Divide each side by 3.	$n=13$ 1st integer
Substitute to get the second number.	$n+1$ 2nd integer
	$13+1$
	24
Substitute to get the third number.	$n+2$ 3rd integer
	$13+2$
	15
Step 6. **Check:**	$13+14+15 \stackrel{?}{=} 42$ $42=42\checkmark$
Step 7. **Answer** the question.	The three consecutive integers are 13, 14, and 15.

> **TRY IT : :** 9.19 Find three consecutive integers whose sum is

> **TRY IT : :** 9.20 Find three consecutive integers whose sum is

LINKS TO LITERACY

The Links to Literacy activities *Math Curse*, *Missing Mittens* and *Among the Odds and Evens* will provide you with another view of the topics covered in this section.

9.1 EXERCISES

Practice Makes Perfect

Use a Problem-solving Strategy for Word Problems

In the following exercises, use the problem-solving strategy for word problems to solve. Answer in complete sentences.

1. Two-thirds of the children in the fourth-grade class are girls. If there are girls, what is the total number of children in the class?

2. Three-fifths of the members of the school choir are women. If there are women, what is the total number of choir members?

3. Zachary has country music CDs, which is one-fifth of his CD collection. How many CDs does Zachary have?

4. One-fourth of the candies in a bag of are red. If there are red candies, how many candies are in the bag?

5. There are girls in a school club. The number of girls is more than twice the number of boys. Find the number of boys in the club.

6. There are Cub Scouts in Troop The number of scouts is more than five times the number of adult leaders. Find the number of adult leaders.

7. Lee is emptying dishes and glasses from the dishwasher. The number of dishes is less than the number of glasses. If there are dishes, what is the number of glasses?

8. The number of puppies in the pet store window is twelve less than the number of dogs in the store. If there are puppies in the window, what is the number of dogs in the store?

9. After months on a diet, Lisa had lost of her original weight. She lost pounds. What was Lisa's original weight?

10. Tricia got a raise on her weekly salary. The raise was per week. What was her original weekly salary?

11. Tim left a tip for a restaurant bill. What percent tip did he leave?

12. Rashid left a tip for a restaurant bill. What percent tip did he leave?

13. Yuki bought a dress on sale for The sale price was of the original price. What was the original price of the dress?

14. Kim bought a pair of shoes on sale for The sale price was of the original price. What was the original price of the shoes?

Solve Number Problems

In the following exercises, solve each number word problem.

15. The sum of a number and eight is Find the number.

16. The sum of a number and nine is Find the number.

17. The difference of a number and twelve is Find the number.

18. The difference of a number and eight is Find the number.

19. The sum of three times a number and eight is Find the number.

20. The sum of twice a number and six is Find the number.

21. The difference of twice a number and seven is Find the number.

22. The difference of four times a number and seven is Find the number.

23. Three times the sum of a number and nine is Find the number.

24. Six times the sum of a number and eight is Find the number.

25. One number is six more than the other. Their sum is forty-two. Find the numbers.

26. One number is five more than the other. Their sum is thirty-three. Find the numbers.

27. The sum of two numbers is twenty. One number is four less than the other. Find the numbers.

28. The sum of two numbers is twenty-seven. One number is seven less than the other. Find the numbers.

29. A number is one more than twice another number. Their sum is negative five. Find the numbers.

30. One number is six more than five times another. Their sum is six. Find the numbers.

31. The sum of two numbers is fourteen. One number is two less than three times the other. Find the numbers.

32. The sum of two numbers is zero. One number is nine less than twice the other. Find the numbers.

33. One number is fourteen less than another. If their sum is increased by seven, the result is Find the numbers.

34. One number is eleven less than another. If their sum is increased by eight, the result is Find the numbers.

35. The sum of two consecutive integers is Find the integers.

36. The sum of two consecutive integers is Find the integers.

37. The sum of two consecutive integers is Find the integers.

38. The sum of two consecutive integers is Find the integers.

39. The sum of three consecutive integers is Find the integers.

40. The sum of three consecutive integers is Find the integers.

41. Find three consecutive integers whose sum is

42. Find three consecutive integers whose sum is

Everyday Math

43. Shopping Patty paid for a purse on sale for off the original price. What was the original price of the purse?

44. Shopping Travis bought a pair of boots on sale for off the original price. He paid for the boots. What was the original price of the boots?

45. Shopping Minh spent on sticker books to give his nephews. Find the cost of each sticker book.

46. Shopping Alicia bought a package of peaches for Find the cost of each peach.

47. Shopping Tom paid for a new refrigerator, including tax. What was the price of the refrigerator before tax?

48. Shopping Kenji paid for a new living room set, including tax. What was the price of the living room set before tax?

Writing Exercises

49. Write a few sentences about your thoughts and opinions of word problems. Are these thoughts positive, negative, or neutral? If they are negative, how might you change your way of thinking in order to do better?

50. When you start to solve a word problem, how do you decide what to let the variable represent?

Self Check

After completing the exercises, use this checklist to evaluate your mastery of the objectives of this section.

I can...	Confidently	With some help	No-I don't get it!
approach word problems with a positive attitude.			
use a problem solving strategy for word problems.			
solve number problems.			

If most of your checks were:

...confidently. Congratulations! You have achieved the objectives in this section. Reflect on the study skills you used so that you can continue to use them. What did you do to become confident of your ability to do these things? Be specific.

...with some help. This must be addressed quickly because topics you do not master become potholes in your road to success.

In math, every topic builds upon previous work. It is important to make sure you have a strong foundation before you move on. Whom can you ask for help? Your fellow classmates and instructor are good resources. Is there a place on campus where math tutors are available? Can your study skills be improved?

...no—I don't get it! This is a warning sign and you must not ignore it. You should get help right away or you will quickly be overwhelmed. See your instructor as soon as you can to discuss your situation. Together you can come up with a plan to get you the help you need.

9.2 Solve Money Applications

Learning Objectives

By the end of this section, you will be able to:

- Solve coin word problems
- Solve ticket and stamp word problems

BE PREPARED : : 9.4 Before you get started, take this readiness quiz.

Multiply:

If you missed this problem, review Example 5.15.

BE PREPARED : : 9.5 Simplify:

If you missed this problem, review Example 7.22.

BE PREPARED : : 9.6 Solve:

If you missed this problem, review Example 8.44.

Solve Coin Word Problems

Imagine taking a handful of coins from your pocket or purse and placing them on your desk. How would you determine the value of that pile of coins?

If you can form a step-by-step plan for finding the total value of the coins, it will help you as you begin solving coin word problems.

One way to bring some order to the mess of coins would be to separate the coins into stacks according to their value. Quarters would go with quarters, dimes with dimes, nickels with nickels, and so on. To get the total value of all the coins, you would add the total value of each pile.

Figure 9.4 To determine the total value of a stack of nickels, multiply the number of nickels times the value of one nickel.(Credit: Darren Hester via ppdigital)

How would you determine the value of each pile? Think about the dime pile—how much is it worth? If you count the number of dimes, you'll know how many you have—the *number* of dimes.

But this does not tell you the *value* of all the dimes. Say you counted dimes, how much are they worth? Each dime is worth —that is the *value* of one dime. To find the total value of the pile of dimes, multiply by to get This is the total value of all dimes.

Finding the Total Value for Coins of the Same Type

For coins of the same type, the total value can be found as follows:

where *number* is the number of coins, *value* is the value of each coin, and *total value* is the total value of all the coins.

You could continue this process for each type of coin, and then you would know the total value of each type of coin. To get the total value of *all* the coins, add the total value of each type of coin.

Let's look at a specific case. Suppose there are quarters, dimes, nickels, and pennies. We'll make a table to organize the information – the type of coin, the number of each, and the value.

Type			
Quarters			
Dimes			
Nickels			
Pennies			

Table 9.1

The total value of all the coins is Notice how Table 9.1 helped us organize all the information. Let's see how this method is used to solve a coin word problem.

EXAMPLE 9.11

Adalberto has in dimes and nickels in his pocket. He has nine more nickels than dimes. How many of each type of coin does he have?

Solution

Step 1. **Read** the problem. Make sure you understand all the words and ideas.

- Determine the types of coins involved.

Think about the strategy we used to find the value of the handful of coins. The first thing you need is to notice what types of coins are involved. Adalberto has dimes and nickels.

- **Create a table** to organize the information.
  - Label the columns 'type', 'number', 'value', 'total value'.
  - List the types of coins.
  - Write in the value of each type of coin.
  - Write in the total value of all the coins.

We can work this problem all in cents or in dollars. Here we will do it in dollars and put in the dollar sign ($) in the table as a reminder.

The value of a dime is and the value of a nickel is The total value of all the coins is

Type			
Dimes			
Nickels			

Step 2. **Identify** what you are looking for.

- We are asked to find the number of dimes and nickels Adalberto has.

Step 3. **Name** what you are looking for.

- Use variable expressions to represent the number of each type of coin.
- Multiply the number times the value to get the total value of each type of coin.
  In this problem you cannot count each type of coin—that is what you are looking for—but you have a clue. There are nine more nickels than dimes. The number of nickels is nine more than the number of dimes.

  Fill in the "number" column to help get everything organized.

Type			
Dimes			
Nickels			

Now we have all the information we need from the problem!

You multiply the number times the value to get the total value of each type of coin. While you do not know the actual number, you do have an expression to represent it.

And so now multiply and write the results in the Total Value column.

Type			
Dimes			
Nickels			

Step 4. **Translate** into an **equation**. Restate the problem in one sentence. Then translate into an equation.

value of the dimes	+	value of the nickels	=	total value of the coins
$0.10d$	+	$0.05(d+9)$	=	2.25

Step 5. **Solve** the equation using good algebra techniques.

Write the equation.	$0.10d + 0.05(d + 9) = 2.25$
Distribute.	$0.10d + 0.05d + 0.45 = 2.25$
Combine like terms.	$0.15d + 0.45 = 2.25$
Subtract 0.45 from each side.	$0.15d = 1.80$
Divide to find the number of dimes.	$d = 12$
The number of nickels is $d + 9$	$d + 9$ $12 + 9$ 21

Step 6. **Check.**

Step 7. **Answer** the question.

If this were a homework exercise, our work might look like this:

Adalberto has $2.25 in dimes and nickels in his pocket. He has nine more nickels than dimes. How many of each type does he have?

Type	Number • Value($)		= Total Value ($)
Dimes	d	0.10	0.10d
Nickels	d + 9	0.05	0.05(d + 9)
			2.25

$0.10d + 0.05d + 0.45 = 2.25$
$0.15d + 0.45 = 2.25$
$0.15d = 1.80$
$d = 12$ dimes

$d + 9$
$12 + 9$
21 nickles

Check:

TRY IT : : 9.21

Michaela has in dimes and nickels in her change purse. She has seven more dimes than nickels. How many coins of each type does she have?

TRY IT : : 9.22

Liliana has in nickels and quarters in her backpack. She has more nickels than quarters. How many coins of each type does she have?

HOW TO : : SOLVE A COIN WORD PROBLEM.

Step 1. **Read** the problem. Make sure you understand all the words and ideas, and create a table to organize the information.

Step 2. **Identify** what you are looking for.

Step 3. **Name** what you are looking for. Choose a variable to represent that quantity.

- Use variable expressions to represent the number of each type of coin and write them in the table.
- Multiply the number times the value to get the total value of each type of coin.

Step 4. **Translate** into an equation. Write the equation by adding the total values of all the types of coins.

Step 5. **Solve** the equation using good algebra techniques.

Step 6. **Check** the answer in the problem and make sure it makes sense.

Step 7. **Answer** the question with a complete sentence.

You may find it helpful to put all the numbers into the table to make sure they check.

Type	Number	Value ($)	Total Value

Table 9.2

EXAMPLE 9.12

Maria has in quarters and pennies in her wallet. She has twice as many pennies as quarters. How many coins of each type does she have?

Solution

Step 1. **Read** the problem.

- Determine the types of coins involved.
  We know that Maria has quarters and pennies.
- Create a table to organize the information.
  - Label the columns type, number, value, total value.
  - List the types of coins.
  - Write in the value of each type of coin.
  - Write in the total value of all the coins.

Type			
Quarters			
Pennies			

Step 2. **Identify** what you are looking for.

Step 3. **Name:** Represent the number of quarters and pennies using variables.

Type			
Quarters			
Pennies			

Multiply the 'number' and the 'value' to get the 'total value' of each type of coin.

Type			
Quarters			
Pennies			

Step 4. **Translate.** Write the equation by adding the 'total value' of all the types of coins.
Step 5. **Solve** the equation.

Write the equation.	$0.25q + 0.01(2q) = 2.43$
Multiply.	$0.25q + 0.02q = 2.43$
Combine like terms.	$0.27q = 2.43$
Divide by 0.27.	$q = 9$ quarters
The number of pennies is $2q$.	$2q$ $2 \cdot 9$ 18 pennies

Step 6. **Check** the answer in the problem.
Maria has quarters and pennies. Does this make

Step 7. **Answer** the question. Maria has nine quarters and eighteen pennies.

TRY IT : : 9.23

Sumanta has in nickels and dimes in her desk drawer. She has twice as many nickels as dimes. How many coins of each type does she have?

 **TRY IT : :** 9.24

Alison has three times as many dimes as quarters in her purse. She has altogether. How many coins of each type does she have?

In the next example, we'll show only the completed table—make sure you understand how to fill it in step by step.

EXAMPLE 9.13

Danny has worth of pennies and nickels in his piggy bank. The number of nickels is two more than ten times the number of pennies. How many nickels and how many pennies does Danny have?

Solution

Step 1: **Read** the problem.	
Determine the types of coins involved. Create a table.	Pennies and nickels
Write in the value of each type of coin.	Pennies are worth Nickels are worth
Step 2: **Identify** what you are looking for.	the number of pennies and nickels
Step 3: **Name.** Represent the number of each type of coin using variables. The number of nickels is defined in terms of the number of pennies, so start with pennies.	Let
The number of nickels is two more than then times the number of pennies.	

Multiply the number and the value to get the total value of each type of coin.

Type			
pennies			
nickels			

Step 4. **Translate:** Write the equation by adding the total value of all the types of coins.

Step 5. **Solve** the equation.

	$0.01p + 0.50p + 0.10 = 2.14$
	$0.51p + 0.10 = 2.14$
	$0.51p = 2.04$
	$p = 4$ pennies
How many nickels?	$10p + 2$
	$10(4) + 2$
	42 nickels

Step 6. **Check.** Is the total value of pennies and nickels equal to

Step 7. **Answer** the question. Danny has pennies and nickels.

> TRY IT : : 9.25

Jesse has worth of quarters and nickels in his pocket. The number of nickels is five more than two times the number of quarters. How many nickels and how many quarters does Jesse have?

> TRY IT : : 9.26

Elaine has in dimes and nickels in her coin jar. The number of dimes that Elaine has is seven less than three times the number of nickels. How many of each coin does Elaine have?

Solve Ticket and Stamp Word Problems

The strategies we used for coin problems can be easily applied to some other kinds of problems too. Problems involving tickets or stamps are very similar to coin problems, for example. Like coins, tickets and stamps have different values; so we can organize the information in tables much like we did for coin problems.

EXAMPLE 9.14

At a school concert, the total value of tickets sold was Student tickets sold for each and adult tickets sold for each. The number of adult tickets sold was less than three times the number of student tickets sold. How many student tickets and how many adult tickets were sold?

Solution

Step 1: **Read the problem.**

- Determine the types of tickets involved.
  There are student tickets and adult tickets.
- Create a table to organize the information.

Type			
Student			
Adult			

Step 2. **Identify** what you are looking for.

Step 3. **Name.** Represent the number of each type of ticket using variables.

Type			
Student			
Adult			

Step 4. **Translate:** Write the equation by adding the total values of each type of ticket.

Step 5. **Solve** the equation.

Substitute to find the number of adults.

$$3s - 5 = \text{number of adults}$$

$$3(47) - 5 = 136 \text{ adults}$$

Step 6. **Check.** There were student tickets at each and adult tickets at each. Is the total value
We find the total value of each type of ticket by multiplying the number of tickets times its value; we then add to get the total value of all the tickets sold.

Step 7. **Answer** the question. They sold student tickets and adult tickets.

TRY IT : : 9.27

The first day of a water polo tournament, the total value of tickets sold was One-day passes sold for and tournament passes sold for The number of tournament passes sold was more than the number of day passes sold. How many day passes and how many tournament passes were sold?

TRY IT : : 9.28

At the movie theater, the total value of tickets sold was Adult tickets sold for each and senior/child tickets sold for each. The number of senior/child tickets sold was less than twice the number of adult tickets sold. How many senior/child tickets and how many adult tickets were sold?

Now we'll do one where we fill in the table all at once.

EXAMPLE 9.15

Monica paid for stamps she needed to mail the invitations to her sister's baby shower. The number of stamps was four more than twice the number of stamps. How many stamps and how many stamps did Monica buy?

Solution

The type of stamps are stamps and stamps. Their names also give the value.

"The number of cent stamps was four more than twice the number of cent stamps."

Type			
stamps			
stamps			

Write the equation from the total values.

Solve the equation.

Monica bought 8 eight-cent stamps.

Find the number of 49-cent stamps she bought by evaluating.

Check.

Monica bought eight stamps and twenty stamps.

TRY IT : : 9.29

Eric paid for stamps so he could mail thank you notes for his wedding gifts. The number of stamps was eight more than twice the number of stamps. How many stamps and how many stamps did Eric buy?

TRY IT : : 9.30

Kailee paid for stamps. The number of stamps was four less than three times the number of stamps. How many stamps and how many stamps did Kailee buy?

9.2 EXERCISES

Practice Makes Perfect

Solve Coin Word Problems

In the following exercises, solve the coin word problems.

51. Jaime has in dimes and nickels. The number of dimes is more than the number of nickels. How many of each coin does he have?

52. Lee has in dimes and nickels. The number of nickels is more than the number of dimes. How many of each coin does he have?

53. Ngo has a collection of dimes and quarters with a total value of The number of dimes is more than the number of quarters. How many of each coin does he have?

54. Connor has a collection of dimes and quarters with a total value of The number of dimes is more than the number of quarters. How many of each coin does he have?

55. Carolyn has in her purse in nickels and dimes. The number of nickels is less than three times the number of dimes. Find the number of each type of coin.

56. Julio has in his pocket in nickels and dimes. The number of dimes is less than twice the number of nickels. Find the number of each type of coin.

57. Chi has in dimes and quarters. The number of dimes is more than three times the number of quarters. How many dimes and nickels does Chi have?

58. Tyler has in dimes and quarters. The number of quarters is more than four times the number of dimes. How many of each coin does he have?

59. A cash box of and bills is worth The number of bills is more than the number of bills. How many of each bill does it contain?

60. Joe's wallet contains and bills worth The number of bills is more than the number of bills. How many of each bill does he have?

61. In a cash drawer there is in and bills. The number of bills is twice the number of bills. How many of each are in the drawer?

62. John has in and bills in his drawer. The number of bills is three times the number of bills. How many of each are in the drawer?

63. Mukul has in quarters, dimes and nickels in his pocket. He has five more dimes than quarters and nine more nickels than quarters. How many of each coin are in his pocket?

64. Vina has in quarters, dimes and nickels in her purse. She has eight more dimes than quarters and six more nickels than quarters. How many of each coin are in her purse?

Solve Ticket and Stamp Word Problems

In the following exercises, solve the ticket and stamp word problems.

65. The play took in one night. The number of $8 adult tickets was less than twice the number of child tickets. How many of each ticket were sold?

66. If the number of child tickets is seventeen less than three times the number of adult tickets and the theater took in how many of each ticket were sold?

67. The movie theater took in one Monday night. The number of child tickets was ten more than twice the number of adult tickets. How many of each were sold?

68. The ball game took in one Saturday. The number of adult tickets was more than twice the number of child tickets. How many of each were sold?

69. Julie went to the post office and bought both stamps and postcards for her office's bills She spent The number of stamps was more than twice the number of postcards. How many of each did she buy?

70. Before he left for college out of state, Jason went to the post office and bought both stamps and postcards and spent The number of stamps was more than twice the number of postcards. How many of each did he buy?

71. Maria spent at the post office. She bought three times as many stamps as stamps. How many of each did she buy?

72. Hector spent at the post office. He bought four times as many stamps as stamps. How many of each did he buy?

73. Hilda has worth of and stock shares. The numbers of shares is more than twice the number of shares. How many of each does she have?

74. Mario invested in and stock shares. The number of shares was less than three times the number of shares. How many of each type of share did he buy?

Everyday Math

75. Parent Volunteer As the treasurer of her daughter's Girl Scout troop, Laney collected money for some girls and adults to go to a camp. Each girl paid and each adult paid The total amount of money collected for camp was If the number of girls is three times the number of adults, how many girls and how many adults paid for camp?

76. Parent Volunteer Laurie was completing the treasurer's report for her son's Boy Scout troop at the end of the school year. She didn't remember how many boys had paid the full-year registration fee and how many had paid a partial-year fee. She knew that the number of boys who paid for a full-year was ten more than the number who paid for a partial-year. If was collected for all the registrations, how many boys had paid the full-year fee and how many had paid the partial-year fee?

Writing Exercises

77. Suppose you have quarters, dimes, and pennies. Explain how you find the total value of all the coins.

78. Do you find it helpful to use a table when solving coin problems? Why or why not?

79. In the table used to solve coin problems, one column is labeled "number" and another column is labeled '"value." What is the difference between the number and the value?

80. What similarities and differences did you see between solving the coin problems and the ticket and stamp problems?

Self Check

After completing the exercises, use this checklist to evaluate your mastery of the objectives of this section.

I can...	Confidently	With some help	No-I don't get it!
solve coin word problems.			
solve ticket and stamp word problems.			

After reviewing this checklist, what will you do to become confident for all objectives?

9.3 Use Properties of Angles, Triangles, and the Pythagorean Theorem

Learning Objectives

By the end of this section, you will be able to:

- Use the properties of angles
- Use the properties of triangles
- Use the Pythagorean Theorem

BE PREPARED : : 9.7 Before you get started, take this readiness quiz.

Solve:

If you missed this problem, review Example 8.6.

BE PREPARED : : 9.8 Solve: — —

If you missed this problem, review Example 6.42.

BE PREPARED : : 9.9 Simplify: $\sqrt{}$

If you missed this problem, review Example 5.72.

So far in this chapter, we have focused on solving word problems, which are similar to many real-world applications of algebra. In the next few sections, we will apply our problem-solving strategies to some common geometry problems.

Use the Properties of Angles

Are you familiar with the phrase 'do a It means to turn so that you face the opposite direction. It comes from the fact that the measure of an angle that makes a straight line is degrees. See Figure 9.5.

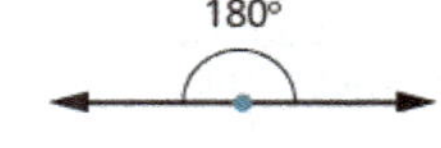

Figure 9.5

An **angle** is formed by two rays that share a common endpoint. Each ray is called a side of the angle and the common endpoint is called the **vertex**. An angle is named by its vertex. In Figure 9.6, is the angle with vertex at point The measure of is written

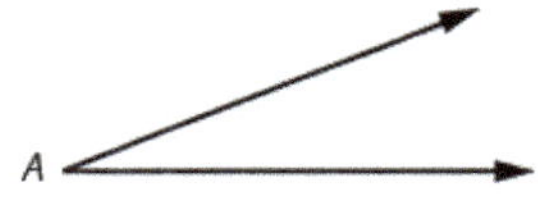

Figure 9.6 is the angle with vertex at

We measure angles in degrees, and use the symbol to represent degrees. We use the abbreviation to for the *measure* of an angle. So if is we would write

If the sum of the measures of two angles is then they are called **supplementary angles**. In Figure 9.7, each pair of angles is supplementary because their measures add to Each angle is the *supplement* of the other.

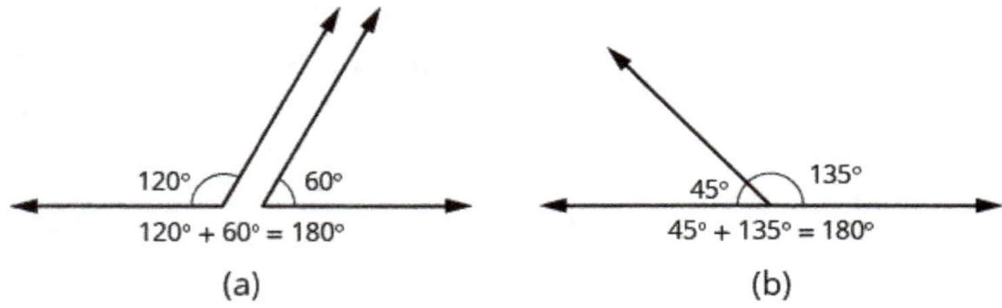

Figure 9.7 The sum of the measures of supplementary angles is

If the sum of the measures of two angles is then the angles are **complementary angles**. In Figure 9.8, each pair of angles is complementary, because their measures add to Each angle is the *complement* of the other.

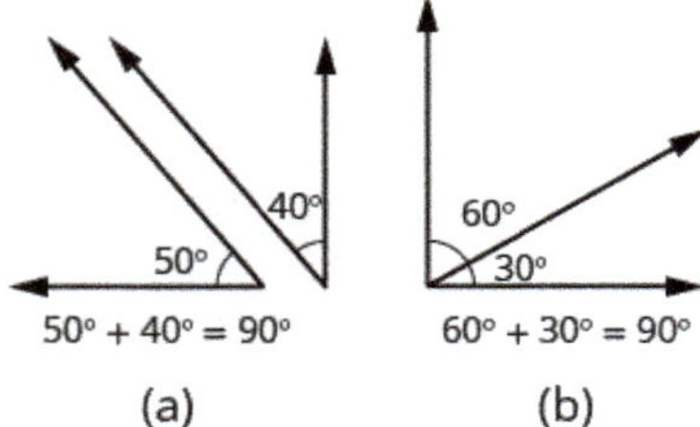

Figure 9.8 The sum of the measures of complementary angles is

Supplementary and Complementary Angles

If the sum of the measures of two angles is then the angles are supplementary.

If and are supplementary, then

If the sum of the measures of two angles is then the angles are complementary.

If and are complementary, then

In this section and the next, you will be introduced to some common geometry formulas. We will adapt our Problem Solving Strategy for Geometry Applications. The geometry formula will name the variables and give us the equation to solve.

In addition, since these applications will all involve geometric shapes, it will be helpful to draw a figure and then label it with the information from the problem. We will include this step in the Problem Solving Strategy for Geometry Applications.

HOW TO : : USE A PROBLEM SOLVING STRATEGY FOR GEOMETRY APPLICATIONS.

Step 1. **Read** the problem and make sure you understand all the words and ideas. Draw a figure and label it with the given information.

Step 2. **Identify** what you are looking for.

Step 3. **Name** what you are looking for and choose a variable to represent it.

Step 4. **Translate** into an equation by writing the appropriate formula or model for the situation. Substitute in the given information.

Step 5. **Solve** the equation using good algebra techniques.

Step 6. **Check** the answer in the problem and make sure it makes sense.

Step 7. **Answer** the question with a complete sentence.

The next example will show how you can use the Problem Solving Strategy for Geometry Applications to answer questions about supplementary and complementary angles.

EXAMPLE 9.16

An angle measures Find its supplement, and its complement.

 **Solution**

Step 1. **Read** the problem. Draw the figure and label it with the given information.	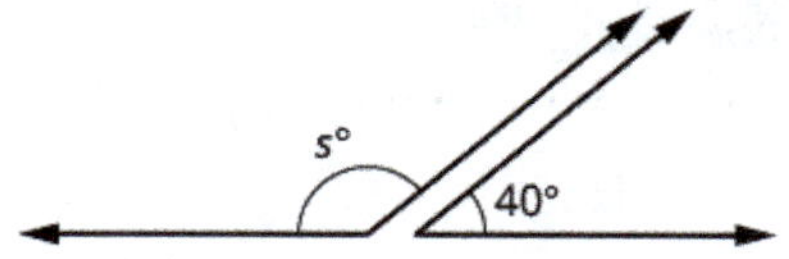
Step 2. **Identify** what you are looking for.	the supplement of a 40° angle.
Step 3. **Name.** Choose a variable to represent it.	let s = the measure of the supplement
Step 4. **Translate.** Write the appropriate formula for the situation and substitute in the given information.	$m\angle A + m\angle B = 180$ $s + 40 = 180$
Step 5. **Solve** the equation.	$s = 140$
Step 6. **Check:** $140 + 40 \stackrel{?}{=} 180$ $180 = 180 ✓$	
Step 7. **Answer** the question.	The supplement of the 40° angle is 140°.

Step 1. **Read** the problem. Draw the figure and label it with the given information.	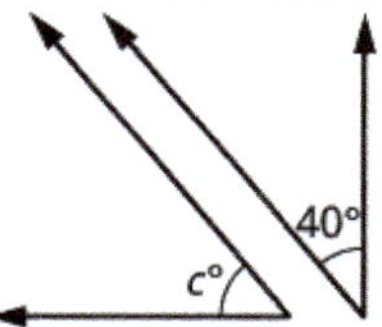
Step 2. **Identify** what you are looking for.	the complement of a 40° angle.
Step 3. **Name.** Choose a variable to represent it.	let c = the measure of the complement
Step 4. **Translate.** Write the appropriate formula for the situation and substitute in the given information.	$m\angle A + m\angle B = 90$
Step 5. **Solve** the equation.	$c + 40 = 90$ $c = 50$
Step 6. **Check:** $50 + 40 \stackrel{?}{=} 90$ $90 = 90 ✓$	
Step 7. **Answer** the question.	The complement of the 40° angle is 50°.

> **TRY IT : :** 9.31 An angle measures Find its: supplement complement.

> **TRY IT : :** 9.32 An angle measures Find its: supplement complement.

Did you notice that the words complementary and supplementary are in alphabetical order just like and are in numerical order?

EXAMPLE 9.17

Two angles are supplementary. The larger angle is more than the smaller angle. Find the measure of both angles.

Solution

Step 1. **Read** the problem. Draw the figure and label it with the given information.	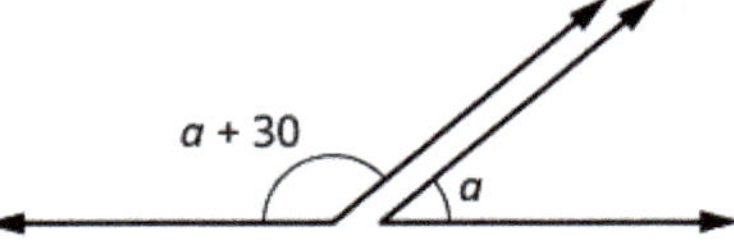
Step 2. **Identify** what you are looking for.	the measures of both angles
Step 3. **Name.** Choose a variable to represent it. The larger angle is 30° more than the smaller angle.	let a = measure of smaller angle $a + 30$ = measure of larger angle
Step 4. **Translate.** Write the appropriate formula and substitute.	$m\angle A + m\angle B = 180$
Step 5. **Solve** the equation.	$(a + 30) + a = 180$ $2a + 30 = 180$ $2a = 150$ $a = 75$ measure of smaller angle $a + 30$ measure of larger angle $75 + 30$ 105
Step 6. **Check:** $m\angle A + m\angle B = 180$ $75 + 105 \stackrel{?}{=} 180$ $180 = 180 \checkmark$	
Step 7. **Answer** the question.	The measures of the angles are 75° and 105°.

> **TRY IT : :** 9.33
>
> Two angles are supplementary. The larger angle is more than the smaller angle. Find the measures of both angles.

TRY IT : : 9.34

Two angles are complementary. The larger angle is more than the smaller angle. Find the measures of both angles.

Use the Properties of Triangles

What do you already know about triangles? Triangle have three sides and three angles. Triangles are named by their vertices. The **triangle** in Figure 9.9 is called read 'triangle '. We label each side with a lower case letter to match the upper case letter of the opposite vertex.

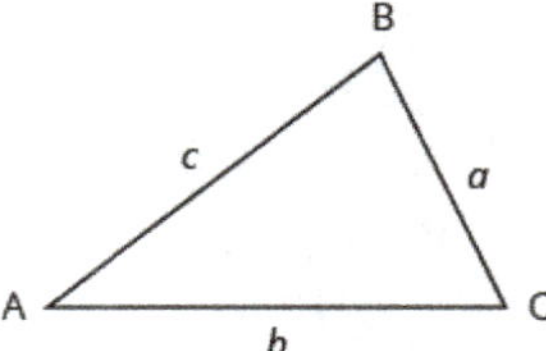

Figure 9.9 has vertices and sides

The three angles of a triangle are related in a special way. The sum of their measures is

Sum of the Measures of the Angles of a Triangle

For any the sum of the measures of the angles is

EXAMPLE 9.18

The measures of two angles of a triangle are and Find the measure of the third angle.

Solution

Step 1. **Read** the problem. Draw the figure and label it with the given information.

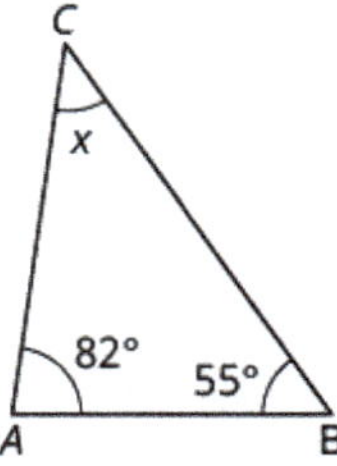

Step 2. **Identify** what you are looking for.	the measure of the third angle in a triangle
Step 3. **Name.** Choose a variable to represent it.	let x = the measure of the angle
Step 4. **Translate.** Write the appropriate formula and substitute.	$m\angle A + m\angle B + m\angle C = 180$
Step 5. **Solve** the equation.	$55 + 82 + x = 180$ $137 + x = 180$ $x = 43$
Step 6. **Check:** $55 + 82 + 43 \stackrel{?}{=} 180$ $180 = 180 \checkmark$	
Step 7. **Answer** the question.	The measure of the third angle is 43 degrees.

> **TRY IT : : 9.35**
>
> The measures of two angles of a triangle are and Find the measure of the third angle.

> **TRY IT : : 9.36** A triangle has angles of and Find the measure of the third angle.

Right Triangles

Some triangles have special names. We will look first at the **right triangle**. A right triangle has one angle, which is often marked with the symbol shown in Figure 9.10.

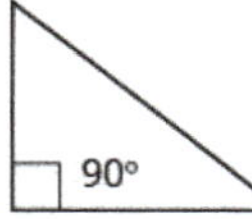

Figure 9.10

If we know that a triangle is a right triangle, we know that one angle measures so we only need the measure of one of the other angles in order to determine the measure of the third angle.

EXAMPLE 9.19

One angle of a right triangle measures What is the measure of the third angle?

Solution

Step 1. **Read** the problem. Draw the figure and label it with the given information.

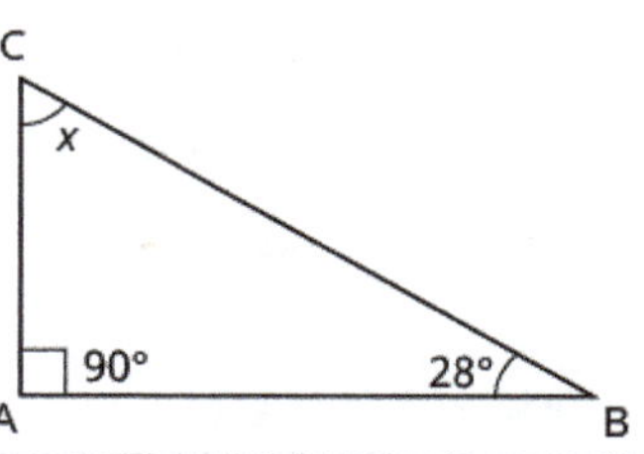

Step 2. **Identify** what you are looking for.	the measure of an angle
Step 3. **Name.** Choose a variable to represent it.	let x = the measure of the angle
Step 4. **Translate.** Write the appropriate formula and substitute.	$m\angle A + m\angle B + m\angle C = 180$
	$x + 90 + 28 = 180$
Step 5. **Solve** the equation.	$x + 118 = 180$
	$x = 62$
Step 6. **Check:** $180 \stackrel{?}{=} 90 + 28 + 62$ $180 = 180\,\checkmark$	
Step 7. **Answer** the question.	The measure of the third angle is 62°.

> **TRY IT : :** 9.37 One angle of a right triangle measures What is the measure of the other angle?

> **TRY IT : :** 9.38 One angle of a right triangle measures What is the measure of the other angle?

In the examples so far, we could draw a figure and label it directly after reading the problem. In the next example, we will have to define one angle in terms of another. So we will wait to draw the figure until we write expressions for all the angles we are looking for.

EXAMPLE 9.20

The measure of one angle of a right triangle is more than the measure of the smallest angle. Find the measures of all three angles.

Solution

Step 1. **Read** the problem.	
Step 2. **Identify** what you are looking for.	the measures of all three angles
Step 3. **Name.** Choose a variable to represent it. Now draw the figure and label it with the given information.	Let a = 1st angle $a + 20$ = 2nd angle 90 = 3rd angle (the right angle) 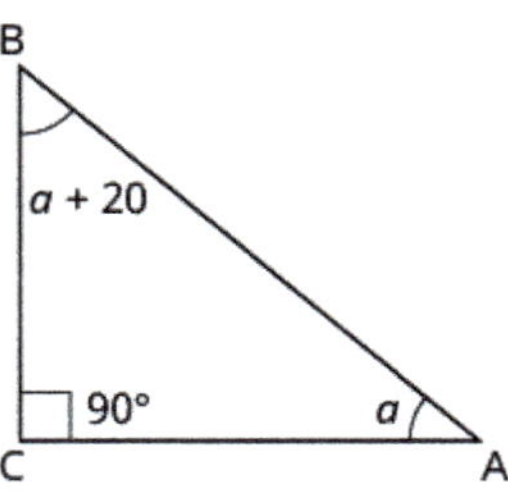
Step 4. **Translate.** Write the appropriate formula and substitute into the formula.	$m\angle A + m\angle B + m\angle C = 180$ $a + (a + 20) + 90 = 180$
Step 5. **Solve** the equation.	$2a + 110 = 180$ $2a = 70$ $a = 35$ first angle $a + 20$ second angle $35 + 20$ 55 90 third angle
Step 6. **Check:** $35 + 55 + 90 \stackrel{?}{=} 180$ $180 = 180\checkmark$	
Step 7. **Answer** the question.	The three angles measure 35°, 55°, and 90°.

TRY IT : : 9.39

The measure of one angle of a right triangle is more than the measure of the smallest angle. Find the measures of all three angles.

TRY IT : : 9.40

The measure of one angle of a right triangle is more than the measure of the smallest angle. Find the measures of all three angles.

Similar Triangles

When we use a map to plan a trip, a sketch to build a bookcase, or a pattern to sew a dress, we are working with similar figures. In geometry, if two figures have exactly the same shape but different sizes, we say they are **similar figures**. One is a scale model of the other. The corresponding sides of the two figures have the same ratio, and all their corresponding angles are have the same measures.

The two triangles in Figure 9.11 are similar. Each side of is four times the length of the corresponding side of and their corresponding angles have equal measures.

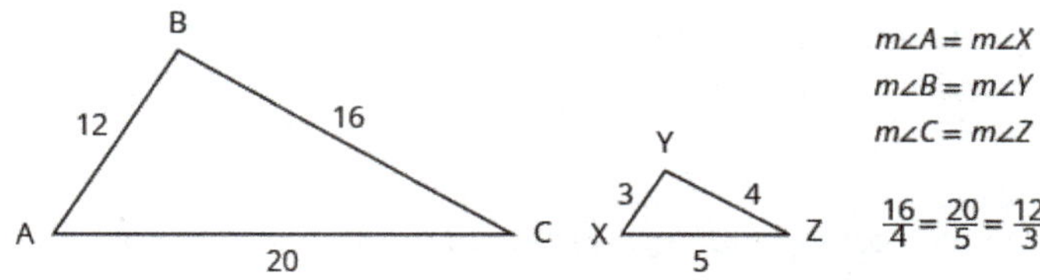

Figure 9.11 and are similar triangles. Their corresponding sides have the same ratio and the corresponding angles have the same measure.

Properties of Similar Triangles

If two triangles are similar, then their corresponding angle measures are equal and their corresponding side lengths are in the same ratio.

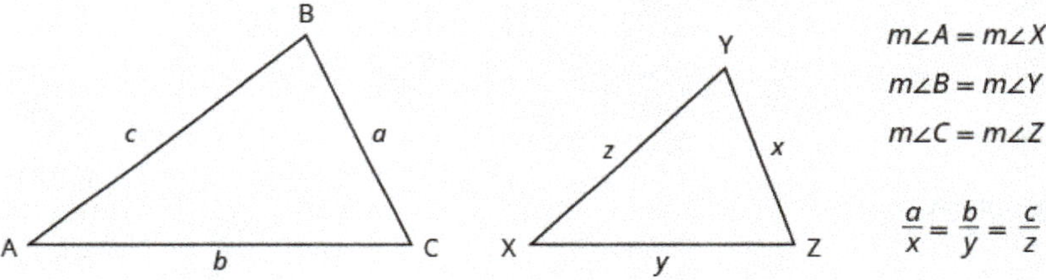

The length of a side of a triangle may be referred to by its endpoints, two vertices of the triangle. For example, in

We will often use this notation when we solve similar triangles because it will help us match up the corresponding side lengths.

EXAMPLE 9.21

and are similar triangles. The lengths of two sides of each triangle are shown. Find the lengths of the third side of each triangle.

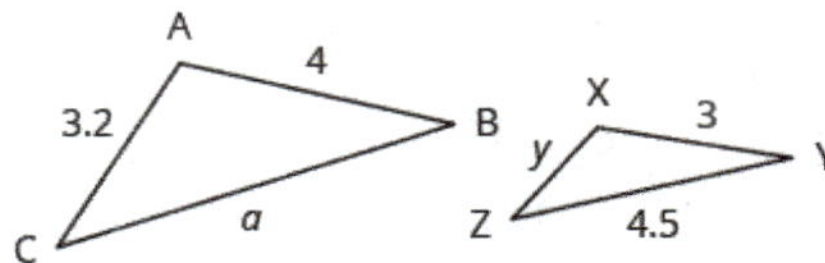

Solution

Step 1. **Read** the problem. Draw the figure and label it with the given information.	The figure is provided.
Step 2. **Identify** what you are looking for.	The length of the sides of similar triangles
Step 3. **Name.** Choose a variable to represent it.	Let a = length of the third side of y = length of the third side
Step 4. **Translate.**	

The triangles are similar, so the corresponding sides are in the same ratio. So

Since the side corresponds to the side , we will use the ratio — to find the other sides.

Be careful to match up corresponding sides correctly.

	To find a:	To find y:
sides of large triangle →	$\frac{AB}{XY} = \frac{BC}{YZ}$	$\frac{AB}{XY} = \frac{AC}{XZ}$
sides of small triangle →	$\frac{4}{3} = \frac{a}{4.5}$	$\frac{4}{3} = \frac{3.2}{y}$

Step 5. **Solve** the equation.	$3a = 4(4.5)$ $3a = 18$ $a = 6$	$4y = 3(3.2)$ $4y = 9.6$ $y = 2.4$

Step 6. **Check:**

$$\frac{4}{3} \stackrel{?}{=} \frac{6}{4.5} \qquad \frac{4}{3} \stackrel{?}{=} \frac{3.2}{2.4}$$

$$4(4.5) \stackrel{?}{=} 6(3) \qquad 4(2.4) \stackrel{?}{=} 3.2(3)$$

$$18 = 18 \checkmark \qquad 9.6 = 9.6 \checkmark$$

Step 7. **Answer** the question.	The third side of is 6 and the third side of is 2.4.

TRY IT : : 9.41 is similar to Find

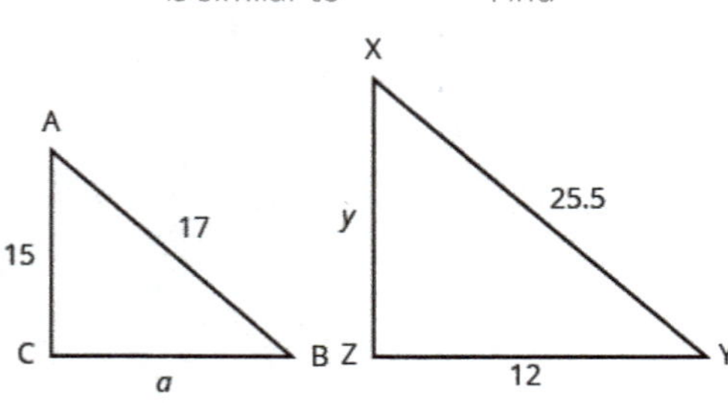

TRY IT : : 9.42 is similar to Find

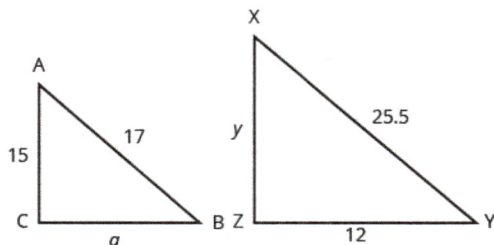

X A 25.5 17 y 15 C B Z 12 Y a

Use the Pythagorean Theorem

The **Pythagorean Theorem** is a special property of right triangles that has been used since ancient times. It is named after the Greek philosopher and mathematician Pythagoras who lived around BCE.

Remember that a right triangle has a angle, which we usually mark with a small square in the corner. The side of the triangle opposite the angle is called the **hypotenuse**, and the other two sides are called the **legs**. See Figure 9.12.

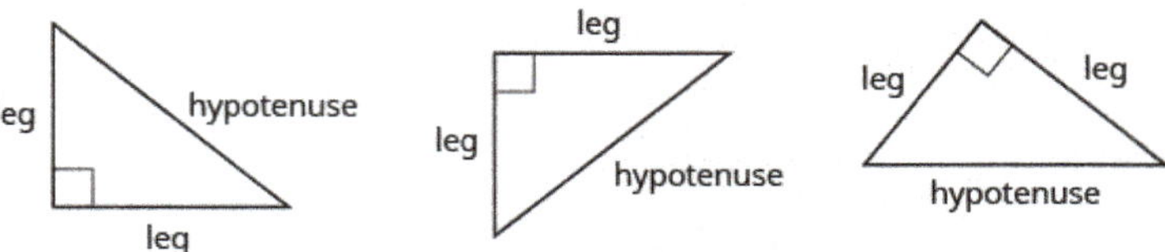

Figure 9.12 In a right triangle, the side opposite the angle is called the hypotenuse and each of the other sides is called a leg.

The Pythagorean Theorem tells how the lengths of the three sides of a right triangle relate to each other. It states that in any right triangle, the sum of the squares of the two legs equals the square of the hypotenuse.

The Pythagorean Theorem

In any right triangle

where is the length of the hypotenuse and are the lengths of the legs.

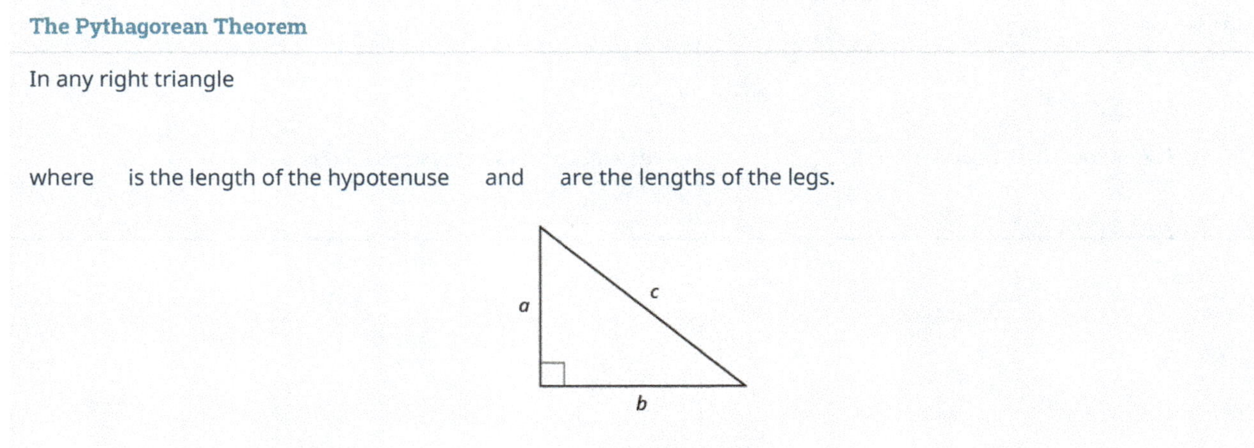

To solve problems that use the Pythagorean Theorem, we will need to find square roots. In Simplify and Use Square Roots we introduced the notation $\sqrt{\ }$ and defined it in this way:

$$\sqrt{\ }$$

For example, we found that $\sqrt{\ }$ is because

We will use this definition of square roots to solve for the length of a side in a right triangle.

EXAMPLE 9.22

Use the Pythagorean Theorem to find the length of the hypotenuse.

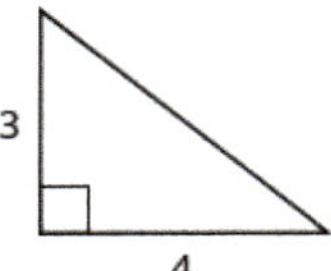

Solution

Step 1. **Read** the problem.	
Step 2. **Identify** what you are looking for.	the length of the hypotenuse of the triangle
Step 3. **Name.** Choose a variable to represent it.	Let 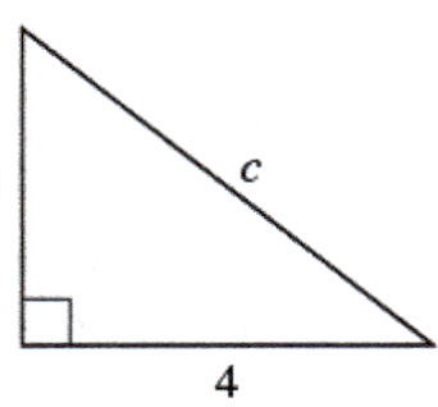
Step 4. **Translate.** Write the appropriate formula. Substitute.	$a^2 + b^2 = c^2$ $3^2 + 4^2 = c^2$
Step 5. **Solve** the equation.	$9 + 16 = c^2$ $25 = c^2$ $\sqrt{25} = c^2$ $5 = c$
Step 6. **Check:** $3^2 + 4^2 = 5^2$ $9 + 16 \stackrel{?}{=} 25$ $25 = 25\checkmark$	
Step 7. **Answer** the question.	The length of the hypotenuse is 5.

TRY IT : : 9.43 Use the Pythagorean Theorem to find the length of the hypotenuse.

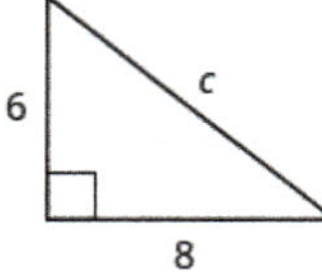

TRY IT : : 9.44 Use the Pythagorean Theorem to find the length of the hypotenuse.

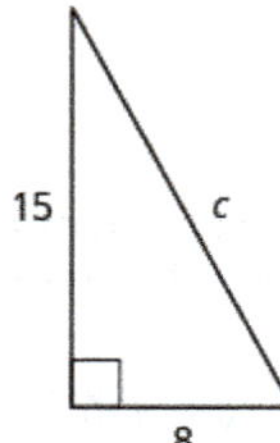

EXAMPLE 9.23

Use the Pythagorean Theorem to find the length of the longer leg.

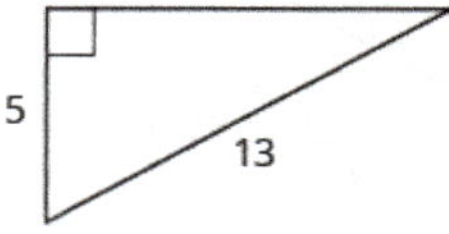

Solution

Step 1. **Read** the problem.	
Step 2. **Identify** what you are looking for.	The length of the leg of the triangle
Step 3. **Name.** Choose a variable to represent it.	Let Label side b 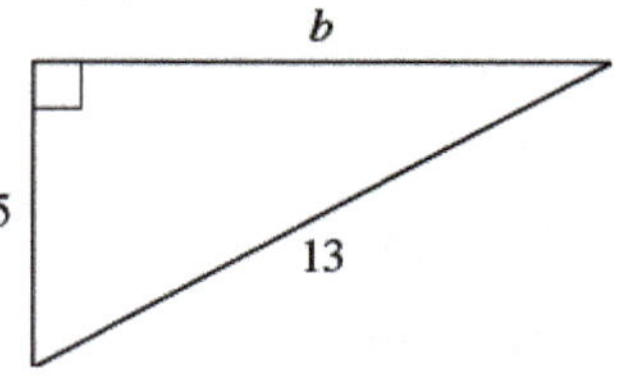
Step 4. **Translate.** Write the appropriate formula. Substitute.	$a^2 + b^2 = c^2$ $5^2 + b^2 = 13^2$
Step 5. **Solve** the equation. Isolate the variable term. Use the definition of the square root. Simplify.	$25 + b^2 = 169$ $b^2 = 144$ $b^2 = \sqrt{144}$ $b = 12$
Step 6. **Check:** $5^2 + 12^2 \stackrel{?}{=} 13^2$ $25 + 144 \stackrel{?}{=} 169$ $169 = 169 \checkmark$	
Step 7. **Answer** the question.	The length of the leg is 12.

 **TRY IT : :** 9.45 Use the Pythagorean Theorem to find the length of the leg.

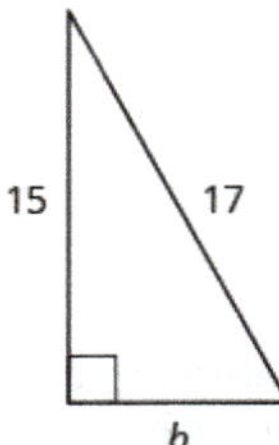

> **TRY IT : :** 9.46 Use the Pythagorean Theorem to find the length of the leg.

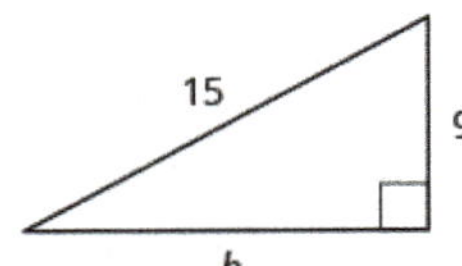

EXAMPLE 9.24

Kelvin is building a gazebo and wants to brace each corner by placing a wooden bracket diagonally as shown. How far below the corner should he fasten the bracket if he wants the distances from the corner to each end of the bracket to be equal? Approximate to the nearest tenth of an inch.

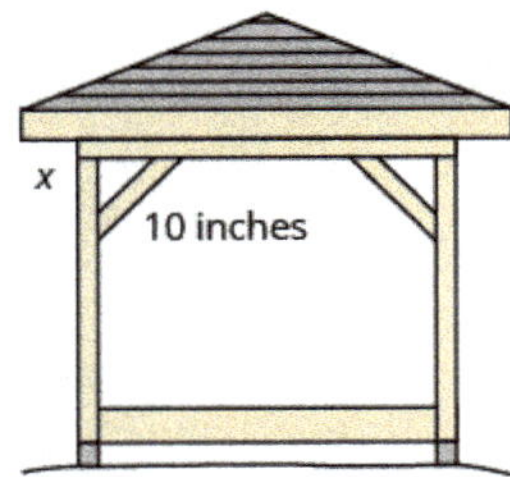

Solution

Step 1. **Read** the problem.	
Step 2. **Identify** what you are looking for.	the distance from the corner that the bracket should be attached
Step 3. **Name.** Choose a variable to represent it.	Let x = the distance from the corner 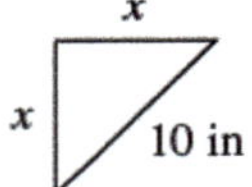
Step 4. **Translate.** Write the appropriate formula. Substitute.	$a^2 + b^2 = c^2$ $x^2 + x^2 = 10^2$
Step 5. **Solve** the equation. Isolate the variable. Use the definition of the square root. Simplify. Approximate to the nearest tenth.	$2x^2 = 100$ $x^2 = 50$ $x = \sqrt{50}$ $b \approx 7.1$
Step 6. **Check:** $a^2 + b^2 = c^2$ $(7.1)^2 + (7.1)^2 \overset{?}{\approx} 10^2$ Yes.	
Step 7. **Answer** the question.	Kelvin should fasten each piece of wood approximately 7.1" from the corner.

TRY IT : : 9.47

John puts the base of a ladder feet from the wall of his house. How far up the wall does the ladder reach?

TRY IT : : 9.48

Randy wants to attach a string of lights to the top of the mast of his sailboat. How far from the base of the mast should he attach the end of the light string?

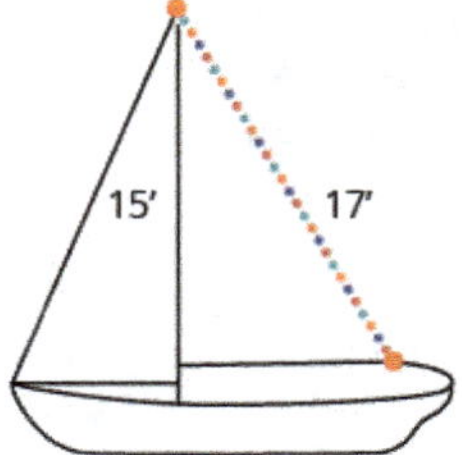

MEDIA : : ACCESS ADDITIONAL ONLINE RESOURCES

- Animation: The Sum of the Interior Angles of a Triangle (http://www.openstax.org/l/24sumintangles)
- Similar Polygons (http://www.openstax.org/l/24simpolygons)
- Example: Determine the Length of the Hypotenuse of a Right Triangle (http://www.openstax.org/l/24hyporighttri)

9.3 EXERCISES

Practice Makes Perfect

Use the Properties of Angles

In the following exercises, find the supplement and the complement of the given angle.

81.

82.

83.

84.

In the following exercises, use the properties of angles to solve.

85. Find the supplement of a angle.

86. Find the complement of a angle.

87. Find the complement of a angle.

88. Find the supplement of a angle.

89. Two angles are supplementary. The larger angle is more than the smaller angle. Find the measures of both angles.

90. Two angles are supplementary. The smaller angle is less than the larger angle. Find the measures of both angles.

91. Two angles are complementary. The smaller angle is less than the larger angle. Find the measures of both angles.

92. Two angles are complementary. The larger angle is more than the smaller angle. Find the measures of both angles.

Use the Properties of Triangles

In the following exercises, solve using properties of triangles.

93. The measures of two angles of a triangle are and Find the measure of the third angle.

94. The measures of two angles of a triangle are and Find the measure of the third angle.

95. The measures of two angles of a triangle are and Find the measure of the third angle.

96. The measures of two angles of a triangle are and Find the measure of the third angle.

97. One angle of a right triangle measures What is the measure of the other angle?

98. One angle of a right triangle measures What is the measure of the other angle?

99. One angle of a right triangle measures What is the measure of the other angle?

100. One angle of a right triangle measures What is the measure of the other angle?

101. The two smaller angles of a right triangle have equal measures. Find the measures of all three angles.

102. The measure of the smallest angle of a right triangle is less than the measure of the other small angle. Find the measures of all three angles.

103. The angles in a triangle are such that the measure of one angle is twice the measure of the smallest angle, while the measure of the third angle is three times the measure of the smallest angle. Find the measures of all three angles.

104. The angles in a triangle are such that the measure of one angle is more than the measure of the smallest angle, while the measure of the third angle is three times the measure of the smallest angle. Find the measures of all three angles.

Find the Length of the Missing Side

In the following exercises, is similar to Find the length of the indicated side.

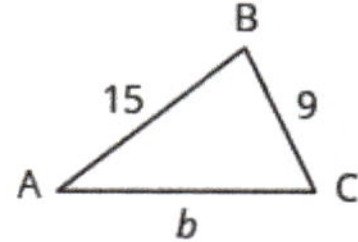

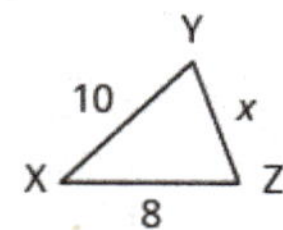

105. side

106. side

On a map, San Francisco, Las Vegas, and Los Angeles form a triangle whose sides are shown in the figure below. The actual distance from Los Angeles to Las Vegas is miles.

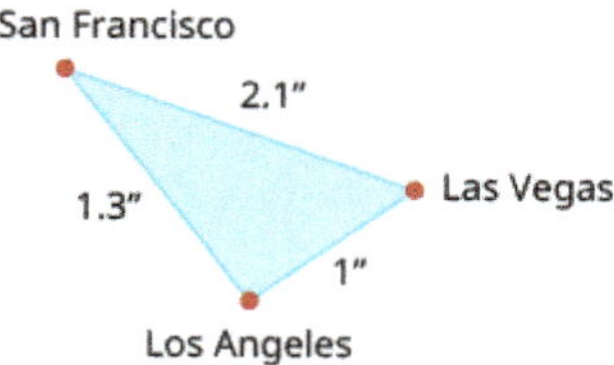

107. Find the distance from Los Angeles to San Francisco.

108. Find the distance from San Francisco to Las Vegas.

Use the Pythagorean Theorem

In the following exercises, use the Pythagorean Theorem to find the length of the hypotenuse.

109.

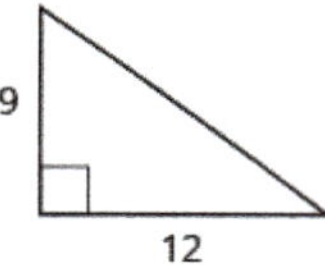

110.

111.

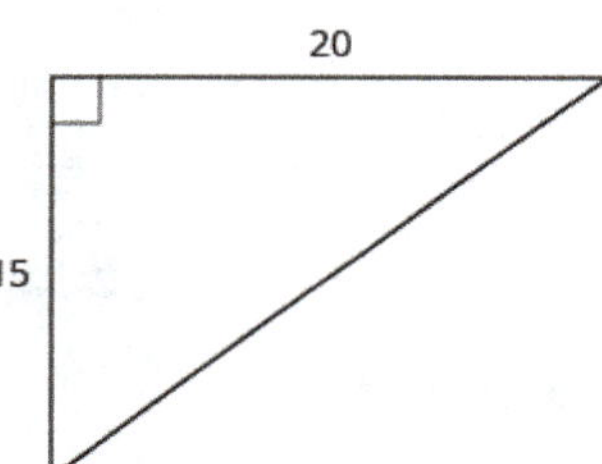

112.

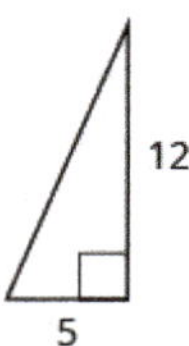

Find the Length of the Missing Side

In the following exercises, use the Pythagorean Theorem to find the length of the missing side. Round to the nearest tenth, if necessary.

113.

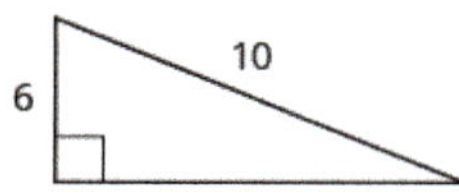

114.

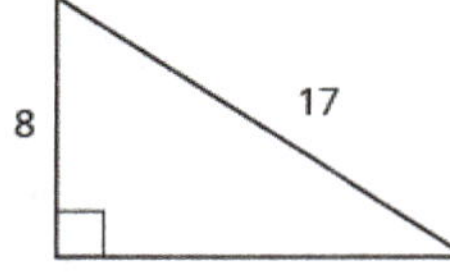

115.

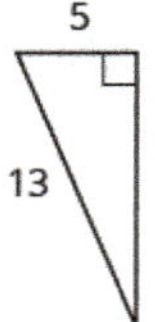

116.

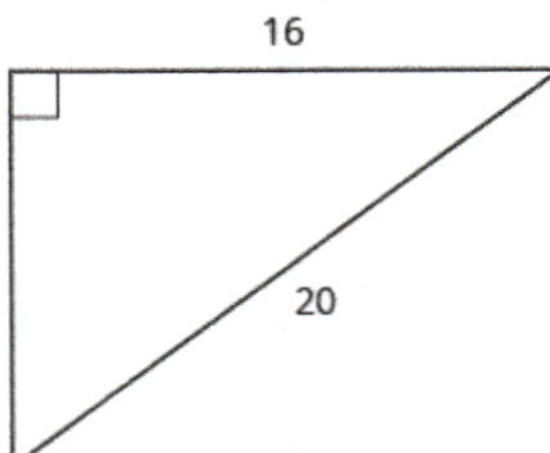

117.

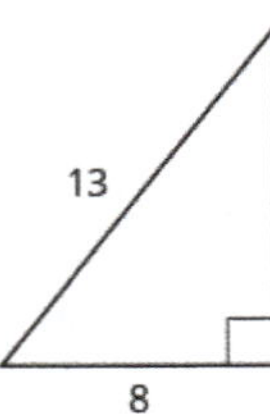

118.

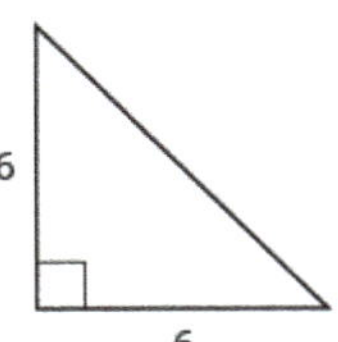

119.

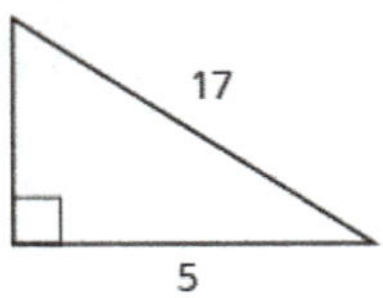

120.

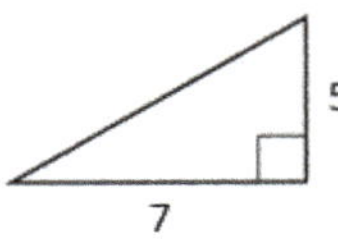

In the following exercises, solve. Approximate to the nearest tenth, if necessary.

121. A string of lights will be attached to the top of a pole for a holiday display. How far from the base of the pole should the end of the string of lights be anchored?

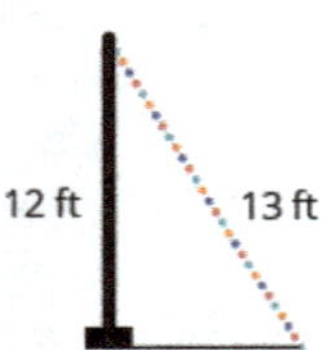

122. Pam wants to put a banner across her garage door to congratulate her son on his college graduation. The garage door is feet high and feet wide. How long should the banner be to fit the garage door?

123. Chi is planning to put a path of paving stones through her flower garden. The flower garden is a square with sides of feet. What will the length of the path be?

124. Brian borrowed a extension ladder to paint his house. If he sets the base of the ladder feet from the house, how far up will the top of the ladder reach?

Everyday Math

125. Building a scale model Joe wants to build a doll house for his daughter. He wants the doll house to look just like his house. His house is feet wide and feet tall at the highest point of the roof. If the dollhouse will be feet wide, how tall will its highest point be?

126. Measurement A city engineer plans to build a footbridge across a lake from point to point as shown in the picture below. To find the length of the footbridge, she draws a right triangle with right angle at She measures the distance from to feet, and from to feet. How long will the bridge be?

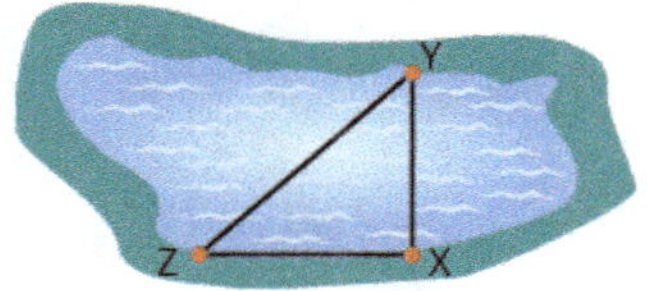

Writing Exercises

127. Write three of the properties of triangles from this section and then explain each in your own words.

128. Explain how the figure below illustrates the Pythagorean Theorem for a triangle with legs of length and

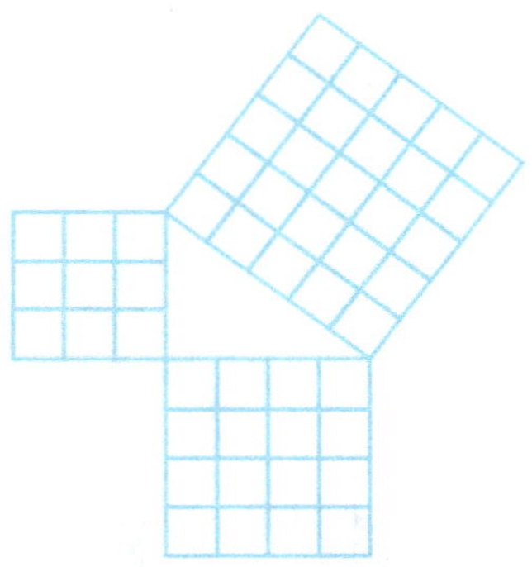

Self Check

After completing the exercises, use this checklist to evaluate your mastery of the objectives of this section.

I can...	Confidently	With some help	No-I don't get it!
use the properties of angles.			
use the properties of triangles.			
use the Pythagorean Theorem.			

What does this checklist tell you about your mastery of this section? What steps will you take to improve?

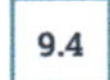 9.4 Use Properties of Rectangles, Triangles, and Trapezoids

Learning Objectives

By the end of this section, you will be able to:

- Understand linear, square, and cubic measure
- Use properties of rectangles
- Use properties of triangles
- Use properties of trapezoids

BE PREPARED : : 9.10

Before you get started, take this readiness quiz.

The length of a rectangle is less than the width. Let represent the width. Write an expression for the length of the rectangle.
If you missed this problem, review Example 2.26.

 **BE PREPARED : :** 9.11

Simplify: —

If you missed this problem, review Example 7.7.

 **BE PREPARED : :** 9.12

Simplify: —

If you missed this problem, review Example 5.36.

In this section, we'll continue working with geometry applications. We will add some more properties of triangles, and we'll learn about the properties of rectangles and trapezoids.

Understand Linear, Square, and Cubic Measure

When you measure your height or the length of a garden hose, you use a ruler or tape measure (Figure 9.13). A tape measure might remind you of a line—you use it for **linear measure**, which measures length. Inch, foot, yard, mile, centimeter and meter are units of linear measure.

Figure 9.13 This tape measure measures inches along the top and centimeters along the bottom.

When you want to know how much tile is needed to cover a floor, or the size of a wall to be painted, you need to know the **area**, a measure of the region needed to cover a surface. Area is measured is **square units**. We often use square inches, square feet, square centimeters, or square miles to measure area. A square centimeter is a square that is one centimeter (cm) on each side. A square inch is a square that is one inch on each side (Figure 9.14).

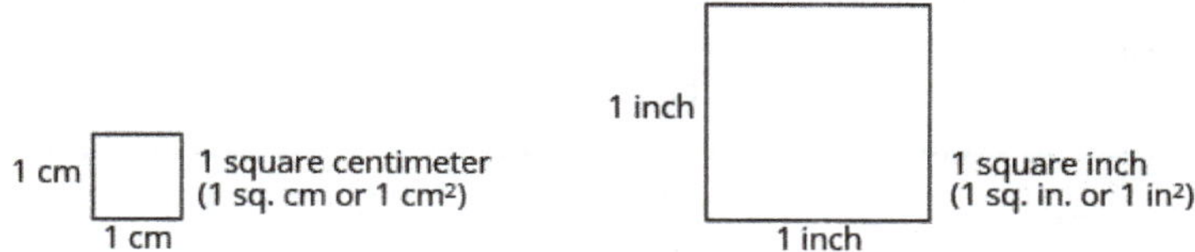

Figure 9.14 Square measures have sides that are each unit in length.

Figure 9.15 shows a rectangular rug that is feet long by feet wide. Each square is foot wide by foot long, or square foot. The rug is made of squares. The area of the rug is square feet.

Figure 9.15 The rug contains six squares of 1 square foot each, so the total area of the rug is 6 square feet.

When you measure how much it takes to fill a container, such as the amount of gasoline that can fit in a tank, or the amount of medicine in a syringe, you are measuring **volume**. Volume is measured in **cubic units** such as cubic inches or cubic centimeters. When measuring the volume of a rectangular solid, you measure how many cubes fill the container. We often use cubic centimeters, cubic inches, and cubic feet. A cubic centimeter is a cube that measures one centimeter on each side, while a cubic inch is a cube that measures one inch on each side (Figure 9.16).

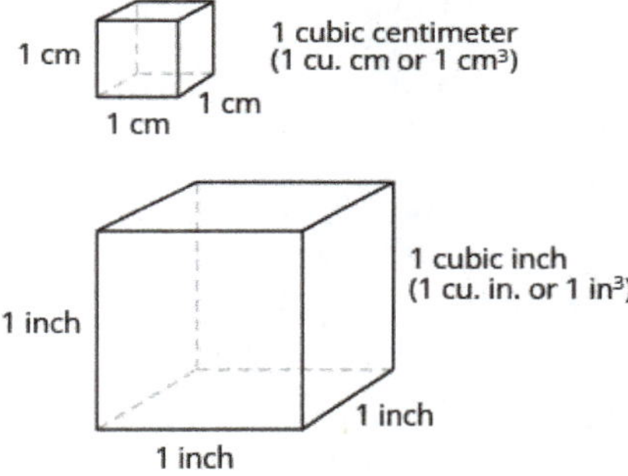

Figure 9.16 Cubic measures have sides that are 1 unit in length.

Suppose the cube in Figure 9.17 measures inches on each side and is cut on the lines shown. How many little cubes does it contain? If we were to take the big cube apart, we would find little cubes, with each one measuring one inch on all sides. So each little cube has a volume of cubic inch, and the volume of the big cube is cubic inches.

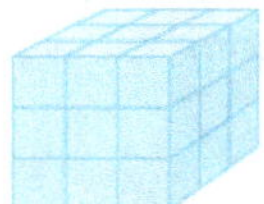

Figure 9.17 A cube that measures 3 inches on each side is made up of 27 one-inch cubes, or 27 cubic inches.

MANIPULATIVE MATHEMATICS

Doing the Manipulative Mathematics activity Visualizing Area and Perimeter will help you develop a better understanding of the difference between the area of a figure and its perimeter.

EXAMPLE 9.25

For each item, state whether you would use linear, square, or cubic measure:

amount of carpeting needed in a room extension cord length amount of sand in a sandbox

length of a curtain rod amount of flour in a canister size of the roof of a doghouse.

Solution

You are measuring how much surface the carpet covers, which is the area.	square measure
You are measuring how long the extension cord is, which is the length.	linear measure
You are measuring the volume of the sand.	cubic measure
You are measuring the length of the curtain rod.	linear measure
You are measuring the volume of the flour.	cubic measure
You are measuring the area of the roof.	square measure

TRY IT : : 9.49

Determine whether you would use linear, square, or cubic measure for each item.

amount of paint in a can height of a tree floor of your bedroom diameter of bike wheel size of a piece of sod amount of water in a swimming pool

TRY IT : : 9.50

Determine whether you would use linear, square, or cubic measure for each item.

volume of a packing box size of patio amount of medicine in a syringe length of a piece of yarn size of housing lot height of a flagpole

Many geometry applications will involve finding the perimeter or the area of a figure. There are also many applications of perimeter and area in everyday life, so it is important to make sure you understand what they each mean.

Picture a room that needs new floor tiles. The tiles come in squares that are a foot on each side—one square foot. How many of those squares are needed to cover the floor? This is the area of the floor.

Next, think about putting new baseboard around the room, once the tiles have been laid. To figure out how many strips are needed, you must know the distance around the room. You would use a tape measure to measure the number of feet around the room. This distance is the perimeter.

Perimeter and Area

The **perimeter** is a measure of the distance around a figure.

The **area** is a measure of the surface covered by a figure.

Figure 9.18 shows a square tile that is inch on each side. If an ant walked around the edge of the tile, it would walk inches. This distance is the perimeter of the tile.

Since the tile is a square that is inch on each side, its area is one square inch. The area of a shape is measured by determining how many square units cover the shape.

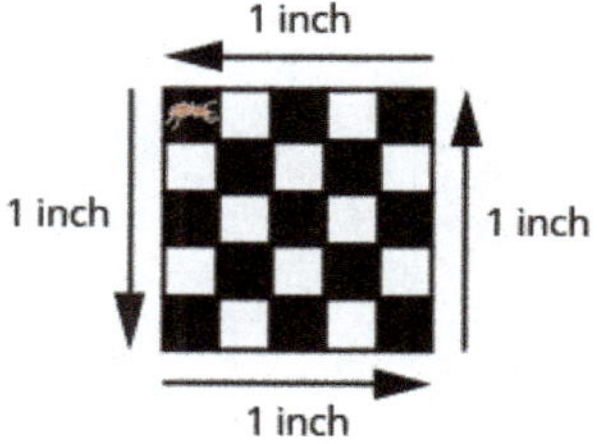

Figure 9.18

When the ant walks completely around the tile on its edge, it is tracing the perimeter of the tile. The area of the tile is 1 square inch.

MANIPULATIVE MATHEMATICS

Doing the Manipulative Mathematics activity Measuring Area and Perimeter will help you develop a better understanding of how to measure the area and perimeter of a figure.

EXAMPLE 9.26

Each of two square tiles is square inch. Two tiles are shown together.

What is the perimeter of the figure? What is the area?

Solution

The perimeter is the distance around the figure. The perimeter is inches.

The area is the surface covered by the figure. There are square inch tiles so the area is square inches.

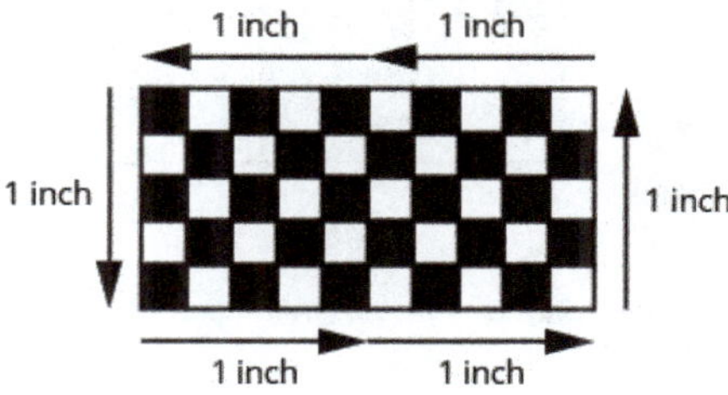

TRY IT : : 9.51 Each box in the figure below is 1 square inch. Find the perimeter and area of the figure:

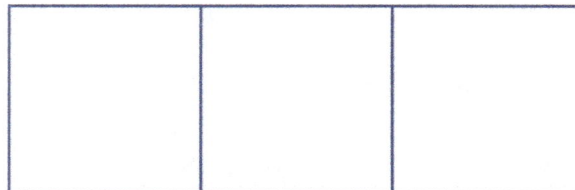

> **TRY IT : :** 9.52 Each box in the figure below is 1 square inch. Find the perimeter and area of the figure:

Use the Properties of Rectangles

A **rectangle** has four sides and four right angles. The opposite sides of a rectangle are the same length. We refer to one side of the rectangle as the length, and the adjacent side as the width, See Figure 9.19.

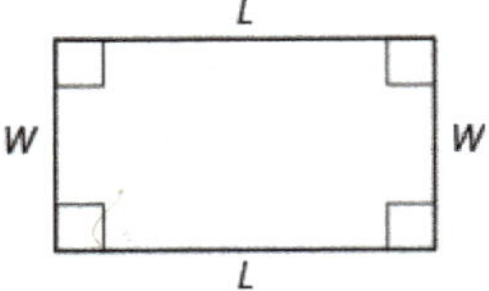

Figure 9.19 A rectangle has four sides, and four right angles. The sides are labeled L for length and W for width.

The perimeter, of the rectangle is the distance around the rectangle. If you started at one corner and walked around the rectangle, you would walk units, or two lengths and two widths. The perimeter then is

What about the area of a rectangle? Remember the rectangular rug from the beginning of this section. It was feet long by feet wide, and its area was square feet. See Figure 9.20. Since we see that the area, is the length, times the width, so the area of a rectangle is

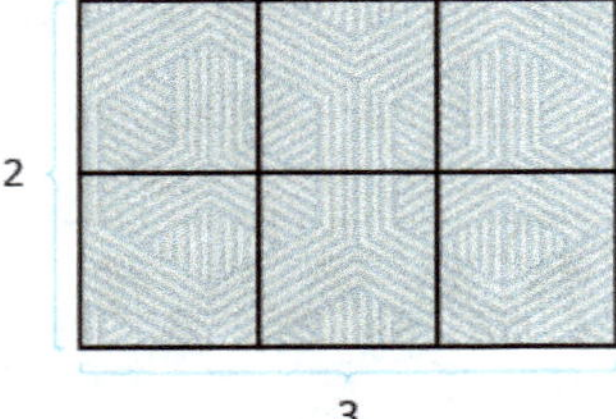

Figure 9.20 The area of this rectangular rug is square feet, its length times its width.

Properties of Rectangles

- Rectangles have four sides and four right angles.
- The lengths of opposite sides are equal.
- The perimeter, of a rectangle is the sum of twice the length and twice the width. See Figure 9.19.
- The area, of a rectangle is the length times the width.

For easy reference as we work the examples in this section, we will restate the Problem Solving Strategy for Geometry Applications here.

HOW TO : : USE A PROBLEM SOLVING STRATEGY FOR GEOMETRY APPLICATIONS

Step 1. **Read** the problem and make sure you understand all the words and ideas. Draw the figure and label it with the given information.

Step 2. **Identify** what you are looking for.

Step 3. **Name** what you are looking for. Choose a variable to represent that quantity.

Step 4. **Translate** into an equation by writing the appropriate formula or model for the situation. Substitute in the given information.

Step 5. **Solve** the equation using good algebra techniques.

Step 6. **Check** the answer in the problem and make sure it makes sense.

Step 7. **Answer** the question with a complete sentence.

EXAMPLE 9.27

The length of a rectangle is meters and the width is meters. Find the perimeter, and the area.

Solution

Step 1. **Read** the problem. Draw the figure and label it with the given information.	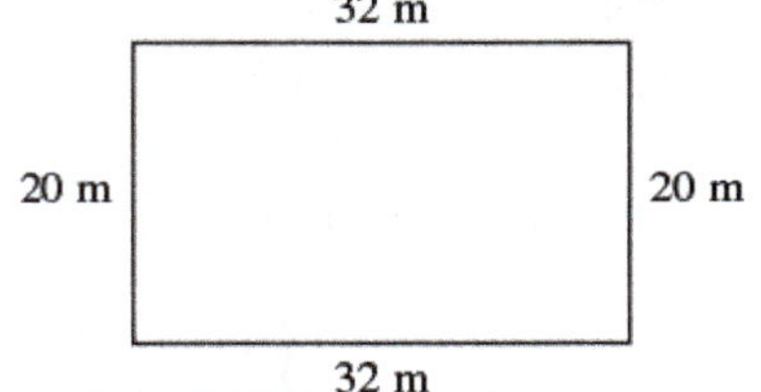
Step 2. **Identify** what you are looking for.	the perimeter of a rectangle
Step 3. **Name.** Choose a variable to represent it.	Let P = the perimeter
Step 4. **Translate.** Write the appropriate formula. Substitute.	$\underbrace{P}\ \underbrace{=}\ \underbrace{2L}\ \underbrace{+}\ \underbrace{2W}$ $P = 2(32) + 2(20)$
Step 5. **Solve** the equation.	$P = 64 + 40$ $P = 104$
Step 6. **Check:** $P \stackrel{?}{=} 104$ $20 + 32 + 20 + 32 \stackrel{?}{=} 104$ $104 = 104$ ✓	
Step 7. **Answer** the question.	The perimeter of the rectangle is 104 meters.

Step 1. **Read** the problem. Draw the figure and label it with the given information.	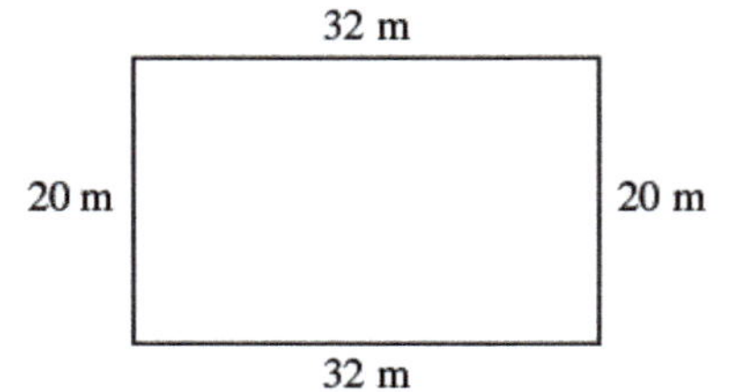
Step 2. **Identify** what you are looking for.	the area of a rectangle
Step 3. **Name.** Choose a variable to represent it.	Let A = the area
Step 4. **Translate.** Write the appropriate formula. Substitute.	$\underbrace{A}_{A} \underbrace{=}_{=} \underbrace{L}_{32\text{ m}} \underbrace{\cdot}_{\cdot} \underbrace{W}_{20\text{ m}}$
Step 5. **Solve** the equation.	$A = 640$
Step 6. **Check:** $A \stackrel{?}{=} 640$ $32 \cdot 20 \stackrel{?}{=} 640$ $640 = 640 ✓$	
Step 7. **Answer** the question.	The area of the rectangle is 60 square meters.

TRY IT : : 9.53

The length of a rectangle is yards and the width is yards. Find the perimeter and the area.

TRY IT : : 9.54

The length of a rectangle is feet and the width is feet. Find the perimeter and the area.

EXAMPLE 9.28

Find the length of a rectangle with perimeter inches and width inches.

Solution

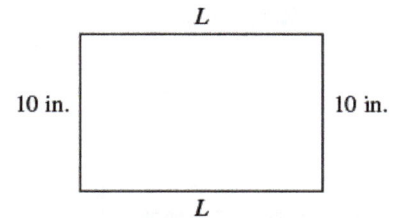

Step 1. **Read** the problem. Draw the figure and label it with the given information.	
Step 2. **Identify** what you are looking for.	the length of the rectangle
Step 3. **Name.** Choose a variable to represent it.	Let L = the length
Step 4. **Translate.** Write the appropriate formula. Substitute.	$\underbrace{P}_{50} \underbrace{=}_{=} \underbrace{2L}_{2L} \underbrace{+}_{+} \underbrace{2W}_{2(10)}$
Step 5. **Solve** the equation.	$50 - 20 = 2L + 20 - 20$ $30 = 2L$ $\frac{30}{2} = \frac{2L}{2}$ $15 = L$
Step 6. **Check:** $P = 50$ $15 + 10 + 15 + 10 \stackrel{?}{=} 50$ $50 = 50 ✓$	
Step 7. **Answer** the question.	The length is 15 inches.

TRY IT : : 9.55 Find the length of a rectangle with a perimeter of inches and width of inches.

TRY IT : : 9.56 Find the length of a rectangle with a perimeter of yards and width of yards.

In the next example, the width is defined in terms of the length. We'll wait to draw the figure until we write an expression for the width so that we can label one side with that expression.

EXAMPLE 9.29

The width of a rectangle is two inches less than the length. The perimeter is inches. Find the length and width.

Solution

Step 1. **Read** the problem.	
Step 2. **Identify** what you are looking for.	the length and width of the rectangle
Step 3. **Name.** Choose a variable to represent it. Now we can draw a figure using these expressions for the length and width.	Since the width is defined in terms of the length, we let L = length. The width is two feet less that the length, so we let $L - 2$ = width 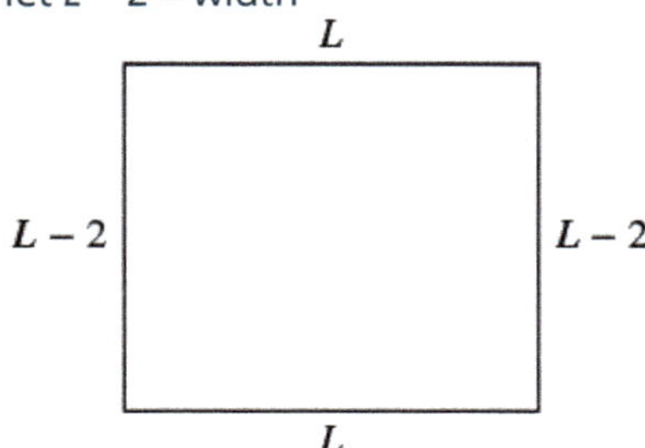
Step 4. **Translate.** Write the appropriate formula. The formula for the perimeter of a rectangle relates all the information. Substitute in the given information.	$\underbrace{P}\ \underbrace{=}\ \underbrace{2L}\ \underbrace{+}\ \underbrace{2W}$ $52 = 2L + 2(L - 2)$
Step 5. **Solve** the equation.	
Combine like terms.	
Add 4 to each side.	
Divide by 4.	
	The length is 14 inches.
Now we need to find the width.	
The width is $L - 2$.	$L - 2$ $14 - 2$ 12 The width is 12 inches.
Step 6. **Check:** Since , this works!	
Step 7. **Answer** the question.	The length is 14 feet and the width is 12 feet.

> **TRY IT : :** 9.57
>
> The width of a rectangle is seven meters less than the length. The perimeter is meters. Find the length and width.

> **TRY IT : :** 9.58
>
> The length of a rectangle is eight feet more than the width. The perimeter is feet. Find the length and width.

EXAMPLE 9.30

The length of a rectangle is four centimeters more than twice the width. The perimeter is centimeters. Find the length and width.

Solution

Step 1. **Read** the problem.	
Step 2. **Identify** what you are looking for.	the length and width
Step 3. **Name.** Choose a variable to represent it.	let W = width The length is four more than twice the width. $2w + 4$ = length 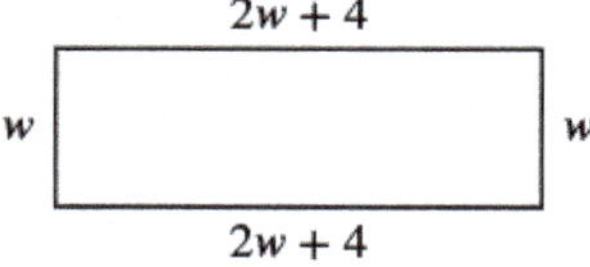
Step 4. **Translate.** Write the appropriate formula and substitute in the given information.	$\underbrace{P}_{32} \; \underbrace{=}_{=} \; \underbrace{2L}_{2(2w+4)} \; \underbrace{+}_{+} \; \underbrace{2W}_{2w}$
Step 5. **Solve** the equation.	$32 = 4w + 8 + 2w$ $32 = 6w + 8$ $24 = 6w$ $4 = w$ width $2w + 4$ length $2(4) + 4$ 12 The length is 12 cm.
Step 6. **Check:** $p = 2L + 2W$ $32 \stackrel{?}{=} 2 \cdot 12 + 2 \cdot 4$ $32 = 32$	
Step 7. **Answer** the question.	The length is 12 cm and the width is 4 cm.

TRY IT : : 9.59

The length of a rectangle is eight more than twice the width. The perimeter is feet. Find the length and width.

TRY IT : : 9.60

The width of a rectangle is six less than twice the length. The perimeter is centimeters. Find the length and width.

EXAMPLE 9.31

The area of a rectangular room is square feet. The length is feet. What is the width?

Solution

Step 1. **Read** the problem.	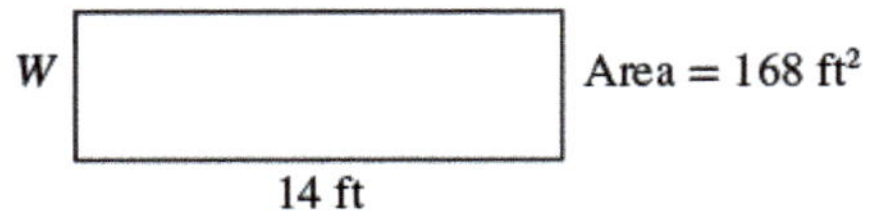
Step 2. **Identify** what you are looking for.	the width of a rectangular room
Step 3. **Name.** Choose a variable to represent it.	Let W = width
Step 4. **Translate.** Write the appropriate formula and substitute in the given information.	$A = LW$ $168 = 14W$
Step 5. **Solve** the equation.	$\frac{168}{14} = \frac{14W}{14}$ $12 = W$
Step 6. **Check:** $A = LW$ $168 \stackrel{?}{=} 14 \cdot 12$ $168 = 168$ ✓	
Step 7. **Answer** the question.	The width of the room is 12 feet.

TRY IT : : 9.61 The area of a rectangle is square feet. The length is feet. What is the width?

TRY IT : : 9.62 The width of a rectangle is meters. The area is square meters. What is the length?

EXAMPLE 9.32

The perimeter of a rectangular swimming pool is feet. The length is feet more than the width. Find the length and width.

Solution

Step 1. **Read** the problem. Draw the figure and label it with the given information.	
Step 2. **Identify** what you are looking for.	the length and width of the pool
Step 3. **Name.** Choose a variable to represent it. The length is 15 feet more than the width.	Let
Step 4.**Translate.** Write the appropriate formula and substitute.	$\underbrace{P}\ \underbrace{=}\ \underbrace{2L}\ \underbrace{+}\ \underbrace{2W}$ $150 = 2(w + 15) + 2w$
Step 5. **Solve** the equation.	$150 = 2w + 30 + 2w$ $150 = 4w + 30$ $120 = 4w$ $30 = w$ the width of the pool $w + 15$ the length of the pool $30 + 15$ 45
Step 6. **Check:** $p = 2L + 2W$ $150 \stackrel{?}{=} 2(45) + 2(30)$ $150 = 150$	
Step 7. **Answer** the question.	The length of the pool is 45 feet and the width is 30 feet.

TRY IT : : 9.63

The perimeter of a rectangular swimming pool is feet. The length is feet more than the width. Find the length and width.

TRY IT : : 9.64

The length of a rectangular garden is yards more than the width. The perimeter is yards. Find the length and width.

Use the Properties of Triangles

We now know how to find the area of a rectangle. We can use this fact to help us visualize the formula for the area of a triangle. In the rectangle in Figure 9.20, we've labeled the length and the width so it's area is

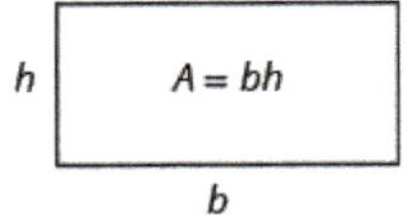

Figure 9.21 The area of a rectangle is the base, times the height,

We can divide this rectangle into two **congruent** triangles (Figure 9.22). Triangles that are congruent have identical side lengths and angles, and so their areas are equal. The area of each triangle is one-half the area of the rectangle, or —

This example helps us see why the formula for the area of a triangle is —

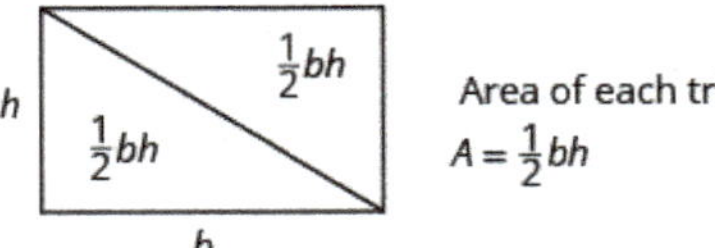

Figure 9.22 A rectangle can be divided into two triangles of equal area. The area of each triangle is one-half the area of the rectangle.

The formula for the area of a triangle is — where is the base and is the height.

To find the area of the triangle, you need to know its base and height. The base is the length of one side of the triangle, usually the side at the bottom. The height is the length of the line that connects the base to the opposite vertex, and makes a angle with the base. Figure 9.23 shows three triangles with the base and height of each marked.

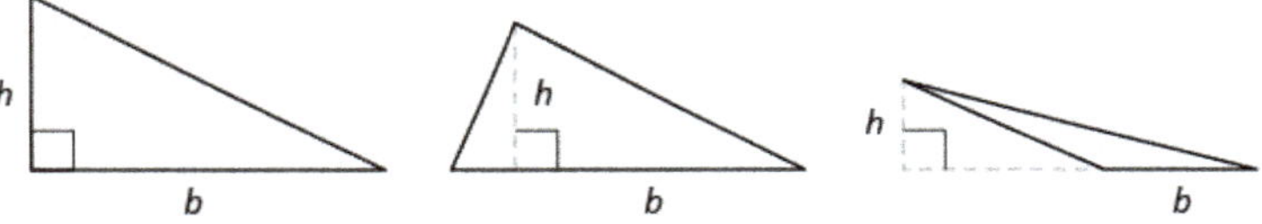

Figure 9.23 The height of a triangle is the length of a line segment that connects the the base to the opposite vertex and makes a angle with the base.

Triangle Properties

For any triangle the sum of the measures of the angles is

The perimeter of a triangle is the sum of the lengths of the sides.

The area of a triangle is one-half the base, times the height,

—

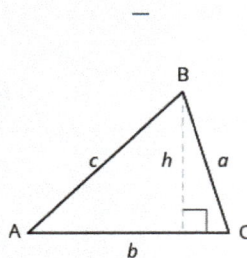

EXAMPLE 9.33

Find the area of a triangle whose base is inches and whose height is inches.

Solution

Step 1. **Read** the problem. Draw the figure and label it with the given information.	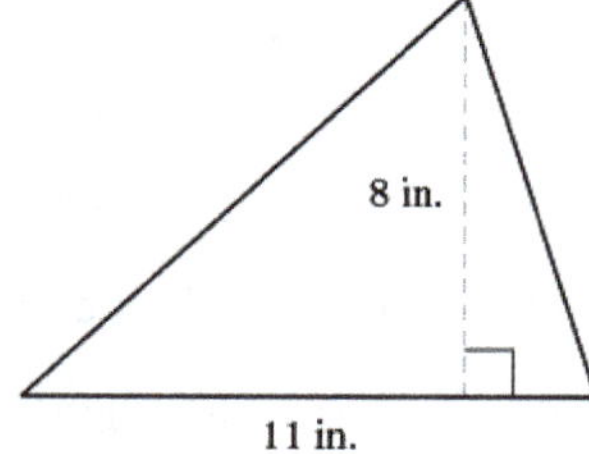
Step 2. **Identify** what you are looking for.	the area of the triangle
Step 3. **Name.** Choose a variable to represent it.	let A = area of the triangle
Step 4.**Translate.** Write the appropriate formula. Substitute.	$\underbrace{A}\ \underbrace{=}\ \underbrace{\frac{1}{2}}\ \underbrace{\cdot}\ \underbrace{b}\ \underbrace{\cdot}\ \underbrace{h}$ $A = \frac{1}{2} \cdot 11 \cdot 8$
Step 5. **Solve** the equation.	$A = 44$ square inches
Step 6. **Check:** $A = \frac{1}{2}bh$ $44 \stackrel{?}{=} \frac{1}{2}(11)8$ $44 = 44 \checkmark$	
Step 7. **Answer** the question.	The area is 44 square inches.

TRY IT : : 9.65 Find the area of a triangle with base inches and height inches.

TRY IT : : 9.66 Find the area of a triangle with base inches and height inches.

EXAMPLE 9.34

The perimeter of a triangular garden is feet. The lengths of two sides are feet and feet. How long is the third side?

Solution

Step 1. **Read** the problem. Draw the figure and label it with the given information.	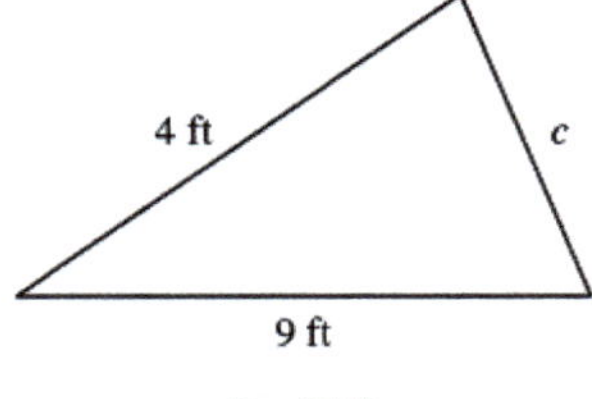 $P = 24$ ft
Step 2. **Identify** what you are looking for.	length of the third side of a triangle
Step 3. **Name.** Choose a variable to represent it.	Let c = the third side
Step 4. **Translate.** Write the appropriate formula. Substitute in the given information.	$\underbrace{P}_{24} = \underbrace{a}_{4} + \underbrace{b}_{9} + \underbrace{c}_{c}$
Step 5. **Solve** the equation.	$24 = 13 + c$ $11 = c$
Step 6. **Check:** $P = a + b + c$ $24 \stackrel{?}{=} 4 + 9 + 11$ $24 = 24 \checkmark$	
Step 7. **Answer** the question.	The third side is 11 feet long.

 **TRY IT : :** 9.67

The perimeter of a triangular garden is feet. The lengths of two sides are feet and feet. How long is the third side?

 **TRY IT : :** 9.68

The lengths of two sides of a triangular window are feet and feet. The perimeter is feet. How long is the third side?

EXAMPLE 9.35

The area of a triangular church window is square meters. The base of the window is meters. What is the window's height?

Solution

Step 1. **Read** the problem. Draw the figure and label it with the given information.	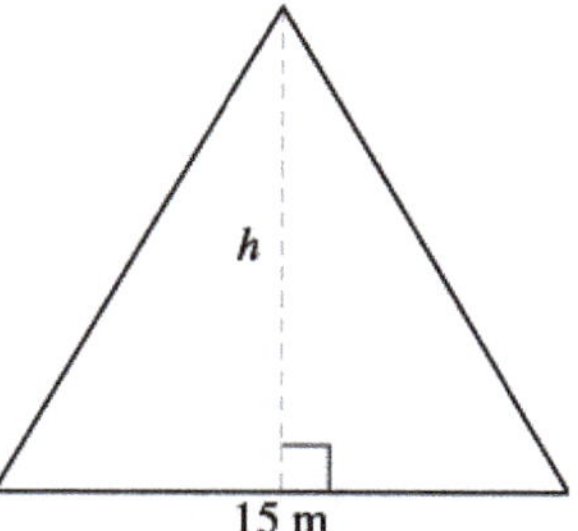
Step 2. **Identify** what you are looking for.	height of a triangle
Step 3. **Name.** Choose a variable to represent it.	Let h = the height
Step 4. **Translate.** Write the appropriate formula. Substitute in the given information.	$A = \frac{1}{2} \cdot b \cdot h$ $90 = \frac{1}{2} \cdot 15 \cdot h$
Step 5. **Solve** the equation.	$90 = \frac{15}{2}h$ $12 = h$
Step 6. **Check:** $A = \frac{1}{2}bh$ $90 \stackrel{?}{=} \frac{1}{2} \cdot 15 \cdot 12$ $90 = 90 \checkmark$	
Step 7. **Answer** the question.	The height of the triangle is 12 meters.

TRY IT : : 9.69

The area of a triangular painting is square inches. The base is inches. What is the height?

TRY IT : : 9.70 A triangular tent door has an area of square feet. The height is feet. What is the base?

Isosceles and Equilateral Triangles

Besides the right triangle, some other triangles have special names. A triangle with two sides of equal length is called an **isosceles triangle**. A triangle that has three sides of equal length is called an **equilateral triangle**. Figure 9.24 shows both types of triangles.

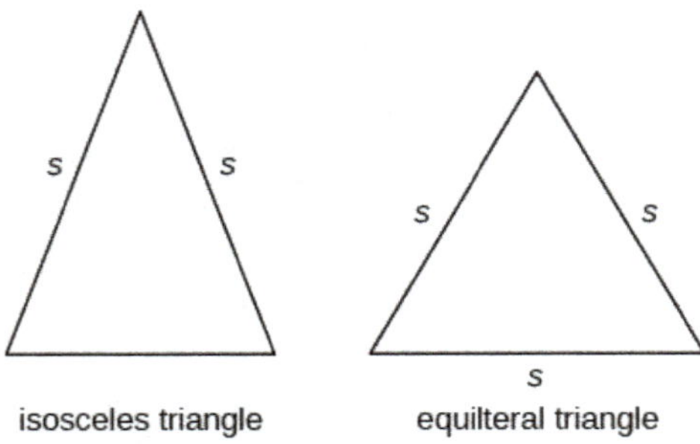

Figure 9.24 In an isosceles triangle, two sides have the same length, and the third side is the base. In an equilateral triangle, all three sides have the same length.

Isosceles and Equilateral Triangles

An **isosceles** triangle has two sides the same length.

An **equilateral** triangle has three sides of equal length.

EXAMPLE 9.36

The perimeter of an equilateral triangle is inches. Find the length of each side.

Solution

Step 1. **Read** the problem. Draw the figure and label it with the given information.	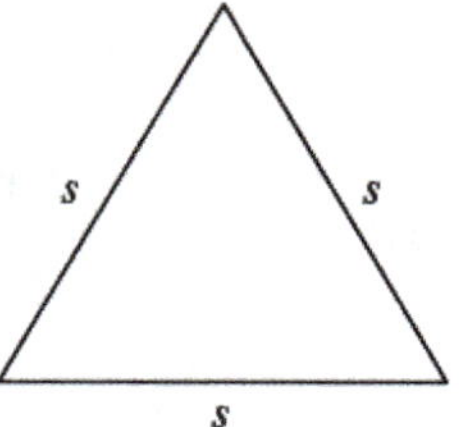 Perimeter = 93 in.
Step 2. **Identify** what you are looking for.	length of the sides of an equilateral triangle
Step 3. **Name.** Choose a variable to represent it.	Let s = length of each side
Step 4. **Translate.** Write the appropriate formula. Substitute.	$\underbrace{P}_{93} \; \underbrace{=}_{=} \; \underbrace{a}_{s} \; \underbrace{+}_{+} \; \underbrace{b}_{s} \; \underbrace{+}_{+} \; \underbrace{c}_{s}$
Step 5. **Solve** the equation.	$93 = 3s$ $31 = s$
Step 6. **Check:** 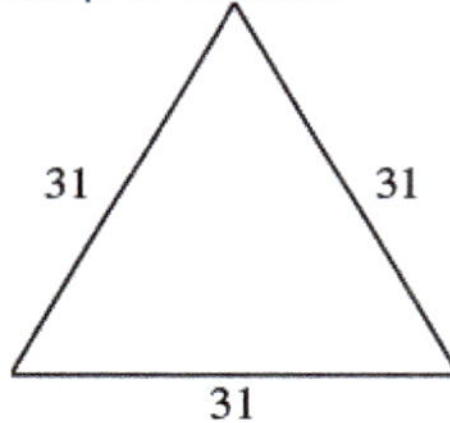 $93 \stackrel{?}{=} 31 + 31 + 31$ $93 = 93 ✓$	
Step 7. **Answer** the question.	Each side is 31 inches.

TRY IT : : 9.71 Find the length of each side of an equilateral triangle with perimeter inches.

TRY IT : : 9.72 Find the length of each side of an equilateral triangle with perimeter centimeters.

EXAMPLE 9.37

Arianna has inches of beading to use as trim around a scarf. The scarf will be an isosceles triangle with a base of inches. How long can she make the two equal sides?

Solution

Step 1. **Read** the problem. Draw the figure and label it with the given information.	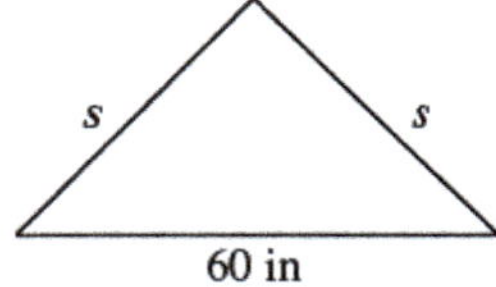 $P = 156$ in.
Step 2. **Identify** what you are looking for.	the lengths of the two equal sides
Step 3. **Name.** Choose a variable to represent it.	Let s = the length of each side
Step 4.**Translate.** Write the appropriate formula. Substitute in the given information.	$\underbrace{P}_{156} \underbrace{=}_{=} \underbrace{a}_{s} \underbrace{+}_{+} \underbrace{b}_{60} \underbrace{+}_{+} \underbrace{c}_{s}$
Step 5. **Solve** the equation.	$156 = 2s + 60$ $96 = 2s$ $48 = s$
Step 6. **Check:** $p = a + b + c$ $156 \stackrel{?}{=} 48 + 60 + 48$ $156 = 156\checkmark$	
Step 7. **Answer** the question.	Arianna can make each of the two equal sides 48 inches long.

TRY IT : : 9.73

A backyard deck is in the shape of an isosceles triangle with a base of feet. The perimeter of the deck is feet. How long is each of the equal sides of the deck?

TRY IT : : 9.74

A boat's sail is an isosceles triangle with base of meters. The perimeter is meters. How long is each of the equal sides of the sail?

Use the Properties of Trapezoids

A **trapezoid** is four-sided figure, a *quadrilateral*, with two sides that are parallel and two sides that are not. The parallel sides are called the bases. We call the length of the smaller base and the length of the bigger base The height, of a trapezoid is the distance between the two bases as shown in Figure 9.25.

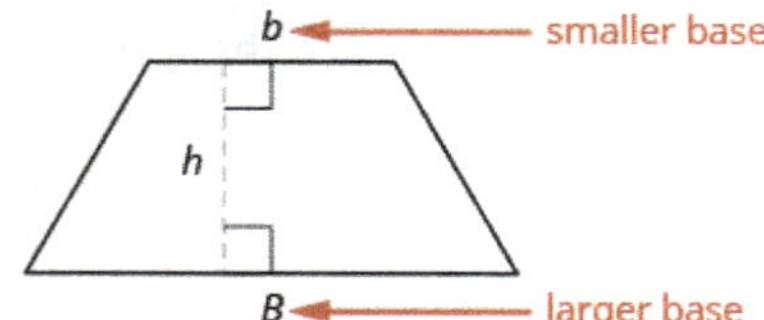

Figure 9.25 A trapezoid has a larger base, and a smaller base, The height is the distance between the bases.

The formula for the area of a trapezoid is:

—

Splitting the trapezoid into two triangles may help us understand the formula. The area of the trapezoid is the sum of the areas of the two triangles. See Figure 9.26.

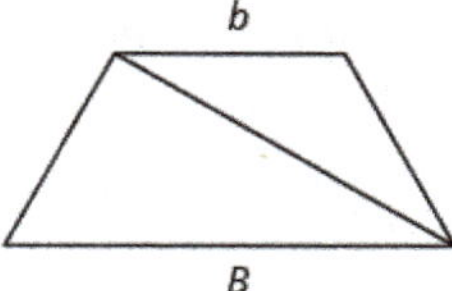

Figure 9.26 Splitting a trapezoid into two triangles may help you understand the formula for its area.

The height of the trapezoid is also the height of each of the two triangles. See Figure 9.27.

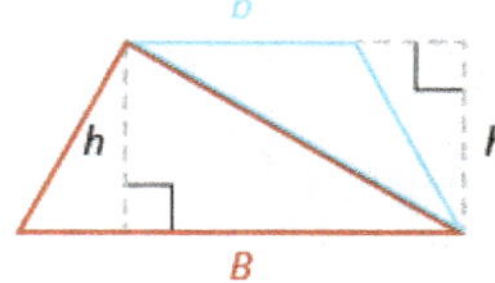

Figure 9.27

The formula for the area of a trapezoid is

$$\text{Area}_{\text{trapezoid}} = \frac{1}{2}h(b + B)$$

If we distribute, we get,

$$\text{Area}_{trapezoid} = \frac{1}{2}bh + \frac{1}{2}Bh$$

$$\text{Area}_{trapezoid} = A_{blue\Delta} + A_{red\Delta}$$

Properties of Trapezoids

- A trapezoid has four sides. See Figure 9.25.
- Two of its sides are parallel and two sides are not.
- The area, of a trapezoid is —) .

EXAMPLE 9.38

Find the area of a trapezoid whose height is 6 inches and whose bases are and inches.

Solution

Step 1. **Read** the problem. Draw the figure and label it with the given information.

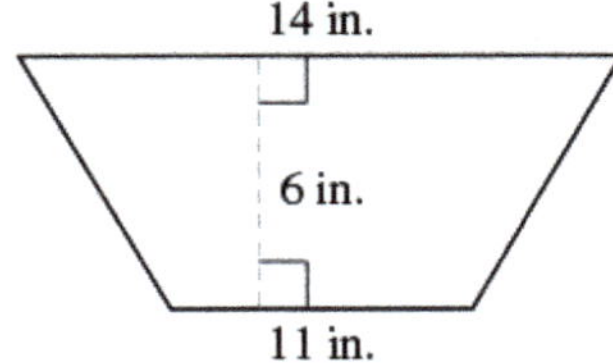

Step 2. **Identify** what you are looking for. the area of the trapezoid

Step 3. **Name.** Choose a variable to represent it. Let

Step 4. **Translate.**
Write the appropriate formula.
Substitute.

$$A = \frac{1}{2} \cdot h \cdot (b + B)$$

$$A = \frac{1}{2} \cdot 6 \cdot (11 + 14)$$

Step 5. **Solve** the equation.

$$A = \frac{1}{2} \cdot 6(25)$$

$$A = 3(25)$$

$$A = 75 \text{ square inches}$$

Step 6. **Check:** Is this answer reasonable?

If we draw a rectangle around the trapezoid that has the same big base and a height its area should be greater than that of the trapezoid.

If we draw a rectangle inside the trapezoid that has the same little base and a height its area should be smaller than that of the trapezoid.

14, 6, 11	14, 6, 11	6
$A_{rectangle} = bh$	$A_{trapezoid} = \frac{1}{2}h(b + B)$	$A_{rectangle} = bh$
$A_{rectangle} = 14 \cdot 6$	$A_{trapezoid} = \frac{1}{2} \cdot 6(11 + 14)$	$A_{rectangle} = 11 \cdot 6$
$A_{rectangle} = 84$ sq. in.	$A_{trapezoid} = 75$ sq. in.	$A_{rectangle} = 66$ sq. in.

The area of the larger rectangle is square inches and the area of the smaller rectangle is square inches. So it makes sense that the area of the trapezoid is between and square inches

Step 7. **Answer** the question. The area of the trapezoid is square inches.

TRY IT : : 9.75 The height of a trapezoid is yards and the bases are and yards. What is the area?

TRY IT : : 9.76

The height of a trapezoid is centimeters and the bases are and centimeters. What is the area?

EXAMPLE 9.39

Find the area of a trapezoid whose height is feet and whose bases are and feet.

Solution

Step 1. **Read** the problem. Draw the figure and label it with the given information.

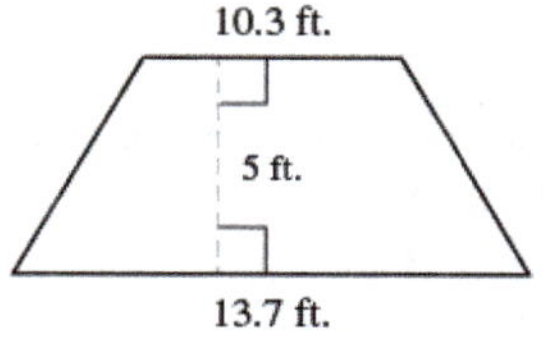

Step 2. **Identify** what you are looking for.	the area of the trapezoid
Step 3. **Name.** Choose a variable to represent it.	Let A = the area
Step 4. **Translate.** Write the appropriate formula. Substitute.	$\underbrace{A}\ \underbrace{=}\ \underbrace{\frac{1}{2}}\ \underbrace{\cdot}\ \underbrace{h}\ \underbrace{\cdot}\ \underbrace{(b+B)}$ $A = \frac{1}{2} \cdot 5 \cdot (10.3 + 13.7)$
Step 5. **Solve** the equation.	$A = \frac{1}{2} \cdot 5(24)$ $A = 12 \cdot 5$ $A = 60$ square feet

Step 6. **Check:** Is this answer reasonable?
The area of the trapezoid should be less than the area of a rectangle with base 13.7 and height 5, but more than the area of a rectangle with base 10.3 and height 5.

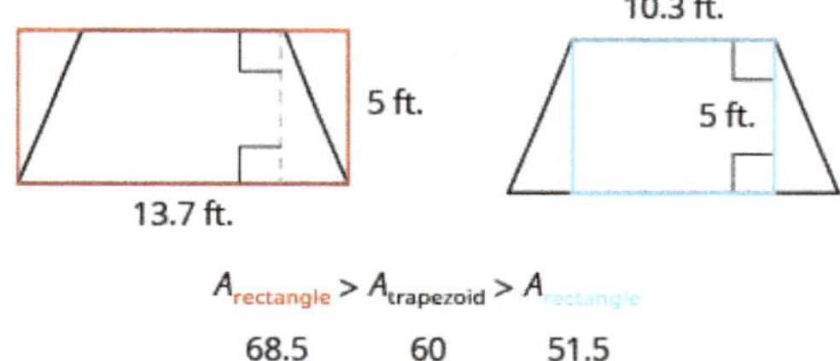

$A_{\text{rectangle}} > A_{\text{trapezoid}} > A_{\text{rectangle}}$

68.5 60 51.5

Step 7. **Answer** the question.	The area of the trapezoid is 60 square feet.

TRY IT : : 9.77

The height of a trapezoid is centimeters and the bases are and centimeters. What is the area?

TRY IT : : 9.78

The height of a trapezoid is meters and the bases are and meters. What is the area?

EXAMPLE 9.40

Vinny has a garden that is shaped like a trapezoid. The trapezoid has a height of yards and the bases are and yards. How many square yards will be available to plant?

Solution

Step 1. **Read** the problem. Draw the figure and label it with the given information.

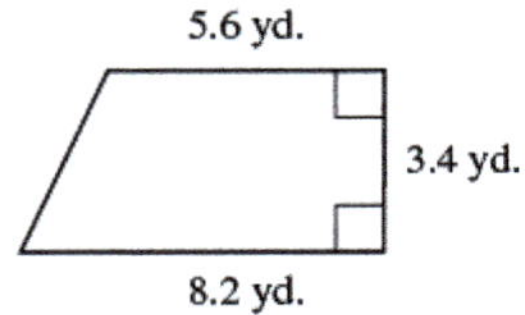

Step 2. **Identify** what you are looking for.	the area of a trapezoid
Step 3. **Name.** Choose a variable to represent it.	Let A = the area
Step 4. **Translate.** Write the appropriate formula. Substitute.	$A = \frac{1}{2} \cdot h \cdot (b + B)$ $A = \frac{1}{2} \cdot 3.4 \cdot (5.6 + 8.2)$
Step 5. **Solve** the equation.	$A = \frac{1}{2}(3.4)(13.8)$ $A = 23.46$ square yards

Step 6. **Check:** Is this answer reasonable?
Yes. The area of the trapezoid is less than the area of a rectangle with a base of 8.2 yd and height 3.4 yd, but more than the area of a rectangle with base 5.6 yd and height 3.4 yd.

$A_{rectangle} = Bh$ $= (8.2)(3.4)$ $= 27.88\text{ yd}^2$	$A_{trapezoid} = \frac{1}{2}(3.4\text{ yd})(5.6 = 8.2)$ $= 23.46\text{ yd}^2$	$A_{rectangle} = bh$ $= (5.6)(3.4)$ $= 19.04\text{ yd}^2$
$A_{rectangle} > A_{trapezoid} > A_{rectangle}$ 27.88 23.46 19.04		

Step 7. **Answer** the question.	Vinny has 23.46 square yards in which he can plant.

TRY IT : : 9.79

Lin wants to sod his lawn, which is shaped like a trapezoid. The bases are yards and yards, and the height is yards. How many square yards of sod does he need?

TRY IT : : 9.80

Kira wants cover his patio with concrete pavers. If the patio is shaped like a trapezoid whose bases are feet and feet and whose height is feet, how many square feet of pavers will he need?

LINKS TO LITERACY

The Links to Literacy activity *Spaghetti and Meatballs for All* will provide you with another view of the topics covered in this section."

MEDIA : : ACCESS ADDITIONAL ONLINE RESOURCES

- Perimeter of a Rectangle (http://www.openstax.org/l/24perirect)
- Area of a Rectangle (http://www.openstax.org/l/24arearect)
- Perimeter and Area Formulas (http://www.openstax.org/l/24periareaform)
- Area of a Triangle (http://www.openstax.org/l/24areatri)
- Area of a Triangle with Fractions (http://www.openstax.org/l/24areatrifract)
- Area of a Trapezoid (http://www.openstax.org/l/24areatrap)

9.4 EXERCISES

Practice Makes Perfect

Understand Linear, Square, and Cubic Measure

In the following exercises, determine whether you would measure each item using linear, square, or cubic units.

129. amount of water in a fish tank

130. length of dental floss

131. living area of an apartment

132. floor space of a bathroom tile

133. height of a doorway

134. capacity of a truck trailer

In the following exercises, find the perimeter and area of each figure. Assume each side of the square is cm.

135.

136.

137.

138.

139.

140.

Use the Properties of Rectangles

In the following exercises, find the perimeter and area of each rectangle.

141. The length of a rectangle is feet and the width is feet.

142. The length of a rectangle is inches and the width is inches.

143. A rectangular room is feet wide by feet long.

144. A driveway is in the shape of a rectangle feet wide by feet long.

In the following exercises, solve.

145. Find the length of a rectangle with perimeter inches and width inches.

146. Find the length of a rectangle with perimeter yards and width of yards.

147. Find the width of a rectangle with perimeter meters and length meters.

148. Find the width of a rectangle with perimeter meters and length meters.

149. The area of a rectangle is square meters. The length is meters. What is the width?

150. The area of a rectangle is square centimeters. The width is centimeters. What is the length?

151. The length of a rectangle is inches more than the width. The perimeter is inches. Find the length and the width.

152. The width of a rectangle is inches more than the length. The perimeter is inches. Find the length and the width.

153. The perimeter of a rectangle is meters. The width of the rectangle is meters less than the length. Find the length and the width of the rectangle.

154. The perimeter of a rectangle is feet. The width is feet less than the length. Find the length and the width.

155. The width of the rectangle is meters less than the length. The perimeter of a rectangle is meters. Find the dimensions of the rectangle.

156. The length of the rectangle is meters less than the width. The perimeter of a rectangle is meters. Find the dimensions of the rectangle.

157. The perimeter of a rectangle of feet. The length of the rectangle is twice the width. Find the length and width of the rectangle.

158. The length of a rectangle is three times the width. The perimeter is feet. Find the length and width of the rectangle.

159. The length of a rectangle is meters less than twice the width. The perimeter is meters. Find the length and width.

160. The length of a rectangle is inches more than twice the width. The perimeter is inches. Find the length and width.

161. The width of a rectangular window is inches. The area is square inches. What is the length?

162. The length of a rectangular poster is inches. The area is square inches. What is the width?

163. The area of a rectangular roof is square meters. The length is meters. What is the width?

164. The area of a rectangular tarp is square feet. The width is feet. What is the length?

165. The perimeter of a rectangular courtyard is feet. The length is feet more than the width. Find the length and the width.

166. The perimeter of a rectangular painting is centimeters. The length is centimeters more than the width. Find the length and the width.

167. The width of a rectangular window is inches less than the height. The perimeter of the doorway is inches. Find the length and the width.

168. The width of a rectangular playground is meters less than the length. The perimeter of the playground is meters. Find the length and the width.

Use the Properties of Triangles

In the following exercises, solve using the properties of triangles.

169. Find the area of a triangle with base inches and height inches.

170. Find the area of a triangle with base centimeters and height centimeters.

171. Find the area of a triangle with base meters and height meters.

172. Find the area of a triangle with base feet and height feet.

173. A triangular flag has base of foot and height of feet. What is its area?

174. A triangular window has base of feet and height of feet. What is its area?

175. If a triangle has sides of feet and feet and the perimeter is feet, how long is the third side?

176. If a triangle has sides of centimeters and centimeters and the perimeter is centimeters, how long is the third side?

177. What is the base of a triangle with an area of square inches and height of inches?

178. What is the height of a triangle with an area of square inches and base of inches?

179. The perimeter of a triangular reflecting pool is yards. The lengths of two sides are yards and yards. How long is the third side?

180. A triangular courtyard has perimeter of meters. The lengths of two sides are meters and meters. How long is the third side?

181. An isosceles triangle has a base of centimeters. If the perimeter is centimeters, find the length of each of the other sides.

182. An isosceles triangle has a base of inches. If the perimeter is inches, find the length of each of the other sides.

183. Find the length of each side of an equilateral triangle with a perimeter of yards.

184. Find the length of each side of an equilateral triangle with a perimeter of meters.

185. The perimeter of an equilateral triangle is meters. Find the length of each side.

186. The perimeter of an equilateral triangle is miles. Find the length of each side.

187. The perimeter of an isosceles triangle is feet. The length of the shortest side is feet. Find the length of the other two sides.

188. The perimeter of an isosceles triangle is inches. The length of the shortest side is inches. Find the length of the other two sides.

189. A dish is in the shape of an equilateral triangle. Each side is inches long. Find the perimeter.

190. A floor tile is in the shape of an equilateral triangle. Each side is feet long. Find the perimeter.

191. A road sign in the shape of an isosceles triangle has a base of inches. If the perimeter is inches, find the length of each of the other sides.

192. A scarf in the shape of an isosceles triangle has a base of meters. If the perimeter is meters, find the length of each of the other sides.

193. The perimeter of a triangle is feet. One side of the triangle is foot longer than the second side. The third side is feet longer than the second side. Find the length of each side.

194. The perimeter of a triangle is feet. One side of the triangle is feet longer than the second side. The third side is feet longer than the second side. Find the length of each side.

195. One side of a triangle is twice the smallest side. The third side is feet more than the shortest side. The perimeter is feet. Find the lengths of all three sides.

196. One side of a triangle is three times the smallest side. The third side is feet more than the shortest side. The perimeter is feet. Find the lengths of all three sides.

Use the Properties of Trapezoids

In the following exercises, solve using the properties of trapezoids.

197. The height of a trapezoid is feet and the bases are and feet. What is the area?

198. The height of a trapezoid is yards and the bases are and yards. What is the area?

199. Find the area of a trapezoid with a height of meters and bases of and meters.

200. Find the area of a trapezoid with a height of inches and bases of and inches.

201. The height of a trapezoid is centimeters and the bases are and centimeters. What is the area?

202. The height of a trapezoid is feet and the bases are and feet. What is the area?

203. Find the area of a trapezoid with a height of meters and bases of and meters.

204. Find the area of a trapezoid with a height of centimeters and bases of and centimeters.

205. Laurel is making a banner shaped like a trapezoid. The height of the banner is feet and the bases are and feet. What is the area of the banner?

206. Niko wants to tile the floor of his bathroom. The floor is shaped like a trapezoid with width feet and lengths feet and feet. What is the area of the floor?

207. Theresa needs a new top for her kitchen counter. The counter is shaped like a trapezoid with width inches and lengths and inches. What is the area of the counter?

208. Elena is knitting a scarf. The scarf will be shaped like a trapezoid with width inches and lengths inches and inches. What is the area of the scarf?

Everyday Math

209. Fence Jose just removed the children's playset from his back yard to make room for a rectangular garden. He wants to put a fence around the garden to keep out the dog. He has a foot roll of fence in his garage that he plans to use. To fit in the backyard, the width of the garden must be feet. How long can he make the other side if he wants to use the entire roll of fence?

210. Gardening Lupita wants to fence in her tomato garden. The garden is rectangular and the length is twice the width. It will take feet of fencing to enclose the garden. Find the length and width of her garden.

211. Fence Christa wants to put a fence around her triangular flowerbed. The sides of the flowerbed are feet, feet, and feet. The fence costs per foot. How much will it cost for Christa to fence in her flowerbed?

212. Painting Caleb wants to paint one wall of his attic. The wall is shaped like a trapezoid with height feet and bases feet and feet. The cost of the painting one square foot of wall is about About how much will it cost for Caleb to paint the attic wall?

Writing Exercises

213. If you need to put tile on your kitchen floor, do you need to know the perimeter or the area of the kitchen? Explain your reasoning.

214. If you need to put a fence around your backyard, do you need to know the perimeter or the area of the backyard? Explain your reasoning.

215. Look at the two figures.

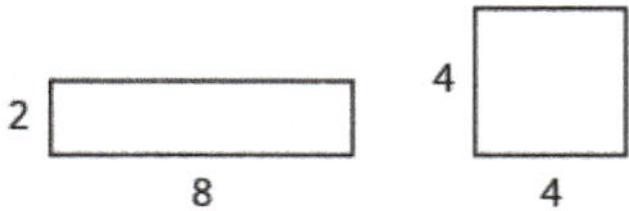

Which figure looks like it has the larger area? Which looks like it has the larger perimeter?

Now calculate the area and perimeter of each figure. Which has the larger area? Which has the larger perimeter?

216. The length of a rectangle is feet more than the width. The area is square feet. Find the length and the width.

Write the equation you would use to solve the problem.

Why can't you solve this equation with the methods you learned in the previous chapter?

Self Check

After completing the exercises, use this checklist to evaluate your mastery of the objectives of this section.

I can...	Confidently	With some help	No-I don't get it!
understand linear, square, and cubic measure.			
use the properties of rectangles.			
use the properties of triangles.			
use the properties of trapezoids.			

On a scale of 1-10, how would you rate your mastery of this section in light of your responses on the checklist? How can you improve this?

9.5 Solve Geometry Applications: Circles and Irregular Figures

Learning Objectives

By the end of this section, you will be able to:

- Use the properties of circles
- Find the area of irregular figures

BE PREPARED : : 9.13 Before you get started, take this readiness quiz.

Evaluate when

If you missed this problem, review Example 2.15.

BE PREPARED : : 9.14

Using for approximate the (a) circumference and (b) the area of a circle with radius inches.

If you missed this problem, review Example 5.39.

BE PREPARED : : 9.15 Simplify — and round to the nearest thousandth.

If you missed this problem, review Example 5.36.

In this section, we'll continue working with geometry applications. We will add several new formulas to our collection of formulas. To help you as you do the examples and exercises in this section, we will show the Problem Solving Strategy for Geometry Applications here.

Problem Solving Strategy for Geometry Applications

Step 1. **Read** the problem and make sure you understand all the words and ideas. Draw the figure and label it with the given information.

Step 2. **Identify** what you are looking for.

Step 3. **Name** what you are looking for. Choose a variable to represent that quantity.

Step 4. **Translate** into an equation by writing the appropriate formula or model for the situation. Substitute in the given information.

Step 5. **Solve** the equation using good algebra techniques.

Step 6. **Check** the answer in the problem and make sure it makes sense.

Step 7. **Answer** the question with a complete sentence.

Use the Properties of Circles

Do you remember the properties of circles from Decimals and Fractions Together? We'll show them here again to refer to as we use them to solve applications.

Properties of Circles

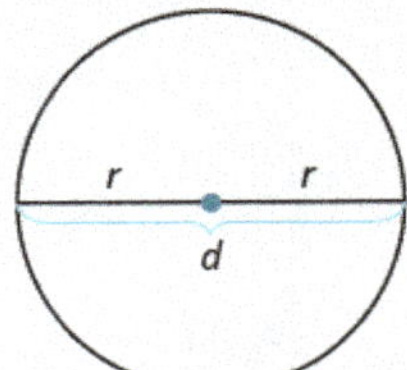

- is the length of the radius

- is the length of the diameter
-
- Circumference is the perimeter of a circle. The formula for **circumference** is
- The formula for area of a circle is

Remember, that we approximate with or — depending on whether the radius of the circle is given as a decimal or a fraction. If you use the key on your calculator to do the calculations in this section, your answers will be slightly different from the answers shown. That is because the key uses more than two decimal places.

EXAMPLE 9.41

A circular sandbox has a radius of feet. Find the circumference and area of the sandbox.

Solution

Step 1. **Read** the problem. Draw the figure and label it with the given information.	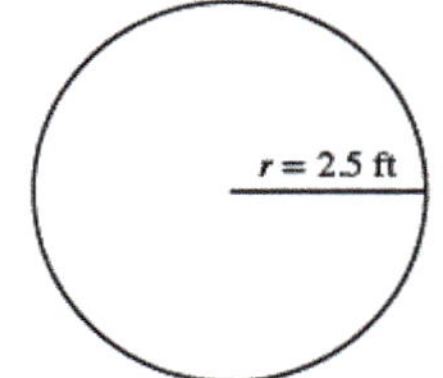
Step 2. **Identify** what you are looking for.	the circumference of the circle
Step 3. **Name.** Choose a variable to represent it.	Let c = circumference of the circle
Step 4. **Translate.** Write the appropriate formula Substitute	
Step 5. **Solve** the equation.	
Step 6. **Check.** Does this answer make sense? Yes. If we draw a square around the circle, its sides would be 5 ft (twice the radius), so its perimeter would be 20 ft. This is slightly more than the circle's circumference, 15.7 ft. 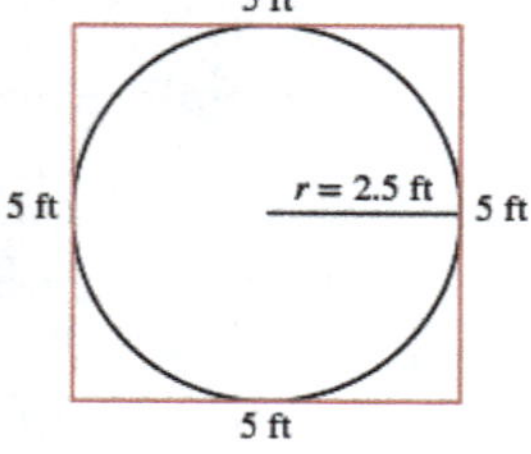	
Step 7. **Answer** the question.	The circumference of the sandbox is 15.7 feet.

Step 1. **Read** the problem. Draw the figure and label it with the given information.	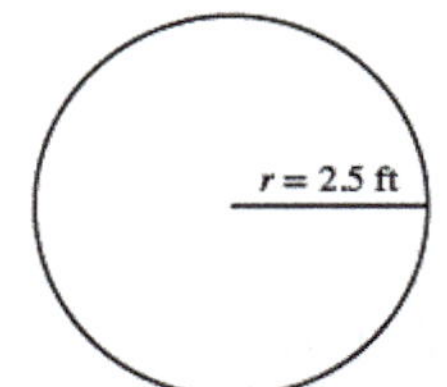
Step 2. **Identify** what you are looking for.	the area of the circle
Step 3. **Name.** Choose a variable to represent it.	Let A = the area of the circle
Step 4. **Translate.** Write the appropriate formula Substitute	
Step 5. **Solve** the equation.	
Step 6. **Check.** Yes. If we draw a square around the circle, its sides would be 5 ft, as shown in part . So the area of the square would be 25 sq. ft. This is slightly more than the circle's area, 19.625 sq. ft.	
Step 7. **Answer** the question.	The area of the circle is 19.625 square feet.

TRY IT : : 9.81 A circular mirror has radius of inches. Find the circumference and area of the mirror.

TRY IT : : 9.82 A circular spa has radius of feet. Find the circumference and area of the spa.

We usually see the formula for circumference in terms of the radius of the circle:

But since the diameter of a circle is two times the radius, we could write the formula for the circumference in terms

We will use this form of the circumference when we're given the length of the diameter instead of the radius.

EXAMPLE 9.42

A circular table has a diameter of four feet. What is the circumference of the table?

Solution

Step 1. **Read** the problem. Draw the figure and label it with the given information.

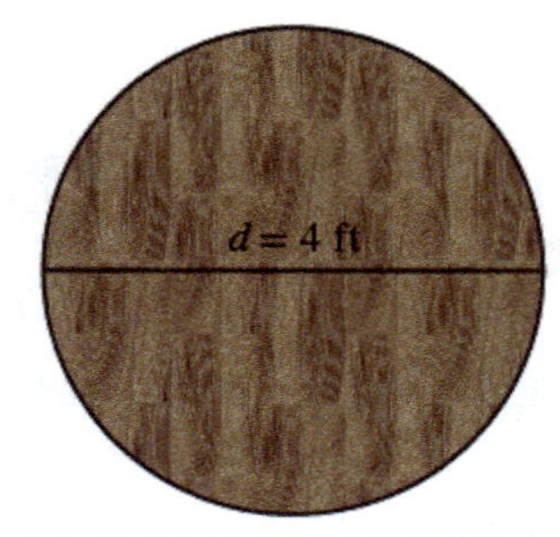

Step 2. **Identify** what you are looking for.	the circumference of the table
Step 3. **Name.** Choose a variable to represent it.	Let c = the circumference of the table
Step 4. **Translate.** Write the appropriate formula for the situation. Substitute.	
Step 5. **Solve** the equation, using 3.14 for	
Step 6. **Check:** If we put a square around the circle, its side would be 4. The perimeter would be 16. It makes sense that the circumference of the circle, 12.56, is a little less than 16.	

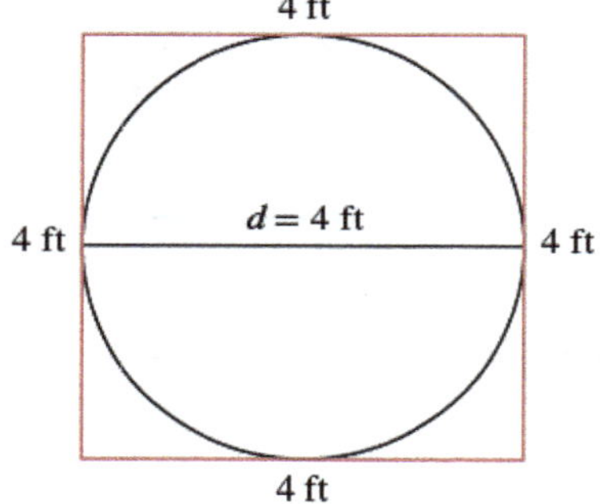

Step 7. **Answer** the question.	The diameter of the table is 12.56 square feet.

TRY IT : : 9.83 Find the circumference of a circular fire pit whose diameter is feet.

TRY IT : : 9.84 If the diameter of a circular trampoline is feet, what is its circumference?

EXAMPLE 9.43

Find the diameter of a circle with a circumference of centimeters.

Solution

Step 1. **Read** the problem. Draw the figure and label it with the given information.	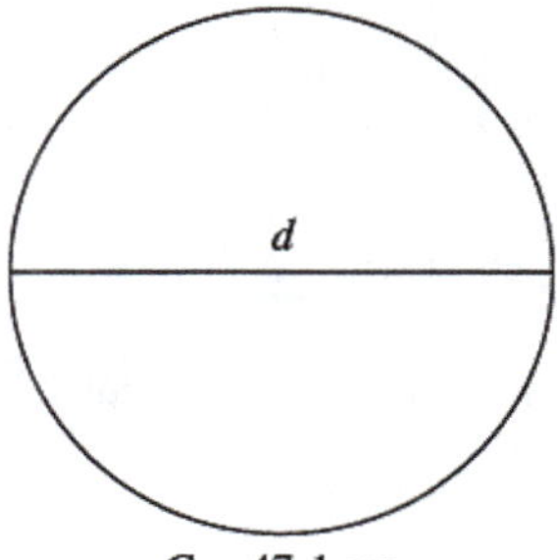
Step 2. **Identify** what you are looking for.	the diameter of the circle
Step 3. **Name.** Choose a variable to represent it.	Let d = the diameter of the circle
Step 4. **Translate.**	
Write the formula. Substitute, using 3.14 to approximate .	$C = \pi d$ $47.1 \approx 3.14d$
Step 5. **Solve.**	$\frac{47.1}{3.14} \approx \frac{3.14d}{3.14}$ $15 \approx d$
Step 6. **Check:** $C = \pi d$	
Step 7. **Answer the question.**	The diameter of the circle is approximately 15 centimeters.

TRY IT : : 9.85 Find the diameter of a circle with circumference of centimeters.

TRY IT : : 9.86 Find the diameter of a circle with circumference of feet.

Find the Area of Irregular Figures

So far, we have found area for rectangles, triangles, trapezoids, and circles. An **irregular figure** is a figure that is not a standard geometric shape. Its area cannot be calculated using any of the standard area formulas. But some irregular figures are made up of two or more standard geometric shapes. To find the area of one of these irregular figures, we can split it into figures whose formulas we know and then add the areas of the figures.

EXAMPLE 9.44

Find the area of the shaded region.

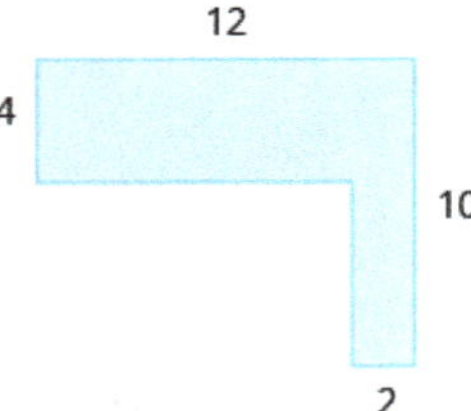

Solution

The given figure is irregular, but we can break it into two rectangles. The area of the shaded region will be the sum of the areas of both rectangles.

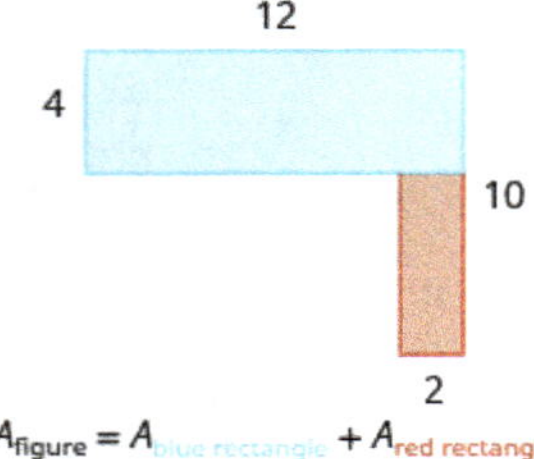

The blue rectangle has a width of and a length of The red rectangle has a width of but its length is not labeled. The right side of the figure is the length of the red rectangle plus the length of the blue rectangle. Since the right side of the blue rectangle is units long, the length of the red rectangle must be units.

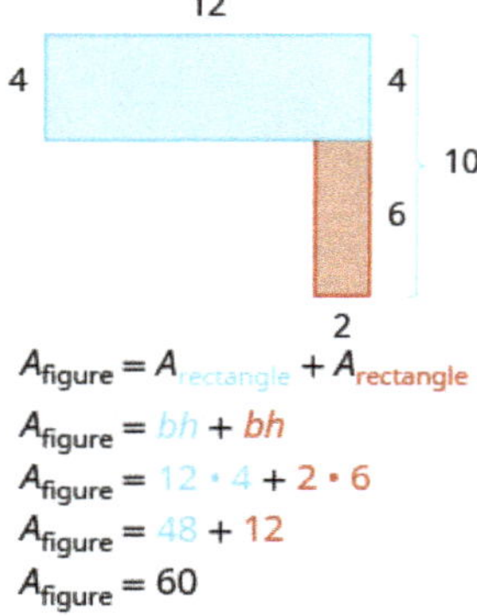

12
4
4
10
6
2
$A_{figure} = A_{rectangle} + A_{rectangle}$
$A_{figure} = bh + bh$
$A_{figure} = 12 \cdot 4 + 2 \cdot 6$
$A_{figure} = 48 + 12$
$A_{figure} = 60$

The area of the figure is square units.

Is there another way to split this figure into two rectangles? Try it, and make sure you get the same area.

TRY IT : : 9.87 Find the area of each shaded region:

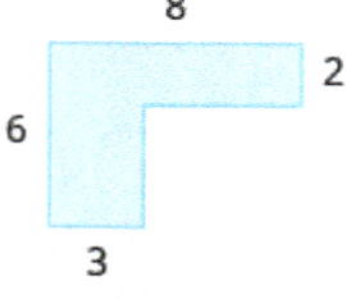

TRY IT : : 9.88 Find the area of each shaded region:

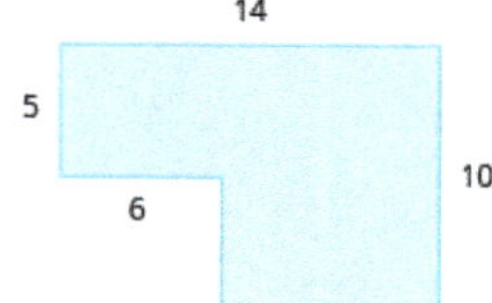

EXAMPLE 9.45

Find the area of the shaded region.

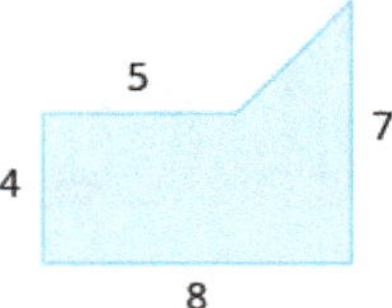

Solution

We can break this irregular figure into a triangle and rectangle. The area of the figure will be the sum of the areas of triangle and rectangle.

The rectangle has a length of units and a width of units.

We need to find the base and height of the triangle.

Since both sides of the rectangle are the vertical side of the triangle is , which is .

The length of the rectangle is so the base of the triangle will be , which is .

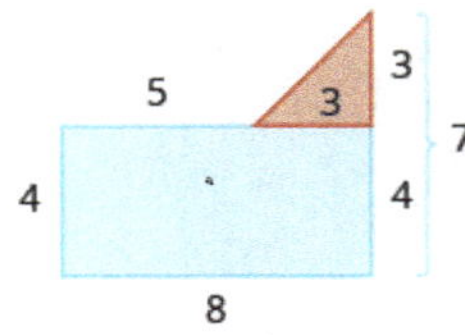

Now we can add the areas to find the area of the irregular figure.

$$A_{\text{figure}} = A_{\text{rectangle}} + A_{\text{triangle}}$$

$$A_{\text{figure}} = lw + \frac{1}{2}bh$$

$$A_{\text{figure}} = 8 \cdot 4 + \frac{1}{2} \cdot 3 \cdot 3$$

$$A_{\text{figure}} = 32 + 4.5$$

$$A_{\text{figure}} = 36.5 \text{ sq. units}$$

The area of the figure is square units.

TRY IT : : 9.89 Find the area of each shaded region.

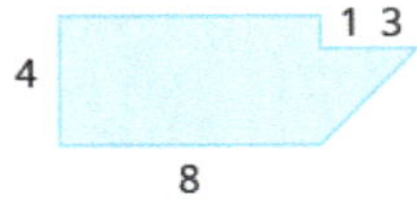

TRY IT : : 9.90 Find the area of each shaded region.

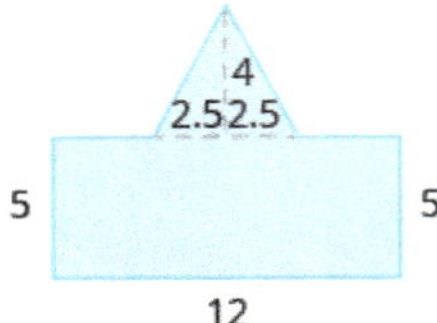

EXAMPLE 9.46

A high school track is shaped like a rectangle with a semi-circle (half a circle) on each end. The rectangle has length meters and width meters. Find the area enclosed by the track. Round your answer to the nearest hundredth.

Solution

We will break the figure into a rectangle and two semi-circles. The area of the figure will be the sum of the areas of the rectangle and the semicircles.

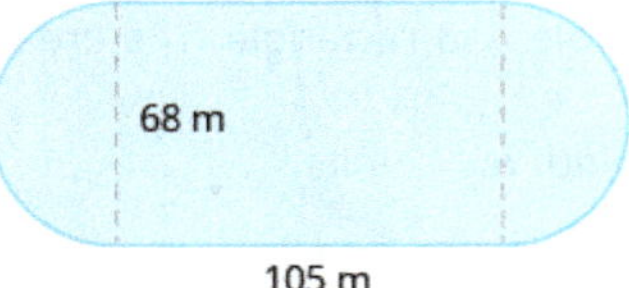

The rectangle has a length of m and a width of m. The semi-circles have a diameter of m, so each has a radius of m.

$$A_{figure} = A_{rectangle} + A_{semicircles}$$

$$A_{figure} = bh + 2\left(\frac{1}{2}\pi \cdot r^2\right)$$

$$A_{figure} \approx 105 \cdot 68 + 2\left(\frac{1}{2} \cdot 3.14 \cdot 34^2\right)$$

$$A_{figure} \approx 7140 + 3629.84$$

$$A_{figure} \approx 10{,}769.84 \text{ square meters}$$

TRY IT : : 9.91 Find the area:

TRY IT : : 9.92 Find the area:

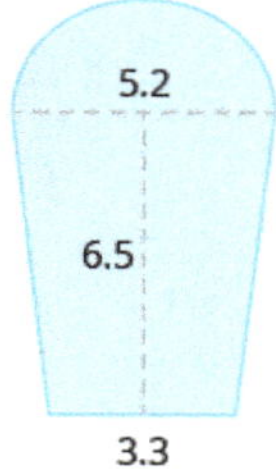

MEDIA : : ACCESS ADDITIONAL ONLINE RESOURCES

- Circumference of a Circle (http://www.openstax.org/l/24circumcircle)
- Area of a Circle (http://www.openstax.org/l/24areacircle)
- Area of an L-shaped polygon (http://www.openstax.org/l/24areaLpoly)
- Area of an L-shaped polygon with Decimals (http://www.openstax.org/l/24areaLpolyd)
- Perimeter Involving a Rectangle and Circle (http://www.openstax.org/l/24perirectcirc)
- Area Involving a Rectangle and Circle (http://www.openstax.org/l/24arearectcirc)

9.5 EXERCISES

Practice Makes Perfect

Use the Properties of Circles

In the following exercises, solve using the properties of circles.

217. The lid of a paint bucket is a circle with radius inches. Find the circumference and area of the lid.

218. An extra-large pizza is a circle with radius inches. Find the circumference and area of the pizza.

219. A farm sprinkler spreads water in a circle with radius of feet. Find the circumference and area of the watered circle.

220. A circular rug has radius of feet. Find the circumference and area of the rug.

221. A reflecting pool is in the shape of a circle with diameter of feet. What is the circumference of the pool?

222. A turntable is a circle with diameter of inches. What is the circumference of the turntable?

223. A circular saw has a diameter of inches. What is the circumference of the saw?

224. A round coin has a diameter of centimeters. What is the circumference of the coin?

225. A barbecue grill is a circle with a diameter of feet. What is the circumference of the grill?

226. The top of a pie tin is a circle with a diameter of inches. What is the circumference of the top?

227. A circle has a circumference of inches. Find the diameter.

228. A circle has a circumference of feet. Find the diameter.

229. A circle has a circumference of meters. Find the diameter.

230. A circle has a circumference of centimeters. Find the diameter.

In the following exercises, find the radius of the circle with given circumference.

231. A circle has a circumference of feet.

232. A circle has a circumference of centimeters.

233. A circle has a circumference of miles.

234. A circle has a circumference of inches.

Find the Area of Irregular Figures

In the following exercises, find the area of the irregular figure. Round your answers to the nearest hundredth.

235.

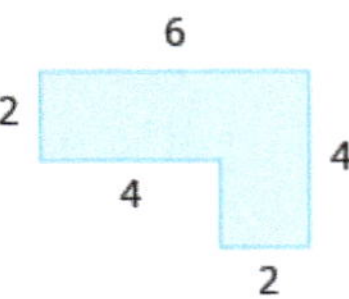

236.

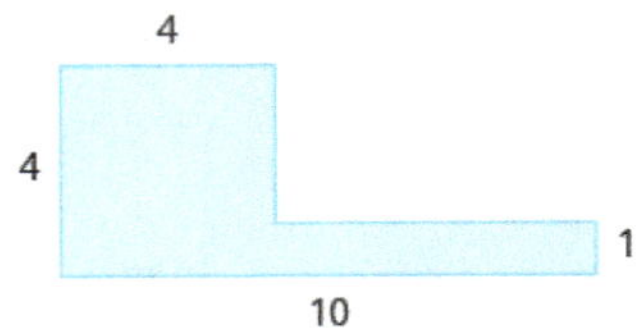

237.

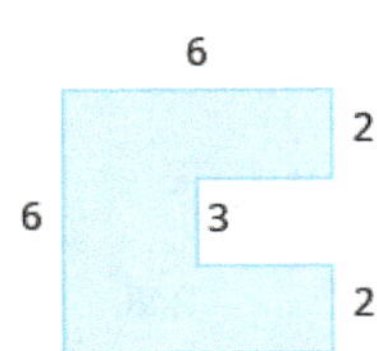

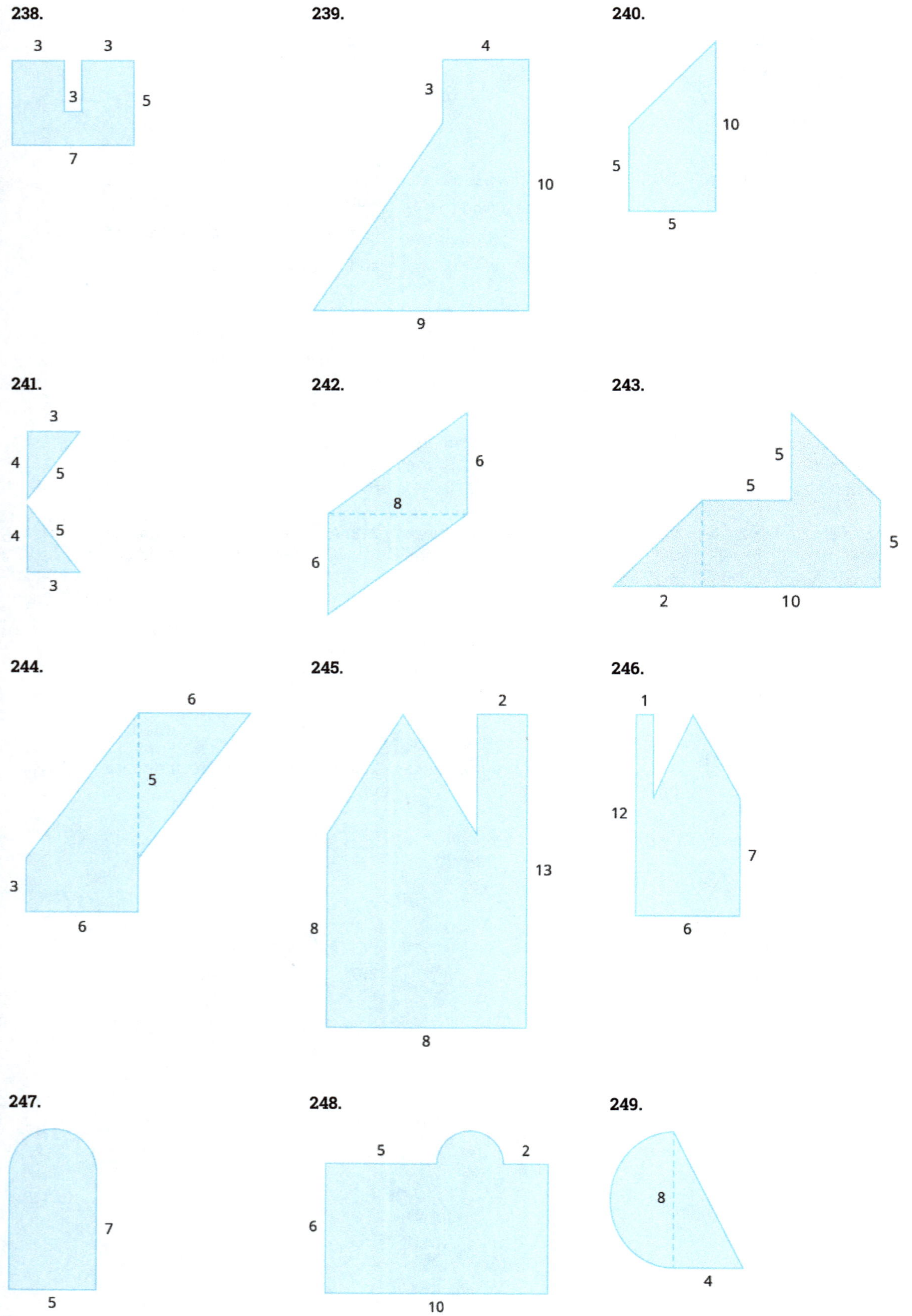
238.
3
3
3
5
7
239.
4
3
10
9
240.
10
5
5
241.
3
4
5
4
5
3
242.
6
8
6
243.
5
5
5
2
10
244.
6
5
3
6
245.
2
13
8
8
246.
1
12
7
6
247.
7
5
248.
5
2
6
10
249.
8
4

250.

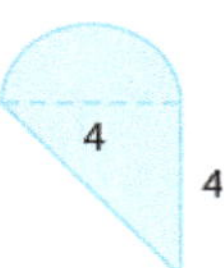

251.

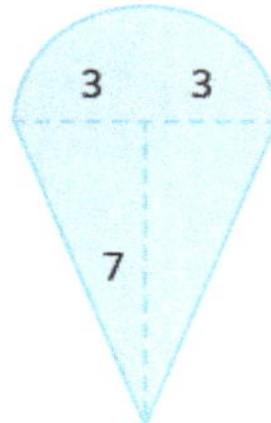

252.

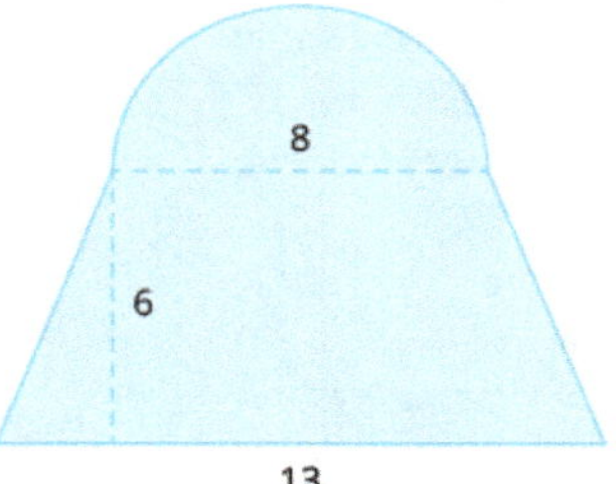

253.

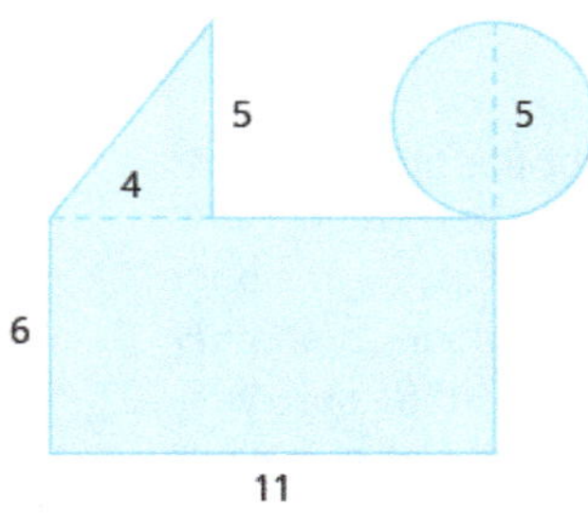

254.

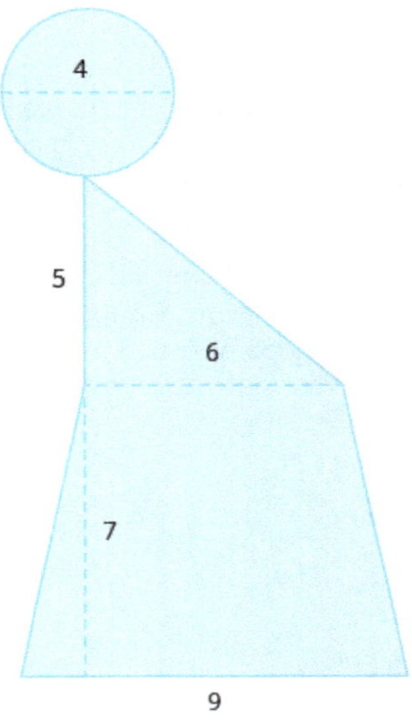

In the following exercises, solve.

255. A city park covers one block plus parts of four more blocks, as shown. The block is a square with sides feet long, and the triangles are isosceles right triangles. Find the area of the park.

256. A gift box will be made from a rectangular piece of cardboard measuring inches by inches, with squares cut out of the corners of the sides, as shown. The sides of the squares are inches. Find the area of the cardboard after the corners are cut out.

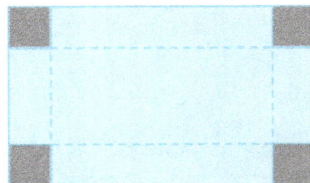

257. Perry needs to put in a new lawn. His lot is a rectangle with a length of feet and a width of feet. The house is rectangular and measures feet by feet. His driveway is rectangular and measures feet by feet, as shown. Find the area of Perry's lawn.

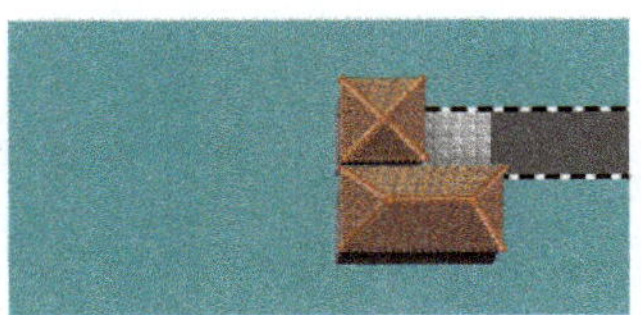

258. Denise is planning to put a deck in her back yard. The deck will be a by rectangle with a semicircle of diameter feet, as shown below. Find the area of the deck.

Everyday Math

259. Area of a Tabletop Yuki bought a drop-leaf kitchen table. The rectangular part of the table is a by rectangle with a semicircle at each end, as shown. Find the area of the table with one leaf up. Find the area of the table with both leaves up.

260. Painting Leora wants to paint the nursery in her house. The nursery is an by rectangle, and the ceiling is feet tall. There is a by door on one wall, a by closet door on another wall, and one by window on the third wall. The fourth wall has no doors or windows. If she will only paint the four walls, and not the ceiling or doors, how many square feet will she need to paint?

Writing Exercises

261. Describe two different ways to find the area of this figure, and then show your work to make sure both ways give the same area.

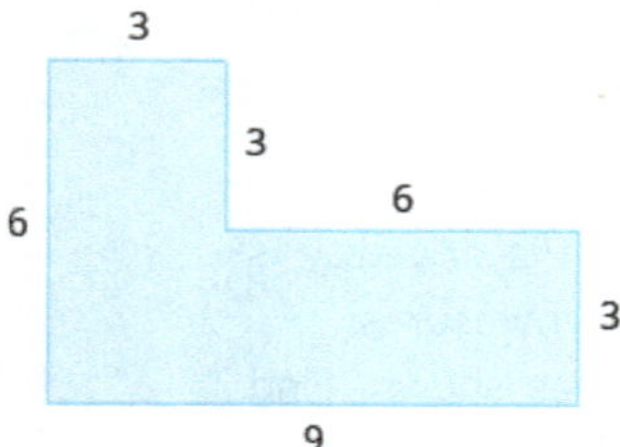

262. A circle has a diameter of feet. Find the area of the circle using for using — for Which calculation to do prefer? Why?

Self Check

After completing the exercises, use this checklist to evaluate your mastery of the objectives of this section.

I can...	Confidently	With some help	No-I don't get it!
use the properties of circles.			
find the area of irregular figures.			

After looking at the checklist, do you think you are well prepared for the next section? Why or why not?

9.6 Solve Geometry Applications: Volume and Surface Area

Learning Objectives

By the end of this section, you will be able to:

- Find volume and surface area of rectangular solids
- Find volume and surface area of spheres
- Find volume and surface area of cylinders
- Find volume of cones

BE PREPARED : : 9.16 Before you get started, take this readiness quiz.

Evaluate when

If you missed this problem, review Example 2.15.

BE PREPARED : : 9.17 Evaluate when

If you missed this problem, review Example 2.16.

BE PREPARED : : 9.18 Find the area of a circle with radius —

If you missed this problem, review Example 5.39.

In this section, we will finish our study of geometry applications. We find the volume and surface area of some three-dimensional figures. Since we will be solving applications, we will once again show our Problem-Solving Strategy for Geometry Applications.

Problem Solving Strategy for Geometry Applications

Step 1. **Read** the problem and make sure you understand all the words and ideas. Draw the figure and label it with the given information.

Step 2. **Identify** what you are looking for.

Step 3. **Name** what you are looking for. Choose a variable to represent that quantity.

Step 4. **Translate** into an equation by writing the appropriate formula or model for the situation. Substitute in the given information.

Step 5. **Solve** the equation using good algebra techniques.

Step 6. **Check** the answer in the problem and make sure it makes sense.

Step 7. **Answer** the question with a complete sentence.

Find Volume and Surface Area of Rectangular Solids

A cheerleading coach is having the squad paint wooden crates with the school colors to stand on at the games. (See Figure 9.28). The amount of paint needed to cover the outside of each box is the **surface area**, a square measure of the total area of all the sides. The amount of space inside the crate is the volume, a cubic measure.

VARSITY
Cheerleader

Figure 9.28 This wooden crate is in the shape of a rectangular solid.

Each crate is in the shape of a **rectangular solid**. Its dimensions are the length, width, and height. The rectangular solid shown in Figure 9.29 has length units, width units, and height units. Can you tell how many cubic units there are altogether? Let's look layer by layer.

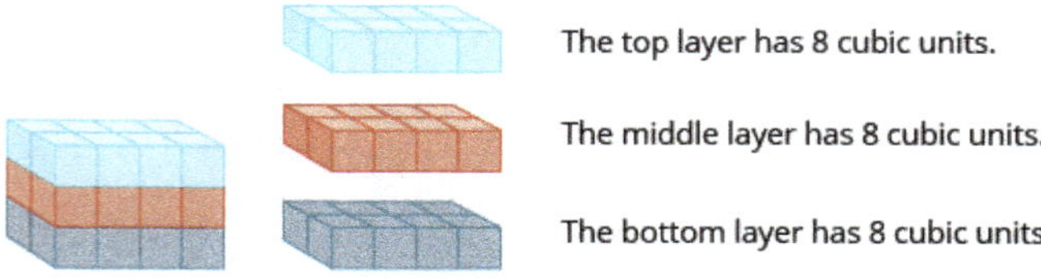
The top layer has 8 cubic units.
The middle layer has 8 cubic units.
The bottom layer has 8 cubic units.

Figure 9.29 Breaking a rectangular solid into layers makes it easier to visualize the number of cubic units it contains. This by by rectangular solid has cubic units.

Altogether there are cubic units. Notice that is the

$$\underbrace{V}_{24} \;\underbrace{=}_{=}\; \underbrace{L}_{4} \;\underbrace{\cdot}_{\cdot}\; \underbrace{W}_{2} \;\underbrace{\cdot}_{\cdot}\; \underbrace{H}_{3}$$

The volume, of any rectangular solid is the product of the length, width, and height.

We could also write the formula for volume of a rectangular solid in terms of the area of the base. The area of the base, is equal to

We can substitute for in the volume formula to get another form of the volume formula.

$$V = L \cdot W \cdot H$$
$$V = (L \cdot W) \cdot H$$
$$V = Bh$$

We now have another version of the volume formula for rectangular solids. Let's see how this works with the rectangular solid we started with. See Figure 9.29.

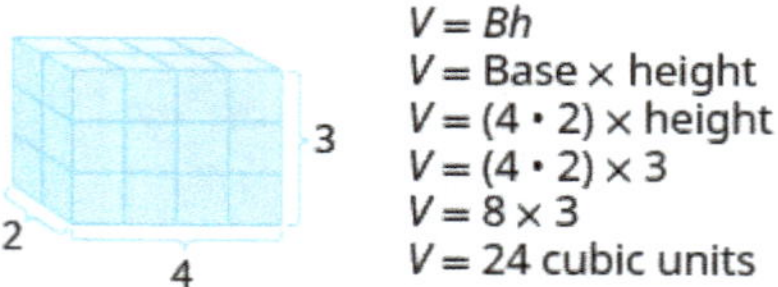
3
2
4
V = Bh
V = Base × height
V = (4 • 2) × height
V = (4 • 2) × 3
V = 8 × 3
V = 24 cubic units

Figure 9.30

To find the *surface area* of a rectangular solid, think about finding the area of each of its faces. How many faces does the rectangular solid above have? You can see three of them.

Notice for each of the three faces you see, there is an identical opposite face that does not show.

The surface area of the rectangular solid shown in Figure 9.30 is square units.

In general, to find the surface area of a rectangular solid, remember that each face is a rectangle, so its area is the product of its length and its width (see Figure 9.31). Find the area of each face that you see and then multiply each area by two to account for the face on the opposite side.

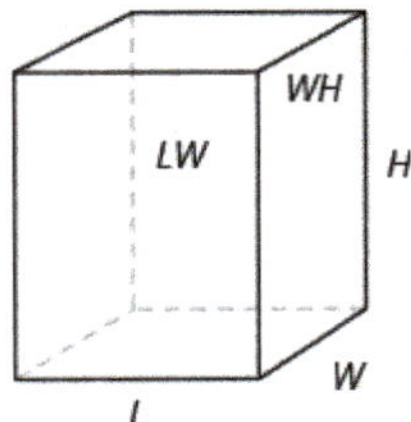

Figure 9.31 For each face of the rectangular solid facing you, there is another face on the opposite side. There are faces in all.

Volume and Surface Area of a Rectangular Solid

For a rectangular solid with length width and height

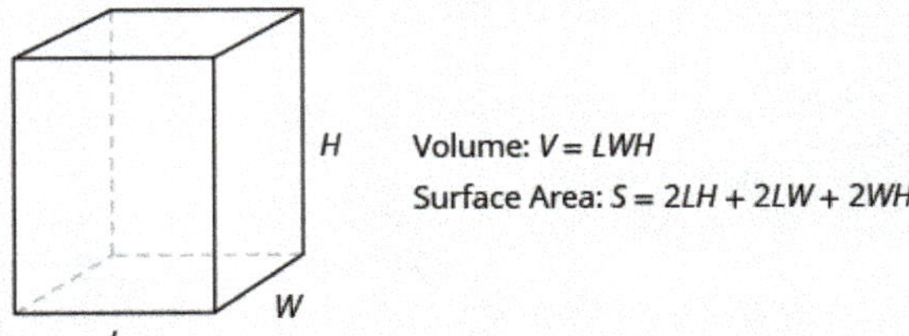

MANIPULATIVE MATHEMATICS

Doing the Manipulative Mathematics activity "Painted Cube" will help you develop a better understanding of volume and surface area.

EXAMPLE 9.47

For a rectangular solid with length cm, height cm, and width cm, find the volume and surface area.

Solution

Step 1 is the same for both and , so we will show it just once.

Step 1. **Read** the problem. Draw the figure and label it with the given information.	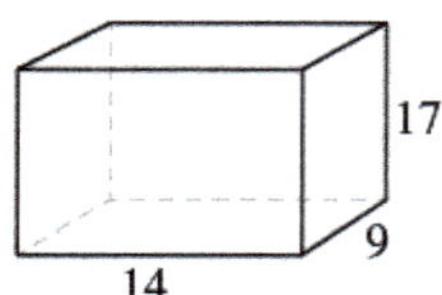

Step 2. **Identify** what you are looking for.	the volume of the rectangular solid
Step 3. **Name.** Choose a variable to represent it.	Let = volume
Step 4. **Translate.** Write the appropriate formula. Substitute.	
Step 5. **Solve** the equation.	
Step 6. **Check** We leave it to you to check your calculations.	
Step 7. **Answer** the question.	The volume is cubic centimeters.

Step 2. **Identify** what you are looking for.	the surface area of the solid
Step 3. **Name.** Choose a variable to represent it.	Let = surface area
Step 4. **Translate.** Write the appropriate formula. Substitute.	
Step 5. **Solve the equation.**	
Step 6. **Check:** Double-check with a calculator.	
Step 7. **Answer** the question.	The surface area is 1,034 square centimeters.

> **TRY IT : :** 9.93
>
> Find the volume and surface area of rectangular solid with the: length feet, width feet, and height feet.

> **TRY IT : :** 9.94
>
> Find the volume and surface area of rectangular solid with the: length feet, width feet, and height feet.

EXAMPLE 9.48

A rectangular crate has a length of inches, width of inches, and height of inches. Find its volume and surface area.

Solution

Step 1 is the same for both and , so we will show it just once.

Step 1. **Read** the problem. Draw the figure and label it with the given information.	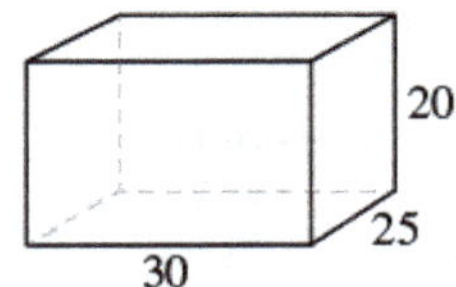

Step 2. **Identify** what you are looking for.	the volume of the crate
Step 3. **Name.** Choose a variable to represent it.	let = volume
Step 4. **Translate.** Write the appropriate formula. Substitute.	
Step 5. **Solve** the equation.	
Step 6. **Check:** Double check your math.	
Step 7. **Answer** the question.	The volume is 15,000 cubic inches.

Step 2. **Identify** what you are looking for.	the surface area of the crate
Step 3. **Name.** Choose a variable to represent it.	let = surface area
Step 4. **Translate.** Write the appropriate formula. Substitute.	
Step 5. **Solve** the equation.	
Step 6. **Check:** Check it yourself!	
Step 7. **Answer** the question.	The surface area is 3,700 square inches.

TRY IT : : 9.95

A rectangular box has length feet, width feet, and height feet. Find its volume and surface area.

TRY IT : : 9.96

A rectangular suitcase has length inches, width inches, and height inches. Find its volume and surface area.

Volume and Surface Area of a Cube

A **cube** is a rectangular solid whose length, width, and height are equal. See Volume and Surface Area of a Cube, below. Substituting, s for the length, width and height into the formulas for volume and surface area of a rectangular solid, we get:

So for a cube, the formulas for volume and surface area are and

Volume and Surface Area of a Cube

For any cube with sides of length

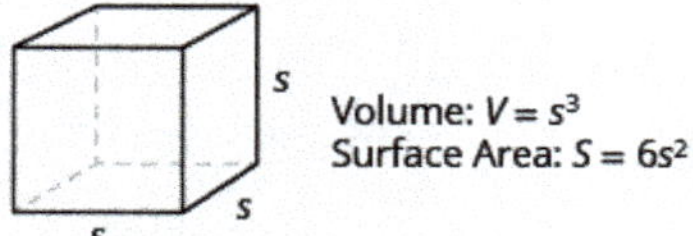

EXAMPLE 9.49

A cube is inches on each side. Find its volume and surface area.

Solution

Step 1 is the same for both and , so we will show it just once.

Step 1. **Read** the problem. Draw the figure and label it with the given information.	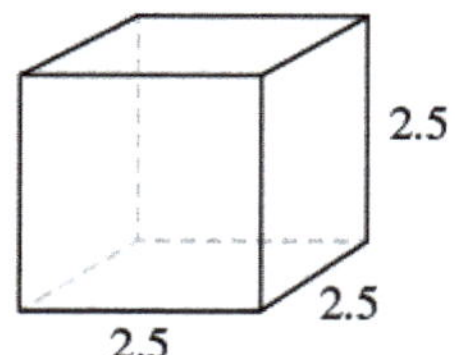

Step 2. **Identify** what you are looking for.	the volume of the cube
Step 3. **Name.** Choose a variable to represent it.	let V = volume
Step 4. **Translate.** Write the appropriate formula.	
Step 5. **Solve.** Substitute and solve.	
Step 6. **Check:** Check your work.	
Step 7. **Answer** the question.	The volume is 15.625 cubic inches.

Step 2. **Identify** what you are looking for.	the surface area of the cube
Step 3. **Name.** Choose a variable to represent it.	let S = surface area
Step 4. **Translate.** Write the appropriate formula.	
Step 5. **Solve.** Substitute and solve.	
Step 6. **Check:** The check is left to you.	
Step 7. **Answer** the question.	The surface area is 37.5 square inches.

> **TRY IT : :** 9.97 For a cube with side 4.5 meters, find the volume and surface area of the cube.

> **TRY IT : :** 9.98 For a cube with side 7.3 yards, find the volume and surface area of the cube.

EXAMPLE 9.50

A notepad cube measures inches on each side. Find its volume and surface area.

Solution

Step 1. **Read** the problem. Draw the figure and label it with the given information.	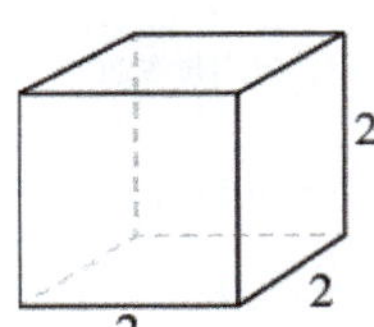
Step 2. **Identify** what you are looking for.	the volume of the cube
Step 3. **Name.** Choose a variable to represent it.	let V = volume
Step 4. **Translate.** Write the appropriate formula.	
Step 5. **Solve** the equation.	
Step 6. **Check:** Check that you did the calculations correctly.	
Step 7. **Answer** the question.	The volume is 8 cubic inches.

Step 2. **Identify** what you are looking for.	the surface area of the cube
Step 3. **Name.** Choose a variable to represent it.	let S = surface area
Step 4. **Translate.** Write the appropriate formula.	
Step 5. **Solve** the equation.	
Step 6. **Check:** The check is left to you.	
Step 7. **Answer** the question.	The surface area is 24 square inches.

TRY IT : : 9.99

A packing box is a cube measuring feet on each side. Find its volume and surface area.

TRY IT : : 9.100

A wall is made up of cube-shaped bricks. Each cube is inches on each side. Find the volume and surface area of each cube.

Find the Volume and Surface Area of Spheres

A **sphere** is the shape of a basketball, like a three-dimensional circle. Just like a circle, the size of a sphere is determined by its radius, which is the distance from the center of the sphere to any point on its surface. The formulas for the volume and surface area of a sphere are given below.

Showing where these formulas come from, like we did for a rectangular solid, is beyond the scope of this course. We will approximate with

Volume and Surface Area of a Sphere

For a sphere with radius

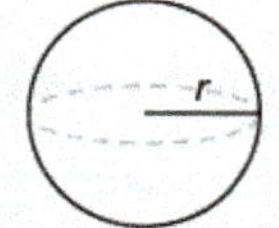

Volume: $V = \frac{4}{3}\pi r^3$
Surface Area: $S = 4\pi r^2$

EXAMPLE 9.51

A sphere has a radius inches. Find its volume and surface area.

Solution

Step 1 is the same for both and , so we will show it just once.

Step 1. **Read** the problem. Draw the figure and label it with the given information.	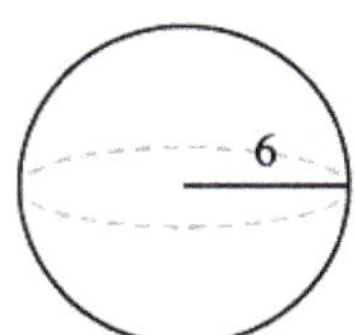

Step 2. **Identify** what you are looking for.	the volume of the sphere
Step 3. **Name.** Choose a variable to represent it.	let V = volume
Step 4. **Translate.** Write the appropriate formula.	–
Step 5. **Solve.**	–
Step 6. **Check:** Double-check your math on a calculator.	
Step 7. **Answer** the question.	The volume is approximately 904.32 cubic inches.

Step 2. **Identify** what you are looking for.	the surface area of the cube
Step 3. **Name.** Choose a variable to represent it.	let S = surface area
Step 4. **Translate.** Write the appropriate formula.	
Step 5. **Solve.**	
Step 6. **Check:** Double-check your math on a calculator	
Step 7. **Answer** the question.	The surface area is approximately 452.16 square inches.

TRY IT : : 9.101 Find the volume and surface area of a sphere with radius 3 centimeters.

TRY IT : : 9.102 Find the volume and surface area of each sphere with a radius of foot

EXAMPLE 9.52

A globe of Earth is in the shape of a sphere with radius centimeters. Find its volume and surface area. Round the answer to the nearest hundredth.

Solution

Step 1. **Read** the problem. Draw a figure with the given information and label it.	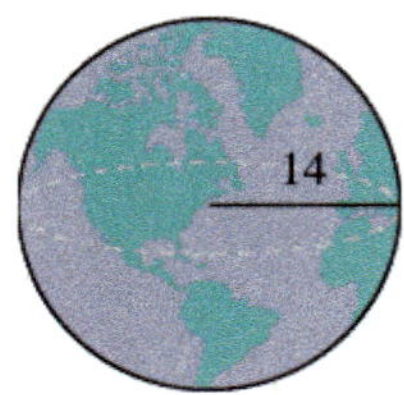

Step 2. **Identify** what you are looking for.	the volume of the sphere
Step 3. **Name.** Choose a variable to represent it.	let V = volume
Step 4. **Translate.** Write the appropriate formula. Substitute. (Use 3.14 for)	
Step 5. **Solve.**	
Step 6. **Check:** We leave it to you to check your calculations.	
Step 7. **Answer** the question.	The volume is approximately 11,488.21 cubic inches.

Step 2. **Identify** what you are looking for.	the surface area of the sphere
Step 3. **Name.** Choose a variable to represent it.	let S = surface area
Step 4. **Translate.** Write the appropriate formula. Substitute. (Use 3.14 for)	
Step 5. **Solve.**	
Step 6. **Check:** We leave it to you to check your calculations.	
Step 7. **Answer** the question.	The surface area is approximately 2461.76 square inches.

TRY IT : : 9.103

A beach ball is in the shape of a sphere with radius of inches. Find its volume and surface area.

TRY IT : : 9.104

A Roman statue depicts Atlas holding a globe with radius of feet. Find the volume and surface area of the globe.

Find the Volume and Surface Area of a Cylinder

If you have ever seen a can of soda, you know what a cylinder looks like. A **cylinder** is a solid figure with two parallel circles of the same size at the top and bottom. The top and bottom of a cylinder are called the bases. The height of a cylinder is the distance between the two bases. For all the cylinders we will work with here, the sides and the height, , will be perpendicular to the bases.

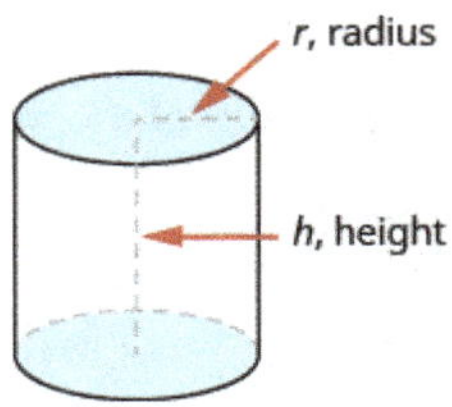

Figure 9.32 A cylinder has two circular bases of equal size. The height is the distance between the bases.

Rectangular solids and cylinders are somewhat similar because they both have two bases and a height. The formula for the volume of a rectangular solid, , can also be used to find the volume of a cylinder.

For the rectangular solid, the area of the base, , is the area of the rectangular base, length × width. For a cylinder, the area of the base, is the area of its circular base, Figure 9.33 compares how the formula is used for rectangular solids and cylinders.

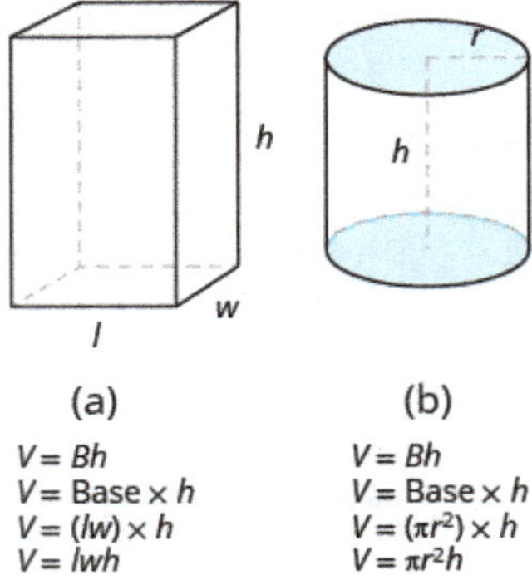

Figure 9.33 Seeing how a cylinder is similar to a rectangular solid may make it easier to understand the formula for the volume of a cylinder.

To understand the formula for the surface area of a cylinder, think of a can of vegetables. It has three surfaces: the top, the bottom, and the piece that forms the sides of the can. If you carefully cut the label off the side of the can and unroll it, you will see that it is a rectangle. See Figure 9.34.

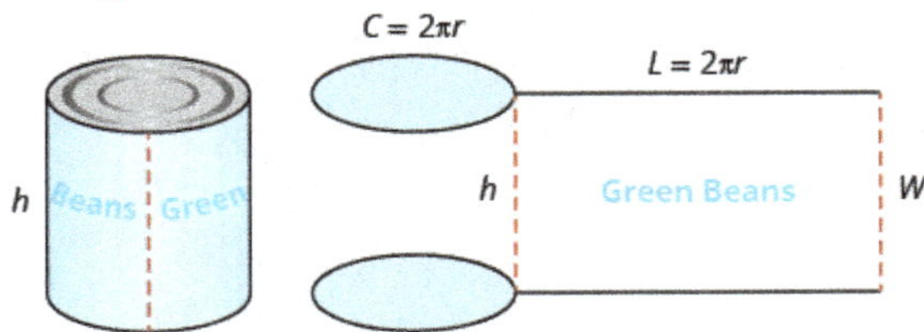

Figure 9.34 By cutting and unrolling the label of a can of vegetables, we can see that the surface of a cylinder is a rectangle. The length of the rectangle is the circumference of the cylinder's base, and the width is the height of the cylinder.

The distance around the edge of the can is the **circumference** of the cylinder's base it is also the length of the rectangular label. The height of the cylinder is the width of the rectangular label. So the area of the label can be represented as

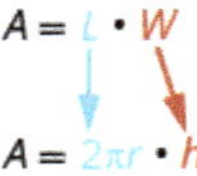

To find the total surface area of the cylinder, we add the areas of the two circles to the area of the rectangle.

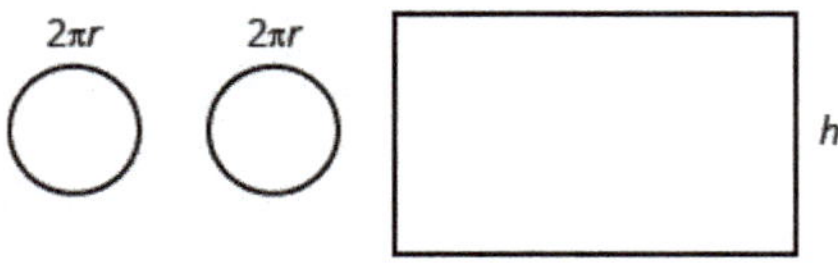

$$S = A_{\text{top circle}} + A_{\text{bottom circle}} + A_{\text{rectangle}}$$
$$S = \pi r^2 + \pi r^2 + 2\pi r \cdot h$$
$$S = 2 \cdot \pi r^2 + 2\pi rh$$
$$S = 2\pi r^2 + 2\pi rh$$

The surface area of a cylinder with radius and height is

Volume and Surface Area of a Cylinder

For a cylinder with radius and height

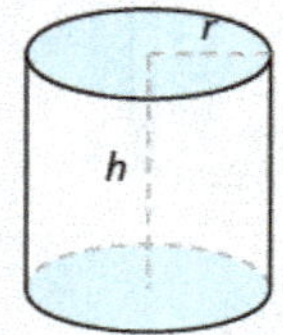

Volume: $V = \pi r^2 h$ or $V = Bh$
Surface Area: $S = 2\pi r^2 + 2\pi rh$

EXAMPLE 9.53

A cylinder has height centimeters and radius centimeters. Find the volume and surface area.

Solution

Step 1. **Read** the problem. Draw the figure and label it with the given information.

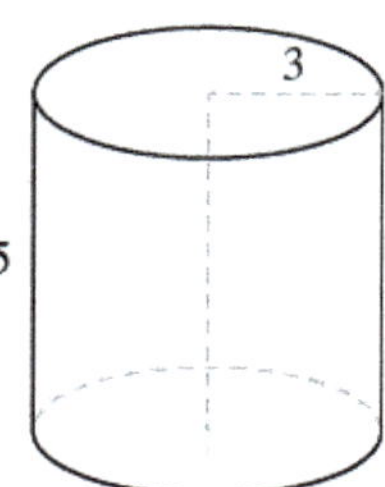

Step 2. **Identify** what you are looking for.	the volume of the cylinder
Step 3. **Name.** Choose a variable to represent it.	let V = volume
Step 4. **Translate.** Write the appropriate formula. Substitute. (Use 3.14 for)	
Step 5. **Solve.**	
Step 6. **Check:** We leave it to you to check your calculations.	
Step 7. **Answer** the question.	The volume is approximately 141.3 cubic inches.

Step 2. **Identify** what you are looking for.	the surface area of the cylinder
Step 3. **Name.** Choose a variable to represent it.	let S = surface area
Step 4. **Translate.** Write the appropriate formula. Substitute. (Use 3.14 for)	
Step 5. **Solve.**	
Step 6. **Check:** We leave it to you to check your calculations.	
Step 7. **Answer** the question.	The surface area is approximately 150.72 square inches.

TRY IT : : 9.105 Find the volume and surface area of the cylinder with radius 4 cm and height 7cm.

TRY IT : : 9.106 Find the volume and surface area of the cylinder with given radius 2 ft and height 8 ft.

EXAMPLE 9.54

Find the volume and surface area of a can of soda. The radius of the base is centimeters and the height is centimeters. Assume the can is shaped exactly like a cylinder.

Solution

Step 1. **Read** the problem. Draw the figure and label it with the given information.	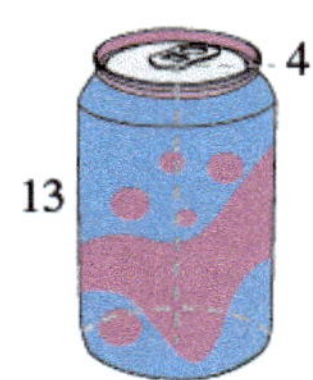

Step 2. **Identify** what you are looking for.	the volume of the cylinder
Step 3. **Name.** Choose a variable to represent it.	let V = volume
Step 4. **Translate.** Write the appropriate formula. Substitute. (Use 3.14 for)	
Step 5. **Solve.**	
Step 6. **Check:** We leave it to you to check.	
Step 7. **Answer** the question.	The volume is approximately 653.12 cubic centimeters.

Step 2. **Identify** what you are looking for.	the surface area of the cylinder
Step 3. **Name.** Choose a variable to represent it.	let S = surface area
Step 4. **Translate.** Write the appropriate formula. Substitute. (Use 3.14 for)	
Step 5. **Solve.**	
Step 6. **Check:** We leave it to you to check your calculations.	
Step 7. **Answer** the question.	The surface area is approximately 427.04 square centimeters.

TRY IT : : 9.107

Find the volume and surface area of a can of paint with radius 8 centimeters and height 19 centimeters. Assume the can is shaped exactly like a cylinder.

TRY IT : : 9.108

Find the volume and surface area of a cylindrical drum with radius 2.7 feet and height 4 feet. Assume the drum is shaped exactly like a cylinder.

Find the Volume of Cones

The first image that many of us have when we hear the word 'cone' is an ice cream cone. There are many other applications of cones (but most are not as tasty as ice cream cones). In this section, we will see how to find the volume of a cone.

In geometry, a **cone** is a solid figure with one circular base and a vertex. The height of a cone is the distance between its base and the vertex.The cones that we will look at in this section will always have the height perpendicular to the base. See Figure 9.35.

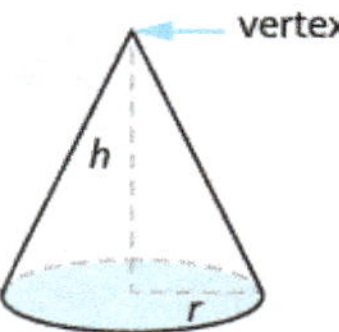

Figure 9.35 The height of a cone is the distance between its base and the vertex.

Earlier in this section, we saw that the volume of a cylinder is We can think of a cone as part of a cylinder. Figure 9.36 shows a cone placed inside a cylinder with the same height and same base. If we compare the volume of the cone and the cylinder, we can see that the volume of the cone is less than that of the cylinder.

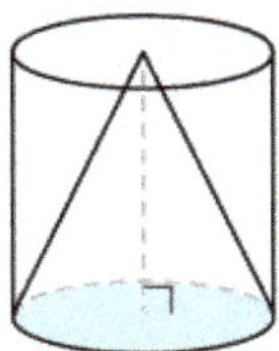

Figure 9.36 The volume of a cone is less than the volume of a cylinder with the same base and height.

In fact, the volume of a cone is exactly one-third of the volume of a cylinder with the same base and height. The volume of a cone is

$$V=\frac{1}{3}Bh$$

Since the base of a cone is a circle, we can substitute the formula of area of a circle, , for to get the formula for volume of a cone.

$$V=\frac{1}{3}\pi r^2 h$$

In this book, we will only find the volume of a cone, and not its surface area.

Volume of a Cone

For a cone with radius and height .

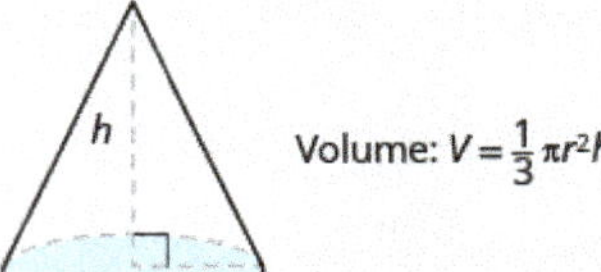

EXAMPLE 9.55

Find the volume of a cone with height inches and radius of its base inches.

Solution

Step 1. **Read** the problem. Draw the figure and label it with the given information.	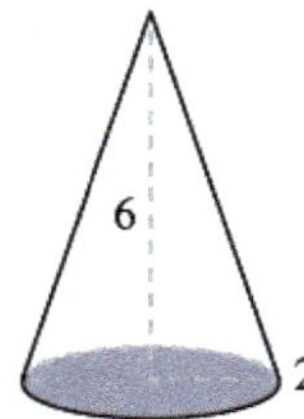
Step 2. **Identify** what you are looking for.	the volume of the cone
Step 3. **Name.** Choose a variable to represent it.	let V = volume
Step 4. **Translate.** Write the appropriate formula. Substitute. (Use 3.14 for)	— —
Step 5. **Solve.**	
Step 6. **Check:** We leave it to you to check your calculations.	
Step 7. **Answer** the question.	The volume is approximately 25.12 cubic inches.

> **TRY IT : : 9.109** Find the volume of a cone with height inches and radius inches

> **TRY IT : : 9.110** Find the volume of a cone with height centimeters and radius centimeters

EXAMPLE 9.56

Marty's favorite gastro pub serves french fries in a paper wrap shaped like a cone. What is the volume of a conic wrap that is inches tall and inches in diameter? Round the answer to the nearest hundredth.

Solution

Step 1. **Read** the problem. Draw the figure and label it with the given information. Notice here that the base is the circle at the top of the cone.	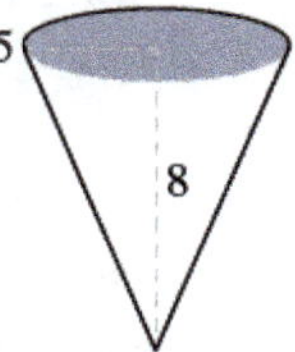
Step 2. **Identify** what you are looking for.	the volume of the cone
Step 3. **Name.** Choose a variable to represent it.	let V = volume
Step 4. **Translate.** Write the appropriate formula. Substitute. (Use 3.14 for , and notice that we were given the distance across the circle, which is its diameter. The radius is 2.5 inches.)	– –
Step 5. **Solve.**	
Step 6. **Check:** We leave it to you to check your calculations.	
Step 7. **Answer** the question.	The volume of the wrap is approximately 52.33 cubic inches.

TRY IT : : 9.111

How many cubic inches of candy will fit in a cone-shaped piñata that is inches long and inches across its base? Round the answer to the nearest hundredth.

TRY IT : : 9.112

What is the volume of a cone-shaped party hat that is inches tall and inches across at the base? Round the answer to the nearest hundredth.

Summary of Geometry Formulas

The following charts summarize all of the formulas covered in this chapter.

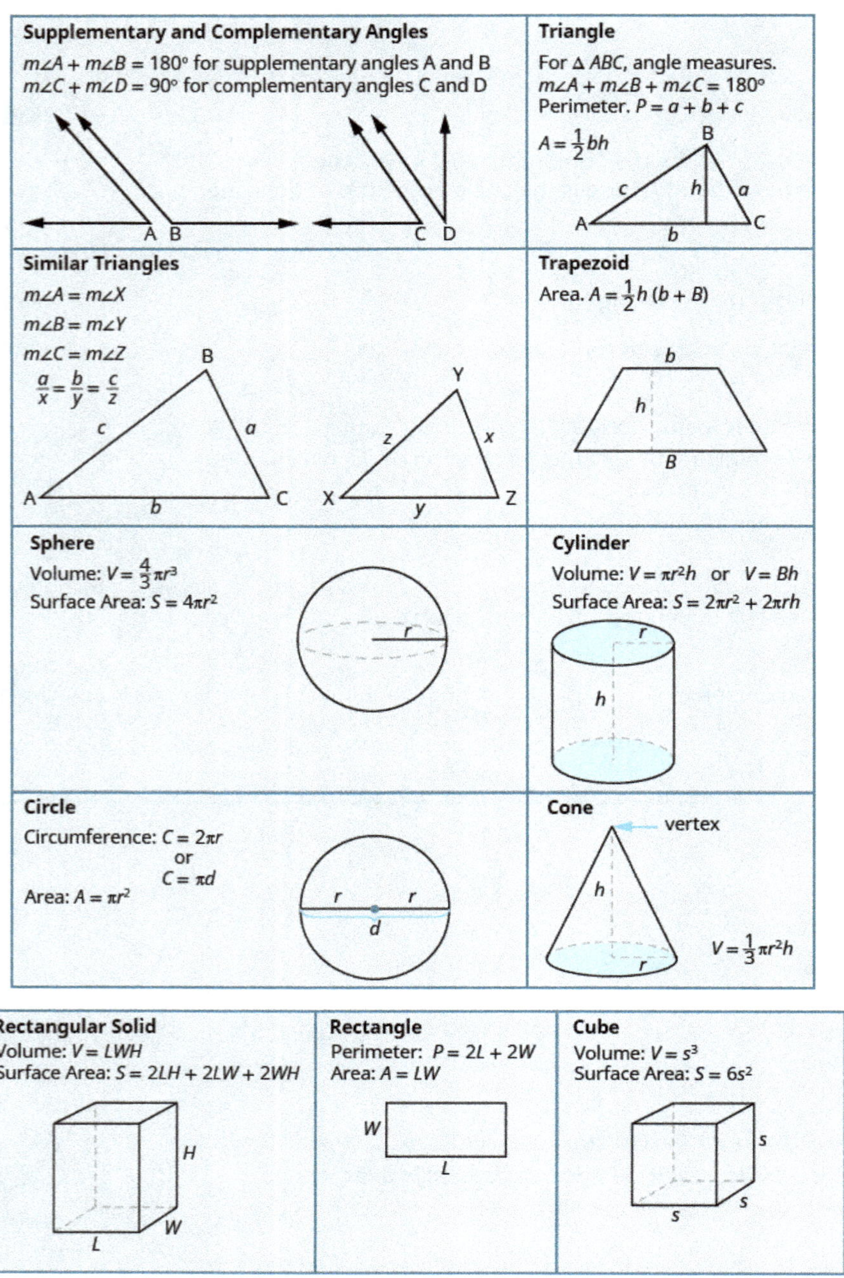

MEDIA : : ACCESS ADDITIONAL ONLINE RESOURCES

- Volume of a Cone (http://openstaxcollege.org/l/24volcone)

9.6 EXERCISES

Practice Makes Perfect

Find Volume and Surface Area of Rectangular Solids

In the following exercises, find the volume and the surface area of the rectangular solid with the given dimensions.

263. length meters, width meters, height meters

264. length feet, width feet, height feet

265. length yards, width yards, height yards

266. length centimeters, width centimeters, height centimeters

In the following exercises, solve.

267. Moving van A rectangular moving van has length feet, width feet, and height feet. Find its volume and surface area.

268. Gift box A rectangular gift box has length inches, width inches, and height inches. Find its volume and surface area.

269. Carton A rectangular carton has length cm, width cm, and height cm. Find its volume and surface area.

270. Shipping container A rectangular shipping container has length feet, width feet, and height feet. Find its volume and surface area.

In the following exercises, find the volume and the surface area of the cube with the given side length.

271. centimeters

272. inches

273. feet

274. meters

In the following exercises, solve.

275. Science center Each side of the cube at the Discovery Science Center in Santa Ana is feet long. Find its volume and surface area.

276. Museum A cube-shaped museum has sides meters long. Find its volume and surface area.

277. Base of statue The base of a statue is a cube with sides meters long. Find its volume and surface area.

278. Tissue box A box of tissues is a cube with sides 4.5 inches long. Find its volume and surface area.

Find the Volume and Surface Area of Spheres

In the following exercises, find the volume and the surface area of the sphere with the given radius. Round answers to the nearest hundredth.

279. centimeters

280. inches

281. feet

282. yards

In the following exercises, solve. Round answers to the nearest hundredth.

283. Exercise ball An exercise ball has a radius of inches. Find its volume and surface area.

284. Balloon ride The Great Park Balloon is a big orange sphere with a radius of feet . Find its volume and surface area.

285. Golf ball A golf ball has a radius of centimeters. Find its volume and surface area.

286. Baseball A baseball has a radius of inches. Find its volume and surface area.

Find the Volume and Surface Area of a Cylinder

In the following exercises, find the volume and the surface area of the cylinder with the given radius and height. Round answers to the nearest hundredth.

287. radius feet, height feet

288. radius centimeters, height centimeters

289. radius meters, height meters

290. radius yards, height yards

In the following exercises, solve. Round answers to the nearest hundredth.

291. Coffee can A can of coffee has a radius of cm and a height of cm. Find its volume and surface area.

292. Snack pack A snack pack of cookies is shaped like a cylinder with radius cm and height cm. Find its volume and surface area.

293. Barber shop pole A cylindrical barber shop pole has a diameter of inches and height of inches. Find its volume and surface area.

294. Architecture A cylindrical column has a diameter of feet and a height of feet. Find its volume and surface area.

Find the Volume of Cones

In the following exercises, find the volume of the cone with the given dimensions. Round answers to the nearest hundredth.

295. height feet and radius feet

296. height inches and radius inches

297. height centimeters and radius cm

298. height meters and radius meters

In the following exercises, solve. Round answers to the nearest hundredth.

299. Teepee What is the volume of a cone-shaped teepee tent that is feet tall and feet across at the base?

300. Popcorn cup What is the volume of a cone-shaped popcorn cup that is inches tall and inches across at the base?

301. Silo What is the volume of a cone-shaped silo that is feet tall and feet across at the base?

302. Sand pile What is the volume of a cone-shaped pile of sand that is meters tall and meters across at the base?

Everyday Math

303. Street light post The post of a street light is shaped like a truncated cone, as shown in the picture below. It is a large cone minus a smaller top cone. The large cone is feet tall with base radius foot. The smaller cone is feet tall with base radius of feet. To the nearest tenth,

find the volume of the large cone.

find the volume of the small cone.

find the volume of the post by subtracting the volume of the small cone from the volume of the large cone.

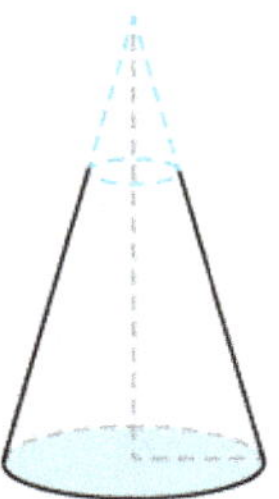

304. Ice cream cones A regular ice cream cone is 4 inches tall and has a diameter of inches. A waffle cone is inches tall and has a diameter of inches. To the nearest hundredth,

find the volume of the regular ice cream cone.

find the volume of the waffle cone.

how much more ice cream fits in the waffle cone compared to the regular cone?

Writing Exercises

305. The formulas for the volume of a cylinder and a cone are similar. Explain how you can remember which formula goes with which shape.

306. Which has a larger volume, a cube of sides of feet or a sphere with a diameter of feet? Explain your reasoning.

Self Check

After completing the exercises, use this checklist to evaluate your mastery of the objectives of this section.

I can...	Confidently	With some help	No-I don't get it!
find volume and surface area of rectangular solids.			
find volume and surface area of spheres.			
find volume and surface area of cylinders.			
find volume of cones.			

After reviewing this checklist, what will you do to become confident for all objectives?

9.7 Solve a Formula for a Specific Variable

Learning Objectives

By the end of this section, you will be able to:

- Use the distance, rate, and time formula
- Solve a formula for a specific variable

BE PREPARED : : 9.19 Before you get started, take this readiness quiz.

Write miles per gallon as a unit rate.

If you missed this problem, review Example 5.65.

BE PREPARED : : 9.20 Solve

If you missed this problem, review Example 8.20.

BE PREPARED : : 9.21

Find the simple interest earned after years on at an interest rate of

If you missed this problem, review Example 6.33.

Use the Distance, Rate, and Time Formula

One formula you'll use often in algebra and in everyday life is the formula for distance traveled by an object moving at a constant speed. The basic idea is probably already familiar to you. Do you know what distance you travel if you drove at a steady rate of miles per hour for hours? (This might happen if you use your car's cruise control while driving on the Interstate.) If you said miles, you already know how to use this formula!

The math to calculate the distance might look like this:

In general, the formula relating distance, rate, and time is

Distance, Rate and Time

For an object moving in at a uniform (constant) rate, the distance traveled, the elapsed time, and the rate are related by the formula

where distance, rate, and time.

Notice that the units we used above for the rate were miles per hour, which we can write as a ratio Then when we multiplied by the time, in hours, the common units 'hour' divided out. The answer was in miles.

EXAMPLE 9.57

Jamal rides his bike at a uniform rate of miles per hour for hours. How much distance has he traveled?

Solution

Step 1. **Read** the problem. You may want to create a mini-chart to summarize the information in the problem.	—
Step 2. **Identify** what you are looking for.	distance traveled
Step 3. **Name.** Choose a variable to represent it.	let d = distance
Step 4. **Translate.** Write the appropriate formula for the situation. Substitute in the given information.	—
Step 5. **Solve** the equation.	
Step 6. **Check:** Does 42 miles make sense? Jamal rides 12 miles in 1 hour, 24 miles in 2 hours, 36 miles in 3 hours, 48 miles in 4 hours, 42 miles in $3\frac{1}{2}$ hours is reasonable	
Step 7. **Answer** the question with a complete sentence.	Jamal rode 42 miles.

TRY IT : : 9.113 Lindsay drove for — hours at miles per hour. How much distance did she travel?

TRY IT : : 9.114 Trinh walked for — hours at miles per hour. How far did she walk?

EXAMPLE 9.58

Rey is planning to drive from his house in San Diego to visit his grandmother in Sacramento, a distance of miles. If he can drive at a steady rate of miles per hour, how many hours will the trip take?

Solution

Step 1. **Read** the problem. Summarize the information in the problem.	
Step 2. **Identify** what you are looking for.	how many hours (time)
Step 3. **Name:** Choose a variable to represent it.	let t = time
Step 4. **Translate.** Write the appropriate formula. Substitute in the given information.	
Step 5. **Solve** the equation.	
Step 6. **Check:** Substitute the numbers into the formula and make sure the result is a true statement.	
Step 7. **Answer** the question with a complete sentence. We know the units of time will be hours because we divided miles by miles per hour.	Rey's trip will take 8 hours.

TRY IT : : 9.115

Lee wants to drive from Phoenix to his brother's apartment in San Francisco, a distance of miles. If he drives at a steady rate of miles per hour, how many hours will the trip take?

TRY IT : : 9.116

Yesenia is miles from Chicago. If she needs to be in Chicago in hours, at what rate does she need to drive?

Solve a Formula for a Specific Variable

In this chapter, you became familiar with some formulas used in geometry. Formulas are also very useful in the sciences and social sciences—fields such as chemistry, physics, biology, psychology, sociology, and criminal justice. Healthcare workers use formulas, too, even for something as routine as dispensing medicine. The widely used spreadsheet program Microsoft Excel™ relies on formulas to do its calculations. Many teachers use spreadsheets to apply formulas to compute student grades. It is important to be familiar with formulas and be able to manipulate them easily.

In Example 9.57 and Example 9.58, we used the formula This formula gives the value of when you substitute in the values of and But in Example 9.58, we had to find the value of We substituted in values of and and then used algebra to solve to If you had to do this often, you might wonder why there isn't a formula that gives the value of when you substitute in the values of and We can get a formula like this by solving the formula for

To solve a formula for a specific variable means to get that variable by itself with a coefficient of on one side of the equation and all the other variables and constants on the other side. We will call this solving an equation for a specific variable *in general.* This process is also called *solving a literal equation*. The result is another formula, made up only of variables. The formula contains letters, or *literals*.

Let's try a few examples, starting with the distance, rate, and time formula we used above.

EXAMPLE 9.59

Solve the formula for

when and in general.

Solution

We'll write the solutions side-by-side so you can see that solving a formula in general uses the same steps as when we have numbers to substitute.

	when $d = 520$ and $r = 65$	in general
Write the forumla.	$d = rt$	$d = rt$
Substitute any given values.	$520 = 65t$	
Divide to isolate t.	$\frac{520}{65} = \frac{65t}{65}$	$\frac{d}{r} = \frac{rt}{r}$
Simplify.	$8 = t$ $t = 8$	$\frac{d}{r} = t$ $t = \frac{d}{r}$

Notice that the solution for is the same as that in Example 9.58. We say the formula — is solved for We can use this version of the formula anytime we are given the distance and rate and need to find the time.

> **TRY IT : :** 9.117 Solve the formula for
>
> when and in general

> **TRY IT : :** 9.118
>
> Solve the formula for
>
> when and in general

We used the formula — in **Use Properties of Rectangles, Triangles, and Trapezoids** to find the area of a triangle when we were given the base and height. In the next example, we will solve this formula for the height.

EXAMPLE 9.60

The formula for area of a triangle is — Solve this formula for

when and in general

Solution

	when $A = 90$ and $b = 15$	in general
Write the forumla.	$A = \frac{1}{2}bh$	$A = \frac{1}{2}bh$
Substitute any given values.	$90 = \frac{1}{2} \cdot 15 \cdot h$	
Clear the fractions.	$2 \cdot 90 = 2 \cdot \frac{1}{2} \cdot 15 \cdot h$	$2 \cdot A = 2 \cdot \frac{1}{2} \cdot b \cdot h$
Simplify.	$180 = 15h$	$2A = bh$
Solve for h.	$12 = h$	$\frac{2A}{b} = h$

We can now find the height of a triangle, if we know the area and the base, by using the formula

—

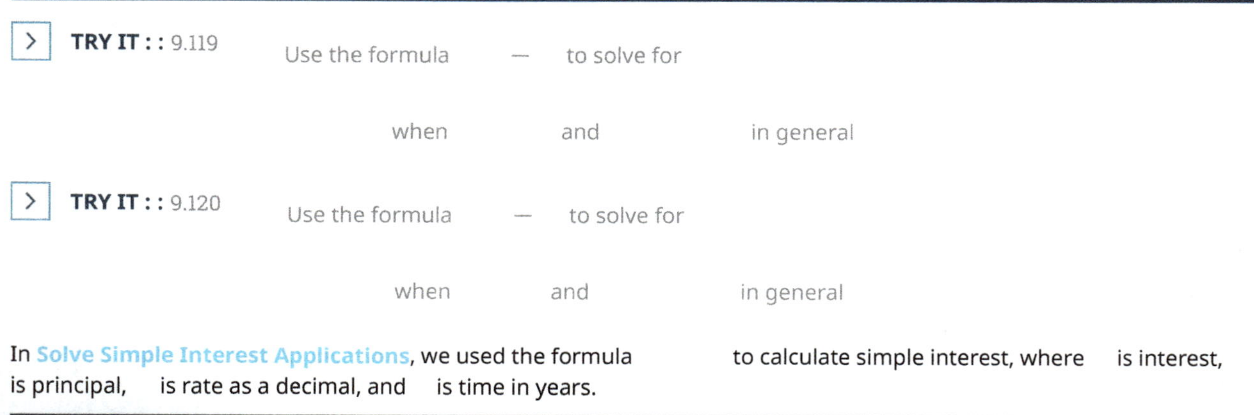

TRY IT : : 9.119 Use the formula — to solve for

when and in general

TRY IT : : 9.120 Use the formula — to solve for

when and in general

In **Solve Simple Interest Applications**, we used the formula to calculate simple interest, where is interest, is principal, is rate as a decimal, and is time in years.

EXAMPLE 9.61

Solve the formula to find the principal,

when in general

Solution

	$I = \$5600, r = 4\%, t = 7$ years	in general
Write the forumla.	$I = Prt$	$I = Prt$
Substitute any given values.	$5600 = P(0.04)(7)$	$I = Prt$
Multiply $r \cdot t$.	$5600 = P(0.28)$	$I = P(rt)$
Divide to isolate P.	$\frac{5600}{0.28} = \frac{P(0.28)}{0.28}$	$\frac{I}{rt} = \frac{P(rt)}{rt}$
Simplify.	$20{,}000 = P$	$\frac{I}{rt} = P$
State the answer.	The principal is \$20,000.	$P = \frac{I}{rt}$

TRY IT : : 9.121 Use the formula

Find when in general

TRY IT : : 9.122 Use the formula

Find when in general

Later in this class, and in future algebra classes, you'll encounter equations that relate two variables, usually and You might be given an equation that is solved for and need to solve it for or vice versa. In the following example, we're given an equation with both and on the same side and we'll solve it for To do this, we will follow the same steps that we used to solve a formula for a specific variable.

EXAMPLE 9.62

Solve the formula for

when in general

Solution

	when $x = 4$	in general
Write the equation.	$3x + 2y = 18$	$3x + 2y = 18$
Substitute any given values.	$3(4) + 2y = 18$	$3x + 2y = 18$
Simplify if possible.	$12 + 2y = 18$	$3x + 2y = 18$
Subtract to isolate the y-term.	$12 - 12 + 2y = 18 - 12$	$3x - 3x + 2y = 18 - 3x$
Simplify.	$2y = 6$	$2y = 18 - 3x$
Divide.	$\frac{2y}{2} = \frac{6}{2}$	$\frac{2y}{2} = \frac{18 - 3x}{2}$
Simplify.	$y = 3$	$y = \frac{18 - 3x}{2}$

TRY IT : : 9.123 Solve the formula for

when in general

TRY IT : : 9.124 Solve the formula for

when in general

In the previous examples, we used the numbers in part (a) as a guide to solving in general in part (b). Do you think you're ready to solve a formula in general without using numbers as a guide?

EXAMPLE 9.63

Solve the formula for

Solution

We will isolate on one side of the equation.

We will isolate *a* on one side of the equation.	
Write the equation.	
Subtract *b* and *c* from both sides to isolate *a*.	$P - b - c = a + b + c - b - c$
Simplify.	

So,

TRY IT : : 9.125 Solve the formula for

TRY IT : : 9.126 Solve the formula for

EXAMPLE 9.64

Solve the equation for

Solution

We will isolate on one side of the equation.

We will isolate *y* on one side of the equation.	
Write the equation.	
Subtract 3*x* from both sides to isolate *y*.	$3x - 3x + y = 10 - 3x$
Simplify.	

TRY IT : : 9.127 Solve the formula for

TRY IT : : 9.128 Solve the formula for

EXAMPLE 9.65

Solve the equation for

Solution

We will isolate on one side of the equation.

We will isolate y on one side of the equation.	
Write the equation.	$6x + 5y = 13$
Subtract to isolate the term with y.	$6x + 5y - 6x = 13 - 6x$
Simplify.	$5y = 13 - 6x$
Divide 5 to make the coefficient 1.	$\frac{5y}{5} = \frac{13 - 6x}{5}$
Simplify.	$y = \frac{13 - 6x}{5}$

> **TRY IT : :** 9.129 Solve the formula for

> **TRY IT : :** 9.130 Solve the formula for

LINKS TO LITERACY

The Links to Literacy activity *What's Faster than a Speeding Cheetah?* will provide you with another view of the topics covered in this section.

MEDIA : : ACCESS ADDITIONAL ONLINE RESOURCES

- Distance=RatexTime (http://www.openstax.org/l/24distratextime)
- Distance, Rate, Time (http://www.openstax.org/l/24distratetime)
- Simple Interest (http://www.openstax.org/l/24simpinterest)
- Solving a Formula for a Specific Variable (http://www.openstax.org/l/24solvespvari)
- Solving a Formula for a Specific Variable (http://www.openstax.org/l/24solvespecivar)

9.7 EXERCISES

Practice Makes Perfect

Use the Distance, Rate, and Time Formula

In the following exercises, solve.

307. Steve drove for — hours at miles per hour. How much distance did he travel?

308. Socorro drove for — hours at miles per hour. How much distance did she travel?

309. Yuki walked for — hours at miles per hour. How far did she walk?

310. Francie rode her bike for — hours at miles per hour. How far did she ride?

311. Connor wants to drive from Tucson to the Grand Canyon, a distance of miles. If he drives at a steady rate of miles per hour, how many hours will the trip take?

312. Megan is taking the bus from New York City to Montreal. The distance is miles and the bus travels at a steady rate of miles per hour. How long will the bus ride be?

313. Aurelia is driving from Miami to Orlando at a rate of miles per hour. The distance is miles. To the nearest tenth of an hour, how long will the trip take?

314. Kareem wants to ride his bike from St. Louis, Missouri to Champaign, Illinois. The distance is miles. If he rides at a steady rate of miles per hour, how many hours will the trip take?

315. Javier is driving to Bangor, Maine, which is miles away from his current location. If he needs to be in Bangor in hours, at what rate does he need to drive?

316. Alejandra is driving to Cincinnati, Ohio, miles away. If she wants to be there in hours, at what rate does she need to drive?

317. Aisha took the train from Spokane to Seattle. The distance is miles, and the trip took hours. What was the speed of the train?

318. Philip got a ride with a friend from Denver to Las Vegas, a distance of miles. If the trip took hours, how fast was the friend driving?

Solve a Formula for a Specific Variable

In the following exercises, use the formula.

319. Solve for when and in general

320. Solve for when and in general

321. Solve for when and in general

322. Solve for when and in general

323. Solve for when and in general

324. Solve for when and in general

325. Solve for when and in general

326. Solve for when and in general.

In the following exercises, use the formula —

327. Solve for

when and

in general

328. Solve for

when and

in general

329. Solve for

when and

in general

330. Solve for

when and

in general

In the following exercises, use the formula

331. Solve for the principal, for:

, ,

in general

332. Solve for the principal, for:

, ,

in general

333. Solve for the time, for:

,

,

in general

334. Solve for the time, for:

, ,

in general

In the following exercises, solve.

335. Solve the formula for

when

in general

336. Solve the formula for

when

in general

337. Solve the formula for

when

in general

338. Solve the formula for

when

in general

339. Solve for

340. Solve for

341. Solve for

342. Solve for

343. Solve the formula for

344. Solve the formula for

345. Solve the formula for

346. Solve the formula for

347. Solve the formula for

348. Solve the formula for

349. Solve the formula for

350. Solve the formula for

351. Solve the formula for

352. Solve the formula for

353. Solve the formula for

354. Solve the formula for

355. Solve the formula for

356. Solve the formula
for

Everyday Math

357. Converting temperature While on a tour in Greece, Tatyana saw that the temperature was Celsius. Solve for in the formula — to find the temperature in Fahrenheit.

358. Converting temperature Yon was visiting the United States and he saw that the temperature in Seattle was Fahrenheit. Solve for in the formula — to find the temperature in Celsius.

Writing Exercises

359. Solve the equation for

when in general

Which solution is easier for you? Explain why.

360. Solve the equation for

when in general

Which solution is easier for you? Explain why.

Self Check

After completing the exercises, use this checklist to evaluate your mastery of the objectives of this section.

I can...	Confidently	With some help	No-I don't get it!
use the distance, rate, and time formula.			
solve a formula for a specific variable.			

Overall, after looking at the checklist, do you think you are well-prepared for the next Chapter? Why or why not?

CHAPTER 9 REVIEW

KEY TERMS

angle An angle is formed by two rays that share a common endpoint. Each ray is called a side of the angle.

area The area is a measure of the surface covered by a figure.

complementary angles If the sum of the measures of two angles is , then they are called complementary angles.

cone A cone is a solid figure with one circular base and a vertex.

cube A cube is a rectangular solid whose length, width, and height are equal.

cylinder A cylinder is a solid figure with two parallel circles of the same size at the top and bottom.

equilateral triangle A triangle with all three sides of equal length is called an equilateral triangle.

hypotenuse The side of the triangle opposite the 90° angle is called the hypotenuse.

irregular figure An irregular figure is a figure that is not a standard geometric shape. Its area cannot be calculated using any of the standard area formulas.

isosceles triangle A triangle with two sides of equal length is called an isosceles triangle.

legs of a right triangle The sides of a right triangle adjacent to the right angle are called the legs.

perimeter The perimeter is a measure of the distance around a figure.

rectangle A rectangle is a geometric figure that has four sides and four right angles.

right triangle A right triangle is a triangle that has one angle.

similar figures In geometry, if two figures have exactly the same shape but different sizes, we say they are similar figures.

supplementary angles If the sum of the measures of two angles is , then they are called supplementary angles.

trapezoid A trapezoid is four-sided figure, a quadrilateral, with two sides that are parallel and two sides that are not.

triangle A triangle is a geometric figure with three sides and three angles.

vertex of an angle When two rays meet to form an angle, the common endpoint is called the vertex of the angle.

KEY CONCEPTS

9.1 Use a Problem Solving Strategy

- **Problem Solving Strategy**

  Step 1. Read the word problem. Make sure you understand all the words and ideas. You may need to read the problem two or more times. If there are words you don't understand, look them up in a dictionary or on the internet.

  Step 2. Identify what you are looking for.

  Step 3. Name what you are looking for. Choose a variable to represent that quantity.

  Step 4. Translate into an equation. It may be helpful to first restate the problem in one sentence before translating.

  Step 5. Solve the equation using good algebra techniques.

  Step 6. Check the answer in the problem. Make sure it makes sense.

  Step 7. Answer the question with a complete sentence.

9.2 Solve Money Applications

- **Finding the Total Value for Coins of the Same Type**
  - For coins of the same type, the total value can be found as follows:

    where number is the number of coins, value is the value of each coin, and total value is the total value of all the coins.
- **Solve a Coin Word Problem**

  Step 1. **Read** the problem. Make sure you understand all the words and ideas, and create a table to organize the information.

Step 2. **Identify** what you are looking for.

Step 3. **Name** what you are looking for. Choose a variable to represent that quantity.

- Use variable expressions to represent the number of each type of coin and write them in the table.
- Multiply the number times the value to get the total value of each type of coin.

Step 4. **Translate** into an equation. Write the equation by adding the total values of all the types of coins.

Step 5. **Solve** the equation using good algebra techniques.

Step 6. **Check** the answer in the problem and make sure it makes sense.

Step 7. **Answer** the question with a complete sentence.

Type			

9.3 Use Properties of Angles, Triangles, and the Pythagorean Theorem

- **Supplementary and Complementary Angles**
  - If the sum of the measures of two angles is 180°, then the angles are supplementary.
  - If and are supplementary, then .
  - If the sum of the measures of two angles is 90°, then the angles are complementary.
  - If and are complementary, then .
- **Solve Geometry Applications**

Step 1. Read the problem and make sure you understand all the words and ideas. Draw a figure and label it with the given information.

Step 2. Identify what you are looking for.

Step 3. Name what you are looking for and choose a variable to represent it.

Step 4. Translate into an equation by writing the appropriate formula or model for the situation. Substitute in the given information.

Step 5. Solve the equation using good algebra techniques.

Step 6. Check the answer in the problem and make sure it makes sense.

Step 7. Answer the question with a complete sentence.

- **Sum of the Measures of the Angles of a Triangle**

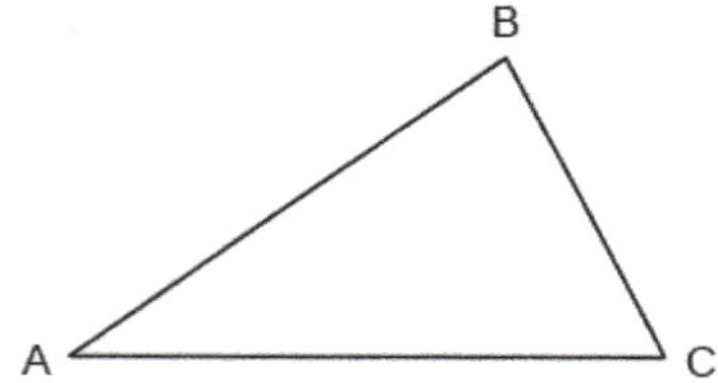

- For any the sum of the measures is 180°

- **Right Triangle**

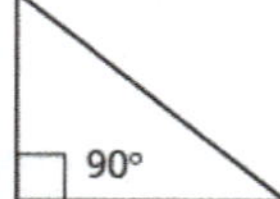

- A right triangle is a triangle that has one 90° angle, which is often marked with a symbol.

- **Properties of Similar Triangles**
    - If two triangles are similar, then their corresponding angle measures are equal and their corresponding side lengths have the same ratio.

9.4 Use Properties of Rectangles, Triangles, and Trapezoids

- **Properties of Rectangles**
    - Rectangles have four sides and four right (90°) angles.
    - The lengths of opposite sides are equal.
    - The perimeter, , of a rectangle is the sum of twice the length and twice the width.
    - The area, , of a rectangle is the length times the width.
- **Triangle Properties**
    - For any triangle , the sum of the measures of the angles is 180°.
    - The perimeter of a triangle is the sum of the lengths of the sides.
    - The area of a triangle is one-half the base, b, times the height, h.
        - —

9.5 Solve Geometry Applications: Circles and Irregular Figures

- **Problem Solving Strategy for Geometry Applications**

Step 1. Read the problem and make sure you understand all the words and ideas. Draw the figure and label it with the given information.

Step 2. Identify what you are looking for.

Step 3. Name what you are looking for. Choose a variable to represent that quantity.

Step 4. Translate into an equation by writing the appropriate formula or model for the situation. Substitute in the given information.

Step 5. Solve the equation using good algebra techniques.

Step 6. Check the answer in the problem and make sure it makes sense.

Step 7. Answer the question with a complete sentence.

- **Properties of Circles**

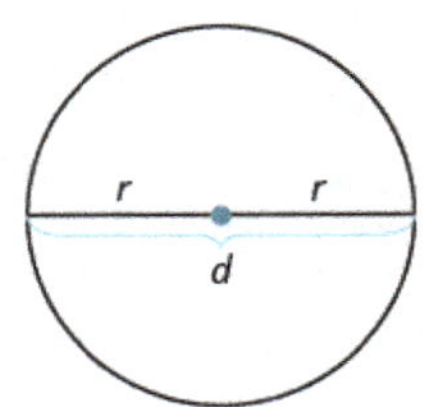

- **Circumference:** or
- **Area:**

9.6 Solve Geometry Applications: Volume and Surface Area

- **Volume and Surface Area of a Rectangular Solid**
  -
  -
- **Volume and Surface Area of a Cube**
  -
  -
- **Volume and Surface Area of a Sphere**
  - —
  -
- **Volume and Surface Area of a Cylinder**
  -
  -
- **Volume of a Cone**
  - For a cone with radius and height :

    Volume: —

9.7 Solve a Formula for a Specific Variable

- **Distance, Rate, and Time**
  -

REVIEW EXERCISES

9.1 Use a Problem Solving Strategy

Approach Word Problems with a Positive Attitude

In the following exercises, solve.

361. How has your attitude towards solving word problems changed as a result of working through this chapter? Explain.

362. Did the Problem Solving Strategy help you solve word problems in this chapter? Explain.

Use a Problem Solving Strategy for Word Problems

In the following exercises, solve using the problem-solving strategy for word problems. Remember to write a complete sentence to answer each question.

363. Three-fourths of the people at a concert are children. If there are children, what is the total number of people at the concert?

364. There are saxophone players in the band. The number of saxophone players is one less than twice the number of tuba players. Find the number of tuba players.

365. Reza was very sick and lost of his original weight. He lost pounds. What was his original weight?

366. Dolores bought a crib on sale for The sale price was of the original price. What was the original price of the crib?

Solve Number Problems

In the following exercises, solve each number word problem.

367. The sum of a number and three is forty-one. Find the number.

368. Twice the difference of a number and ten is fifty-four. Find the number.

369. One number is nine less than another. Their sum is twenty-seven. Find the numbers.

370. The sum of two consecutive integers is Find the numbers.

9.2 Solve Money Applications

Solve Coin Word Problems

In the following exercises, solve each coin word problem.

371. Francie has in dimes and quarters. The number of dimes is more than the number of quarters. How many of each coin does she have?

372. Scott has in pennies and nickels. The number of pennies is times the number of nickels. How many of each coin does he have?

373. Paulette has in and bills. The number of bills is one less than twice the number of bills. How many of each does she have?

374. Lenny has in pennies, dimes, and quarters. The number of pennies is more than the number of dimes. The number of quarters is twice the number of dimes. How many of each coin does he have?

Solve Ticket and Stamp Word Problems

In the following exercises, solve each ticket or stamp word problem.

375. A church luncheon made Adult tickets cost each and children's tickets cost each. The number of children was more than twice the number of adults. How many of each ticket were sold?

376. Tickets for a basketball game cost for students and for adults. The number of students was less than times the number of adults. The total amount of money from ticket sales was How many of each ticket were sold?

377. Ana spent buying stamps. The number of stamps she bought was more than the number of stamps. How many of each did she buy?

378. Yumi spent buying stamps. The number of stamps she bought was less than times the number of stamps. How many of each did she buy?

9.3 Use Properties of Angles, Triangles, and the Pythagorean Theorem

Use Properties of Angles

In the following exercises, solve using properties of angles.

379. What is the supplement of a angle?

380. What is the complement of a angle?

381. Two angles are complementary. The smaller angle is less than the larger angle. Find the measures of both angles.

382. Two angles are supplementary. The larger angle is more than the smaller angle. Find the measures of both angles.

Use Properties of Triangles

In the following exercises, solve using properties of triangles.

383. The measures of two angles of a triangle are and degrees. Find the measure of the third angle.

384. One angle of a right triangle measures degrees. What is the measure of the other small angle?

385. One angle of a triangle is more than the smallest angle. The largest angle is the sum of the other angles. Find the measures of all three angles.

386. One angle of a triangle is twice the measure of the smallest angle. The third angle is more than the measure of the smallest angle. Find the measures of all three angles.

In the following exercises, is similar to Find the length of the indicated side.

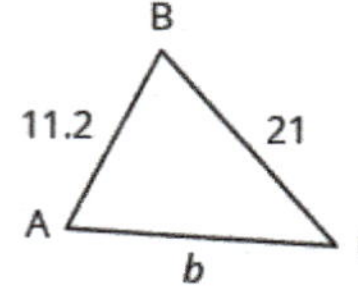

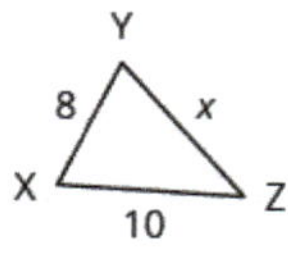

387. side

388. side

Use the Pythagorean Theorem

In the following exercises, use the Pythagorean Theorem to find the length of the missing side. Round to the nearest tenth, if necessary.

389.

390.

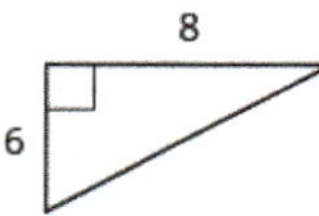

391.

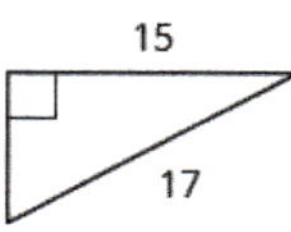

392.

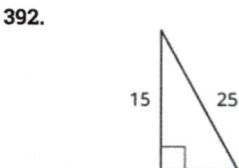

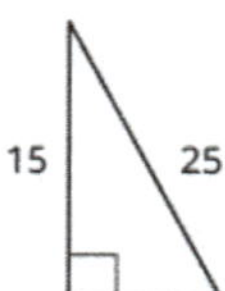

393.

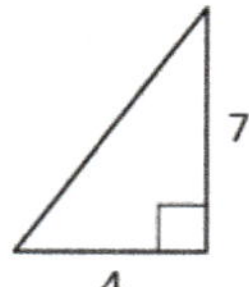

394.

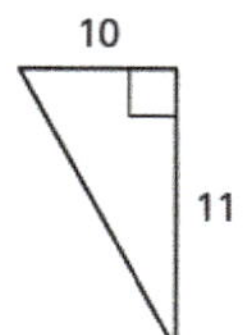

In the following exercises, solve. Approximate to the nearest tenth, if necessary.

395. Sergio needs to attach a wire to hold the antenna to the roof of his house, as shown in the figure. The antenna is feet tall and Sergio has feet of wire. How far from the base of the antenna can he attach the wire?

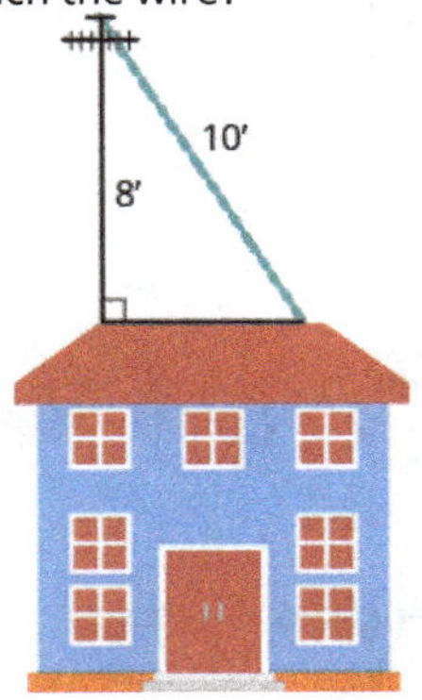

396. Seong is building shelving in his garage. The shelves are inches wide and inches tall. He wants to put a diagonal brace across the back to stabilize the shelves, as shown. How long should the brace be?

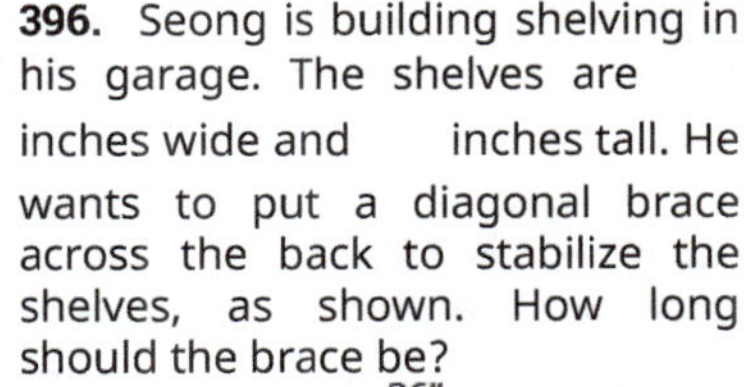

9.4 Use Properties of Rectangles, Triangles, and Trapezoids

Understand Linear, Square, Cubic Measure

In the following exercises, would you measure each item using linear, square, or cubic measure?

397. amount of sand in a sandbag

398. height of a tree

399. size of a patio

400. length of a highway

In the following exercises, find

the perimeter the area of each figure

401.

402.

Use Properties of Rectangles

In the following exercises, find the perimeter area of each rectangle

403. The length of a rectangle is meters and the width is meters.

404. The length of a rectangle is feet and the width is feet.

405. A sidewalk in front of Kathy's house is in the shape of a rectangle feet wide by feet long.

406. A rectangular room is feet wide by feet long.

In the following exercises, solve.

407. Find the length of a rectangle with perimeter of centimeters and width of centimeters.

408. Find the width of a rectangle with perimeter and length

409. The area of a rectangle is square meters. The length is meters. What is the width?

410. The width of a rectangle is centimeters. The area is square centimeters. What is the length?

411. The length of a rectangle is centimeters more than the width. The perimeter is centimeters. Find the length and the width.

412. The width of a rectangle is more than twice the length. The perimeter is inches. Find the length and the width.

Use Properties of Triangles

In the following exercises, solve using the properties of triangles.

413. Find the area of a triangle with base inches and height inches.

414. Find the area of a triangle with base centimeters and height centimeters.

415. A triangular road sign has base inches and height inches. What is its area?

416. If a triangular courtyard has sides feet and feet and the perimeter is feet, how long is the third side?

417. A tile in the shape of an isosceles triangle has a base of inches. If the perimeter is inches, find the length of each of the other sides.

418. Find the length of each side of an equilateral triangle with perimeter of yards.

419. The perimeter of a triangle is feet. One side of the triangle is feet longer than the shortest side. The third side is feet longer than the shortest side. Find the length of each side.

420. One side of a triangle is three times the smallest side. The third side is feet more than the shortest side. The perimeter is feet. Find the lengths of all three sides.

Use Properties of Trapezoids

In the following exercises, solve using the properties of trapezoids.

421. The height of a trapezoid is feet and the bases are and feet. What is the area?

422. The height of a trapezoid is yards and the bases are and yards. What is the area?

423. Find the area of the trapezoid with height meters and bases and meters.

424. A flag is shaped like a trapezoid with height centimeters and the bases are and centimeters. What is the area of the flag?

9.5 Solve Geometry Applications: Circles and Irregular Figures

Use Properties of Circles

In the following exercises, solve using the properties of circles. Round answers to the nearest hundredth.

425. A circular mosaic has radius meters. Find the

circumference

area of the mosaic

426. A circular fountain has radius feet. Find the

circumference

area of the fountain

427. Find the diameter of a circle with circumference inches.

428. Find the radius of a circle with circumference centimeters

Find the Area of Irregular Figures

In the following exercises, find the area of each shaded region.

429.

8
3
5
3

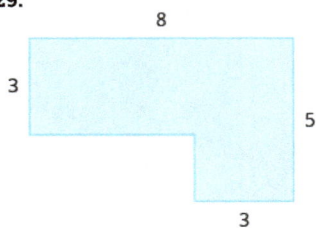

430.

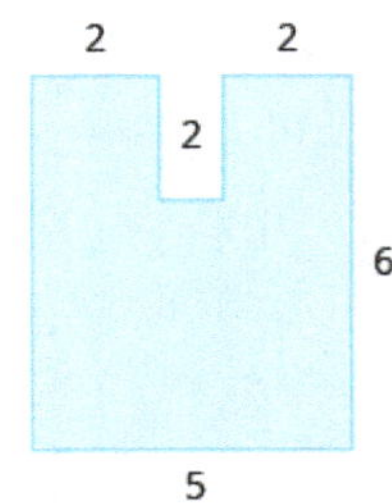

431.

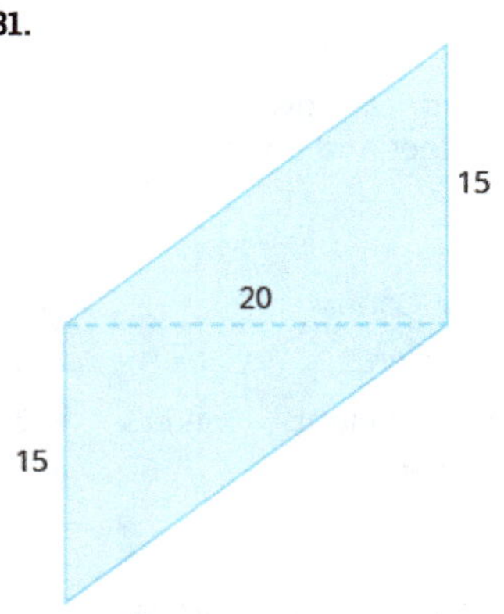

432.

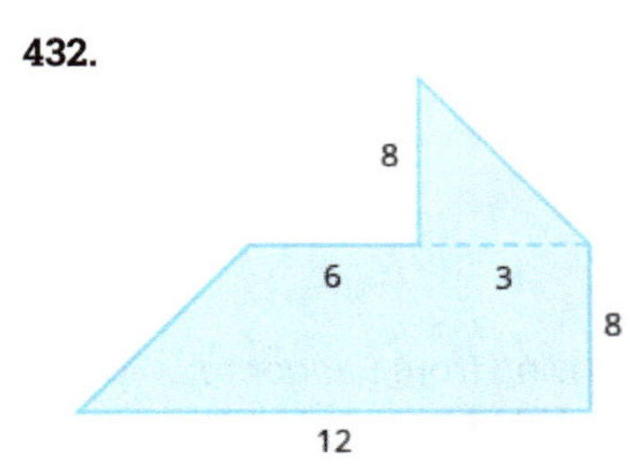

433.

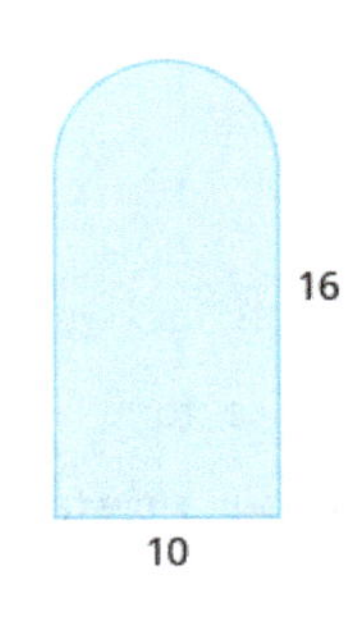

434.

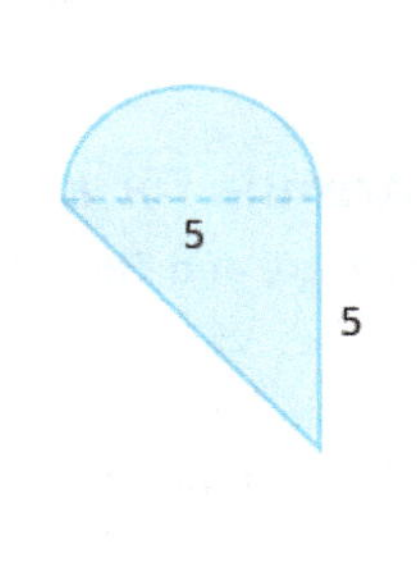

9.6 Solve Geometry Applications: Volume and Surface Area

Find Volume and Surface Area of Rectangular Solids

In the following exercises, find the

volume surface area of the rectangular solid

435. a rectangular solid with length centimeters, width centimeters, and height centimeters

436. a cube with sides that are feet long

437. a cube of tofu with sides inches

438. a rectangular carton with length inches, width inches, and height inches

Find Volume and Surface Area of Spheres

In the following exercises, find the

volume surface area of the sphere.

439. a sphere with radius yards

440. a sphere with radius meters

441. a baseball with radius inches

442. a soccer ball with radius centimeters

Find Volume and Surface Area of Cylinders

In the following exercises, find the

volume surface area of the cylinder

443. a cylinder with radius yards and height yards

444. a cylinder with diameter inches and height inches

445. a juice can with diameter centimeters and height centimeters

446. a cylindrical pylon with diameter feet and height feet

Find Volume of Cones

In the following exercises, find the volume of the cone.

447. a cone with height meters and radius meter

448. a cone with height feet and radius feet

449. a cone-shaped water cup with diameter inches and height inches

450. a cone-shaped pile of gravel with diameter yards and height yards

9.7 Solve a Formula for a Specific Variable

Use the Distance, Rate, and Time Formula

In the following exercises, solve using the formula for distance, rate, and time.

451. A plane flew hours at miles per hour. What distance was covered?

452. Gus rode his bike for — hours at miles per hour. How far did he ride?

453. Jack is driving from Bangor to Portland at a rate of miles per hour. The distance is miles. To the nearest tenth of an hour, how long will the trip take?

454. Jasmine took the bus from Pittsburgh to Philadelphia. The distance is miles and the trip took hours. What was the speed of the bus?

Solve a Formula for a Specific Variable

In the following exercises, use the formula

455. Solve for

when and

in general

456. Solve for

when and

in general

In the following exercises, use the formula —

457. Solve for

when and

in general

458. Solve for

when and

in general

In the following exercises, use the formula

459. Solve for the principal, for:

, ,

in general

460. Solve for the time, for:

,

,

in general

In the following exercises, solve.

461. Solve the formula for

when

in general

462. Solve the formula for

when

in general

463. Solve for

464. Solve for

465. Solve the formula for

466. Solve the formula for

467. Solve the formula for

468. Solve the formula for

469. Describe how you have used two topics from this chapter in your life outside of math class during the past month.

PRACTICE TEST

470. Four-fifths of the people on a hike are children. If there are children, what is the total number of people on the hike?

471. The sum of and twice a number is Find the number.

472. One number is less than another number. Their sum is Find the numbers.

473. Bonita has in dimes and quarters in her pocket. If she has more dimes than quarters, how many of each coin does she have?

474. At a concert, in tickets were sold. Adult tickets were each and children's tickets were each. If the number of adult tickets was fewer than twice the number of children's tickets, how many of each kind were sold?

475. Find the complement of a angle.

476. The measure of one angle of a triangle is twice the measure of the smallest angle. The measure of the third angle is more than the measure of the smallest angle. Find the measures of all three angles.

477. The perimeter of an equilateral triangle is feet. Find the length of each side.

478. is similar to Find the length of side

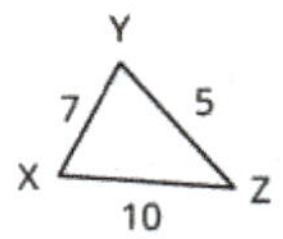

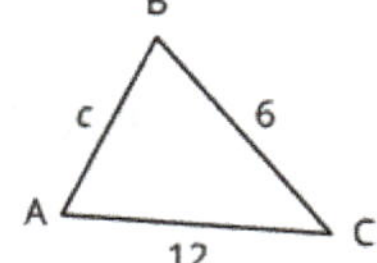

479. Find the length of the missing side. Round to the nearest tenth, if necessary.

480. Find the length of the missing side. Round to the nearest tenth, if necessary.

481. A baseball diamond is shaped like a square with sides feet long. How far is it from home plate to second base, as shown?

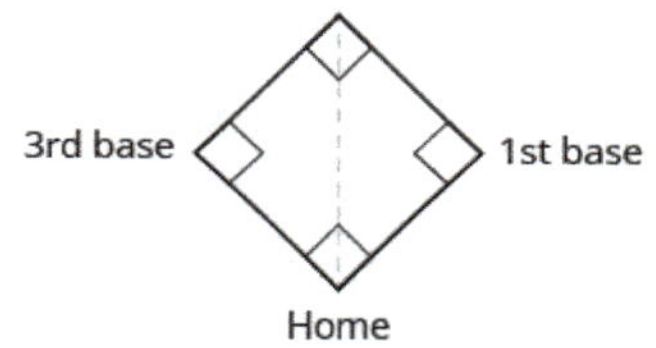

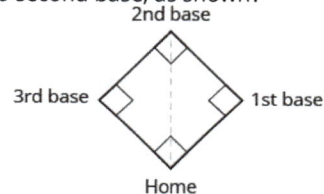

482. The length of a rectangle is feet more than five times the width. The perimeter is feet. Find the dimensions of the rectangle.

483. A triangular poster has base centimeters and height centimeters. Find the area of the poster.

484. A trapezoid has height inches and bases inches and inches. Find the area of the trapezoid.

485. A circular pool has diameter inches. What is its circumference? Round to the nearest *tenth.*

486. Find the area of the shaded region. Round to the nearest tenth.

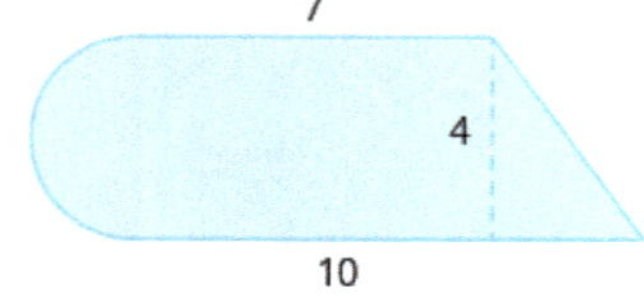

487. Find the volume of a rectangular room with width feet, length feet, and height feet.

488. A coffee can is shaped like a cylinder with height inches and radius inches. Find (a) the surface area and (b) the volume of the can. Round to the nearest tenth.

489. A traffic cone has height centimeters. The radius of the base is centimeters. Find the volume of the cone. Round to the nearest tenth.

490. Leon drove from his house in Cincinnati to his sister's house in Cleveland. He drove at a uniform rate of miles per hour and the trip took hours. What was the distance?

491. The Catalina Express takes — hours to travel from Long Beach to Catalina Island, a distance of miles. To the nearest tenth, what is the speed of the boat?

492. Use the formula to solve for the principal, for:

years

in general

493. Solve the formula — for

when and

in general

494. Solve for

Figure 10.1 The paths of rockets are calculated using polynomials. (credit: NASA, Public Domain)

Chapter Outline

Introduction

Expressions known as polynomials are used widely in algebra. Applications of these expressions are essential to many careers, including economists, engineers, and scientists. In this chapter, we will find out what polynomials are and how to manipulate them through basic mathematical operations.

10.1 Add and Subtract Polynomials

Learning Objectives

By the end of this section, you will be able to:

- Identify polynomials, monomials, binomials, and trinomials
- Determine the degree of polynomials
- Add and subtract monomials
- Add and subtract polynomials
- Evaluate a polynomial for a given value

BE PREPARED : : 10.1 Before you get started, take this readiness quiz.

Simplify:

If you missed this problem, review Example 2.22.

BE PREPARED : : 10.2 Subtract:

If you missed this problem, review Example 7.29.

BE PREPARED : : 10.3 Evaluate: when
If you missed this problem, review Example 2.18.

Identify Polynomials, Monomials, Binomials, and Trinomials

In Evaluate, Simplify, and Translate Expressions, you learned that a term is a constant or the product of a constant and one or more variables. When it is of the form where is a constant and is a whole number, it is called a monomial. A monomial, or a sum and/or difference of monomials, is called a polynomial.

Polynomials

polynomial—A monomial, or two or more monomials, combined by addition or subtraction
monomial—A polynomial with exactly one term
binomial— A polynomial with exactly two terms
trinomial—A polynomial with exactly three terms

Notice the roots:

- *poly-* means many
- *mono-* means one
- *bi-* means two
- *tri-* means three

Here are some examples of polynomials:

Polynomial			
Monomial			
Binomial			
Trinomial			

Notice that every monomial, binomial, and trinomial is also a polynomial. They are special members of the family of polynomials and so they have special names. We use the words 'monomial', 'binomial', and 'trinomial' when referring to these special polynomials and just call all the rest 'polynomials'.

EXAMPLE 10.1

Determine whether each polynomial is a monomial, binomial, trinomial, or other polynomial:

Solution

Polynomial	Number of terms	Type
	3	Trinomial
	1	Monomial
	5	Polynomial
	2	Binomial
	1	Monomial

TRY IT : : 10.1 Determine whether each polynomial is a monomial, binomial, trinomial, or other polynomial.

TRY IT : : 10.2 Determine whether each polynomial is a monomial, binomial, trinomial, or other polynomial.

Determine the Degree of Polynomials

In this section, we will work with polynomials that have only one variable in each term. The degree of a polynomial and the degree of its terms are determined by the exponents of the variable.

A monomial that has no variable, just a constant, is a special case. The degree of a constant is —it has no variable.

Degree of a Polynomial

The **degree of a term** is the exponent of its variable.

The degree of a constant is

The degree of a polynomial is the highest degree of all its terms.

Let's see how this works by looking at several polynomials. We'll take it step by step, starting with monomials, and then progressing to polynomials with more terms.

Remember: Any base written without an exponent has an implied exponent of

Monomials	5	$4b^2$	$-9x^3$	-18
Degree	0	2	3	0
Binomial	$b+1$	$3a-7$	y^2-9	$17x^3+14x^2$
Degree of each term	1 0	1 0	2 0	3 2
Degree of polynomial	1	1	2	3
Trinomial	x^2-5x+6	$4y^2-7y+2$	$5a^4-3a^3+a$	x^4+2x^2-5
Degree of each term	2 1 0	2 1 0	4 3 1	4 2 0
Degree of polynomial	2	2	4	4
Polynomial	$b+1$	$4y^2-7y+2$	$4x^4+x^3+8x^2-9x+1$	
Degree of each term	1 0	2 1 0	4 3 2 1 0	
Degree of polynomial	1	2	4	

EXAMPLE 10.2

Find the degree of the following polynomials:

Solution

The exponent of is one.	The degree is 1.
The highest degree of all the terms is 3.	The degree is 3
The degree of a constant is 0.	The degree is 0.
The highest degree of all the terms is 2.	The degree is 2.
The highest degree of all the terms is 1.	The degree is 1.

TRY IT : : 10.3 Find the degree of the following polynomials:

TRY IT : : 10.4 Find the degree of the following polynomials:

Working with polynomials is easier when you list the terms in descending order of degrees. When a polynomial is written this way, it is said to be in **standard form**. Look back at the polynomials in Example 10.2. Notice that they are all written in standard form. Get in the habit of writing the term with the highest degree first.

Add and Subtract Monomials

In The Language of Algebra, you simplified expressions by combining like terms. Adding and subtracting monomials is the same as combining like terms. Like terms must have the same variable with the same exponent. Recall that when combining like terms only the coefficients are combined, never the exponents.

EXAMPLE 10.3

Add:

Solution

Combine like terms.

TRY IT : : 10.5 Add:

TRY IT : : 10.6 Add:

EXAMPLE 10.4

Subtract:

Solution

Combine like terms.

TRY IT : : 10.7 Subtract:

TRY IT : : 10.8 Subtract:

EXAMPLE 10.5

Simplify:

Solution

Combine like terms.	

Remember, and are not like terms. The variables are not the same.

TRY IT : : 10.9 Add:

TRY IT : : 10.10 Add:

Add and Subtract Polynomials

Adding and subtracting polynomials can be thought of as just adding and subtracting like terms. Look for like terms—those with the same variables with the same exponent. The Commutative Property allows us to rearrange the terms to put like terms together. It may also be helpful to underline, circle, or box like terms.

EXAMPLE 10.6

Find the sum:

Solution

	$(4x^2 - 5x + 1) + (3x^2 - 8x - 9)$
Identify like terms.	$4x^2 - 5x + 1 + 3x^2 - 8x - 9$
Rearrange to get the like terms together.	$4x^2 + 3x^2 - 5x - 8x + 1 - 9$
Combine like terms.	$7x^2 - 13x - 8$

TRY IT : : 10.11 Find the sum:

TRY IT : : 10.12 Find the sum:

Parentheses are grouping symbols. When we add polynomials as we did in Example 10.6, we can rewrite the expression without parentheses and then combine like terms. But when we subtract polynomials, we must be very careful with the signs.

EXAMPLE 10.7

Find the difference:

Solution

	$(7u^2 - 5u + 3) - (4u^2 - 2)$
Distribute and identify like terms.	$7u^2 - 5u + \boxed{3} - 4u^2 + \boxed{2}$
Rearrange the terms.	$7u^2 - 4u^2 - 5u + 3 + \boxed{2}$
Combine like terms.	$3u^2 - 5u + 5$

TRY IT : : 10.13 Find the difference:

TRY IT : : 10.14 Find the difference:

EXAMPLE 10.8

Subtract: from

Solution

	Subtract $(m^2 - 3m + 8)$ from $(9m^2 - 7m + 4)$
Distribute and identify like terms.	$9m^2 - 7m + \boxed{4} - m^2 + 3m - \boxed{8}$
Rearrange the terms.	$9m^2 - m^2 - 7m + 3m + \boxed{4} - \boxed{8}$
Combine like terms.	$8m^2 - 4m - 4$

TRY IT : : 10.15 Subtract: from

TRY IT : : 10.16 Subtract: from

Evaluate a Polynomial for a Given Value

In The Language of Algebra we evaluated expressions. Since polynomials are expressions, we'll follow the same procedures to evaluate polynomials—substitute the given value for the variable into the polynomial, and then simplify.

EXAMPLE 10.9

Evaluate when

Solution

Substitute 3 for	
Simplify the expression with the exponent.	
Multiply.	
Simplify.	

Substitute −1 for	
Simplify the expression with the exponent.	
Multiply.	
Simplify.	

TRY IT : : 10.17 Evaluate: when

TRY IT : : 10.18 Evaluate: when

EXAMPLE 10.10

The polynomial gives the height of an object seconds after it is dropped from a foot tall bridge. Find the height after seconds.

Solution

	$-16t^2 + 300$
Substitute 3 for	$-16(3)^2 + 300$
Simplify the expression with the exponent.	$-16 \cdot 9 + 300$
Multiply.	$-144 + 300$
Simplify.	156

TRY IT : : 10.19

The polynomial gives the height, in feet, of a ball seconds after it is tossed into the air, from an initial height of feet. Find the height after seconds.

TRY IT : : 10.20

The polynomial gives the height, in feet, of a ball seconds after it is tossed into the air, from an initial height of feet. Find the height after seconds.

MEDIA : : ACCESS ADDITIONAL ONLINE RESOURCES

- Adding Polynomials (http://openstaxcollege.org/l/24addpolynomi)
- Subtracting Polynomials (http://openstaxcollege.org/l/24subtractpoly)

10.1 EXERCISES

Practice Makes Perfect

Identify Polynomials, Monomials, Binomials and Trinomials

In the following exercises, determine if each of the polynomials is a monomial, binomial, trinomial, or other polynomial.

1.

2.

3.

4.

5.

6.

7.

8.

Determine the Degree of Polynomials

In the following exercises, determine the degree of each polynomial.

9.

10.

11.

12.

13.

14.

Add and Subtract Monomials

In the following exercises, add or subtract the monomials.

15.

16.

17.

18.

19.

20.

21. Add:

22. Add:

23.

24.

25. Subtract

26. Subtract

Add and Subtract Polynomials

In the following exercises, add or subtract the polynomials.

27.

28.

29.

30.

31.

32.

33.

34.

35.

36.

37.

38.

39. Find the sum of and

40. Find the sum of and

41. Subtract from

42. Subtract from

43. Find the difference of and

44. Find the difference of and

Evaluate a Polynomial for a Given Value

In the following exercises, evaluate each polynomial for the given value.

45.

46.

47.

48.

49. A window washer drops a squeegee from a platform feet high. The polynomial gives the height of the squeegee seconds after it was dropped. Find the height after seconds.

50. A manufacturer of microwave ovens has found that the revenue received from selling microwaves at a cost of p dollars each is given by the polynomial Find the revenue received when dollars.

Everyday Math

51. Fuel Efficiency The fuel efficiency (in miles per gallon) of a bus going at a speed of miles per hour is given by the polynomial —— — Find the fuel efficiency when

52. Stopping Distance The number of feet it takes for a car traveling at miles per hour to stop on dry, level concrete is given by the polynomial Find the stopping distance when

Writing Exercises

53. Using your own words, explain the difference between a monomial, a binomial, and a trinomial.

54. Eloise thinks the sum is What is wrong with her reasoning?

Self Check

After completing the exercises, use this checklist to evaluate your mastery of the objectives of this section.

I can...	Confidently	With some help	No-I don't get it!
identify polynomials, monomials, binomials, and trinomials.			
determine the degree of polynomials.			
add and subtract monomials.			
add and subtract polynomials.			
evaluate a polynomial for a given value.			

If most of your checks were:

...confidently. Congratulations! You have achieved the objectives in this section. Reflect on the study skills you used so that you can continue to use them. What did you do to become confident of your ability to do these things? Be specific.

...with some help. This must be addressed quickly because topics you do not master become potholes in your road to success. In math, every topic builds upon previous work. It is important to make sure you have a strong foundation before you move on. Whom can you ask for help? Your fellow classmates and instructor are good resources. Is there a place on campus where math

tutors are available? Can your study skills be improved?

...no—I don't get it! This is a warning sign and you must not ignore it. You should get help right away or you will quickly be overwhelmed. See your instructor as soon as you can to discuss your situation. Together you can come up with a plan to get you the help you need.

10.2 Use Multiplication Properties of Exponents

Learning Objectives

By the end of this section, you will be able to:

- Simplify expressions with exponents
- Simplify expressions using the Product Property of Exponents
- Simplify expressions using the Power Property of Exponents
- Simplify expressions using the Product to a Power Property
- Simplify expressions by applying several properties
- Multiply monomials

BE PREPARED : : 10.4 Before you get started, take this readiness quiz.

Simplify: — —

If you missed the problem, review Example 4.25.

BE PREPARED : : 10.5 Simplify:

If you missed the problem, review Example 3.52.

Simplify Expressions with Exponents

Remember that an exponent indicates repeated multiplication of the same quantity. For example, means to multiply four factors of so means This format is known as **exponential notation**.

Exponential Notation

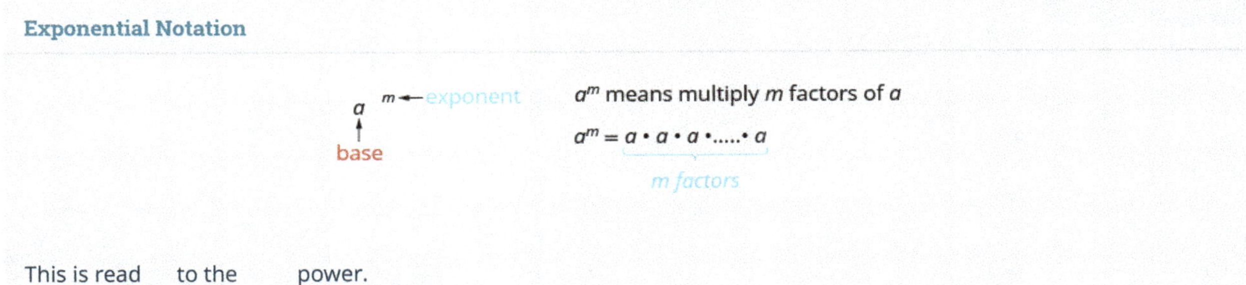

This is read to the power.

In the expression the exponent tells us how many times we use the base as a factor.

$$7^3 \qquad (-8)^5$$

$$\underbrace{7 \cdot 7 \cdot 7}_{3\ factors} \qquad \underbrace{(-8)(-8)(-8)(-8)(-8)}_{5\ factors}$$

Before we begin working with variable expressions containing exponents, let's simplify a few expressions involving only numbers.

EXAMPLE 10.11

Simplify:

Solution

Multiply 3 factors of 5.
Simplify.

Multiply 1 factor of 9.

TRY IT : : 10.21 Simplify:

TRY IT : : 10.22 Simplify:

EXAMPLE 10.12

Simplify:

–

Solution

	–
Multiply two factors.	– –
Simplify.	—

Multiply two factors.

Simplify.

TRY IT : : 10.23 Simplify:

—

TRY IT : : 10.24 Simplify:

—

EXAMPLE 10.13

Simplify:

Solution

Multiply four factors of −3.

Simplify.

Multiply two factors.

Simplify.

Notice the similarities and differences in parts and . Why are the answers different? In part the parentheses tell us to raise the (−3) to the 4th power. In part we raise only the 3 to the 4th power and then find the opposite.

TRY IT : : 10.25 Simplify:

> **TRY IT : :** 10.26 Simplify:

Simplify Expressions Using the Product Property of Exponents

You have seen that when you combine like terms by adding and subtracting, you need to have the same base with the same exponent. But when you multiply and divide, the exponents may be different, and sometimes the bases may be different, too. We'll derive the properties of exponents by looking for patterns in several examples. All the exponent properties hold true for any real numbers, but right now we will only use whole number exponents.

First, we will look at an example that leads to the Product Property.

	$x^2 \cdot x^3$
What does this mean? How many factors altogether?	$\underbrace{x \cdot x}_{2\ factors} \cdot \underbrace{x \cdot x \cdot x}_{3\ factors}$ *5 factors*
So, we have	x^5
Notice that 5 is the sum of the exponents, 2 and 3.	$x^2 \cdot x^3$ is x^{2+3}, or x^5
We write:	

The base stayed the same and we added the exponents. This leads to the Product Property for Exponents.

Product Property of Exponents

If is a real number and are counting numbers, then

To multiply with like bases, add the exponents.

An example with numbers helps to verify this property.

EXAMPLE 10.14

Simplify:

Solution

Use the product property,	x^{5+7}
Simplify.	

TRY IT : : 10.27 Simplify:

TRY IT : : 10.28 Simplify:

EXAMPLE 10.15

Simplify:

Solution

Rewrite,	
Use the product property,	b^{4+1}
Simplify.	

TRY IT : : 10.29 Simplify:

TRY IT : : 10.30 Simplify:

EXAMPLE 10.16

Simplify:

Solution

Use the product property,	2^{7+9}
Simplify.	

TRY IT : : 10.31 Simplify:

TRY IT : : 10.32 Simplify:

EXAMPLE 10.17

Simplify:

Solution

Notice, the bases are the same, so add the exponents.	y^{17+23}
Simplify.	

TRY IT : : 10.33 Simplify:

TRY IT : : 10.34 Simplify:

We can extend the Product Property of Exponents to more than two factors.

EXAMPLE 10.18

Simplify:

Solution

Add the exponents, since the bases are the same.	x^{3+4+2}
Simplify.	

TRY IT : : 10.35 Simplify:

TRY IT : : 10.36 Simplify:

Simplify Expressions Using the Power Property of Exponents

Now let's look at an exponential expression that contains a power raised to a power. See if you can discover a general property.

	$(x^2)^3$
	$x^2 \cdot x^2 \cdot x^2$
What does this mean? How many factors altogether?	$\underbrace{x \cdot x}_{2\ factors} \cdot \underbrace{x \cdot x}_{2\ factors} \cdot \underbrace{x \cdot x}_{2\ factors}$ 6 *factors*
So, we have	x^6
Notice that 6 is the product of the exponents, 2 and 3.	$(x^2)^3$ is $x^{2 \cdot 3}$ or x^6
We write:	

We multiplied the exponents. This leads to the Power Property for Exponents.

Power Property of Exponents

If is a real number and are whole numbers, then

To raise a power to a power, multiply the exponents.

An example with numbers helps to verify this property.

EXAMPLE 10.19

Simplify:

Solution

Use the Power Property,	$x^{5 \cdot 7}$
Simplify.	

Use the Power Property,	$3^{6 \cdot 8}$
Simplify.	

TRY IT : : 10.37 Simplify:

TRY IT : : 10.38 Simplify:

Simplify Expressions Using the Product to a Power Property

We will now look at an expression containing a product that is raised to a power. Look for a pattern.

What does this mean?
We group the like factors together.
How many factors of 2 and of
Notice that each factor was raised to the power.
We write:

The exponent applies to each of the factors. This leads to the Product to a Power Property for Exponents.

Product to a Power Property of Exponents

If and are real numbers and is a whole number, then

To raise a product to a power, raise each factor to that power.

An example with numbers helps to verify this property:

EXAMPLE 10.20

Simplify:

Solution

Use the Power of a Product Property,	$(-11)^2 x^2$
Simplify.	

TRY IT : : 10.39 Simplify:

TRY IT : : 10.40 Simplify:

EXAMPLE 10.21

Simplify:

Solution

Raise each factor to the third power.	$3^3 x^3 y^3$
Simplify.	

TRY IT : : 10.41 Simplify:

TRY IT : : 10.42 Simplify:

Simplify Expressions by Applying Several Properties

We now have three properties for multiplying expressions with exponents. Let's summarize them and then we'll do some examples that use more than one of the properties.

Properties of Exponents

If are real numbers and are whole numbers, then

EXAMPLE 10.22

Simplify:

Solution

Use the Power Property.
Add the exponents.

TRY IT : : 10.43 Simplify:

TRY IT : : 10.44 Simplify:

EXAMPLE 10.23

Simplify:

Solution

Take each factor to the second power.
Use the Power Property.

TRY IT : : 10.45 Simplify:

TRY IT : : 10.46 Simplify:

EXAMPLE 10.24

Simplify:

Solution

Raise to the second power.
Simplify.
Use the Commutative Property.
Multiply the constants and add the exponents.

Notice that in the first monomial, the exponent was outside the parentheses and it applied to both factors inside. In the second monomial, the exponent was inside the parentheses and so it only applied to the *n*.

TRY IT : : 10.47 Simplify:

TRY IT : : 10.48 Simplify:

EXAMPLE 10.25

Simplify:

Solution

Use the Power of a Product Property.	
Use the Power Property.	
Use the Commutative Property.	
Multiply the constants and add the exponents for each variable.	

TRY IT : : 10.49 Simplify:

TRY IT : : 10.50 Simplify:

Multiply Monomials

Since a monomial is an algebraic expression, we can use the properties for simplifying expressions with exponents to multiply the monomials.

EXAMPLE 10.26

Multiply:

Solution

Use the Commutative Property to rearrange the factors.	
Multiply.	

TRY IT : : 10.51 Multiply:

TRY IT : : 10.52 Multiply:

EXAMPLE 10.27

Multiply: —

Solution

—

Use the Commutative Property to rearrange the factors.	—
Multiply.	

TRY IT : : 10.53 Multiply: —

TRY IT : : 10.54 Multiply: —

MEDIA : : ACCESS ADDITIONAL ONLINE RESOURCES

- Exponent Properties (http://www.openstax.org/l/24expproperties)
- Exponent Properties 2 (http://www.openstax.org/l/expproperties2)

10.2 EXERCISES

Practice Makes Perfect

Simplify Expressions with Exponents

In the following exercises, simplify each expression with exponents.

55.

56.

57. —

58. —

59.

60.

61.

62.

63.

64.

65.

66.

67. —

68. —

69.

70.

Simplify Expressions Using the Product Property of Exponents

In the following exercises, simplify each expression using the Product Property of Exponents.

71.

72.

73.

74.

75.

76.

77.

78.

79.

80.

81.

82.

Simplify Expressions Using the Power Property of Exponents

In the following exercises, simplify each expression using the Power Property of Exponents.

83.

84.

85.

86.

87.

88.

89.

90.

91.

92.

93.

94.

Simplify Expressions Using the Product to a Power Property

In the following exercises, simplify each expression using the Product to a Power Property.

95.

96.

97.

98.

99.

100.

101.

102.

Simplify Expressions by Applying Several Properties

In the following exercises, simplify each expression.

103.

104.

105.

106.

107.

108.

109.

110.

111.

112.

113.

114.

115.

116.

117. —

118. —

119.

120.

121.

122.

123. —

124. —

125.

126.

Multiply Monomials

In the following exercises, multiply the following monomials.

127.

128.

129.

130.

131. —

132. —

133.

134.

135. —

136. —

137. — —

138. — —

Everyday Math

139. Email Janet emails a joke to six of her friends and tells them to forward it to six of their friends, who forward it to six of their friends, and so on. The number of people who receive the email on the second round is on the third round is as shown in the table. How many people will receive the email on the eighth round? Simplify the expression to show the number of people who receive the email.

Round	Number of people

140. Salary Raul's boss gives him a raise every year on his birthday. This means that each year, Raul's salary is times his last year's salary. If his original salary was , his salary after year was after years was after years was as shown in the table below. What will Raul's salary be after years? Simplify the expression, to show Raul's salary in dollars.

Year	Salary

Writing Exercises

141. Use the Product Property for Exponents to explain why

142. Explain why but

143. Jorge thinks — is What is wrong with his reasoning?

144. Explain why is and not

Self Check

After completing the exercises, use this checklist to evaluate your mastery of the objectives of this section.

I can...	Confidently	With some help	No-I don't get it!
simplify expressions with exponents.			
simplify expressions using the Product Property for Exponents.			
simplify expressions using the Power Property for Exponents.			
simplify expressions using the Product to a Power Property.			
simplify expressions by applying several properties.			
multiply monomials.			

After reviewing this checklist, what will you do to become confident for all objectives?

10.3 Multiply Polynomials

Learning Objectives

By the end of this section, you will be able to:

- Multiply a polynomial by a monomial
- Multiply a binomial by a binomial
- Multiply a trinomial by a binomial

BE PREPARED : : 10.6 Before you get started, take this readiness quiz.

Distribute:

If you missed the problem, review Example 7.17.

BE PREPARED : : 10.7 Distribute:

If you missed the problem, review Example 7.26.

BE PREPARED : : 10.8 Combine like terms:

If you missed the problem, review Example 2.21.

Multiply a Polynomial by a Monomial

In Distributive Property you learned to use the Distributive Property to simplify expressions such as You multiplied both terms in the parentheses, by to get With this chapter's new vocabulary, you can say you were multiplying a binomial, by a monomial, Multiplying a binomial by a monomial is nothing new for you!

EXAMPLE 10.28

Multiply:

Solution

Distribute. 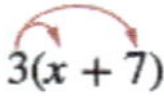

Simplify.

 **TRY IT : :** 10.55 Multiply:

 **TRY IT : :** 10.56 Multiply:

EXAMPLE 10.29

Multiply:

Solution

	$x(x-8)$
Distribute.	$x(x-8)$
	x^2-8x
Simplify.	x^2-8x

TRY IT : : 10.57 Multiply:

TRY IT : : 10.58 Multiply:

EXAMPLE 10.30

Multiply:

Solution

	$10x(4x+y)$
Distribute.	$10x(4x+y)$
	$10x\cdot4x+10x\cdot y$
Simplify.	$40x^2+10xy$

TRY IT : : 10.59 Multiply:

TRY IT : : 10.60 Multiply:

Multiplying a monomial by a trinomial works in much the same way.

EXAMPLE 10.31

Multiply:

Solution

Distribute.	$-2x(5x^2 + 7x - 3)$
Simplify.	

TRY IT : : 10.61 Multiply:

TRY IT : : 10.62 Multiply:

EXAMPLE 10.32

Multiply:

Solution

Distribute.	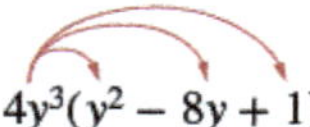
Simplify.	

TRY IT : : 10.63 Multiply:

TRY IT : : 10.64 Multiply:

Now we will have the monomial as the second factor.

EXAMPLE 10.33

Multiply:

Solution

Distribute.	$(x + 3)p$
Simplify.	

TRY IT : : 10.65 Multiply:

TRY IT : : 10.66 Multiply:

Multiply a Binomial by a Binomial

Just like there are different ways to represent multiplication of numbers, there are several methods that can be used to multiply a binomial times a binomial.

Using the Distributive Property

We will start by using the Distributive Property. Look again at Example 10.33.

	$(x + 3)p$
We distributed the to get	$xp + 3p$
What if we have instead of ? Think of the $(x + 7)$ as the p above.	$(x + 3)(x + 7)$
Distribute .	$x(x + 7) + 3(x + 7)$
Distribute again.	
Combine like terms.	

Notice that before combining like terms, we had four terms. We multiplied the two terms of the first binomial by the two terms of the second binomial—four multiplications.

Be careful to distinguish between a sum and a product.

EXAMPLE 10.34

Multiply:

Solution

	$(x+6)(x+8)$
Distribute	$x(x+8)+6(x+8)$
Distribute again.	
Simplify.	

TRY IT : : 10.67 Multiply:

TRY IT : : 10.68 Multiply:

Now we'll see how to multiply binomials where the variable has a coefficient.

EXAMPLE 10.35

Multiply:

Solution

Distribute.	$2x(3x+4)+9(3x+4)$
Distribute again.	
Simplify.	

TRY IT : : 10.69 Multiply:

TRY IT : : 10.70 Multiply:

In the previous examples, the binomials were sums. When there are differences, we pay special attention to make sure the signs of the product are correct.

EXAMPLE 10.36

Multiply:

Solution

Distribute.	$4y(6y-5)+3(6y-5)$
Distribute again.	
Simplify.	

TRY IT : : 10.71 Multiply:

TRY IT : : 10.72 Multiply:

Up to this point, the product of two binomials has been a trinomial. This is not always the case.

EXAMPLE 10.37

Multiply:

Solution

	$(x+2)(x-y)$
Distribute.	$x(x-y)+2(x-y)$
Distribute again.	$x^2-xy+2x-2y$
Simplify.	There are no like terms to combine.

TRY IT : : 10.73 Multiply:

TRY IT : : 10.74 Multiply:

Using the FOIL Method

Remember that when you multiply a binomial by a binomial you get four terms. Sometimes you can combine like terms to get a trinomial, but sometimes there are no like terms to combine. Let's look at the last example again and pay particular attention to how we got the four terms.

Where did the first term, come from?

It is the product of the **first** terms in

First $(x+2)(x-y)$

The next term, is the product of the two **outer** terms.

$(x+2)(x-y)$ Outer

The third term, is the product of the two **inner** terms.

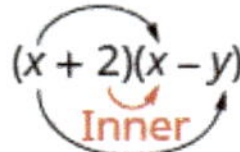

And the last term, came from multiplying the two **last** terms.

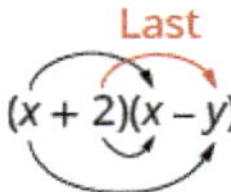

We abbreviate "First, Outer, Inner, Last" as FOIL. The letters stand for 'First, Outer, Inner, Last'. The word FOIL is easy to remember and ensures we find all four products. We might say we use the FOIL method to multiply two binomials.

first last first last

$(a+b)\quad(c+d)$

inner

outer

Let's look at again. Now we will work through an example where we use the FOIL pattern to multiply two binomials.

Distributive Property	FOIL
$(x+3)(x+7)$	$(x+3)(x+7)$
$x(x+7)+3(x+7)$	
$x^2+7x+3x+21$ F O I L	$x^2+7x+3x+21$ F O I L
$x^2+10x+21$	$x^2+10x+21$

EXAMPLE 10.38

Multiply using the FOIL method:

Solution

Step 1: Multiply the **First** terms.	$(x+6)(x+9)$	$x^2+\_\_+\_\_+\_\_$ F O I L
Step 2: Multiply the **Outer** terms.	$(x+6)(x+9)$	$x^2+9x+\_\_+\_\_$ F O I L
Step 3: Multiply the **Inner** terms.	$(x+6)(x+9)$	$x^2+9x+6x+\_\_$ F O I L
Step 4: Multiply the **Last** terms.	$(x+6)(x+9)$	$x^2+9x+6x+54$ F O I L
Step 5: Combine like terms, when possible.		$x^2+15x+54$

TRY IT : : 10.75 Multiply using the FOIL method:

TRY IT : : 10.76 Multiply using the FOIL method:

We summarize the steps of the FOIL method below. The FOIL method only applies to multiplying binomials, not other polynomials!

HOW TO : : USE THE FOIL METHOD FOR MULTIPLYING TWO BINOMIALS.

Step 1. Multiply the **First** terms.
Step 2. Multiply the **Outer** terms.
Step 3. Multiply the **Inner** terms.
Step 4. Multiply the **Last** terms.
Step 5. Combine like terms, when possible.

first last first last
$(a+b)\quad(c+d)$
inner
outer

EXAMPLE 10.39

Multiply:

Solution

Step 1: Multiply the **First** terms.	$(y-8)(y+6)$	$y^2+\_+\_+\_$ F O I L
Step 2: Multiply the **Outer** terms.	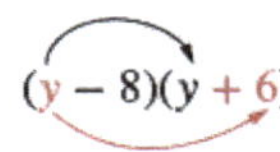	$y^2+6y+\_+\_$ F O I L
Step 3: Multiply the **Inner** terms.	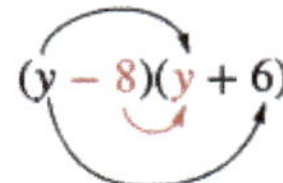	$y^2+6y-8y+\_$ F O I L
Step 4: Multiply the **Last** terms.	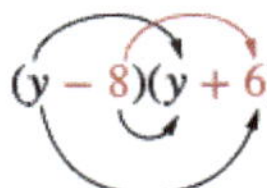	$y^2+6y-8y-48$ F O I L
Step 5: Combine like terms		$y^2-2y-48$

TRY IT : : 10.77 Multiply:

TRY IT : : 10.78 Multiply:

EXAMPLE 10.40

Multiply:

Solution

		$(2a+3)(3a-1)$
		$(2a+3)(3a-1)$
Multiply the **First** terms.	$2a \cdot 3a$	$\underset{F}{6a^2} + \underset{O}{\_} + \underset{I}{\_} + \underset{L}{\_}$
Multiply the **Outer** terms.	$2a \cdot (-1)$	$\underset{F}{6a^2} \underset{O}{- 2a} + \underset{I}{\_} + \underset{L}{\_}$
Multiply the **Inner** terms.	$3 \cdot 3a$	$\underset{F}{6a^2} \underset{O}{- 2a} \underset{I}{+ 9a} + \underset{L}{\_}$
Multiply the **Last** terms.	$3 \cdot (-1)$	$\underset{F}{6a^2} \underset{O}{- 2a} \underset{I}{+ 9a} \underset{L}{- 3}$
Combine like terms.		$6a^2 + 7a - 3$

> TRY IT : : 10.79 Multiply:

> TRY IT : : 10.80 Multiply:

EXAMPLE 10.41

Multiply:

Solution

	$(5x-y)(2x-7)$
	$(5x-y)(2x-7)$
Multiply the **First** terms.	$\underset{F}{10x^2} + \underset{O}{\_} + \underset{I}{\_} + \underset{L}{\_}$
Multiply the **Outer** terms.	$\underset{F}{10x^2} \underset{O}{- 35x} + \underset{I}{\_} + \underset{L}{\_}$
Multiply the **Inner** terms.	$\underset{F}{10x^2} \underset{O}{- 35x} \underset{I}{- 2xy} + \underset{L}{\_}$
Multiply the **Last** terms.	$\underset{F}{10x^2} \underset{O}{- 35x} \underset{I}{- 2xy} \underset{L}{+ 7y}$
Combine like terms. There are none.	$10x^2 - 35x - 2xy + 7y$

> TRY IT : : 10.81 Multiply:

TRY IT : : 10.82 Multiply:

Using the Vertical Method

The FOIL method is usually the quickest method for multiplying two binomials, but it works *only* for binomials. You can use the Distributive Property to find the product of any two polynomials. Another method that works for all polynomials is the Vertical Method. It is very much like the method you use to multiply whole numbers. Look carefully at this example of multiplying two-digit numbers.

23	
×46	
138	partial product
92	partial product
1058	product

You start by multiplying by to get

Then you multiply by lining up the partial product in the correct columns.

Last, you add the partial products.

Now we'll apply this same method to multiply two binomials.

EXAMPLE 10.42

Multiply using the vertical method:

Solution

It does not matter which binomial goes on the top. Line up the columns when you multiply as we did when we multiplied

	$2x - 7$ $\times\ 5x - 1$	
Multiply by .	$-2x + 7$	partial product
Multiply by .	$10x^2 - 35x$	partial product
Add like terms.	$10x^2 - 37x + 7$	product

Notice the partial products are the same as the terms in the FOIL method.

$$(5x - 1)(2x - 7)$$
$$10x^2 - 35x - 2x + 7$$
$$10x^2 - 37x + 7$$

$$\begin{array}{r} 2x - 7 \\ \times\ 5x - 1 \\ \hline -2x + 7 \\ 10x^2 - 35x \\ \hline 10x^2 - 37x + 7 \end{array}$$

TRY IT : : 10.83 Multiply using the vertical method:

TRY IT : : 10.84 Multiply using the vertical method:

We have now used three methods for multiplying binomials. Be sure to practice each method, and try to decide which one you prefer. The three methods are listed here to help you remember them.

Multiplying Two Binomials

To multiply binomials, use the:

- Distributive Property
- FOIL Method
- Vertical Method

Remember, FOIL only works when multiplying two binomials.

Multiply a Trinomial by a Binomial

We have multiplied monomials by monomials, monomials by polynomials, and binomials by binomials. Now we're ready to multiply a trinomial by a binomial. Remember, the FOIL method will not work in this case, but we can use either the Distributive Property or the Vertical Method. We first look at an example using the Distributive Property.

EXAMPLE 10.43

Multiply using the Distributive Property:

Solution

	$(x+3)(2x^2-5x+8)$
Distribute.	$x(2x^2-5x+8)+3(2x^2-5x+8)$
Multiply.	
Combine like terms.	

TRY IT : : 10.85 Multiply using the Distributive Property:

TRY IT : : 10.86 Multiply using the Distributive Property:

Now let's do this same multiplication using the Vertical Method.

EXAMPLE 10.44

Multiply using the Vertical Method:

Solution

It is easier to put the polynomial with fewer terms on the bottom because we get fewer partial products this way.

		$2x^2-5x+8$
		$\times\quad x+3$
Multiply	by 3.	$6x^2-15x+24$
Multiply	by .	$2x^3-5x^2+8x$
Add like terms.		$2x^3+x^2-7x+24$

TRY IT : : 10.87 Multiply using the Vertical Method:

TRY IT : : 10.88 Multiply using the Vertical Method:

MEDIA : : ACCESS ADDITIONAL ONLINE RESOURCES

- Multiply Monomials (http://www.openstax.org/l/24multmonomials)
- Multiply Polynomials (http://www.openstax.org/l/24multpolyns)
- Multiply Polynomials 2 (http://www.openstax.org/l/24multpolyns2)
- Multiply Polynomials Review (http://www.openstax.org/l/24multpolynsrev)
- Multiply Polynomials Using the Distributive Property (http://www.openstax.org/l/24multpolynsdis)
- Multiply Binomials (http://www.openstax.org/l/24multbinomials)

10.3 EXERCISES

Practice Makes Perfect

Multiply a Polynomial by a Monomial

In the following exercises, multiply.

145. 146. 147.

148. 149. 150.

151. 152. 153.

154. 155. 156.

157. 158. 159.

160. 161. 162.

163. 164. 165.

166. 167. 168.

169. 170. 171.

172. 173. 174.

175. 176. 177.

178.

Multiply a Binomial by a Binomial

In the following exercises, multiply the following binomials using: the Distributive Property the FOIL method the Vertical method

179. 180. 181.

182.

In the following exercises, multiply the following binomials. Use any method.

183. 184. 185.

186. 187. 188.

189. 190. 191.

192. 193. 194.

195. 196. 197.

198.

199.

200.

201.

202.

203.

204.

205.

206.

Multiply a Trinomial by a Binomial

In the following exercises, multiply using the Distributive Property and the Vertical Method.

207.

208.

209.

210.

In the following exercises, multiply. Use either method.

211.

212.

213.

214.

Everyday Math

215. Mental math You can use binomial multiplication to multiply numbers without a calculator. Say you need to multiply times Think of as and as

Multiply by the FOIL method.

Multiply without using a calculator.

Which way is easier for you? Why?

216. Mental math You can use binomial multiplication to multiply numbers without a calculator. Say you need to multiply times Think of as and as

Multiply by the FOIL method.

Multiply without using a calculator.

Which way is easier for you? Why?

Writing Exercises

217. Which method do you prefer to use when multiplying two binomials—the Distributive Property, the FOIL method, or the Vertical Method? Why?

218. Which method do you prefer to use when multiplying a trinomial by a binomial—the Distributive Property or the Vertical Method? Why?

Self Check

After completing the exercises, use this checklist to evaluate your mastery of the objectives of this section.

I can...	Confidently	With some help	No-I don't get it!
multiply a polynomial by a monomial.			
multiply a binomial by a binomial.			
multiply a trinomial by a binomial.			

What does this checklist tell you about your mastery of this section? What steps will you take to improve?

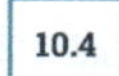

Divide Monomials

Learning Objectives

By the end of this section, you will be able to:

- Simplify expressions using the Quotient Property of Exponents
- Simplify expressions with zero exponents
- Simplify expressions using the Quotient to a Power Property
- Simplify expressions by applying several properties
- Divide monomials

BE PREPARED : : 10.9

Before you get started, take this readiness quiz.

Simplify: —

If you missed the problem, review Example 4.19.

BE PREPARED : : 10.10

Simplify:

If you missed the problem, review Example 10.23.

BE PREPARED : : 10.11

Simplify: —

If you missed the problem, review Example 4.23.

Simplify Expressions Using the Quotient Property of Exponents

Earlier in this chapter, we developed the properties of exponents for multiplication. We summarize these properties here.

Summary of Exponent Properties for Multiplication

If are real numbers and are whole numbers, then

Now we will look at the exponent properties for division. A quick memory refresher may help before we get started. In Fractions you learned that fractions may be simplified by dividing out common factors from the numerator and denominator using the Equivalent Fractions Property. This property will also help us work with algebraic fractions—which are also quotients.

Equivalent Fractions Property

If are whole numbers where then

— — — —

As before, we'll try to discover a property by looking at some examples.

Notice that in each case the bases were the same and we subtracted the exponents.

- When the larger exponent was in the numerator, we were left with factors in the numerator and in the denominator, which we simplified.
- When the larger exponent was in the denominator, we were left with factors in the denominator, and in the numerator, which could not be simplified.

We write:

Quotient Property of Exponents

If is a real number, and are whole numbers, then

A couple of examples with numbers may help to verify this property.

When we work with numbers and the exponent is less than or equal to we will apply the exponent. When the exponent is greater than , we leave the answer in exponential form.

EXAMPLE 10.45

Simplify:

Solution

To simplify an expression with a quotient, we need to first compare the exponents in the numerator and denominator.

Since 10 > 8, there are more factors of in the numerator.	
Use the quotient property with .	x^{10-8}
Simplify.	

Since 9 > 2, there are more factors of 2 in the numerator.	
Use the quotient property with	2^{9-2}
Simplify.	

Notice that when the larger exponent is in the numerator, we are left with factors in the numerator.

TRY IT : : 10.89 Simplify:

TRY IT : : 10.90 Simplify:

EXAMPLE 10.46

Simplify:

Solution

To simplify an expression with a quotient, we need to first compare the exponents in the numerator and denominator.

Since 15 > 10, there are more factors of in the denominator.	—
Use the quotient property with — ——	$\frac{1}{b^{15-10}}$
Simplify.	—

Since 5 > 3, there are more factors of 3 in the denominator.	—
Use the quotient property with — ——	$\frac{1}{3^{5-3}}$
Simplify.	—
Apply the exponent.	—

Notice that when the larger exponent is in the denominator, we are left with factors in the denominator and in the numerator.

TRY IT : : 10.91 Simplify:

— —

TRY IT : : 10.92 Simplify:

— —

EXAMPLE 10.47

Simplify:

— —

Solution

Since 9 > 5, there are more 's in the denominator and so we will end up with factors in the denominator.	—
Use the Quotient Property for — —	$\frac{1}{a^{9-5}}$
Simplify.	—

Notice there are more factors of in the numerator, since 11 > 7. So we will end up with factors in the numerator.	—
Use the Quotient Property for —	x^{11-7}
Simplify.	

> **TRY IT : : 10.93** Simplify:
>
> — —

> **TRY IT : : 10.94** Simplify:
>
> — —

Simplify Expressions with Zero Exponents

A special case of the Quotient Property is when the exponents of the numerator and denominator are equal, such as an expression like — From earlier work with fractions, we know that

— — —

In words, a number divided by itself is So — for any ( ), since any number divided by itself is

The Quotient Property of Exponents shows us how to simplify — when and when by subtracting exponents. What if ?

Now we will simplify — in two ways to lead us to the definition of the **zero exponent**.

Consider first — which we know is

	—
Write 8 as .	—
Subtract exponents.	
Simplify.	

In general, for $a \neq 0$:

$\frac{a^m}{a^m}$	$\frac{a^m}{a^m}$
a^{m-m}	$\frac{\overbrace{\not a \cdot \not a \cdot \ldots \cdot \not a}^{m \text{ factors}}}{\underbrace{\not a \cdot \not a \cdot \ldots \cdot \not a}_{m \text{ factors}}}$
a^0	1

We see —— simplifies to a and to . So .

Zero Exponent

If is a non-zero number, then

Any nonzero number raised to the zero power is

In this text, we assume any variable that we raise to the zero power is not zero.

EXAMPLE 10.48

Simplify:

Solution

The definition says any non-zero number raised to the zero power is

Use the definition of the zero exponent.	1

Use the definition of the zero exponent.	1

TRY IT : : 10.95 Simplify:

TRY IT : : 10.96 Simplify:

Now that we have defined the zero exponent, we can expand all the Properties of Exponents to include whole number exponents.

What about raising an expression to the zero power? Let's look at We can use the product to a power rule to rewrite this expression.

Use the Product to a Power Rule.
Use the Zero Exponent Property.
Simplify. 1

This tells us that any non-zero expression raised to the zero power is one.

EXAMPLE 10.49

Simplify:

Solution

Use the definition of the zero exponent.	1

TRY IT : : 10.97 Simplify:

TRY IT : : 10.98 Simplify: —

EXAMPLE 10.50

Simplify:

Solution

The product is raised to the zero power.
Use the definition of the zero exponent.

Notice that only the variable is being raised to the zero power.
Use the definition of the zero exponent.
Simplify.

TRY IT : : 10.99 Simplify:

TRY IT : : 10.100 Simplify:

Simplify Expressions Using the Quotient to a Power Property

Now we will look at an example that will lead us to the Quotient to a Power Property.

This means	
Multiply the fractions.	
Write with exponents.	

Notice that the exponent applies to both the numerator and the denominator.

We see that is

This leads to the Quotient to a Power Property for Exponents.

Quotient to a Power Property of Exponents

If and are real numbers, and is a counting number, then

To raise a fraction to a power, raise the numerator and denominator to that power.

An example with numbers may help you understand this property:

EXAMPLE 10.51

Simplify:

Solution

	$\left(\frac{5}{8}\right)^2$
Use the Quotient to a Power Property,	$\frac{5^2}{8^2}$
Simplify.	$\frac{25}{64}$

	$\left(\frac{x}{3}\right)^4$
Use the Quotient to a Power Property,	$\frac{x^4}{3^4}$
Simplify.	$\frac{x^4}{81}$

	$\left(\frac{y}{m}\right)^3$
Raise the numerator and denominator to the third power.	$\frac{y^3}{m^3}$

TRY IT : : 10.101 Simplify:

TRY IT : : 10.102 Simplify:

Simplify Expressions by Applying Several Properties

We'll now summarize all the properties of exponents so they are all together to refer to as we simplify expressions using several properties. Notice that they are now defined for whole number exponents.

Summary of Exponent Properties

If are real numbers and are whole numbers, then

EXAMPLE 10.52

Simplify:

Solution

Multiply the exponents in the numerator, using the Power Property.	
Subtract the exponents.	

TRY IT : : 10.103

Simplify:

TRY IT : : 10.104

Simplify:

EXAMPLE 10.53

Simplify:

Solution

Multiply the exponents in the numerator, using the Power Property.

Subtract the exponents.

Zero power property

TRY IT : : 10.105

Simplify:

TRY IT : : 10.106

Simplify:

EXAMPLE 10.54

Simplify:

Solution

Remember parentheses come before exponents, and the bases are the same so we can simplify inside the parentheses. Subtract the exponents.

Simplify.

Multiply the exponents.

TRY IT : : 10.107

Simplify:

TRY IT : : 10.108

Simplify:

EXAMPLE 10.55

Simplify:

Solution

Here we cannot simplify inside the parentheses first, since the bases are not the same.

Raise the numerator and denominator to the third power using the Quotient to a Power Property,	
Use the Power Property,	

> **TRY IT : : 10.109**
> Simplify:

> **TRY IT : : 10.110**
> Simplify:

EXAMPLE 10.56

Simplify:

Solution

Raise the numerator and denominator to the fourth power using the Quotient to a Power Property.	
Raise each factor to the fourth power, using the Power to a Power Property.	
Use the Power Property and simplify.	

> **TRY IT : : 10.111**
> Simplify:

TRY IT : : 10.112

Simplify:

EXAMPLE 10.57

Simplify:

Solution

Use the Power Property.	
Add the exponents in the numerator, using the Product Property.	
Use the Quotient Property.	

TRY IT : : 10.113

Simplify:

TRY IT : : 10.114

Simplify:

Divide Monomials

We have now seen all the properties of exponents. We'll use them to divide monomials. Later, you'll use them to divide polynomials.

EXAMPLE 10.58

Find the quotient:

Solution

Rewrite as a fraction.	
Use fraction multiplication to separate the number part from the variable part.	
Use the Quotient Property.	

TRY IT : : 10.115 Find the quotient:

TRY IT : : 10.116 Find the quotient:

When we divide monomials with more than one variable, we write one fraction for each variable.

EXAMPLE 10.59

Find the quotient:

Solution

Use fraction multiplication.	
Simplify and use the Quotient Property.	
Multiply.	

TRY IT : : 10.117 Find the quotient:

TRY IT : : 10.118 Find the quotient:

EXAMPLE 10.60

Find the quotient:

Solution

Use fraction multiplication.	
Simplify and use the Quotient Property.	
Multiply.	

> **TRY IT : :** 10.119 Find the quotient:

> **TRY IT : :** 10.120 Find the quotient:

Once you become familiar with the process and have practiced it step by step several times, you may be able to simplify a fraction in one step.

EXAMPLE 10.61

Find the quotient:

Solution

Simplify and use the Quotient Property.	

Be very careful to simplify by dividing out a common factor, and to simplify the variables by subtracting their exponents.

> **TRY IT : :** 10.121 Find the quotient:

> **TRY IT : :** 10.122 Find the quotient:

In all examples so far, there was no work to do in the numerator or denominator before simplifying the fraction. In the next example, we'll first find the product of two monomials in the numerator before we simplify the fraction.

EXAMPLE 10.62

Find the quotient:

Solution

Remember, the fraction bar is a grouping symbol. We will simplify the numerator first.

Simplify the numerator.	
Simplify, using the Quotient Rule.	

TRY IT : : 10.123

Find the quotient:

TRY IT : : 10.124

Find the quotient:

MEDIA : : ACCESS ADDITIONAL ONLINE RESOURCES

- Simplify a Quotient (http://openstaxcollege.org/l/24simpquot)
- Zero Exponent (http://openstaxcollege.org/l/zeroexponent)
- Quotient Rule (http://openstaxcollege.org/l/24quotientrule)
- Polynomial Division (http://openstaxcollege.org/l/24polyndivision)
- Polynomial Division 2 (http://openstaxcollege.org/l/24polydivision2)

10.4 EXERCISES

Practice Makes Perfect

Simplify Expressions Using the Quotient Property of Exponents

In the following exercises, simplify.

219. —

220. —

221. —

222. —

223. —

224. —

225. —

226. —

227. —

228. —

229. —

230. —

Simplify Expressions with Zero Exponents

In the following exercises, simplify.

231.

232.

233.

234.

235.

236.

237.

238.

239.

240.

241.

242.

243.

244.

Simplify Expressions Using the Quotient to a Power Property

In the following exercises, simplify.

245. —

246. —

247. —

248. —

249. —

250. —

251. —

252. —

Simplify Expressions by Applying Several Properties

In the following exercises, simplify.

253. ——

254. ——

255. ——

256. ——

257. ——

258. ——

259. ——

260. ——

261. —

262. —

263. ——

264. ——

265. ——

266. ——

267. ——

268. ——

269. —

270. —

271. ——

272. ——

273. ——

274. ——

275. ——

276. ——

Divide Monomials

In the following exercises, divide the monomials.

277.

278.

279.

280.

281. ——

282. ——

283. ——

284. ——

285. ——

286. ——

287. ——

288. ——

289. ——

290. ——

291. ——

292. ——

293. ——

294. ——

295. ——

Mixed Practice

296.

297.

298.

299.

300.

— —

301.

— —

302.

303.

304. —— ——

305. —— ——

306. —— ——

307. —— ——

308. —— ——

309. —— ——

Everyday Math

310. Memory One megabyte is approximately bytes. One gigabyte is approximately bytes. How many megabytes are in one gigabyte?

311. Memory One megabyte is approximately bytes. One terabyte is approximately bytes. How many megabytes are in one terabyte?

Writing Exercises

312. Vic thinks the quotient —— simplifies to What is wrong with his reasoning?

313. Mai simplifies the quotient — by writing What is wrong with her reasoning?

314. When Dimple simplified and she got the same answer. Explain how using the Order of Operations correctly gives different answers.

315. Roxie thinks simplifies to What would you say to convince Roxie she is wrong?

Self Check

After completing the exercises, use this checklist to evaluate your mastery of the objectives of this section.

I can...	Confidently	With some help	No-I don't get it!
simplify expressions using the Quotient Property for Exponents.			
simplify expressions with zero exponents.			
simplify expressions using the Quotient to a Power Property.			
simplify expressions by applying several properties.			
divide monomials.			

On a scale of 1–10, how would you rate your mastery of this section in light of your responses on the checklist? How can you improve this?

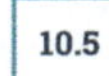

Integer Exponents and Scientific Notation

Learning Objectives

By the end of this section, you will be able to:

- Use the definition of a negative exponent
- Simplify expressions with integer exponents
- Convert from decimal notation to scientific notation
- Convert scientific notation to decimal form
- Multiply and divide using scientific notation

BE PREPARED : : 10.12 Before you get started, take this readiness quiz.

What is the place value of the in the number

If you missed this problem, review Example 1.3.

BE PREPARED : : 10.13 Name the decimal

If you missed this problem, review Exercise 5.0.

BE PREPARED : : 10.14 Subtract:

If you missed this problem, review Example 3.37.

Use the Definition of a Negative Exponent

The Quotient Property of Exponents, introduced in Divide Monomials, had two forms depending on whether the exponent in the numerator or denominator was larger.

Quotient Property of Exponents

If is a real number, and are whole numbers, then

What if we just subtract exponents, regardless of which is larger? Let's consider

We subtract the exponent in the denominator from the exponent in the numerator.

We can also simplify by dividing out common factors:

$$\frac{x \cdot x}{x \cdot x \cdot x \cdot x \cdot x}$$

$$\frac{1}{x^3}$$

This implies that and it leads us to the definition of a **negative exponent**.

Negative Exponent

If is a positive integer and then —

The negative exponent tells us to re-write the expression by taking the reciprocal of the base and then changing the sign of the exponent. Any expression that has negative exponents is not considered to be in simplest form. We will use the definition of a negative exponent and other properties of exponents to write an expression with only positive exponents.

EXAMPLE 10.63

Simplify:

Solution

Use the definition of a negative exponent,	— —
Simplify.	—

Use the definition of a negative exponent,	— —
Simplify.	—

TRY IT : : 10.125 Simplify:

TRY IT : : 10.126 Simplify:

When simplifying any expression with exponents, we must be careful to correctly identify the base that is raised to each exponent.

EXAMPLE 10.64

Simplify:

Solution

The negative in the exponent does not affect the sign of the base.

The exponent applies to the base, .	
Take the reciprocal of the base and change the sign of the exponent.	
Simplify.	

The expression means "find the opposite of ". The exponent applies only to the base, 3.	
Rewrite as a product with −1.	
Take the reciprocal of the base and change the sign of the exponent.	
Simplify.	

TRY IT : : 10.127 Simplify:

TRY IT : : 10.128 Simplify:

We must be careful to follow the order of operations. In the next example, parts and look similar, but we get different results.

EXAMPLE 10.65

Simplify:

Solution

Remember to always follow the order of operations.

Do exponents before multiplication.	
Use	
Simplify.	

Simplify inside the parentheses first.	
Use —	—
Simplify.	–

TRY IT : : 10.129 Simplify:

TRY IT : : 10.130 Simplify:

When a variable is raised to a negative exponent, we apply the definition the same way we did with numbers.

EXAMPLE 10.66

Simplify:

Solution

Use the definition of a negative exponent,	— —

TRY IT : : 10.131 Simplify:

TRY IT : : 10.132 Simplify:

When there is a product and an exponent we have to be careful to apply the exponent to the correct quantity. According to the order of operations, expressions in parentheses are simplified before exponents are applied. We'll see how this works in the next example.

EXAMPLE 10.67

Simplify:

Solution

Notice the exponent applies to just the base .	
Take the reciprocal of and change the sign of the exponent.	—
Simplify.	—

Here the parentheses make the exponent apply to the base .	
Take the reciprocal of and change the sign of the exponent.	—
Simplify.	—

The base is . Take the reciprocal of and change the sign of the exponent.	—
Simplify.	—
Use — —	—

TRY IT : : 10.133 Simplify:

TRY IT : : 10.134 Simplify:

Now that we have defined negative exponents, the Quotient Property of Exponents needs only one form, —

where and *m* and *n* are integers.

When the exponent in the denominator is larger than the exponent in the numerator, the exponent of the quotient will be negative. If the result gives us a negative exponent, we will rewrite it by using the definition of negative exponents, —

Simplify Expressions with Integer Exponents

All the exponent properties we developed earlier in this chapter with whole number exponents apply to integer exponents, too. We restate them here for reference.

Summary of Exponent Properties

If are real numbers and are integers, then

—

— —

—

EXAMPLE 10.68

Simplify:

Solution

Use the Product Property,

Simplify.

The bases are the same, so add the exponents.

Simplify.

Use the definition of a negative exponent, — —

The bases are the same, so add the exponents.	
Simplify.	
Use the definition of a negative exponent,	

> TRY IT : : 10.135 Simplify:

> TRY IT : : 10.136 Simplify:

In the next two examples, we'll start by using the Commutative Property to group the same variables together. This makes it easier to identify the like bases before using the Product Property of Exponents.

EXAMPLE 10.69

Simplify:

Solution

Use the Commutative Property to get like bases together.	
Add the exponents for each base.	
Take reciprocals and change the signs of the exponents.	
Simplify.	

> TRY IT : : 10.137 Simplify:

> TRY IT : : 10.138 Simplify:

If the monomials have numerical coefficients, we multiply the coefficients, just as we did in Integer Exponents and Scientific Notation.

EXAMPLE 10.70

Simplify:

Solution

Rewrite with the like bases together.	
Simplify.	
Use the definition of a negative exponent,	
Simplify.	

TRY IT : : 10.139 Simplify:

TRY IT : : 10.140 Simplify:

In the next two examples, we'll use the Power Property and the Product to a Power Property.

EXAMPLE 10.71

Simplify:

Solution

Use the Product to a Power Property,	
Simplify.	
Rewrite with a positive exponent.	

TRY IT : : 10.141 Simplify:

TRY IT : : 10.142 Simplify:

EXAMPLE 10.72

Simplify:

Solution

Use the Product to a Power Property,	
Simplify and multiply the exponents of using the Power Property,	
Rewrite by using the definition of a negative exponent, —	—
Simplify	—

> **TRY IT : :** 10.143 Simplify:

> **TRY IT : :** 10.144 Simplify:

To simplify a fraction, we use the Quotient Property.

EXAMPLE 10.73

Simplify: —

Solution

	$\frac{r^5}{r^{-4}}$
Use the Quotient Property, — .	$r^{5-(-4)}$
Be careful to subtract 5 − (−4).	
Simplify.	r^9

> **TRY IT : :** 10.145 Simplify: —

> **TRY IT : :** 10.146 Simplify: —

Convert from Decimal Notation to Scientific Notation

Remember working with place value for whole numbers and decimals? Our number system is based on powers of We use tens, hundreds, thousands, and so on. Our decimal numbers are also based on powers of tens—tenths, hundredths, thousandths, and so on.

Consider the numbers and We know that means and means —— If we write the as a power of ten in exponential form, we can rewrite these numbers in this way:

When a number is written as a product of two numbers, where the first factor is a number greater than or equal to one but less than and the second factor is a power of written in exponential form, it is said to be in *scientific notation.*

Scientific Notation

A number is expressed in **scientific notation** when it is of the form

where and and is an integer.

It is customary in scientific notation to use as the multiplication sign, even though we avoid using this sign elsewhere in algebra.

Scientific notation is a useful way of writing very large or very small numbers. It is used often in the sciences to make calculations easier.

If we look at what happened to the decimal point, we can see a method to easily convert from decimal notation to scientific notation.

$4000. = 4 \times 10^3$	$0.004 = 4 \times 10^{-3}$
$4000. = 4 \times 10^3$	$0.004 = 4 \times 10^{-3}$
Moved the decimal point 3 places to the left.	Moved the decimal point 3 places to the right.

In both cases, the decimal was moved places to get the first factor, by itself.

- The power of is positive when the number is larger than
- The power of is negative when the number is between and

EXAMPLE 10.74

Write in scientific notation.

Solution

Step 1: Move the decimal point so that the first factor is greater than or equal to 1 but less than 10.	37000.
Step 2: Count the number of decimal places, , that the decimal point was moved.	3.70000 4 places
Step 3: Write the number as a product with a power of 10.	
If the original number is: • greater than 1, the power of 10 will be . • between 0 and 1, the power of 10 will be	
Step 4: Check.	
is 10,000 and 10,000 times 3.7 will be 37,000.	

TRY IT : : 10.147 Write in scientific notation:

TRY IT : : 10.148 Write in scientific notation:

HOW TO : : CONVERT FROM DECIMAL NOTATION TO SCIENTIFIC NOTATION.

Step 1. Move the decimal point so that the first factor is greater than or equal to but less than

Step 2. Count the number of decimal places, that the decimal point was moved.

Step 3. Write the number as a product with a power of

- If the original number is:
  - greater than the power of will be
  - between and the power of will be

Step 4. Check.

EXAMPLE 10.75

Write in scientific notation:

Solution

	0.0052
Move the decimal point to get 5.2, a number between 1 and 10.	0.0052
Count the number of decimal places the point was moved.	3 places
Write as a product with a power of 10.	
Check your answer:	

> **TRY IT : :** 10.149 Write in scientific notation:

> **TRY IT : :** 10.150 Write in scientific notation:

Convert Scientific Notation to Decimal Form

How can we convert from scientific notation to decimal form? Let's look at two numbers written in scientific notation and see.

If we look at the location of the decimal point, we can see an easy method to convert a number from scientific notation to decimal form.

$$9.12 \times 10^4 = 91{,}200 \qquad 9.12 \times 10^{-4} = 0.000912$$

$$9.12\_\_ \times 10^4 = 91{,}200 \qquad \_\_\_9.12 \times 10^{-4} = 0.000912$$

In both cases the decimal point moved 4 places. When the exponent was positive, the decimal moved to the right. When the exponent was negative, the decimal point moved to the left.

EXAMPLE 10.76

Convert to decimal form:

Solution

Step 1: Determine the exponent, , on the factor 10.	
Step 2: Move the decimal point places, adding zeros if needed.	6,200.
• If the exponent is positive, move the decimal point places to the right. • If the exponent is negative, move the decimal point places to the left.	6,200
Step 3: Check to see if your answer makes sense.	
is 1000 and 1000 times 6.2 will be 6,200.	

TRY IT : : 10.151 Convert to decimal form:

TRY IT : : 10.152 Convert to decimal form:

HOW TO : : CONVERT SCIENTIFIC NOTATION TO DECIMAL FORM.

Step 1. Determine the exponent, on the factor

Step 2. Move the decimal places, adding zeros if needed.

- If the exponent is positive, move the decimal point places to the right.
- If the exponent is negative, move the decimal point places to the left.

Step 3. Check.

EXAMPLE 10.77

Convert to decimal form:

Solution

Determine the exponent , on the factor 10.	The exponent is −2.
Move the decimal point 2 places to the left.	− 8.9
Add zeros as needed for placeholders.	0.089

The Check is left to you.

TRY IT : : 10.153 Convert to decimal form:

TRY IT : : 10.154 Convert to decimal form:

Multiply and Divide Using Scientific Notation

We use the Properties of Exponents to multiply and divide numbers in scientific notation.

EXAMPLE 10.78

Multiply. Write answers in decimal form:

Solution

Use the Commutative Property to rearrange the factors.

Multiply 4 by 2 and use the Product Property to multiply by .

Change to decimal form by moving the decimal two places left.

TRY IT : : 10.155 Multiply. Write answers in decimal form:

TRY IT : : 10.156 Multiply. Write answers in decimal form:

EXAMPLE 10.79

Divide. Write answers in decimal form:

Solution

Separate the factors.

Divide 9 by 3 and use the Quotient Property to divide by .

Change to decimal form by moving the decimal five places right.

TRY IT : : 10.157 Divide. Write answers in decimal form:

TRY IT : : 10.158 Divide. Write answers in decimal form:

MEDIA : : ACCESS ADDITIONAL ONLINE RESOURCES

- Negative Exponents (http://www.openstax.org/l/24negexponents)
- Examples of Simplifying Expressions with Negative Exponents (http://www.openstax.org/l/24simpexprnegex)
- Scientific Notation (http://www.openstax.org/l/24scnotation)

10.5 EXERCISES

Practice Makes Perfect

Use the Definition of a Negative Exponent

In the following exercises, simplify.

316.

317.

318.

319.

320.

321.

322.

323.

324.

325.

326.

327.

328.

329.

330.

331.

332.

333.

334.

335.

336.

337.

338.

339.

340.

341.

342.

343.

Simplify Expressions with Integer Exponents

In the following exercises, simplify.

344.

345.

346.

347.

348.

349.

350.

351.

352.

353.

354.

355.

356.

357.

358.

359.

360.

361.

362.

363.

364.

365.

366.

367.

368.

369.

370.

371.

372.

373.

374.

375.

376. ——

377. ——

378. ——

379. ——

380. ——

381. —

382. ——

383. ——

Convert from Decimal Notation to Scientific Notation

In the following exercises, write each number in scientific notation.

384. 45,000

385. 280,000

386. 8,750,000

387. 1,290,000

388. 0.036

389. 0.041

390. 0.00000924

391. 0.0000103

392. The population of the United States on July 4, 2010 was almost

393. The population of the world on July 4, 2010 was more than

394. The average width of a human hair is centimeters.

395. The probability of winning the Megamillions lottery is about

Convert Scientific Notation to Decimal Form

In the following exercises, convert each number to decimal form.

396.

397.

398.

399.

400.

401.

402.

403.

404. In 2010, the number of Facebook users each day who changed their status to 'engaged' was

405. At the start of 2012, the US federal budget had a deficit of more than

406. The concentration of carbon dioxide in the atmosphere is

407. The width of a proton is of the width of an atom.

Multiply and Divide Using Scientific Notation

In the following exercises, multiply or divide and write your answer in decimal form.

408.

409.

410.

411.

412. ________

413. ________

414. ________

415. ________

Everyday Math

416. Calories In May 2010 the Food and Beverage Manufacturers pledged to reduce their products by trillion calories by the end of 2015.

Write trillion in decimal notation.

Write trillion in scientific notation.

417. Length of a year The difference between the calendar year and the astronomical year is day.

Write this number in scientific notation.

How many years does it take for the difference to become 1 day?

418. Calculator display Many calculators automatically show answers in scientific notation if there are more digits than can fit in the calculator's display. To find the probability of getting a particular 5-card hand from a deck of cards, Mario divided by and saw the answer Write the number in decimal notation.

419. Calculator display Many calculators automatically show answers in scientific notation if there are more digits than can fit in the calculator's display. To find the number of ways Barbara could make a collage with of her favorite photographs, she multiplied Her calculator gave the answer Write the number in decimal notation.

Writing Exercises

420.

Explain the meaning of the exponent in the expression

Explain the meaning of the exponent in the expression

421. When you convert a number from decimal notation to scientific notation, how do you know if the exponent will be positive or negative?

Self Check

After completing the exercises, use this checklist to evaluate your mastery of the objectives of this section.

I can...	Confidently	With some help	No-I don't get it!
use the definition of a negative exponent.			
simplify expressions with integer exponents.			
convert from decimal notation to scientific notation.			
convert scientific notation to decimal form.			
multiply and divide using scientific notation.			

After looking at the checklist, do you think you are well prepared for the next section? Why or why not?

10.6 Introduction to Factoring Polynomials

Learning Objectives

By the end of this section, you will be able to:

- Find the greatest common factor of two or more expressions
- Factor the greatest common factor from a polynomial

BE PREPARED : : 10.15 Before you get started, take this readiness quiz.

Factor into primes.

If you missed this problem, review Example 2.48.

BE PREPARED : : 10.16 Multiply:

If you missed this problem, review Example 7.25.

BE PREPARED : : 10.17 Multiply:

If you missed this problem, review Example 10.32.

Find the Greatest Common Factor of Two or More Expressions

Earlier we multiplied factors together to get a product. Now, we will be reversing this process; we will start with a product and then break it down into its factors. Splitting a product into factors is called factoring.

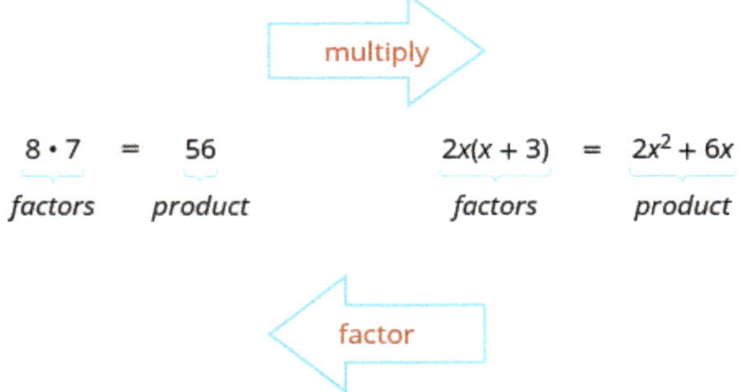

In The Language of Algebra we factored numbers to find the least common multiple (LCM) of two or more numbers. Now we will factor expressions and find the *greatest common factor* of two or more expressions. The method we use is similar to what we used to find the LCM.

Greatest Common Factor

The greatest common factor (GCF) of two or more expressions is the largest expression that is a factor of all the expressions.

First we will find the greatest common factor of two numbers.

EXAMPLE 10.80

Find the greatest common factor of and

Solution

Step 1: Factor each coefficient into primes. Write all variables with exponents in expanded form.	Factor 24 and 36.	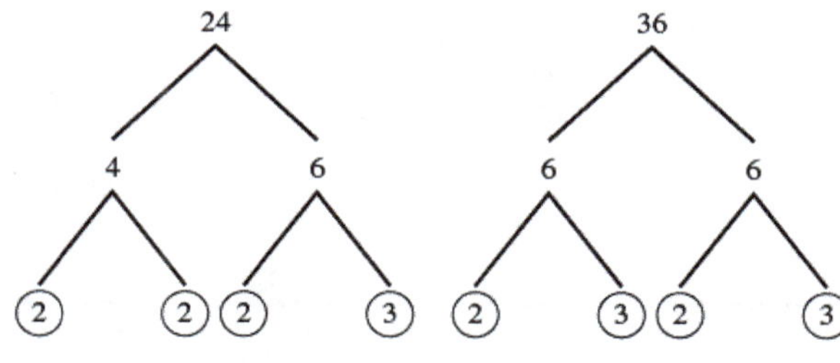
Step 2: List all factors--matching common factors in a column.		$24 = 2 \cdot 2 \cdot 2 \cdot 3$ $36 = 2 \cdot 2 \cdot \quad 3 \cdot 3$
In each column, circle the common factors.	Circle the 2, 2, and 3 that are shared by both numbers.	$24 = (2) \cdot (2) \cdot 2 \cdot (3)$ $36 = (2) \cdot (2) \cdot \quad (3) \cdot 3$ $\text{GCF} = 2 \cdot 2 \cdot \quad 3$ $\text{GCF} = 12$
Step 3: Bring down the common factors that all expressions share.	Bring down the 2, 2, 3 and then multiply.	
Step 4: Multiply the factors.		The GCF of 24 and 36 is 12.

Notice that since the GCF is a factor of both numbers, and can be written as multiples of

 **TRY IT : :** 10.159 Find the greatest common factor:

 **TRY IT : :** 10.160 Find the greatest common factor:

In the previous example, we found the greatest common factor of constants. The greatest common factor of an algebraic expression can contain variables raised to powers along with coefficients. We summarize the steps we use to find the greatest common factor.

HOW TO : : FIND THE GREATEST COMMON FACTOR.

Step 1. Factor each coefficient into primes. Write all variables with exponents in expanded form.

Step 2. List all factors—matching common factors in a column. In each column, circle the common factors.

Step 3. Bring down the common factors that all expressions share.

Step 4. Multiply the factors.

EXAMPLE 10.81

Find the greatest common factor of

Solution

Factor each number into primes. Circle the common factors in each column. Bring down the common factors.	$5x = 5 \cdot x$ $15 = 3 \cdot 5 \cdot$ $\text{GCF} = 5$
	The GCF of 5x and 15 is 5.

TRY IT : : 10.161 Find the greatest common factor:

TRY IT : : 10.162 Find the greatest common factor:

In the examples so far, the greatest common factor was a constant. In the next two examples we will get variables in the greatest common factor.

EXAMPLE 10.82

Find the greatest common factor of and

Solution

Factor each coefficient into primes and write the variables with exponents in expanded form. Circle the common factors in each column. Bring down the common factors. Multiply the factors.	$12x^2 = 2 \cdot 2 \cdot 3 \cdot x \cdot x$ $18x^3 = 2 \cdot 3 \cdot 3 \cdot x \cdot x \cdot x$ $\text{GCF} = 2 \cdot 3 \cdot x \cdot x$ $\text{GCF} = 6x^2$

TRY IT : : 10.163 Find the greatest common factor:

TRY IT : : 10.164 Find the greatest common factor:

EXAMPLE 10.83

Find the greatest common factor of

Solution

Factor each coefficient into primes and write the variables with exponents in expanded form. Circle the common factors in each column. Bring down the common factors. Multiply the factors.	$14x^3 = 2 \cdot 7 \cdot x \cdot x \cdot x$ $8x^2 = 2 \cdot 2 \cdot 2 \cdot x \cdot x$ $10x = 2 \cdot 5 \cdot x$ $\text{GCF} = 2 \cdot x$ $\text{GCF} = 2x$

TRY IT : : 10.165 Find the greatest common factor:

TRY IT : : 10.166 Find the greatest common factor:

Factor the Greatest Common Factor from a Polynomial

Just like in arithmetic, where it is sometimes useful to represent a number in factored form (for example, as in algebra it can be useful to represent a polynomial in factored form. One way to do this is by finding the greatest common factor of all the terms. Remember that you can multiply a polynomial by a monomial as follows:

Here, we will start with a product, like and end with its factors, To do this we apply the Distributive Property "in reverse".

Distributive Property

If are real numbers, then

The form on the left is used to multiply. The form on the right is used to factor.

So how do we use the Distributive Property to factor a polynomial? We find the GCF of all the terms and write the polynomial as a product!

EXAMPLE 10.84

Factor:

Solution

Step 1: Find the GCF of all the terms of the polynomial.	Find the GCF of 2x and 14.	$2x = 2 \cdot x$ $14 = 2 \cdot 7$ $\text{GCF} = 2$
Step 2: Rewrite each term as a product using the GCF.	Rewrite 2x and 14 as products of their GCF, 2.	$2x + 14$ $2 \cdot x + 2 \cdot 7$
Step 3: Use the Distributive Property 'in reverse' to factor the expression.		
Step 4: Check by multiplying the factors.		Check: $2(x + 7)$ $2 \cdot x + 2 \cdot 7$ $2x + 14$ ✓

 **TRY IT : : 10.167** Factor:

 **TRY IT : : 10.168** Factor:

Notice that in Example 10.84, we used the word *factor* as both a noun and a verb:

HOW TO : : FACTOR THE GREATEST COMMON FACTOR FROM A POLYNOMIAL.

Step 1. Find the GCF of all the terms of the polynomial.
Step 2. Rewrite each term as a product using the GCF.
Step 3. Use the Distributive Property ‘in reverse’ to factor the expression.
Step 4. Check by multiplying the factors.

EXAMPLE 10.85

Factor:

Solution

Find the GCF of $3a$, 3.	$3a = 3 \cdot a$ $3 = 3$ $\text{GCF} = 3$
	$3a + 3$
Rewrite each term as a product using the GCF.	$3 \cdot a + 3 \cdot 1$
Use the Distributive Property 'in reverse' to factor the GCF.	$3(a + 1)$
Check by multiplying the factors to get the original polynomial.	
$3(a + 1)$ $3 \cdot a + 3 \cdot 1$ $3a + 3$ ✓	

TRY IT : : 10.169 Factor:

TRY IT : : 10.170 Factor:

The expressions in the next example have several factors in common. Remember to write the GCF as the product of all the common factors.

EXAMPLE 10.86

Factor:

Solution

Find the GCF of 12*x* and 60.	$\begin{aligned} 12x &= 2 \cdot 2 \cdot 3 \cdot x \\ 60 &= 2 \cdot 2 \cdot 3 \cdot 5 \\ \hline \text{GCF} &= 2 \cdot 2 \cdot 3 \\ \text{GCF} &= 12 \end{aligned}$
	$12x - 60$
Rewrite each term as a product using the GCF.	$12 \cdot x - 12 \cdot 5$
Factor the GCF.	$12(x - 5)$
Check by multiplying the factors.	

$$12(x - 5)$$
$$12 \cdot x - 12 \cdot 5$$
$$12x - 60 \checkmark$$

TRY IT : : 10.171 Factor:

TRY IT : : 10.172 Factor:

Now we'll factor the greatest common factor from a trinomial. We start by finding the GCF of all three terms.

EXAMPLE 10.87

Factor:

Solution

Find the GCF of $3y^2$, 6*y*, and 9.	$\begin{aligned} 3y^2 &= 3 \cdot y \cdot y \\ 6y &= 2 \cdot 3 \cdot y \\ 9 &= 3 \cdot 3 \\ \hline \text{GCF} &= 3 \end{aligned}$
	$3y^2 + 6y + 9$
Rewrite each term as a product using the GCF.	$3 \cdot y^2 + 3 \cdot 2y + 3 \cdot 3$
Factor the GCF.	$3(y^2 + 2y + 3)$
Check by multiplying.	

$$3(y^2 + 2y + 3)$$
$$3 \cdot y^2 + 3 \cdot 2y + 3 \cdot 3$$
$$3y^2 + 6y + 9 \checkmark$$

TRY IT : : 10.173 Factor:

TRY IT : : 10.174 Factor:

In the next example, we factor a variable from a binomial.

EXAMPLE 10.88

Factor:

Solution

Find the GCF of and and the math that goes with it.	$6x^2 = 2 \cdot 3 \cdot x \cdot x$ $5x = 5 \cdot x$ $\text{GCF} = x$
Rewrite each term as a product.	$x \cdot 6x + x \cdot 5$
Factor the GCF.	
Check by multiplying.	

TRY IT : : 10.175 Factor:

TRY IT : : 10.176 Factor:

When there are several common factors, as we'll see in the next two examples, good organization and neat work helps!

EXAMPLE 10.89

Factor:

Solution

Find the GCF of $4x^3$ and $20x^2$.	$4x^3 = 2 \cdot 2 \cdot x \cdot x \cdot x$ $20x^2 = 2 \cdot 2 \cdot 5 \cdot x \cdot x$ $\text{GCF} = 2 \cdot 2 \cdot x \cdot x$ $\text{GCF} = 4x^2$	
		$4x^3 - 20x^2$
Rewrite each term.		$4x^2 \cdot x - 4x^2 \cdot 5$
Factor the GCF.		$4x^2(x - 5)$
Check.	$4x^2(x - 5)$ $4x^2 \cdot x - 4x^2 \cdot 5$ $4x^3 - 20x^2$ ✓	

TRY IT : : 10.177 Factor:

TRY IT : : 10.178 Factor:

EXAMPLE 10.90

Factor:

Solution

Find the GCF of and	$\begin{array}{l} 21y^2 = 3 \cdot 7 \cdot y \cdot y \\ 35y = 5 \cdot 7 \cdot y \cdot \\ \hline \text{GCF} = 7 \cdot y \\ \text{GCF} = 7y \end{array}$
	$21y^2 + 35y$
Rewrite each term.	$7y \cdot 3y + 7y \cdot 5$
Factor the GCF.	$7y(3y + 5)$

TRY IT : : 10.179 Factor:

TRY IT : : 10.180 Factor:

EXAMPLE 10.91

Factor:

Solution

Previously, we found the GCF of to be

Rewrite each term using the GCF, 2x.	$2x \cdot 7x^2 + 2x \cdot 4x - 2x \cdot 5$
Factor the GCF.	
Check. $\begin{array}{c} 2x\,(7x^2 + 4x - 5) \\ 2x \cdot 7x^2 + 2x \cdot 4x - 2x \cdot 5 \\ 14x^3 + 8x^2 - 10x\ ✓ \end{array}$	

TRY IT : : 10.181 Factor:

TRY IT : : 10.182 Factor:

When the leading coefficient, the coefficient of the first term, is negative, we factor the negative out as part of the GCF.

EXAMPLE 10.92

Factor:

Solution

When the leading coefficient is negative, the GCF will be negative. Ignoring the signs of the terms, we first find the GCF of 9*y* and 27 is 9.	$9y = 3 \cdot 3 \cdot y$ $27 = 3 \cdot 3 \cdot 3$ $\text{GCF} = 3 \cdot 3$ $\text{GCF} = 9$
Since the expression −9y−27 has a negative leading coefficient, we use −9 as the GCF.	
Rewrite each term using the GCF.	$-9 \cdot y + (-9) \cdot 3$
Factor the GCF.	
Check. $-9(y+3)$ $-9 \cdot y + (-9) \cdot 3$ $-9y - 27$ ✓	

TRY IT : : 10.183 Factor:

TRY IT : : 10.184 Factor:

Pay close attention to the signs of the terms in the next example.

EXAMPLE 10.93

Factor:

Solution

The leading coefficient is negative, so the GCF will be negative.	
	$4a^2 = 2 \cdot 2 \cdot a \cdot a$ $16a = 2 \cdot 2 \cdot 2 \cdot 2 \cdot a$ $\text{GCF} = 2 \cdot 2 \cdot a$ $\text{GCF} = 4a$
Since the leading coefficient is negative, the GCF is negative, −4*a*.	
Rewrite each term.	$-4a \cdot a - (-4a) \cdot 4$
Factor the GCF.	
Check on your own by multiplying.	

TRY IT : : 10.185 Factor:

TRY IT : : 10.186 Factor:

MEDIA : : ACCESS ADDITIONAL ONLINE RESOURCES

- Factor GCF (http://www.openstax.org/l/24factorgcf)
- Factor a Binomial (http://www.openstax.org/l/24factorbinomi)
- Identify GCF (http://www.openstax.org/l/24identifygcf)

10.6 EXERCISES

Practice Makes Perfect

Find the Greatest Common Factor of Two or More Expressions

In the following exercises, find the greatest common factor.

422. 423. 424.

425. 426. 427.

428. 429. 430.

431. 432. 433.

434. 435. 436.

437. 438. 439.

440. 441.

Factor the Greatest Common Factor from a Polynomial

In the following exercises, factor the greatest common factor from each polynomial.

442. 443. 444.

445. 446. 447.

448. 449. 450.

451. 452. 453.

454. 455. 456.

457. 458. 459.

460. 461. 462.

463. 464. 465.

466. 467. 468.

469. 470. 471.

472. 473. 474.

475. 476. 477.

478. **479.** **480.**

481. **482.** **483.**

484. **485.** **486.**

487. **488.** **489.**

Everyday Math

490. Revenue A manufacturer of microwave ovens has found that the revenue received from selling microwaves a cost of dollars each is given by the polynomial Factor the greatest common factor from this polynomial.

491. Height of a baseball The height of a baseball hit with velocity feet/second at feet above ground level is with the number of seconds since it was hit. Factor the greatest common factor from this polynomial.

Writing Exercises

492. The greatest common factor of and is Explain what this means.

493. What is the GCF of , , and ? Write a general rule that tells how to find the GCF of , , and .

Self Check

After completing the exercises, use this checklist to evaluate your mastery of the objectives of this section.

I can...	Confidently	With some help	No-I don't get it!
find the greatest common factor of two or more expressions.			
factor the greatest common factor from a polynomial.			

Overall, after looking at the checklist, do you think you are well-prepared for the next Chapter? Why or why not?

CHAPTER 10 REVIEW

KEY TERMS

binomial A binomial is a polynomial with exactly two terms.

degree of a constant The degree of a constant is .

degree of a polynomial The degree of a polynomial is the highest degree of all its terms.

degree of a term The degree of a term of a polynomial is the exponent of its variable.

greatest common factor The greatest common factor (GCF) of two or more expressions is the largest expression that is a factor of all the expressions.

monomial A term of the form , where is a constant and is a whole number, is called a monomial.

negative exponent If is a positive integer and , then —.

polynomial A polynomial is a monomial, or two or more monomials, combined by addition or subtraction.

scientific notation A number expressed in scientific notation when it is of the form where and and is an integer.

trinomial A trinomial is a polynomial with exactly three terms.

zero exponent If is a non-zero number, then . Any nonzero number raised to the zero power is 1.

KEY CONCEPTS

10.2 Use Multiplication Properties of Exponents

- **Exponential Notation**

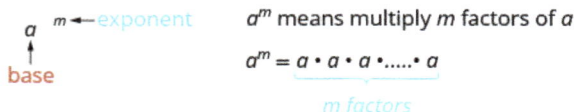

a^m means multiply m factors of a

$a^m = a \cdot a \cdot a \cdot \ldots \cdot a$

This is read to the power.

- **Product Property of Exponents**
    - If is a real number and are counting numbers, then
    - To multiply with like bases, add the exponents.
- **Power Property for Exponents**
    - If is a real number and are counting numbers, then
- **Product to a Power Property for Exponents**
    - If and are real numbers and is a whole number, then

10.3 Multiply Polynomials

- **Use the FOIL method for multiplying two binomials.**

Step 1. Multiply the **First** terms.	*first last first last*
Step 2. Multiply the **Outer** terms.	$(a + b) \quad (c + d)$
Step 3. Multiply the **Inner** terms.	*inner*
Step 4. Multiply the **Last** terms.	*outer*
Step 5. Combine like terms, when possible.	

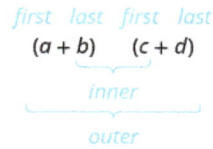

- **Multiplying Two Binomials:** To multiply binomials, use the:
    - Distributive Property
    - FOIL Method
    - Vertical Method
- **Multiplying a Trinomial by a Binomial:** To multiply a trinomial by a binomial, use the:
    - Distributive Property
    - Vertical Method

10.4 Divide Monomials

- **Equivalent Fractions Property**
    - If are whole numbers where then
- **Zero Exponent**
    - If is a non-zero number, then
    - Any nonzero number raised to the zero power is
- **Quotient Property for Exponents**
    - If is a real number, and are whole numbers, then
- **Quotient to a Power Property for Exponents**
    - If and are real numbers, and is a counting number, then
    - To raise a fraction to a power, raise the numerator and denominator to that power.

10.5 Integer Exponents and Scientific Notation

- **Summary of Exponent Properties**
    - If are real numbers and are integers, then

- **Convert from Decimal Notation to Scientific Notation:** To convert a decimal to scientific notation:
  Step 1. Move the decimal point so that the first factor is greater than or equal to 1 but less than 10.
  Step 2. Count the number of decimal places, , that the decimal point was moved.
  Step 3. Write the number as a product with a power of 10.
    - If the original number is greater than 1, the power of 10 will be .
    - If the original number is between 0 and 1, the power of 10 will be .

  Step 4. Check.
- **Convert Scientific Notation to Decimal Form:** To convert scientific notation to decimal form:
  Step 1. Determine the exponent, , on the factor 10.
  Step 2. Move the decimal places, adding zeros if needed.
    - If the exponent is positive, move the decimal point places to the right.
    - If the exponent is negative, move the decimal point places to the left.

  Step 3. Check.

10.6 Introduction to Factoring Polynomials

- **Find the greatest common factor.**
  Step 1. Factor each coefficient into primes. Write all variables with exponents in expanded form.
  Step 2. List all factors—matching common factors in a column. In each column, circle the common factors.
  Step 3. Bring down the common factors that all expressions share.
  Step 4. Multiply the factors.
- **Distributive Property**
  - If , , are real numbers, then
    and
- **Factor the greatest common factor from a polynomial.**
  Step 1. Find the GCF of all the terms of the polynomial.
  Step 2. Rewrite each term as a product using the GCF.
  Step 3. Use the Distributive Property 'in reverse' to factor the expression.
  Step 4. Check by multiplying the factors.

REVIEW EXERCISES

10.1 Add and Subtract Polynomials

Identify Polynomials, Monomials, Binomials and Trinomials

In the following exercises, determine if each of the following polynomials is a monomial, binomial, trinomial, or other polynomial.

494.

495.

496.

497.

Determine the Degree of Polynomials

In the following exercises, determine the degree of each polynomial.

498.

499.

500.

501.

Add and Subtract Monomials

In the following exercises, add or subtract the monomials.

502.

503.

504. Add

505. Subtract from

Add and Subtract Polynomials

In the following exercises, add or subtract the polynomials.

506.

507.

508.

509.

510. Find the sum of and

511. Find the difference of and

Evaluate a Polynomial for a Given Value of the Variable

In the following exercises, evaluate each polynomial for the given value.

512. — when

513. — when

514. — when

515. — when

516. — when

517. — when

518. A pair of glasses is dropped off a bridge feet above a river. The polynomial gives the height of the glasses seconds after they were dropped. Find the height of the glasses when

519. The fuel efficiency (in miles per gallon) of a bus going at a speed of miles per hour is given by the polynomial —— — Find the fuel efficiency when mph.

10.2 Use Multiplication Properties of Exponents

Simplify Expressions with Exponents

In the following exercises, simplify.

520.

521. —

522.

523.

Simplify Expressions Using the Product Property of Exponents

In the following exercises, simplify each expression.

524.

525.

526.

527.

Simplify Expressions Using the Power Property of Exponents

In the following exercises, simplify each expression.

528.

529.

530.

531.

Simplify Expressions Using the Product to a Power Property

In the following exercises, simplify each expression.

532.

533.

534.

535.

Simplify Expressions by Applying Several Properties

In the following exercises, simplify each expression.

536.

537.

538.

539.

Multiply Monomials

In the following exercises, multiply the monomials.

540.

541. —

542.

543. — —

10.3 Multiply Polynomials

Multiply a Polynomial by a Monomial

In the following exercises, multiply.

544.

545.

546.

547.

Multiply a Binomial by a Binomial

In the following exercises, multiply the binomials using various methods.

548.

549.

550.

551.

552.

553.

554.

555.

556.

557.

558.

559.

Multiply a Trinomial by a Binomial

In the following exercises, multiply using any method.

560.

561.

562.

563.

10.4 Divide Monomials

Simplify Expressions Using the Quotient Property of Exponents

In the following exercises, simplify.

564. —

565. —

566. —

567. —

Simplify Expressions with Zero Exponents

In the following exercises, simplify.

568.

569.

570.

571.

Simplify Expressions Using the Quotient to a Power Property

In the following exercises, simplify.

572. —

573. —

574. —

575. —

Simplify Expressions by Applying Several Properties

In the following exercises, simplify.

576. —

577. —

578. —

579. —

580. —

581. —

Divide Monomials

In the following exercises, divide the monomials.

582.

583.

584. —

585. —

586. —

587. —

588. ______

589. ______ ______

10.5 Integer Exponents and Scientific Notation

Use the Definition of a Negative Exponent

In the following exercises, simplify.

590.

591.

592.

593.

Simplify Expressions with Integer Exponents

In the following exercises, simplify.

594.

595.

596.

597.

598.

599. ___

600. ___

601. ___

Convert from Decimal Notation to Scientific Notation

In the following exercises, write each number in scientific notation.

602.

603.

604. The thickness of a piece of paper is about millimeter.

605. According to www.cleanair.com, U.S. businesses use about tons of paper per year.

Convert Scientific Notation to Decimal Form

In the following exercises, convert each number to decimal form.

606.

607.

608.

609.

Multiply and Divide Using Scientific Notation

In the following exercises, multiply and write your answer in decimal form.

610.

611.

612. ______

613. ______

10.6 Introduction to Factoring Polynomials

Find the Greatest Common Factor of Two or More Expressions

In the following exercises, find the greatest common factor.

614.

615.

616.

617.

Factor the Greatest Common Factor from a Polynomial

In the following exercises, factor the greatest common factor from each polynomial.

618.

619.

620.

621.

622.

623.

624.

625.

PRACTICE TEST

626. For the polynomial

Is it a monomial, binomial, or trinomial?

What is its degree?

In the following exercises, simplify each expression.

627.

628.

629. —

630.

631.

632.

633.

634.

635.

636.

637.

638.

639. —

640. ———

641.

642. ———

643. ———————

644.

645.

646.

647.

648. ——

In the following exercises, factor the greatest common factor from each polynomial.

649.

650.

651. According to www.cleanair.org, the amount of trash generated in the US in one year averages out to pounds of trash per person. Write this number in scientific notation.

652. Convert to decimal form.

In the following exercises, simplify, and write your answer in decimal form.

653.

654. ————

655. A hiker drops a pebble from a bridge feet above a canyon. The polynomial gives the height of the pebble seconds a after it was dropped. Find the height when

Figure 11.1 Cyclists speed toward the finish line. (credit: ewan traveler, Flickr)

Chapter Outline

Introduction

Which cyclist will win the race? What will the winning time be? How many seconds will separate the winner from the runner-up? One way to summarize the information from the race is by creating a graph. In this chapter, we will discuss the basic concepts of graphing. The applications of graphing go far beyond races. They are used to present information in almost every field, including healthcare, business, and entertainment.

11.1 Use the Rectangular Coordinate System

Learning Objectives

By the end of this section, you will be able to:

- Plot points on a rectangular coordinate system
- Identify points on a graph
- Verify solutions to an equation in two variables
- Complete a table of solutions to a linear equation
- Find solutions to linear equations in two variables

BE PREPARED : : 11.1 Before you get started, take this readiness quiz.

Evaluate: when

If you missed this problem, review Example 3.23.

BE PREPARED : : 11.2 Evaluate: when

If you missed this problem, review Example 3.56.

BE PREPARED : : 11.3 Solve for

If you missed this problem, review Example 8.20.

Plot Points on a Rectangular Coordinate System

Many maps, such as the Campus Map shown in Figure 11.2, use a grid system to identify locations. Do you see the numbers and across the top and bottom of the map and the letters A, B, C, and D along the sides? Every location on the map can be identified by a number and a letter.

For example, the Student Center is in section 2B. It is located in the grid section above the number and next to the letter B. In which grid section is the Stadium? The Stadium is in section 4D.

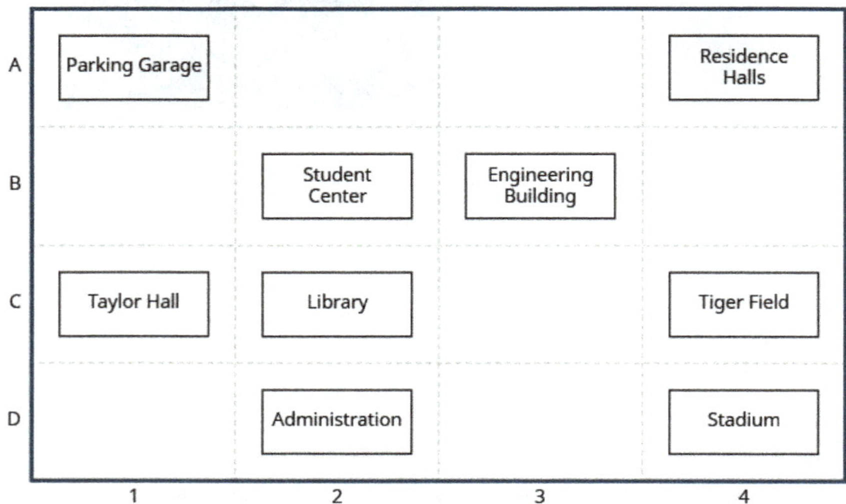

Figure 11.2

EXAMPLE 11.1

Use the map in Figure 11.2.

Find the grid section of the Residence Halls. What is located in grid section 4C?

Solution

Read the number below the Residence Halls, and the letter to the side, A. So the Residence Halls are in grid section 4A.

Find across the bottom of the map and C along the side. Look below the and next to the C. Tiger Field is in grid section 4C.

TRY IT : : 11.1 Use the map in Figure 11.2.

Find the grid section of Taylor Hall. What is located in section 3B?

TRY IT : : 11.2 Use the map in Figure 11.2.

Find the grid section of the Parking Garage. What is located in section 2C?

Just as maps use a grid system to identify locations, a grid system is used in algebra to show a relationship between two variables in a rectangular coordinate system. To create a rectangular coordinate system, start with a horizontal number line. Show both positive and negative numbers as you did before, using a convenient scale unit. This horizontal number line is called the ***x*-axis**.

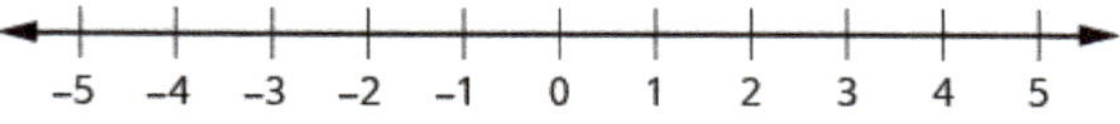

Now, make a vertical number line passing through the at Put the positive numbers above and the negative numbers below See Figure 11.3. This vertical line is called the ***y*-axis**.

Vertical grid lines pass through the integers marked on the Horizontal grid lines pass through the integers marked on the The resulting grid is the rectangular coordinate system.

The rectangular coordinate system is also called the plane, the coordinate plane, or the Cartesian coordinate system (since it was developed by a mathematician named René Descartes.)

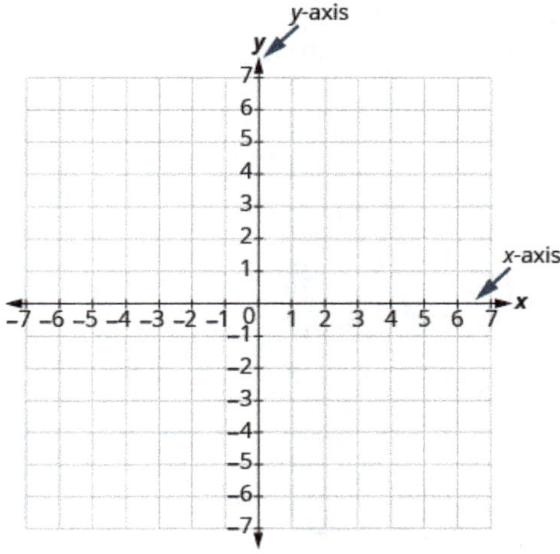

Figure 11.3 The rectangular coordinate system.

The and the form the rectangular coordinate system. These axes divide a plane into four areas, called **quadrants**. The quadrants are identified by Roman numerals, beginning on the upper right and proceeding counterclockwise. See Figure 11.4.

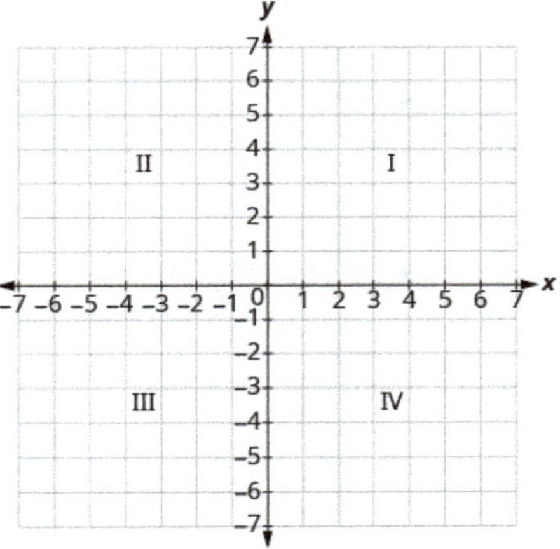

Figure 11.4 The four quadrants of the rectangular coordinate system

In the rectangular coordinate system, every point is represented by an **ordered pair**. The first number in the ordered pair is the *x*-coordinate of the point, and the second number is the *y*-coordinate of the point.

Ordered Pair

An ordered pair, gives the coordinates of a point in a rectangular coordinate system.

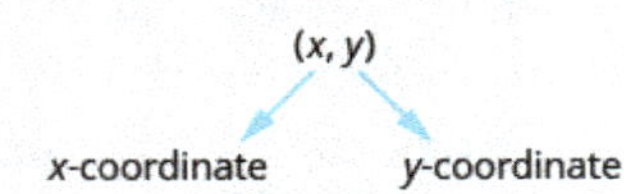

So how do the coordinates of a point help you locate a point on the plane?

Let's try locating the point . In this ordered pair, the -coordinate is and the -coordinate is .

We start by locating the value, on the Then we lightly sketch a vertical line through as shown in Figure 11.5.

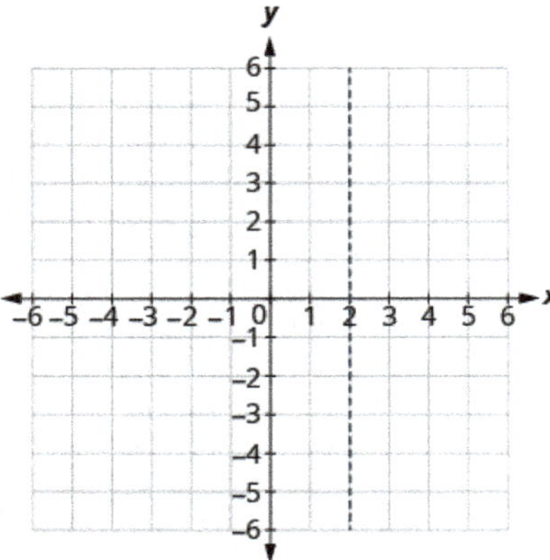

Figure 11.5

Now we locate the value, on the -axis and sketch a horizontal line through . The point where these two lines meet is the point with coordinates We plot the point there, as shown in Figure 11.6.

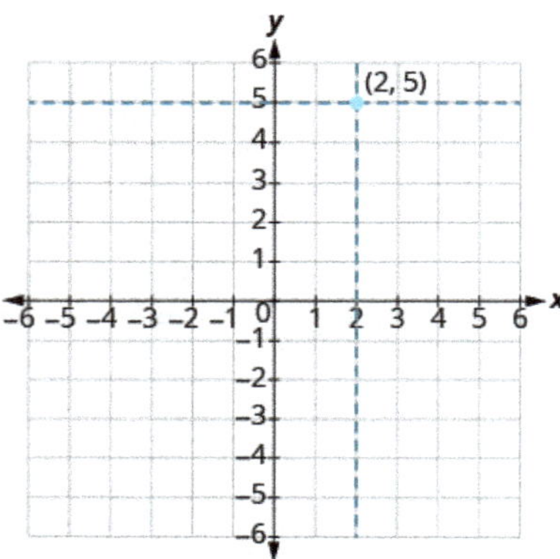

Figure 11.6

EXAMPLE 11.2

Plot and in the same rectangular coordinate system.

Solution

The coordinate values are the same for both points, but the and values are reversed. Let's begin with point The is so find on the and sketch a vertical line through The is so we find on the and sketch a horizontal line through Where the two lines meet, we plot the point

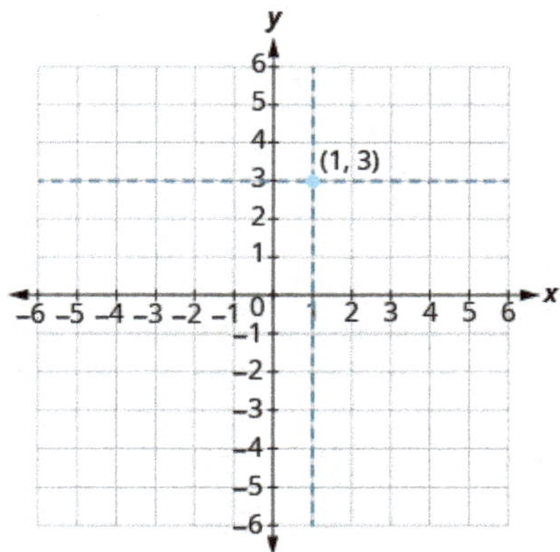

To plot the point we start by locating on the and sketch a vertical line through Then we find on the and sketch a horizontal line through Where the two lines meet, we plot the point

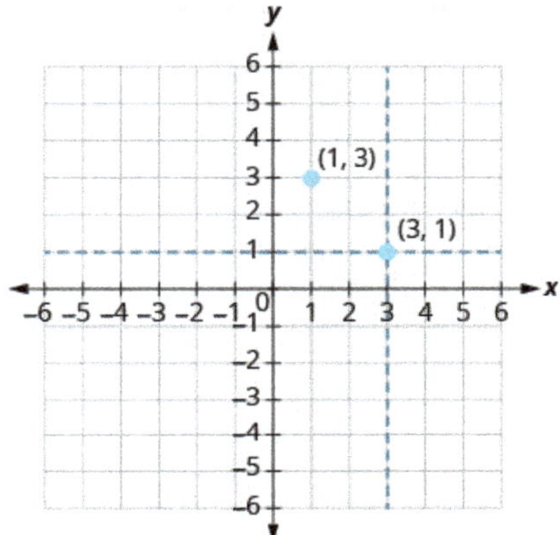

Notice that the order of the coordinates does matter, so, is not the same point as

TRY IT : : 11.3 Plot each point on the same rectangular coordinate system:

TRY IT : : 11.4 Plot each point on the same rectangular coordinate system:

EXAMPLE 11.3

Plot each point in the rectangular coordinate system and identify the quadrant in which the point is located:

—

Solution

The first number of the coordinate pair is the and the second number is the

Since the point is in Quadrant II.

Since the point is in Quadrant III.

Since the point is in Quadrant IV.

Since — the point — is in Quadrant I. It may be helpful to write — as the mixed number, — or decimal, Then we know that the point is halfway between and on the

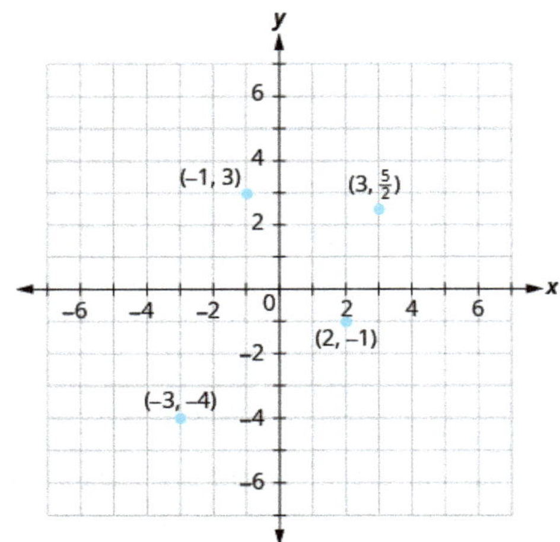

> **TRY IT : :** 11.5

Plot each point on a rectangular coordinate system and identify the quadrant in which the point is located.

—

> **TRY IT : :** 11.6

Plot each point on a rectangular coordinate system and identify the quadrant in which the point is located.

—

How do the signs affect the location of the points?

EXAMPLE 11.4

Plot each point:

Solution

As we locate the and the we must be careful with the signs.

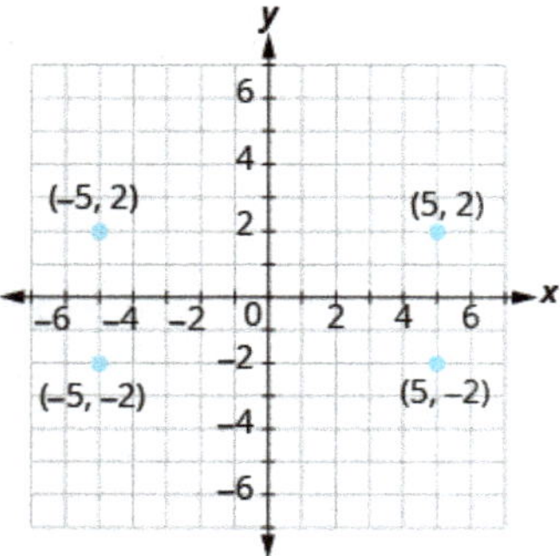

> **TRY IT : :** 11.7 Plot each point:

TRY IT : : 11.8 Plot each point:

You may have noticed some patterns as you graphed the points in the two previous examples.

For each point in Quadrant IV, what do you notice about the signs of the coordinates?

What about the signs of the coordinates of the points in the third quadrant? The second quadrant? The first quadrant?

Can you tell just by looking at the coordinates in which quadrant the point (−2, 5) is located? In which quadrant is (2, −5) located?

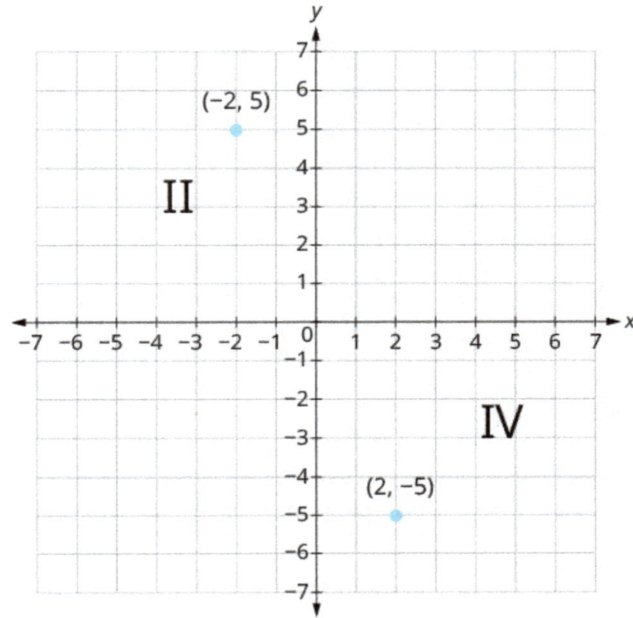

We can summarize sign patterns of the quadrants as follows. Also see Figure 11.7.

Quadrant I	Quadrant II	Quadrant III	Quadrant IV
(x,y)	(x,y)	(x,y)	(x,y)
(+,+)	(−,+)	(−,−)	(+,−)

Table 11.1

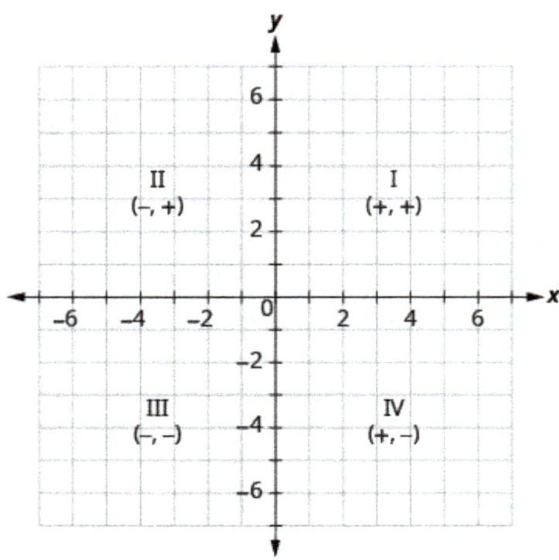

Figure 11.7

What if one coordinate is zero? Where is the point located? Where is the point located? The point is on the *y*-axis and the point is on the *x*-axis.

Points on the Axes

Points with a equal to are on the and have coordinates

Points with an equal to are on the and have coordinates

What is the ordered pair of the point where the axes cross? At that point both coordinates are zero, so its ordered pair is . The point has a special name. It is called the *origin*.

The Origin

The point is called the **origin**. It is the point where the *x*-axis and *y*-axis intersect.

EXAMPLE 11.5

Plot each point on a coordinate grid:

Solution

Since the point whose coordinates are is on the

Since the point whose coordinates are is on the

Since the point whose coordinates are is on the

Since and the point whose coordinates are is the origin.

Since the point whose coordinates are is on the

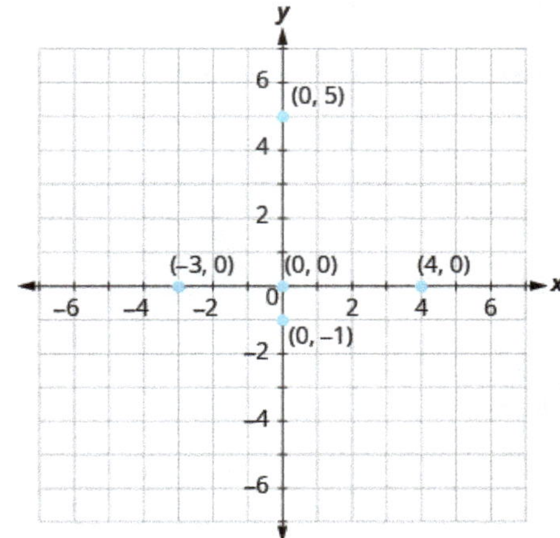

> TRY IT : : 11.9 Plot each point on a coordinate grid:

> TRY IT : : 11.10 Plot each point on a coordinate grid:

Identify Points on a Graph

In algebra, being able to identify the coordinates of a point shown on a graph is just as important as being able to plot points. To identify the *x*-coordinate of a point on a graph, read the number on the *x*-axis directly above or below the point. To identify the *y*-coordinate of a point, read the number on the *y*-axis directly to the left or right of the point. Remember, to write the ordered pair using the correct order

EXAMPLE 11.6

Name the ordered pair of each point shown:

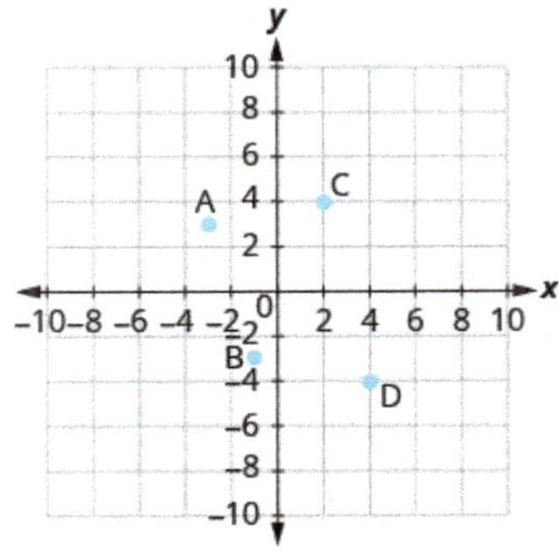

Solution

Point A is above on the so the of the point is The point is to the left of on the so the of the point is The coordinates of the point are

Point B is below on the so the of the point is The point is to the left of on the so the of the point is The coordinates of the point are

Point C is above on the so the of the point is The point is to the right of on the so the of the point is The coordinates of the point are

Point D is below on the so the of the point is The point is to the right of on the so the of the point is The coordinates of the point are

> **TRY IT : :** 11.11 Name the ordered pair of each point shown:

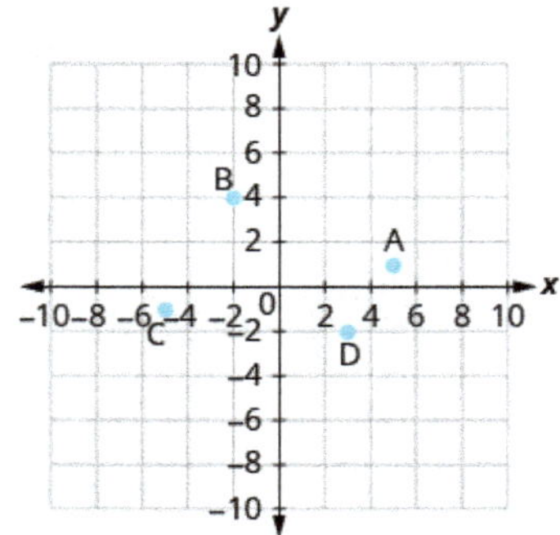

> **TRY IT : :** 11.12 Name the ordered pair of each point shown:

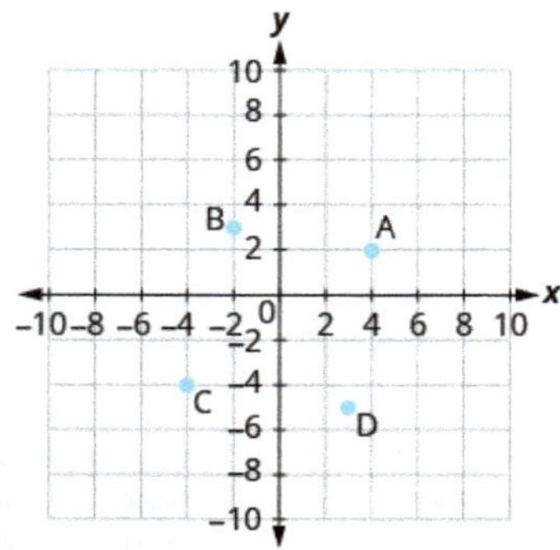

EXAMPLE 11.7

Name the ordered pair of each point shown:

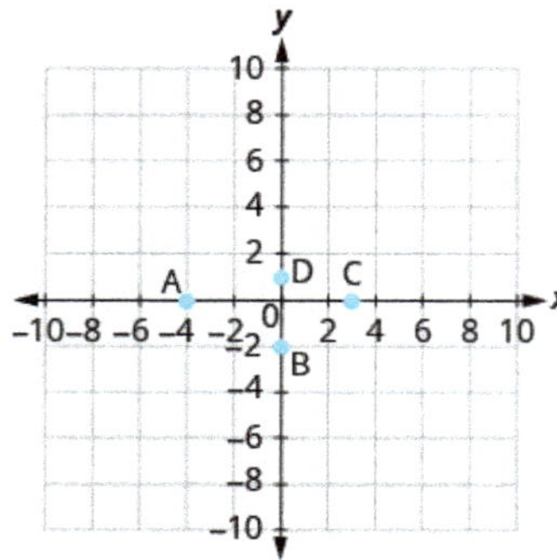

Solution

Point A is on the *x*-axis at .	The coordinates of point A are .
Point B is on the *y*-axis at	The coordinates of point B are .
Point C is on the *x*-axis at .	The coordinates of point C are .
Point D is on the *y*-axis at .	The coordinates of point D are .

TRY IT : : 11.13 Name the ordered pair of each point shown:

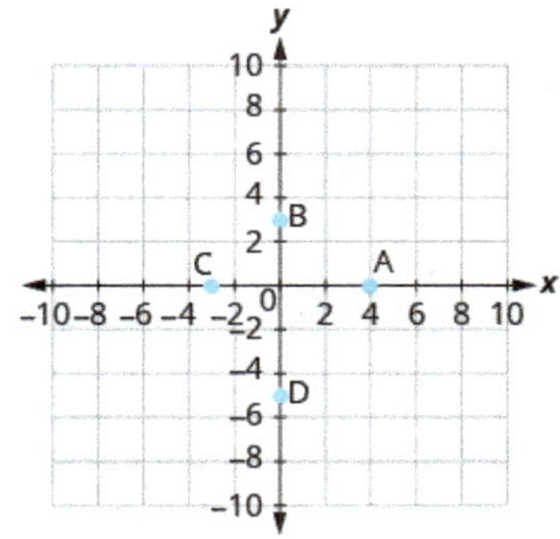

TRY IT : : 11.14 Name the ordered pair of each point shown:

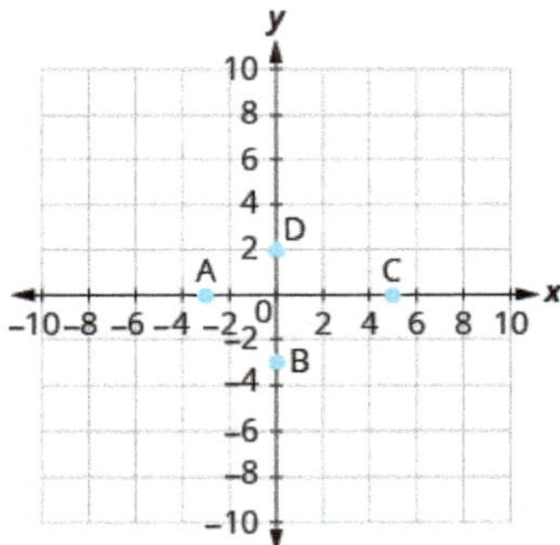

Verify Solutions to an Equation in Two Variables

All the equations we solved so far have been equations with one variable. In almost every case, when we solved the equation we got exactly one solution. The process of solving an equation ended with a statement such as Then we checked the solution by substituting back into the equation.

Here's an example of a linear equation in one variable, and its one solution.

But equations can have more than one variable. Equations with two variables can be written in the general form An equation of this form is called a linear equation in two variables.

Linear Equation

An equation of the form where are not both zero, is called a linear equation in two variables.

Notice that the word "line" is in linear.

Here is an example of a linear equation in two variables, and

$$Ax + By = C$$

$$x + 4y = 8$$

$$A = 1, B = 4, C = 8$$

Is a linear equation? It does not appear to be in the form But we could rewrite it in this form.

	$y = -5x + 1$
Add to both sides.	$y + 5x = -5x + 1 + 5x$
Simplify.	$y + 5x = 1$
Use the Commutative Property to put it in	$Ax + By = C$ $5x + \quad y = 1$

By rewriting as we can see that it is a linear equation in two variables because it can be written in the form

Linear equations in two variables have infinitely many solutions. For every number that is substituted for there is a corresponding value. This pair of values is a **solution to the linear equation** and is represented by the ordered pair When we substitute these values of and into the equation, the result is a true statement because the value on the left side is equal to the value on the right side.

Solution to a Linear Equation in Two Variables

An ordered pair is a solution to the linear equation if the equation is a true statement when the and of the ordered pair are substituted into the equation.

EXAMPLE 11.8

Determine which ordered pairs are solutions of the equation

Solution

Substitute the from each ordered pair into the equation and determine if the result is a true statement.

$x = 0, y = 2$	$x = 2, y = -4$	$x = -4, y = 3$
$x + 4y = 8$	$x + 4y = 8$	$x + 4y = 8$
$0 + 4 \cdot 2 \stackrel{?}{=} 8$	$2 + 4(-4) \stackrel{?}{=} 8$	$-4 + 4 \cdot 3 \stackrel{?}{=} 8$
$0 + 8 \stackrel{?}{=} 8$	$2 + (-16) \stackrel{?}{=} 8$	$-4 + 12 \stackrel{?}{=} 8$
$8 = 8 \checkmark$	$-14 \neq 8$	$8 = 8 \checkmark$
is a solution.	is not a solution.	is a solution.

TRY IT : : 11.15 Determine which ordered pairs are solutions to the given equation:

TRY IT : : 11.16 Determine which ordered pairs are solutions to the given equation:

EXAMPLE 11.9

Determine which ordered pairs are solutions of the equation.

Solution

Substitute the and from each ordered pair into the equation and determine if it results in a true statement.

$x = 0, y = -1$	$x = 1, y = 4$	$x = -2, y = -7$
$y = 5x - 1$	$y = 5x - 1$	$y = 5x - 1$
$-1 \stackrel{?}{=} 5(0) - 1$	$4 \stackrel{?}{=} 5(1) - 1$	$-7 \stackrel{?}{=} 5(-2) - 1$
$-1 \stackrel{?}{=} 0 - 1$	$4 \stackrel{?}{=} 5 - 1$	$-7 \stackrel{?}{=} -10 - 1$
$-1 = -1$ ✓	$4 = 4$ ✓	$-7 \neq -11$
is a solution.	is a solution.	is not a solution.

TRY IT : : 11.17 Determine which ordered pairs are solutions of the given equation:

TRY IT : : 11.18 Determine which ordered pairs are solutions of the given equation:

Complete a Table of Solutions to a Linear Equation

In the previous examples, we substituted the of a given ordered pair to determine whether or not it was a solution to a linear equation. But how do we find the ordered pairs if they are not given? One way is to choose a value for and then solve the equation for Or, choose a value for and then solve for

We'll start by looking at the solutions to the equation we found in Example 11.9. We can summarize this information in a table of solutions.

y x		
x	y	x y

To find a third solution, we'll let and solve for

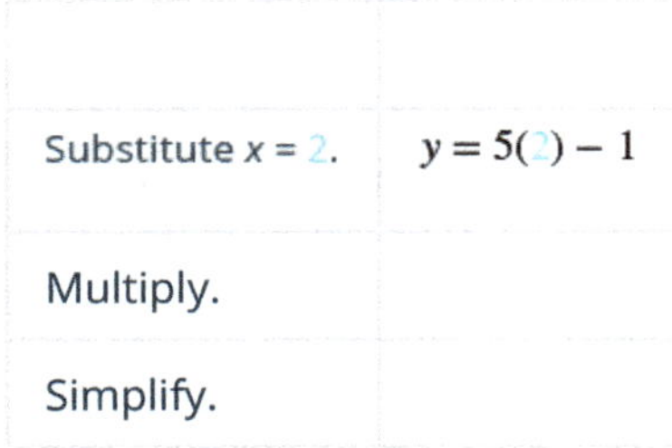

Substitute $x = 2$.	$y = 5(2) - 1$
Multiply.	
Simplify.	

The ordered pair is a solution to . We will add it to the table.

y x		
x	y	x y

We can find more solutions to the equation by substituting any value of or any value of and solving the resulting equation to get another ordered pair that is a solution. There are an infinite number of solutions for this equation.

EXAMPLE 11.10

Complete the table to find three solutions to the equation

y x		
x	y	x y

Solution

Substitute and into

$x = 0$	$x = -1$	$x = 2$
$y = 4 \cdot 0 - 2$	$y = 4(-1) - 2$	$y = 4 \cdot 2 - 2$

The results are summarized in the table.

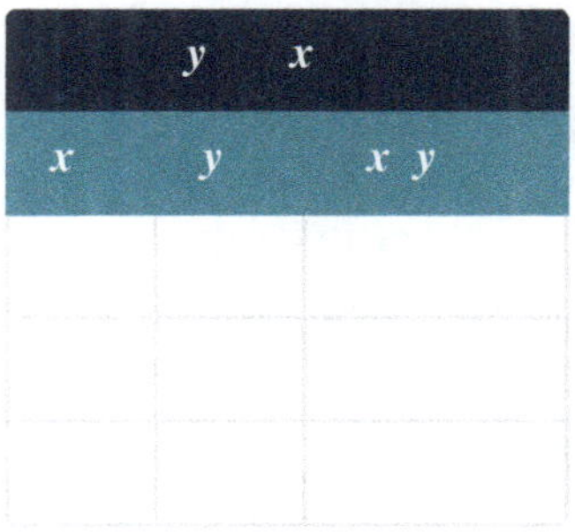

y x		
x	y	x y

> **TRY IT : : 11.19** Complete the table to find three solutions to the equation:

y x		
x	y	x y

> **TRY IT : : 11.20** Complete the table to find three solutions to the equation:

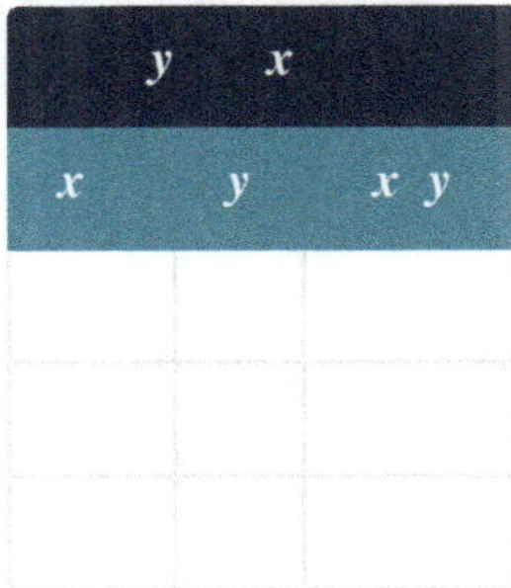

y x		
x	y	x y

EXAMPLE 11.11

Complete the table to find three solutions to the equation

x y		
x	y	x y

Solution

$x = 0$	$y = 0$	$y = 5$
$5x - 4y = 20$	$5x - 4y = 20$	$5x - 4y = 20$
$5 \cdot 0 - 4y = 20$	$5x - 4 \cdot 0 = 20$	$5x - 4 \cdot 5 = 20$
$0 - 4y = 20$	$5x - 0 = 20$	$5x - 20 = 20$
$-4y = 20$	$5x = 20$	$5x = 40$
$y = -5$	$x = 4$	$x = 8$
$(0, -5)$	$(4, 0)$	$(8, 5)$

The results are summarized in the table.

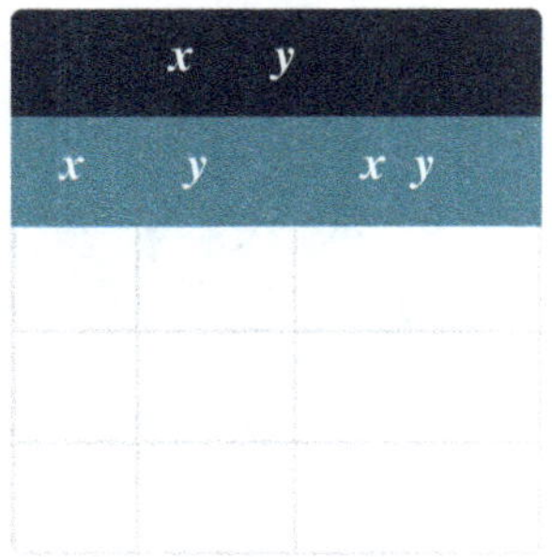

x y		
x	y	x y

> **TRY IT : :** 11.21 Complete the table to find three solutions to the equation:

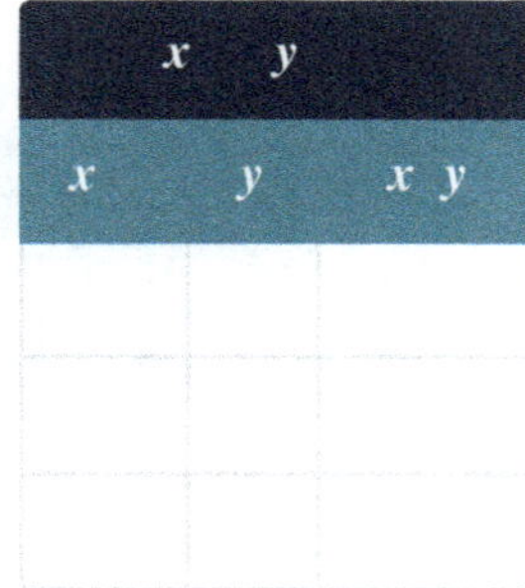

x y		
x	y	x y

TRY IT : : 11.22 Complete the table to find three solutions to the equation:

Find Solutions to Linear Equations in Two Variables

To find a solution to a linear equation, we can choose any number we want to substitute into the equation for either or We could choose or any other value we want. But it's a good idea to choose a number that's easy to work with. We'll usually choose as one of our values.

EXAMPLE 11.12

Find a solution to the equation

Solution

Step 1: Choose any value for one of the variables in the equation.		We can substitute any value we want for or any value for Let's pick What is the value of if ?
Step 2: Substitute that value into the equation. Solve for the other variable.	Substitute for Simplify. Divide both sides by 2.	$3x + 2y = 6$ $3 \cdot 0 + 2y = 6$ $0 + 2y = 6$ $2y = 6$ $y = 3$
Step 3: Write the solution as an ordered pair.	So, when	This solution is represented by the ordered pair
Step 4: Check.	Substitute $x = 0, y = 3$ into the equation $3x + 2y = 6$. Is the result a true equation? Yes!	$3x + 2y = 6$ $3 \cdot 0 + 2 \cdot 3 \stackrel{?}{=} 6$ $0 + 6 \stackrel{?}{=} 6$ $6 = 6 \checkmark$

TRY IT : : 11.23 Find a solution to the equation:

TRY IT : : 11.24 Find a solution to the equation:

We said that linear equations in two variables have infinitely many solutions, and we've just found one of them. Let's find some other solutions to the equation

EXAMPLE 11.13

Find three more solutions to the equation

Solution

To find solutions to choose a value for or Remember, we can choose any value we want for or Here we chose for and and for

	$y = 0$	$x = 1$	$y = -3$
	$3x + 2y = 6$	$3x + 2y = 6$	$3x + 2y = 6$
Substitute it into the equation.	$3x + 2(0) = 6$	$3(1) + 2y = 6$	$3x + 2(-3) = 6$
Simplify. Solve.	$3x + 0 = 6$ $3x = 6$	$3 + 2y = 6$ $2y = 3$	$3x - 6 = 6$ $3x = 12$
	$x = 2$	$y = \frac{3}{2}$	$x = 4$
Write the ordered pair.		—	

Check your answers.

—

$3x + 2y = 6$	$3x + 2y = 6$	$3x + 2y = 6$
$3 \cdot 2 + 2 \cdot 0 \stackrel{?}{=} 6$	$3 \cdot 1 + 2 \cdot \frac{3}{2} \stackrel{?}{=} 6$	$3 \cdot 4 + 2(-3) \stackrel{?}{=} 6$
$6 + 0 \stackrel{?}{=} 6$	$3 + 3 \stackrel{?}{=} 6$	$12 + (-6) \stackrel{?}{=} 6$
$6 = 6$ ✓	$6 = 6$ ✓	$6 = 6$ ✓

So — and are all solutions to the equation In the previous example, we found that is a solution, too. We can list these solutions in a table.

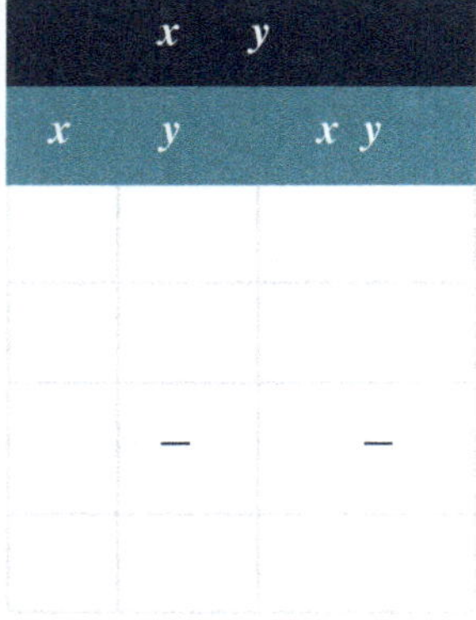

x y		
x	y	x y
	—	—

TRY IT : : 11.25 Find three solutions to the equation:

TRY IT : : 11.26 Find three solutions to the equation:

Let's find some solutions to another equation now.

EXAMPLE 11.14

Find three solutions to the equation

Solution

	$x - 4y = 8$	$x - 4y = 8$	$x - 4y = 8$
Choose a value for or	$x = 0$	$y = 0$	$y = 3$
Substitute it into the equation.	$0 - 4y = 8$	$x - 4 \cdot 0 = 8$	$x - 4 \cdot 3 = 8$
Solve.	$-4y = 8$ $y = -2$	$x - 0 = 8$ $x = 8$	$x - 12 = 8$ $x = 20$
Write the ordered pair.			

So and are three solutions to the equation

x y		
x	y	x y

Remember, there are an infinite number of solutions to each linear equation. Any point you find is a solution if it makes the equation true.

TRY IT : : 11.27 Find three solutions to the equation:

TRY IT : : 11.28 Find three solutions to the equation:

MEDIA : : ACCESS ADDITIONAL ONLINE RESOURCES

- Plotting Points (http://openstaxcollege.org/l/24plotpoints)
- Identifying Quadrants (http://openstaxcollege.org/l/24quadrants)
- Verifying Solution to Linear Equation (http://openstaxcollege.org/l/24verlineq)

11.1 EXERCISES

Practice Makes Perfect

Plot Points on a Rectangular Coordinate System

In the following exercises, plot each point on a coordinate grid.

1.

2.

3.

4.

5.

6.

7.

In the following exercises, plot each point on a coordinate grid and identify the quadrant in which the point is located.

8.

—

9.

—

10.

—

In the following exercises, plot each point on a coordinate grid.

11.

12.

13.

1.
2.
3.
4.

Identify Points on a Graph

In the following exercises, name the ordered pair of each point shown.

14.

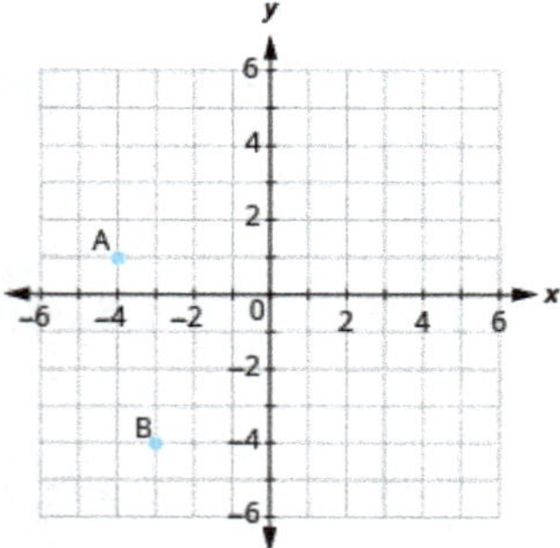

15.

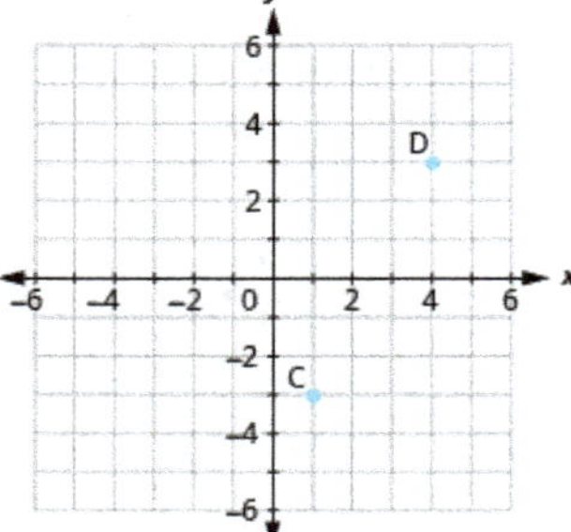

16.

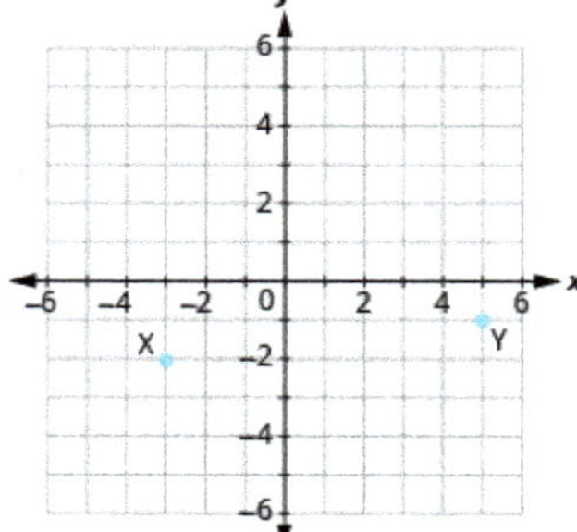

17.

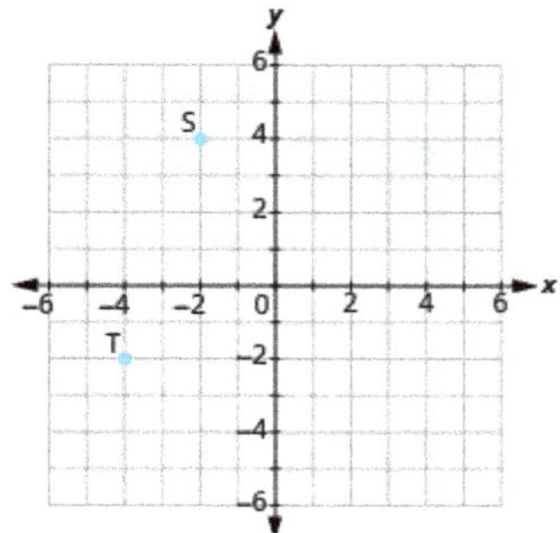

18.

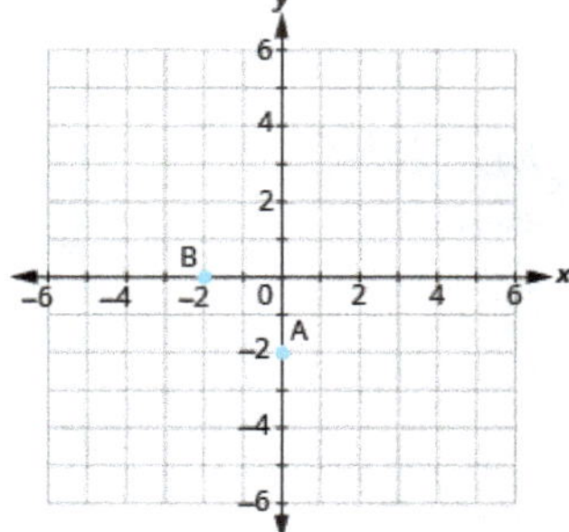

19.

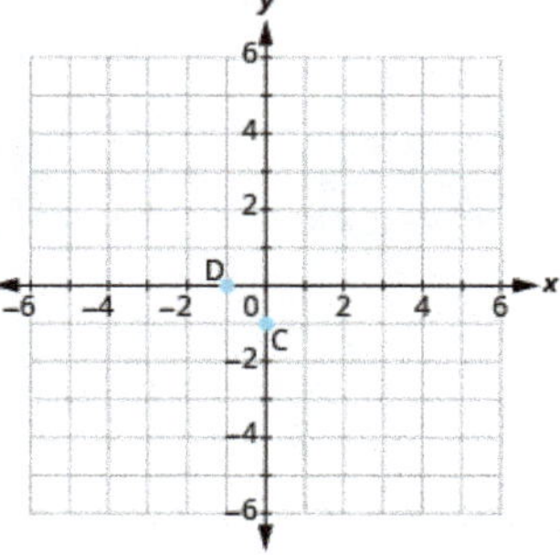

20.

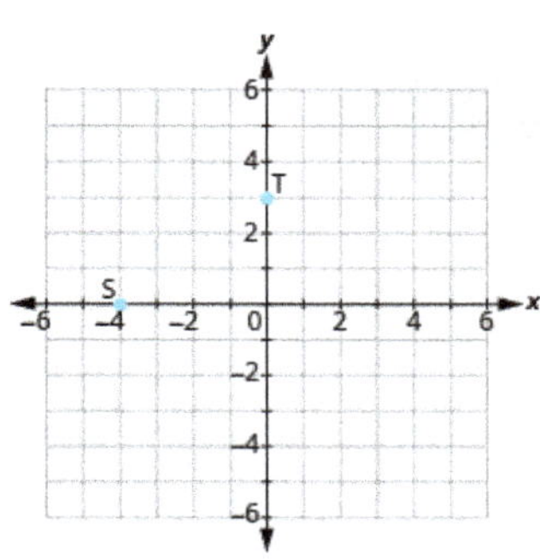

Verify Solutions to an Equation in Two Variables

In the following exercises, determine which ordered pairs are solutions to the given equation.

21.

22.

23.

24.

25.

—

26.

—

27. —

28. —

Find Solutions to Linear Equations in Two Variables

In the following exercises, complete the table to find solutions to each linear equation.

29.

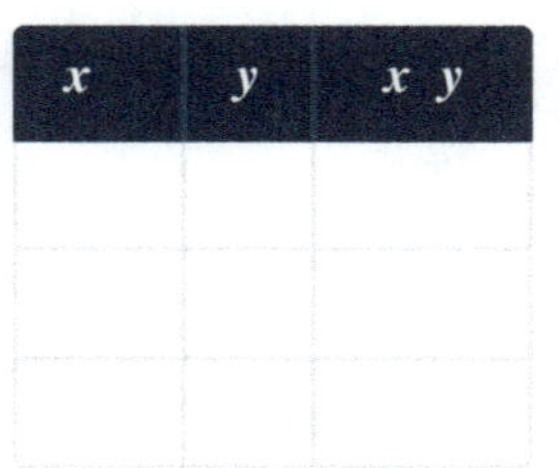

x	y	(x, y)

30.

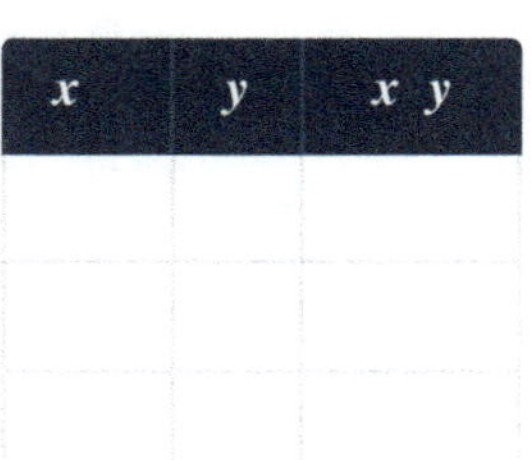

x	y	(x, y)

31.

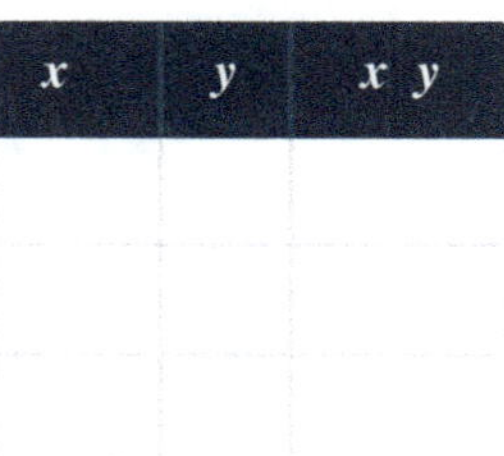

x	y	(x, y)

32. —

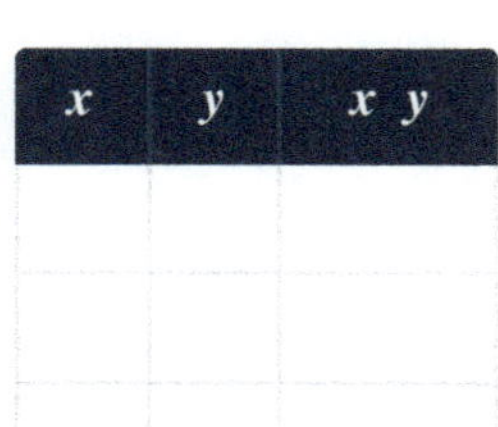

x	y	(x, y)

33. —

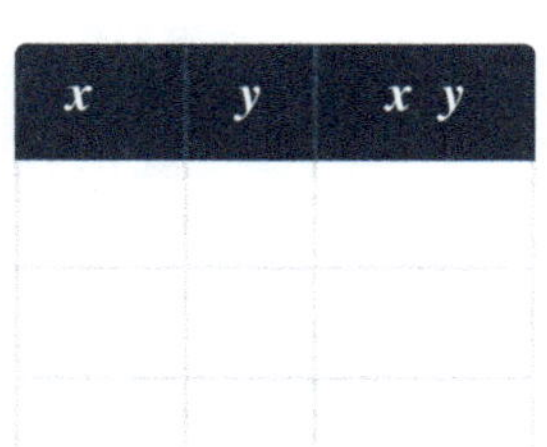

x	y	(x, y)

34.

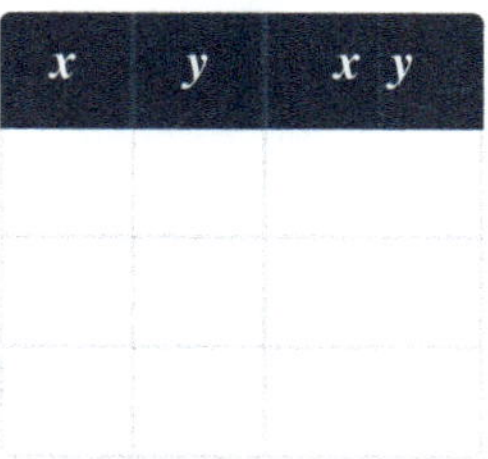

x	y	(x, y)

Everyday Math

35. Weight of a baby Mackenzie recorded her baby's weight every two months. The baby's age, in months, and weight, in pounds, are listed in the table, and shown as an ordered pair in the third column.

Plot the points on a coordinate grid.

Why is only Quadrant I needed?

36. Weight of a child Latresha recorded her son's height and weight every year. His height, in inches, and weight, in pounds, are listed in the table, and shown as an ordered pair in the third column.

Plot the points on a coordinate grid.

Why is only Quadrant I needed?

Writing Exercises

37. Have you ever used a map with a rectangular coordinate system? Describe the map and how you used it.

38. How do you determine if an ordered pair is a solution to a given equation?

Self Check

After completing the exercises, use this checklist to evaluate your mastery of the objectives of this section.

I can...	Confidently	With some help	No-I don't get it!
plot points on a rectangular coordinate system.			
identify points on a graph.			
verify solutions to an equation in two variables.			
complete a table of solutions to a linear equation.			
find solutions to linear equations in two variables.			

If most of your checks were:

...confidently. Congratulations! You have achieved the objectives in this section. Reflect on the study skills you used so that you can continue to use them. What did you do to become confident of your ability to do these things? Be specific.

...with some help. This must be addressed quickly because topics you do not master become potholes in your road to success. In math, every topic builds upon previous work. It is important to make sure you have a strong foundation before you move on. Whom can you ask for help? Your fellow classmates and instructor are good resources. Is there a place on campus where math tutors are available? Can your study skills be improved?

...no—I don't get it! This is a warning sign and you must not ignore it. You should get help right away or you will quickly be overwhelmed. See your instructor as soon as you can to discuss your situation. Together you can come up with a plan to get you the help you need.

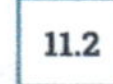

11.2 Graphing Linear Equations

Learning Objectives

By the end of this section, you will be able to:

- Recognize the relation between the solutions of an equation and its graph
- Graph a linear equation by plotting points
- Graph vertical and horizontal lines

BE PREPARED : : 11.4 Before you get started, take this readiness quiz.

Evaluate: when

If you missed this problem, review Example 3.56.

BE PREPARED : : 11.5 Solve the formula: for

If you missed this problem, review Example 9.62.

BE PREPARED : : 11.6 Simplify: —

If you missed this problem, review Example 4.28.

Recognize the Relation Between the Solutions of an Equation and its Graph

In Use the Rectangular Coordinate System, we found a few solutions to the equation . They are listed in the table below. So, the ordered pairs , , — , , are some solutions to the equation . We can plot these solutions in the rectangular coordinate system as shown on the graph at right.

$3x + 2y = 6$		
x	y	(x, y)
0	3	(0, 3)
2	0	(2, 0)
1	$\frac{3}{2}$	$\left(1, \frac{3}{2}\right)$
4	–3	(4, –3)

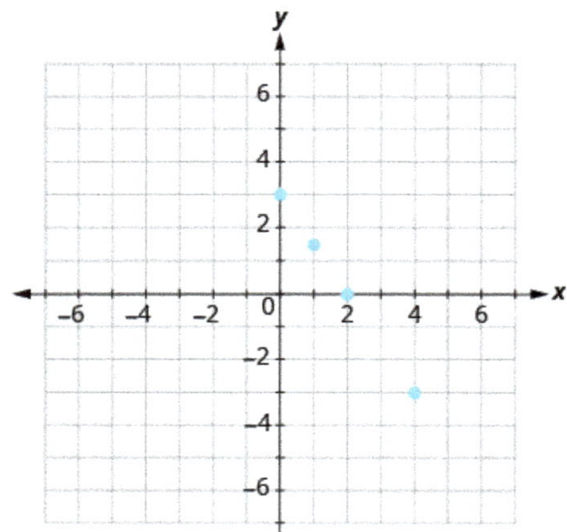

Notice how the points line up perfectly? We connect the points with a straight line to get the graph of the equation . Notice the arrows on the ends of each side of the line. These arrows indicate the line continues.

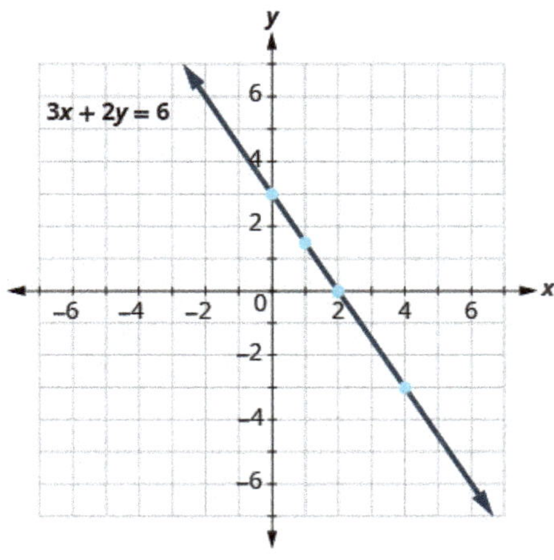

Every point on the line is a solution of the equation. Also, every solution of this equation is a point on this line. Points not on the line are *not* solutions!

Notice that the point whose coordinates are is on the line shown in Figure 11.8. If you substitute and into the equation, you find that it is a solution to the equation.

Test $(-2, 6)$:

$$3x + 2y = 6$$
$$3(-2) + 2(6) \stackrel{?}{=} 6$$
$$-6 + 12 \stackrel{?}{=} 6$$
$$6 = 6 \checkmark$$

So $(-2, 6)$ is a solution to the equation.

What about $(4, 1)$?

$$3x + 2y = 6$$
$$3 \cdot 4 + 2 \cdot 1 \stackrel{?}{=} 6$$
$$12 + 2 \stackrel{?}{=} 6$$
$$14 \neq 6$$

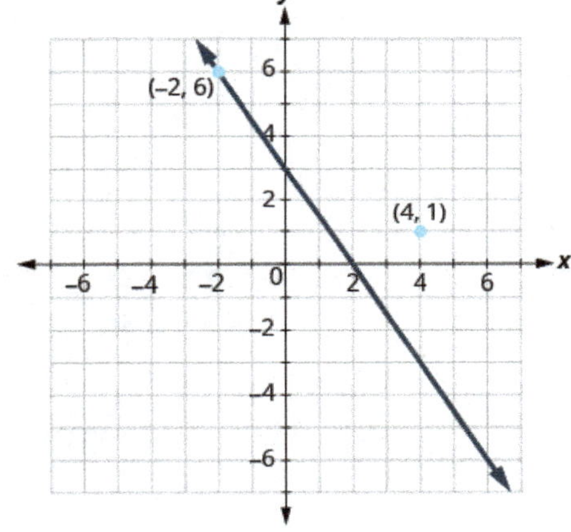

Figure 11.8

So is not a solution to the equation . Therefore the point is not on the line.

This is an example of the saying," A picture is worth a thousand words." The line shows you all the solutions to the equation. Every point on the line is a solution of the equation. And, every solution of this equation is on this line. This line is called the *graph* of the equation .

Graph of a Linear Equation

The graph of a linear equation is a straight line.

- Every point on the line is a solution of the equation.
- Every solution of this equation is a point on this line.

EXAMPLE 11.15

The graph of is shown below.

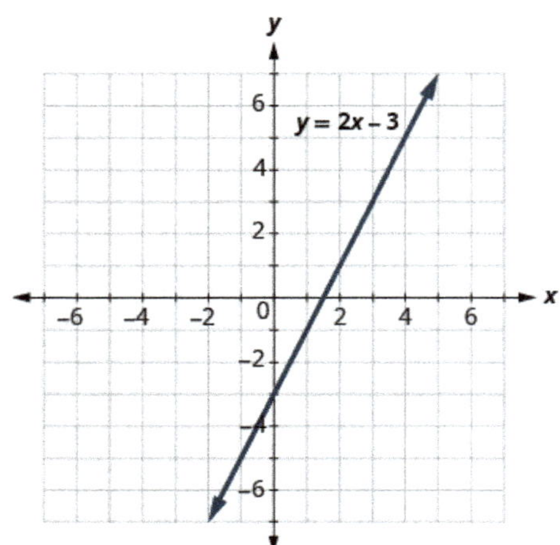

For each ordered pair decide

Is the ordered pair a solution to the equation? Is the point on the line?

(a) (b) (c) (d)

Solution

Substitute the - and -values into the equation to check if the ordered pair is a solution to the equation.

(a) $(0, -3)$

$y = 2x - 3$

$-3 \stackrel{?}{=} 2(0) - 3$

$-3 = -3$ ✓

$(0, -3)$ is a solution.

(b) $(3, 3)$

$y = 2x - 3$

$3 \stackrel{?}{=} 2(3) - 3$

$3 = 3$ ✓

$(3, 3)$ is a solution.

(c) $(2, -3)$

$y = 2x - 3$

$-3 \stackrel{?}{=} 2(2) - 3$

$-3 \neq 1$

$(2, -3)$ is not a solution.

(d) $(-1, -5)$

$y = 2x - 3$

$-5 \stackrel{?}{=} 2(-1) - 3$

$-5 = -5$ ✓

$(-1, -5)$ is a solution.

Plot the points A: B: C: and D: .

The points , , and are on the line , and the point is not on the line.

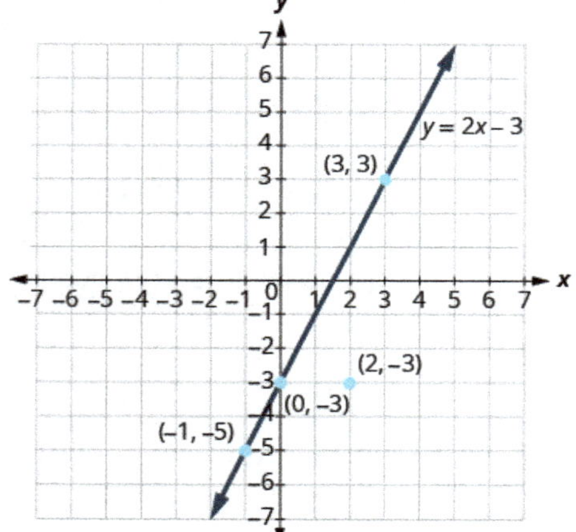

The points which are solutions to are on the line, but the point which is not a solution is not on the line.

TRY IT : : 11.29 The graph of is shown.

For each ordered pair, decide

is the ordered pair a solution to the equation? is the point on the line?

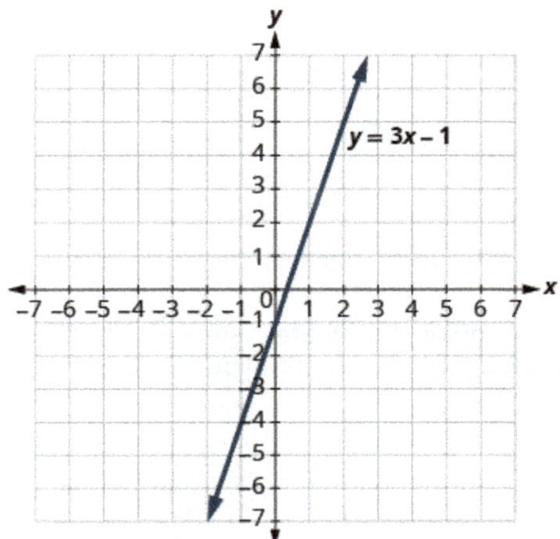

1.

2.

3.

4.

Graph a Linear Equation by Plotting Points

There are several methods that can be used to graph a linear equation. The method we used at the start of this section to graph is called plotting points, or the Point-Plotting Method.

Let's graph the equation by plotting points.

We start by finding three points that are solutions to the equation. We can choose any value for or and then solve for the other variable.

Since is isolated on the left side of the equation, it is easier to choose values for We will use and for for this example. We substitute each value of into the equation and solve for

$x = -2$	$x = 0$	$x = 1$
$y = 2x + 1$	$y = 2x + 1$	$y = 2x + 1$
$y = 2(-2) + 1$	$y = 2(0) + 1$	$y = 2(1) + 1$
$y = -4 + 1$	$y = 0 + 1$	$y = 2 + 1$
$y = -3$	$y = 1$	$y = 3$
$(-2, -3)$	$(0, 1)$	$(1, 3)$

We can organize the solutions in a table. See Table 11.2.

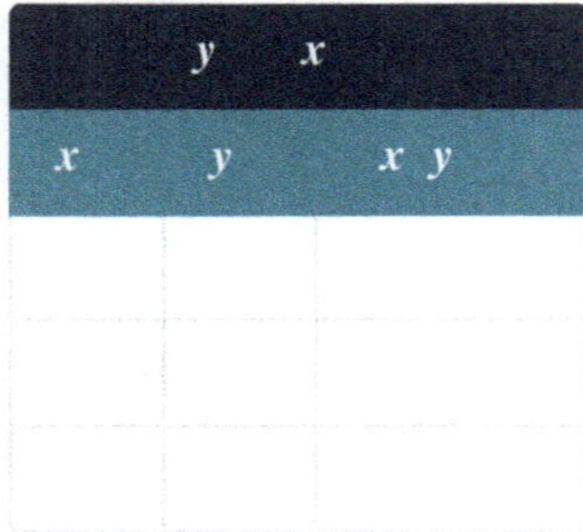

y x		
x	y	x y

Table 11.2

Now we plot the points on a rectangular coordinate system. Check that the points line up. If they did not line up, it would mean we made a mistake and should double-check all our work. See Figure 11.9.

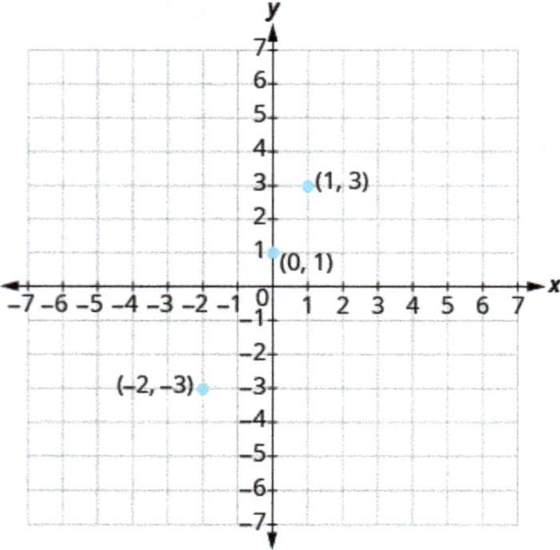

Figure 11.9

Draw the line through the three points. Extend the line to fill the grid and put arrows on both ends of the line. The line is the graph of

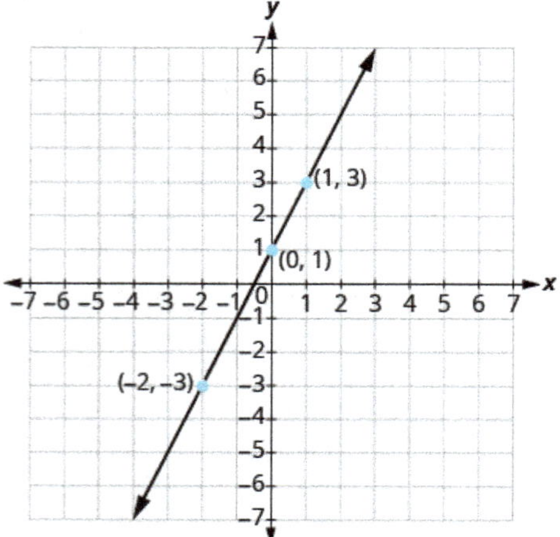

Figure 11.10

HOW TO : : GRAPH A LINEAR EQUATION BY PLOTTING POINTS.

Step 1. Find three points whose coordinates are solutions to the equation. Organize them in a table.

Step 2. Plot the points on a rectangular coordinate system. Check that the points line up. If they do not, carefully check your work.

Step 3. Draw the line through the points. Extend the line to fill the grid and put arrows on both ends of the line.

It is true that it only takes two points to determine a line, but it is a good habit to use three points. If you plot only two points and one of them is incorrect, you can still draw a line but it will not represent the solutions to the equation. It will be the wrong line. If you use three points, and one is incorrect, the points will not line up. This tells you something is wrong and you need to check your work. See Figure 11.11.

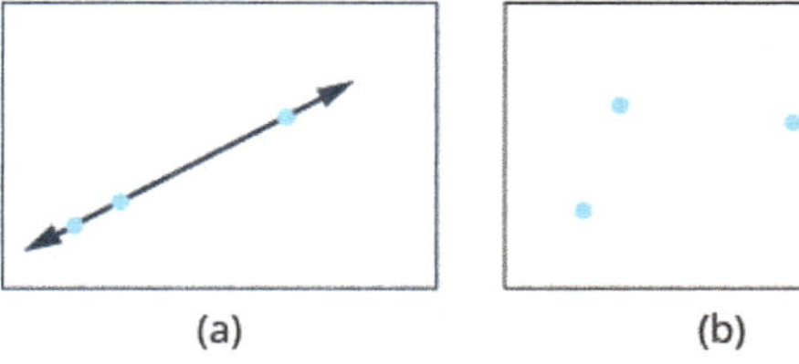

Figure 11.11 Look at the difference between (a) and (b). All three points in (a) line up so we can draw one line through them. The three points in (b) do not line up. We cannot draw a single straight line through all three points.

EXAMPLE 11.16

Graph the equation

Solution

Find three points that are solutions to the equation. It's easier to choose values for and solve for Do you see why?

$x = 0$	$x = 1$	$x = -2$
$y = -3x$	$y = -3x$	$y = -3x$
$y = -3(0)$	$y = -3(1)$	$y = -3(-2)$
$y = 0$	$y = -3$	$y = 6$
$(0, 0)$	$(1, -3)$	$(-2, 6)$

List the points in a table.

Plot the points, check that they line up, and draw the line as shown.

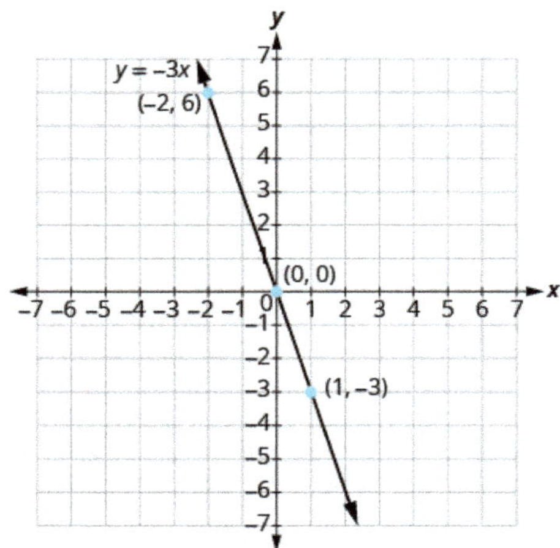

> **TRY IT : :** 11.30 Graph the equation by plotting points:

> **TRY IT : :** 11.31 Graph the equation by plotting points:

When an equation includes a fraction as the coefficient of we can substitute any numbers for But the math is easier if we make 'good' choices for the values of This way we will avoid fraction answers, which are hard to graph precisely.

EXAMPLE 11.17

Graph the equation —

Solution

Find three points that are solutions to the equation. Since this equation has the fraction — as a coefficient of we will choose values of carefully. We will use zero as one choice and multiples of for the other choices.

$x = 0$	$x = 2$	$x = 4$
$y = \frac{1}{2}x + 3$	$y = \frac{1}{2}x + 3$	$y = \frac{1}{2}x + 3$
$y = \frac{1}{2}(0) + 3$	$y = \frac{1}{2}(2) + 3$	$y = \frac{1}{2}(4) + 3$
$y = 3$	$y = 4$	$y = 5$
(0, 3)	(2, 4)	(4, 5)

The points are shown in the table.

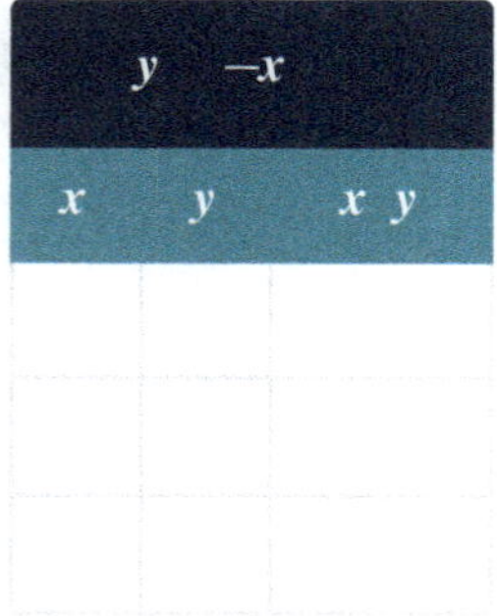

y —x		
x	y	x y

Plot the points, check that they line up, and draw the line as shown.

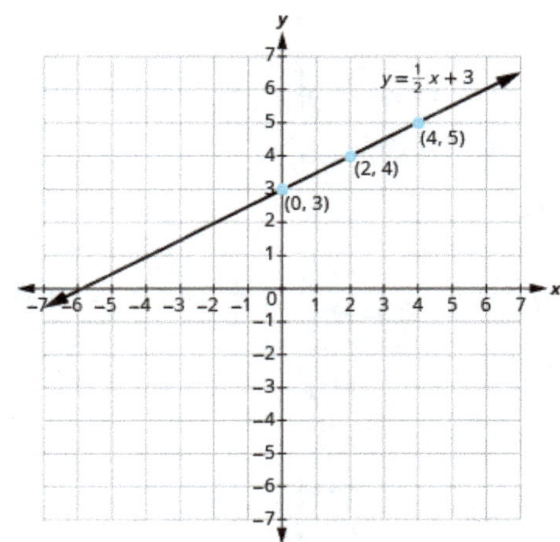

TRY IT : : 11.32 Graph the equation: —

TRY IT : : 11.33 Graph the equation: —

So far, all the equations we graphed had given in terms of Now we'll graph an equation with and on the same side.

EXAMPLE 11.18

Graph the equation

Solution

Find three points that are solutions to the equation. Remember, you can start with *any* value of or

$x = 0$	$x = 1$	$x = 4$
$x + y = 5$	$x + y = 5$	$x + y = 5$
$0 + y = 5$	$1 + y = 5$	$4 + y = 5$
$y = 5$	$y = 4$	$y = 1$
(0, 5)	(1, 4)	(4, 1)

We list the points in a table.

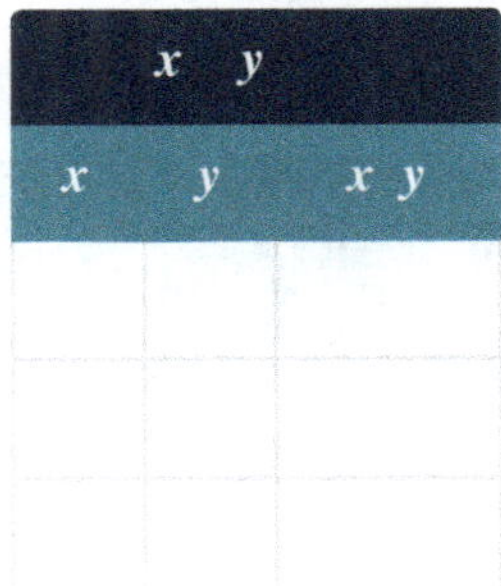

x	y	
x	y	x y

Then plot the points, check that they line up, and draw the line.

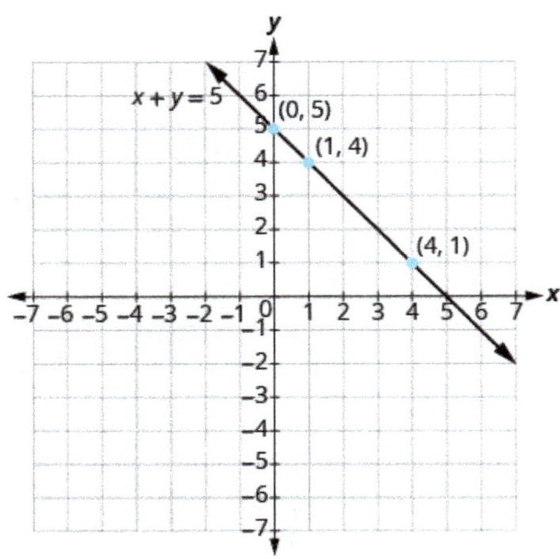

TRY IT : : 11.34 Graph the equation:

TRY IT : : 11.35 Graph the equation:

In the previous example, the three points we found were easy to graph. But this is not always the case. Let's see what happens in the equation If is what is the value of

$$2x + y = 3$$
$$2x + 0 = 3$$
$$2x = 3$$
$$x = \frac{3}{2}$$

The solution is the point — This point has a fraction for the -coordinate. While we could graph this point, it is hard to be precise graphing fractions. Remember in the example — we carefully chose values for so as not to graph fractions at all. If we solve the equation for it will be easier to find three solutions to the equation.

Now we can choose values for that will give coordinates that are integers. The solutions for and are shown.

	y	x
x	y	x y

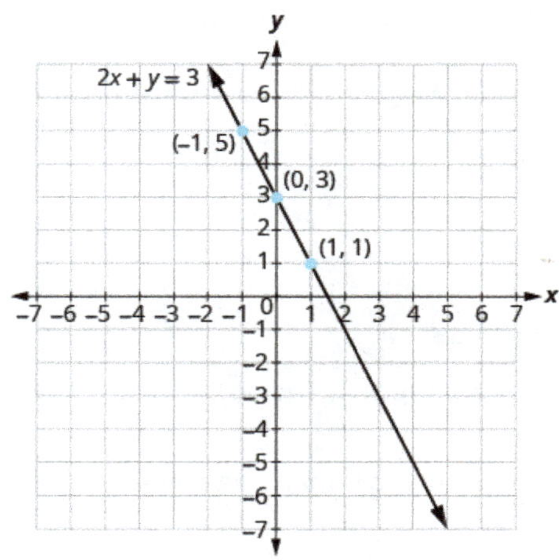

EXAMPLE 11.19

Graph the equation

 **Solution**

Find three points that are solutions to the equation.

First, solve the equation for

We'll let be and to find three points. The ordered pairs are shown in the table. Plot the points, check that they line up, and draw the line.

	y	x
x	y	x y

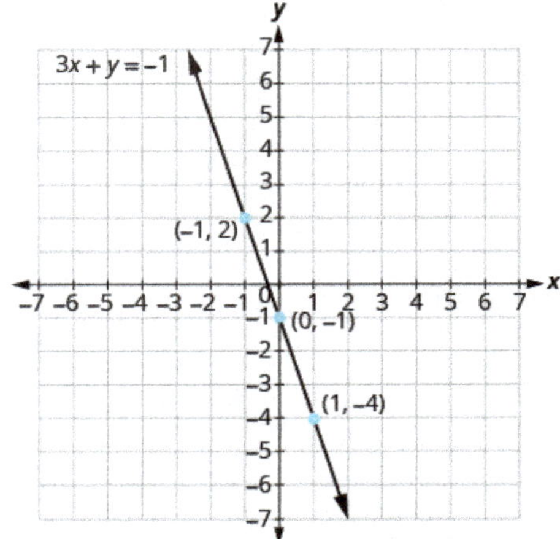

If you can choose any three points to graph a line, how will you know if your graph matches the one shown in the answers in the book? If the points where the graphs cross the and -axes are the same, the graphs match.

> **TRY IT : :** 11.36 Graph each equation:

> **TRY IT : :** 11.37 Graph each equation:

Graph Vertical and Horizontal Lines

Can we graph an equation with only one variable? Just and no or just without an How will we make a table of values to get the points to plot?

Let's consider the equation The equation says that is always equal to so its value does not depend on No matter what is, the value of is always

To make a table of solutions, we write for all the values. Then choose any values for Since does not depend on you can chose any numbers you like. But to fit the size of our coordinate graph, we'll use and for the -coordinates as shown in the table.

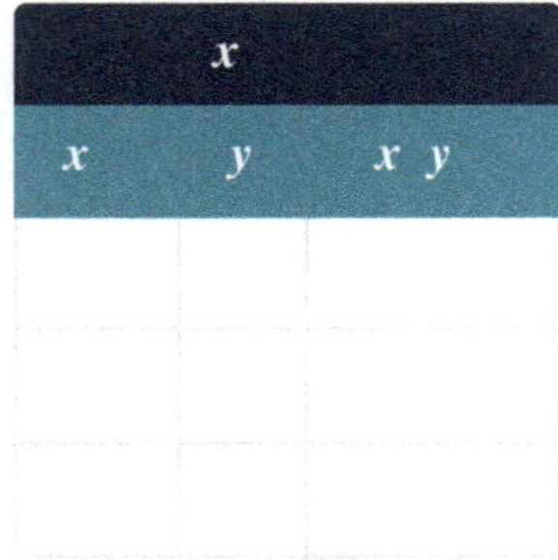

x		
x	y	x y

Then plot the points and connect them with a straight line. Notice in Figure 11.12 that the graph is a **vertical line**.

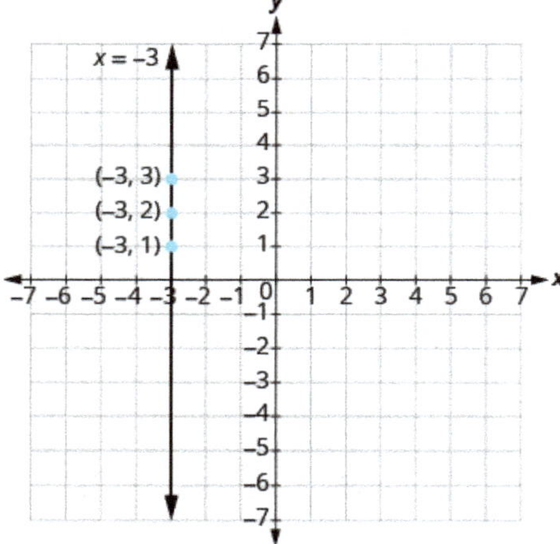

Figure 11.12

Vertical Line

A vertical line is the graph of an equation that can be written in the form

The line passes through the -axis at .

EXAMPLE 11.20

Graph the equation What type of line does it form?

Solution

The equation has only variable, and is always equal to We make a table where is always and we put in any values for

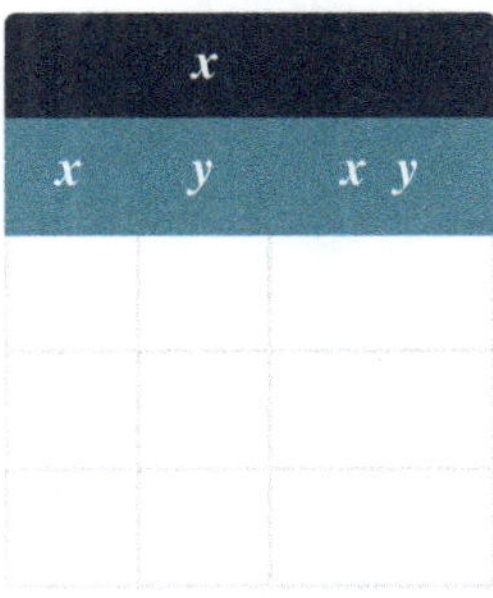

x		
x	y	x y

Plot the points and connect them as shown.

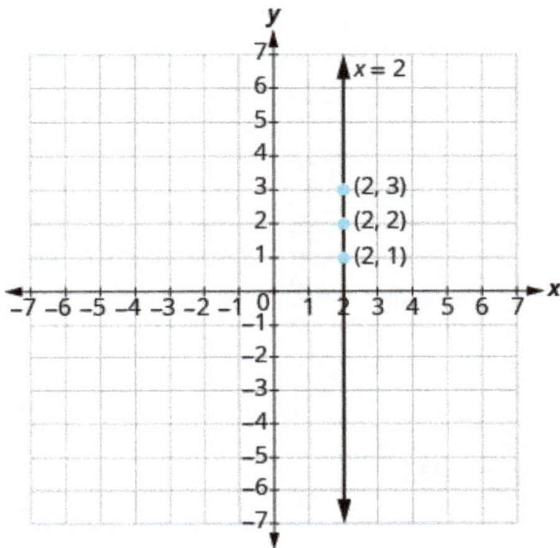

The graph is a vertical line passing through the -axis at

TRY IT : : 11.38 Graph the equation:

TRY IT : : 11.39 Graph the equation:

What if the equation has but no ? Let's graph the equation This time the -value is a constant, so in this equation does not depend on

To make a table of solutions, write for all the values and then choose any values for

We'll use and for the -values.

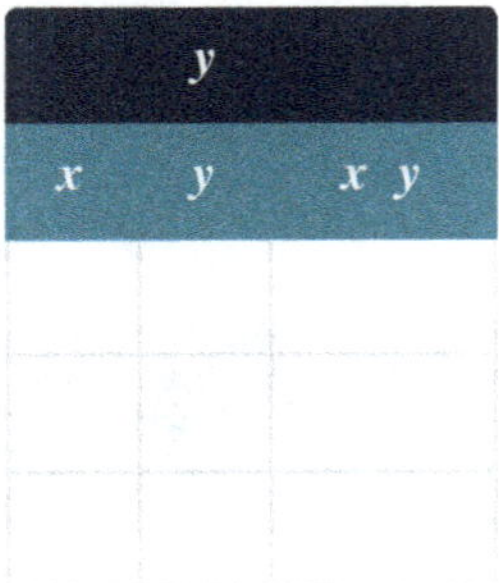

y		
x	y	x y

Plot the points and connect them, as shown in Figure 11.13. This graph is a **horizontal line** passing through the at

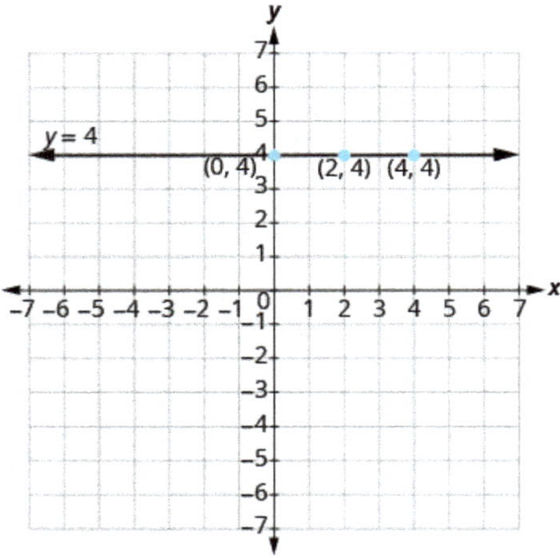

Figure 11.13

Horizontal Line

A horizontal line is the graph of an equation that can be written in the form

The line passes through the at

EXAMPLE 11.21

Graph the equation

Solution

The equation has only variable, The value of is constant. All the ordered pairs in the table have the same -coordinate, . We choose and as values for

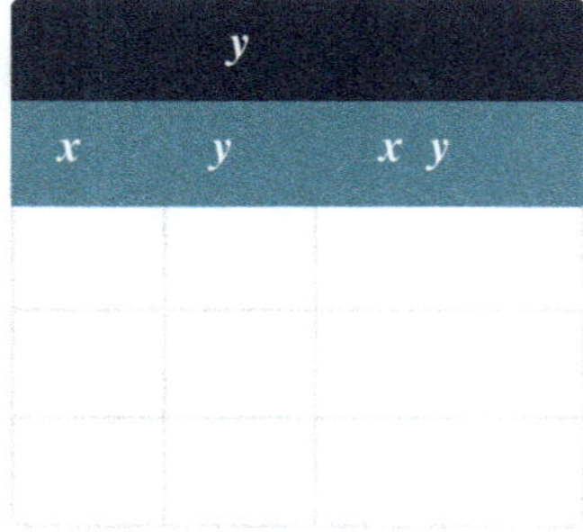

y		
x	y	x y

The graph is a horizontal line passing through the -axis at as shown.

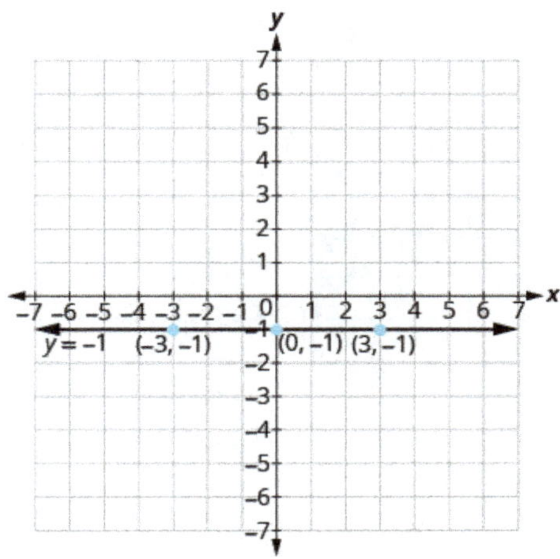

TRY IT : : 11.40 Graph the equation:

TRY IT : : 11.41 Graph the equation:

The equations for vertical and horizontal lines look very similar to equations like What is the difference between the equations and

The equation has both and The value of depends on the value of The changes according to the value of

The equation has only one variable. The value of is constant. The is always It does not depend on the value of

$y = 4x$		
x	y	(x, y)
0	0	(0, 0)
1	4	(1, 4)
2	8	(2, 8)

$y = 4$		
x	y	(x, y)
0	4	(0, 4)
1	4	(1, 4)
2	4	(2, 4)

The graph shows both equations.

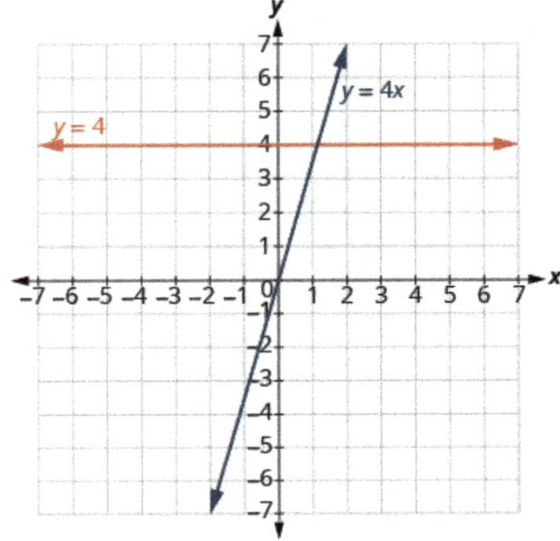

Notice that the equation gives a slanted line whereas gives a horizontal line.

EXAMPLE 11.22

Graph and in the same rectangular coordinate system.

Solution

Find three solutions for each equation. Notice that the first equation has the variable while the second does not. Solutions for both equations are listed.

$y = -3x$		
x	y	(x, y)
0	0	(0, 0)
1	−3	(1, −3)
2	−6	(2, −6)

$y = -3$		
x	y	(x, y)
0	−3	(0, −3)
1	−3	(1, −3)
2	−3	(2, −3)

The graph shows both equations.

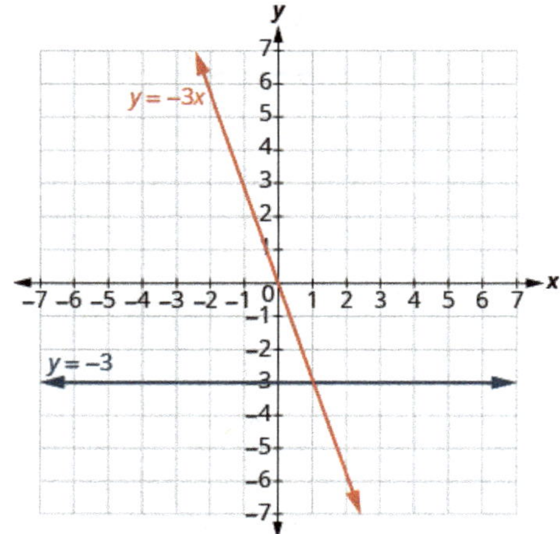

TRY IT : : 11.42 Graph the equations in the same rectangular coordinate system: and

TRY IT : : 11.43 Graph the equations in the same rectangular coordinate system: and

MEDIA : : ACCESS ADDITIONAL ONLINE RESOURCES

- Use a Table of Values (http://www.openstax.org/l/24tabofval)
- Graph a Linear Equation Involving Fractions (http://www.openstax.org/l/24graphlineq)
- Graph Horizontal and Vertical Lines (http://www.openstax.org/l/24graphhorvert)

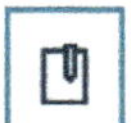

11.2 EXERCISES

Practice Makes Perfect

Recognize the Relation Between the Solutions of an Equation and its Graph

In each of the following exercises, an equation and its graph is shown. For each ordered pair, decide

is the ordered pair a solution to the equation? is the point on the line?

39.

40.

41. —

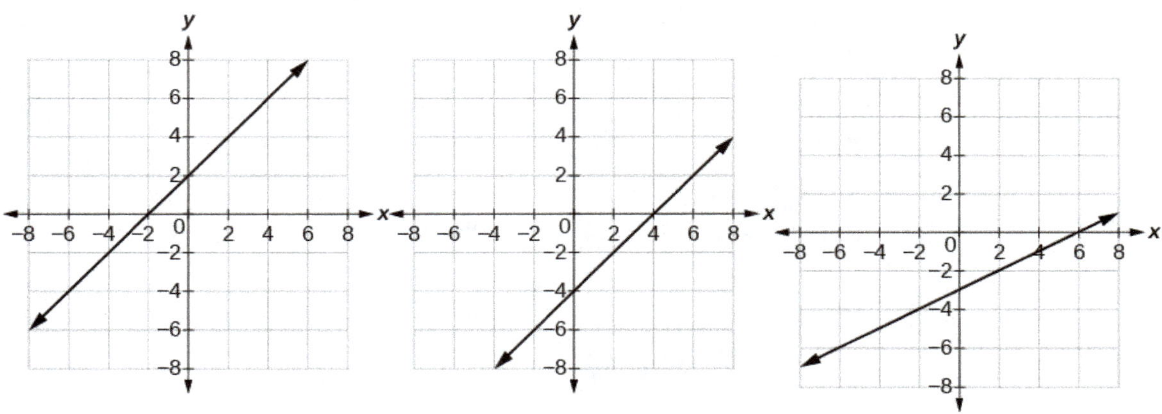

39.
1.
2.
3.
4.

40.
1.
2.
3.
4.

41.
1.
2.
3.
4.

42. —

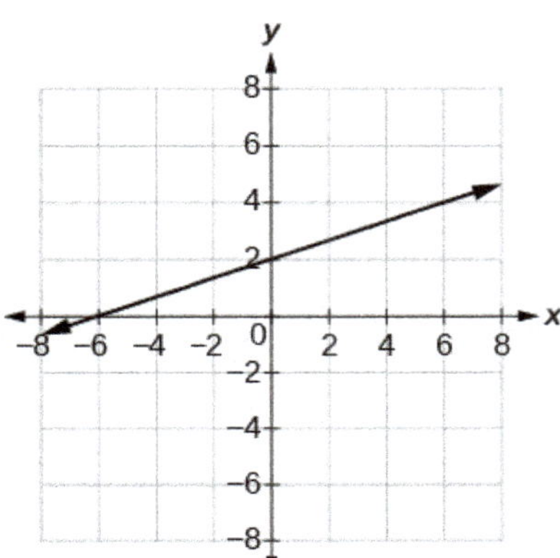

1.
2.
3.
4.

Graph a Linear Equation by Plotting Points

In the following exercises, graph by plotting points.

43. **44.** **45.**

46. **47.** **48.**

49. **50.** **51.**

52. **53.** **54.**

55. — **56.** — **57.** —

58. — **59.** — **60.** —

61. — **62.** — **63.**

64. **65.** **66.**

67. **68.** **69.**

70. **71.** **72.**

73. **74.** **75.**

76. **77.** **78.**

79. **80.** **81.** —

82. —

Graph Vertical and Horizontal lines

In the following exercises, graph the vertical and horizontal lines.

83. **84.** **85.**

86. **87.** **88.**

89. **90.** **91.** —

92. —

In the following exercises, graph each pair of equations in the same rectangular coordinate system.

93. — and — **94.** — and — **95.** and

96. and

Mixed Practice

In the following exercises, graph each equation.

97.

98.

99. —

100. —

101.

102.

103.

104.

105.

106.

107.

108.

109.

110.

111.

112.

Everyday Math

113. Motor home cost The Robinsons rented a motor home for one week to go on vacation. It cost them plus per mile to rent the motor home, so the linear equation gives the cost, for driving miles. Calculate the rental cost for driving miles, and then graph the line.

114. Weekly earning At the art gallery where he works, Salvador gets paid per week plus of the sales he makes, so the equation gives the amount he earns for selling dollars of artwork. Calculate the amount Salvador earns for selling and then graph the line.

Writing Exercises

115. Explain how you would choose three to make a table to graph the line —

116. What is the difference between the equations of a vertical and a horizontal line?

Self Check

After completing the exercises, use this checklist to evaluate your mastery of the objectives of this section.

I can...	Confidently	With some help	No-I don't get it!
graph a linear equation by plotting points.			
graph vertical and horizontal lines.			

After reviewing this checklist, what will you do to become confident for all objectives?

11.3 Graphing with Intercepts

Learning Objectives

By the end of this section, you will be able to:

- Identify the intercepts on a graph
- Find the intercepts from an equation of a line
- Graph a line using the intercepts
- Choose the most convenient method to graph a line

BE PREPARED : : 11.7 Before you get started, take this readiness quiz.

Solve: for when

If you missed this problem, review Example 9.62.

BE PREPARED : : 11.8 Is the point on the or

If you missed this problem, review Example 11.5.

BE PREPARED : : 11.9 Which ordered pairs are solutions to the equation

If you missed this problem, review Example 11.8.

Identify the Intercepts on a Graph

Every linear equation has a unique line that represents all the solutions of the equation. When graphing a line by plotting points, each person who graphs the line can choose any three points, so two people graphing the line might use different sets of points.

At first glance, their two lines might appear different since they would have different points labeled. But if all the work was done correctly, the lines will be exactly the same line. One way to recognize that they are indeed the same line is to focus on where the line crosses the axes. Each of these points is called an **intercept of the line**.

Intercepts of a Line

Each of the points at which a line crosses the and the is called an intercept of the line.

Let's look at the graph of the lines shown in Figure 11.14.

a) $2x + y = 6$

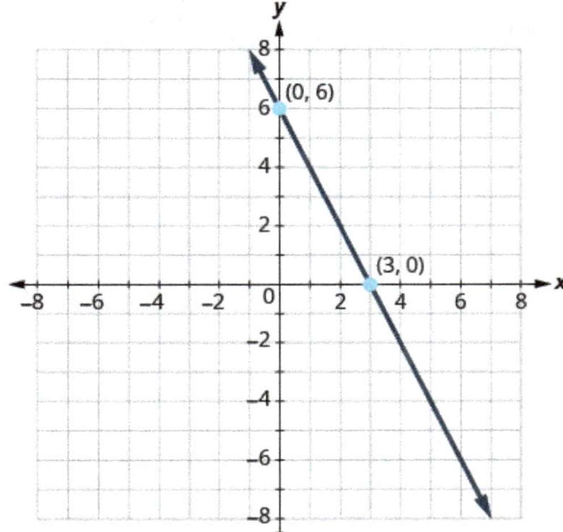

b) $3x - 4y = 12$

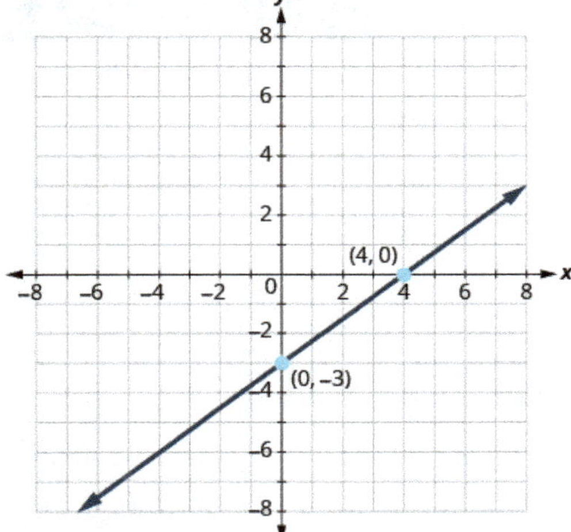

c) $x - y = 5$

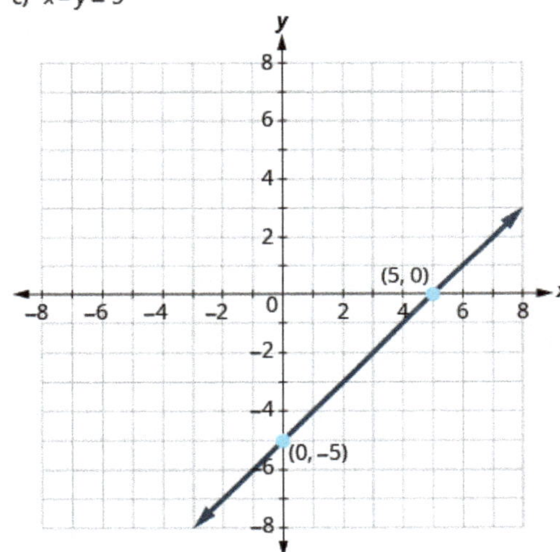

d) $y = -2x$

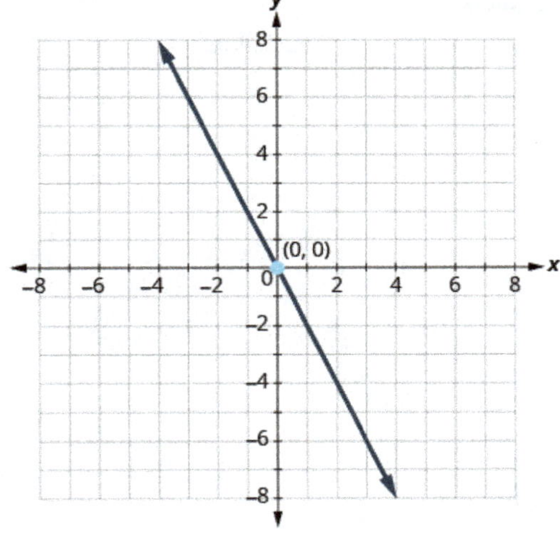

Figure 11.14

First, notice where each of these lines crosses the *x*- axis:

Figure:	The line crosses the x-axis at:	Ordered pair of this point
42	3	(3,0)
43	4	(4,0)
44	5	(5,0)
45	0	(0,0)

Table 11.3

Do you see a pattern?

For each row, the *y*- coordinate of the point where the line crosses the *x*- axis is zero. The point where the line crosses the *x*- axis has the form ; and is called the *x-intercept* of the line. The **x-** intercept occurs when y is zero.

Now, let's look at the points where these lines cross the y-axis.

Figure:	The line crosses the y-axis at:	Ordered pair for this point
42	6	(0,6)
43	-3	(0,-3)
44	-5	(0,-5)
45	0	(0,0)

Table 11.4

x- intercept and *y*- intercept of a line

The is the point, where the graph crosses the The occurs when is zero.

The is the point, where the graph crosses the

The occurs when is zero.

EXAMPLE 11.23

Find the of each line:

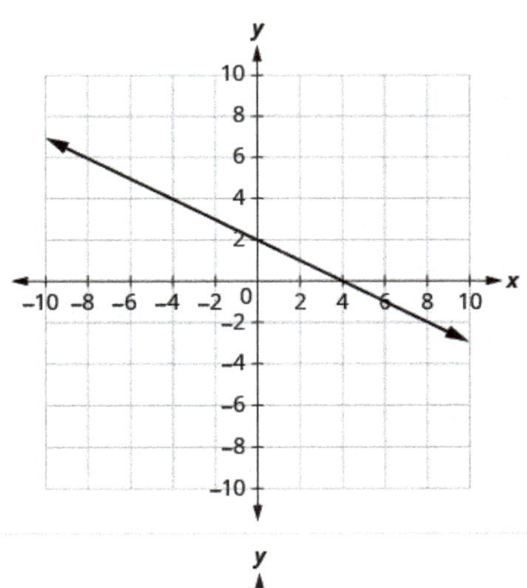

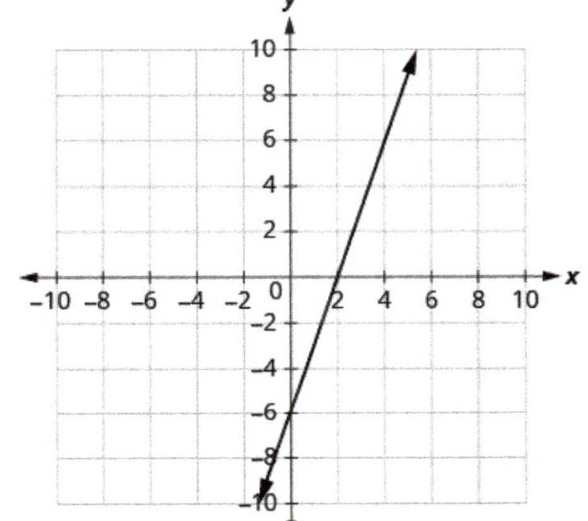

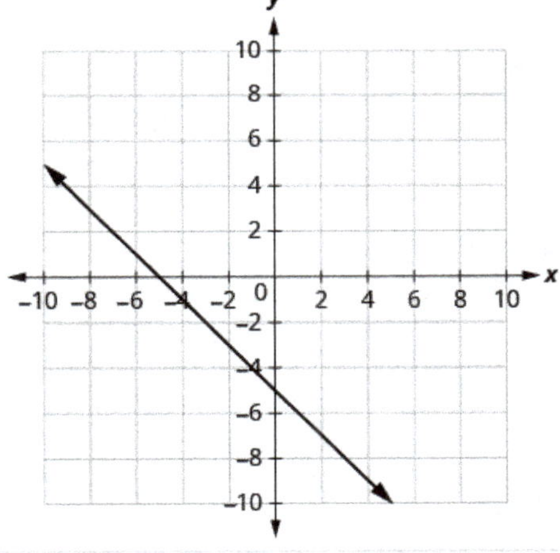

Solution

The graph crosses the *x*-axis at the point (4, 0).	The *x*-intercept is (4, 0).
The graph crosses the *y*-axis at the point (0, 2).	The *x*-intercept is (0, 2).

The graph crosses the x-axis at the point (2, 0).	The x-intercept is (2, 0)
The graph crosses the y-axis at the point (0, −6).	The y-intercept is (0, −6).

The graph crosses the x-axis at the point (−5, 0).	The x-intercept is (−5, 0).
The graph crosses the y-axis at the point (0, −5).	The y-intercept is (0, −5).

> **TRY IT : :** 11.44 Find the and of the graph:

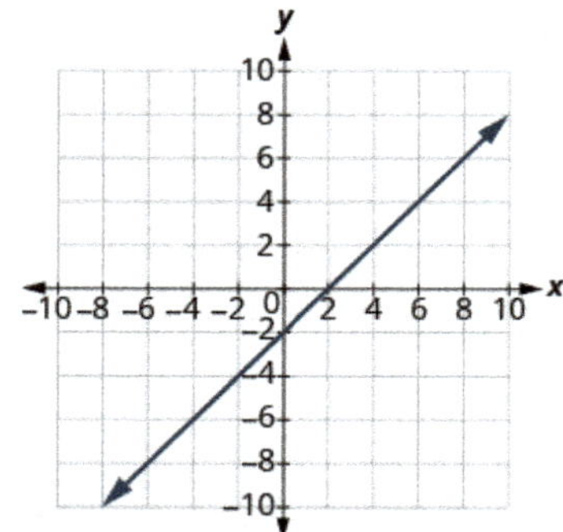

> **TRY IT : :** 11.45 Find the and of the graph:

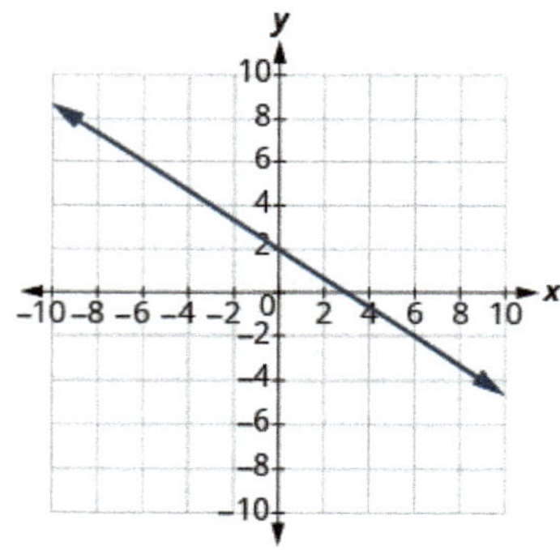

Find the Intercepts from an Equation of a Line

Recognizing that the occurs when is zero and that the occurs when is zero gives us a method to find the intercepts of a line from its equation. To find the let and solve for To find the let and solve for

Find the x and y from the Equation of a Line

Use the equation to find:

- the x-intercept of the line, let and solve for x.
- the y-intercept of the line, let and solve for y.

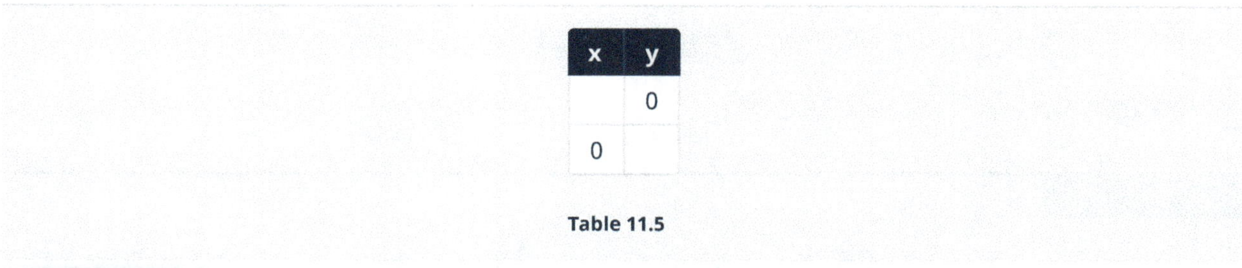

x	y
	0
0	

Table 11.5

EXAMPLE 11.24

Find the intercepts of

Solution

We'll fill in Figure 11.15.

$2x + y = 6$		
x	y	
	0	x-intercept
0		y-intercept

Figure 11.15

To find the x- intercept, let :

	$2x + y = 6$
Substitute 0 for y.	$2x + 0 = 6$
Add.	$2x = 6$
Divide by 2.	$x = 3$
The x-intercept is (3, 0).	

To find the y- intercept, let :

	$2x + y = 6$
Substitute 0 for x.	$2 \cdot 0 + y = 6$
Multiply.	$0 + y = 6$
Add.	$y = 6$
The y-intercept is (0, 6).	

$2x + y = 6$	
x	y
3	0
0	6

Figure 11.16

The intercepts are the points and .

> **TRY IT : :** 11.46 Find the intercepts:

> **TRY IT : :** 11.47 Find the intercepts:

EXAMPLE 11.25

Find the intercepts of

Solution

To find the let

Substitute 0 for
Multiply.
Subtract.
Divide by 4.

The is

To find the let

Substitute 0 for
Multiply.
Simplify.
Divide by −3.

The is

The intercepts are the points and

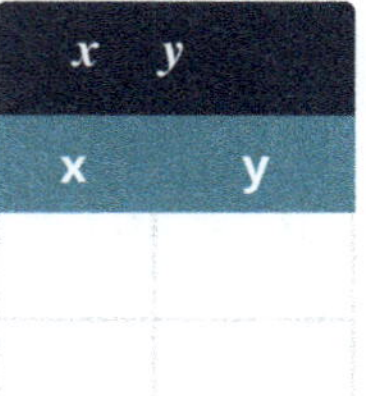

x	y
x	**y**

> **TRY IT : :** 11.48 Find the intercepts of the line:

> **TRY IT : :** 11.49 Find the intercepts of the line:

Graph a Line Using the Intercepts

To graph a linear equation by plotting points, you can use the intercepts as two of your three points. Find the two intercepts, and then a third point to ensure accuracy, and draw the line. This method is often the quickest way to graph a line.

EXAMPLE 11.26

Graph using intercepts.

First, find the Let

The is

Now find the Let

The is

Find a third point. We'll use

A third solution to the equation is

Summarize the three points in a table and then plot them on a graph.

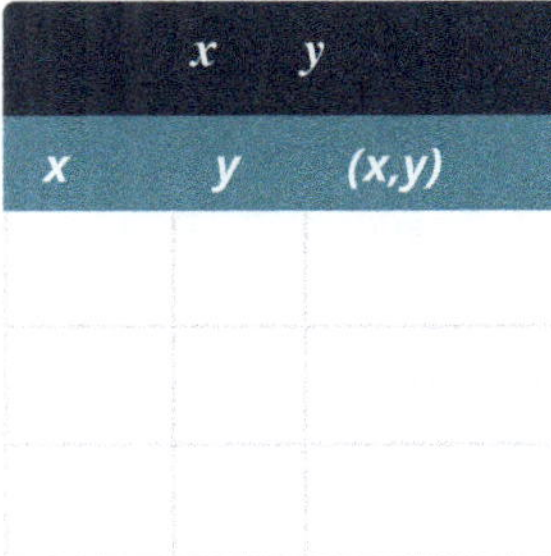

	x	y
x	y	(x,y)

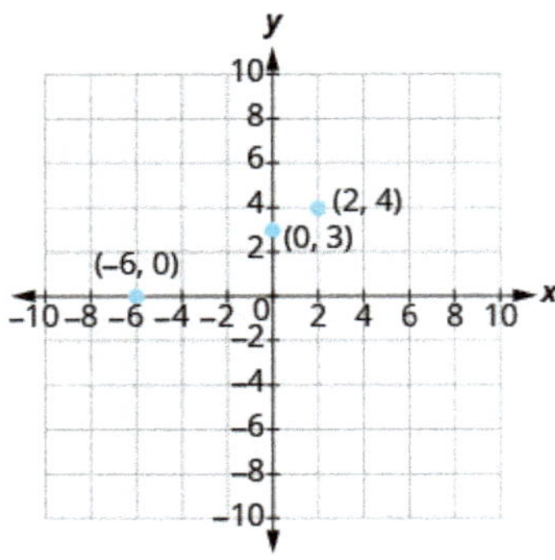

Do the points line up? Yes, so draw line through the points.

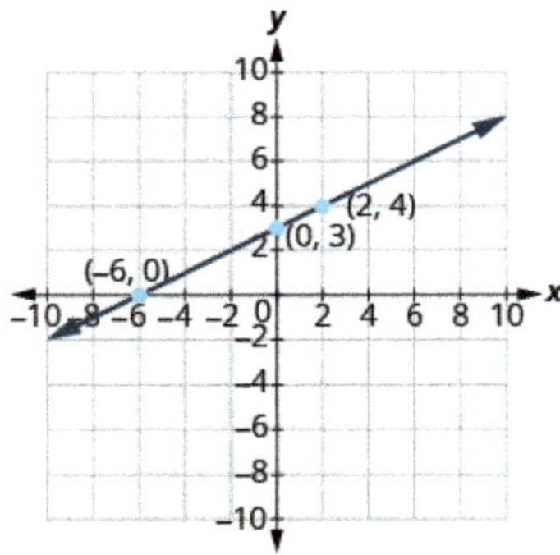

TRY IT : : 11.50 Graph the line using the intercepts:

TRY IT : : 11.51 Graph the line using the intercepts:

HOW TO : : GRAPH A LINE USING THE INTERCEPTS.

Step 1. Find the and of the line.

- Let and solve for
- Let and solve for

Step 2. Find a third solution to the equation.

Step 3. Plot the three points and then check that they line up.

Step 4. Draw the line.

EXAMPLE 11.27

Graph using intercepts.

Solution

Find the intercepts and a third point.

x-intercept, let $y=0$	y-intercept, let $x=0$	third point, let $y=4$
$4x-3y=12$	$4x-3y=12$	$4x-3y=12$
$4x-3(0)=12$	$4(0)-3y=12$	$4x-3(4)=12$
$4x=12$	$-3y=12$	$4x-12=12$
$x=3$	$y=-4$	$4x=24$
		$x=6$

We list the points and show the graph.

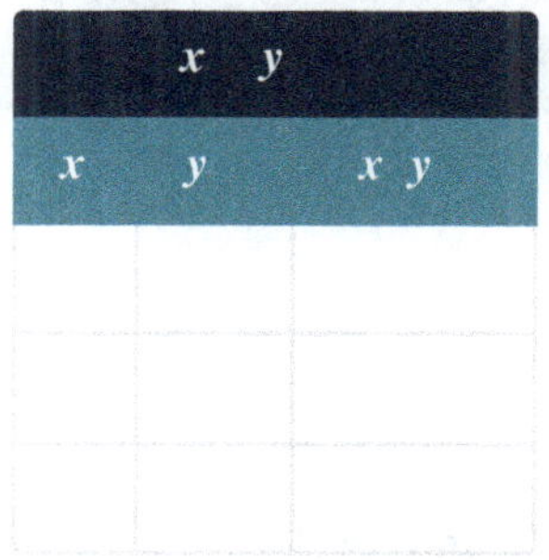

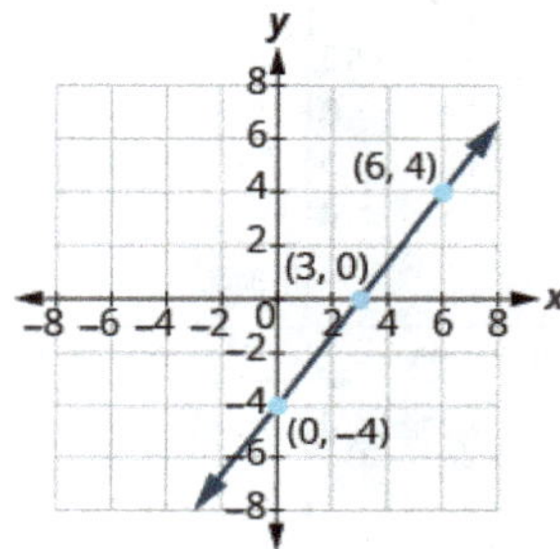

TRY IT : : 11.52 Graph the line using the intercepts:

TRY IT : : 11.53 Graph the line using the intercepts:

EXAMPLE 11.28

Graph using the intercepts.

Solution

x-intercept; Let $y=0$.	y-intercept; Let $x=0$.
$y=5x$	$y=5x$
$0=5x$	$y=5(0)$
$0=x$	$y=0$
$x=0$	The y-intercept is (0, 0).
The x-intercept is (0, 0).	

This line has only one intercept! It is the point

To ensure accuracy, we need to plot three points. Since the intercepts are the same point, we need two more points to graph the line. As always, we can choose any values for so we'll let be and

$$\begin{array}{ll} x = 1 & x = -1 \\ y = 5x & y = 5x \\ y = 5(1) & y = 5(-1) \\ y = 5 & y = -5 \\ (1, 5) & (-1, -5) \end{array}$$

Organize the points in a table.

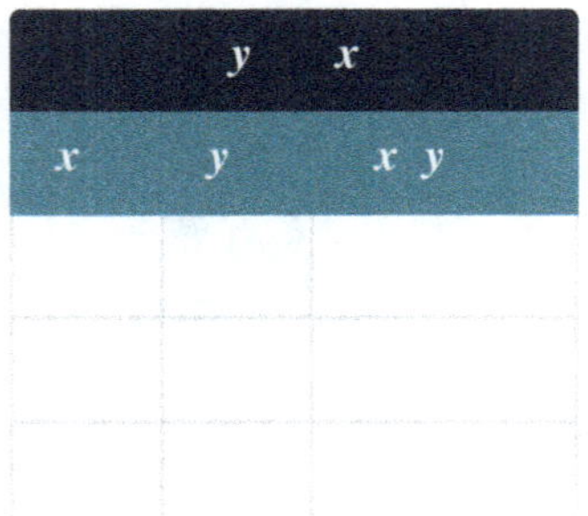

	y	x
x	y	x y

Plot the three points, check that they line up, and draw the line.

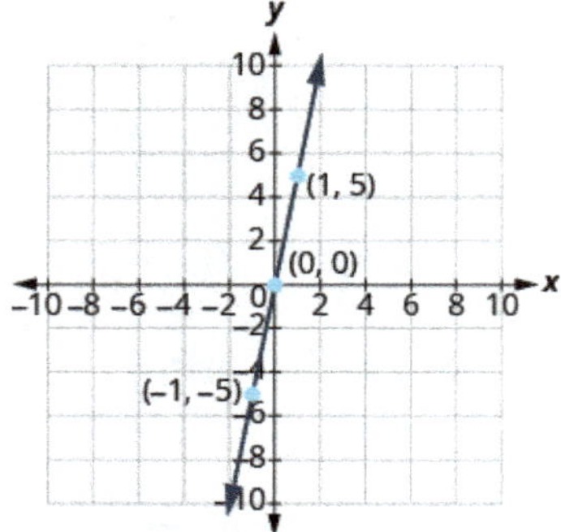

> TRY IT : : 11.54 Graph using the intercepts:

> TRY IT : : 11.55 Graph using the intercepts:

Choose the Most Convenient Method to Graph a Line

While we could graph any linear equation by plotting points, it may not always be the most convenient method. This table shows six of equations we've graphed in this chapter, and the methods we used to graph them.

	Equation	Method
#1		Plotting points
#2	—	Plotting points
#3		Vertical line
#4		Horizontal line
#5		Intercepts
#6		Intercepts

What is it about the form of equation that can help us choose the most convenient method to graph its line?

Notice that in equations #1 and #2, *y* is isolated on one side of the equation, and its coefficient is 1. We found points by substituting values for *x* on the right side of the equation and then simplifying to get the corresponding *y*- values.

Equations #3 and #4 each have just one variable. Remember, in this kind of equation the value of that one variable is constant; it does not depend on the value of the other variable. Equations of this form have graphs that are vertical or horizontal lines.

In equations #5 and #6, both *x* and *y* are on the same side of the equation. These two equations are of the form . We substituted and to find the *x*- and *y*- intercepts, and then found a third point by choosing a value for *x* or *y*.

This leads to the following strategy for choosing the most convenient method to graph a line.

HOW TO : : CHOOSE THE MOST CONVENIENT METHOD TO GRAPH A LINE.

Step 1. If the equation has only one variable. It is a vertical or horizontal line.

- is a vertical line passing through the at
- is a horizontal line passing through the at

Step 2. If is isolated on one side of the equation. Graph by plotting points.

- Choose any three values for and then solve for the corresponding values.

Step 3. If the equation is of the form find the intercepts.

- Find the and intercepts and then a third point.

EXAMPLE 11.29

Identify the most convenient method to graph each line:

—

Solution

This equation has only one variable, Its graph is a horizontal line crossing the at

This equation is of the form Find the intercepts and one more point.

There is only one variable, The graph is a vertical line crossing the at

—

Since is isolated on the left side of the equation, it will be easiest to graph this line by plotting three points.

TRY IT : : 11.56 Identify the most convenient method to graph each line:

—

TRY IT : : 11.57 Identify the most convenient method to graph each line:

—

MEDIA : : ACCESS ADDITIONAL ONLINE RESOURCES

- **Graph by Finding Intercepts (http://www.openstax.org/l/24findinter)**
- **Use Intercepts to Graph (http://www.openstax.org/l/24useintercept)**
- **State the Intercepts from a Graph (http://www.openstax.org/l/24statintercept)**

11.3 EXERCISES

Practice Makes Perfect

Identify the Intercepts on a Graph

In the following exercises, find the and intercepts.

117.

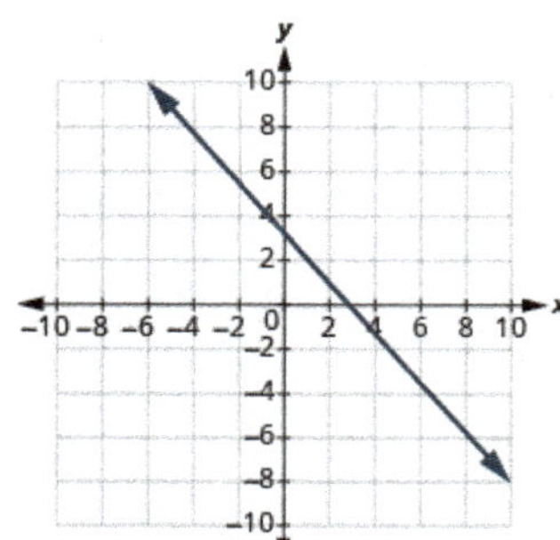

118.

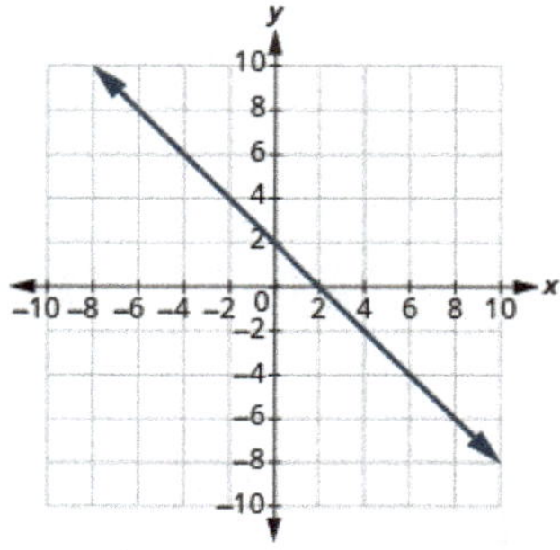

119.

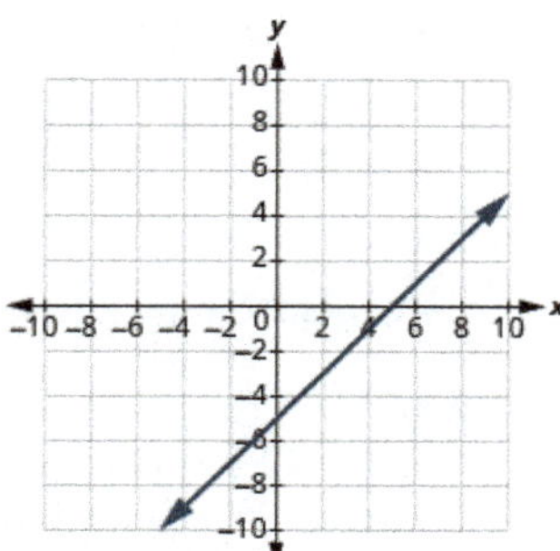

120.

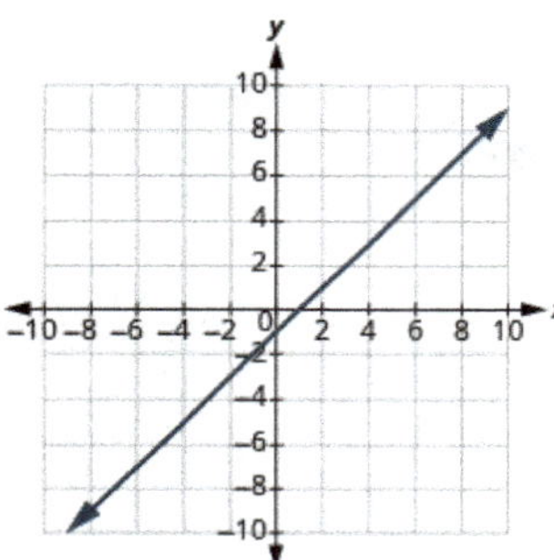

121.

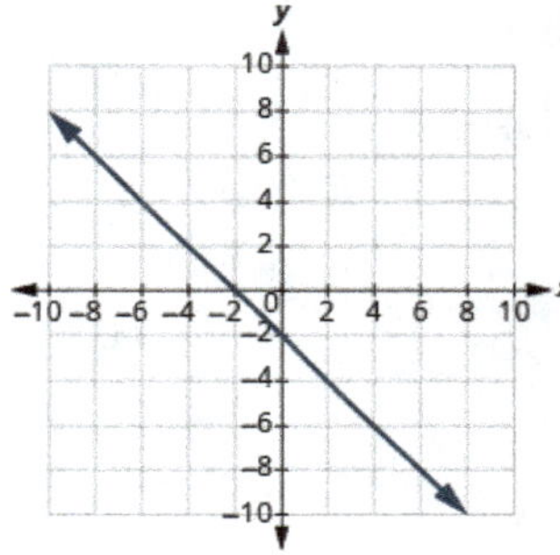

122.

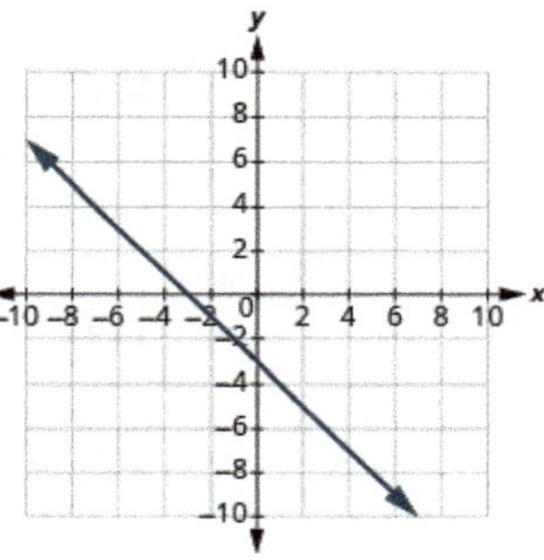

123.

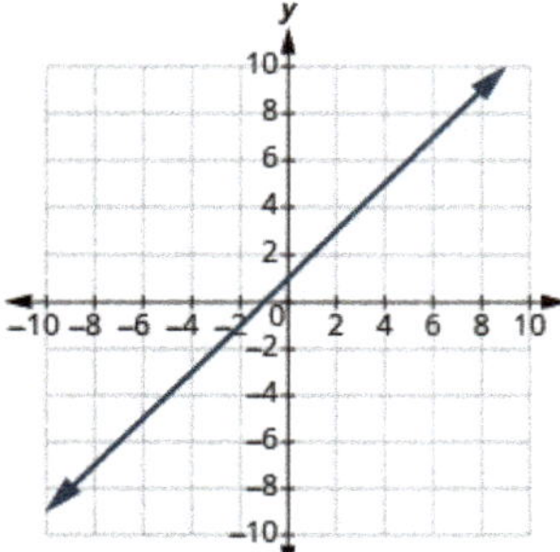

124.

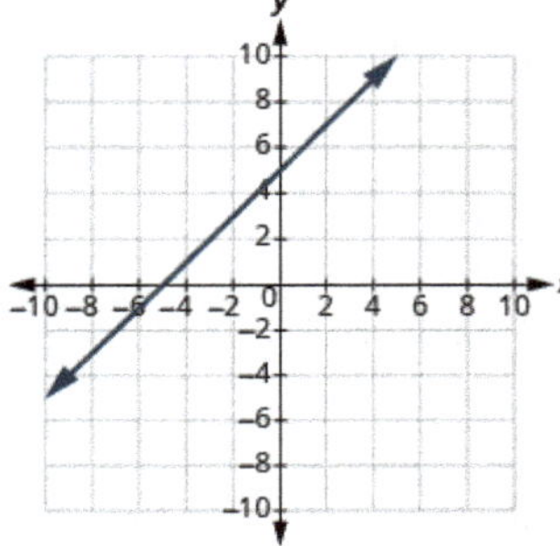

125.

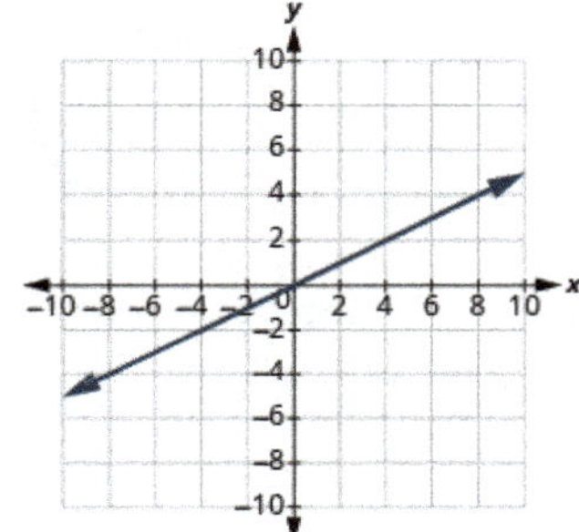

126.

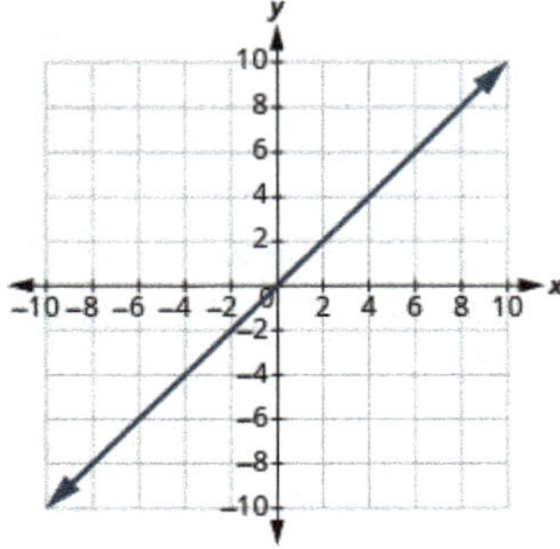

Find the x and y Intercepts from an Equation of a Line

In the following exercises, find the intercepts.

127. **128.** **129.**

130. **131.** **132.**

133. **134.** **135.**

136. **137.** **138.**

139. **140.** **141.**

142. **143.** **144.**

145. **146.** **147.** —

148. — **149.** — **150.** —

151. **152.** **153.**

154.

Graph a Line Using the Intercepts

In the following exercises, graph using the intercepts.

155. **156.** **157.**

158. **159.** **160.**

161. **162.** **163.**

164. **165.** **166.**

167. **168.** **169.**

170. **171.** **172.**

173. **174.** **175.**

176. **177.** **178.**

179. **180.**

Choose the Most Convenient Method to Graph a Line

In the following exercises, identify the most convenient method to graph each line.

181. **182.** **183.**

184. **185.** **186.**

187. **188.** **189.** —

190. — **191.** **192.**

193. **194.** **195.** —

196. —

Everyday Math

197. Road trip Damien is driving from Chicago to Denver, a distance of miles. The on the graph below shows the time in hours since Damien left Chicago. The represents the distance he has left to drive.

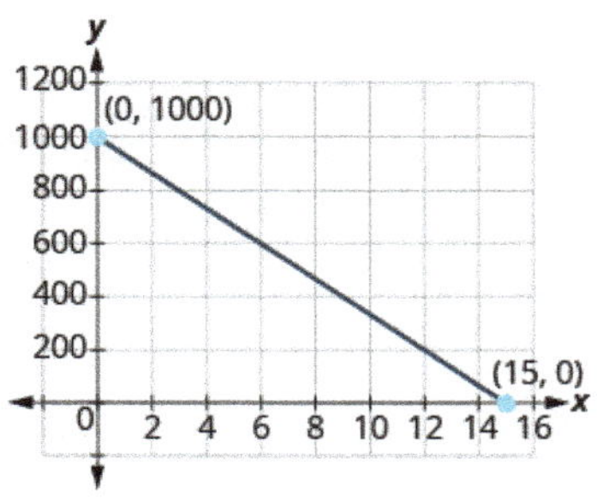

Find the and intercepts

Explain what the and intercepts mean for Damien.

198. Road trip Ozzie filled up the gas tank of his truck and went on a road trip. The on the graph shows the number of miles Ozzie drove since filling up. The represents the number of gallons of gas in the truck's gas tank.

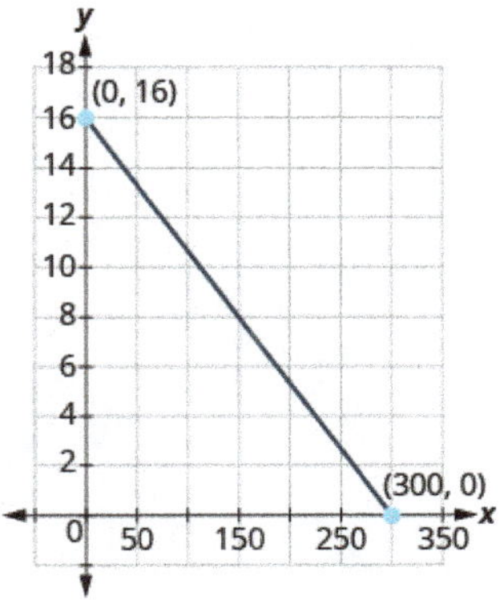

Find the and intercepts.

Explain what the and intercepts mean for Ozzie.

Writing Exercises

199. How do you find the of the graph of

200. How do you find the of the graph of

201. Do you prefer to graph the equation by plotting points or intercepts? Why?

202. Do you prefer to graph the equation — by plotting points or intercepts? Why?

Self Check

After completing the exercises, use this checklist to evaluate your mastery of the objectives of this section.

I can...	Confidently	With some help	No-I don't get it!
identify the intercepts on a graph.			
find the intercepts from an equation of a line.			
graph a line using the intercepts.			
choose the most convenient method to graph a line.			

What does this checklist tell you about your mastery of this section? What steps will you take to improve?

11.4 Understand Slope of a Line

Learning Objectives

BE PREPARED : : 11.10

Before you get started, take this readiness quiz.

Simplify: ——

If you missed this problem, review Example 4.49.

BE PREPARED : : 11.11

Divide: — —

If you missed this problem, review Example 7.37.

BE PREPARED : : 11.12

Simplify: — — —

If you missed this problem, review Example 4.47.

As we've been graphing linear equations, we've seen that some lines slant up as they go from left to right and some lines slant down. Some lines are very steep and some lines are flatter. What determines whether a line slants up or down, and if its slant is steep or flat?

The steepness of the slant of a line is called the **slope of the line**. The concept of slope has many applications in the real world. The pitch of a roof and the grade of a highway or wheelchair ramp are just some examples in which you literally see slopes. And when you ride a bicycle, you feel the slope as you pump uphill or coast downhill.

Use Geoboards to Model Slope

In this section, we will explore the concepts of slope.

Using rubber bands on a geoboard gives a concrete way to model lines on a coordinate grid. By stretching a rubber band between two pegs on a geoboard, we can discover how to find the slope of a line. And when you ride a bicycle, you feel the slope as you pump uphill or coast downhill.

MANIPULATIVE MATHEMATICS

Doing the Manipulative Mathematics activity "Exploring Slope" will help you develop a better understanding of the slope of a line.

We'll start by stretching a rubber band between two pegs to make a line as shown in Figure 11.17.

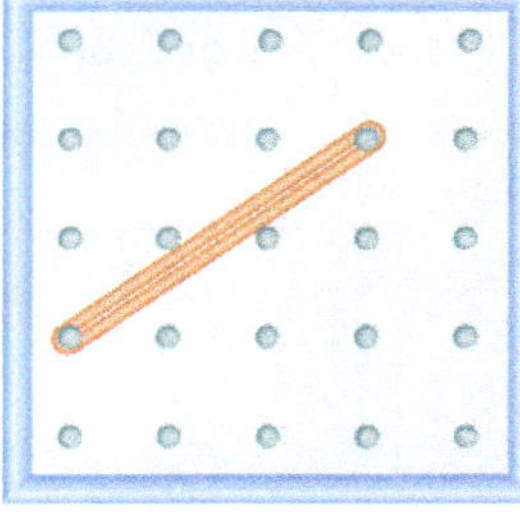

Figure 11.17

Does it look like a line?

Now we stretch one part of the rubber band straight up from the left peg and around a third peg to make the sides of a right triangle as shown in Figure 11.18. We carefully make a angle around the third peg, so that one side is vertical and the other is horizontal.

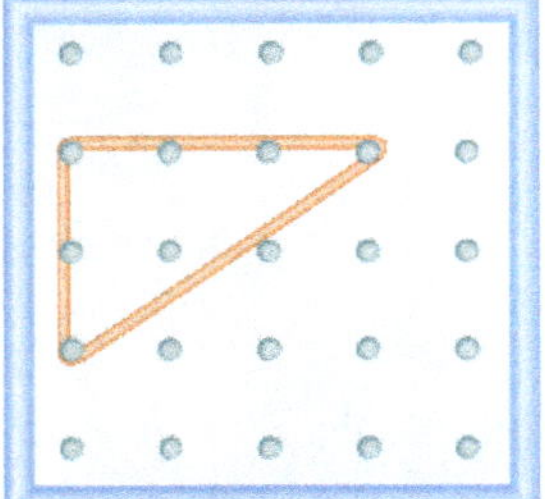

Figure 11.18

To find the slope of the line, we measure the distance along the vertical and horizontal legs of the triangle. The vertical distance is called the ***rise*** and the horizontal distance is called the ***run***, as shown in Figure 11.19.

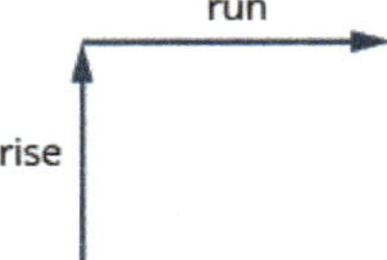

Figure 11.19

To help remember the terms, it may help to think of the images shown in Figure 11.20.

It goes straight up,
as if along the y-axis.
RISE ↕
A jogger runs straight across,
as if along the x-axis.
RUN ↔

Figure 11.20

On our geoboard, the rise is units because the rubber band goes up spaces on the vertical leg. See Figure 11.21.

What is the run? Be sure to count the spaces between the pegs rather than the pegs themselves! The rubber band goes across spaces on the horizontal leg, so the run is units.

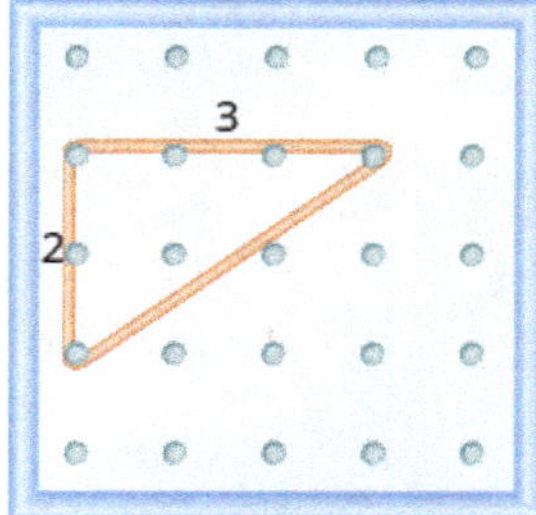

Figure 11.21

The slope of a line is the ratio of the rise to the run. So the slope of our line is — In mathematics, the slope is always represented by the letter

Slope of a line

The slope of a line is ——

The **rise** measures the vertical change and the **run** measures the horizontal change.

What is the slope of the line on the geoboard in Figure 11.21?

——

—

—

When we work with geoboards, it is a good idea to get in the habit of starting at a peg on the left and connecting to a peg to the right. Then we stretch the rubber band to form a right triangle.

If we start by going up the rise is positive, and if we stretch it down the rise is negative. We will count the run from left to right, just like you read this paragraph, so the run will be positive.

Since the slope formula has rise over run, it may be easier to always count out the rise first and then the run.

EXAMPLE 11.30

What is the slope of the line on the geoboard shown?

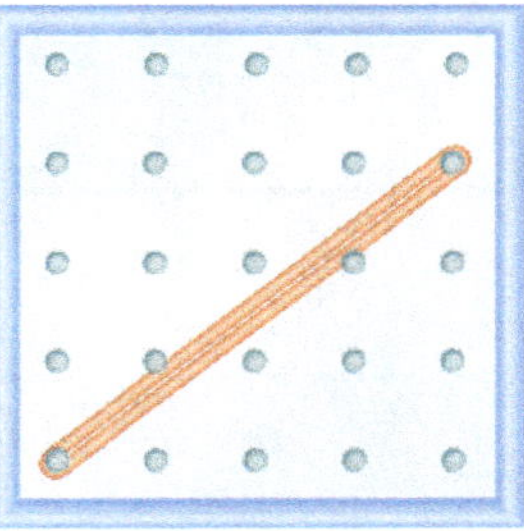

Solution

Use the definition of slope.

——

Start at the left peg and make a right triangle by stretching the rubber band up and to the right to reach the second peg. Count the rise and the run as shown.

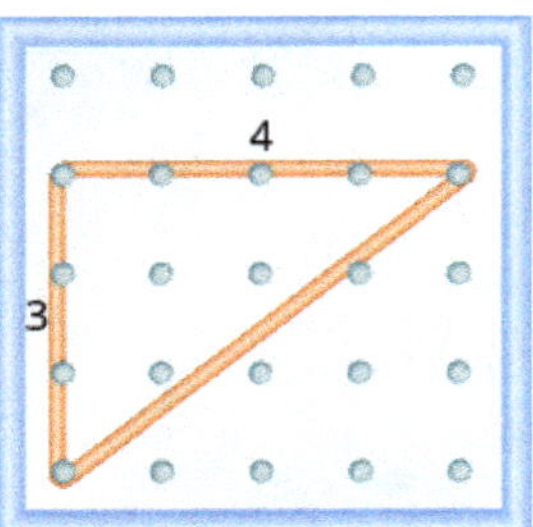

TRY IT : : 11.58 What is the slope of the line on the geoboard shown?

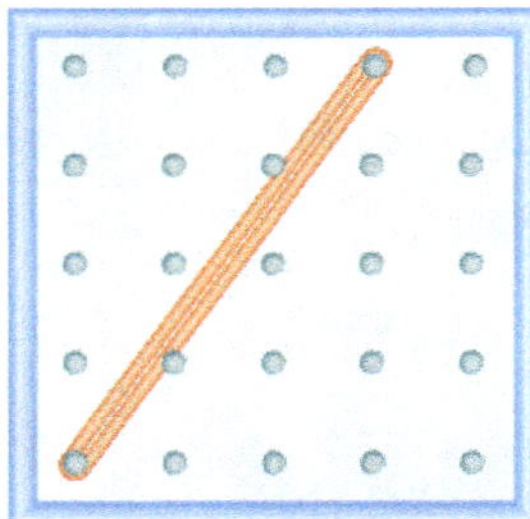

TRY IT : : 11.59 What is the slope of the line on the geoboard shown?

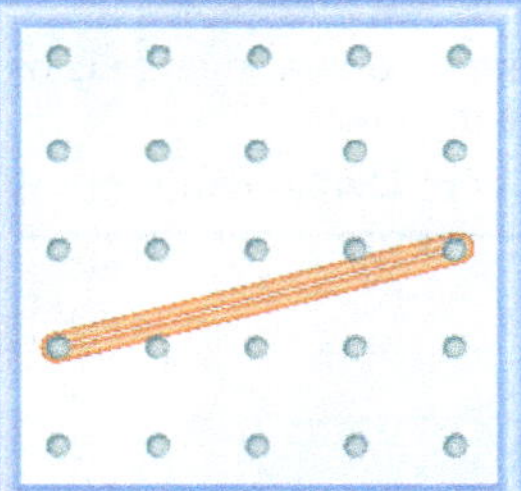

EXAMPLE 11.31

What is the slope of the line on the geoboard shown?

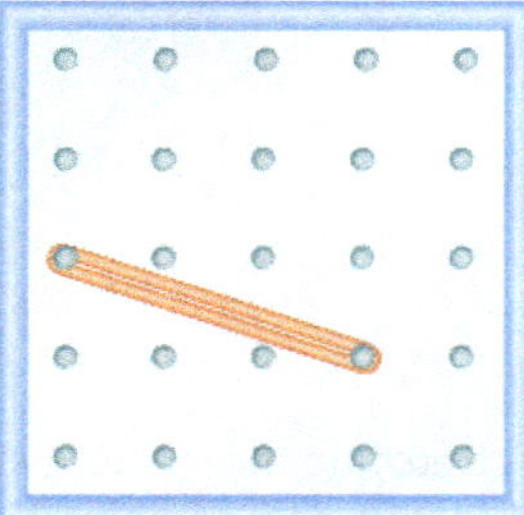

Solution

Use the definition of slope.

Start at the left peg and make a right triangle by stretching the rubber band to the peg on the right. This time we need to stretch the rubber band down to make the vertical leg, so the rise is negative.

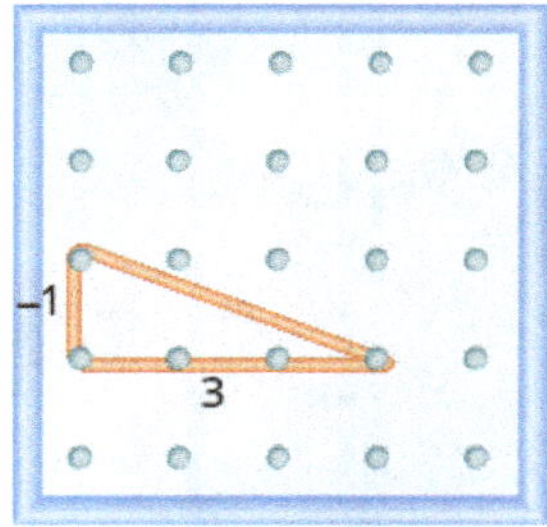

> **TRY IT : : 11.60** What is the slope of the line on the geoboard?

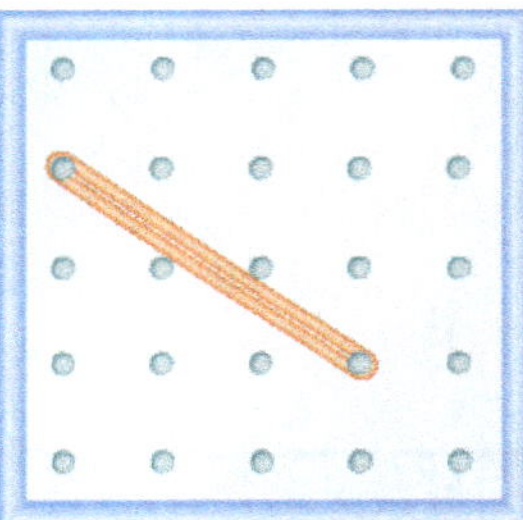

> **TRY IT : : 11.61** What is the slope of the line on the geoboard?

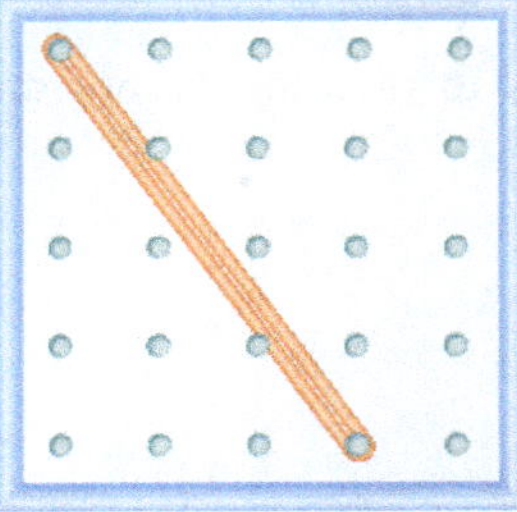

Notice that in the first example, the slope is positive and in the second example the slope is negative. Do you notice any difference in the two lines shown in Figure 11.22.

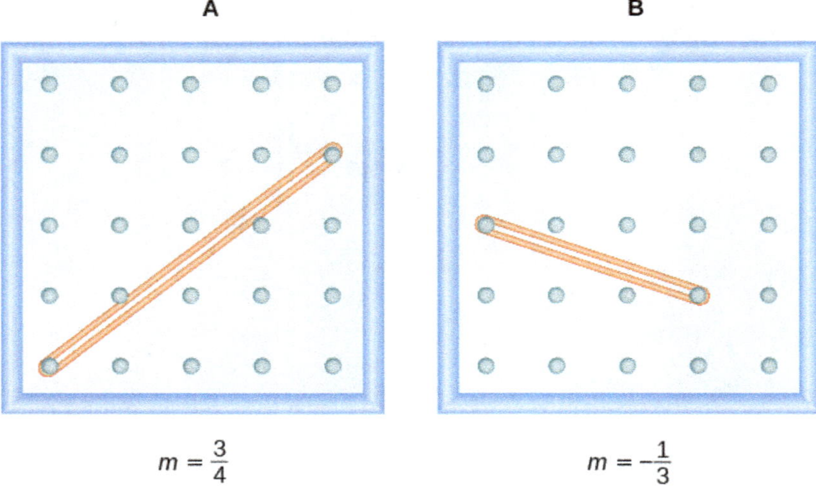

Figure 11.22

As you read from left to right, the line in Figure A, is going up; it has positive slope. The line Figure B is going down; it has negative slope.

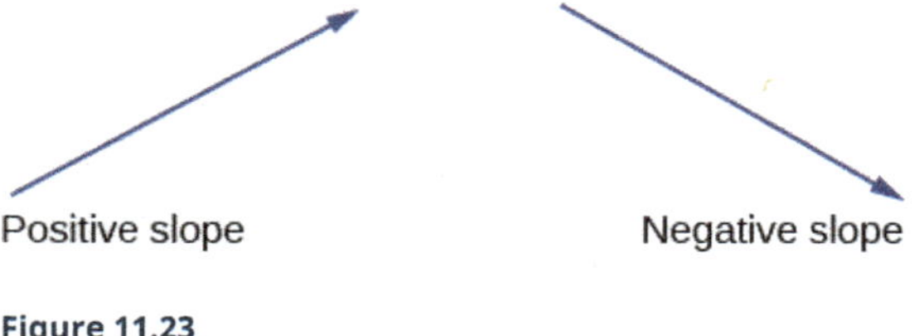

Figure 11.23

EXAMPLE 11.32

Use a geoboard to model a line with slope —

Solution

To model a line with a specific slope on a geoboard, we need to know the rise and the run.

Use the slope formula.	——
Replace with —.	— ——

So, the rise is unit and the run is units.

Start at a peg in the lower left of the geoboard. Stretch the rubber band up unit, and then right units.

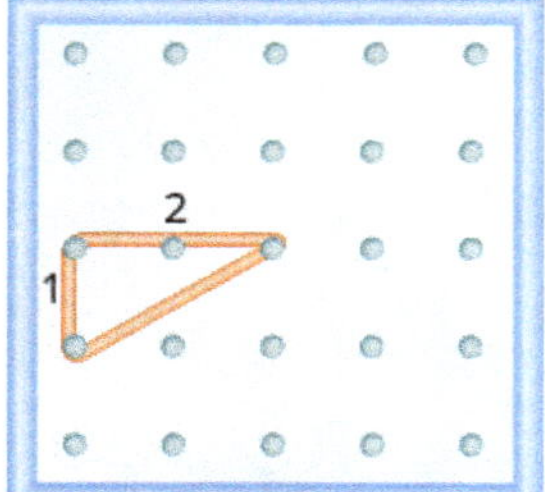

The hypotenuse of the right triangle formed by the rubber band represents a line with a slope of —

TRY IT : : 11.62 Use a geoboard to model a line with the given slope: —

TRY IT : : 11.63 Use a geoboard to model a line with the given slope: —

EXAMPLE 11.33

Use a geoboard to model a line with slope ——

Solution

Use the slope formula.	——
Replace with —.	— ——

So, the rise is and the run is

Since the rise is negative, we choose a starting peg on the upper left that will give us room to count down. We stretch the rubber band down unit, then to the right units.

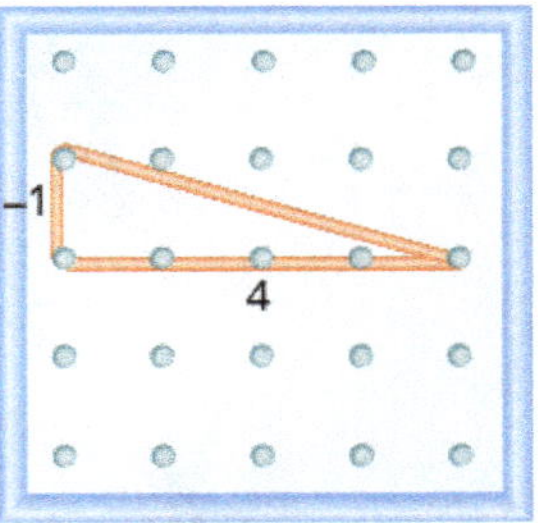

The hypotenuse of the right triangle formed by the rubber band represents a line whose slope is —

TRY IT : : 11.64 Use a geoboard to model a line with the given slope: ——

TRY IT : : 11.65 Use a geoboard to model a line with the given slope: ——

Find the Slope of a Line from its Graph

Now we'll look at some graphs on a coordinate grid to find their slopes. The method will be very similar to what we just modeled on our geoboards.

MANIPULATIVE MATHEMATICS

Doing the Manipulative Mathematics activity "Slope of Lines Between Two Points" will help you develop a better understanding of how to find the slope of a line from its graph.

To find the slope, we must count out the rise and the run. But where do we start?

We locate any two points on the line. We try to choose points with coordinates that are integers to make our calculations easier. We then start with the point on the left and sketch a right triangle, so we can count the rise and run.

EXAMPLE 11.34

Find the slope of the line shown:

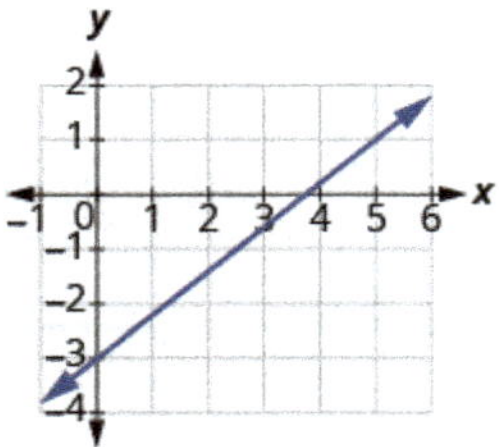

Solution

Locate two points on the graph, choosing points whose coordinates are integers. We will use and

Starting with the point on the left, sketch a right triangle, going from the first point to the second point,

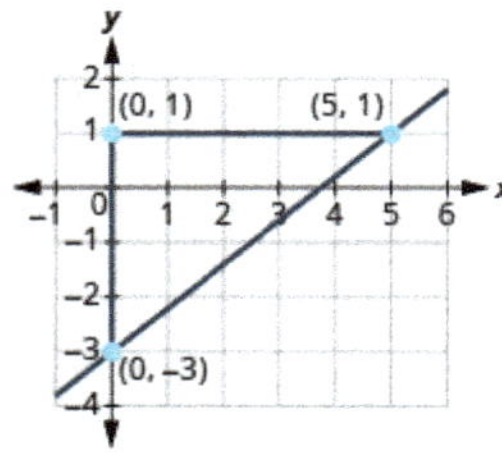

Count the rise on the vertical leg of the triangle.	The rise is 4 units.
Count the run on the horizontal leg.	The run is 5 units.
Use the slope formula.	——
Substitute the values of the rise and run.	—
	The slope of the line is —.

Notice that the slope is positive since the line slants upward from left to right.

TRY IT : : 11.66 Find the slope of the line:

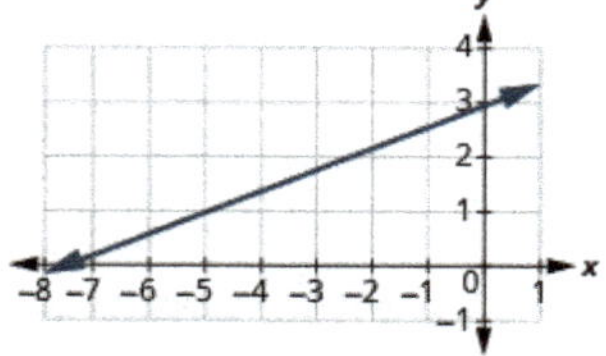

TRY IT : : 11.67 Find the slope of the line:

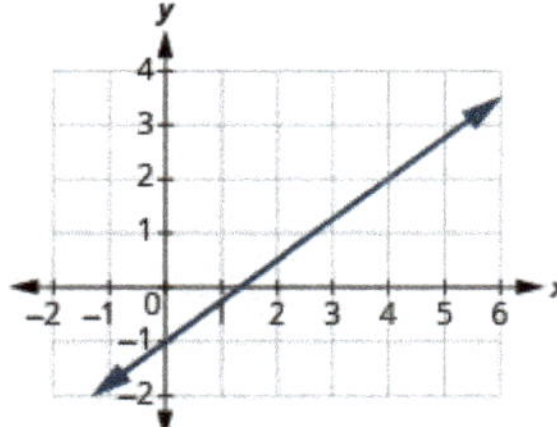

HOW TO : : FIND THE SLOPE FROM A GRAPH.

Step 1. Locate two points on the line whose coordinates are integers.

Step 2. Starting with the point on the left, sketch a right triangle, going from the first point to the second point.

Step 3. Count the rise and the run on the legs of the triangle.

Step 4. Take the ratio of rise to run to find the slope. $m = \frac{\text{rise}}{\text{run}}$

EXAMPLE 11.35

Find the slope of the line shown:

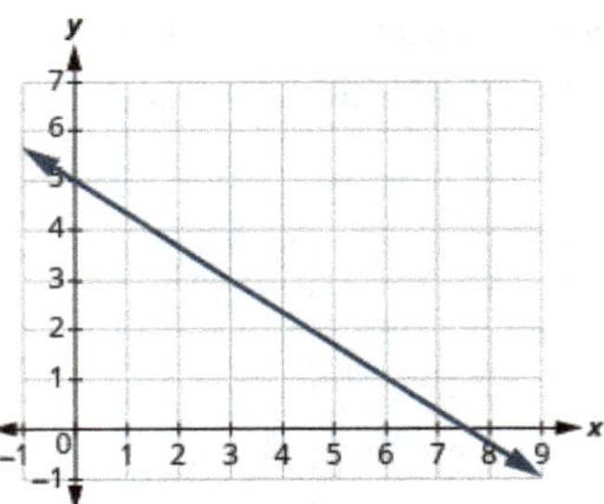

Solution

Locate two points on the graph. Look for points with coordinates that are integers. We can choose any points, but we will use (0, 5) and (3, 3). Starting with the point on the left, sketch a right triangle, going from the first point to the second point.

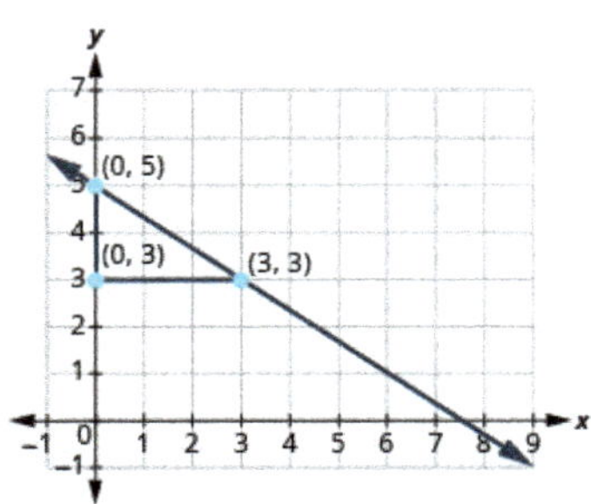

Count the rise – it is negative.	The rise is −2.
Count the run.	The run is 3.
Use the slope formula.	
Substitute the values of the rise and run.	
Simplify.	
	The slope of the line is

Notice that the slope is negative since the line slants downward from left to right.

What if we had chosen different points? Let’s find the slope of the line again, this time using different points. We will use the points and

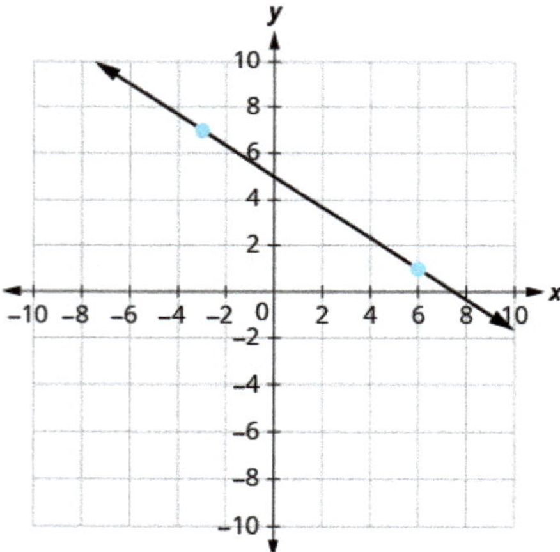

Starting at sketch a right triangle to

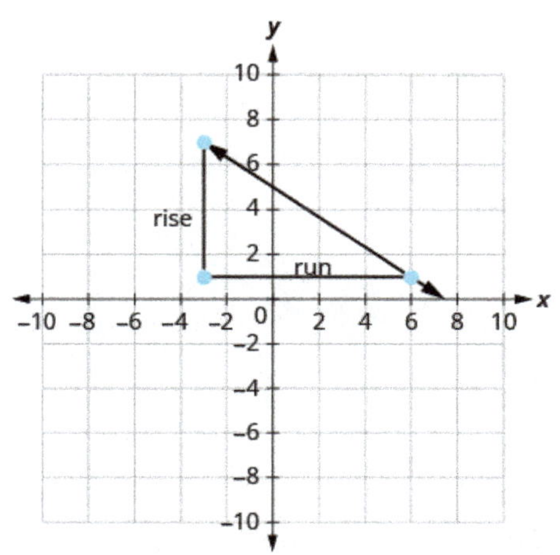

Count the rise.	The rise is −6.
Count the run.	The run is 9.
Use the slope formula.	—
Substitute the values of the rise and run.	—
Simplify the fraction.	–
	The slope of the line is –

It does not matter which points you use—the slope of the line is always the same. The slope of a line is constant!

TRY IT : : 11.68 Find the slope of the line:

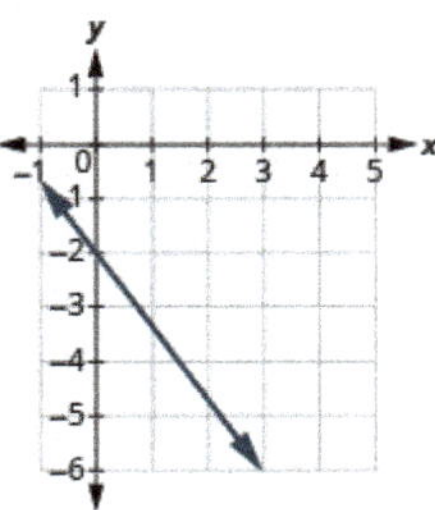

TRY IT : : 11.69 Find the slope of the line:

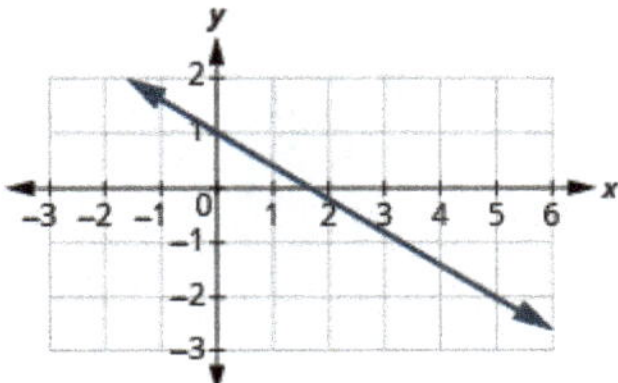

The lines in the previous examples had -intercepts with integer values, so it was convenient to use the *y*-intercept as one of the points we used to find the slope. In the next example, the -intercept is a fraction. The calculations are easier if we use two points with integer coordinates.

EXAMPLE 11.36

Find the slope of the line shown:

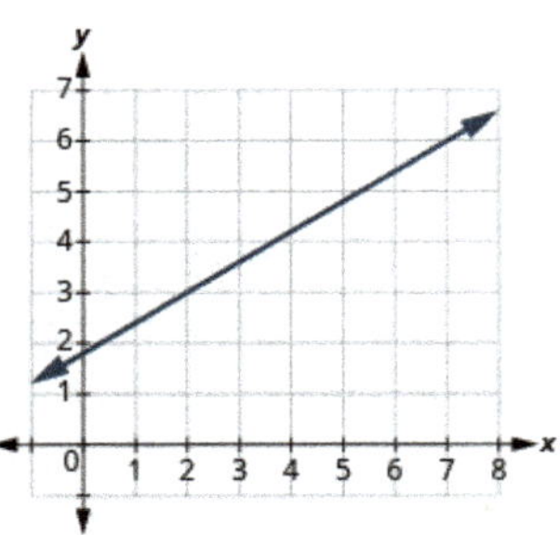

Solution

Locate two points on the graph whose coordinates are integers.	and
Which point is on the left?	
Starting at , sketch a right angle to as shown below.	

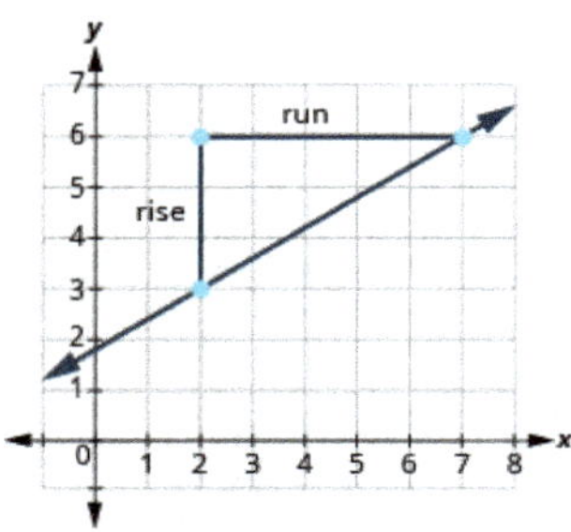

Count the rise.	The rise is 3.
Count the run.	The run is 5.
Use the slope formula.	—
Substitute the values of the rise and run.	–
	The slope of the line is –

TRY IT : : 11.70 Find the slope of the line:

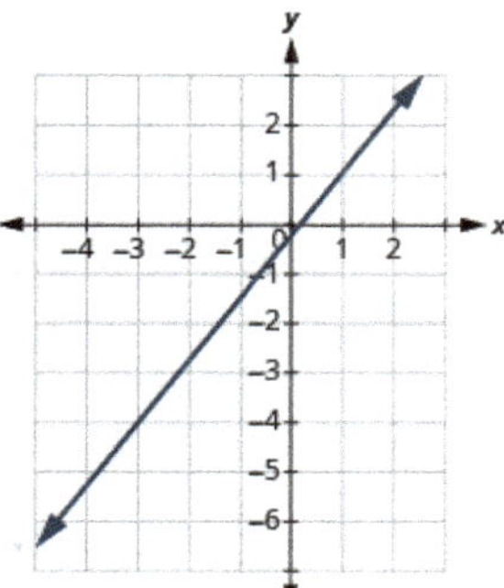

TRY IT : : 11.71 Find the slope of the line:

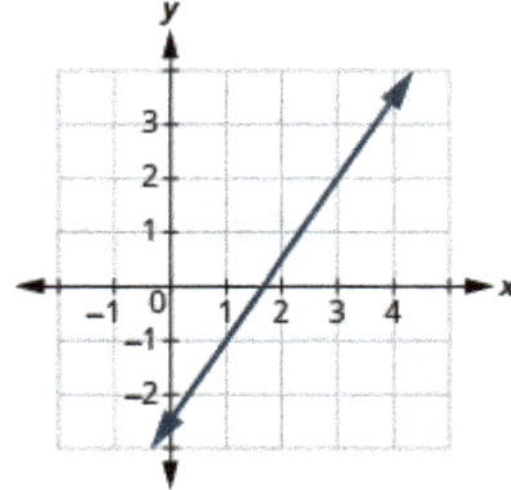

Find the Slope of Horizontal and Vertical Lines

Do you remember what was special about horizontal and vertical lines? Their equations had just one variable.

- horizontal line all the -coordinates are the same.
- vertical line all the -coordinates are the same.

So how do we find the slope of the horizontal line One approach would be to graph the horizontal line, find two points on it, and count the rise and the run. Let's see what happens in Figure 11.24. We'll use the two points and to count the rise and run.

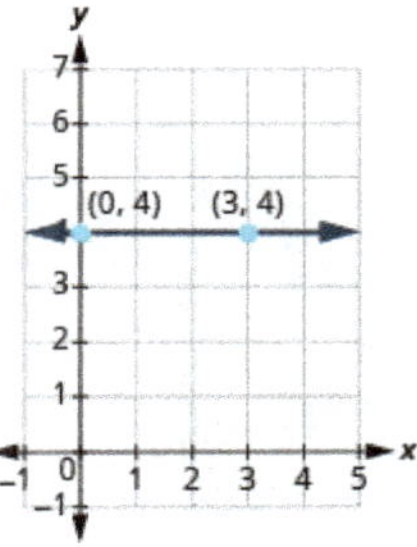

Figure 11.24

What is the rise?	The rise is 0.
What is the run?	The run is 3.
What is the slope?	——
	—

The slope of the horizontal line is

All horizontal lines have slope . When the -coordinates are the same, the rise is .

Slope of a Horizontal Line

The slope of a horizontal line, is

Now we'll consider a vertical line, such as the line , shown in Figure 11.25. We'll use the two points and to count the rise and run.

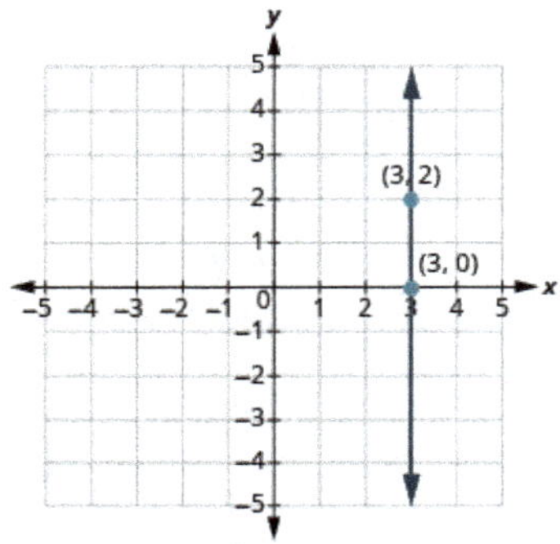

Figure 11.25

What is the rise?	The rise is 2.
What is the run?	The run is 0.
What is the slope?	——
	—

But we can't divide by Division by is undefined. So we say that the slope of the vertical line is undefined. The slope of all vertical lines is undefined, because the run is

Slope of a Vertical Line

The slope of a vertical line, is undefined.

EXAMPLE 11.37

Find the slope of each line:

Solution

This is a vertical line, so its slope is undefined.

This is a horizontal line, so its slope is

TRY IT : : 11.72 Find the slope of the line:

TRY IT : : 11.73 Find the slope of the line:

Quick Guide to the Slopes of Lines

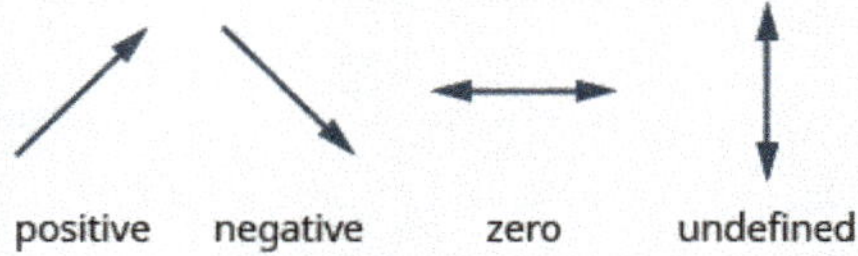

Use the Slope Formula to find the Slope of a Line between Two Points

Sometimes we need to find the slope of a line between two points and we might not have a graph to count out the rise and the run. We could plot the points on grid paper, then count out the rise and the run, but there is a way to find the slope without graphing.

Before we get to it, we need to introduce some new algebraic notation. We have seen that an ordered pair gives the coordinates of a point. But when we work with slopes, we use two points. How can the same symbol be used to represent two different points?

Mathematicians use subscripts to distinguish between the points. A subscript is a small number written to the right of, and a little lower than, a variable.

-
-

We will use to identify the first point and to identify the second point. If we had more than two points, we could use and so on.

To see how the rise and run relate to the coordinates of the two points, let's take another look at the slope of the line between the points and in Figure 11.26.

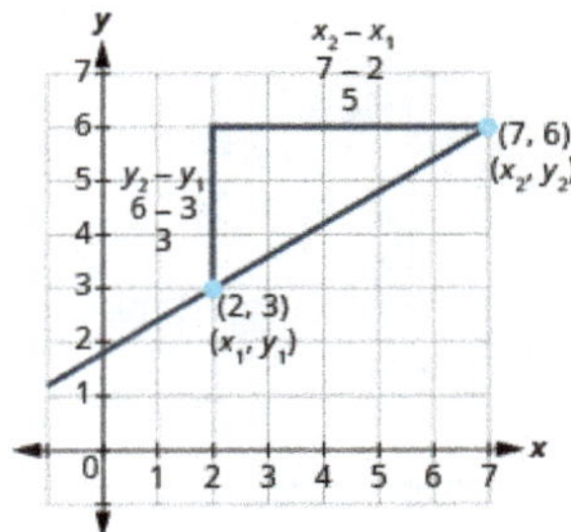

Figure 11.26

Since we have two points, we will use subscript notation.

On the graph, we counted the rise of The rise can also be found by subtracting the of the points.

We counted a run of The run can also be found by subtracting the

We know	
So	
We rewrite the rise and run by putting in the coordinates.	
But 6 is the -coordinate of the second point, and 3 is the -coordinate of the first point . So we can rewrite the rise using subscript notation.	
Also 7 is the -coordinate of the second point, and 2 is the -coordinate of the first point . So we rewrite the run using subscript notation.	

We've shown that is really another version of We can use this formula to find the slope of a line when we have two points on the line.

Slope Formula

The slope of the line between two points and is

Say the formula to yourself to help you remember it:

MANIPULATIVE MATHEMATICS

Doing the Manipulative Mathematics activity "Slope of Lines Between Two Points" will help you develop a better understanding of how to find the slope of a line between two points.

EXAMPLE 11.38

Find the slope of the line between the points and

Solution

We'll call point #1 and point #2.	
Use the slope formula.	
Substitute the values in the slope formula:	
of the second point minus of the first point	
of the second point minus of the first point	
Simplify the numerator and the denominator.	

Let's confirm this by counting out the slope on the graph.

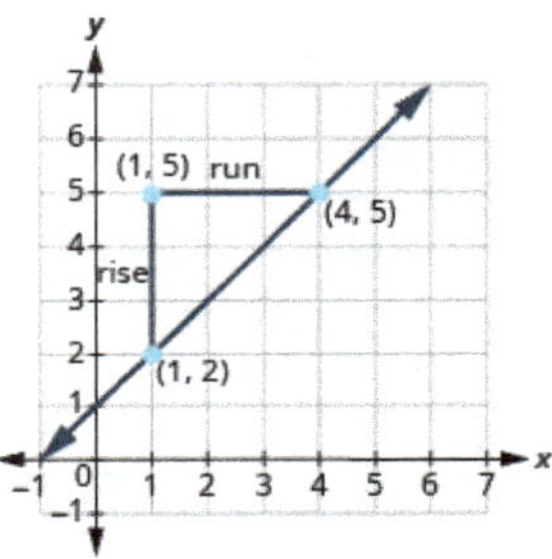

The rise is and the run is so

> **TRY IT : : 11.74** Find the slope of the line through the given points: and

> **TRY IT : : 11.75** Find the slope of the line through the given points: and

How do we know which point to call #1 and which to call #2? Let's find the slope again, this time switching the names of the points to see what happens. Since we will now be counting the run from right to left, it will be negative.

We'll call point #1 and point #2.	
Use the slope formula.	
Substitute the values in the slope formula:	
of the second point minus of the first point	
of the second point minus of the first point	
Simplify the numerator and the denominator.	

The slope is the same no matter which order we use the points.

EXAMPLE 11.39

Find the slope of the line through the points and

Solution

We'll call point #1 and point #2.	
Use the slope formula.	
Substitute the values	
of the second point minus of the first point	
of the second point minus of the first point	
Simplify.	

Let's confirm this on the graph shown.

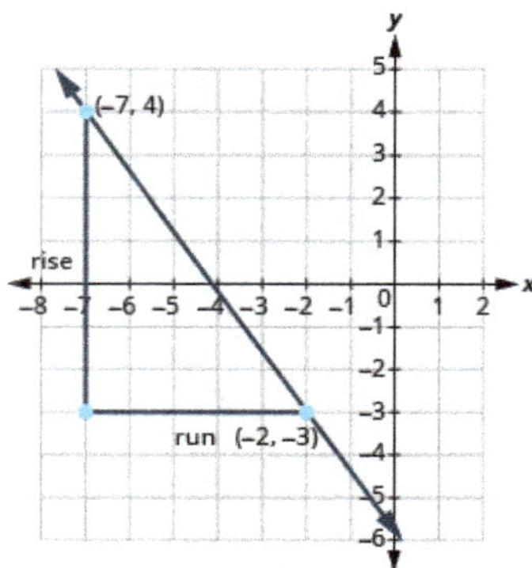

> **TRY IT : : 11.76** Find the slope of the line through the pair of points: and

> **TRY IT : : 11.77** Find the slope of the line through the pair of points: and

Graph a Line Given a Point and the Slope

In this chapter, we graphed lines by plotting points, by using intercepts, and by recognizing horizontal and vertical lines.

Another method we can use to graph lines is the point-slope method. Sometimes, we will be given one point and the slope of the line, instead of its equation. When this happens, we use the definition of slope to draw the graph of the line.

EXAMPLE 11.40

Graph the line passing through the point whose slope is

Solution

Plot the given point,

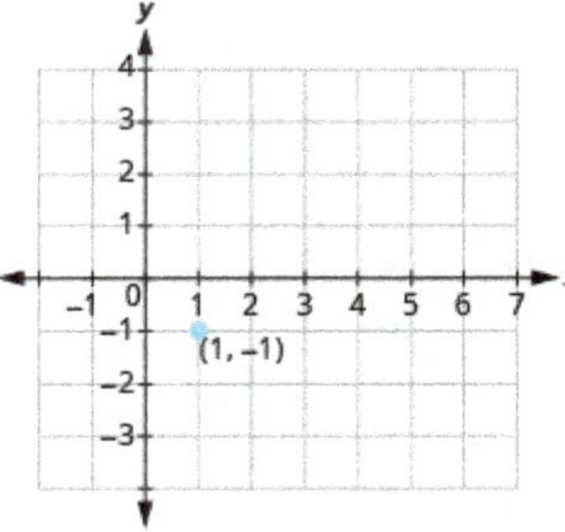

Use the slope formula to identify the rise and the run.

Starting at the point we plotted, count out the rise and run to mark the second point. We count units up and units right.

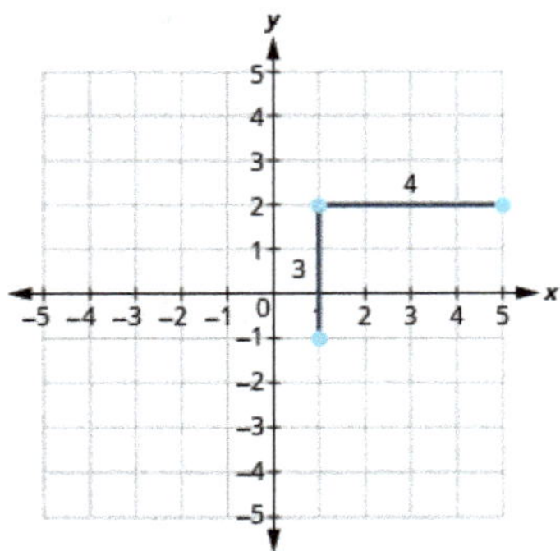

Then we connect the points with a line and draw arrows at the ends to show it continues.

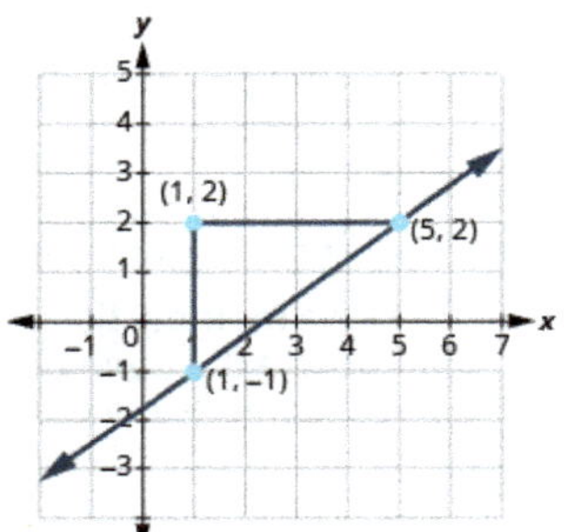

We can check our line by starting at any point and counting up and to the right We should get to another point on the line.

TRY IT : : 11.78 Graph the line passing through the point with the given slope:

TRY IT : : 11.79 Graph the line passing through the point with the given slope:

HOW TO : : GRAPH A LINE GIVEN A POINT AND A SLOPE.

Step 1. Plot the given point.
Step 2. Use the slope formula to identify the rise and the run.
Step 3. Starting at the given point, count out the rise and run to mark the second point.
Step 4. Connect the points with a line.

EXAMPLE 11.41

Graph the line with -intercept and slope —

Solution

Plot the given point, the -intercept

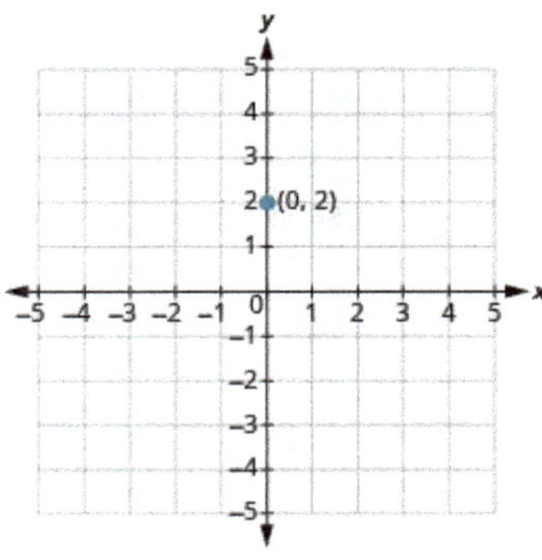

Use the slope formula —— to identify the rise and the run.

Starting at count the rise and the run and mark the second point.

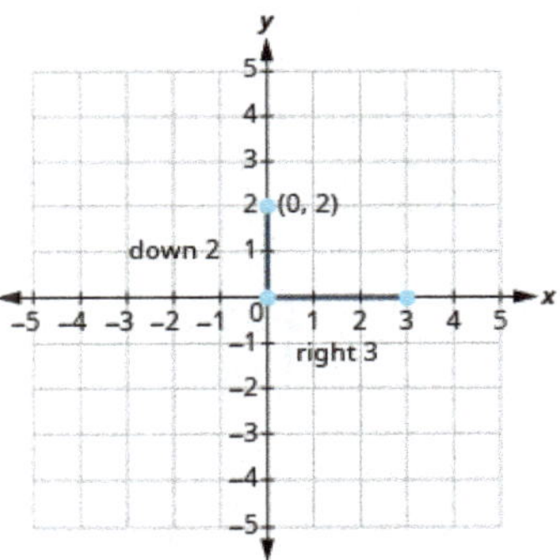

Connect the points with a line.

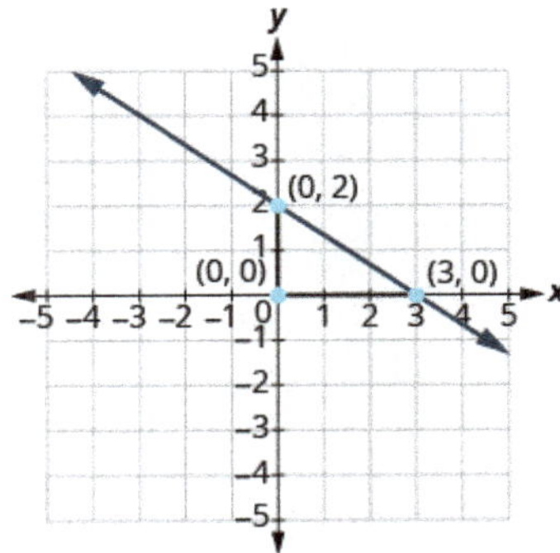

TRY IT : : 11.80 Graph the line with the given intercept and slope:

-intercept —

TRY IT : : 11.81 Graph the line with the given intercept and slope:

-intercept —

EXAMPLE 11.42

Graph the line passing through the point whose slope is

 **Solution**

Plot the given point.

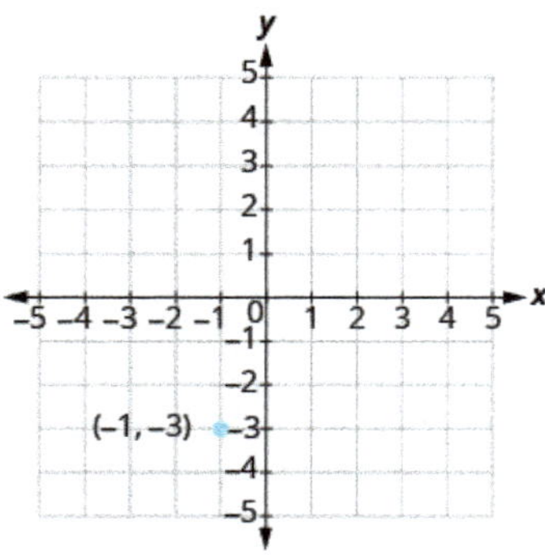

Identify the rise and the run.

Write 4 as a fraction. —— –

Count the rise and run.

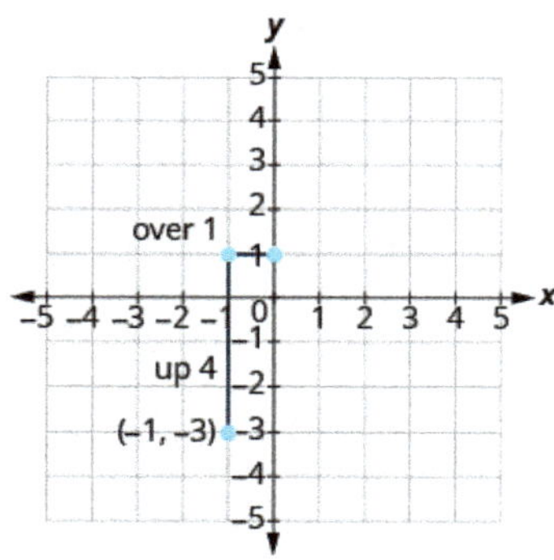

Mark the second point. Connect the two points with a line.

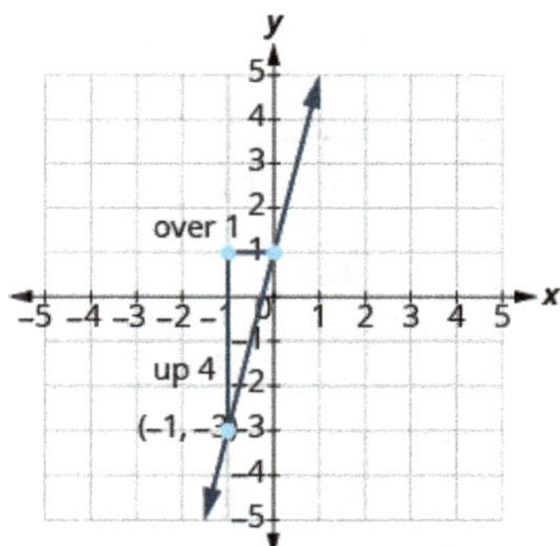

TRY IT : : 11.82 Graph the line passing through the point and with slope

TRY IT : : 11.83 Graph the line passing through the point and with slope

Solve Slope Applications

At the beginning of this section, we said there are many applications of slope in the real world. Let's look at a few now.

EXAMPLE 11.43

The pitch of a building's roof is the slope of the roof. Knowing the pitch is important in climates where there is heavy snowfall. If the roof is too flat, the weight of the snow may cause it to collapse. What is the slope of the roof shown?

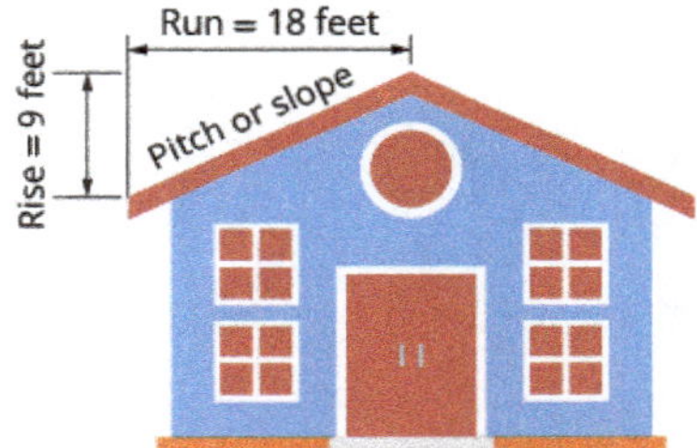

Solution

Use the slope formula.	—
Substitute the values for rise and run.	—
Simplify.	–
	The slope of the roof is –.

TRY IT : : 11.84 Find the slope given rise and run: A roof with a rise and run

TRY IT : : 11.85 Find the slope given rise and run: A roof with a rise and run

Have you ever thought about the sewage pipes going from your house to the street? Their slope is an important factor in how they take waste away from your house.

EXAMPLE 11.44

Sewage pipes must slope down — inch per foot in order to drain properly. What is the required slope?

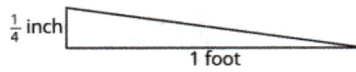

$\frac{1}{4}$ inch

1 foot

Solution

Use the slope formula.	—
	—
	—
Convert 1 foot to 12 inches.	—
Simplify.	—
The slope of the pipe is	—

TRY IT : : 11.86 Find the slope of the pipe: The pipe slopes down — inch per foot.

TRY IT : : 11.87 Find the slope of the pipe: The pipe slopes down — inch per yard.

MEDIA : : ACCESS ADDITIONAL ONLINE RESOURCES

- Determine Positive slope from a Graph (http://www.openstax.org/l/24posslope)
- Determine Negative slope from a Graph (http://www.openstax.org/l/24negslope)
- Determine Slope from Two Points (http://www.openstax.org/l/24slopetwopts)

11.4 EXERCISES

Practice Makes Perfect

Use Geoboards to Model Slope

In the following exercises, find the slope modeled on each geoboard.

203.

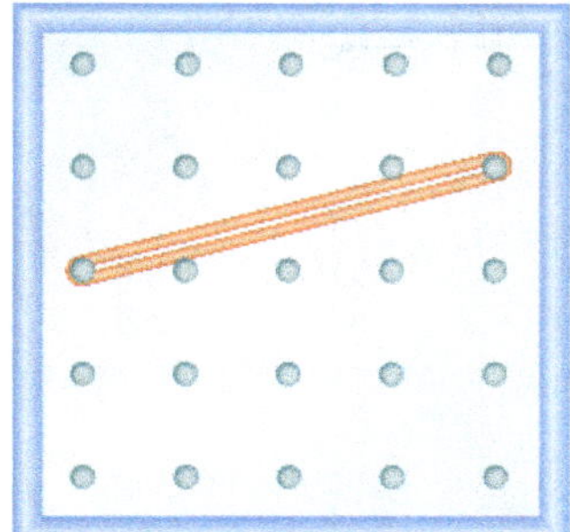

204.

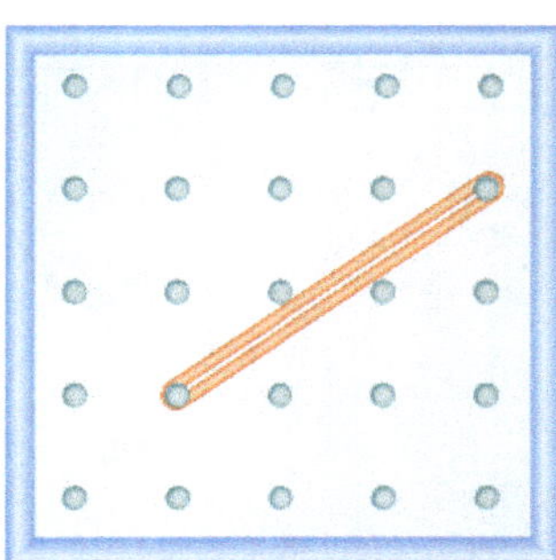

205.

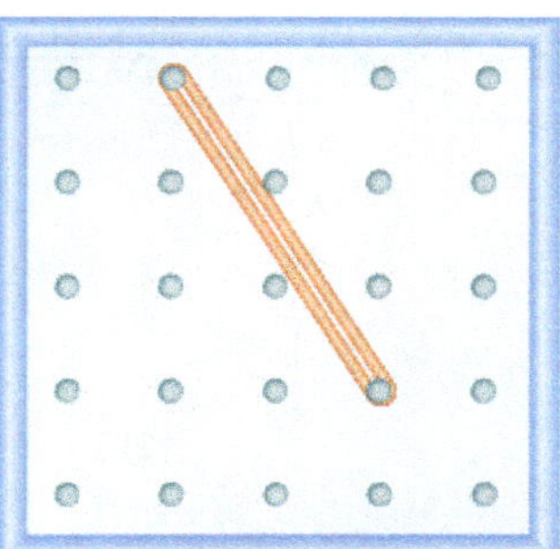

206.

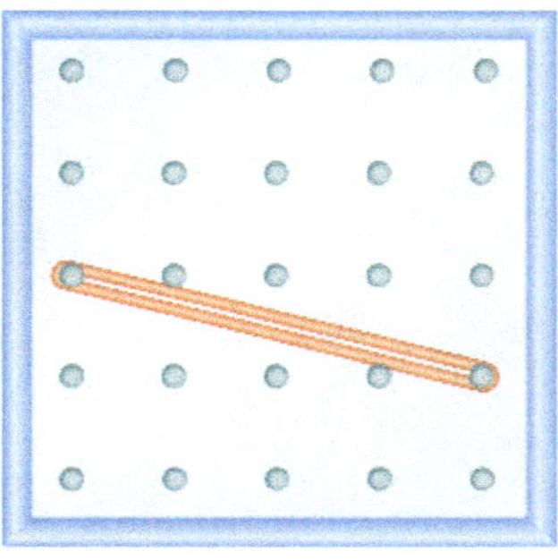

In the following exercises, model each slope. Draw a picture to show your results.

207. —

208. —

209. —

210. —

211. —

212. —

213. —

214. —

Find the Slope of a Line from its Graph

In the following exercises, find the slope of each line shown.

215.

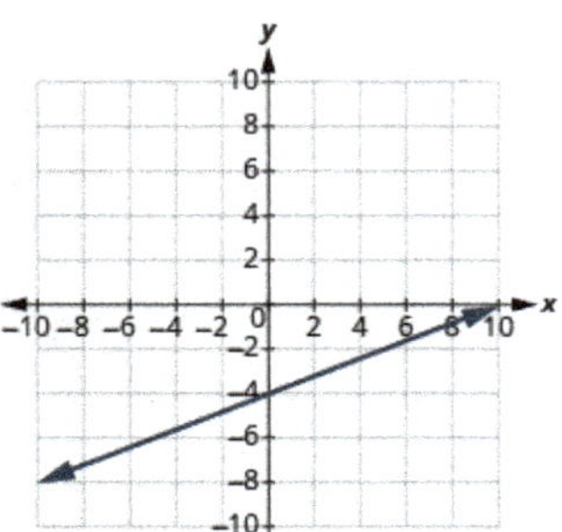

216.

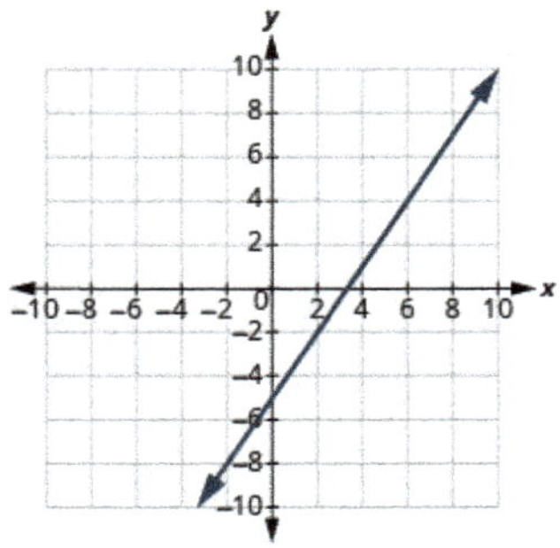

217.

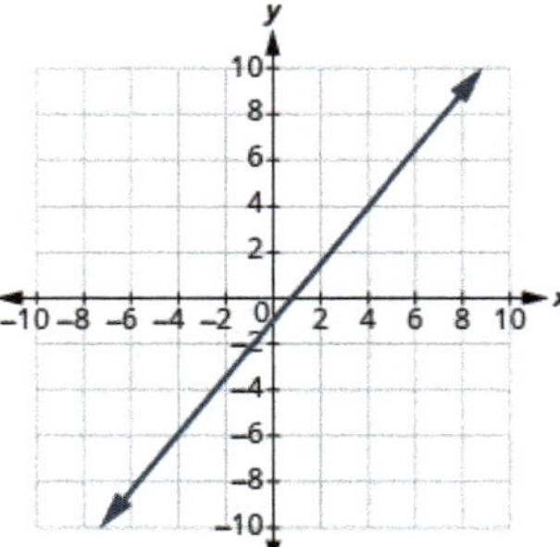

218.

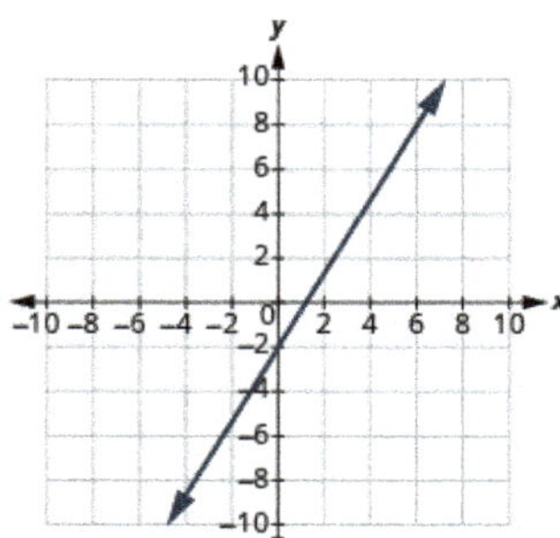

219.

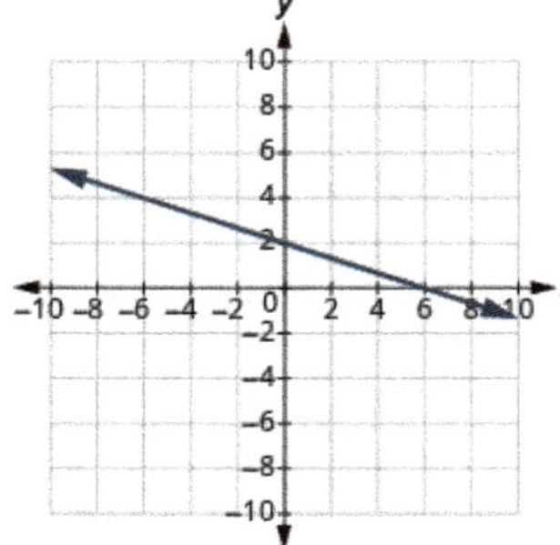

220.

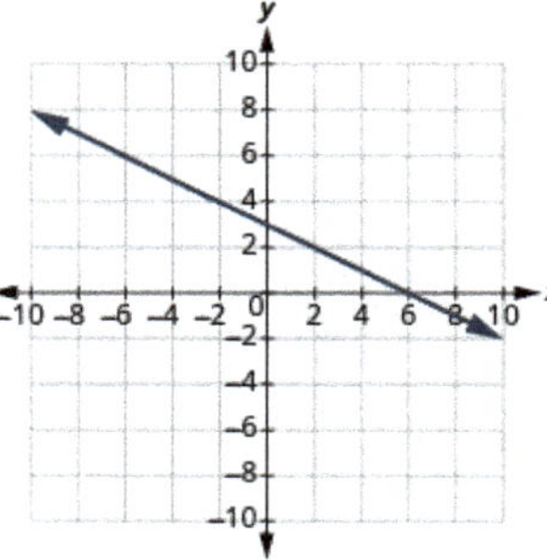

221.

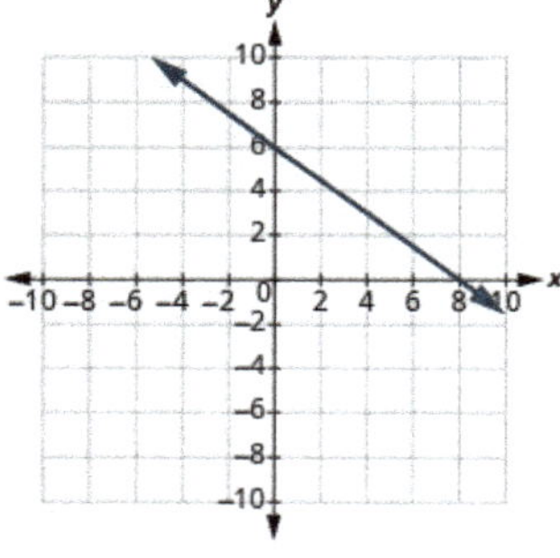

222.

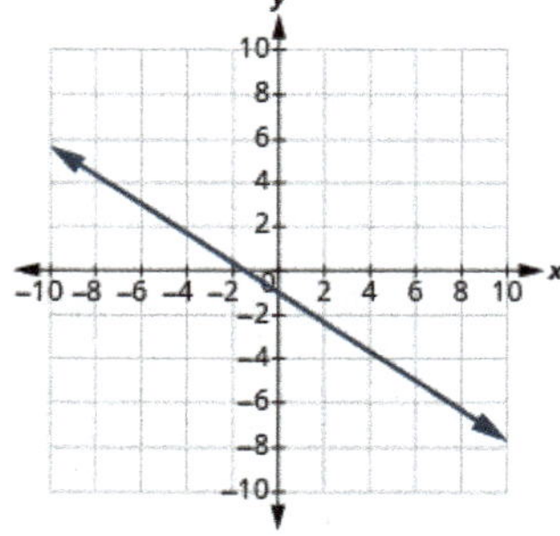

223.

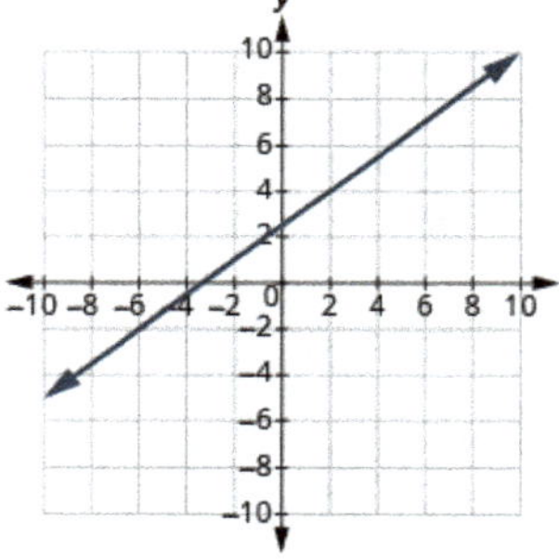

224.

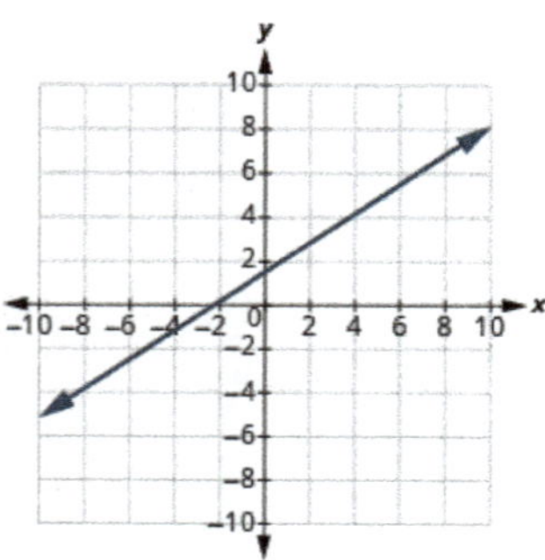

225.

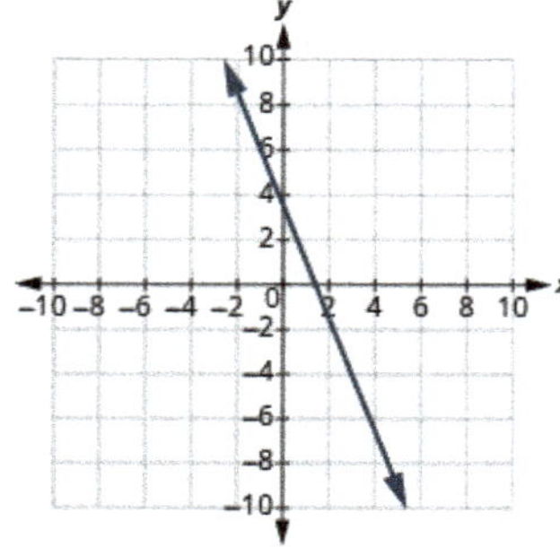

226.

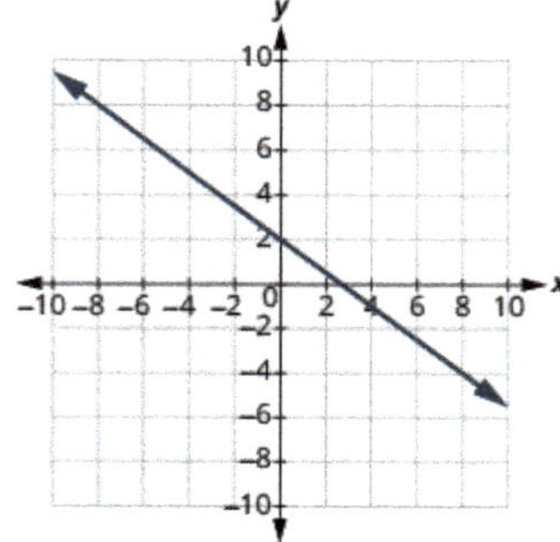

227.

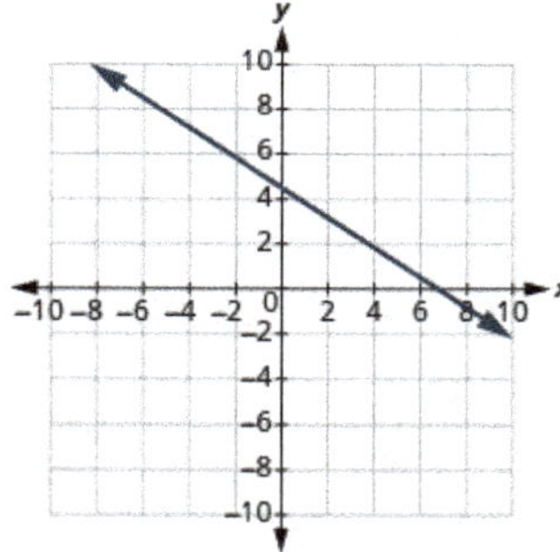

228.

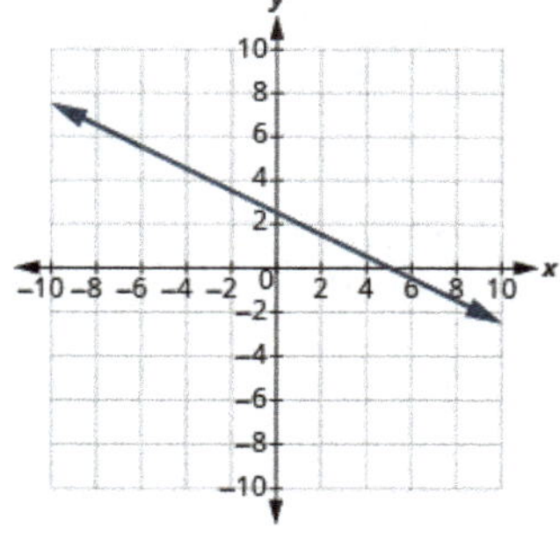

229.

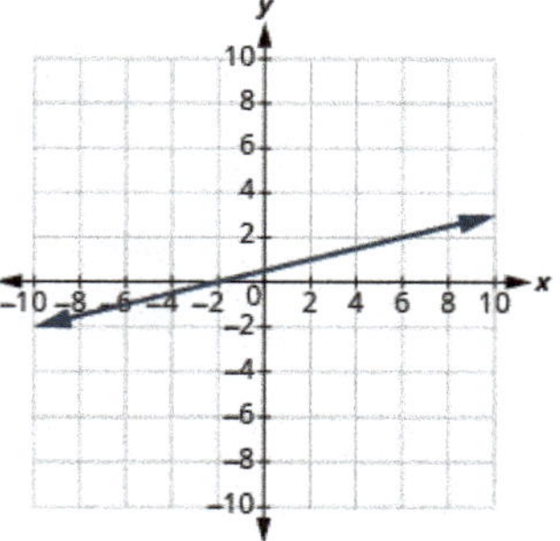

230.

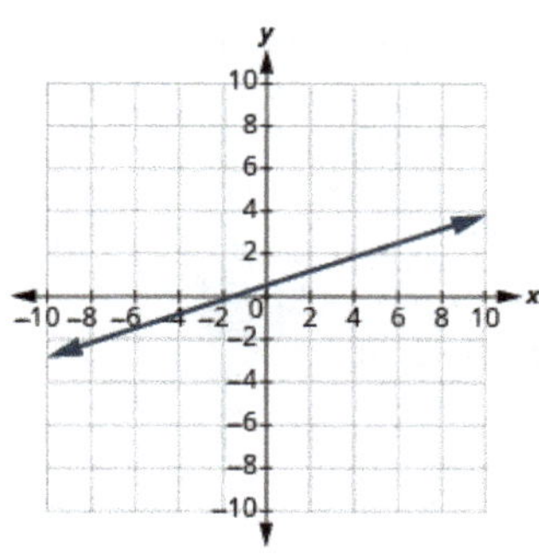

Find the Slope of Horizontal and Vertical Lines

In the following exercises, find the slope of each line.

231.

232.

233.

234.

235.

236.

237.

238.

Use the Slope Formula to find the Slope of a Line between Two Points

In the following exercises, use the slope formula to find the slope of the line between each pair of points.

239.

240.

241.

242.

243.

244.

245.

246.

247.

248.

249.

250.

Graph a Line Given a Point and the Slope

In the following exercises, graph the line given a point and the slope.

251. —

252. —

253. —

254. —

255. —

256. —

257. —

258. —

259. —

260. —

261. —

262. —

263.

264.

265.

266.

Solve Slope Applications

In the following exercises, solve these slope applications.

267. Slope of a roof A fairly easy way to determine the slope is to take a level and set it on one end on the roof surface. Then take a tape measure or ruler, and measure from the other end of the level down to the roof surface. You can use these measurements to calculate the slope of the roof. What is the slope of the roof in this picture?

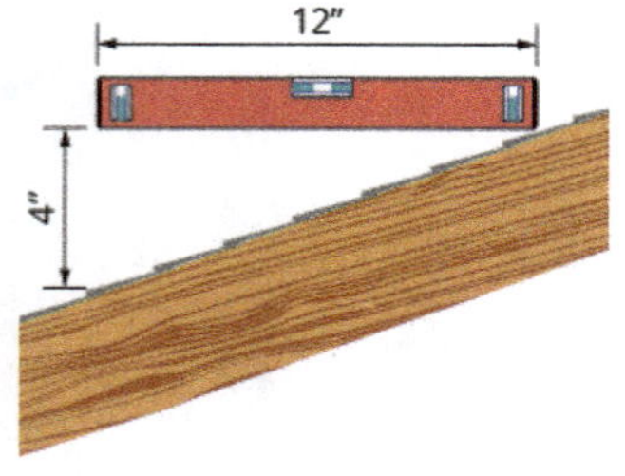

268. What is the slope of the roof shown?

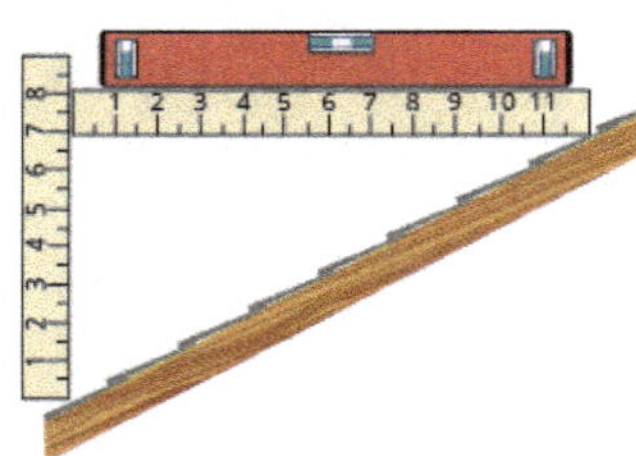

269. Road grade A local road has a grade of The grade of a road is its slope expressed as a percent.

Find the slope of the road as a fraction and then simplify the fraction.

What rise and run would reflect this slope or grade?

270. Highway grade A local road rises feet for every feet of highway.

What is the slope of the highway?

The grade of a highway is its slope expressed as a percent. What is the grade of this highway?

Everyday Math

271. Wheelchair ramp The rules for wheelchair ramps require a maximum inch rise for a inch run.

What run must the ramp have to accommodate a rise to the door?

Draw a model of this ramp.

272. Wheelchair ramp A rise for a run makes it easier for the wheelchair rider to ascend the ramp.

What run must the ramp have to easily accommodate a rise to the door?

Draw a model of this ramp.

Writing Exercises

273. What does the sign of the slope tell you about a line?

274. How does the graph of a line with slope —

differ from the graph of a line with slope

275. Why is the slope of a vertical line undefined?

276. Explain how you can graph a line given a point and its slope.

Self Check

After completing the exercises, use this checklist to evaluate your mastery of the objectives of this section.

I can...	Confidently	With some help	No-I don't get it!
use geoboards to model slope.			
find the slope of a line from its graph.			
find the slope of horizontal and vertical lines.			
use the slope formula to find the slope of a line between two points.			
graph a line given a point and the slope.			
solve slope applications.			

On a scale of 1–10, how would you rate your mastery of this section in light of your responses on the checklist? How can you improve this?

CHAPTER 11 REVIEW

KEY TERMS

horizontal line A horizontal line is the graph of an equation that can be written in the form . The line passes through the *y*-axis at

intercepts of a line Each of the points at which a line crosses the *x*-axis and the *y*-axis is called an intercept of the line.

linear equation An equation of the form where are not both zero, is called a linear equation in two variables.

ordered pair An ordered pair gives the coordinates of a point in a rectangular coordinate system. The first number is the -coordinate. The second number is the -coordinate.

origin The point is called the origin. It is the point where the -axis and -axis intersect.

quadrants The -axis and -axis divide a rectangular coordinate system into four areas, called quadrants.

slope of a line The slope of a line is ——. The rise measures the vertical change and the run measures the horizontal change.

solution to a linear equation in two variables An ordered pair is a solution to the linear equation , if the equation is a true statement when the *x*- and *y*-values of the ordered pair are substituted into the equation.

vertical line A vertical line is the graph of an equation that can be written in the form . The line passes through the *x*-axis at

***x*-axis** The *x*-axis is the horizontal axis in a rectangular coordinate system.

***y*-axis** The *y*-axis is the vertical axis on a rectangular coordinate system.

KEY CONCEPTS

11.1 Use the Rectangular Coordinate System

- **Sign Patterns of the Quadrants**

Quadrant I	Quadrant II	Quadrant III	Quadrant IV
(x,y)	(x,y)	(x,y)	(x,y)
$(+,+)$	$(-,+)$	$(-,-)$	$(+,-)$

- **Coordinates of Zero**
  - Points with a *y*-coordinate equal to 0 are on the *x*-axis, and have coordinates $(a, 0)$.
  - Points with a *x*-coordinate equal to 0 are on the *y*-axis, and have coordinates $(0, b)$.
  - The point (0, 0) is called the origin. It is the point where the *x*-axis and *y*-axis intersect.

11.2 Graphing Linear Equations

- **Graph a linear equation by plotting points.**

  Step 1. Find three points whose coordinates are solutions to the equation. Organize them in a table.

  Step 2. Plot the points on a rectangular coordinate system. Check that the points line up. If they do not, carefully check your work.

  Step 3. Draw the line through the points. Extend the line to fill the grid and put arrows on both ends of the line.

- **Graph of a Linear Equation:**The graph of a linear equation is a straight line.

- Every point on the line is a solution of the equation.
- Every solution of this equation is a point on this line.

11.3 Graphing with Intercepts

- **Intercepts**
  - The *x*-intercept is the point, , where the graph crosses the *x*-axis. The *x*-intercept occurs when y is zero.
  - The *y*-intercept is the point, , where the graph crosses the *y*-axis. The *y*-intercept occurs when x is zero.
  - The *x*-intercept occurs when y is zero.
  - The *y*-intercept occurs when x is zero.
- **Find the *x* and *y* intercepts from the equation of a line**
  - To find the *x*-intercept of the line, let and solve for *x*.
  - To find the *y*-intercept of the line, let and solve for *y*.

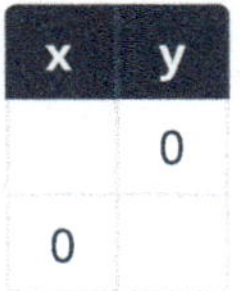

x	y
	0
0	

- **Graph a line using the intercepts**

  Step 1. Find the *x*- and *y*- intercepts of the line.
  - Let and solve for *x*.
  - Let and solve for *y*.

  Step 2. Find a third solution to the equation.

  Step 3. Plot the three points and then check that they line up.

  Step 4. Draw the line.
- **Choose the most convenient method to graph a line**

  Step 1. Determine if the equation has only one variable. Then it is a vertical or horizontal line.
  is a vertical line passing through the *x*-axis at *a*.
  is a horizontal line passing through the *y*-axis at *b*.

  Step 2. Determine if *y* is isolated on one side of the equation. The graph by plotting points.
  Choose any three values for *x* and then solve for the corresponding *y*- values.

  Step 3. Determine if the equation is of the form , find the intercepts.
  Find the *x*- and *y*- intercepts and then a third point.

11.4 Understand Slope of a Line

- **Find the slope from a graph**

  Step 1. Locate two points on the line whose coordinates are integers.

  Step 2. Starting with the point on the left, sketch a right triangle, going from the first point to the second point.

  Step 3. Count the rise and the run on the legs of the triangle.

  Step 4. Take the ratio of rise to run to find the slope,
- **Slope of a Horizontal Line**
  - The slope of a horizontal line, , is 0.
- **Slope of a Vertical Line**
  - The slope of a vertical line, , is undefined.

- **Slope Formula**
    - The slope of the line between two points and is

- **Graph a line given a point and a slope.**

  Step 1. Plot the given point.

  Step 2. Use the slope formula to identify the rise and the run.

  Step 3. Starting at the given point, count out the rise and run to mark the second point.

  Step 4. Connect the points with a line.

REVIEW EXERCISES

11.1 Use the Rectangular Coordinate System

Plot Points in a Rectangular Coordinate System

In the following exercises, plot each point in a rectangular coordinate system.

277.

278.

In the following exercises, plot each point in a rectangular coordinate system and identify the quadrant in which the point is located.

279.

280.

Identify Points on a Graph

In the following exercises, name the ordered pair of each point shown in the rectangular coordinate system.

281.

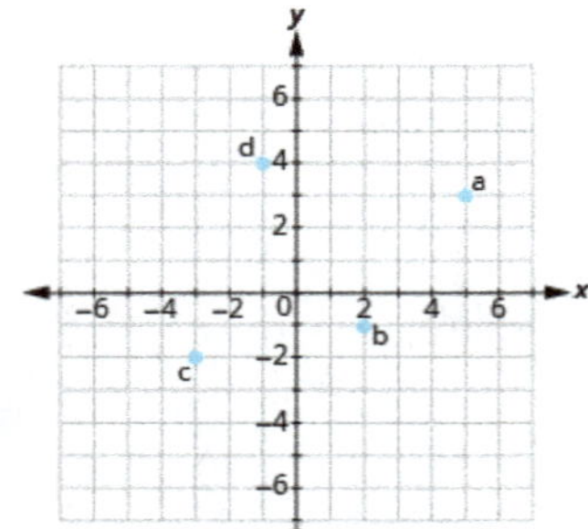

282.

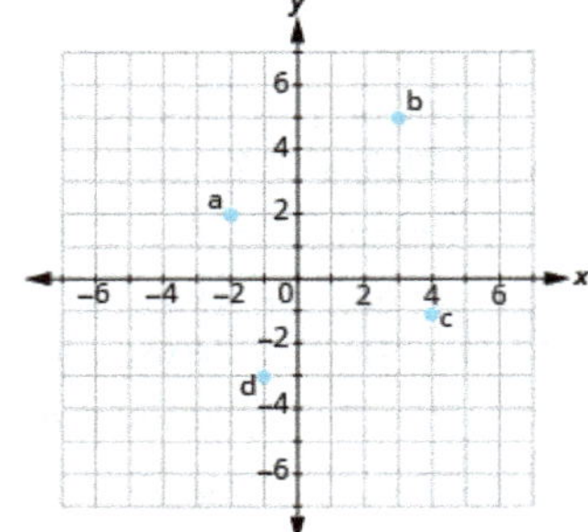

283.

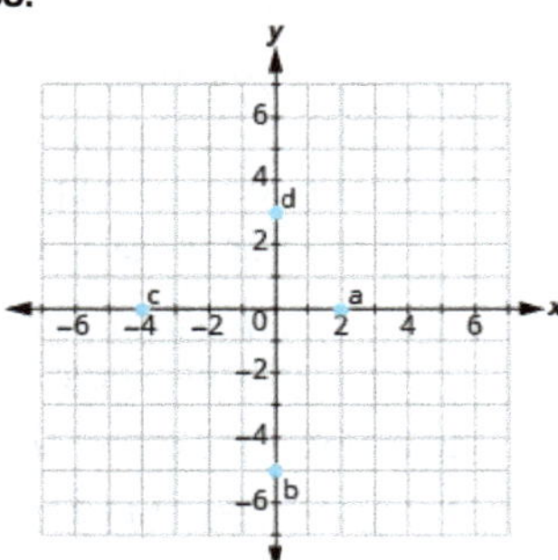

284.

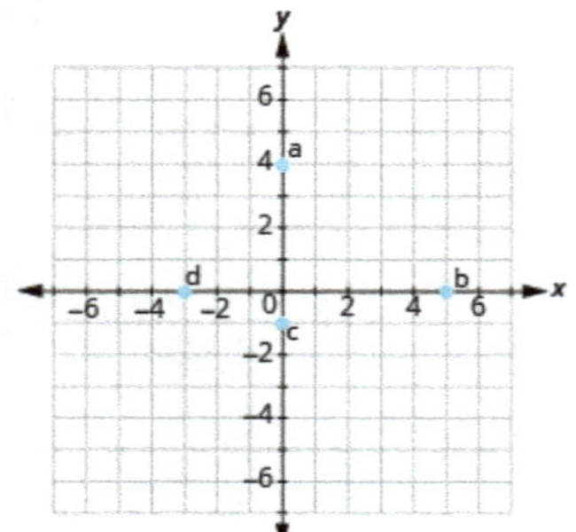

Verify Solutions to an Equation in Two Variables

In the following exercises, find the ordered pairs that are solutions to the given equation.

285.

286.

—

Complete a Table of Solutions to a Linear Equation in Two Variables

In the following exercises, complete the table to find solutions to each linear equation.

287.

x	y	x y

288. —

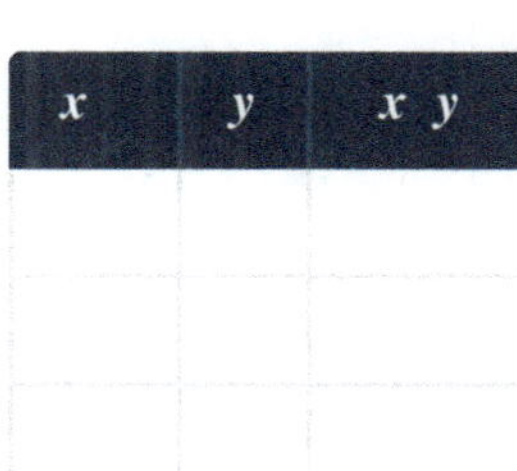

289.

x	y	x y

290.

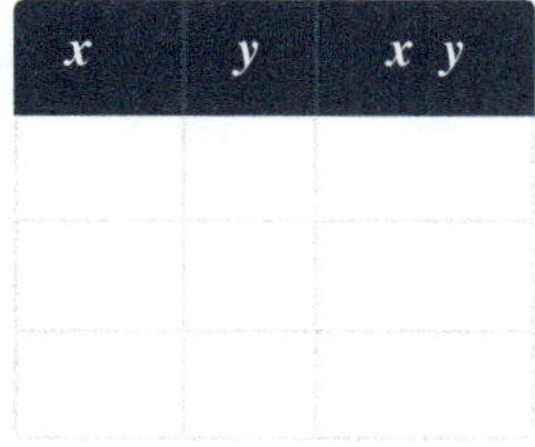

Find Solutions to a Linear Equation in Two Variables

In the following exercises, find three solutions to each linear equation.

291.

292.

293.

294.

11.2 Graphing Linear Equations

Recognize the Relation Between the Solutions of an Equation and its Graph

In each of the following exercises, an equation and its graph is shown. For each ordered pair, decide

if the ordered pair is a solution to the equation. if the point is on the line.

295.

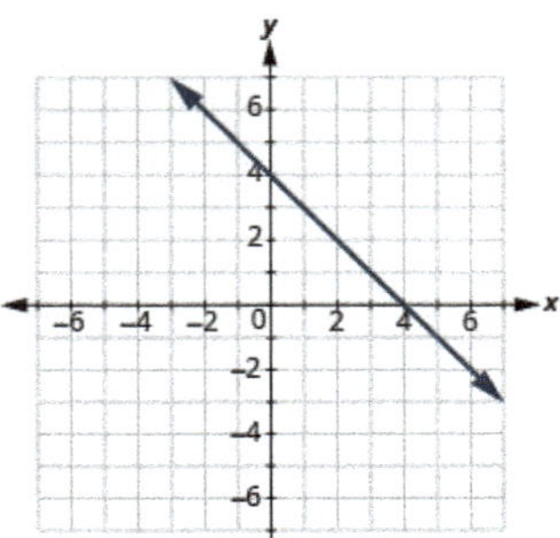

1.
2.
3.
4.

296. –

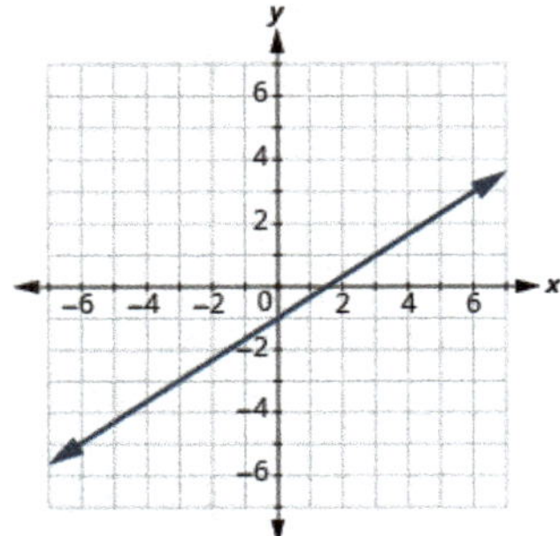

1.
2.
3.
4.

Graph a Linear Equation by Plotting Points

In the following exercises, graph by plotting points.

297.

298.

299.

Graph Vertical and Horizontal lines

In the following exercises, graph the vertical or horizontal lines.

300.

301.

11.3 Graphing with Intercepts

Identify the Intercepts on a Graph

In the following exercises, find the and

302.

303.

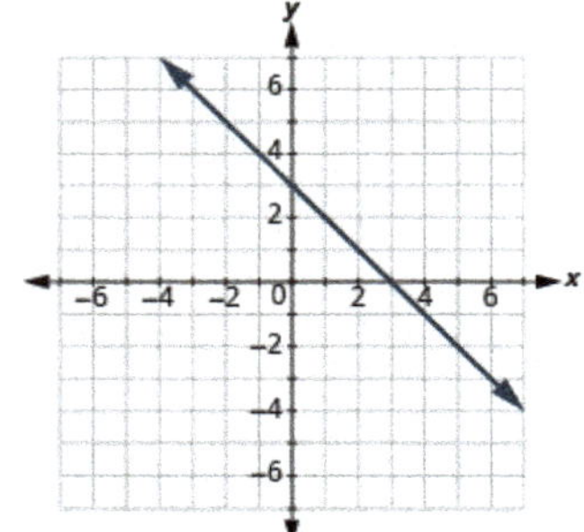

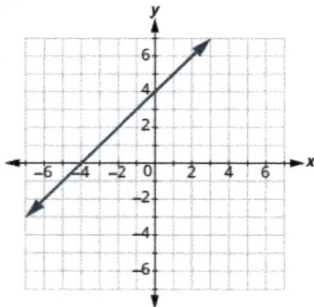

Find the Intercepts from an Equation of a Line

In the following exercises, find the intercepts.

304.

305.

306. –

307.

Graph a Line Using the Intercepts

In the following exercises, graph using the intercepts.

308.

309.

Choose the Most Convenient Method to Graph a Line

In the following exercises, identify the most convenient method to graph each line.

310.

311.

312.

313.

314. —

315. —

11.4 Understand Slope of a Line

Use Geoboards to Model Slope

In the following exercises, find the slope modeled on each geoboard.

316.

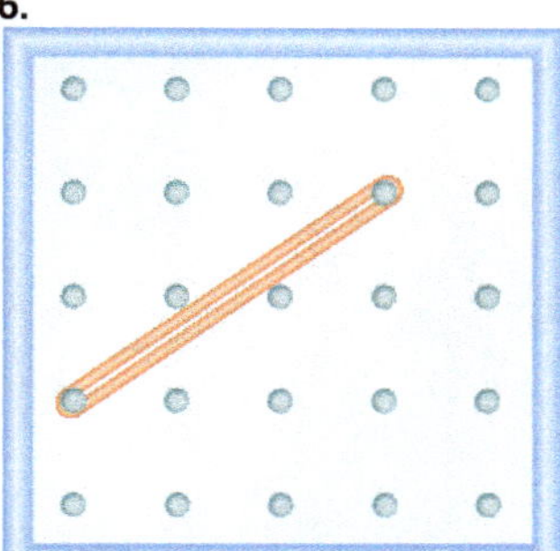

317.

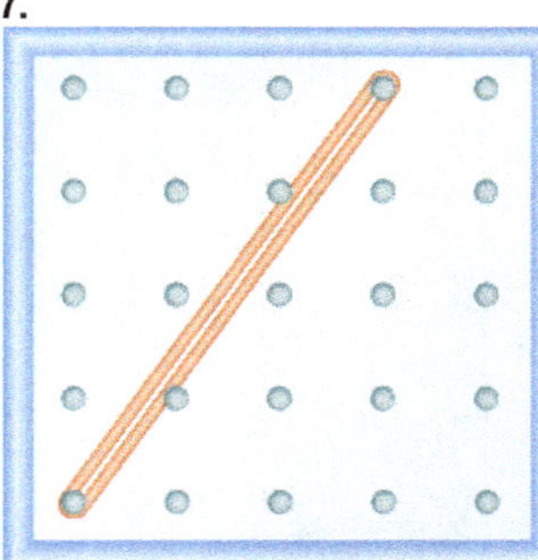

318.

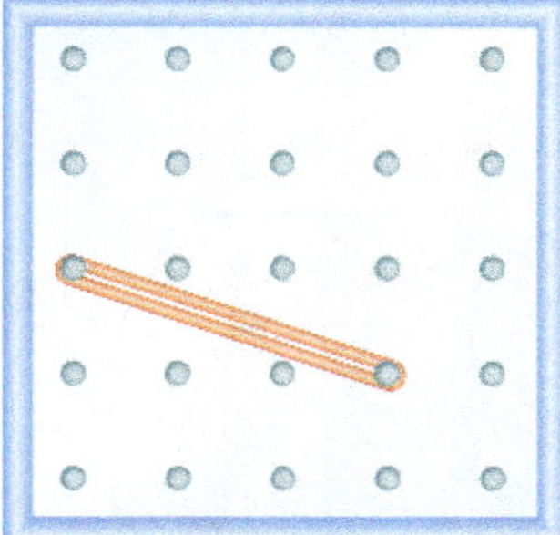

319.

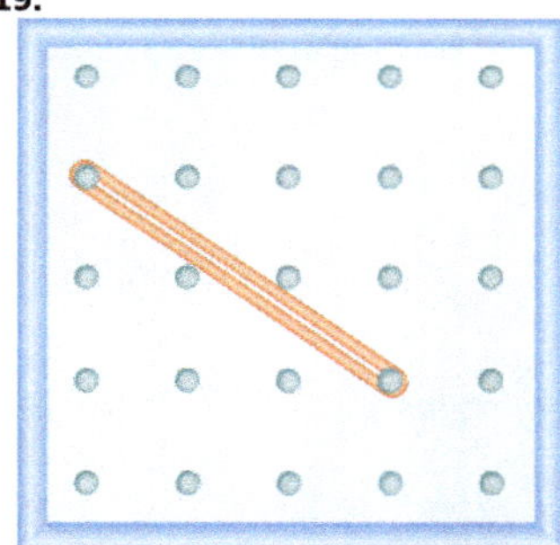

In the following exercises, model each slope. Draw a picture to show your results.

320. —

321. —

322. —

323. —

Find the Slope of a Line from its Graph

In the following exercises, find the slope of each line shown.

324.

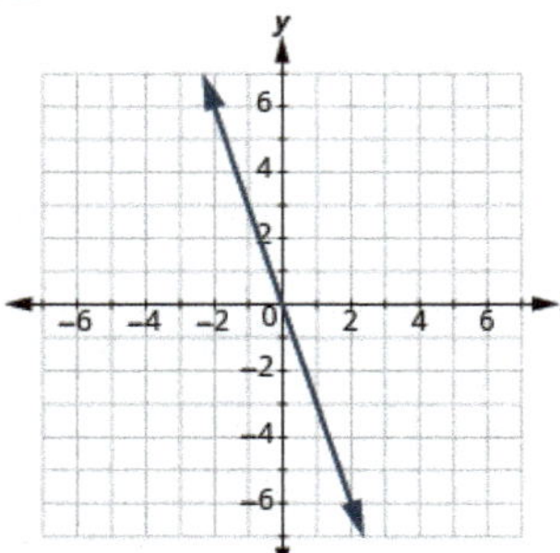

325.

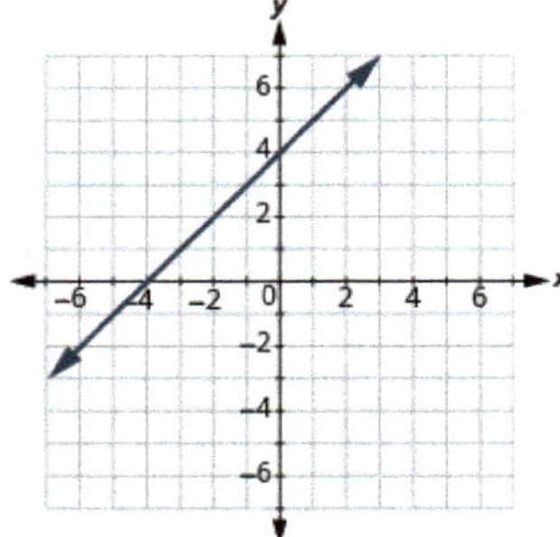

326.

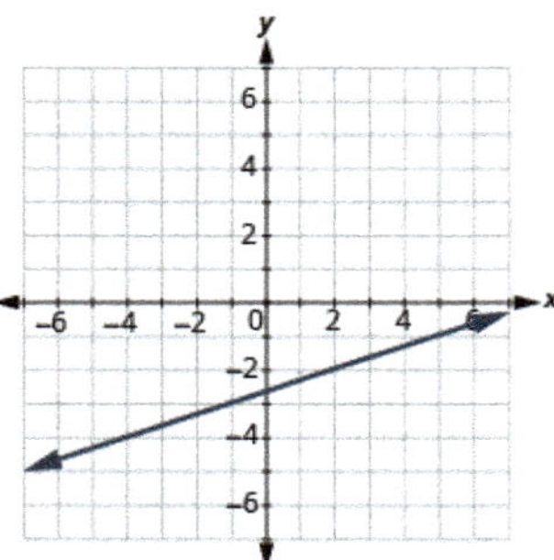

327.

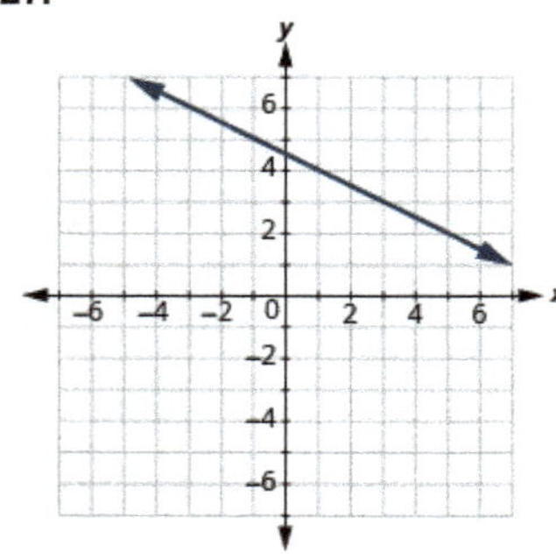

Find the Slope of Horizontal and Vertical Lines

In the following exercises, find the slope of each line.

328.

329.

330.

331.

Use the Slope Formula to find the Slope of a Line between Two Points

In the following exercises, use the slope formula to find the slope of the line between each pair of points.

332.

333.

334.

335.

Graph a Line Given a Point and the Slope

In the following exercises, graph the line given a point and the slope.

336. —

337. —

Solve Slope Applications

In the following exercise, solve the slope application.

338. A roof has rise feet and run feet. What is its slope?

PRACTICE TEST

339. Plot and label these points:

340. Name the ordered pair for each point shown.

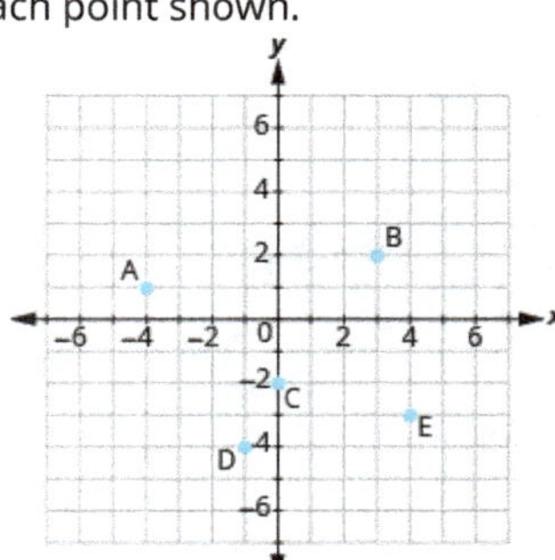

341. Find the and on the line shown.

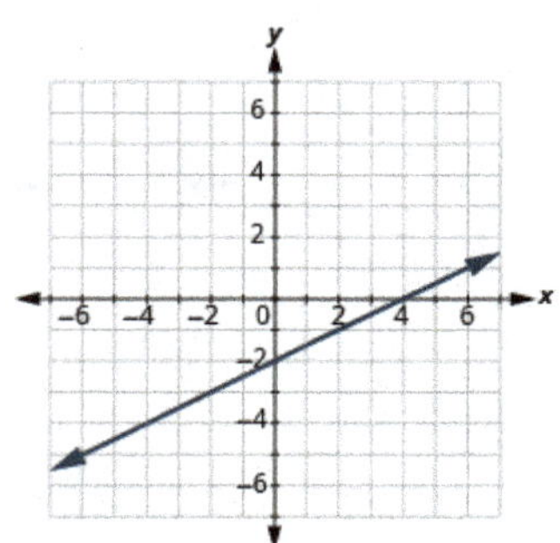

342. Find the and of the equation

343. Is a solution to the equation How do you know?

344. Complete the table to find four solutions to the equation

345. Complete the table to find three solutions to the equation

In the following exercises, find three solutions to each equation and then graph each line.

346.

347.

In the following exercises, find the slope of each line.

348.

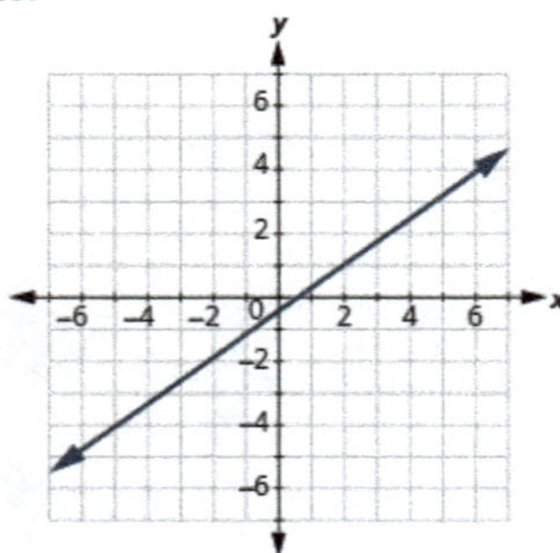

349.

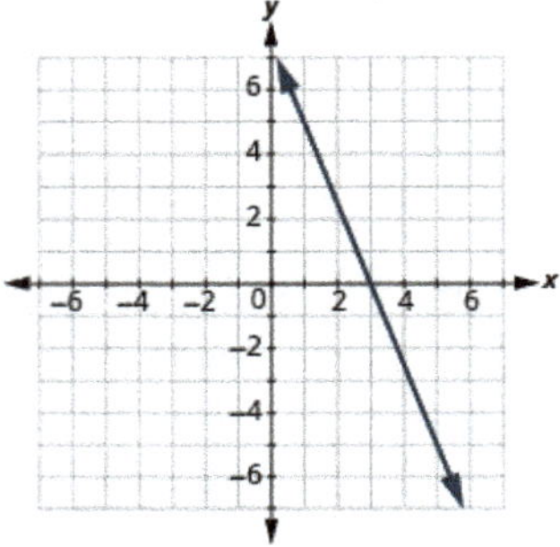

350. Use the slope formula to find the slope of the line between and

351. Find the slope of the line

352. Graph the line passing through with slope —

353. A bicycle route climbs feet for feet of horizontal distance. What is the slope of the route?

A CUMULATIVE REVIEW

Note: Answers to the Cumulative Review can be found in the Supplemental Resources. Please visit http://openstaxcollege.org to view an updated list of the Learning Resources for this title and how to access them.

Chapter 1 Whole Numbers

No exercises.

Chapter 2 The Language of Algebra

Simplify:

Solve:

Translate into an algebraic expression.

less than the product of and

Translate into an algebraic equation and solve.

Twice the difference of and gives

Find all the factors of

Find the prime factorization of

Find the least common multiple of and

Chapter 3 Integers

Simplify:

Translate into an algebraic expression or equation.

The sum of and increased by

The product of

The quotient of and the sum of

The product of and is

Solve:

Chapter 4 Fractions

Locate the numbers on a number line. — — —

Simplify:

——

— ——

— —

Chapter 5 Decimals

Simplify:

Write in order from smallest to largest:

Solve:

Using as the estimate for pi, approximate the (a) circumference and (b) area of a circle whose radius is inches.

Find the mean of the numbers,

Find the median of the numbers,

Identify the mode of the numbers,

Find the unit price of one t-shirt if they are sold at for

Chapter 6 Percents

Convert to (a) a fraction and (b) a decimal.

Translate and solve.

is of what number?

The nutrition label on a package of granola bars says that each granola bar has calories, and calories are from fat. What percent of the total calories is from fat?

Elliot received commission when he sold a painting at the art gallery where he works. What was the rate of commission?

Nandita bought a set of towels on sale for The original price of the towels was What was the discount rate?

Alan invested in a friend's business. In years the friend paid him the plus interest. What was the rate of interest?

Solve:

Chapter 7 The Properties of Real Numbers

List the (a) whole numbers, (b) integers, (c) rational numbers, (d) irrational numbers, (e) real numbers

Simplify:

A playground is feet wide. Convert the width to yards.

Every day last week Amit recorded the number of minutes he spent reading. The recorded number of minutes he read each day was How many hours did Amit spend reading last week?

June walked kilometers. Convert this length to miles knowing mile is kilometer.

Chapter 8 Solve Linear Equations

Solve:

Translate and solve.

Four less than is

Chapter 9 Math Models and Geometry

One number is less than another. Their sum is negative twenty-two. Find the numbers.

The sum of two consecutive integers is Find the numbers.

Wilma has in dimes and quarters. The number of dimes is less than the number of quarters. How many of each coin does she have?

Two angles are supplementary. The larger angle is more than the smaller angle. Find the measurements of both angles.

One angle of a triangle is more than the smallest angle. The largest angle is the sum of the other angles. Find the measurements of all three angles.

Erik needs to attach a wire to hold the antenna to the roof of his house, as shown in the figure. The antenna is feet tall and Erik has feet of wire. How far from the base of the antenna can he attach the wire?

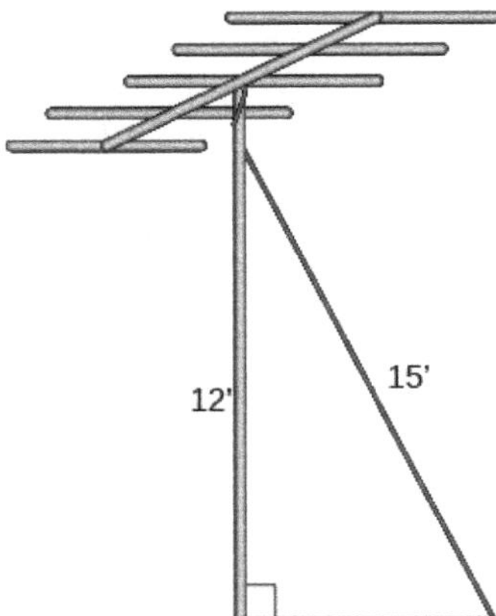

The width of a rectangle is less than the length. The perimeter is inches. Find the length and the width.

Find the (a) volume and (b) surface area of a rectangular carton with length inches, width inches, and height inches.

Chapter 10 Polynomials

Simplify:

Write in scientific notation:

Factor the greatest common factor from the polynomial.

Chapter 11 Graphs

Graph:

—

Find the intercepts.

Graph using the intercepts.

Find the slope of the line shown.

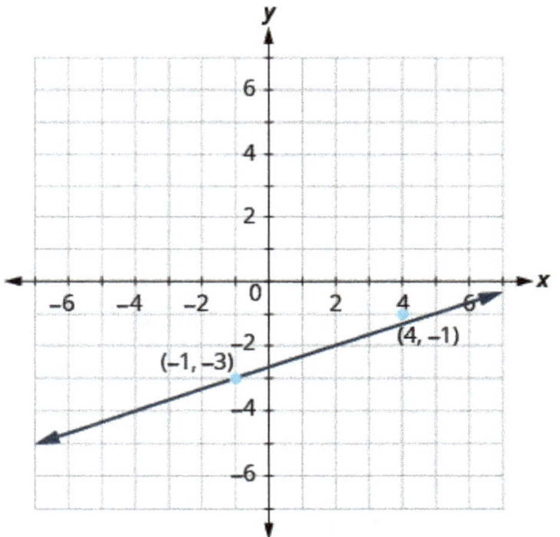

Use the slope formula to find the slope of the line between the points

Graph the line passing through the point and with slope —

B POWERS AND ROOTS TABLES

n	n	$\sqrt{n}$	n	$\sqrt{n}$

Table B1

n	n	$\sqrt{n}$	n	$\sqrt{n}$

Table B1

n	n	$\sqrt{n}$	n	$\sqrt{n}$

Table B1

n	n	$\sqrt{n}$	n	$\sqrt{n}$

Table B1

C GEOMETRIC FORMULAS

Name	Shape	Formulas
Rectangle	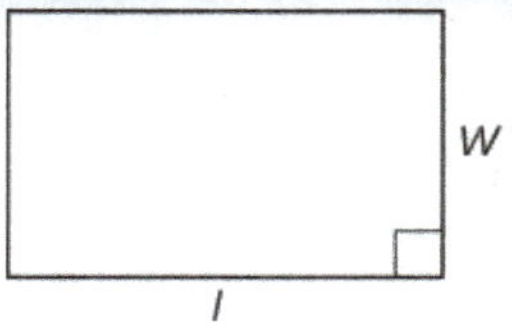	
Square	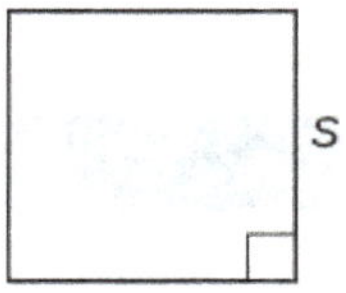	
Triangle	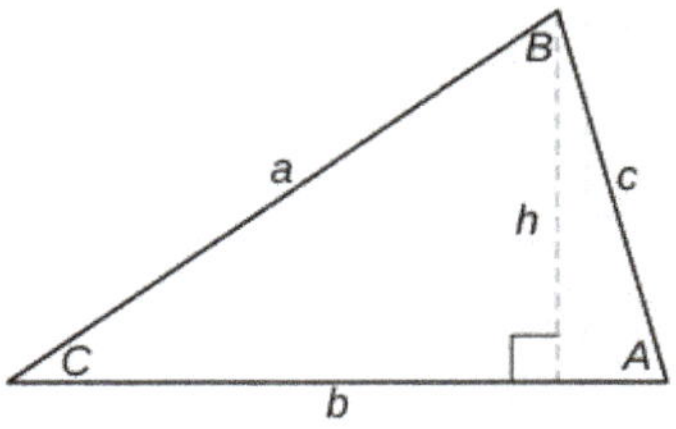	—
Right Triangle	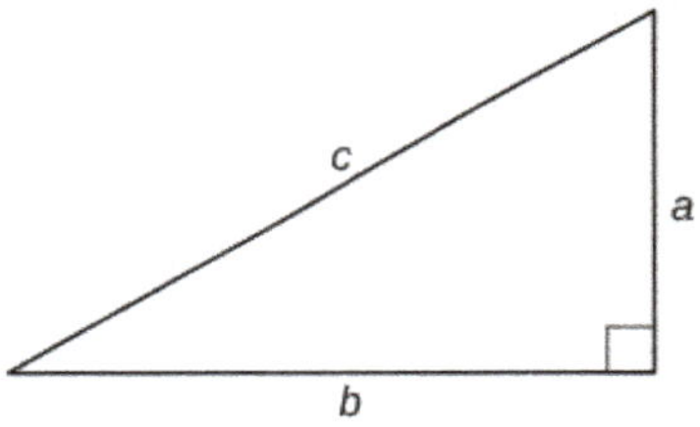	—
Circle	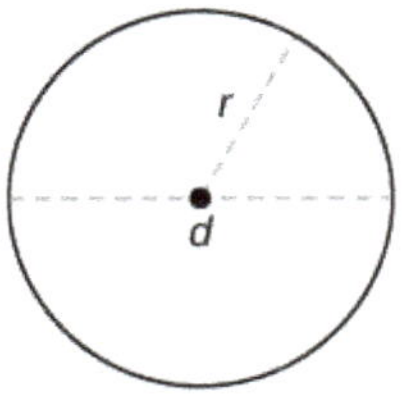	
Parellelogram	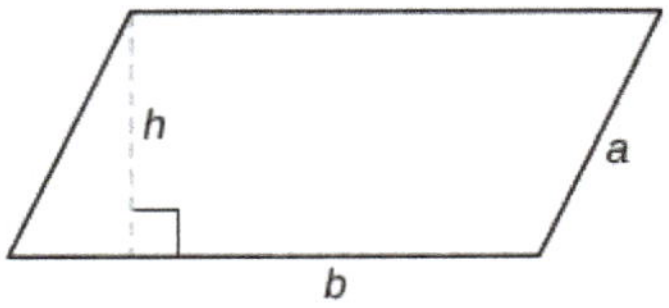	

Table C1 2 Dimensions

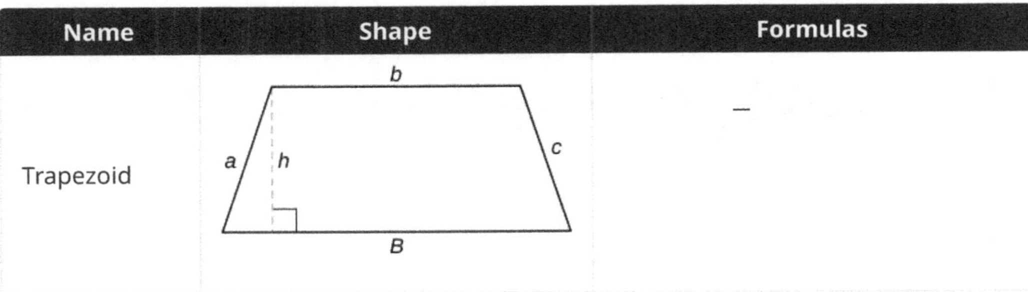

Name	Shape	Formulas
Trapezoid		

Table C1 2 Dimensions

Name	Shape	Formulas
Rectangular Solid	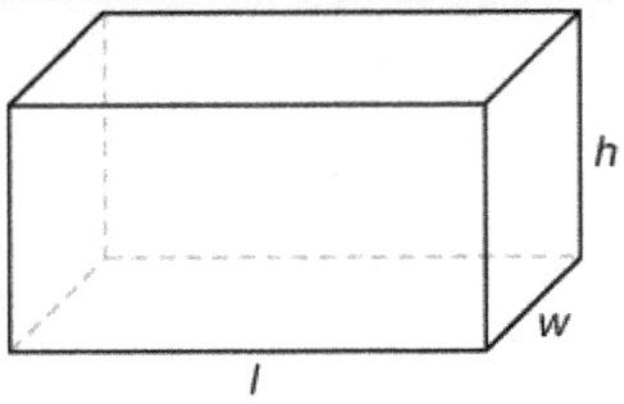	
Cube	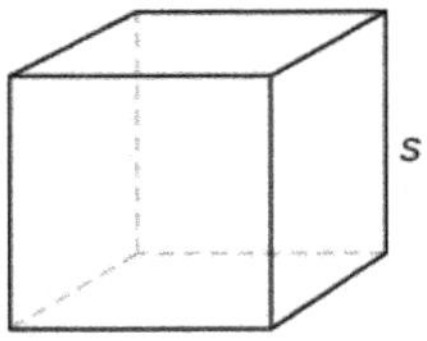	
Cone	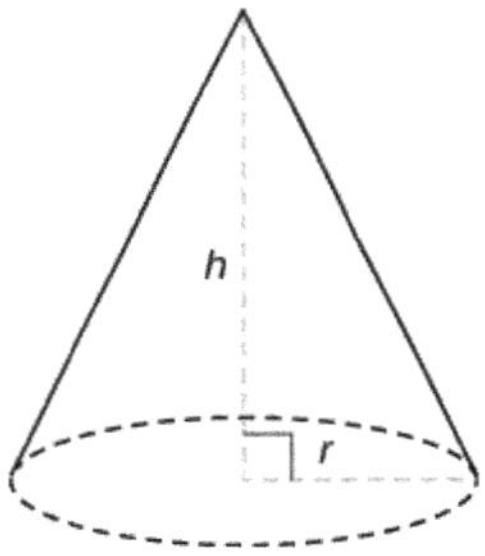	
Sphere	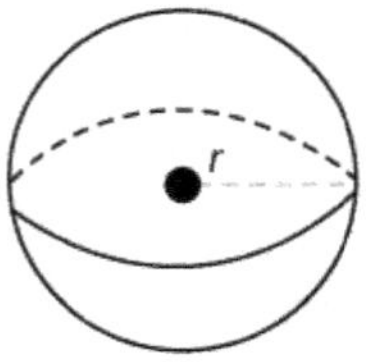	

Table C2 3 Dimensions

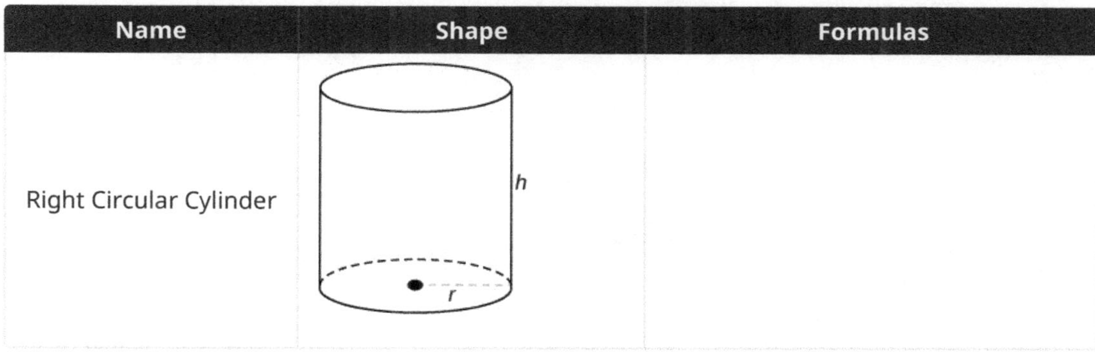

Name	Shape	Formulas
Right Circular Cylinder		

Table C2 3 Dimensions

ANSWER KEY

Chapter 1

Be Prepared

1.1.

1.2.

1.3.

1.4.

1.5.

1.6.

1.7.

1.8.

1.9.

Try It

1.1.

2, 9, 241, 376

0, 2, 9, 241, 376

1.2.

7, 13, 201

0, 7, 13, 201

1.3. 176

1.4. 237

1.5.

ten millions

tens

hundred thousands

millions

ones

1.6.

billions

ten thousands

tens

hundred thousands

hundred millions

1.7. nine trillion, two hundred fifty-eight billion, one hundred thirty-seven million, nine hundred four thousand, sixty-one

1.8. seventeen trillion, eight hundred sixty-four billion, three hundred twenty-five million, six hundred nineteen thousand, four

1.9. three hundred sixteen million, one hundred twenty-eight thousand, eight hundred thirty nine

1.10. thirty one million, five hundred thirty-six thousand

1.11. 53,809,051

1.12. 2,022,714,466

1.13. 34,000,000 miles

1.14. 204,000,000 pounds

1.15. 160

1.16. 880

1.17. 17,900

1.18. 5,000

1.19. 64,000

1.20. 156,000

1.21.

eight plus four; the sum of eight and four

eighteen plus eleven; the sum of eighteen and eleven

1.22.

twenty-one plus sixteen; the sum of twenty-one and sixteen

one hundred plus two hundred; the sum of one hundred and two hundred

1.23.

$3 + 6 = 9$

1.24.

$5 + 1 = 6$

1.25.

$5 + 7 = 12$

1.26.

$6 + 8 = 14$

1.27.

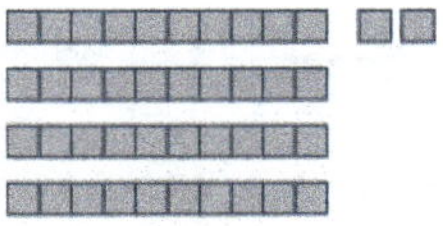

$15 + 27 = 42$

1.28.

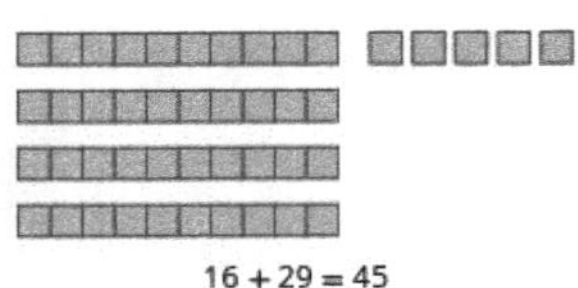

1.29.

1.30.

1.31.

1.32.

1.33.

1.34.

1.35.

1.36.

1.37.

1.38.

1.39.

1.40.

1.41.

1.42.

1.43. Translate: ; Simplify:

1.44. Translate: ; Simplify:

1.45. Translate: ; Simplify

1.46. Translate 37 + 69; Simplify 106

1.47. He rode 140 miles.

1.48. The total number is 720 students.

1.49. The perimeter is 30 inches.

1.50. The perimeter is 36 inches.

1.51.

twelve minus four; the difference of twelve and four

twenty-nine minus eleven; the difference of twenty-nine and eleven

1.52.

eleven minus two; the difference of eleven and two

twenty-nine minus twelve; the difference of twenty-nine and twelve

1.53.

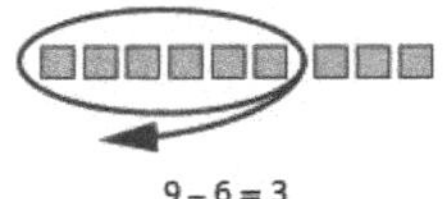

1.54.

6 − 1 = 5

1.55.

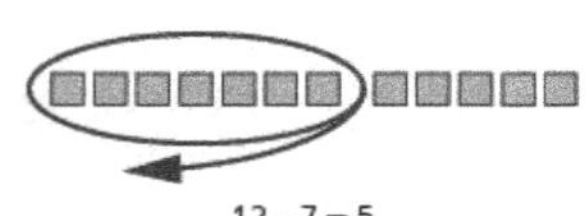

1.56.

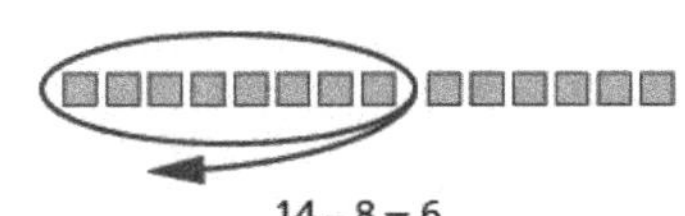

1.57.

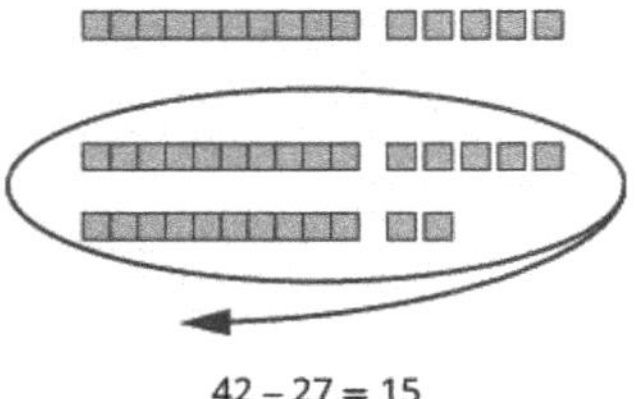

1.58.

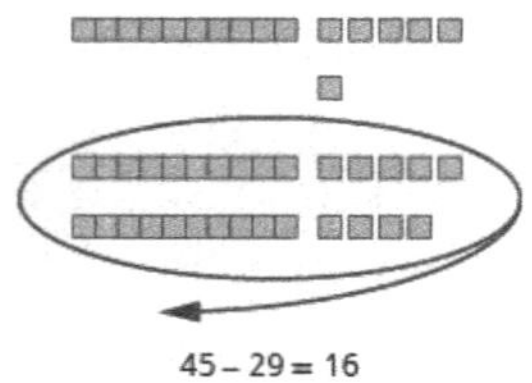

1.59. 7 − 0 = 7; 7 + 0 = 7

1.60. 6 − 2 = 4; 2 + 4 = 6

1.61. 86 − 54 = 32 because 54 + 32 = 86

1.62. 99 − 74 = 25 because 74 + 25 = 99

1.63. 93 − 58 = 35 because 58 + 35 = 93

1.64. 81 − 39 = 42 because 42 + 39 = 81

1.65. 439 − 52 = 387 because 387 + 52 = 439

1.66. 318 − 75 = 243 because 243 + 75 = 318

1.67. 832 − 376 = 456 because 456 + 376 = 832

1.68. 847 − 578 = 269 because 269 + 578 = 847

1.69. 4,585 − 697 = 3,888 because 3,888 + 697 = 4,585

1.70. 5,637 − 899 = 4,738 because 4,738 + 899 = 5,637

1.71.

14 − 9 = 5 37 − 21 = 16

1.72.

11 − 6 = 5 67 − 18 = 49

1.73. The difference is 19 degrees Fahrenheit.

1.74. The difference is 17 degrees Fahrenheit.

1.75. The difference is $149.

1.76. The difference is $136.

1.77.

eight times seven ; the product of eight and seven

eighteen times eleven ; the product of eighteen and eleven

1.78.

thirteen times seven ; the product of thirteen and seven

five times sixteen; the product of five and sixteen

1.79.

1.80.

1.81.

1.82.

1.83.

1.84.

1.85. 54 and 54; both are the same.

1.86. 48 and 48; both are the same.

1.87.

1.88.

1.89.

1.90.

1.91. 3,354

1.92. 3,776

1.93.

540 5,400

1.94.

750 7,500

1.95. 127,995

1.96. 653,462

1.97. 365,462

1.98. 504,108

1.99. $13 \cdot 28$; 364

1.100. $47 \cdot 14$; 658

1.101. 2(167); 334

1.102. 2(258); 516

1.103. Valia donated 144 water bottles.

1.104. Vanessa bought 80 hot dogs.

1.105. Erin needs 28 dahlias.

1.106. There are 36 women in the choir.

1.107. Jane needs 320 tiles.

1.108. Yousef needs 1,080 tiles.

1.109. The area of the rug is 40 square feet.

1.110. The area of the driveway is 900 square feet

1.111.

eighty-four divided by seven; the quotient of eighty-four and seven

eighteen divided by six; the quotient of eighteen and six.

twenty-four divided by eight; the quotient of twenty-four and eight

1.112.

seventy-two divided by nine; the quotient of seventy-two and nine

twenty-one divided by three; the quotient of twenty-one and three

fifty-four divided by six; the quotient of fifty-four and six

1.113.

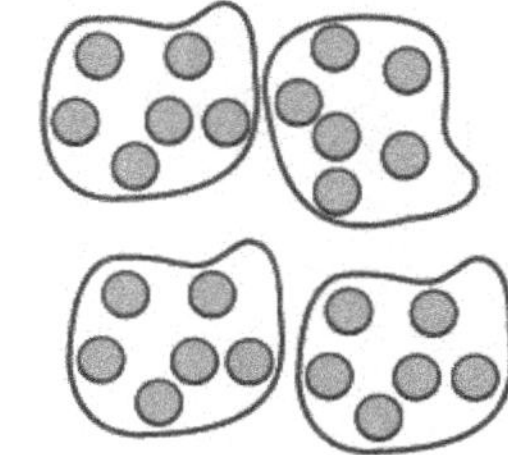

1.114.

1.115. 9 3

1.116. 4 5

1.117.

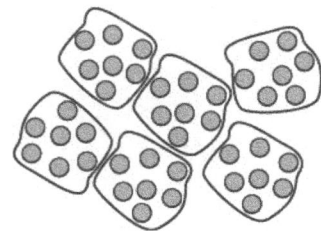

1 27

1.118.

16 4

1.119. 0 undefined

1.120. 0 undefined

1.121. 659

1.122. 679

1.123. 861

1.124. 651

1.125. 704

1.126. 809

1.127. 476 with a remainder of 4

1.128. 539 with a remainder of 7

1.129. 114 R11

1.130. 121 R9

1.131. 307 R49

1.132. 308 R77

1.133. 91 ÷ 13; 7

1.134. 52 ÷ 13; 4

1.135. Marcus can fill 15 cups.

1.136. Andrea can make 9 bows.

Section Exercises

1.

5, 125 5, 125

3.

50, 221 50, 221

5. 561

7. 407

9.

thousands hundreds

tens ten thousands

hundred thousands

11.

hundred thousands

millions thousands

tens ten thousands

13. One thousand, seventy-eight

15. Three hundred sixty-four thousand, five hundred ten

17. Five million, eight hundred forty-six thousand, one hundred three

19. Thirty seven million, eight hundred eighty-nine thousand, five

21. Fourteen thousand, four hundred ten

23. Six hundred thirteen thousand, two hundred

25. Two million, six hundred seventeen thousand, one hundred seventy-six

27. Twenty three million, eight hundred sixty-seven thousand

29. One billion, three hundred seventy-seven million, five hundred eighty-three thousand, one hundred fifty-six

31. 412

33. 35,975

35. 11,044,167

37. 3,226,512,017

39. 7,173,000,000

41. 39,000,000,000,000

43.

390 2,930

45.

13,700 391,800

47.

1,490 1,500

49.

51. Twenty four thousand, four hundred ninety-three dollars

53.

$24,490 $24,500

$24,000 $20,000

55.

57. Answers may vary. The whole numbers are the counting numbers with the inclusion of zero.

59. five plus two; the sum of 5 and 2.

61. thirteen plus eighteen; the sum of 13 and 18.

63. two hundred fourteen plus six hundred forty-two; the sum of 214 and 642

65.

67.

69.

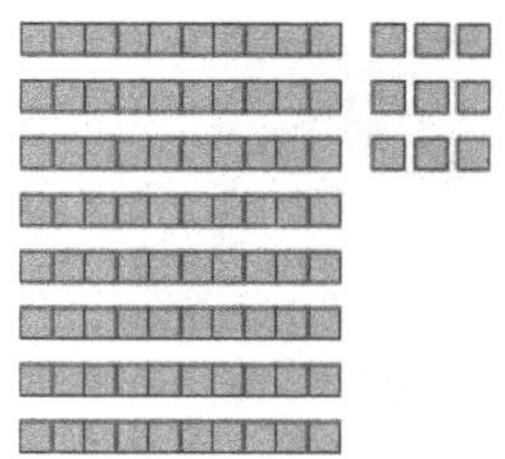

71.

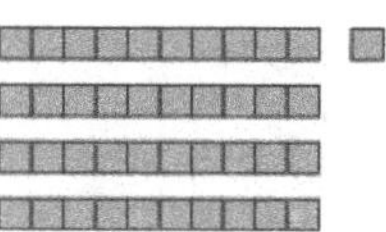

73.

+	0	1	2	3	4	5	6	7	8	9
0	0	1	2	3	4	5	6	7	8	9
1	1	2	3	4	5	6	7	8	9	10
2	2	3	4	5	6	7	8	9	10	11
3	3	4	5	6	7	8	9	10	11	12
4	4	5	6	7	8	9	10	11	12	13
5	5	6	7	8	9	10	11	12	13	14
6	6	7	8	9	10	11	12	13	14	15
7	7	8	9	10	11	12	13	14	15	16
8	8	9	10	11	12	13	14	15	16	17
9	9	10	11	12	13	14	15	16	17	18

75.

+	3	4	5	6	7	8	9
6	9	10	11	12	13	14	15
7	10	11	12	13	14	15	16
8	11	12	13	14	15	16	17
9	12	13	14	15	16	17	18

77.

+	5	6	7	8	9
5	10	11	12	13	14
6	11	12	13	14	15
7	12	13	14	15	16
8	13	14	15	16	17
9	14	15	16	17	18

79.

13 13

81.

83.

85.

87.

89.

91.

93.

95.

97.

99.

101.

103.

105.

107.

109.

111.

113.

115.

117.

119. The total cost was $1,875.

121. Ethan rode 138 miles.

123. The total square footage in the rooms is 1,167 square feet.

125. Natalie's total salary is $237,186.

127. The perimeter of the figure is 44 inches.

129. The perimeter of the figure is 56 meters.

131. The perimeter of the figure is 71 yards.

133. The perimeter of the figure is 62 feet.

135. The total number of calories was 640.

137. Yes, he scored 406 points.

138. Yes, the total weight is 1091 pounds.

139. Answers will vary.

141. fifteen minus nine; the difference of fifteen and nine

143. forty-two minus thirty-five; the difference of forty-two and thirty-five

145. hundred seventy-five minus three hundred fifty; the difference of six hundred seventy-five and three hundred fifty

147.

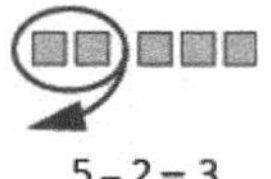

149.

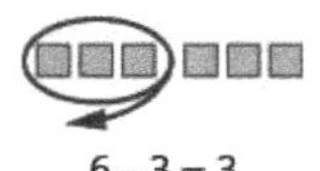

151.

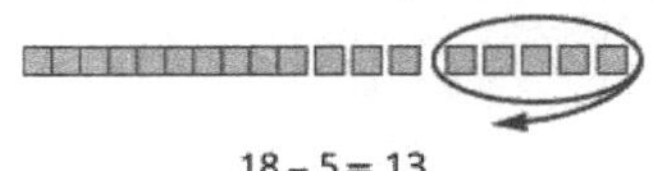

153.

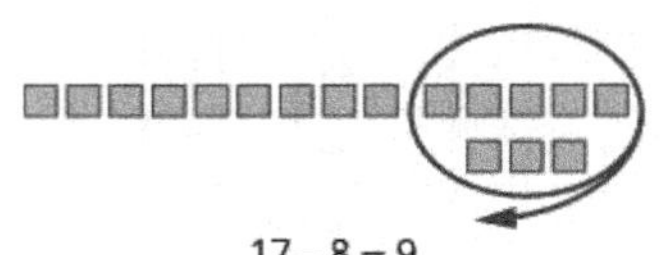

155.

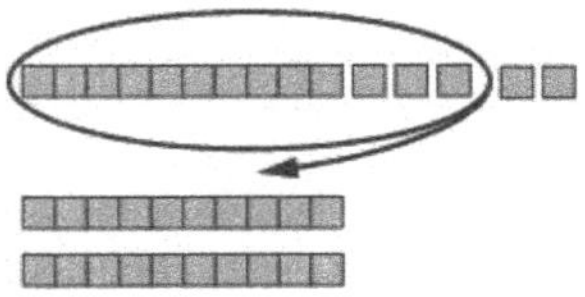

35 − 13 = 22

157.

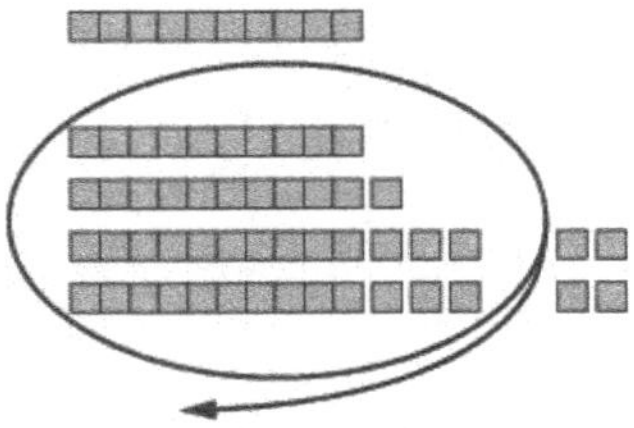

61 − 47 = 14

159. 5

161. 8
163. 22
165. 33
167. 123
169. 1,321
171. 28
173. 222
175. 346
177. 3,519
179. 1,186
181. 34,668
183. 10 − 3; 7
185. 15 − 4; 11
187. 9 − 6; 3
189. 75 − 28; 47
191. 45 − 20; 25
193. 92 − 67; 25
195. 16 − 12; 4
197. 61 − 38; 23
199. 29
201. 72
203. 1,060
205. 22
207. 75 + 35; 110
209. 41 − 13; 28
211. 100 − 76; 24
213. The difference between the high and low temperature was 17 degrees

215. The difference between the third grade and second grade was 13 children.

217. The difference between the regular price and sale price is $251.

219. John needs to save $155 more.

221. 157 miles

223. Answers may vary.

225. four times seven; the product of four and seven

227. five times twelve; the product of five and twelve

229. ten times twenty-five; the product of ten and twenty-five

231. forty-two times thirty-three; the product of forty-two and thirty-three

233.

3 × 6 = 18

235.

5 × 9 = 45

237.

×	0	1	2	3	4	5	6	7	8	9
0	0	0	0	0	0	0	0	0	0	0
1	0	1	2	3	4	5	6	7	8	9
2	0	2	4	6	8	10	12	14	16	18
3	0	3	6	9	12	15	18	21	24	27
4	0	4	8	12	16	20	24	28	32	36
5	0	5	10	15	20	25	30	35	40	45
6	0	6	12	18	24	30	36	42	48	54
7	0	7	14	21	28	35	42	49	56	63
8	0	8	16	24	32	40	48	56	64	72
9	0	9	18	27	36	45	54	63	72	81

239.

×	3	4	5	6	7	8	9
4	12	16	20	24	28	32	36
5	15	20	25	30	35	40	45
6	18	24	30	36	42	48	54
7	21	28	35	42	49	56	63
8	24	32	40	48	56	64	72
9	27	36	45	54	63	72	81

241.

×	3	4	5	6	7	8	9
6	18	24	30	36	42	48	54
7	21	28	35	42	49	56	63
8	24	32	40	48	56	64	72
9	27	36	45	54	63	72	81

243.

×	5	6	7	8	9
5	25	30	35	40	45
6	30	36	42	48	54
7	35	42	49	56	63
8	40	48	56	64	72
9	45	54	63	72	81

245. 0
247. 0
249. 43
251. 28
253. 240,055
255.

42 42

257. 395
259. 1,650
261. 23,947
263. 1,976
265. 7,008
267. 2,295
269. 230
271. 3,600
273. 88,000
275. 50,000,000
277. 34,333
279. 422,506
281. 804,285
283. 26,624
285. 245,340
287. 1,476,417
289. $18 \cdot 33$; 594
291. 51(67); 3,417
293. 2(249); 498
295. 10(375); 3,750
297. 1,406
299. 148
301. 11,000
303. 11,424
305. 0
307. 15,383
309. $50 - 18$; 32
311. 2(35); 70
313. $20 + 980$; 1,000
315. 12(875); 10,500
317. $89 - 74$; 15
319. $3{,}075 + 95$; 3,170
321. $814 - 366$; 448
323. Tim brought 54 cans of soda to the party.
325. There were 308 students.
327. Rey donated 180 t-shirts.
329. Stephanie should use 20 cups of fruit juice.
331. There are 100 senators in the U.S. senate.
333. The area of the wall is 117 square feet.
335. The area of the room is 1,428 square feet.
337. The area of the court is 4,700 square feet.
339. Javier's portfolio gained $3,600.
341. Answers will vary.
343. fifty-four divided by nine; the quotient of fifty-four and nine
345. thirty-two divided by eight; the quotient of thirty-two and eight
347. forty-eight divided by six; the quotient of forty-eight and six
349. sixty-three divided by seven; the quotient of sixty-three and seven
351.

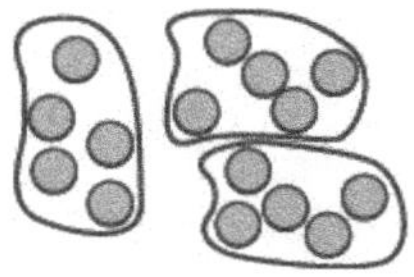

353.

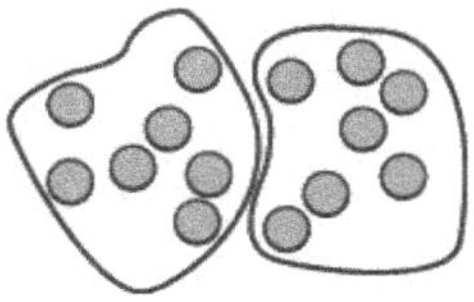

355.

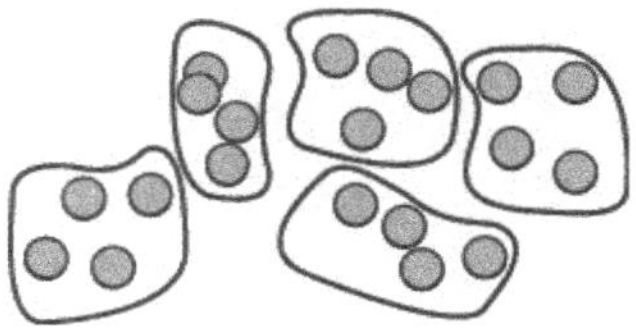

357.

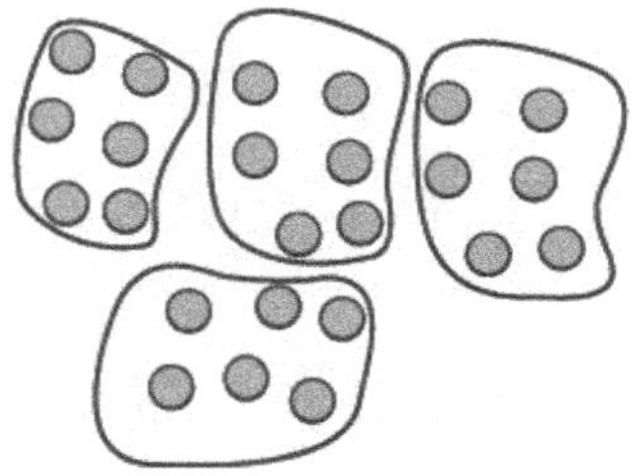

359. 9
361. 9
363. 7
365. 9
367. 9
369. 5
371. 1
373. 1
375. 23
377. 19
379. 0
381. undefined
383. undefined
385. 0
387. 24
389. 12
391. 93
393. 132
395. 871
397. 7,831
399. 2,403
401. 901
403. 704
405. 10,209
407. 352 R6
409. 6,913 R1
411. 86,234 R4
413. 43,338 R2
415. 382 R5
417. 849
419. 96
421. 1,986 R17
423. 3,060
425. 72
427. 1,060
429. 35
431. $45 \div 15$; 3
433. $288 \div 24$; 12
435. Ric can fill 32 bags.
437. There are 25 groups.
439. Marta can wrap 16 cakes from 1 roll.
441. The difference is 28 miles per gallon.
443. They will need 5 vans for the field trip
445. Bill hiked 50 miles
447. LaVonne treated 48 patients last week.

449. Jenna uses 26 pairs of contact lenses, but there is 1 day left over, so she needs 27 pairs for 365 days.

451. Answers may vary. Using multiplication facts can help you check your answers once you've finished division.

Review Exercises

453.

2, 99 0, 2, 99

455.

4, 90 0, 4, 90

457.

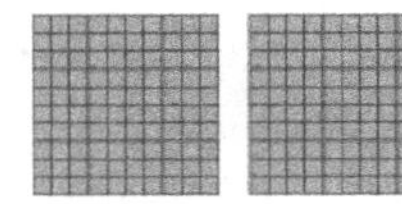

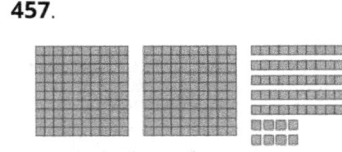

Place Value	Digit	Total Value
hundreds	2	200
tens	5	50
ones	8	8
		258

459.

tens

hundred thousands

ones ten thousands

thousands

461. Five thousand two hundred eighty

463. Five million twelve thousand five hundred eighty-two

466. 15,253

467. 340,912,061

469. 410

471. 3,560

473. 39,000

475. 81,500

477. four plus three; the sum of four and three

479. five hundred seventy-one plus six hundred twenty-nine; the sum of five hundred seventy-one and six hundred twenty-nine

481.

483.

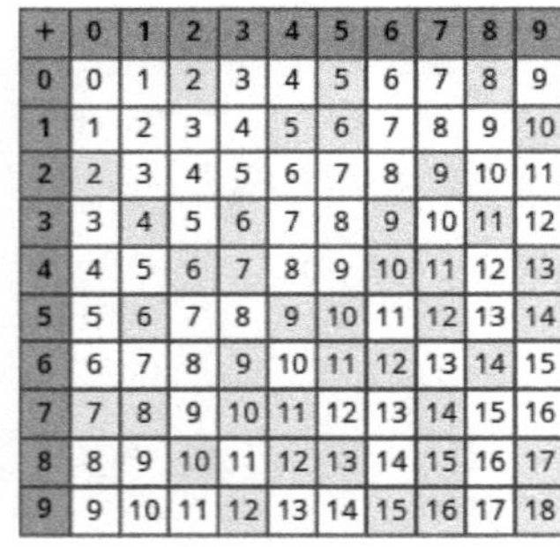

+	0	1	2	3	4	5	6	7	8	9
0	0	1	2	3	4	5	6	7	8	9
1	1	2	3	4	5	6	7	8	9	10
2	2	3	4	5	6	7	8	9	10	11
3	3	4	5	6	7	8	9	10	11	12
4	4	5	6	7	8	9	10	11	12	13
5	5	6	7	8	9	10	11	12	13	14
6	6	7	8	9	10	11	12	13	14	15
7	7	8	9	10	11	12	13	14	15	16
8	8	9	10	11	12	13	14	15	16	17
9	9	10	11	12	13	14	15	16	17	18

485.

19 19

487.

13 13

489. 79

491. 154

493. 19,827

495. 30 + 12; 42

497. 39 + 25; 64

499. $76

501. 46 feet

503. fourteen minus five; the difference of fourteen and five

505. three hundred fifty-one minus two hundred forty-nine; the difference between three hundred fifty-one and two hundred forty-nine

507.

509. 3

511. 14

513. 23

515. 322

517. 380

519. 9,985

521. 19 − 13; 6

523. 74 − 8; 66

525. 58 degrees Fahrenheit

527. eight times five the product of eight and five

529. ten times ninety-five; the product of ten and ninety-five

531.

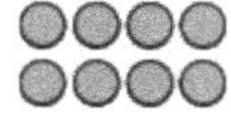

533.

×	0	1	2	3	4	5	6	7	8	9
0	0	0	0	0	0	0	0	0	0	0
1	0	1	2	3	4	5	6	7	8	9
2	0	2	4	6	8	10	12	14	16	18
3	0	3	6	9	12	15	18	21	24	27
4	0	4	8	12	16	20	24	28	32	36
5	0	5	10	15	20	25	30	35	40	45
6	0	6	12	18	24	30	36	42	48	54
7	0	7	14	21	28	35	42	49	56	63
8	0	8	16	24	32	40	48	56	64	72
9	0	9	18	27	36	45	54	63	72	81

535. 0

537. 99

539.

28 28

541. 27,783

543. 640

545. 79,866

547. 1,873,608

549. 15(28); 420

551. 2(575); 1,150

553. 48 marigolds

555. 1,800 seats

557. fifty-four divided by nine; the quotient of fifty-four and nine

559. seventy-two divided by eight; the quotient of seventy-two and eight

561.

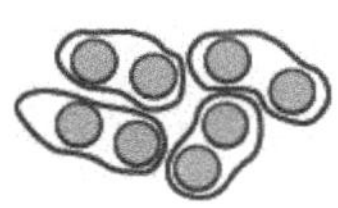

563. 7

565. 13

567. 97

569. undefined

571. 638

573. 300 R5

575. 64 ÷ 16; 4

577. 9 baskets

Practice Test

579.

4, 87 0, 4, 87

581.

613 55,208

583. 68

585. 17

587. 32

589. 0

591. 0

593. 66

595. 3,325

597. 490

599. 442

601. 11

603. 16 + 58; 74

605. 32 − 18; 14

607. 2(524); 1,048

609. 300 − 50; 250

611. Stan had $344 left.

613. Clayton walked 30 blocks.

Chapter 2

Be Prepared

2.1.

2.2.

2.3.

2.4. expression

2.5.

2.6.

2.7.

2.8.

2.9.

2.10. and

2.11.

2.12.

2.13. prime

2.13.

Try It

2.1.

18 plus 11; the sum of eighteen and eleven

27 times 9; the product of twenty-seven and nine

84 divided by 7; the quotient of eighty-four and seven

p minus q; the difference of p and q

2.2.

47 minus 19; the difference of forty-seven and nineteen

72 divided by 9; the quotient of seventy-two and nine

m plus n; the sum of m and n

13 times 7; the product of thirteen and seven

2.3.

fourteen is less than or equal to twenty-seven

nineteen minus two is not equal to eight

twelve is greater than four divided by two

x minus seven is less than one

2.4.

nineteen is greater than or equal to fifteen

seven is equal to twelve minus five

fifteen divided by three is less than eight

y minus three is greater than six

2.5.

$>$ $<$

2.6.

$<$ $>$

2.7.

equation expression

2.8.

expression equation

2.9. 41^5

2.10. 7^9

2.11.

$4 \cdot 4 \cdot 4 \cdot 4 \cdot 4 \cdot 4 \cdot 4 \cdot 4$

$a \cdot a \cdot a \cdot a \cdot a \cdot a \cdot a$

2.12.

$8 \cdot 8 \cdot 8 \cdot 8 \cdot 8 \cdot 8 \cdot 8 \cdot 8$

$b \cdot b \cdot b \cdot b \cdot b \cdot b$

2.13.

125 1

2.14.

49 0

2.15.

2 14

2.16.

35 99

2.17. 18

2.18. 9

2.19. 16

2.20. 23

2.21. 86

2.22. 1

2.23. 81

2.24. 75

2.25.

10 19

2.26.

4 12

2.27.

13 5

2.28.

8 16

2.29. 64

2.30. 216

2.31. 64

2.32. 81

2.33. 33

2.34. 10

2.35. 40

2.36. 9

2.37. The terms are $4x$, $3b$, and 2. The coefficients are 4, 3, and 2.

2.38. The terms are $9a$, $13a^2$, and a^3, The coefficients are 9, 13, and 1.

2.39. 9 and 15; $2x^3$ and $8x^3$; y^2 and $11y^2$

2.40. $4x^3$ and $6x^3$; $8x^2$ and $3x^2$; 19 and 24

2.41. $16x + 17$

2.42. $17y + 7$

2.43. $4x^2 + 14x$

2.44. $12y^2 + 15y$

2.45.

$47 - 41$ $5x \div 2$

2.46.

$17 + 19$ $7x$

2.47.

$x + 11$ $11a - 14$

2.48.

$j + 19$ $2x - 21$

2.49.

$4(p + q)$ $4p + q$

2.50.

$2x - 8$ $2(x - 8)$

2.51. $w - 5$

2.52. $l + 2$
2.53. $6q - 7$
2.54. $4n + 8$
2.55. no
2.56. yes
2.57. yes
2.58. yes
2.59. $x + 1 = 7; x = 6$
2.60. $x + 3 = 4; x = 1$
2.61. $x = 13$
2.62. $x = 5$
2.63. $y = 28$
2.64. $y = 46$
2.65. $x = 22$
2.66. $y = 4$
2.67. $a = 37$
2.68. $n = 41$
2.69. $7 + 6 = 13$
2.70. $8 + 6 = 14$
2.71. $6 \cdot 9 = 54$
2.72. $21 \cdot 3 = 63$
2.73. $2(x - 5) = 30$
2.74. $2(y - 4) = 16$
2.75. $x + 7 = 37; x = 30$
2.76. $y + 11 = 28; y = 17$
2.77. $z - 17 = 37; z = 54$
2.78. $x - 19 = 45; x = 64$
2.79.

yes no

2.80.

no yes

2.81.

yes no

2.82.

no yes

2.83.

no yes

2.84.

yes no

2.85.

yes no

2.86.

no yes

2.87. Divisible by 2, 3, 5, and 10
2.88. Divisible by 2 and 3, not 5 or 10.
2.89. Divisible by 2, 3, not 5 or 10.
2.90. Divisible by 3 and 5.
2.91. 1, 2, 3, 4, 6, 8, 12, 16, 24, 32, 48, 96
2.92. 1, 2, 4, 5, 8, 10, 16, 20, 40, 80
2.93. composite
2.94. prime
2.95. $2 \cdot 2 \cdot 2 \cdot 2 \cdot 5$, or $2^4 \cdot 5$
2.96. $2 \cdot 2 \cdot 3 \cdot 5$, or $2^2 \cdot 3 \cdot 5$
2.97. $2 \cdot 3 \cdot 3 \cdot 7$, or $2 \cdot 3^2 \cdot 7$
2.98. $2 \cdot 3 \cdot 7 \cdot 7$, or $2 \cdot 3 \cdot 7^2$
2.99. $2 \cdot 2 \cdot 2 \cdot 2 \cdot 5$, or $2^4 \cdot 5$
2.100. $2 \cdot 2 \cdot 3 \cdot 5$, or $2^2 \cdot 3 \cdot 5$
2.101. $2 \cdot 3 \cdot 3 \cdot 7$, or $2 \cdot 3^2 \cdot 7$
2.102. $2 \cdot 3 \cdot 7 \cdot 7$, or $2 \cdot 3 \cdot 7^2$
2.103. 36
2.104. 72
2.105. 60
2.106. 105
2.107. 440
2.108. 360

Section Exercises

1. 16 minus 9, the difference of sixteen and nine
3. 5 times 6, the product of five and six
5. 28 divided by 4, the quotient of twenty-eight and four
7. x plus 8, the sum of x and eight
9. 2 times 7, the product of two and seven
11. fourteen is less than twenty-one
13. thirty-six is greater than or equal to nineteen
15. 3 times n equals 24, the product of three and n equals twenty-four
17. y minus 1 is greater than 6, the difference of y and one is greater than six
19. 2 is less than or equal to 18 divided by 6; 2 is less than or equal to the quotient of eighteen and six
21. a is not equal to 7 times 4, a is not equal to the product of seven and four
23. equation
25. expression
27. expression
29. equation
31. 3^7
33. x^5
35. 5x5x5
37. 2x2x2x2x2x2x2x2
39.

43 55

41. 5
43. 34
45. 58
47. 6
49. 13
51. 4
53. 35
55. 10
57. 41
59. 81
61. 149
63. 50
69. 22
71. 26
73. 144
75. 32
77. 27
79. 21
81. 41
83. 9
84. 225
85. 73
87. 54
89. $15x^2, 6x, 2$
91. $10y^3, y, 2$
93. 8
95. 5
97. x^3 and $8x^3$; 14 and 5
99. $16ab$ and $4ab$; $16b^2$ and $9b^2$
101. $13x$
103. $26a$
105. $7c$
107. $12x + 8$
109. $10u + 3$
111. $12p + 10$
113. $22a + 1$
115. $17x^2 + 20x + 16$
117. $8 + 12$
119. $14 - 9$
121. $9 \cdot 7$

123. $36 \div 9$
125. $x - 4$
127. $6y$
129. $8x + 3x$
131. $y \div 3$
133. $8(y - 9)$
135. $5(x + y)$
137. $b + 15$
139. $b - 4$
141. $2n - 7$
143. He will pay $750. His insurance company will pay $1350.
147.

yes no

149.

no yes

151.

yes no

153.

no yes

155.

no yes

157.

no yes

159. $x + 2 = 5$; $x = 3$
161. $x + 3 = 6$; $x = 3$
163. $a = 16$
165. $p = 5$
167. $r = 24$
169. $x = 7$
171. $p = 69$
173. $d = 67$
175. $y = 22$
177. $u = 30$
179. $f = 178$
181. $n = 32$
183. $p = 48$
185. $y = 467$
187. $8 + 9 = 17$
189. $23 - 19 = 4$
191. $3 \cdot 9 = 27$
193. $54 \div 6 = 9$
195. $2(n - 10) = 52$
197. $3y + 10 = 100$
199. $p + 5 = 21$; $p = 16$
201. $r + 18 = 73$; $r = 55$
203. $d - 30 = 52$; $d = 82$
205. $u - 12 = 89$; $u = 101$
207. $c - 325 = 799$; $c = 1124$
209. $1300
211. $460
215. 2, 4, 6, 8, 10 12, 14, 16, 18, 20, 22, 24, 26, 28, 30, 32, 34, 36, 38, 40, 42, 44, 46, 48
217. 4, 8, 12, 16, 20, 24, 28, 32, 36, 40, 44, 48
219. 6, 12, 18, 24, 30, 36, 42, 48
221. 8, 16, 24, 32, 40, 48
223. 10, 20, 30, 40
225. Divisible by 2, 3, 4, 6
227. Divisible by 3, 5
229. Divisible by 2, 3, 4, 6
231. Divisible by 2, 3, 4, 5, 6, 10
233. Divisible by 2, 4
235. Divisible by 3, 5
237. Divisible by 2, 5, 10
239. Divisible by 2, 5, 10
241. Divisible by 3, 5
243. 1, 2, 3, 4, 6, 9, 12, 18, 36
245. 1, 2, 3, 4, 5, 6, 10, 12, 15, 20, 30, 60
247. 1, 2, 3, 4, 6, 8, 9, 12, 16, 18, 24, 36, 48, 72,144
249. 1, 2, 3, 4, 6, 7, 12, 14, 21, 28, 42, 49, 84, 98, 147, 196, 294, 588
251. prime
253. composite
255. prime
257. composite
259. composite
261. composite
263.

Weeks after graduation	Total number of dollars Frank put in the account	Simplified Total
0	100	100
1	100 + 15	115
2	100 + 15 • 2	130
3	100 + 15 • 3	145
4	100 + 15 • 4	160
5	100 + 15 • 5	175
6	100 + 15 • 6	190
20	100 + 15 • 20	400
x	100 + 15 • x	100 + 15x

267. $2 \cdot 43$
269. $2 \cdot 2 \cdot 3 \cdot 11$
271. $3 \cdot 3 \cdot 7 \cdot 11$
273. $5 \cdot 23$
275. $3 \cdot 3 \cdot 5 \cdot 5 \cdot 11$
277. $2 \cdot 2 \cdot 2 \cdot 7$
279. $2 \cdot 2 \cdot 2 \cdot 3 \cdot 7$
281. $17 \cdot 23$
283. $2 \cdot 2 \cdot 2 \cdot 2 \cdot 3 \cdot 3 \cdot 3$
285. $2 \cdot 2 \cdot 2 \cdot 2 \cdot 3 \cdot 3 \cdot 3 \cdot 5$
287. $2 \cdot 3 \cdot 5 \cdot 5$
289. $3 \cdot 5 \cdot 5 \cdot 7$
291. $2 \cdot 2 \cdot 3 \cdot 3$
293. $2 \cdot 5 \cdot 5 \cdot 7$
295. 24
297. 30
299. 120
301. 300
303. 24
305. 120
307. 420
309. 42
311. 120
313. 40

Review Exercises

317. 3 times 8, the product of three and eight.
319. 24 divided by 6, the quotient of twenty-four and six.
321. 50 is greater than or equal to 47

323. The sum of n and 4 is equal to 13

325. equation

327. expression

329. 2^3

331. x^6

333. $8 \cdot 8 \cdot 8 \cdot 8$

335. $y \cdot y \cdot y \cdot y \cdot y$

337. 81

339. 128

341. 20

343. 18

345. 74

347. 31

349. 58

351. 26

353. $12n^2, 3n, 1$

355. 6

357. 3 and 4; $3x$ and x

359. $24a$

361. $14x$

363. $12n + 11$

365. $10y^2 + 2y + 3$

367. $x - 6$

369. $3n \cdot 9$

371. $5(y + 1)$

373. $c + 3$

375.

yes no

377.

yes no

379.

no yes

381. $x + 3 = 5$; $x = 2$

383. $c = 6$

385. $x = 11$

387. $y = 23$

389. $p = 34$

391. $7 + 33 = 44$

393. $4 \cdot 8 = 32$

395. $2(n - 3) = 76$

397. $x + 8 = 35$; $x = 27$

399. $q - 18 = 57$; $q = 75$

401. $h = 42$

403. $z = 33$

405. $q = 8$

407. $v = 56$

409. 3, 6, 9, 12, 15, 18, 21, 24, 27, 30, 33, 36, 39, 42, 45, 48

411. 8, 16, 24, 32, 40, 48

413. 2, 3, 6

415. 2, 3, 5, 6, 10

417. 1, 2, 3, 5, 6, 10, 15, 30

419. 1, 2, 3, 4, 5, 6, 9, 10, 12, 15, 18, 20, 30, 36, 45, 60, 90, 180

421. prime

423. composite

425. $2 \cdot 2 \cdot 3 \cdot 7$

427. $2 \cdot 5 \cdot 5 \cdot 7$

429. 45

431. 175

433. Answers will vary

Practice Test

435. 15 minus x, the difference of fifteen and x.

437. equation

439.

n^6 $3 \cdot 3 \cdot 3 \cdot 3 \cdot 3 = 243$

441. 36

443. 5

445. 45

447. 125

449. 36

451. $x + 5$

453. $3(a - b)$

455. $n = 31$

457. $y - 15 = 32$; $y = 47$

459. 4, 8, 12, 16, 20, 24, 28, 32, 36, 40, 44, 48

461. $2^3 \cdot 3^3 \cdot 5$

Chapter 3

Be Prepared

3.1.

3.2.

3.3.

3.3.

3.3.

3.4.

3.5.

3.6.

3.7.

3.8.

3.9.

3.10.

3.11.

3.12.

Try It

3.1.

c b a

−4 −3 −2 −1 0 1 2 3 4

3.2.

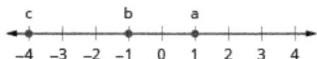

a c b

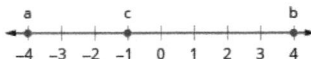

−4 −3 −2 −1 0 1 2 3 4

3.3.

> < > >

3.4.

< > < >

3.5.

−4 3

3.6.

−8 5

3.7. 1

3.8. 5

3.9.

−4 4

3.10.

−11 11

3.11.

12 −28

3.12.

9 −37

3.13.

17 39 −22

−11

3.14.

23 21 −37

−49

3.15.

> > < =

3.16.

> = > <

3.17.

3 18

3.18.

11 63

3.19. 2

3.20. 3

3.21. 16

3.22. 9

3.23.

−9 15 −20

11−(−4)

3.24.

19 −22 −9

−8−(−5)

3.25. 5 yards

3.26. −30 feet

3.27.

6

3.28.

7

3.29.

-6

3.30.

-7

3.31.

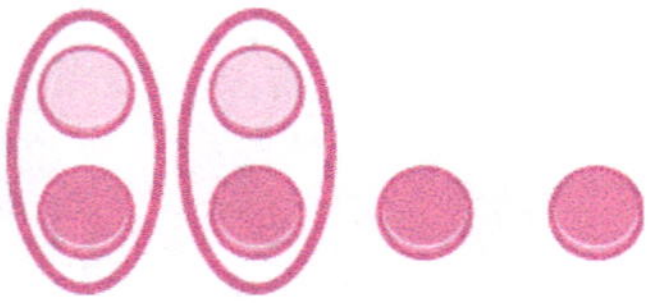

-2

3.32.

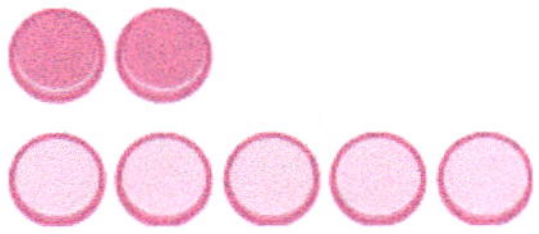

-3

3.33.

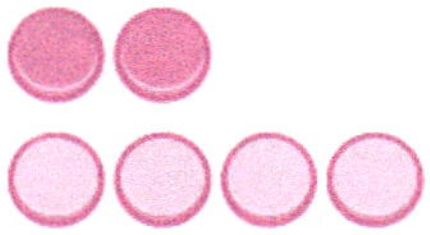

2

3.34.

3

3.35.

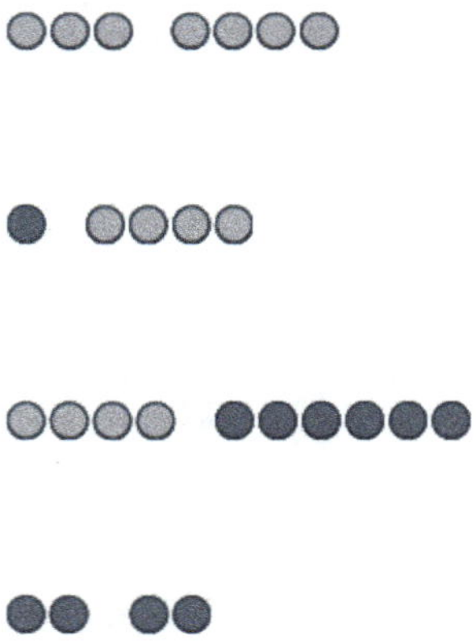

3.36.

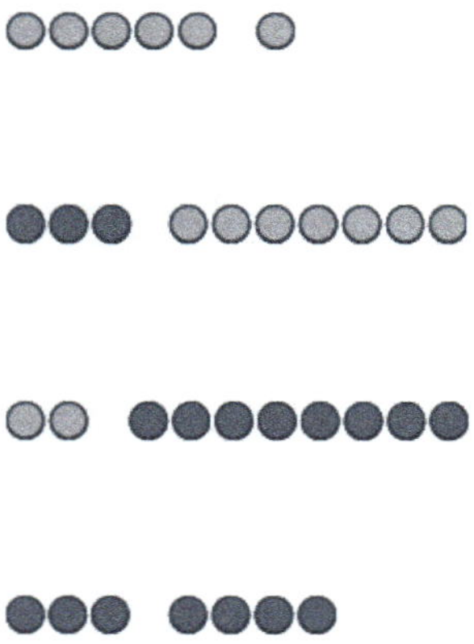

3.37.

−17 57

3.38.

−46 26

3.39. −50

3.40. −70

3.41. 13

3.42. 0

3.43.

2 −12

3.44.

2 −1

3.45.

−6 10

3.46.

−1 17

3.47. 9

3.48. 13

3.49. 196

3.50. 8

3.51. $-7 + 4 = -3$

3.52. $-8 + (-6) = -14$

3.53. $[9 + (-16)] + 4 = -3$

3.54. $[-8 + (-12)] + 7 = -13$

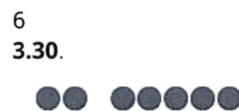

3.55. 4 degrees Celsius

3.56. −33 feet

3.57. 14-yard line

3.58. 37-yard line

3.59.

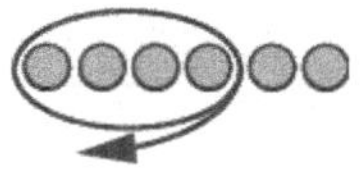

2

3.60.

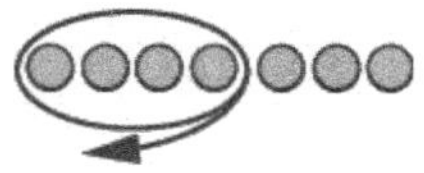

3

3.61.

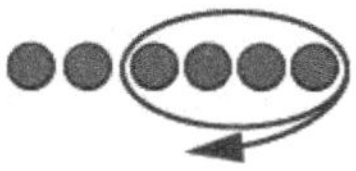

−2

3.62.

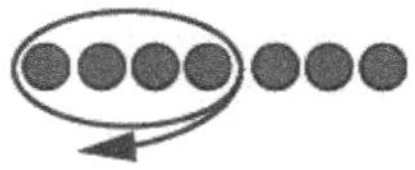

−3

3.63.

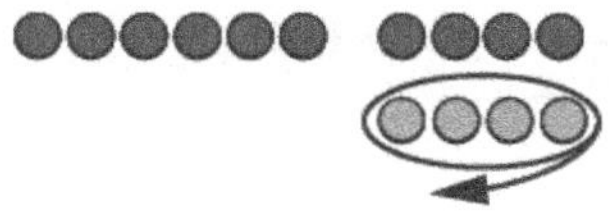

−10

3.64.

−11

3.65.

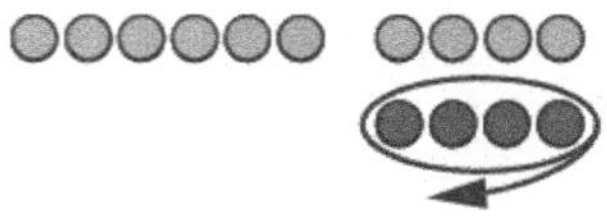

10

3.66.

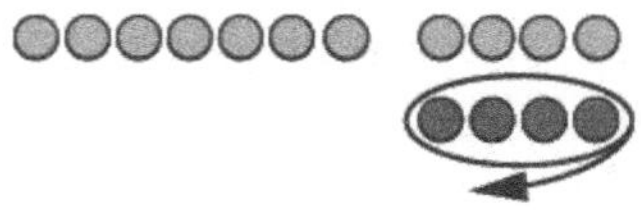

11

3.67.

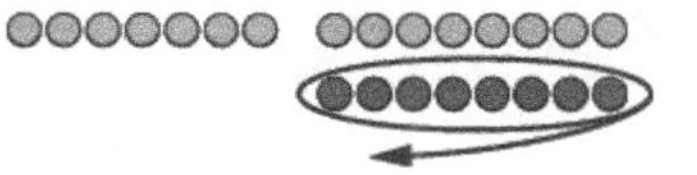

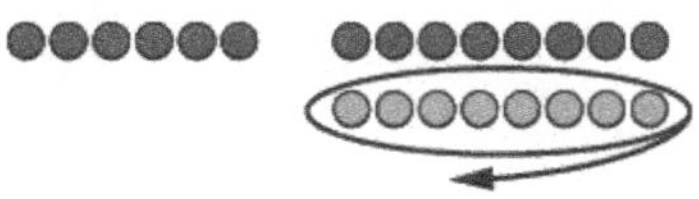

3.68.

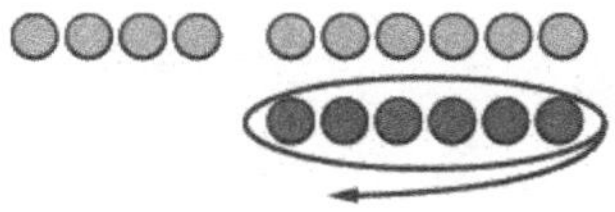

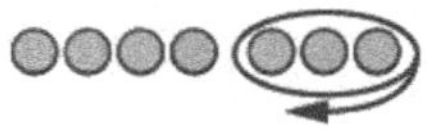

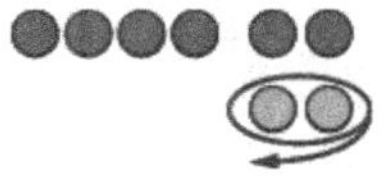

3.69.

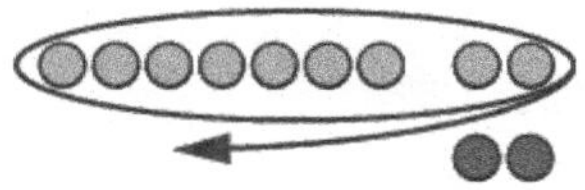

−2

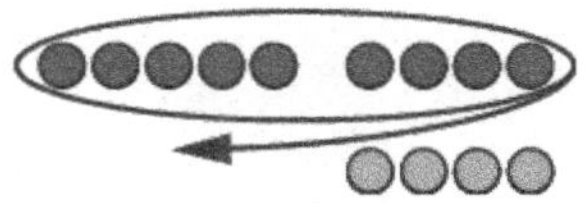

4

3.70.

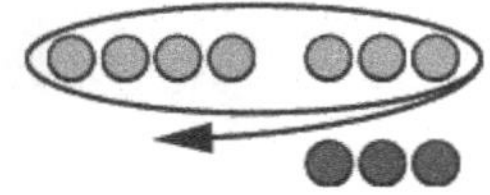

−3

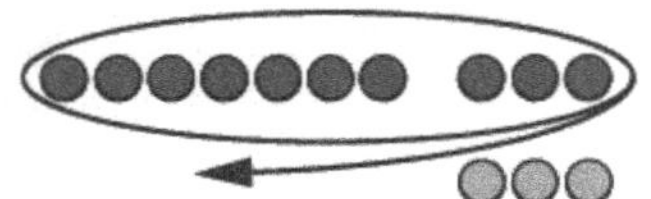

3

3.71.

8, 8 −18, −18

3.72.

8, 8 −22, −22

3.73.

19, 19 −4, −4

3.74.

23, 23 3, 3

3.75. −29

3.76. −26

3.77. 3

3.78. 13

3.79. −69

3.80. −47

3.81.

−2 −15

3.82.

−2 −7

3.83.

−2 36

3.84.

−19 9

3.85.

$-14 - (-23) = 37$

$-17 - 21 = -38$

3.86.

$11 - (-19) = 30$

$-11 - 18 = -29$

3.87. 45 degrees Fahrenheit

3.88. 9 degrees Fahrenheit

3.89. 10,103 feet

3.90. 233 feet

3.91.

$48 −$2 $18

3.92.

−$54 No, −$5

3.93.

−48 28 −63

60

3.94.

−56 54 −28

39

3.95.

−9 17

3.96.

−8 16

3.97.

−7 39

3.98.

−9 23

3.99.

−6 36

3.100.

−28 52

3.101. −63

3.102. −84

3.103.

81 −81

3.104.

49 −49

3.105. 29

3.106. 52

3.107. 4

3.108. 9

3.109. 21

3.110. 6

3.111. 39

3.112. 13

3.113. −8

3.114. 19

3.115. $-5\,(12) = -60$

3.116. $8\,(-13) = -104$

3.117. $-63 \div -9 = 7$

3.118. $-72 \div -9 = 8$

3.119.

no no yes

3.120.

no no yes

3.121. −4

3.122. −19

3.123. −6

3.124. −4

3.125. $4x = 12$; $x = 3$

3.126. $3x = 6$; $x = 2$

3.127. 7

3.128. 11

3.129. −12

3.130. −9

3.131. $x + 7 = -2$; $x = -9$

3.132. $y + 11 = 2$; $y = -9$

3.133. $p - 2 = -4$; $p = -2$

3.134. $q - 7 = -3$; $q = 4$

3.135. $132 = -12y$; $y = -11$

3.136. $117 = -13z$; $z = -9$

Section Exercises

1.

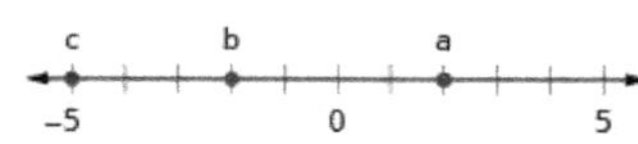

3.

5.

> < < >

7.

< > < >

9.

−2 6

11.

8 −1

13. 4

15. 15

17.

−3 3

19.

−12; 12

21.

7 25 0

23.

32 18 16

25.

28 15

27.

−19 −33

29.

< =

31.

> >

33. 4

35. 56

37. 0

39. 8

41. 80

43.

−8 −(−6), or 6 −3

4−(−3)

45.

−20 −(−5), or 5

−12 18−(−7)

47. −6 degrees

49. −40 feet

51. −12 yards

53. $3

55. +1

57.

20,320 feet −282 feet

59.

$540 million

−$27 billion

61. Sample answer: I have experienced negative temperatures.

63.

11

65.

-9

67.

-2

69.

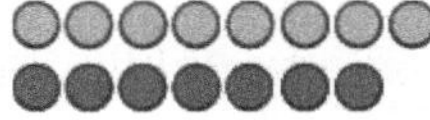

1

71. -80

73. 32

75. -135

77. 0

79. -22

81. 108

83. -4

85. 29

87.
−18 −87

89.
−47 16

91.
−4 10

93.
−13 5

95. -8

97. 10

99. 64

101. 121

103. -14 + 5 = -9

105. -2 + 8 = 6

107. -15 + (-10) = -25

109. [-1 + (-12)] + 6 = -7

111. [10 + (-19)] + 4 = -5

113. 7°F

115. -$118

117. -8 pounds

119. 25-yard line

121. 20 calories

123. -32

125. Sample answer: In the first case, there are more negatives so the sum is negative. In the second case, there are more positives so the sum is positive.

127.

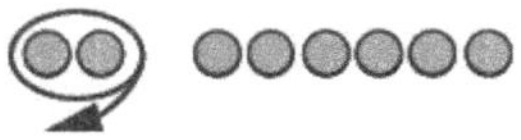

6

129.

−4

131.

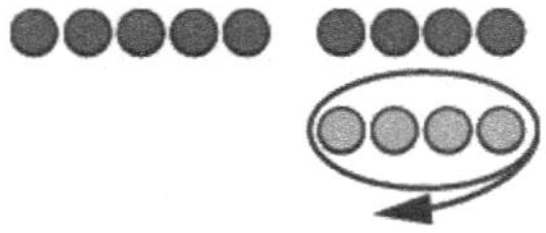

−9

133.

12

135.

9 9

137.

16 16

139.

17 17

141.

45 45

143. 27

145. 29

147. −39

149. −48

151. −42

153. −59

155. −51

157. 9

159. −2

161. −2

163. 22

165. 53

167. −20

169. 0

171. 4

173. 6

175. 5.2

177. −11

179.

−3 −9

181.

3 7

183. −8

185. −192

187.

$3 - (-10) = 13$

$45 - (-20) = 65$

189.

$-6 - 9 = -15$

$-16 - (-12) = -4$

191.

$-17 - 8 = -25$

$-24 - 37 = -61$

193.

$6 - 21 = -15$

$-19 - 31 = -50$

195. −10°

197. 96°

199. 21-yard line

201. $65

203. −$40

205. $26

207. 13°

209. Sample answer: On a number line, 9 is 15 units away from −6.

211. −32

213. −35

215. 36

217. −63

219. −6

221. 14

223. −4

225. −8

227. 13

229. −12

231. −49

233. −47

235. 43

237. −125

239. 64

241. −16

243. 90

245. −88

247. 9

249. 41

251. −5

253. −9

255. −29

257. 5

259.

1 33

261.

−5 25

263. 11

265. 21

267. 38

269. −56

271. $-3 \cdot 15 = -45$

273. $-60 \div (-20) = 3$

275. ———

277. $-10(p-q)$

279. −$3,600

281. Sample answer: Multiplying two integers with the same sign results in a positive product. Multiplying two integers with different signs results in a negative product.

283. Sample answer: In the first expression the base is positive and after you raise it to the power you should take the opposite. Then in the second expression the base is negative so you simply raise it to the power.

285.

no no yes

287.

no no yes

289. $n=-7$

291. $p=-17$

293. $u=-4$

295. $h=6$

297. $x=-16$

299. $r=-14$

301. $3x=6$; $x=2$

303. $2x=8$; $x=4$

305. $x=9$

307. $c=-8$

309. $p=3$

311. $q=-12$

313. $x=20$

315. $z=0$

317. $n+4=1$; $n=-3$

319. $8+p=-3$; $p=-11$

321. $a-3=-14$; $a=-11$

323. $-42=-7x$; $x=6$

325. $-15f=75$; $f=-5$

327. $-6+c=4$; $c=10$

329. $m-9=-4$; $m=5$

331.

$x=8$ $x=5$

333.

$p=-9$ $p=30$

335. $a=20$

337. $m=7$

339. $u=-52$

341. $r=-9$

343. $d=5$

345. $x=-42$

347. 17 cookies

349. Sample answer: It is helpful because it shows how the counters can be divided among the envelopes.

351. Sample answer: The operation used in the equation is multiplication. The inverse of multiplication is division, not addition.

Review Exercises

353.

355.

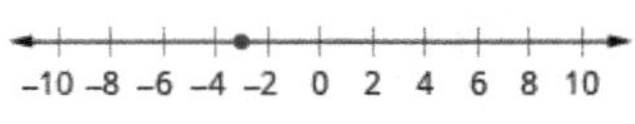

357.

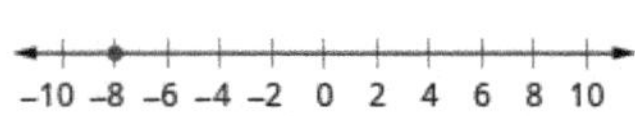

359. <

361. >

363. >

365. −6

367. 4

369.

−8 8

371.

−32 32

373. 21

375. 36

377. 0

379. 14

381. −33

383. <

385. =

387. 55; −55

389. 7

391. 54

393. −1

395. −16

397. −3

399. −10°

401. 10

403. 1

405. 96

407. −50

409. −1

411. 21

413.

3 −16

415. −27

417. $-8+2=-6$

419. $10+[-5+(-6)]=-1$

421. 16 degrees

423.

5

425.

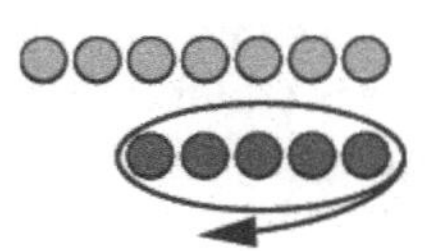

7

427. 8

429. −38

431. −58

433. −1

435.

−2 −11

437. 41

439. −12 − 5 = −17

441. −2 degrees

443. −36

445. 121

447. −7

449. −8

451. −45

453. −9

455. −81

457. 54

459. 4

461. −66

463. −58

465. −12(6) = −72

467.

no yes no

469. −12

471. −7

473. $3x = 9$; $x = 3$

475. 9

477. 4

479. $-6y = -42$; $y = 7$

481. $m + 4 = -48$; $m = -52$

483. Answers will vary.

Practice Test

485.

< >

487.

7 −8

489. 5

491. −27

493. 11

495. 54

497. −8

499. 22

501. 39

503. 34

505. −7 − (−4) = −3

507. 4°F

509. $n = -1$

511. $r = 6$

513. $y - 8 = -32$; $y = -24$

Chapter 4

Be Prepared

4.1.

4.2.

4.3.

4.4.

4.5. Answers may vary. Acceptable answers include

— — — etc.

4.6.

4.7. —

4.8. —

4.9.

4.10.

4.11. —

4.12. — —

4.13.

4.14.

4.15. —

4.16. —

4.17.

4.18.

4.19.

Try It

4.1.

— —

4.2.

— —

4.3.

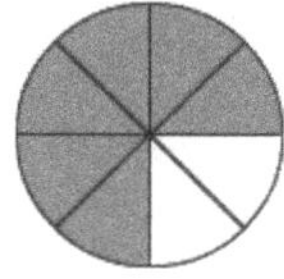

4.4. 

4.5.

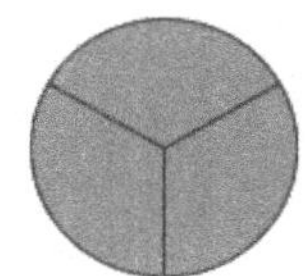

4.6.

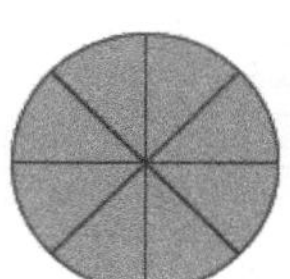

4.7.

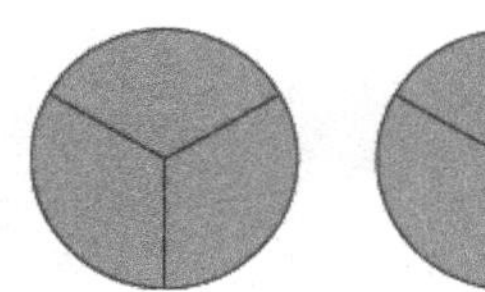

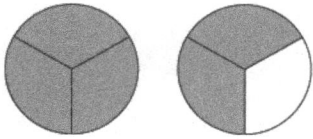

4.8.

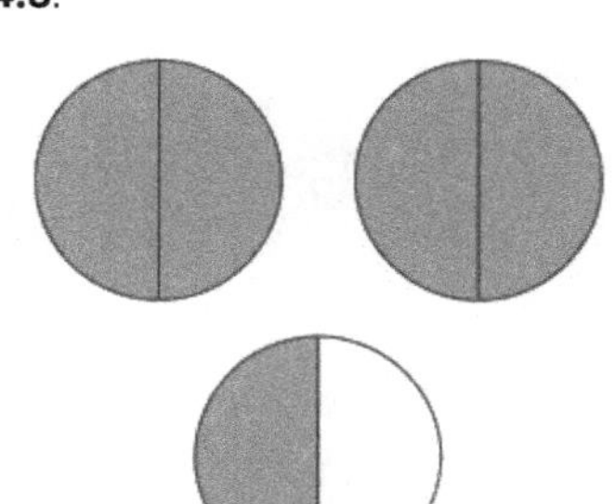

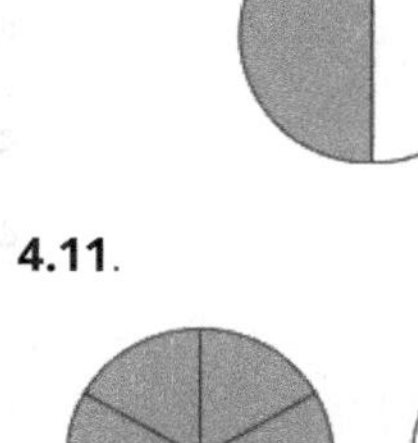

4.9.

4.10.

4.11.

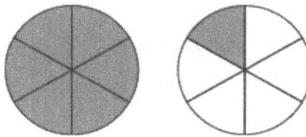

4.12.

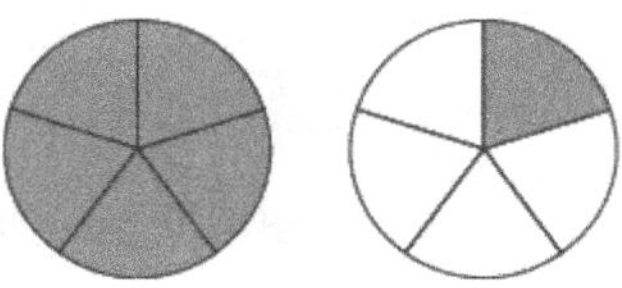

4.13.

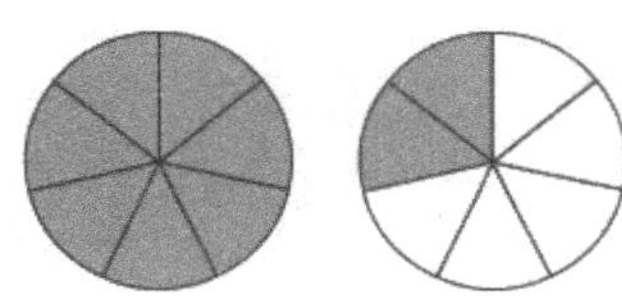

4.14.

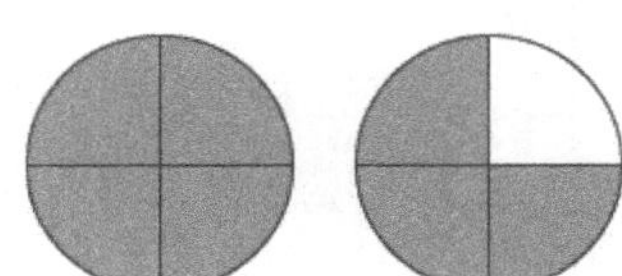

4.15.

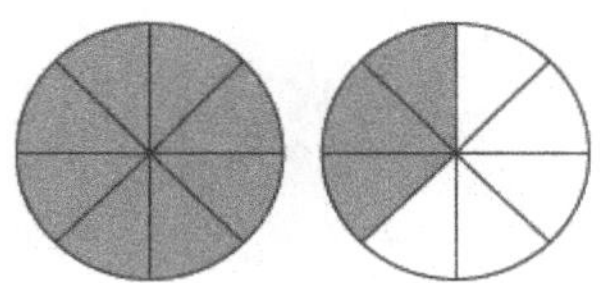

4.16.

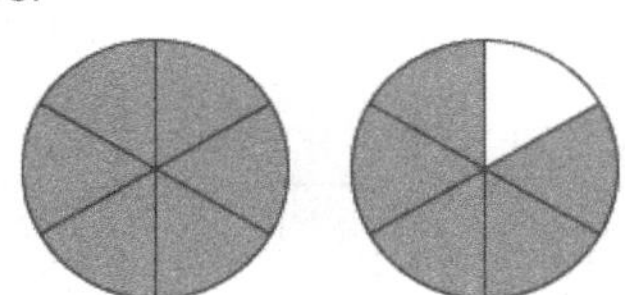

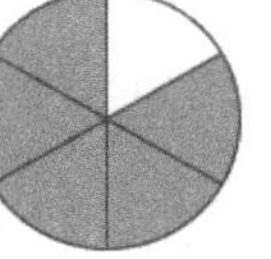

4.17.

4.18.

4.19. 

4.20.

4.21.

4.22. 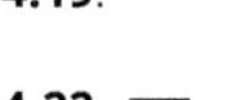

4.23.

4.24.

4.25. 2

4.26. 3

4.27. Correct answers include

4.28. Correct answers include

4.29. 

4.30.

4.31.

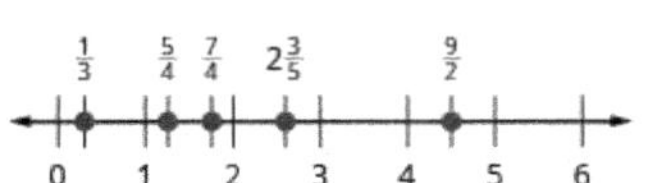

4.32.

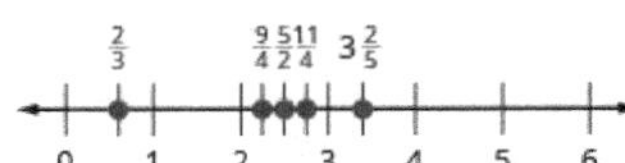

4.33.

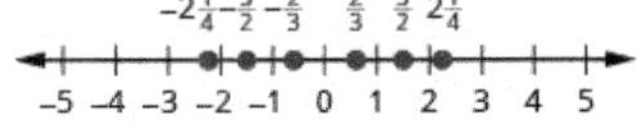

4.34. 

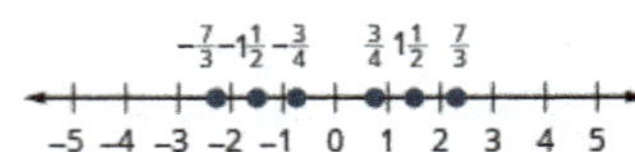

4.35.

\> > < <

4.36.

\> < > <

4.37. —

4.38. —

4.39. —

4.40. —

4.41. —

4.42. —

4.43. —

4.44. —

4.45. —

4.46. —

4.47. —

4.48. —

4.49. —

4.50. —

4.51. —

4.52. —

4.53. —

4.54. —

4.55.

9 $-33a$

4.56.

—

4.57.

— —

—

4.58.

— —

—

4.59.

Number	Opposite	Absolute Value	Reciprocal
$-\frac{5}{8}$	$\frac{5}{8}$	$\frac{5}{8}$	$-\frac{8}{5}$
$\frac{1}{4}$	$-\frac{1}{4}$	$\frac{1}{4}$	4
$\frac{8}{3}$	$-\frac{8}{3}$	$\frac{8}{3}$	$\frac{3}{8}$
−8	8	8	$-\frac{1}{8}$

4.60.

Number	Opposite	Absolute Value	Reciprocal
$-\frac{4}{7}$	$\frac{4}{7}$	$\frac{4}{7}$	$-\frac{7}{4}$
$\frac{1}{8}$	$-\frac{1}{8}$	$\frac{1}{8}$	8
$\frac{9}{4}$	$-\frac{9}{4}$	$\frac{9}{4}$	$\frac{4}{9}$
−1	1	1	−1

4.61. 2

$\frac{1}{3}$
$\frac{1}{6}$ $\frac{1}{6}$

4.62. 2

$\frac{1}{2}$
$\frac{1}{4}$ $\frac{1}{4}$

4.63. 6

1 1
$\frac{1}{3}$ $\frac{1}{3}$ $\frac{1}{3}$ $\frac{1}{3}$ $\frac{1}{3}$ $\frac{1}{3}$

4.64. 6

1 1 1
$\frac{1}{2}$ $\frac{1}{2}$ $\frac{1}{2}$ $\frac{1}{2}$ $\frac{1}{2}$ $\frac{1}{2}$

4.65. —

4.66. —

4.67. —

4.68. —

4.69. —

4.70. —

4.71. —

4.72. —

4.73. 2

4.74. —

4.75. −15

4.76. —

4.77. —

4.78. —

4.79. 2

4.80. —

4.81. —

4.82. —

4.83. —

4.84. —

4.85. —

4.86. —

4.87. —

4.88. —

4.89. —

4.90. —

4.91.

4.92.

4.93.

4.94.

4.95.

4.96.

4.97.

4.98.

4.99.

4.100.

4.101. 4

4.102. 2

4.103.

4.104.

4.105.

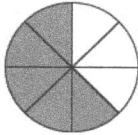

4.106. 1

4.107.

4.108.

4.109.

4.110.

4.111.

4.112.

4.113.

4.114.

4.115. , models may differ.

4.116. , models may differ

4.117.

4.118.

4.119.

4.120.

4.121.

4.122.

4.123. -1

4.124.

4.125. 60

4.126. 15

4.127. 96

4.128. 224

4.129.

4.130.

4.131.

4.132.

4.133.

4.134.

4.135.

4.136.

4.137.

4.138.

4.139.

4.140.

4.141.

4.142.

4.143.

4.144.

4.145.

4.146.

4.147.

4.148.

4.149.

4.150. 272

4.151. 2

4.152.

4.153.

4.154.

4.155.

4.156.

4.157.

4.158.

4.159.

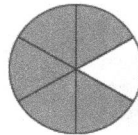

4.160. —

4.161. 5

4.162. 5

4.163. —

4.164. —

4.165. —

4.166. —

4.167. —

4.168. —

4.169. —

4.170. —

4.171. —

4.172. —

4.173. —

4.174. —

4.175. —

4.176. —

4.177. —

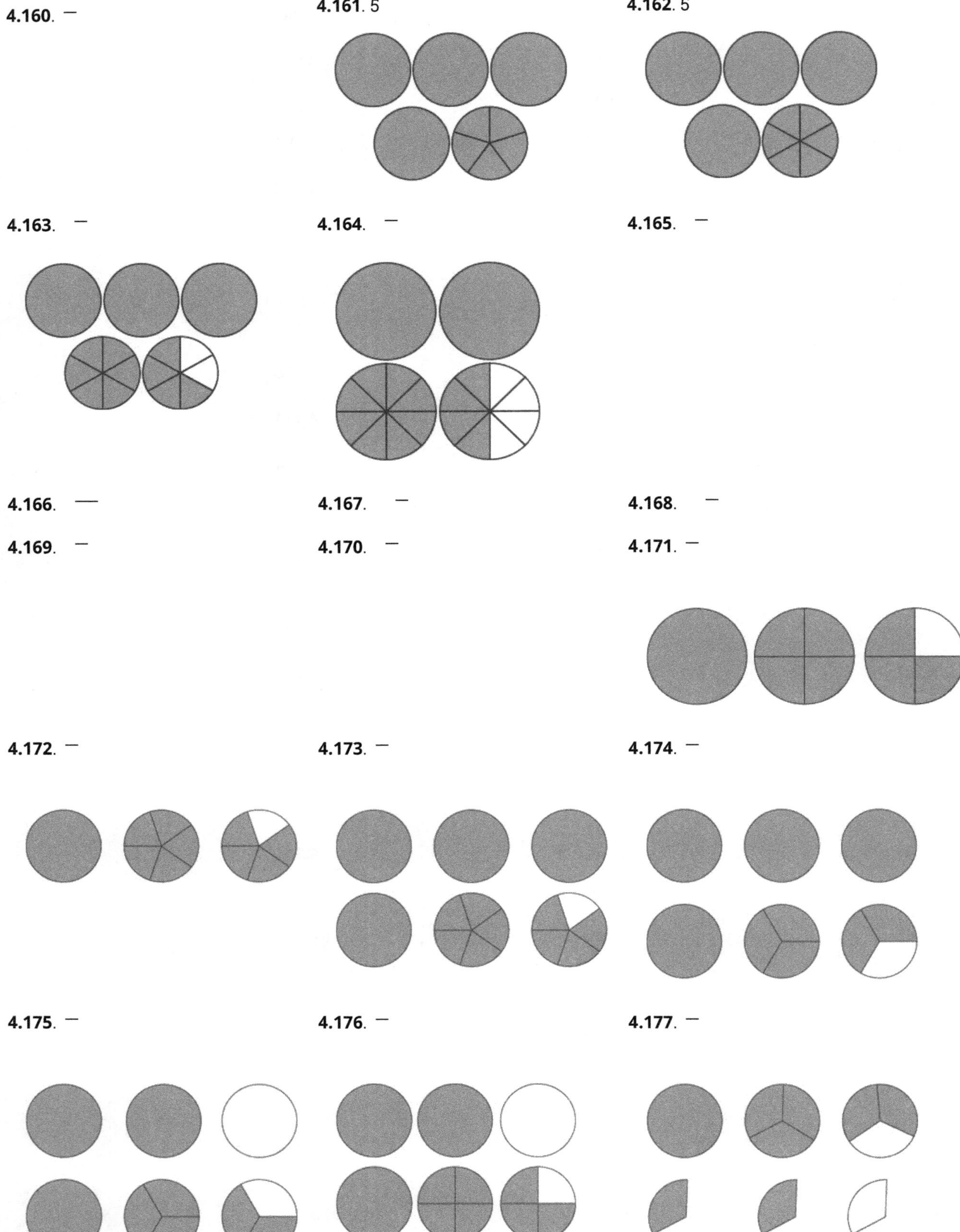

4.178. —

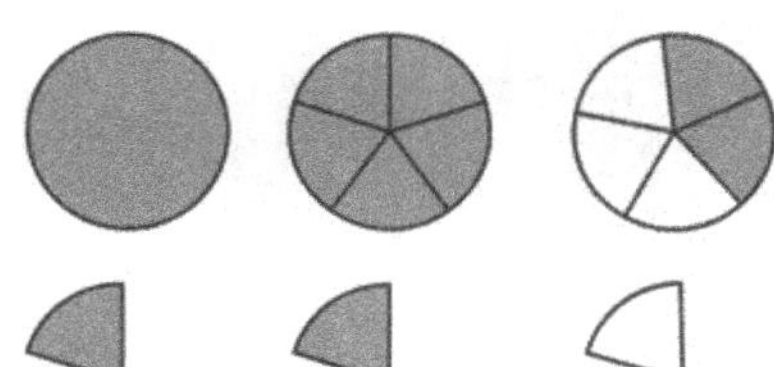

4.179. —

4.180. —

4.181. —

4.182. —

4.183. —

4.184. —

4.185. —

4.186. —

4.187. —

4.188. —

4.189.
no yes no

4.190.
no no yes

4.191. —

4.192. —

4.193. -1

4.194. —

4.195. —

4.196. —

4.197. -125

4.198. -243

4.199. 245

4.200. 132

4.201. -48

4.202. 23

4.203. 35

4.204. 18

4.205. -91

4.206. -108

4.207. —

4.208. —

4.209. —

4.210. —

4.211. —

4.212. —

4.213. — — —

4.214. — — —

4.215. — — —

4.216. — — —

Section Exercises

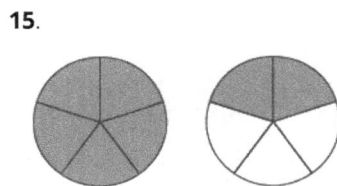

1.
— — — —

3.

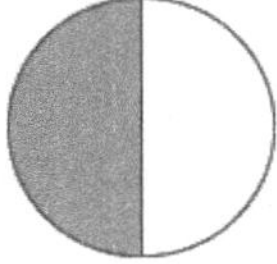

5.

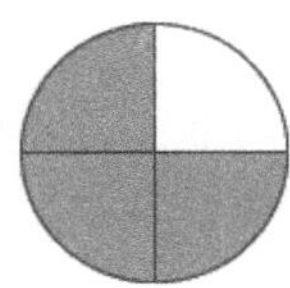

7.

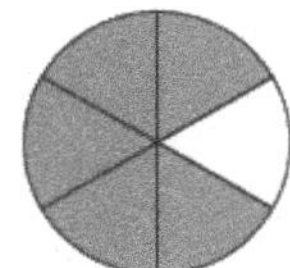

9.

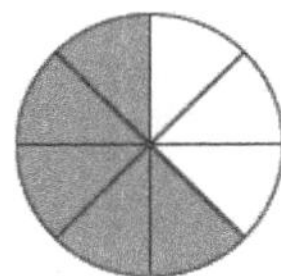

11.

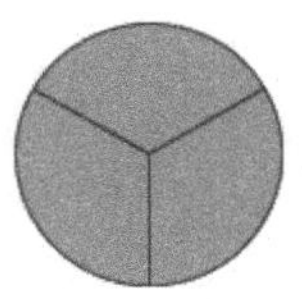

13.

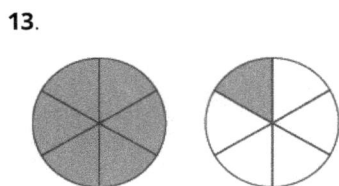

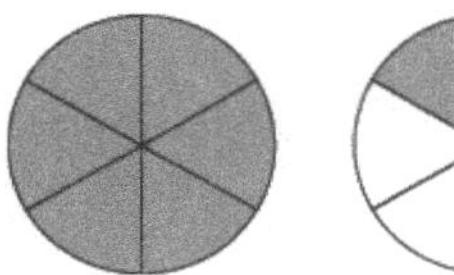

15.

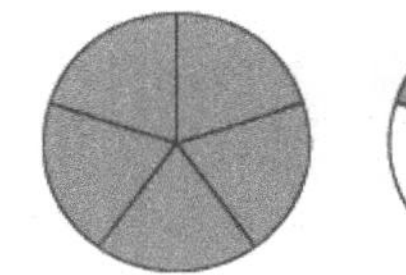

17.
— — — —
— —

19. — — — —

21.

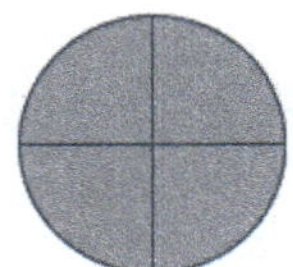

23.

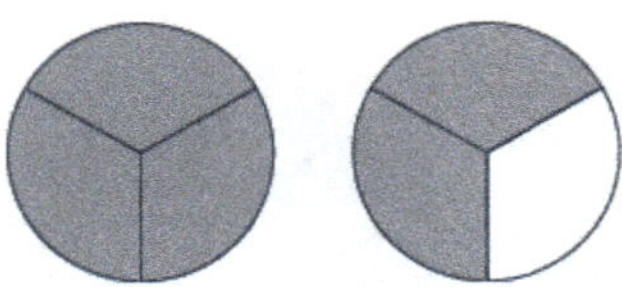

25.

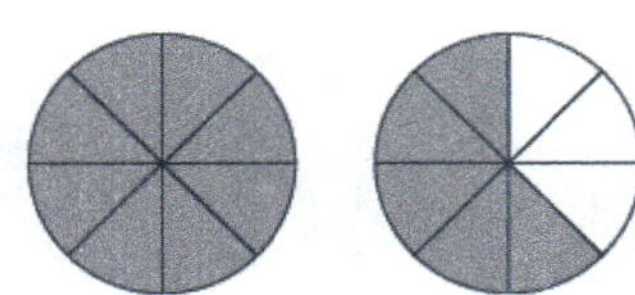

27.

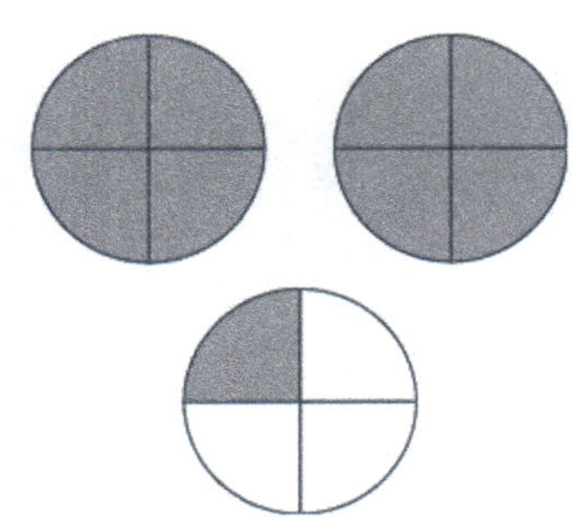

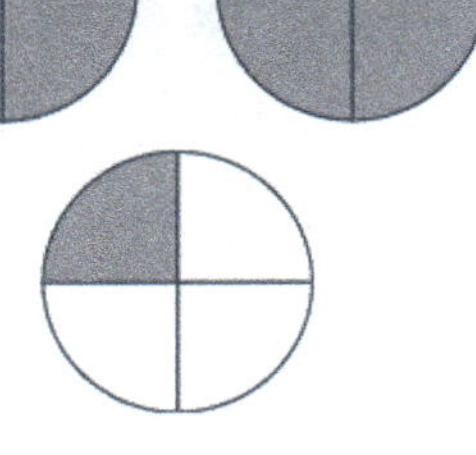

29. —

31. —

33. —

35. —

37. —

39. —

41. —

43. —

45. 4

47. 9

49. 9

51. Answers may vary. Correct answers include — — —

53. Answers may vary. Correct answers include — — —

55. Answers may vary. Correct answers include — — —

57.

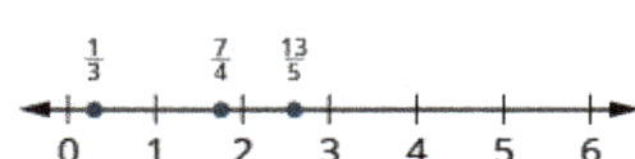

59.

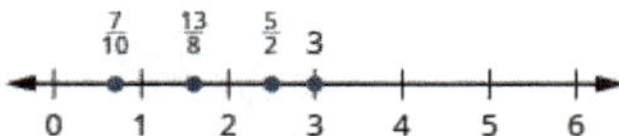

61.

$-1\frac{3}{5}$ $1\frac{3}{4}$

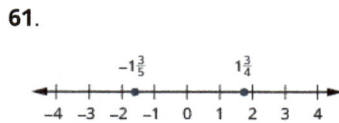

–4 –3 –2 –1 0 1 2 3 4

63.

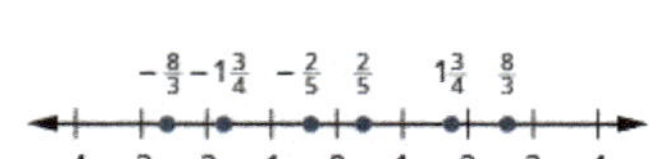

65. <

67. >

69. <

71. <

73.

2 —

75. Answers will vary.

77. —

79. —

81. —

83. —

85. —

87. —

89. —

91. —

93. —

95. —

97.

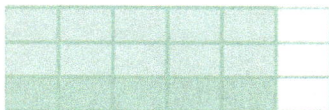

99.

101.

103.

105.

107.

109.

111.

113.

115. $9n$

117. $7p$

119. −34

121.

123.

125.

127.

129.

131.

133.

135. 1

136.

Number	Opposite	Absolute Value	Reciprocal
$-\frac{4}{7}$	$\frac{4}{7}$	$\frac{4}{7}$	$-\frac{7}{4}$
$\frac{1}{8}$	$-\frac{1}{8}$	$\frac{1}{8}$	8
$\frac{9}{4}$	$-\frac{9}{4}$	$\frac{9}{4}$	$\frac{4}{9}$
−1	1	1	−1

137.

Number	Opposite	Absolute Value	Reciprocal
$-\frac{3}{13}$	$\frac{3}{13}$	$\frac{3}{13}$	$-\frac{13}{3}$
$\frac{9}{14}$	$-\frac{9}{14}$	$\frac{9}{14}$	$\frac{14}{9}$
$\frac{15}{7}$	$-\frac{15}{7}$	$\frac{15}{7}$	$\frac{7}{15}$
−9	9	9	$-\frac{1}{9}$

139. 4

$\frac{1}{2}$

$\frac{1}{8}$ $\frac{1}{8}$ $\frac{1}{8}$ $\frac{1}{8}$

141. 12

1 1 1

$\frac{1}{4}$ $\frac{1}{4}$ $\frac{1}{4}$ $\frac{1}{4}$ $\frac{1}{4}$ $\frac{1}{4}$ $\frac{1}{4}$ $\frac{1}{4}$ $\frac{1}{4}$ $\frac{1}{4}$ $\frac{1}{4}$

143. 4

145.

147.

149. 1

151.

153.

155.

157.

159.

161. −12

163.

165. 9

167.

169.

Answers will vary.

171.

173. Answers will vary.

175. Answers will vary.

177.

179.

181.

183.

185.

187.

189. 2

191. 5

193.

195.

197.

199.

201.

203.

205.

207.

209.

211. 28

213.

215.

217.

219.

221.

223.

225.

227.

229.

231. 2

233.

235.

237.

239.

241. -10

243. -2

245.

247.

249.

251. Answers will vary.

253. Answers will vary.

255.

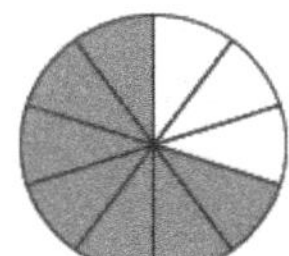

257.

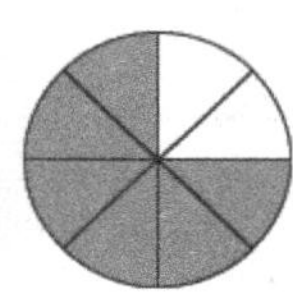

259.

261.

263.

265.

267.

269.

271.

273.

275.

277.

279.

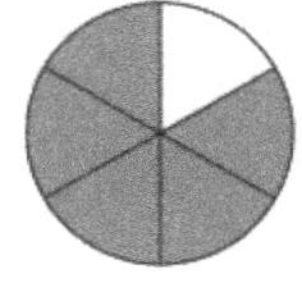

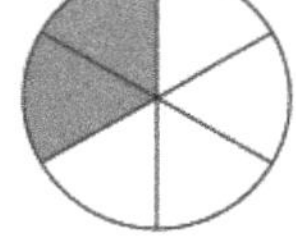

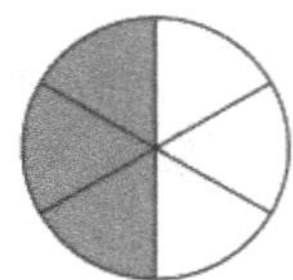

281.

283.

285.

287.

289.

291.

293.

295.

297.

299.

301.

303.

305.

307.

309.

311.

313.

315. No, adding up the number of pieces gives —, which is greater than 1. (Answers may vary.)

317. 20

319. 48

321. 240

323. 245

325. 60

327.

329.

331.

333.

335.

337.

339.

341.

343.

345.

347.

349.

351.

353.

355.

357.

359.

361.

363.

365.

367.

369.

371.

373.

375.

377.

381.

383.

385.

387.

389.

391.

393. 32

395.

397.

399.

401.

403.

405. -9

407.

409.

411.

413. 1

415.

417.

419.

421.

423.

425.

427.

429. -2

431. 3

433. —

435. Answers will vary.

437.

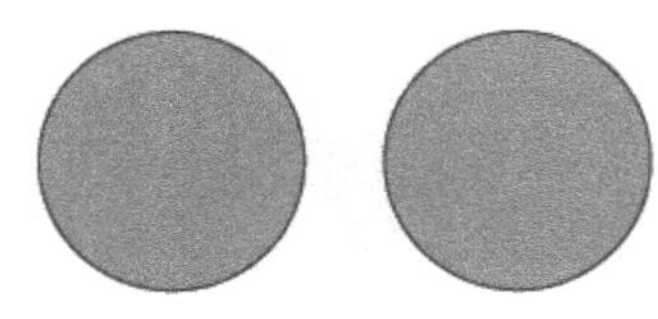

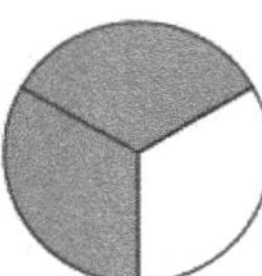

—

439.

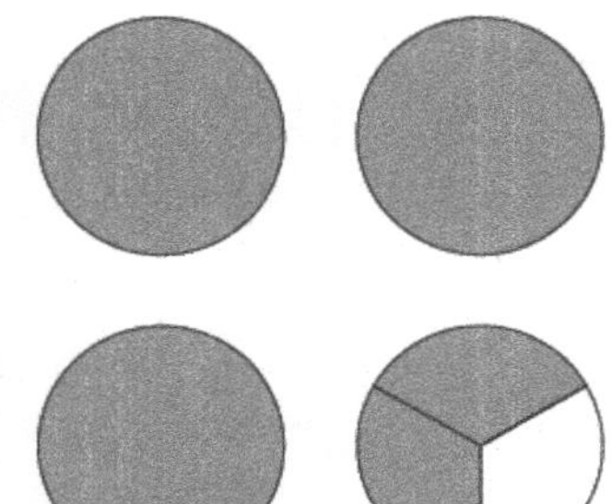

—

441. —

443. 11

445. —

447. —

449.

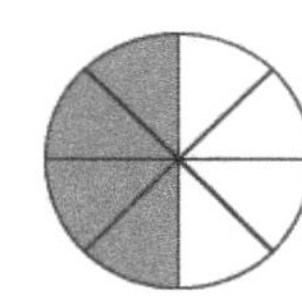

—

451. —

453. —

455. —

457. —

459. —

461. —

463. —

465. —

467. —

469. —

471. 23

473. —

475. —

477. —

479. —

481. —

483. —

485. —

487. —

489. —

491. —

493. —

495. Answers will vary.

497. Answers will vary.

499.
no yes no

501.
no yes no

503. —

505. —

507. $c = -1$

509. $z = -1$

511.

513.

515.

517.

519. $b = -27$

521. $x = -256$

523. $q = 160$

525. $s = 45$

527. $y = -42$

529.

531. $p = 100$

533. $m = -16$

535. $b = -21$

537. $v = 36$

539. $y = 0$

541.

543.

545.

547.

549.

551.

553.

555.

557.

559.

561.

563.

565.

567.

569.

571. 30 inches

573. Answers will vary.

Review Exercises

575.

577.

579.

581.

583. Answers may vary.

585.

$-\frac{7}{2}$ $-1\frac{1}{3}$ $-\frac{1}{4}$ $\frac{1}{4}$ $1\frac{1}{3}$ $\frac{7}{2}$

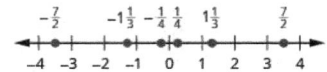

–4 –3 –2 –1 0 1 2 3 4

587. >

589.

591.

593.

595.

597. 6

599.

601. −4

603. 4

605.

607.

609.

611. 8

613.

615.

617. 22

619.

621.

623.

625.

627.

629.

631.

633. 15

635. 60

637.

639.

641.

643.

645.

647.

649.

651.

653.

655.

657.

659.

661.

no yes no

663.

665.

667. $z = -23$

669.

671.

Practice Test

673.

675.

677.

679. $16u$

681. −2

683.

685. 5

687.

689. 13

691.

693. −1

695.

697. 3

699.

701.

703. $c = -27$

Chapter 5

Be Prepared

5.1. Four million, nine hundred twenty-six thousand, fifteen

5.2.

5.3.

5.4.

5.5.

5.6.

5.7.

5.8.

5.9.

5.10.

5.11.

5.12.

5.13.

5.14.

5.15.

5.16.

5.17.

5.18.

5.19.

5.20.

5.21.

Try It

5.1.

six and seven tenths

nineteen and fifty-eight hundredths

eighteen thousandths

negative two and fifty-three thousandths

5.2.

five and eight tenths

three and fifty-seven hundredths

five thousandths

negative thirteen and four hundred sixty-one thousandths

5.3. 13.68

5.4. 5.894

5.5. 0.058

5.6. 0.067

5.7.

5.8.

5.9.

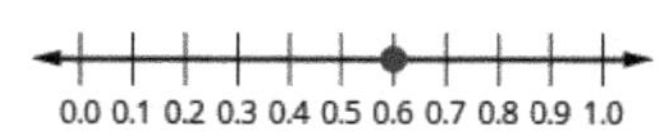

5.10.

0.0 0.1 0.2 0.3 0.4 0.5 0.6 0.7 0.8 0.9 1.0

5.11.

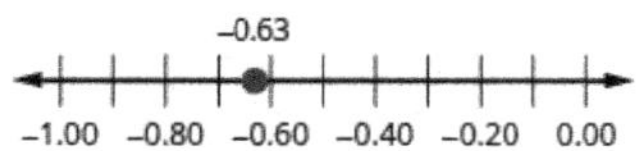

5.12.

−0.25

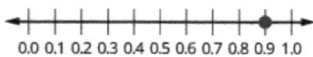

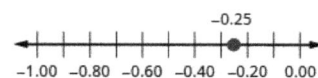

−1.00 −0.80 −0.60 −0.40 −0.20 0.00

5.13.

> >

5.14.

< <

5.15. >

5.16. >

5.17. 1.05

5.18. 9.17

5.19.

6.58 6.6 7

5.20.

15.218 15.22 15.2

5.21. 17.6

5.22. 33.11

5.23. 16.49

5.24. 23.593

5.25. 0.42

5.26. 12.58

5.27. −1.53

5.28. −3.58

5.29. 27.4815

5.30. 87.6148

5.31. −13.427

5.32. −38.122

5.33. 0.00348

5.34. 0.00603

5.35.

25.8 258 2,580

5.36.

142 1,420 14,200

5.37. 0.07

5.38. 0.08

5.39. $0.19

5.40. $0.42

5.41. −0.39

5.42. −0.4

5.43. 587.3

5.44. 34.25

5.45. 200

5.46. 350

5.47. $16.52

5.48. $65.97

5.49. $41.28

5.50. $48.75

5.51. $15.47

5.52. $14.25

5.53. $5.97

5.54. $43.00

5.55. 0.25

5.56. 0.375

5.57. −2.25

5.58. −5.5

5.59.

5.60.

5.61. 5.275

5.62. 6.35

5.63. >

5.64. <

5.65. <

5.66. >

5.67. — —

5.68. — —

5.69.

−183.2 4.5

5.70.

−776.25 2.2

5.71. 11.16

5.72. 1.51

5.73.

314 in. 7850 sq. in.

5.74.

628 ft. 31,400 sq. ft.

5.75.

325.304 cm

8425.3736 sq. cm

5.76.

165.792 m

2188.4544 sq. m

5.77.

— —

5.78.

— —

5.79.

no yes no

5.80.

yes no no

5.81. $y = -8$

5.82. $y = -8.4$

5.83. $a = 1.07$

5.84. $n = 0.83$

5.85. $b = -12$

5.86. $c = -8$

5.87. $c = 11.7$

5.88. $b = 6.48$

5.89. $y - 4.9 = 2.8; y = 7.7$

5.90. $z - 5.7 = 3.4; z = 9.1$

5.91. $-4.3x = 12.04; x = -2.8$

5.92. $-3.1m = 26.66; m = -8.6$

5.93. —

5.94. —

5.95. $j + 3.8 = 2.6; j = -1.2$

5.96. $k + 4.7 = 0.3; k = -4.4$

5.97. 8.5

5.98. 9

5.99. 21.5 years years

5.100. 10

5.101. $7.19

5.102. $39.62

5.103. 43

5.104. 30

5.105. 8.5

5.106. 20.5

5.107. 2

5.108. 5

5.109. 21

5.110. 5

5.111.

—

5.112.

—

5.113.

—

5.114.

—

5.115.

— —

5.116.

— —

5.117.

— —

5.118.

— —

5.119. —

5.120. —

5.121.

5.122.

5.123.

5.124.

5.125.

5.126.

5.127. $18.00/hour

5.128. $19.00/hour

5.129. 23.5 mpg

5.130. 28 mpg

5.131. $0.29/box

5.132. $0.53/bottle

5.133. Brand A costs $0.11 per bag. Brand B costs $0.13 per bag. Brand A is the better buy.

5.134. Brand C costs $0.07 per ounce. Brand D costs $0.09 per ounce. Brand C is the better buy.

5.135.

689 mi/h hours

y parents/22 students

$$d$/9 min

5.136.

m mi/9 h

x students/8 buses

$$y$/40 h

5.137.

6 13

5.138.

4 14

5.139.

−2 −15

5.140.

−9 −8

5.141.

not a real number −9

5.142.

not a real number −11

5.143.

7 5

5.144.

17 23

5.145. $\sqrt{}$

5.146. $\sqrt{}$

5.147. ≈ 3.32

5.148. ≈ 3.61

5.149. y

5.150. m

5.151. $8x$

5.152. $13y$

5.153. $-11y$

5.154. $-10p$

5.155. $10ab$

5.156. $15mn$

5.157. 19.2 feet

5.158. 52 centimeters

5.159. 9 seconds

5.160. 3.5 seconds

5.161. 42.7 mph

5.162. 54.1 mph

Section Exercises

1. five and five tenths

3. five and one hundredth

5. eight and seventy-one hundredths

7. two thousandths

9. three hundred eighty-one thousandths

11. negative seventeen and nine tenths

13. 8.03

15. 29.81

17. 0.7

19. 0.001

21. 0.029

23. −11.0009

25. 13.0395

27.

29.

31.

33.

35.

37.

39.

41.

43.

45.

47.

49.

51.

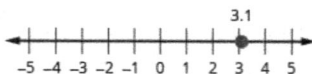

0.8

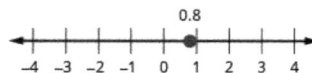

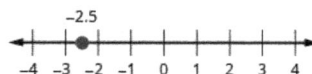

−4 −3 −2 −1 0 1 2 3 4

53.

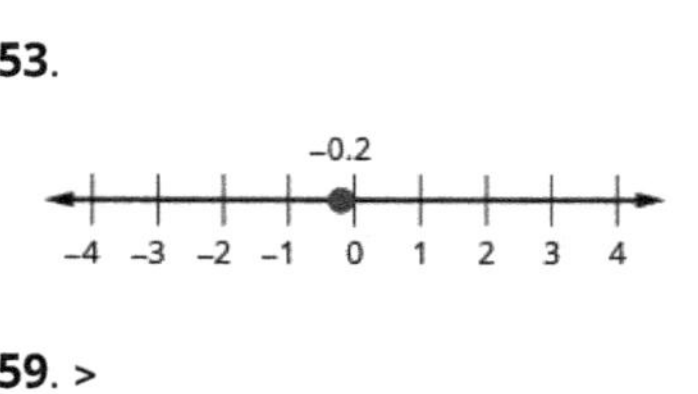

55.

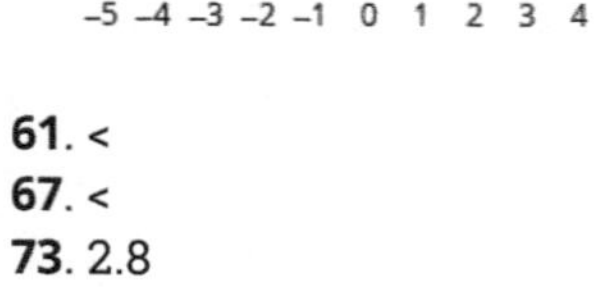

57.

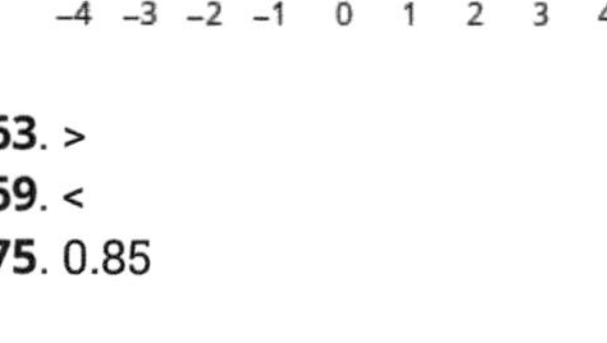

59. >

61. <

63. >

65. >

67. <

69. <

71. 0.7

73. 2.8

75. 0.85

77. 5.79

79. 0.30

81. 4.10

83.

5.78 5.8 6

85.

63.48 63.5 63

87.

$58,966 $59,000

$60,000

89.

$142.19 $142

91. Answers will vary.

93. Tim had the faster time. 12.3 is less than 12.32, so Tim had the faster time.

95. 24.48

97. 170.88

99. −9.23

101. 49.73

103. −40.91

105. −7.22

107. −13.5

109. 35.8

111. −27.5

113. 15.73

115. 42.51

117. 102.212

119. 51.31

121. −4.89

123. 0.12

125. 0.144

127. 42.008

129. 26.7528

131. −11.653

133. 337.8914

135. 2.2302

137. 1.305

139. 92.4

141. 55,200

143. 0.03

145. 0.19

147. $0.71

149. $2.44

151. 3

153. −4.8

155. 35

157. 2.08

159. 150

161. 20

163. 19.2

165. 12.09

167. 32.706

169. $48.60

171. 20

173. 2

175. $17.80

177. $24.89

179. $29.06

181. $3.19

183. 181.7 pounds

185. $15.00

187. $296.00

189. $12.75

191.

$3 $2 $1.50

$1.20 $1

193. $18.64

195. $259.45

197.

$243.57 $79.35

199. The difference: 0.03 seconds. Three hundredths of a second.

201. 0.4

203. −0.375

205. 0.85

207. 2.75

209. −12.4

211.

213.

215.

217. 7

219. 3.025

221. 10.58

223. <

225. >

227. <

229. <

231. >

233. >

235.

237.

239.

241.

243. −187

245. 295.12

247. 6.15

249. 20.2

251. 107.11

253. 449

255. 9.14

257. −0.23

259. −3.25

261. 16.29

263. 632.045

265. −5.742

267.

31.4 in 78.5 sq.in.

269.

56.52.ft. 254.34 sq.ft.

271.

288.88 cm

6644.24 sq.cm

273.

116.808 m

1086.3144 sq.m

275.

277.

279.

281. $56.66

283. Answers will vary.

285.

no no yes

287.

no yes no

289. $y = 2.8$
291. $f = -0.85$
293. $a = -7.9$
295. $c = -4.65$
297. $n = 4.4$
299. $x = -3.5$
301. $j = -4.68$
303. $m = -1.42$
305. $x = 7$
307. $c = -5$
309. $p = 3$
311. $q = -80$
313. $x = 20$
315. $z = 2.7$
317. $a = -8$
319. $x = -0.28$
321. $p = 8.25$
323. $r = 7.2$
325. $x = -6$
327. $p = -10$
329. $m = 8$
331. —
333. —
335. —
337. —
339. $a = -0.8$
341. $r = -1.45$
343. $h = 24$
345.
347. $-6.2x = -4.96$; 0.8
349. —
351. $n + (-7.3) = 2.4$; 9.7
353. $104
355. Answers will vary.
357. 4
359. 35
361. 245.5
363. 11.65
365. $18.84
367. 19.5 minutes
369. 0.329
371. 21
373. 45
375. 4.5
377. 99.65
379. $6.50
381. 40.5 months
383. 2
385. 22
387. 2 children
389. 11 units
391. —
393. —
395. —
397. —
399.

$285.47 $275.63

$236.25

401. Answers will vary.
403. —
405. —
407. —
409. —
411. —
413. —
415. —
417. —
419. —
421. —
423. —
425. —
427. —
429. —
431. —
433. —
435. —
437. —
439. 11.67 calories/ounce
441. 2.73 lbs./sq. in.
443. 69.71 mph
445. $14.88/hour
447. 32 mpg
449. 2.69 lbs./week
451. 92 beats/minute
453. 8,000
455. $1.09/bar
457. $1.33/pair
459. $0.48/pack
461. $0.60/disc
463. $1.29/box
465. The 50.7-ounce size costs $0.138 per ounce. The 33.8-ounce size costs $0.142 per ounce. The 50.7-ounce size is the better buy.
467. The 18-ounce size costs $0.222 per ounce. The 14-ounce size costs $0.235 per ounce. The 18-ounce size is a better buy.
469. The regular bottle costs $0.075 per ounce. The squeeze bottle costs $0.069 per ounce. The squeeze bottle is a better buy.
471. The half-pound block costs $6.78/lb, so the 1-lb. block is a better buy.
473. —
475. —
477. —
479. —

481. 15.2 students per teacher

483.

72 calories/ounce

3.87 grams of fat/ounce

5.73 grams carbs/ounce

3.33 grams protein/ounce

485. Answers will vary.

487. Kathryn should swim for approximately 16.35 minutes. Explanations will vary.

489. 6

491. 8

493. −2

495. −1

497. not a real number

499. not a real number

501. 5

503. 7

505. √

507. √

509. 4.36

511. 7.28

513. y

515. $7x$

517. $-8a$

519. $12xy$

521. 8.7 feet

523. 8 seconds

525. 15.8 seconds

527. 72 mph

529. 53.0 mph

531. 45 inches

533. Answers will vary. 9^2 reads: "nine squared" and means nine times itself. The expression √ reads: "the square root of nine" which gives us the number such that if it were multiplied by itself would give you the number inside of the square root.

Review Exercises

535. three hundred seventy-five thousandths

537. five and twenty-four hundredths

539. negative four and nine hundredths

541. 0.09

543. 10.035

545. −0.05

547. —

549. —

551. <

553. <

555.

12.53 12.5 13

557.

5.90 5.9 6

559. 24.67

561. 24.831

563. −2.37

565. −1.6

567. 15,400

569. 0.18

571. 4

573. 200

575. $28.22

577. $1.79

579. 0.875

581. −5.25

583.

585. >

587. >

589. >

591. — –

593. 6.03

595. 1.975

597. −0.22

599.

21.98 ft. 38.465 sq.ft.

601.

34.54 cm 94.985 sq.cm

603.

no yes

605.

no yes

607. $h = -3.51$

609. $p = 2.65$

611. $j = 3.72$

613. $x = -4$

615. $a = -7.2$

617. $s = 25$

619. $-5.9x = -3.54$; $x = 0.6$

621. $m + (-4.03) = 6.8$; $m = 10.83$

623. $269.10

625. $40.94

627. 24.5

629. 16 clients

631. 2

633. –

635. –

637. –

639. –

641. –

643. ——

645. ——

647. 12 pounds/sq.in.

649. \$17.50/hour
651. \$0.42
653. \$1.65
655. \$0.11, \$0.12; 60 tablets for \$6.49
657.
659.
661. 12
663. −9
665. not a real number
667. 17
669.
671. 7.55
673. $8b$
675. $15mn$
677. $7y$
679. $11cd$
681. 5.5 feet
683. 72 mph

Practice Test

685.
687. 16.7 16.75 17
689. 18.42
691. 0.192
693. −0.08
695. 2
697. 200
699. 1.975
701. −1.2
703. 8.8
705. \$26.45
707. 52.1 51.5 55
709. The unit prices are \$0.172 per ounce for 64 ounces, and \$0.177 per ounce for 48 ounces; 64 ounces is the better buy.
711. $12n$
713. 15 feet

Chapter 6

Be Prepared

6.1.
6.2.
6.3.
6.4.
6.5.
6.6.
6.7.
6.8.
6.9.
6.10.
6.11.
6.12.
6.13.

Try It

6.1.
6.2.
6.3.
6.4.
6.5.
6.6.
6.7.
6.8.
6.9. 0.09 0.87
6.10. 0.03 0.91
6.11. 1.15 0.235
6.12. 1.23 0.168
6.13.
6.14.
6.15.
6.16.
6.17. 1% 17%
6.18. 4% 41%
6.19. 175% 8.25%
6.20. 225% 9.25%
6.21. 62.5% 275% 340%
6.22. 87.5% 225% 160%
6.23. 42.9%
6.24. 57.1%

6.25.

6.26.

6.27. 36

6.28. 33

6.29. 117

6.30. 126

6.31. 68

6.32. 64

6.33. $26

6.34. $36

6.35. 75%

6.36. 80%

6.37. 125%

6.38. 175%

6.39. $14.67

6.40. $2.16

6.41. 24.1 grams

6.42. 2,375 mg

6.43. 26%

6.44. 37%

6.45. 8.8%

6.46. 50%

6.47. 6.3%

6.48. 10%

6.49.
$45.25 $769.25

6.50.
$20.50 $270.50

6.51. 9%

6.52. 7.5%

6.53. $273

6.54. $394.20

6.55. 4%

6.56. 5.5%

6.57. $450

6.58. $82

6.59.
$11.60 $17.40

6.60.
$256.75 $138.25

6.61.
$184.80 33%

6.62.
$60 15%

6.63.
$600 $1,800

6.64.
$2,975 $11,475

6.65. $160

6.66. $56

6.67. $2,750

6.68. $3,560

6.69. 4.5%

6.70. 6.5%

6.71. $142.50

6.72. $7,020

6.73. 6%

6.74. 4%

6.75. $11,450

6.76. $9,600

6.77. $57.00

6.78. $143.50

6.79.

6.80.

6.81.
no yes

6.82.
no no

6.83. 77

6.84. 104

6.85. 65

6.86. 24

6.87. -7

6.88. -9

6.89. 12 ml

6.90. 180 mg

6.91. 300

6.92. 5 pieces

6.93. 590 Euros

6.94. 56,460 yen

6.95.

6.96.

6.97.

6.98.

6.99.

6.100.

6.101.

6.102.

6.103.

6.104.

6.105.

6.106.

6.107.

6.108.

Section Exercises

1.

3.

5.

7.

9.

11.

13.

15.

17.

19.

21.

23.

25. 0.05

27. 0.01

29. 0.63

31. 0.4
33. 1.15
35. 1.5
37. 0.214
39. 0.078
41.

—

43.

—

45.

—

47.

—

49. 1%
51. 18%
53. 135%
55. 300%
57. 0.9%
59. 8.75%
61. 150%
63. 225.4%
65. 25%
67. 37.5%
69. 175%
71. 680%
73. —
75.
77. 42.9%
79.
81. 25%
83. 35%
85.

Fraction	Decimal	Percent
—		
—		
—		
—		
—		

87.

—

89. —

91. 80%; 0.8
93. — — —
95. The Szetos sold their home for five times what they paid 30 years ago.
97. 54
99. 26.88
101. 162.5
103. 18,000
105. 112
107. 108
109. $35
111. $940
113. 30%
115. 36%
117. 150%
119. 175%
121. $11.88
123. $259.80
125. 24.2 grams
127. 2,407 mg
129. 45%
131. 25%
133. 13.2%
135. 125%
137. 72.7%
139. 2.5%
141. 11%
143. 5.5%
145. 21.2%
147. The original number should be greater than 44.80% is less than 100%, so when 80% is converted to a decimal and multiplied to the base in the percent equation, the resulting amount of 44 is less. 44 is only the larger number in cases where the percent is greater than 100%.
149. Alex should have packed half as many shorts and twice as many shirts.
151.

$4.20 $$88.20

153.

$9.68 $138.68

155.

$17.13 $267.13

157.

$61.45 $1,260.45

159. 6.5%

161. 6.85%

163. $20.25

165. $975

167. $859.25

169. 3%

171. 16%

173. 15.5%

175. $139

176. $81

177. $125

179.

$26.97 $17.98

181.

$128.37 $260.63

183.

$332.48 $617.47

185.

$576 30%

187.

$53.25 15%

189.

$370 43.5%

191.

$7.20 $23.20

193.

$0.20 $0.80

195.

$258.75 $373.75

197.

$131.25 $126.25

25% off first, then $20 off

199.

Priam is correct. The original price is 100%. Since the discount rate was 40%, the sale price was 60% of the original price.

Yes.

201. $180

203. $14,000

205. 6.3%

207. $90

209. $579.96

211. $14,167

213. $3,280

215. $860

217. $24,679.91

219. 4%

221. 5.5%

223. $116

225. $4,836

227. 3%

229. 3.75%

231. $35,000

233. $3,345

235. $332.10

237. $195.00

239. Answers will vary.

241. Answers will vary.

243. — —

245. — —

247. — —

249. — —

251. — —

253. — —

255. yes

257. no

259. no

261. yes

263. $x = 49$

265. $z = 7$

267. $a = 9$

269. $p = -11$

271. $a = 7$

273. $c = 2$

275. $j = 0.6$

277. $m = 4$

279. 9 ml

281. 114 beats/minute. Carol has not met her target heart rate.

283. 159 cal

285. —

287. $252.50

289. 0.8 Euros

291. 48 quarters

293. 19 gallons, $58.71

295. 12.8 hours

297. 4 bags

299. — —

301. — —

303. — —

305. — —

307. — —

309. — —

311. — —

313. — —

315. — —

317. — —

319. — —

321. — —

323. — —

325. — —

327. He must add 20 oz of water to obtain a final solution of 32 oz.

329. Answers will vary.

Review Exercises

331. —

333. —

335. —

337.
339. 0.06
341. 1.28
343.
345.
347. 4%
349. 282%
351. 0.3%
353. 75%
355. 362.5%
357. 40%
359. 161
361. 240
363. 25
365. 68%
367. $16.70
369. 28.4%
371. 4%
373. $45 $795
375. 7.25%
377. $11,400
379. 15%
381. $45
383. $25.47 $59.43
385. $13 26%
387. $0.48 $1.28
389. $450
391. $4400
393. $900
395. 2.5%
397.
399.
401. yes
403. no
405. 20
407. 4
409. 12 ml
411. 340 calories
413. 13 gallons
415.
417.
419.
421.

Practice Test

423.
425.
427.
429. 25%
431. 40
433. 11.9%
435. 4%
437. $70
439. 6.25%
441. −52

Chapter 7

Be Prepared

7.1.
7.2.
7.3.
7.4.
7.5.
7.6.
7.7.
7.8.
7.9.
7.10.
7.11.
7.12.
7.13.
7.14.
7.15.

Try It

7.1.
7.2.
7.3. rational rational irrational
7.4. rational rational irrational
7.5. rational irrational
7.6. irrational rational

7.7.

Number	Whole	Integer	Rational
−3		✓	✓
$-\sqrt{2}$			
0.3			✓
$\frac{9}{5}$			✓
4	✓	✓	✓
$\sqrt{49}$	✓	✓	✓

Number	Irrational	Real
−3		✓
$-\sqrt{2}$	✓	✓
0.3		✓
$\frac{9}{5}$		✓
4		✓
$\sqrt{49}$		✓

7.8.

Number	Whole	Integer	Rational
$-\sqrt{25}$		✓	✓
$-\frac{3}{8}$			✓
−1		✓	✓
6	✓	✓	✓
$\sqrt{121}$	✓	✓	✓
2.041975...			

Number	Irrational	Real
$-\sqrt{25}$		✓
$-\frac{3}{8}$		✓
−1		✓
6		✓
$\sqrt{121}$		✓
2.041975...	✓	✓

7.9.

$-4 + 7 = 7 + (-4)$

$6 \cdot 12 = 12 \cdot 6$

7.10.

$14 + (-2) = -2 + 14$

$3(-5) = (-5)3$

7.11.

— —

7.12.

— —

7.13. $8(4x) = (8 \cdot 4)x = 32x$

7.14. $-9(7y) = (-9 \cdot 7)y = -63y$

7.15.

0.84 0.84

7.16.

0.975 0.975

7.17.

24 24

7.18.

15 15

7.19. $-48a$

7.20. $-92x$

7.21. —

7.22. —

7.23. —

7.24. —

7.25. $15.58c$

7.26. $17.79d$

7.27. 10.53

7.28. 70.4

7.29. $24y$

7.30. $60z$

7.31. $32r - s$

7.32. $41m + 6n$

7.33. $4x + 8$

7.34. $6x + 42$

7.35. $27y + 72$

7.36. $25w + 45$

7.37. $7x - 42$

7.38. $8x - 40$

7.39. —

7.40. —

7.41. $5y + 3$

7.42. $4n + 9$

7.43. $70 + 15p$

7.44. $4 + 35d$

7.45. $rs - 2r$

7.46. $yz - 8y$

7.47. $xp + 2p$

7.48. $yq + 4q$

7.49. $-18m - 15$

7.50. $-48n - 66$

7.51. $-10 + 15a$

7.52. $-56 + 105y$

7.53. $-z + 11$

7.54. $-x + 4$

7.55. $-3x + 3$

7.56. $2x - 20$

7.57. $5x - 66$

7.58. $7x - 13$

7.59.

120 120

7.60.

126 126

7.61.

30 30

7.62.

3 3

7.63.

−32 −32

7.64.

−45 −45

7.65.

identity property of addition

identity property of multiplication

7.66.

identity property of multiplication

identity property of addition

7.67.

7.68.

7.69.

7.70.

7.71.

0 0 0

7.72.

0 0 0

7.73.

0 0 0

7.74.

0 0 0

7.75.

undefined undefined

7.76.

undefined undefined

7.77. 9

7.78. −18

7.79. p

7.80. r

7.81. 0

7.82. 0

7.83. undefined

7.84. undefined

7.85. $20y + 50$

7.86. $12z + 16$

7.87. 2.5 feet

7.88. 54 feet

7.89. 8600 pounds

7.90. 102,000,000 pounds

7.91. 440,000,000 yards

7.92. 151,200 minutes

7.93. 16 cups

7.94. 48 teaspoons

7.95. 9 lbs. 8 oz

7.96. 21 ft. 6 in.

7.97. 4 lbs. 8 oz.

7.98. 11 gal. 2 qts.

7.99. 5000 m

7.100. 250 cm

7.101. 2.8 kilograms

7.102. 4.5 kilograms

7.103.

0.00725 kL 6300 mL

7.104.

35,000 L 410 cL

7.105. 83 cm

7.106. 1.04 m

7.107. 2 L

7.108. 2.4 kg

7.109. 2.12 quarts

7.110. 3.8 liters

7.111. 19,328 ft

7.112. 3,470 mi

7.113. 15°C

7.114. 5°C

7.115. 59°F

7.116. 50°F

Section Exercises

1.

3.

5. Rational: .
Irrational:

7. Rational: , . Irrational:

9.

rational irrational

11.

irrational rational

13.

Number	Whole	Integer	Rational
−8		✓	✓
0	✓	✓	✓
1.95286...			
$\frac{12}{5}$			✓
$\sqrt{36}$	✓	✓	✓
9	✓	✓	✓

Number	Irrational	Real
−8		✓
0		✓
1.95286...	✓	✓
$\frac{12}{5}$		✓
$\sqrt{36}$		✓
9		✓

15.

Number	Whole	Integer	Rational
$-\sqrt{100}$		✓	✓
−7		✓	✓
$-\frac{8}{5}$			✓
−1		✓	✓
0.77			✓
$3\frac{1}{4}$			✓

Number	Irrational	Real
$-\sqrt{100}$		✓
−7		✓
$-\frac{8}{5}$		✓
−1		✓
0.77		✓
$3\frac{1}{4}$		✓

17.

4

Teachers cannot be divided

It would result in a lower number.

19. Answers will vary.
21. $7 + 6 = 6 + 7$
23. $7(-13) = (-13)7$
25. $(-12)(-18) = (-18)(-12)$
27. $-15 + 7 = 7 + (-15)$
29. $y + 1 = 1 + y$
31. $-3m = m(-3)$
33. $(21 + 14) + 9 = 21 + (14 + 9)$
35. $(14 \cdot 6) \cdot 9 = 14(6 \cdot 9)$
37. $(-2 + 6) + 7 = -2 + (6 + 7)$
39. — —
41. $4(7x) = (4 \cdot 7)x$
43. $(17 + y) + 33 = 17 + (y + 33)$
45.
0.97 0.97
47.
2.375 2.375
49.
21 21
51.
−8 −8
53. 23
55. —
57. —
59. —
61. −176
63. —
65. —
67. $9.89d$
69. 36
71. 29.193
73. $72w$
75. $-46n$
77. $12q$
79. $42u + 30v$
81. $-57p + (-10q)$
83. —
85. $7.41m + 6.57n$
87.
$975 $700 $1675
$185 $270 $1220
89. Answers will vary.
91. $3a + 27$
93. $27w + 63$
95. $7y - 91$
97. $35u - 20$
99. —
101. —
103. $3x - 4$
105. $2 + 9s$
107. $uv - 10u$
109. $ax + 7x$
111. $-3a - 33$
113. $-81a - 36$
115. $-4q + 28$
117. $-42x + 48$
119. $-q - 11$
121. $-5p + 4$
123. $8u + 4$
125. $-4x + 10$
127. $-42n + 39$
129. $-r + 15$
131. $-c + 6$
133. $3y + 1$
135. $47u + 60$
137. $24x + 4$
139. $6n - 72$
141. $17n + 76$
143.
56 56
145.
— —
147.
−89 −89
149.
−525 −525
151.
160 160
153.
−1.03 −1.03
155.
$3(4 - 0.03) = 11.91$
$1.42
157. Answers will vary.
159. identity property of multiplication
161. identity property of addition
163. —
165. —
167. —
169. —
171. —
173. −2
175. 0
177. 0
179. undefined
181. 0
183. 0
185. undefined
187. 16
189. $31s$
191. p
193. $2n$
195. 0
197. 0
199. 0
201. undefined
203. undefined
205. undefined
207. $20q - 35$
209. $225h + 360$
211.
8 hours 8
associative property of multiplication
213. Answers will vary.
215. 24 inches

217. 3.75 feet
219. 58 inches
221. 90 feet
223. 0.98 miles
225. 300,000 pounds
227. 110 tons
229. 162 hours
231. 8100 seconds
233. 256 tablespoons
235. 128 ounces
237. 111 ounces
239. 31.25 gallons
241. 4 lbs. 1 oz.
243. 5 weeks and 1 day
245. 9 ft 2 in
247. 8 yards
249. 8000 meters
251. 245 centimeters
253. 10.911 kilometers
255. 3000 milligrams
257. 25,000 milligrams
259. 3800 grams
261. 0.3 liters
263. 65 centimeters
265. 1.82 kilograms
267. 16.8 grams
269. 42,000 milligrams
271. 106.7 centimeters
273. 8.2 meters
275. 40,900 kilograms
277. 5.2 feet
279. 33 pounds
281. 30.2 liters
283. 25°C
285. −10°C
287. −15.6°C
289. 48.9°C
291. 77°F
293. 5°F
295. 46.4°F
297. 60.8°F
299. 110 reflectors

Review Exercises

303.
305.
307.

309.
irrational rational

311.
none

313. $-14 \cdot 5 = 5(-14)$

315. $a + 8 = 8 + a$
317. $(22 + 7) + 3 = 22 + (7 + 3)$
319.

321.
5.39 5.39

323.
13 13

325.

327. −60
329. $5.98\ d$
331. $25\ q$
333. $34\ m + (-25\ n)$
335. $9y - 36$
337. $56a + 96$
339. $yp + 10p$
341. $-4x + 68$
343.
9 9

345.
36 36

347. identity property of addition
349. identity property of multiplication
351. −19.4
353.
355.
357.
359. 0
361. 0
363. $n + 7$
365. 34
367. $54x - 84$
369. 3.5 feet
371. 15 yards
373. 9000 pounds
375. 64 tablespoons
377. 1.9 gallons
379. 7 hours 10 minutes
381. 3 yards, 12 inches
383. 8.85 kilometers
385. 13,000 milligrams
387. 0.65 liters
389. 855 milliliters
391. 10,000 milligrams
393. 25.6 meters
395. 171.6 pounds
397. 11.4 kilograms
399. −5°C
401. 17.8°C
403. 23°F
405. 75.2°F

Practice Test

407.
409. $x \cdot 14 = 14 \cdot x$
411. $(8 \cdot 2) \cdot 5 = 8 \cdot (2 \cdot 5)$

413. — —

415. $15y$

417. —

419. $30y - 4z$

421. $30x - 24$

423. $2a + 3$

425. $66p - 2$

427. 0

429. 0

431. .276 grams

433. 9.317 miles

435. $95°F$

Chapter 8

Be Prepared

8.1.

8.2.

8.3. yes

8.4.

8.5. —

8.6.

8.7.

8.8.

8.9.

8.10.

8.11.

8.12.

Try It

8.1. no

8.2. no

8.3. $x = -16$

8.4. $x = -20$

8.5. $n = -1$

8.6. $x = -4$

8.7. —

8.8. —

8.9. $b = 6.4$

8.10. $c = 14$

8.11. $y = 15$

8.12. $z = 2$

8.13. $p = 5$

8.14. $q = -16$

8.15. $h = -1$

8.16. $x = 1$

8.17. $x + 11 = 41; x = 30$

8.18. $y - 12 = 51; y = 63$

8.19. $4x - 3x = 14; x = 14$

8.20. $7a - 6a = -8; a = -8$

8.21. $a + 6 = 13$; Athena weighs 7 pounds.

8.22. $26 + h = 68$; Henry has 42 books.

8.23. $19,875 = s - 1025$; the sticker price is $20,900.

8.24. $7.75 = n - 3.25$; the price at night is $11.00.

8.25. $y = -16$

8.26. $z = -13$

8.27. $b = 144$

8.28. $c = 128$

8.29. $k = -8$

8.30. $g = -3$

8.31. $n = 35$

8.32. $y = 18$

8.33. $x = 2$

8.34. $n = -5$

8.35. $c = -3$

8.36. —

8.37. $n = -6$

8.38. $n = -5$

8.39. $x = -4$

8.40. $a = -8$

8.41. $y = 5$

8.42. $m = 9$

8.43. $n = 10$

8.44. $c = 1$

8.45. $p = -7$

8.46. $m = -3$

8.47. $j = 2$

8.48. $h = 1$

8.49. $x = -1$

8.50. $y = 4$

8.51. $q = 1$

8.52. $n = 1$

8.53. $a = -5$

8.54. $k = -6$

8.55. $x = 10$

8.56. $y = -3$

8.57. $x = -5$

8.58. $y = -5$

8.59. $x = 4$

8.60. $y = 1$

8.61. $y = -6$

8.62. $z = 8$

8.63. $a = 2$

8.64. $n = 2$

8.65. —

8.66. —

8.67. $p = -2$

8.68. $q = -8$

8.69. $u = 2$

8.70. $x = 4$

8.71. $n = 1$

8.72. $m = -1$

8.73. —

8.74. $y = 3$

8.75. $v = 40$

8.76. $u = -12$

8.77. $a = -2$

8.78. $c = -2$

8.79. $p = -4$

8.80. $q = 2$

8.81. $n = 2$

8.82. $m = -1$

8.83. $x = 20$

8.84. $x = 10$

8.85. $h = 12$

8.86. $k = -1$

8.87. $n = 9$

8.88. $d = 16$

Section Exercises

1. yes

3. no

5. $x = 5$

7.
9. $p = -11.7$
11. $a = 10$
13.
15. $y = 13.8$
17. $x = -27$
19.
21.
23. $m = 17$
25. $x = 8$
27. $n = -20$
29. $y = 2$
31. $w = -1.7$
33. $x = -2$
35. $m = -4$
37. $k = 6$
39. $c = -41$
41. $y = 28$
43. $x + (-5) = 33; x = 38$
45. $y - 3 = -19; y = -16$
47. $p + 8 = 52; p = 44$
49. $5c - 4c = 60; c = 60$
51.
53. $-9m + 10m = -25; m = -25$
55. Let p equal the number of pages read in the Psychology book. $41 + p = 54$. Jeff read 13 pages in his Psychology book.
57. Let d equal the daughter's age. d = 12 - 5. Eva's daughter's age is 7 years old.
59. 21 pounds
61. 100.5 degrees
63. $121.19
65.
67. Answers will vary.
69. $p = 9$
71. $x = 3$
73. $y = -6$
75. $m = 7$
77. $a = 15$
79. $x = 0$
81. $z = 28$
83. $c = 36$
85. $q = -48$
87. $p = 80$
89. $r = 25$
91. $x = -32$
93.
95. $y = -1$
97. $m = -5$
99.
101. $q = 24$
103. $p = 56$
105. 6 children
107. $1.08
109. 42 yards
111. Answer will vary.
113. $x = 6$
115. $y = 6$
117. $m = -8$
119. $b = -8$
121. $x = -4$
123. $q = -2$
125. $k = -11$
127. $x = 9$
129. $b = -3$
131. $z = 3$
133.
135. $r = -2$
137. $x = 19$
139. $f = 7$
141. $q = -5$
143. $c = -4$
145. $x = 2$
147. $y = 4$
149. $m = -6$
151. $a = 7$
153. $a = -40$
155. $p = 15$
157. $z = 3.46$
159. $w = 60$
161. $x = 23$
163. $y = 9$
165. $x = 6$
167. $y = 3$
169. $n = -2$
171. $p = -1$
173. $x = 5$
175. $v = 1$
177. $m = 0.25$
179. $t = -9$
181. $b = 2$
183. $m = 6$
185.
187. $r = 3$
189. $y = -4$
191. $n = 2$
193. $x = 34$
195. $s = 10$
197.
199. 30 feet
201. 8 nickels
203. Answers will vary.
205. Answers will vary.
207. Answers will vary.
209. $x = -1$
211. $y = -1$
213.
215. $x = 4$
217. $m = 20$
219. $x = -3$
221.
223. $x = 1$
225. $b = 12$
227. $x = 1$
229. $p = -41$
231.
233. $y = 10$
235. $j = 2$
237. $x = 18$
239. $x = 18$
241. $x = 20$
243. $n = 9$
245. $d = 8$
247. $q = 11$
249. $d = 18$
251. Answers will vary.
253. Answers will vary.

Review Examples

255. yes
257. no
259. 12
261. –
263. $u = 17$
265. —
267. $n = 44$
269. —
271. $y = 4$
273. $n = -8$
275. $-6 + m = 25$; $m = 31$
277. $s = 11 - 3$; 8 years old
279. $c - 46.25 = 9.75$; \$56.00
281. $x = 9$
283. $p = 21$
285. $n = 108$
287. $x = 48$
289. $m = 4$
291. $x = 15$
293. $r = 3$
295. $p = 5$
297. $x = -22$
299. $y = -13$
301. $k = -5$
303. $x = 6$
305. $u = -7$
307. $x = -2$
309. $s = -22$
311. $y = 12$
313. $r = 38$
315. $y = 26$
317. $n = 2$
319. —
321. $x = 5$
323. $p = -20$

Practice Test

325.
yes no
327. $c = 16$
329. $x = -5$
331. $x = 9$
333. $y = 4$
335. $m = 6$
337. $d = -32$
339. $x = -2$
341. –
343. $2x - 4 = 16$; $x = 10$

Chapter 9

Be Prepared

9.1.
9.2.
9.3.
9.4.
9.5.
9.6.
9.7.
9.8.
9.9.
9.10.
9.11.
9.12.
9.13.
9.14. (a) ; (b)
.
9.15.
9.16.
9.17.
9.18. —
9.19. ———
9.20.
9.21.

Try It

9.1. \$180
9.2. 40
9.3. 2
9.4. 7
9.5. \$950
9.6. \$4,200
9.7. 25
9.8. 4
9.9. 3
9.10. 6
9.11. 9, 15
9.12. 27, 31
9.13. −8, −15
9.14. −29, 11
9.15. −4, 0
9.16. −2, −3
9.17. 47, 48
9.18. −15, −16
9.19. 31, 32, 33
9.20. −11, −12, −13
9.21. 9 nickels, 16 dimes
9.22. 17 nickels, 5 quarters
9.23. 42 nickels, 21 dimes
9.24. 51 dimes, 17 quarters
9.25. 41 nickels, 18 quarters
9.26. 22 nickels, 59 dimes
9.27. 330 day passes, 367 tournament passes
9.28. 112 adult tickets, 199 senior/child tickets
9.29. 32 at 49 cents, 12 at 8 cents
9.30. 26 at 49 cents, 10 at 21 cents
9.31.
155° 65°
9.32.
103° 13°
9.33. 40°, 140°

9.34. 25°, 65°
9.35. 21°
9.36. 56°
9.37. 34°
9.38. 45°
9.39. 20°, 70°, 90°
9.40. 30°, 60°, 90°
9.41. 8
9.42. 22.5
9.43. 10
9.44. 17
9.45. 8
9.46. 12
9.47. 12 feet
9.48. 8 feet
9.49.

cubic linear

square linear

square cubic

9.50.

cubic square

cubic linear

square linear

9.51.

8 inches 3 sq. inches

9.52.

8 centimeters

4 sq. centimeters

9.53.

340 yd 6000 sq. yd

9.54.

220 ft 2976 sq. ft

9.55. 15 in.
9.56. 9 yd
9.57. 18 m, 11 m
9.58. 11 ft , 19 ft
9.59. 8 ft, 24 ft
9.60. 5 cm, 4 cm
9.61. 26 ft
9.62. 29 m
9.63. 30 ft, 70 ft
9.64. 60 yd, 90 yd
9.65. 13 sq. in.
9.66. 49 sq. in.
9.67. 8 ft
9.68. 6 ft
9.69. 14 in.
9.70. 6 ft
9.71. 13 in.
9.72. 17 cm
9.73. 14 ft
9.74. 7 m
9.75. 161 sq. yd
9.76. 225 sq. cm
9.77. 42 sq. cm
9.78. 63 sq. m
9.79. 40.25 sq. yd
9.80. 240 sq. ft
9.81.

31.4 in. 78.5 sq. in.

9.82.

28.26 ft 63.585 sq. ft

9.83. 17.27 ft
9.84. 37.68 ft
9.85. 30 cm
9.86. 110 ft
9.87. 28 sq. units
9.88. 110 sq. units
9.89. 36.5 sq. units
9.90. 70 sq. units
9.91. 103.2 sq. units
9.92. 38.24 sq. units
9.93.

792 cu. ft 518 sq. ft

9.94.

1,440 cu. ft 792 sq. ft

9.95.

216 cu. ft 228 sq. ft

9.96.

2,772 cu. in.

1,264 sq. in.

9.97.

91.125 cu. m

121.5 sq. m

9.98.

389.017 cu. yd.

319.74 sq. yd.

9.99.

64 cu. ft 96 sq. ft

9.100.

4,096 cu. in.

1536 sq. in.

9.101.

113.04 cu. cm

113.04 sq. cm

9.102.

4.19 cu. ft 12.56 sq. ft

9.103.

3052.08 cu. in.

1017.36 sq. in.

9.104.

14.13 cu. ft 28.26 sq. ft

9.105.

351.68 cu. cm

276.32 sq. cm

9.106.

100.48 cu. ft

125.6 sq. ft

9.107.

3,818.24 cu. cm

1,356.48 sq. cm

9.108.

91.5624 cu. ft

113.6052 sq. ft

9.109. 65.94 cu. in.
9.110. 235.5 cu. cm
9.111. 678.24 cu. in.
9.112. 128.2 cu. in.
9.113. 330 mi
9.114. 7 mi
9.115. 11 hours
9.116. 56 mph
9.117.

9.118.

9.119.

9.120.

9.121.

9.122.

9.123.

9.124.

9.125. $b = P - a - c$

9.126. $c = P - a - b$

9.127. $y = 11 - 7x$

9.128. $y = 8 - 11x$

9.129.

9.130.

Section Exercises

1. There are 30 children in the class.
3. Zachary has 125 CDs.
5. There are 6 boys in the club.
7. There are 17 glasses.
9. Lisa's original weight was 175 pounds.
11. 18%
13. The original price was $120.
15. 4
17. 15
19. 5
21. 12
23. −5
25. 18, 24
27. 8, 12
29. −2, −3
31. 4, 10
33. 32, 46
35. 38, 39
37. −11, −12
39. 25, 26, 27
41. −11, −12, −13
43. The original price was $45.
45. Each sticker book cost $1.25.
47. The price of the refrigerator before tax was $1,080.
49. Answers will vary.
51. 8 nickels, 22 dimes
53. 15 dimes, 8 quarters
55. 12 dimes and 27 nickels
57. 63 dimes, 20 quarters
59. 10 of the $1 bills, 7 of the $5 bills
61. 10 of the $10 bills, 5 of the $5 bills
63. 16 nickels, 12 dimes, 7 quarters
65. 30 child tickets, 50 adult tickets
67. 110 child tickets, 50 adult tickets
69. 40 postcards, 100 stamps
71. 30 at 49 cents, 10 at 21 cents
73. 15 at $10 shares, 5 at $12 shares
75. 9 girls, 3 adults
77. Answers will vary.
79. Answers will vary.
81.
127° 37°
83.
151° 61°
85. 45°
87. 62.5°
89. 62°, 118°
91. 62°, 28°
93. 56°
95. 44°
97. 57°
99. 67.5°
101. 45°, 45°, 90°
103. 30°, 60°, 90°
105. 12
107. 351 miles
109. 15
111. 25
113. 8
115. 12
117. 10.2
119. 8
121. 5 feet
123. 14.1 feet
125. 2.9 feet
127. Answers will vary.
129. cubic
131. square
133. linear
135.
10 cm 4 sq. cm
137.
8 cm 3 sq. cm
139.
10 cm 5 sq. cm
141.
260 ft 3825 sq. ft
143.
58 ft 210 sq. ft
145. 24 inches
147. 27 meters
149. 23 m
151. 7 in., 16 in.
153. 17 m, 12 m
155. 13.5 m, 12.8 m
157. 25 ft, 50 ft
159. 7 m, 11 m
161. 26 in.
163. 55 m
165. 35 ft, 45 ft
167. 76 in., 36 in.
169. 30 sq. in.
171. 25.315 sq. m
173. 0.75 sq. ft

175. 8 ft
177. 23 in.
179. 11 ft
181. 28 cm
183. 17 ft
185. 6 m
187. 15 ft
189. 24 in.
191. 27.5 in.
193. 12 ft, 13 ft, 14 ft
195. 3 ft, 6 ft, 8 ft
197. 144 sq. ft
199. 2805 sq. m
201. 231 sq. cm
203. 28.56 sq. m
205. 13.5 sq. ft
207. 1036 sq. in.
209. 15 ft
211. $24
213. Answers will vary.
215. Answers will vary.
217.
43.96 in. 153.86 sq. in.
219.
53.38 ft 226.865 sq. ft
221. 62.8 ft
223. 37.68 in.
225. 6.908 ft
227. 52 in.
229. 5.5 m
231. 24 ft
233. 6.5 mi
235. 16 sq. units
237. 30 sq. units
239. 57.5 sq. units
241. 12 sq. units
243. 67.5 sq. units
245. 89 sq. units
247. 44.81 sq. units
249. 41.12 sq. units
251. 35.13 sq. units
253. 95.625 sq. units
255. 187,500 sq. ft
257. 9400 sq. ft
259.
6.5325 sq. ft
10.065 sq. ft
261. Answers will vary.
263.
9 cu. m 27 sq. m
265.
17.64 cu. yd.
41.58 sq. yd.
267.
1,024 cu. ft 640 sq. ft
269.
3,350.49 cu. cm
1,622.42 sq. cm
271.
125 cu. cm 150 sq. cm
273.
1124.864 cu. ft.
648.96 sq. ft
275.
262,144 cu. ft
24,576 sq. ft
277.
21.952 cu. m
47.04 sq. m
279.
113.04 cu. cm
113.04 sq. cm
281.
1,766.25 cu. ft
706.5 sq. ft
283.
14,130 cu. in.
2,826 sq. in.
285.
381.51 cu. cm
254.34 sq. cm
287.
254.34 cu. ft
226.08 sq. ft
289.
29.673 cu. m
53.694 sq. m
291.
1,020.5 cu. cm
565.2 sq. cm
293.
678.24 cu. in.
508.68 sq. in.
295. 37.68 cu. ft
297. 324.47 cu. cm
299. 261.67 cu. ft
301. 64,108.33 cu. ft
303.
31.4 cu. ft 2.6 cu. ft
28.8 cu. ft
305. Answers will vary.
307. 612 mi
309. 7 mi
311. 6.5 hours
313. 3.6 hours
315. 60 mph
317. 80 mph
319.
321.
323.
325.
327.
329.
331.
333.
335.

337.

$y = 13$ $y = 7 - 3x$

339.

$b = 90 - a$ $a = 90 - b$

341. $a = 180 - b - c$

343. $y = 15 - 8x$

345. $y = -6 + 4x$

347.

349. $y = 4 + x$

351.

353.

355.

357. 104° F

359. Answers will vary

Review Exercises

361. Answers will vary.

363. There are 116 people at the concert.

365. His original weight was 180 pounds.

367. 38

369. 18, 9

371. 16 dimes, 11 quarters

373. 6 of \$5 bills, 11 of \$10 bills

375. 35 adults, 82 children

377. 3 of 26 -cent stamps, 8 of 41 -cent stamps

379. 132°

381. 33°, 57°

383. 73°

385. 30°, 60°, 90°

387. 15

389. 26

391. 8

393. 8.1

395. 6 feet

397. cubic

399. square

401.

8 units 3 sq. units

403.

140 m 1176 sq. m

405.

98 ft. 180 sq. ft.

407. 25 cm

409. 62 m

411. 24.5 cm., 12.5 cm.

413. 135 sq. in.

415. 600 sq. in.

417. 7 in., 7 in.

419. 17 ft., 20 ft., 22 ft.

421. 100 sq. ft.

423. 675 sq. m

425.

18.84 m 28.26 sq. m

427. 48 in.

429. 30 sq. units

431. 300 sq. units

433. 199.25 sq. units

435.

630 cu. cm 496 sq. cm

437.

15.625 cu. in.

37.5 sq. in.

439.

267.95 cu. yd.

200.96 sq. yd.

441.

12.76 cu. in.

26.41 sq. in.

443.

75.36 cu. yd.

100.48 sq. yd.

445.

753.6 cu. cm

477.28 sq. cm

447. 5.233 cu. m

449. 4.599 cu. in.

451. 1520 miles

453. 1.6 hours

455.

457.

459.

461.

463. $a = 90 - b$

465. $y = 17 - 4x$

467.

Practice Test

471. −16

473. 7 quarters, 12 dimes

475. 38°

477. 48.3

479. 10

481. 127.3 ft

483. 2200 square centimeters

485. 282.6 inches

487. 1440

489. 31,400 cubic inches

491. 14.7 miles per hour

493.

Chapter 10

Be Prepared

10.1.
10.2.
10.3.
10.4. —
10.5.
10.6.
10.7.
10.8.
10.9. –
10.10.
10.11. –
10.12. ten thousand
10.13. twelve ten-thousandths
10.14.
10.15.
10.16.
10.17.

Try It

10.1.

monomial polynomial
trinomial binomial
monomial

10.2.

binomial trinomial
polynomial monomial
monomial

10.3.

1 1 4 2
0

10.4.

0 2 4 5
3

10.5. $17x^2$
10.6. $-3y^2$
10.7. $14n$
10.8. $-2a^3$
10.9. $-2x^2 + 3y^2$
10.10. $-2a^2 + b^2$
10.11. $4x^2 - 8x + 10$
10.12. $11y^2 + 9y - 5$
10.13. $3y^2 + 3y + 3$
10.14. $3u^2 - u + 2$
10.15. $4n^2 + 12n$
10.16. $5a^2 + 8a + 8$
10.17.

13 3

10.18.

114 −2

10.19. 4 feet
10.20. 20 feet
10.21.

64 11

10.22.

81 21

10.23.

—

10.24.

—

10.25.

16 −16

10.26.

64 −64

10.27. x^{15}
10.28. x^{16}
10.29. p^{10}
10.30. m^8
10.31. 6^{10}
10.32. 9^{15}
10.33. y^{43}
10.34. z^{39}
10.35. x^{21}
10.36. y^{15}
10.37.

x^{28} 7^{32}

10.38.

x^{54} 8^{42}

10.39. $196x^2$
10.40. $144a^2$
10.41. $256x^4y^4$
10.42. $216x^3y^3$
10.43. x^{40}
10.44. y^{42}
10.45. $-512x^{12}y^{21}$
10.46. $81a^{20}b^{24}$
10.47. $98n^{14}$
10.48. $48m^5$
10.49. $64u^{18}v^{22}$
10.50. $675x^7y^{18}$
10.51. $-56x^{11}$
10.52. $54y^9$
10.53. $12m^5n^6$
10.54. $12p^{11}q^8$
10.55. $6x + 48$
10.56. $2y + 24$
10.57. $y^2 - 9y$
10.58. $p^2 - 13p$
10.59. $8x^2 + 24xy$
10.60. $18r^2 + 3rs$
10.61. $-32y^3 - 20y^2 + 36y$
10.62. $-54x^3 - 6x^2 + 6x$
10.63. $12x^4 - 9x^3 + 27x^2$
10.64. $24y^4 - 16y^3 - 32y^2$
10.65. $xp + 8p$
10.66. $ap + 4p$
10.67. $x^2 + 17x + 72$
10.68. $a^2 + 9a + 20$
10.69. $20x^2 + 51x + 27$
10.70. $80m^2 + 142m + 63$
10.71. $56y^2 - 13y - 3$
10.72. $15x^2 - 14x - 16$
10.73. $x^2 - xy + 5x - 5y$
10.74. $x^2 - x + 2xy - 2y$
10.75. $x^2 + 15x + 56$
10.76. $y^2 + 16y + 28$
10.77. $y^2 + 5y - 24$
10.78. $q^2 + q - 20$
10.79. $20a^2 + 37a - 18$
10.80. $49x^2 - 28x - 32$
10.81. $12x^2 - 60x - xy + 5y$

10.82. $12a^2 - 54a - 2ab + 9b$

10.83. $12m^2 - 55m + 63$

10.84. $42n^2 - 47n + 10$

10.85. $y^3 - 8y^2 + 9y - 2$

10.86. $3x^3 + 2x^2 - 3x + 10$

10.87. $y^3 - 8y^2 + 9y - 2$

10.88. $3x^3 + 2x^2 - 3x + 10$

10.89.
x^3 7^9

10.90.
y^6 8^8

10.91.

10.92.

10.93.

10.94.

10.95.
1 1

10.96.
1 1

10.97. 1

10.98. 1

10.99.
1 $7x^2$

10.100.
$-23x^2$ 1

10.101.

10.102.

10.103. a^{11}

10.104. b^{19}

10.105. k^2

10.106.

10.107. f^{12}

10.108.

10.109.

10.110.

10.111.

10.112.

10.113.

10.114. $9x^5$

10.115. $7x^4$

10.116. $16y^3$

10.117.

10.118.

10.119.

10.120.

10.121.

10.122.

10.123. $2xy^2$

10.124. $-4ab^5$

10.125.

10.126.

10.127.

10.128.

10.129.

10.130.

10.131.

10.132.

10.133.

10.134.

10.135.

10.136.

10.137.

10.138.

10.139.

10.140.

10.141.

10.142.

10.143.

10.144.

10.145. x^{11}

10.146. y^{13}

10.147. 9.6×10^4

10.148. 4.83×10^4

10.149. 7.8×10^{-3}

10.150. 1.29×10^{-2}

10.151. 1,300

10.152. 92,500

10.153. 0.00012

10.154. 0.075

10.155. 0.06

10.156. 0.009

10.157. 400,000
10.158. 20,000
10.159. 18
10.160. 16
10.161. 7
10.162. 11
10.163. $8x^2$
10.164. $9y^3$
10.165. $3x$
10.166. $5m^2$
10.167. $4(x+3)$
10.168. $6(a+4)$
10.169. $9(a+1)$
10.170. $11(x+1)$
10.171. $11(x-4)$
10.172. $13(y-4)$
10.173. $4(y^2+2y+3)$
10.174. $6(x^2+7x-2)$
10.175. $x(9x+7)$
10.176. $a(5a-12)$
10.177. $2x^2(x+6)$
10.178. $3y^2(2y-5)$
10.179. $9y(2y+7)$
10.180. $8k(4k+7)$
10.181. $6y(3y^2-y-4)$
10.182. $4x(4x^2+2x-3)$
10.183. $-5(y+7)$
10.184. $-8(2z+7)$
10.185. $-7a(a-3)$
10.186. $-x(6x-1)$

Section Exercises

1. binomial
3. trinomial
5. polynomial
7. monomial
9. 5
11. 1
13. 0
15. $15x^2$
17. $-8u$
19. $5a+7b$
21. $-4a-3b$
23. $16x$
25. $-17x^6$
27. $12y^2+4y+8$
29. $-3x^2+17x-1$
31. $4a^2-7a-11$
33. $4m^2-10m+2$
35. $11z+8$
37. $12s^2-16s+9$
39. $2p^3+p^2+9p+10$
41. x^2+3x+4
43. $11w-66$
45.

187 40 2

47.

−104 4 40

49. 19 feet
51. 10 mpg
53. Answers will vary.
55. 1,024
57. —
59. 0.008
61. 625
63. −625
65. −10,000
67. —
69. −0.25
71. x^9
73. a^5
75. 3^{14}
77. z^6
79. x^{a+2}
81. y^{a+b}
83. u^8
85. y^{20}
87. 10^{12}
89. x^{90}
91. x^{2y}
93. 5^{xy}
95. $25a^2$
97. $-216m^3$
99. $16r^2s^2$
101. $256x^4y^4z^4$
103. x^{14}
105. a^{36}
107. $45x^3$
109. $200a^5$
111. $8m^{18}$
113. $1{,}000x^6y^3$
115. $16a^{12}b^8$
117. —
119. $1{,}024a^{10}$
121. $25{,}000p^{24}$
123. $x^{18}y^{18}$
125. $144m^8n^{22}$
127. $-60x^6$
129. $72u^7$
131. $4r^{11}$
133. $36a^5b^7$
135. $8x^2y^5$
137. —
139. 1,679,616
141. Answers will vary.
143. Answers will vary.
145. $4x+40$
147. $15r-360$
149. $-3m-33$
151. $-8z+40$
153. u^2+5u
155. n^3-3n^2
157. $12x^2-120x$
159. $-27a^2-45a$
161. $24x^2+6xy$
163. $55p^2-25pq$
165. $3v^2+30v+75$
167. $8n^3-8n^2+2n$
169. $-8y^3-16y^2+120y$
171. $5q^5-10q^4+30q^3$
173. $-12z^4-48z^3+4z^2$
175. $2y^2-9y$
177. $8w-48$
179. $x^2+10x+24$
181. $n^2+9n-36$
183. $y^2+11y+24$
185. $a^2+22a+96$
187. $u^2-14u+45$
189. $z^2-32z+220$
191. $x^2+3x-28$
193. $v^2+7v-60$
195. $6n^2+11n+5$
197. $20m^2-88m-9$
199. $16c^2-1$
201. $15u^2-82u+112$
203. $2a^2+5ab+3b^2$
205. $5x^2-20x-xy+4y$
207. $u^3+7u^2+14u+8$
209. $3a^3+31a^2+5a-50$
211. $y^3-16y^2+69y-54$
213. $2x^3-9x^2-17x-6$
215.

195 195

Answers will vary.

217. Answers will vary.

219. 4^6

221. x^9

223. r^4

225. —

227. —

229. —

231. 1

233. 1

235. -1

237.

1 10

239.

1 $-27x^5$

241.

1 15

243. 7

245. —

247. —

249. —

251. —

253. x^3

255. u^2

257. —

259. 1

261. —

263. a^{14}

265. y^3

267. —

269. —

271. 1

273. —

275. $3x^8$

277. $8b^6$

279. —

281. $2x$

283. —

285. —

287. —

289. —

291. —

293. —

295. $5u^4v^3$

297.

299.

301.

—

303.

305.

307.

309.

311. 1,000,000

313. Answers will vary.

315. Answers will vary.

317. —

319. —

321. —

323. —

325. —

327. —

329.

— —

331.

— —

333.

— —

335.

— —

337. —

339. —

341.

— — —

343.

— — —

345. r^3

347. —

349. —

351. —

353. —

355.
357.
359.
361.
363.
365.
367.
369.
371. m^6
373.
375.
377. b^8
379. m^7
381.
383. p^3

385. 2.8×10^5
387. 1.29×10^6
389. 4.1×10^{-2}
391. 1.03×10^{-5}
393. 6.85×10^9
395. 5.7×10^{-9}
397. 830
399. 16,000,000,000
401. 0.028
403. 0.0000000615
405. $15,000,000,000,000
407. 0.00001
409. 0.003
411. 0.00000735
413. 20,000,000
415. 50,000,000
417. 1.25×10^{-4} 8,000
419. 11,441,304,000

421. Answers will vary.
423. 15
425. 25
427. 4
429. 5
431. $3x$
433. $12p^3$
435. $2a$
437. $10y$
439. $5x^3$
441. $7b^2$
443. $5(y + 3)$
445. $4(b - 5)$
447. $7(x - 1)$
449. $3(n^2 + 7n + 4)$
451. $6(q^2 + 5q + 7)$
453. $c(9c + 22)$
455. $x(17x + 7)$
457. $q(4q + 7)$
459. $3r(r + 9)$
461. $10u(3u - 1)$
463. $b(a + 8)$
465. $11y(5 - y^3)$
467. $15c^2(3c - 1)$
469. $6c(c^2 - d^2)$
471. $24x^2(2x + 3)$
473. $18a^3(8a^3 + 5)$
475. $10(y^2 + 5y + 4)$
477. $12(u^2 - 3u - 9)$
479. $5p^2(p^2 - 4p - 3)$
481. $8c^3(c^2 + 5c - 7)$
483. $-7(p + 12)$
485. $-6b(3b + 11)$
487. $-8a^2(a - 4)$
489. $-9b^3(b^2 - 7)$
491. $-4(4t^2 - 20t - 1)$
493. Answers will vary.

Review Exercises

494. trinomial
496. binomial
498. 2
500. 0
502. $15p$
504. $-3n^5$
506. $10a^2 + 4a - 1$
508. $6y^2 - 3y + 3$
510. $8q^3 + q^2 + 6q - 29$
512. 995
514. 2,955
516. −163
518. 64 feet
520. 216
522. 0.25
524. p^{13}
526. a^6
528. y^{12}
530. 3^{10}
532. $64n^2$
534. $256a^8b^8$
536. $27a^{15}$
538. x^{21}
540. $-54p^5$
542. $56x^3y^{11}$
544. $70 - 7x$
546. $-625y^4 + 5y$
548. $a^2 + 7a + 10$
550. $6x^2 - 19x - 7$
552. $n^2 + 9n + 8$
554. $5u^2 + 37u - 24$
556. $p^2 + 11p + 28$
558. $27c^2 - 3c - 4$
560. $x^3 - 2x^2 - 24x - 21$
562. $m^3 - m^2 - 72m - 180$
564. 2^6 *or* 64
566.
568. 1
570. 1
572.
574.
576. a^2
578.
580.
582. $9p^9$
584.
586.
588.

590. —

592. —

594. x^6

596. ——

598. k^6

600. b^{10}

602. 5.3×10^6

604. 9.7×10^{-2} millimeter

606. 29,000

608. 0.375

610. 6,000

612. 30,000,000,000

614. 5

616. $4x^2$

618. $8(2u - 3)$

620. $6p(p + 1)$

622. $-9a^3(a^2 + 1)$

624. $5(y^2 - 11y + 9)$

Practice Test

626.

trinomial 4

628. $6x^2 - 3x + 11$

630. n^5

632. $-48x^5y^9$

634. $s^2 + 17s + 72$

636. $55a^2 - 41a + 6$

638. $24a^2 + 34ab - 45b^2$

640. x^{14}

642. —

644. $3y^2 - 7x$

646. —

648. x^9

650. $-6x(x + 5)$

652. 0.000525

654. 3×10^5

Chapter 11

Be Prepared

11.1.

11.2.

11.3.

11.4.

11.5. ———

11.6.

11.7.

11.8.

11.9. b

11.10. —

11.11.

11.12.

Try It

11.1.

1C

Engineering Building

11.2.

1A Library

11.3.

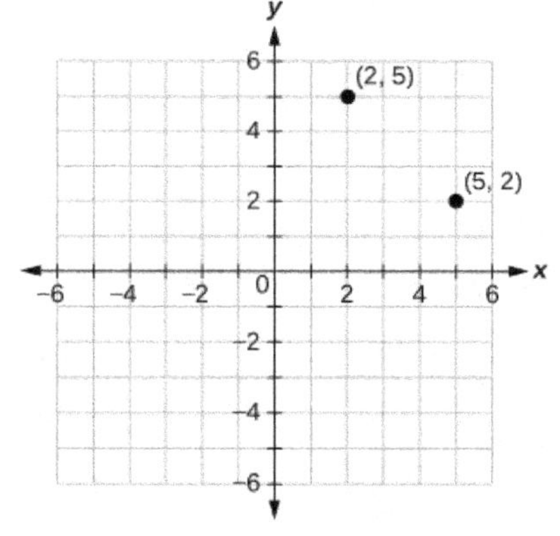

11.4.

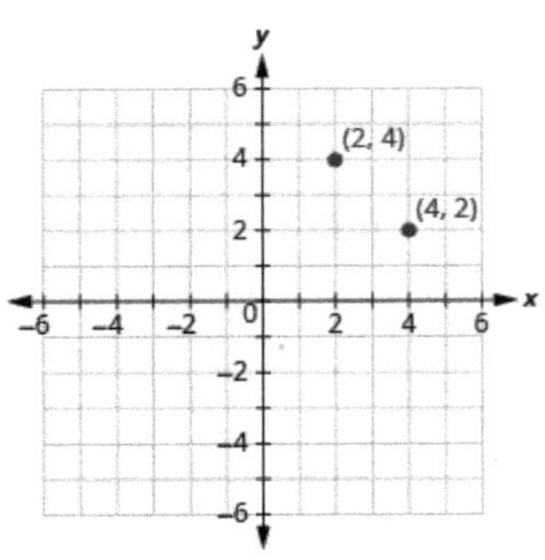

11.5. (a) Quadrant II, (b) Quadrant III, (c) Quadrant IV, (d) Quadrant II

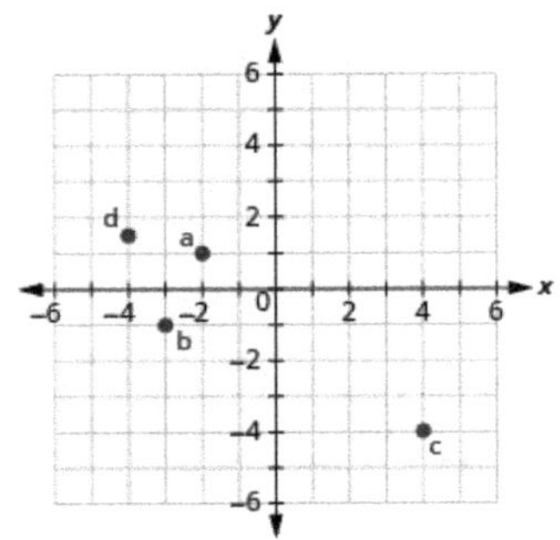

11.6. (a) Quadrant II, (b) Quadrant II, (c) Quadrant IV, (d) Quadrant II

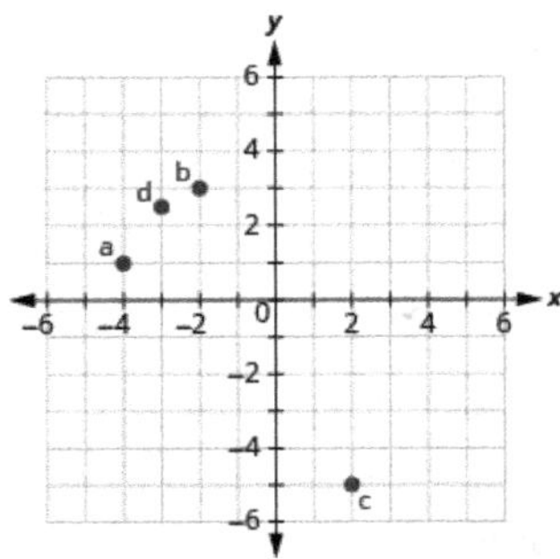

11.7.

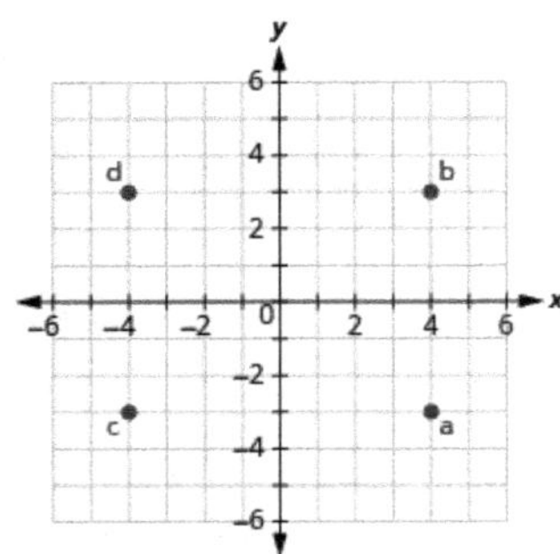

11.8.

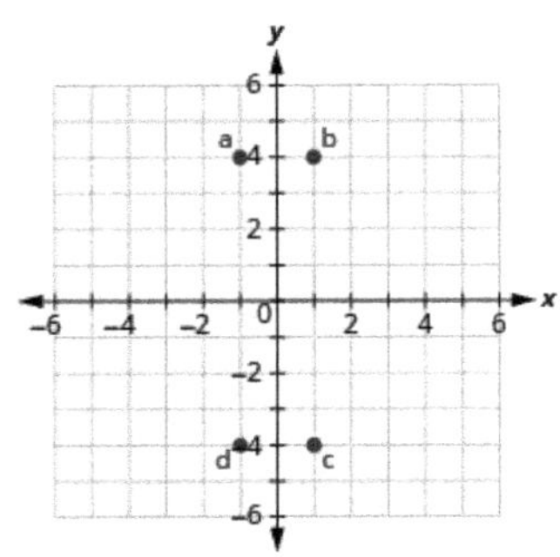

11.9.

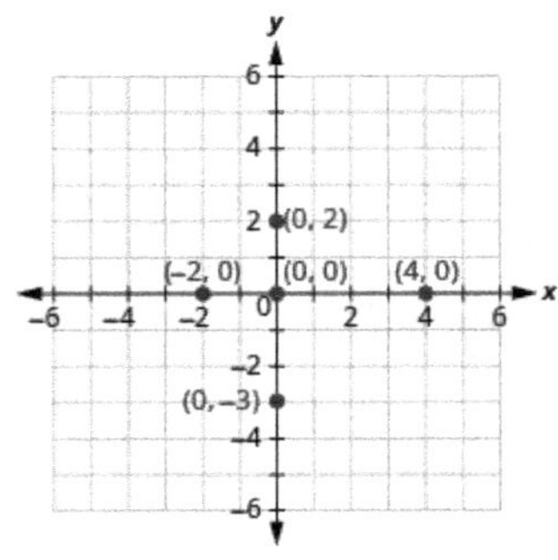

11.10.

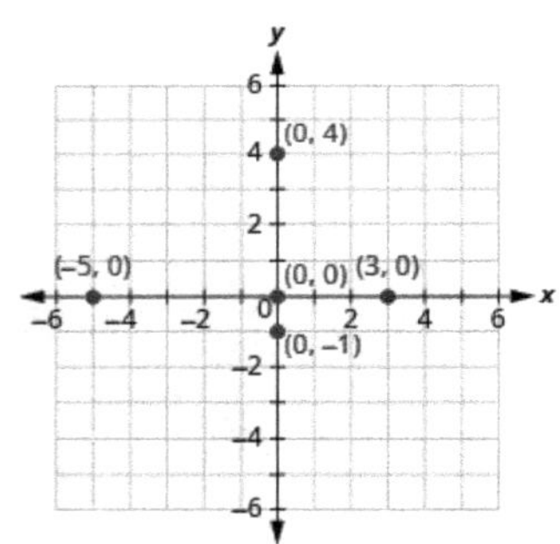

11.11.

1. A: (5,1)
2. B: (−2,4)
3. C: (−5,−1)
4. D: (3,−2)

11.12.

1. A: (4,2)
2. B: (−2,3)
3. C: (−4,−4)
4. D: (3,−5)

11.13.

1. A: (4,0)
2. B: (0,3)
3. C: (−3,0)
4. D: (0,−5)

11.14.

1. A: (−3,0)
2. B: (0,−3)
3. C: (5,0)
4. D: (0,2)

11.15. ,

11.16. ,

11.17.

11.18. ,

11.19.

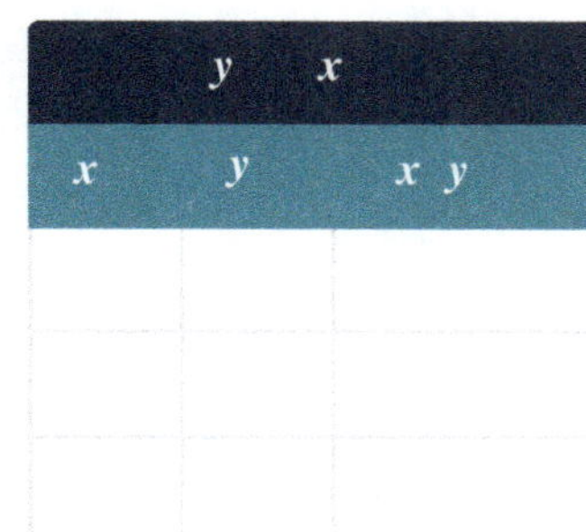

11.20.

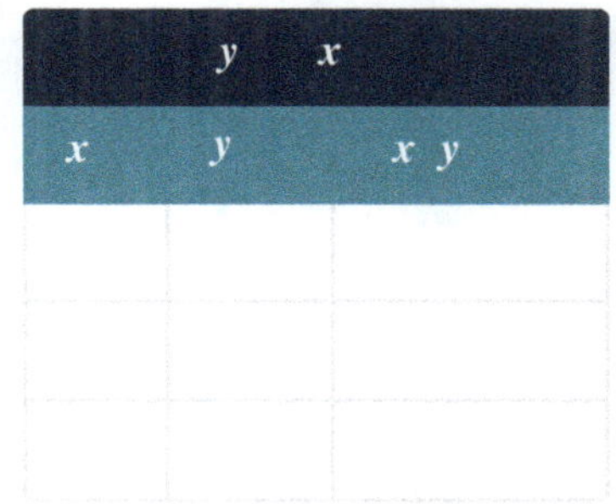

11.21.

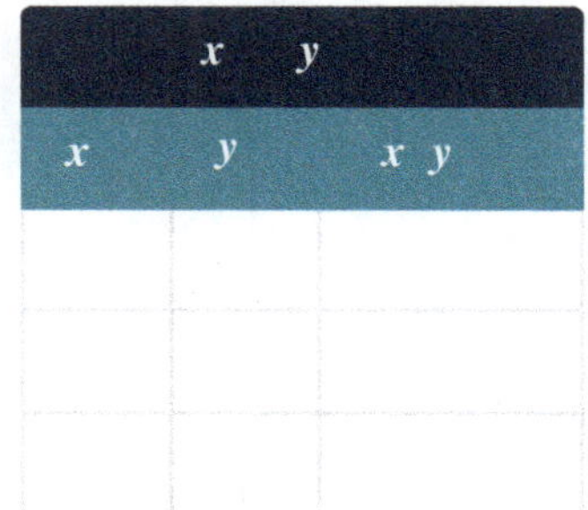

11.22.

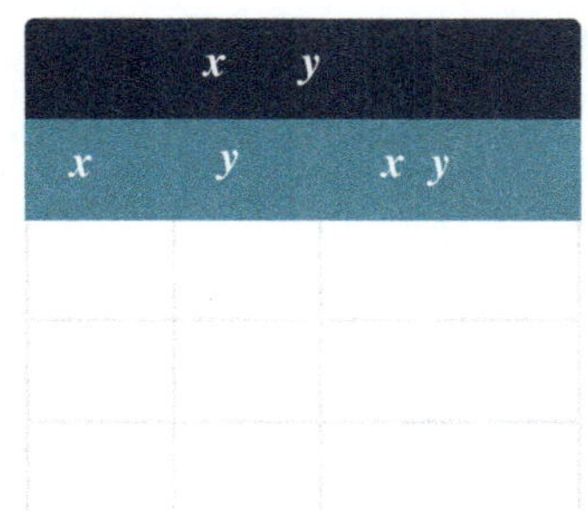

11.23. Answers will vary.

11.24. Answers will vary.

11.25. Answers will vary.

11.26. Answers will vary.

11.27. Answers will vary.

11.28. Answers will vary.

11.29.

1. yes yes
2. no no
3. no no
4. yes yes

11.30.

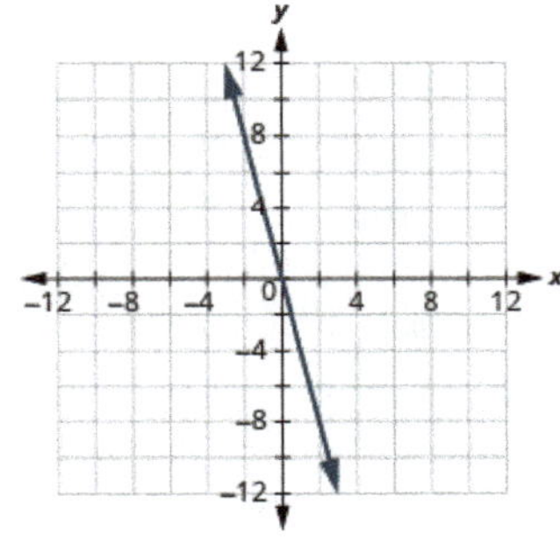

11.31.

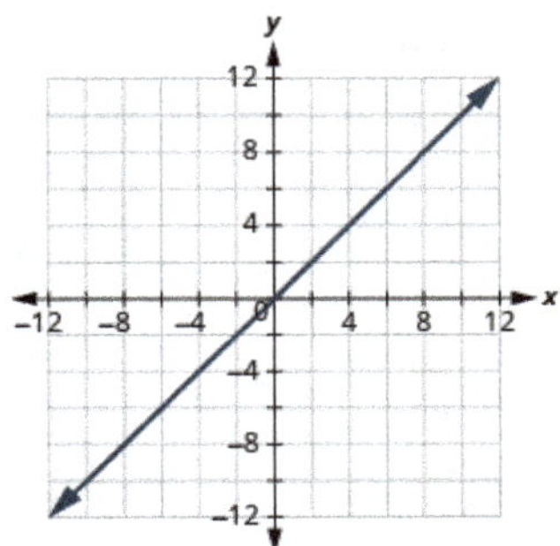

11.32.

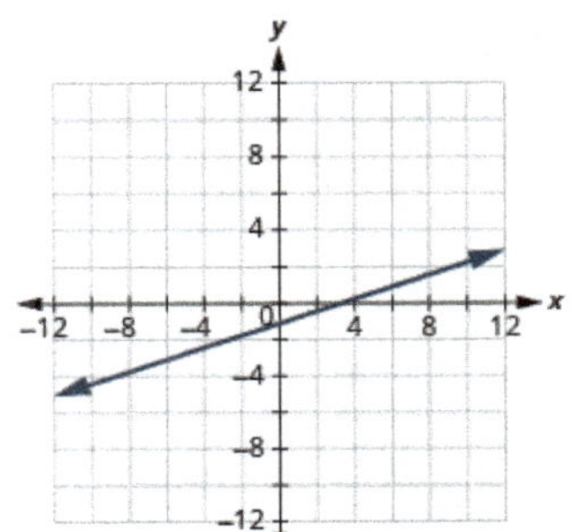

11.33.

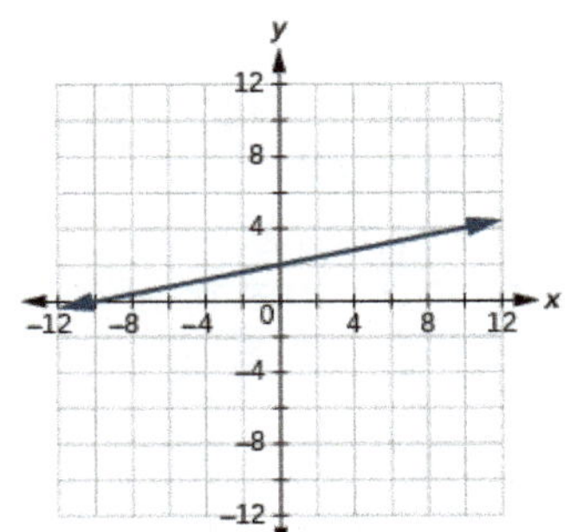

11.34.

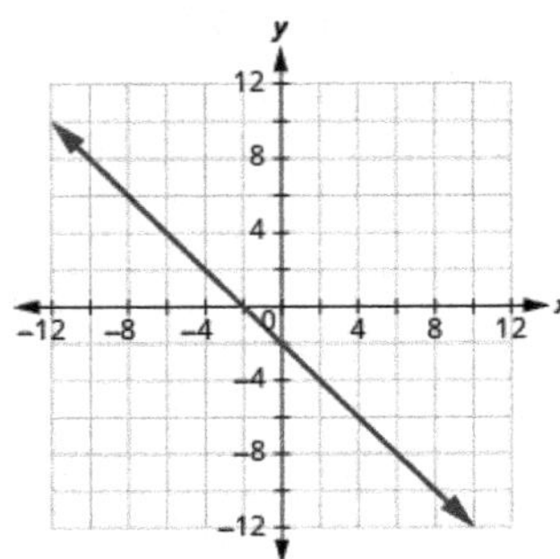

11.35.

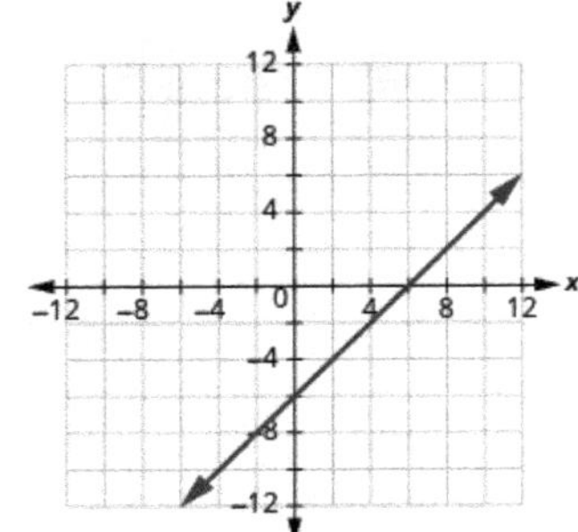

11.36.

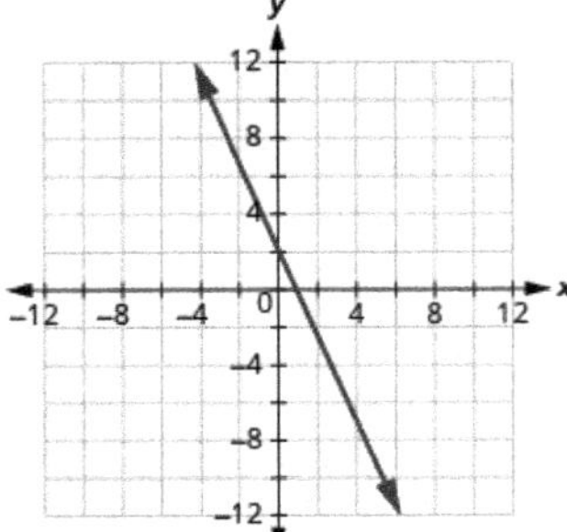

11.37.

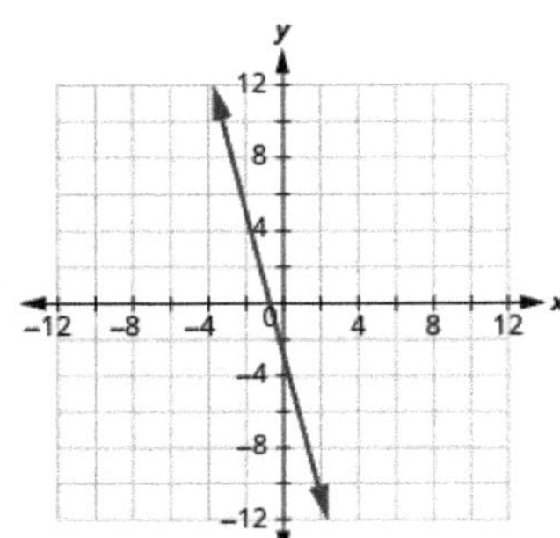

11.38.

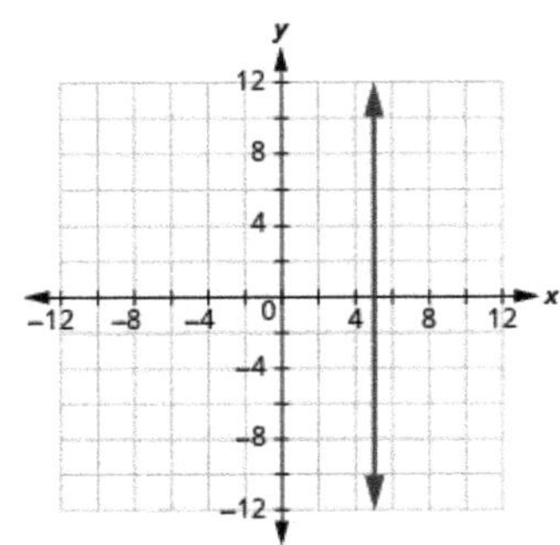

11.39.

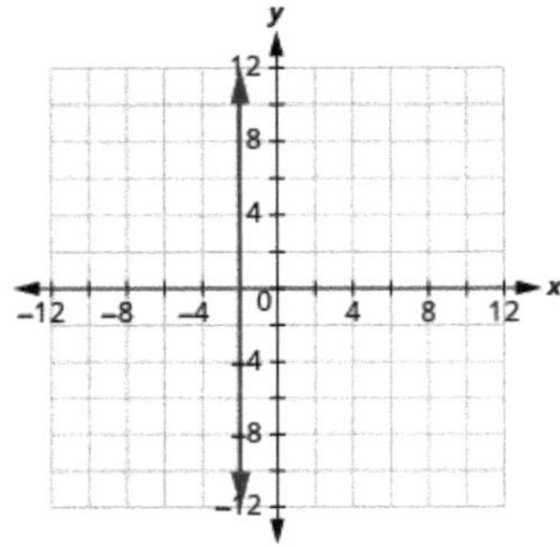

11.40.

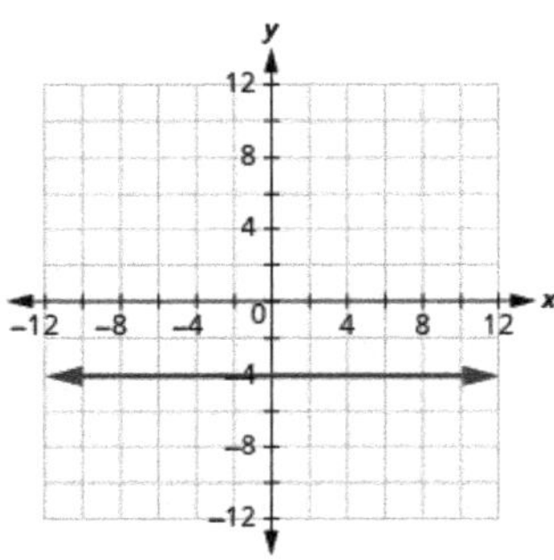

11.41.

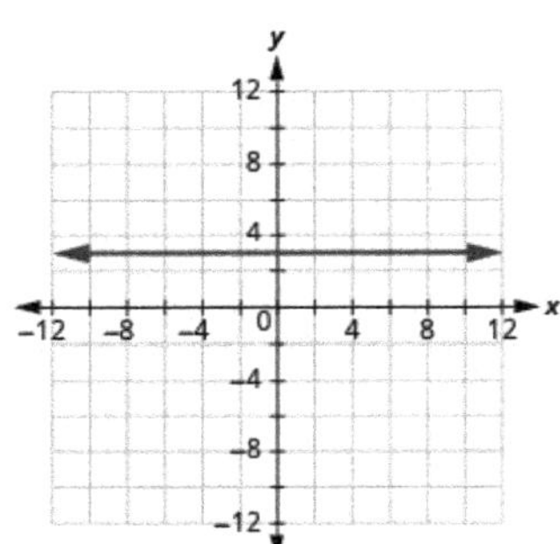

11.42.

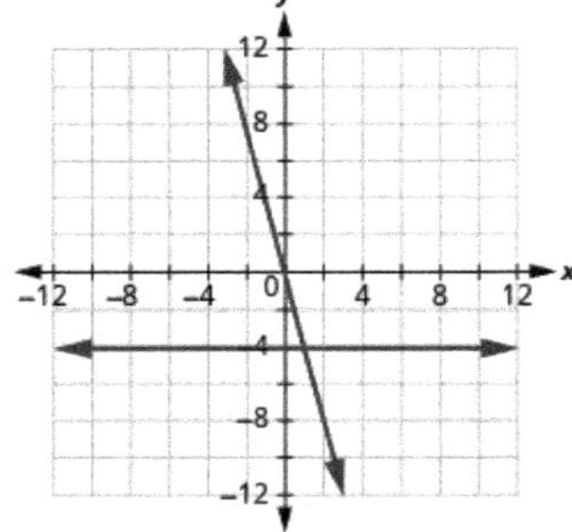

11.43.

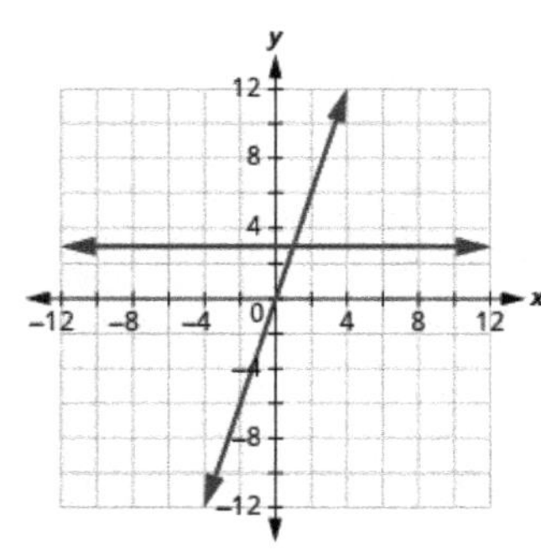

11.44. x-intercept (2,0): y-intercept (0,−2)

11.45. x-intercept (3,0); y-intercept (0,2)

11.46. and

11.47. and

11.48. x-intercept (4,0); y-intercept: (0,−3)

11.49. x-intercept (4,0); y-intercept: (0,−2)

11.50.

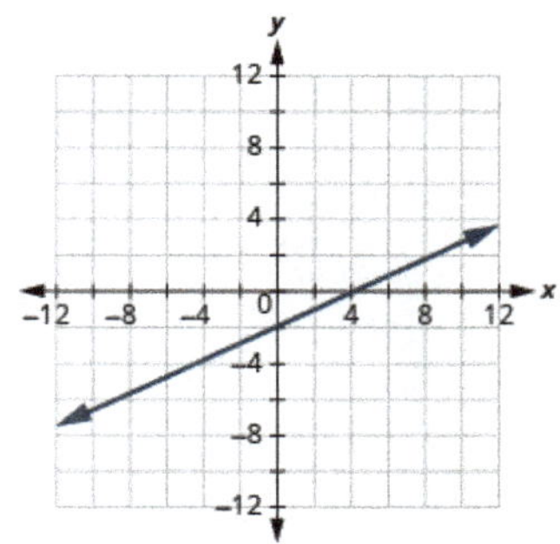

11.51.

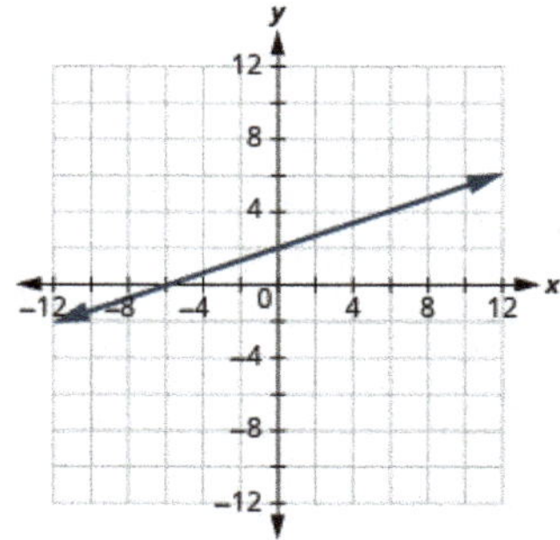

11.52.

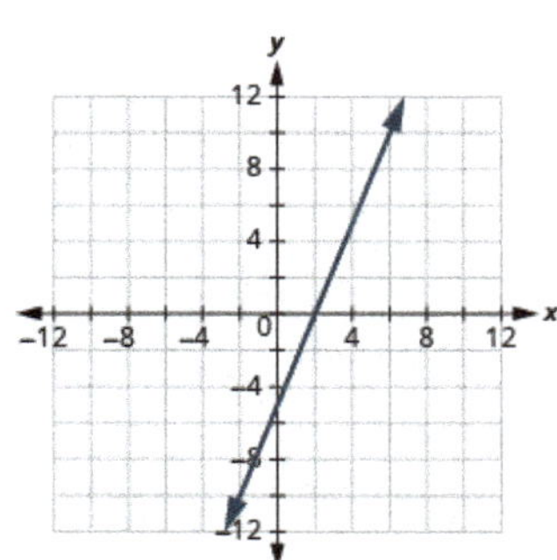

11.53.

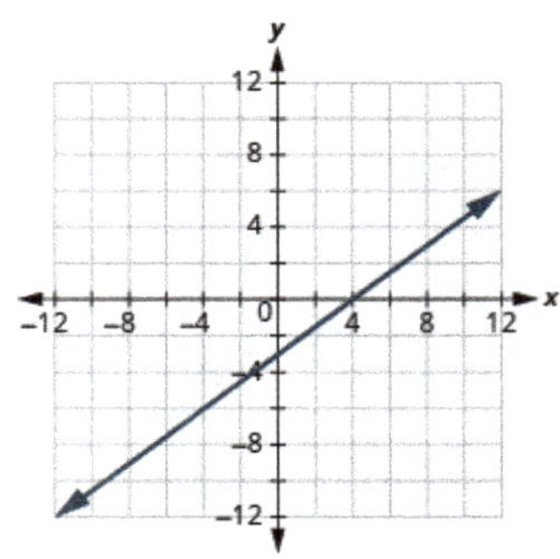

11.54.

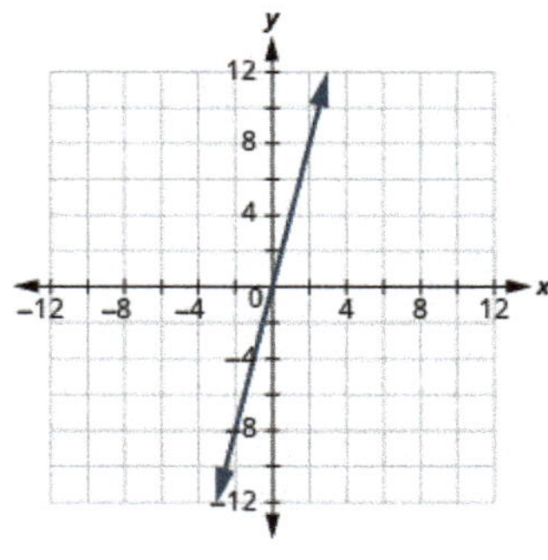

11.55.

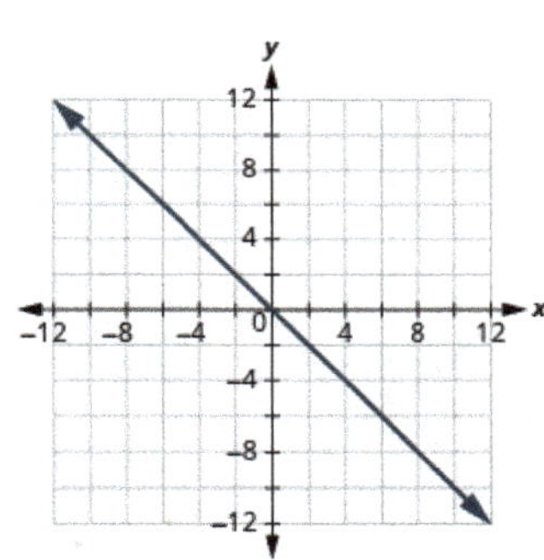

11.56.

intercepts

horizontal line

plotting points

vertical line

11.57.

vertical line

plotting points

horizontal line

intercepts

11.58. —

11.59. —

11.60. —

11.61. —

11.62.

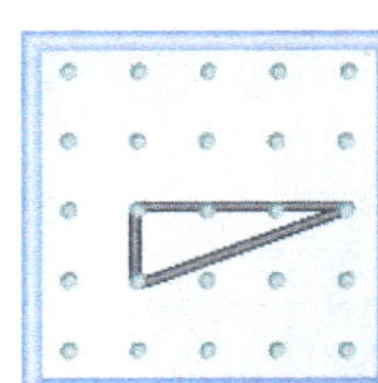

11.63.

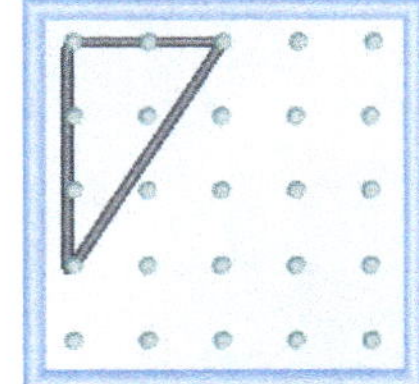

11.64.

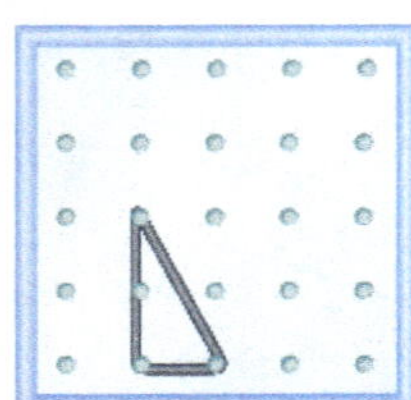

11.65.

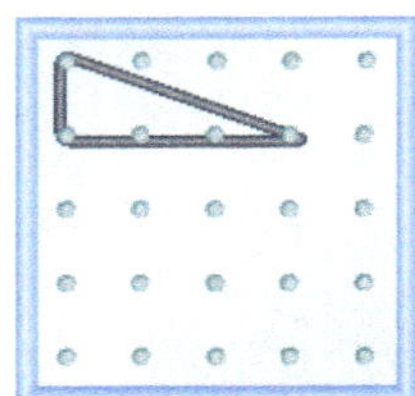

11.66. —

11.67. —

11.68. —

11.69. —

11.70. —

11.71. —

11.72. undefined

11.73. 0

11.74. 1

11.75. 1

11.76. -1

11.77. 10

11.78.

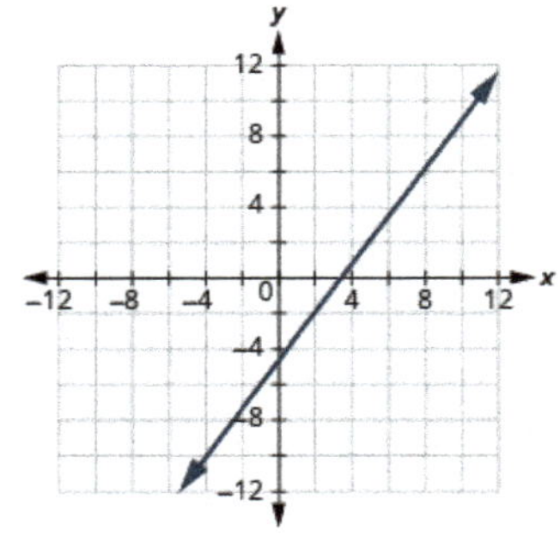

11.79.

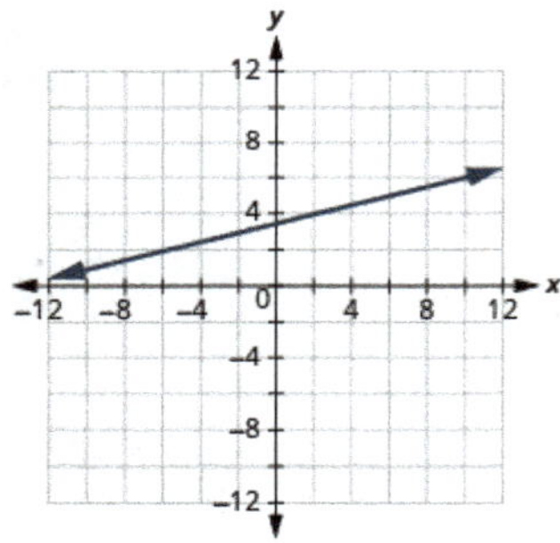

11.80.

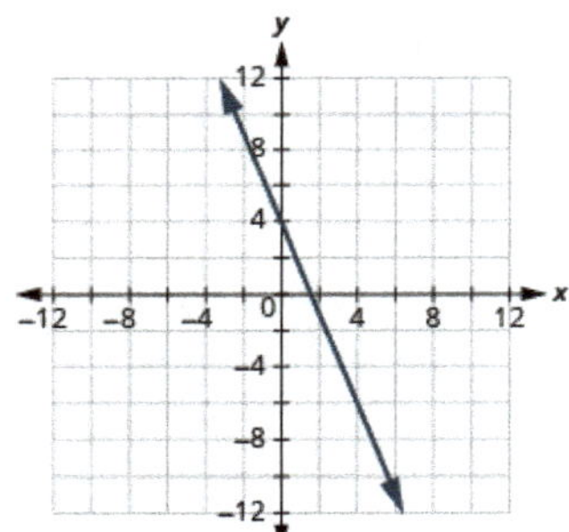

11.81.

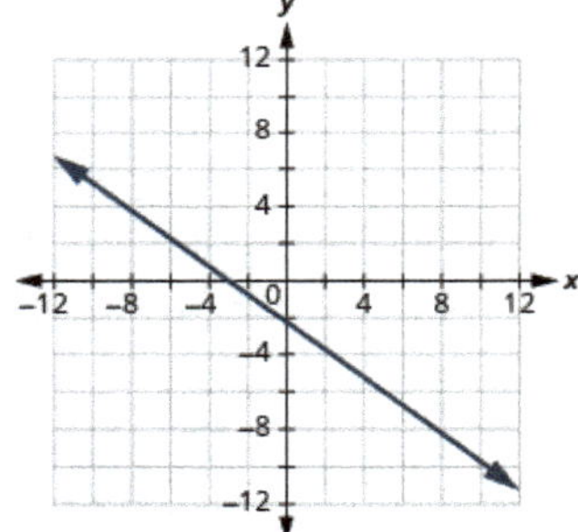

11.82.

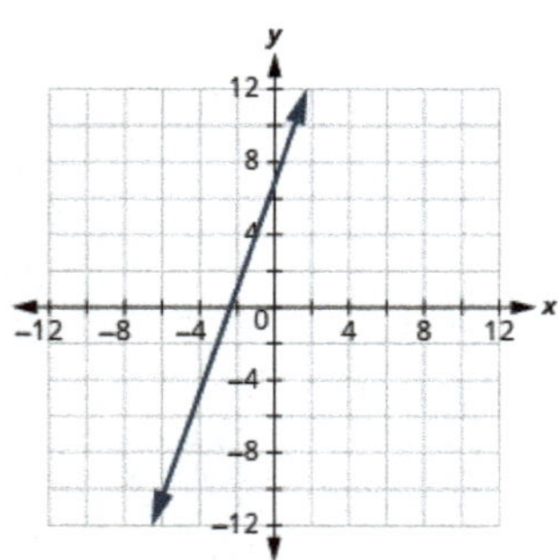

11.83.

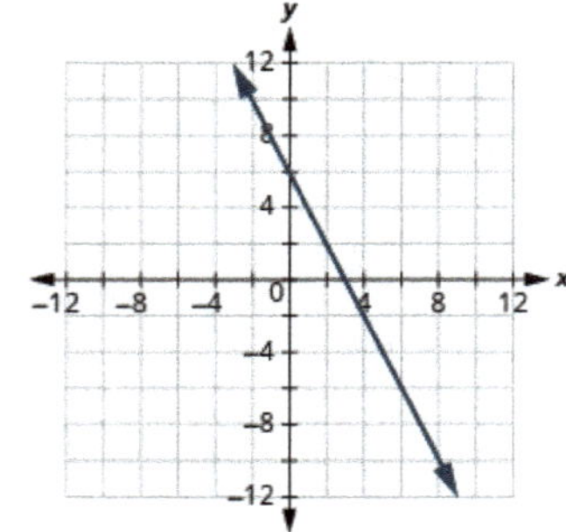

11.84. —

11.85. —

11.86. —

11.87. —

Section Exercises

1.

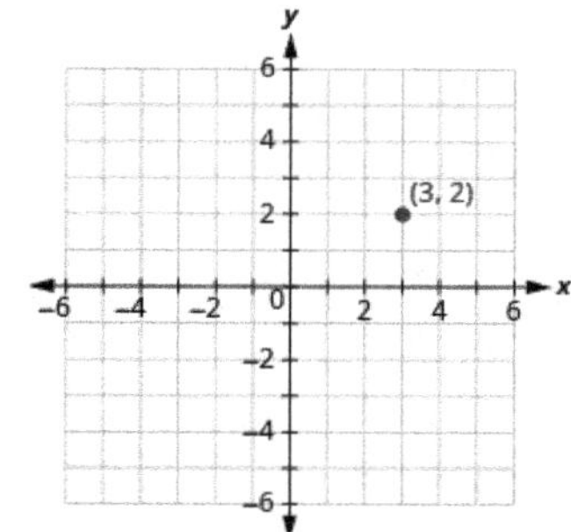

3.

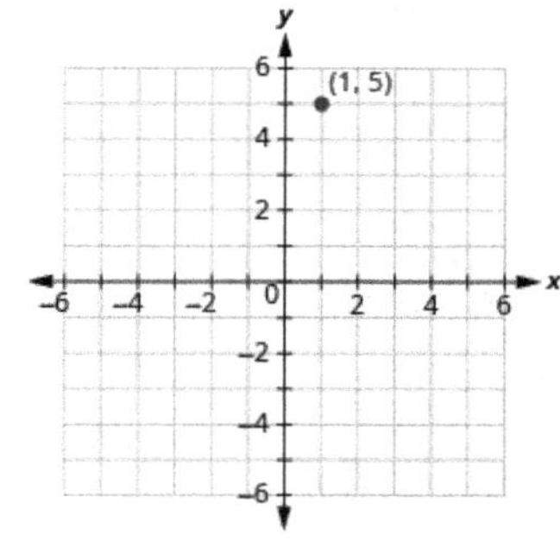

5.

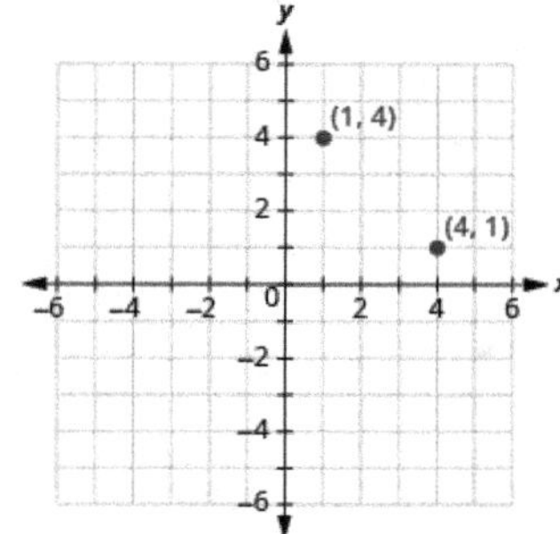

7.

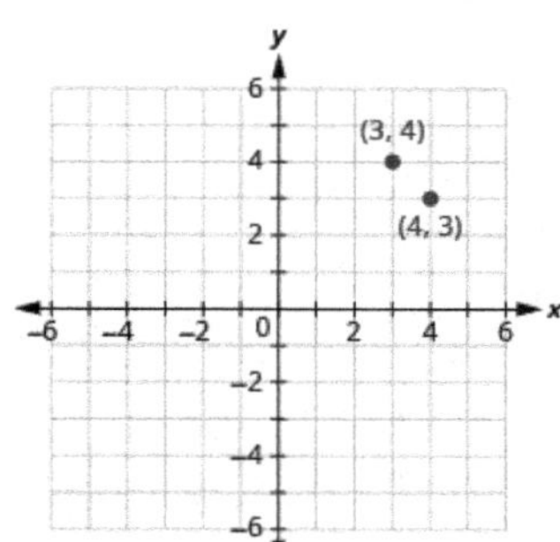

9.

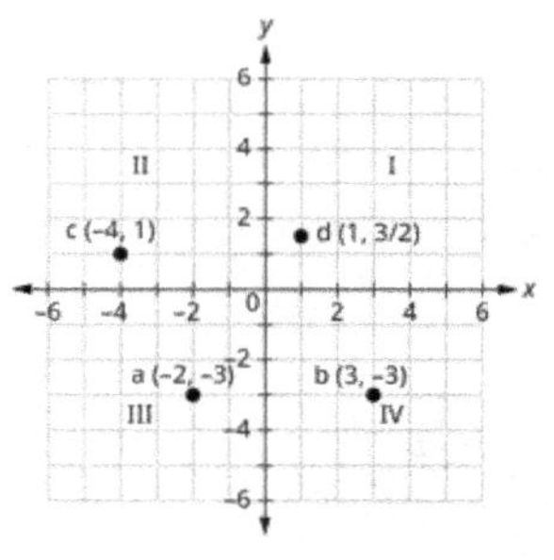

11.

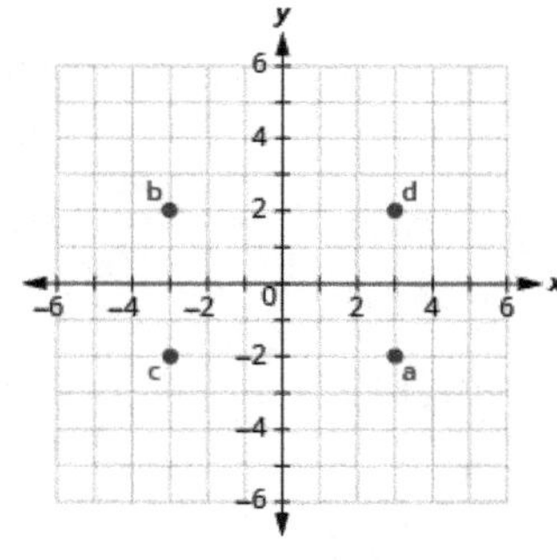

13.

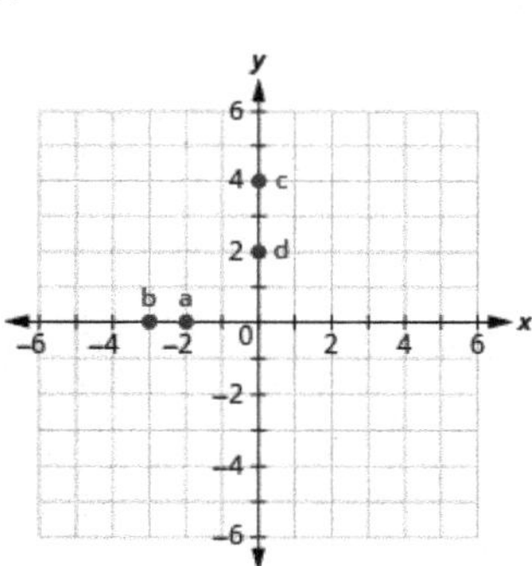

15. C(1, -3) D(4, 3)

17. S(-2, 4) T(-4, -2)

19. C(0, -1) D(-1, 0)

21. ,

23. ,

25. ,

27. ,

29.

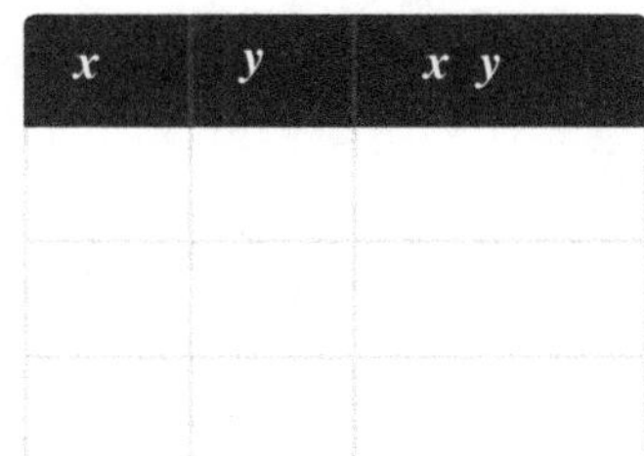

x	y	x y

31.

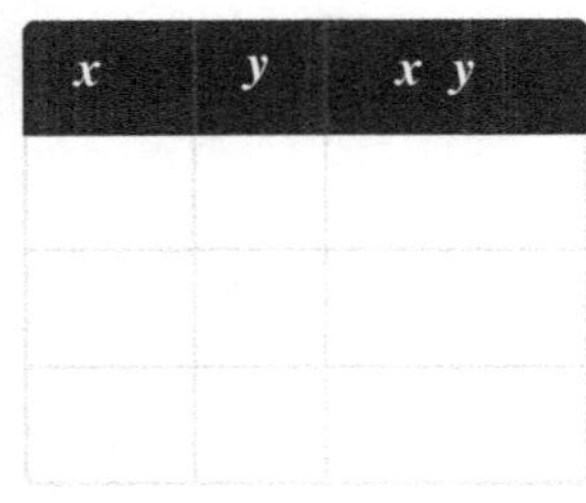

x	y	x y

33.

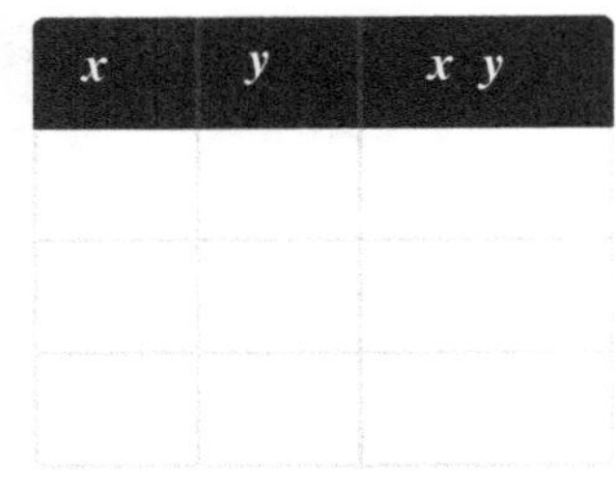

x	y	x y

35.

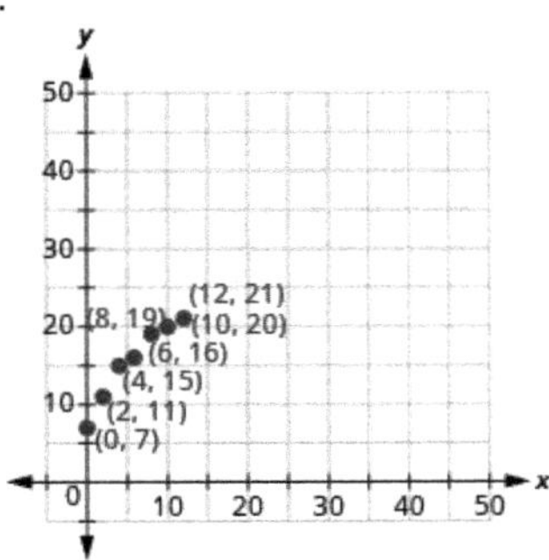

Age and weight are only positive.

37. Answers may vary.

39.

1. yes yes
2. no no
3. yes yes
4. no no

41.

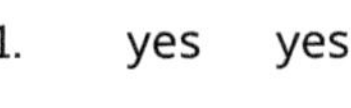

1. yes yes
2. yes yes
3. yes yes
4. no no

43.

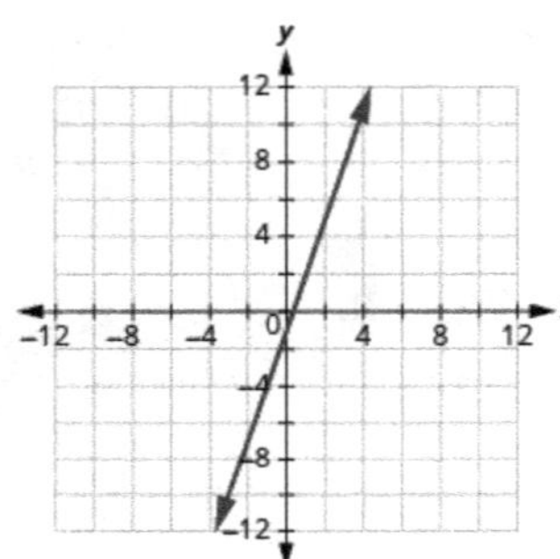

45.

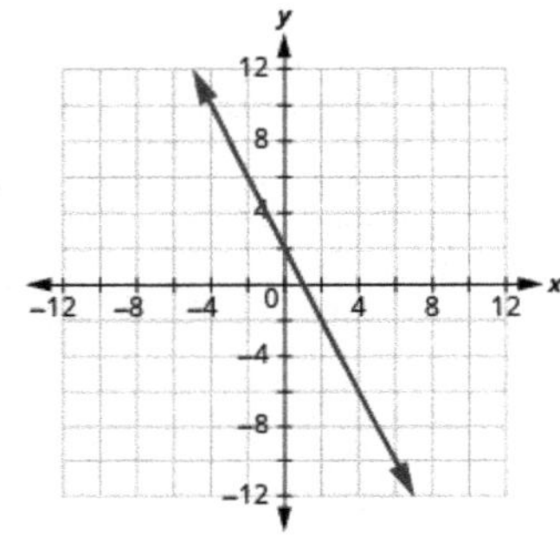

47.

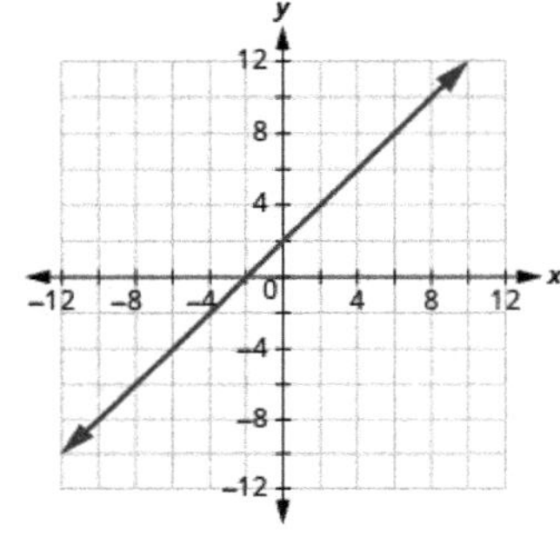

49.

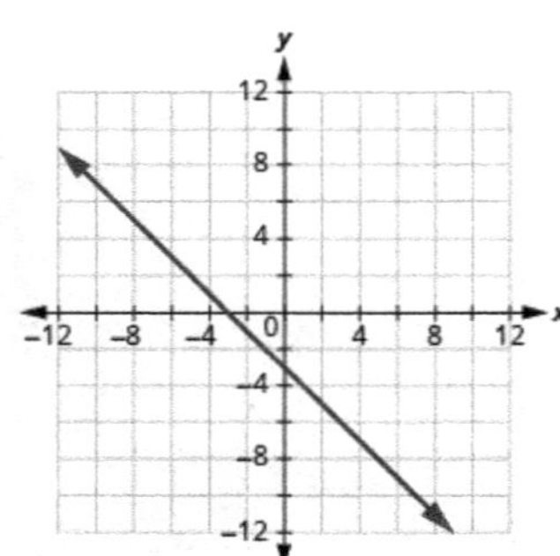

51.

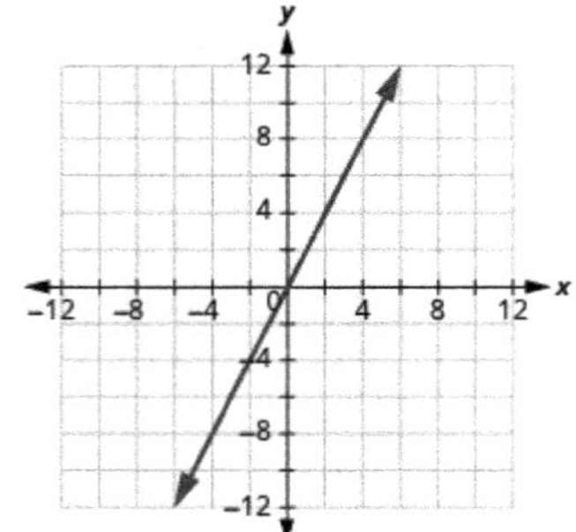

53.

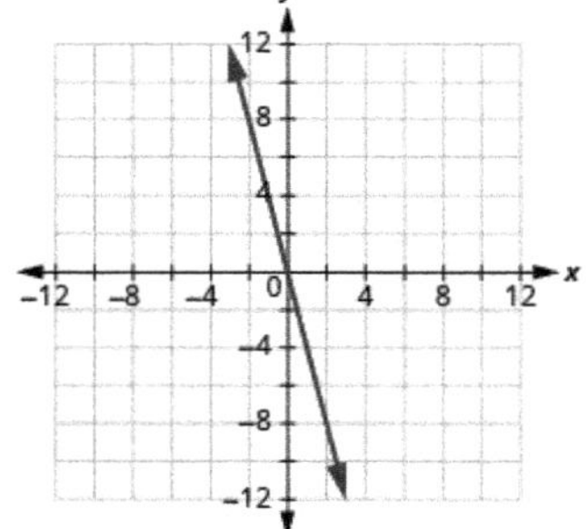

55.

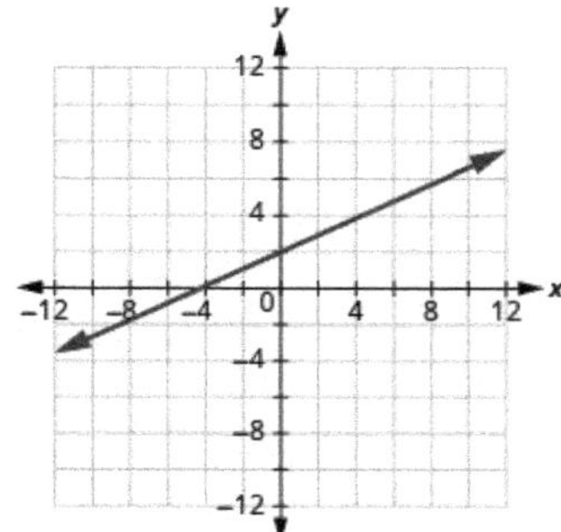

57.

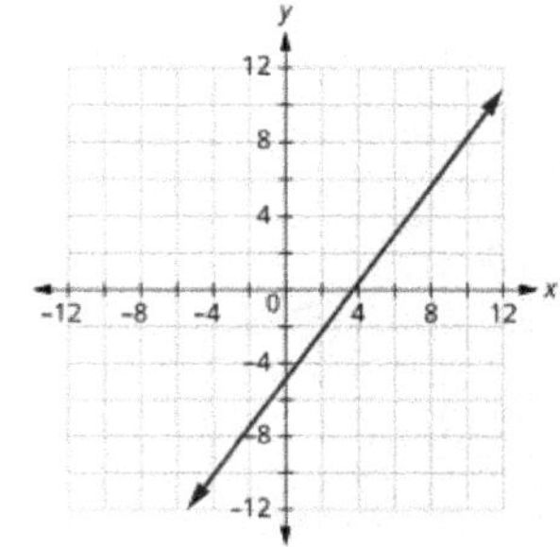

59.

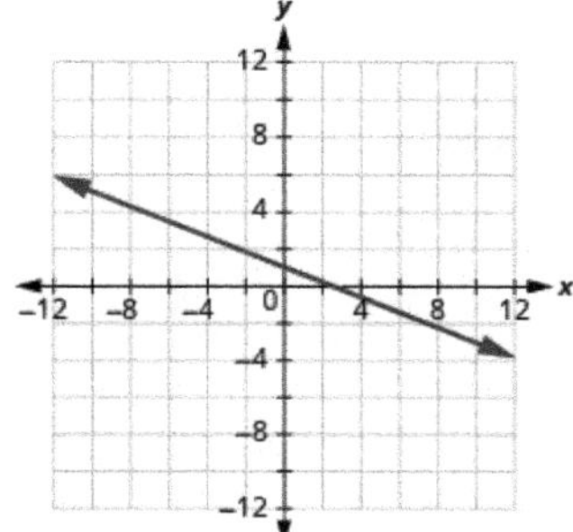

61.

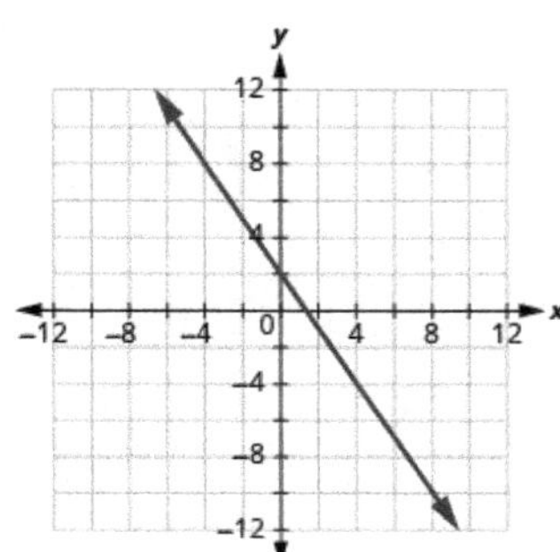

63.

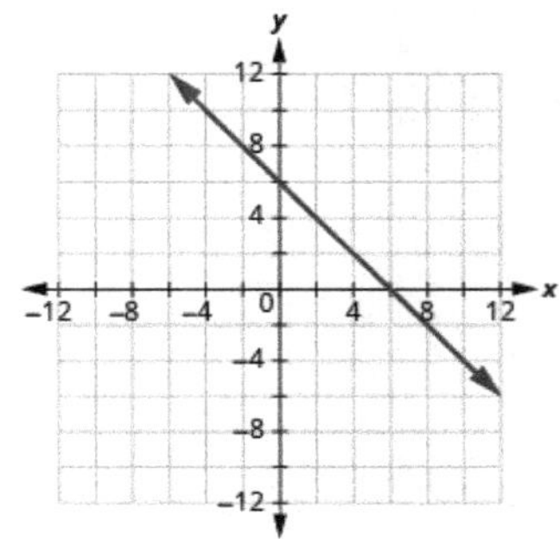

65.

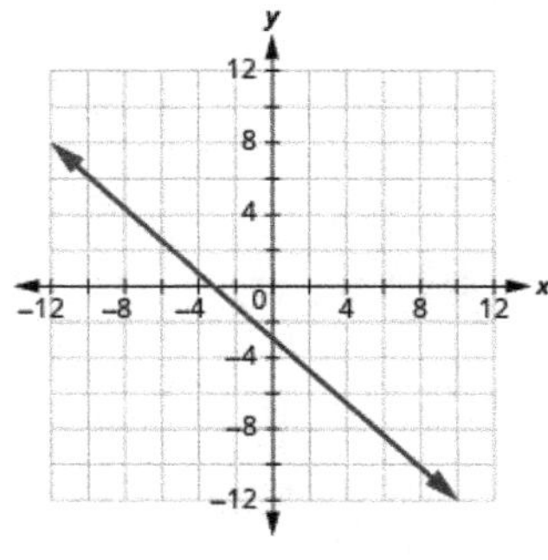

67.

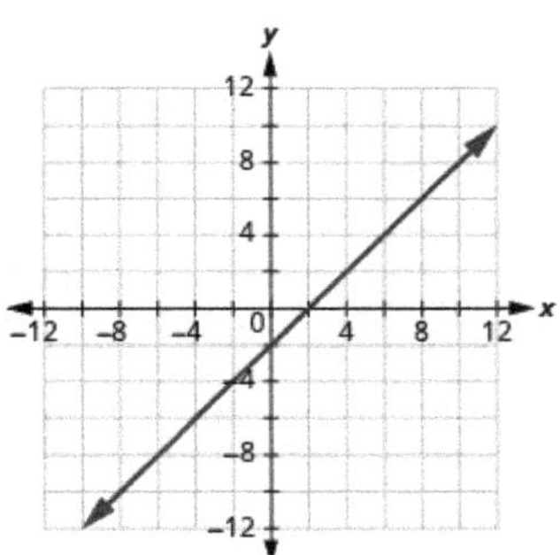

69.

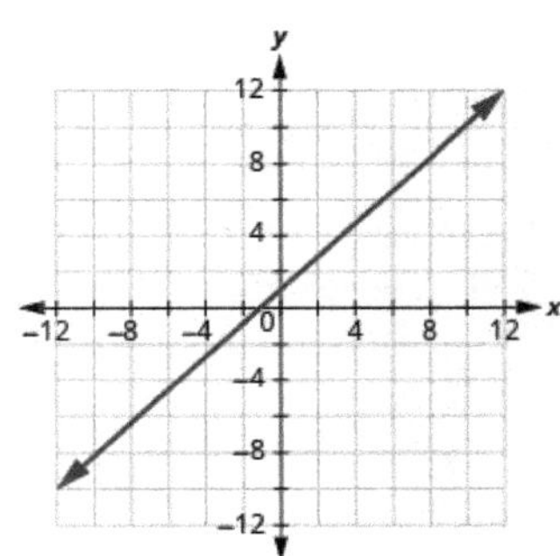

71.

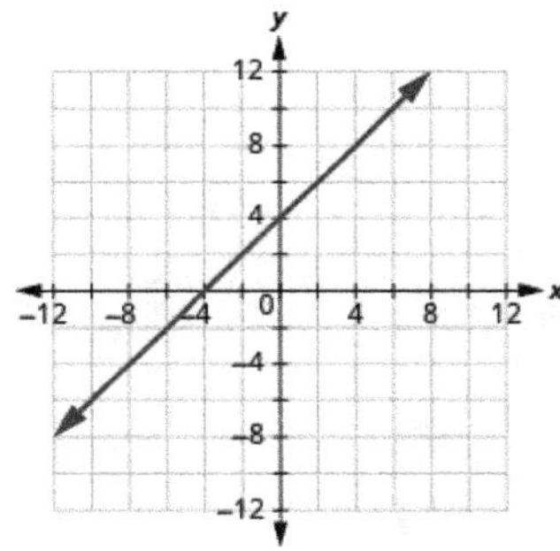

73.

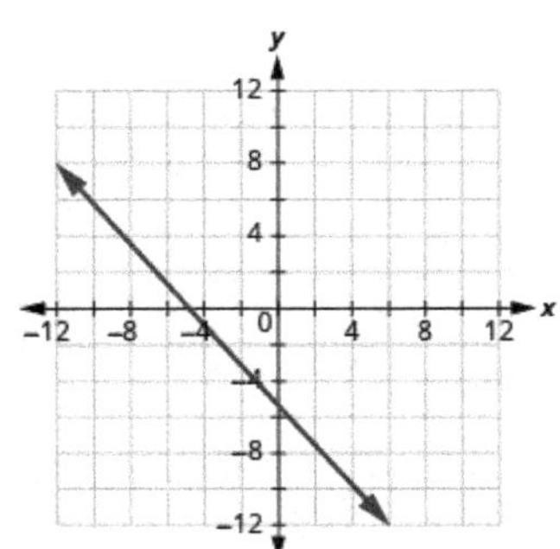

75.

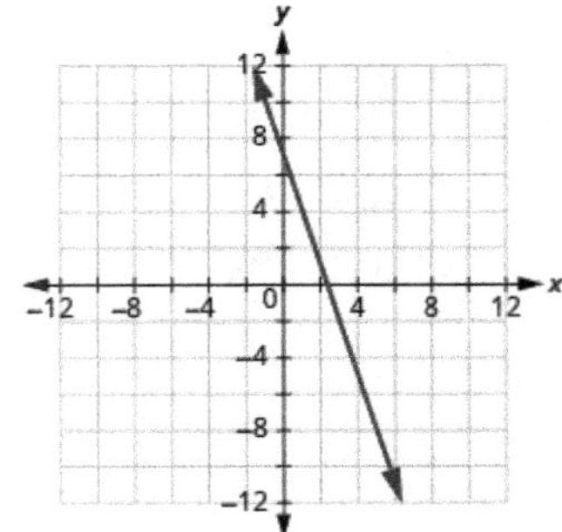

77.

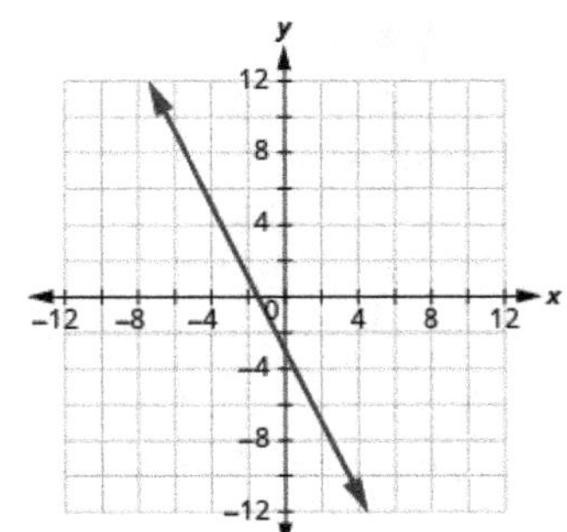

79.

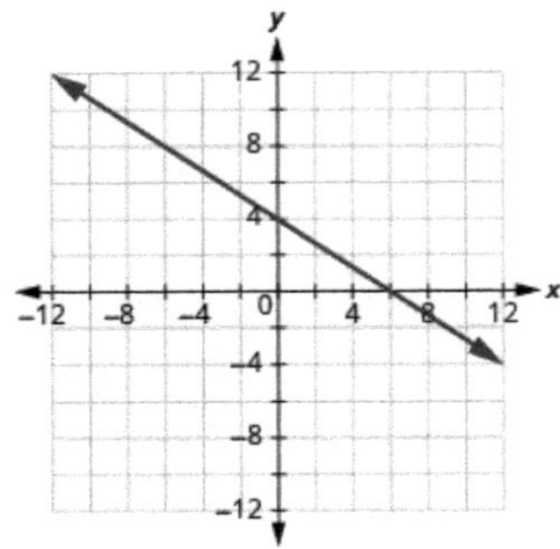

81.

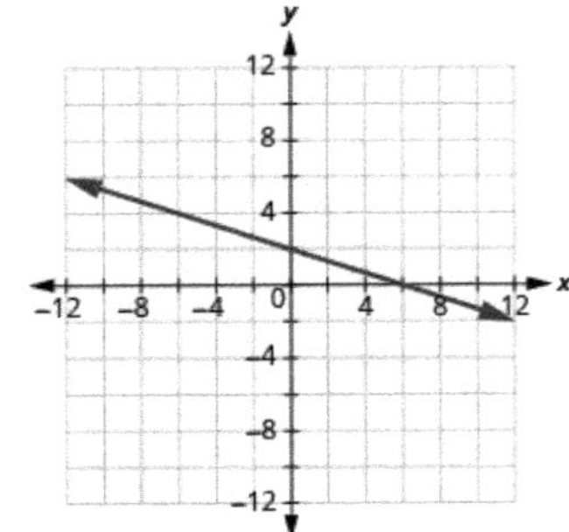

83.

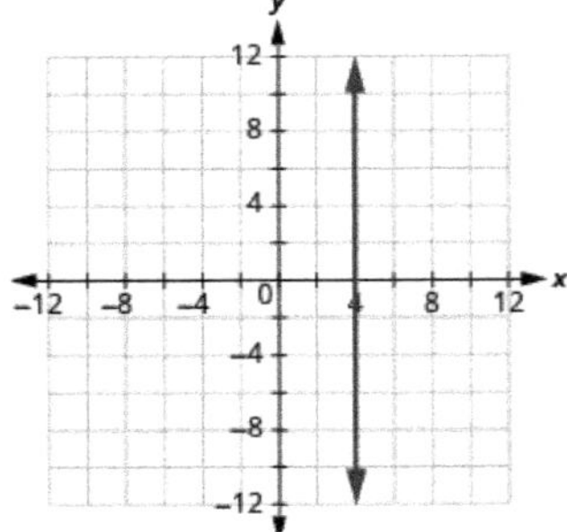

85.

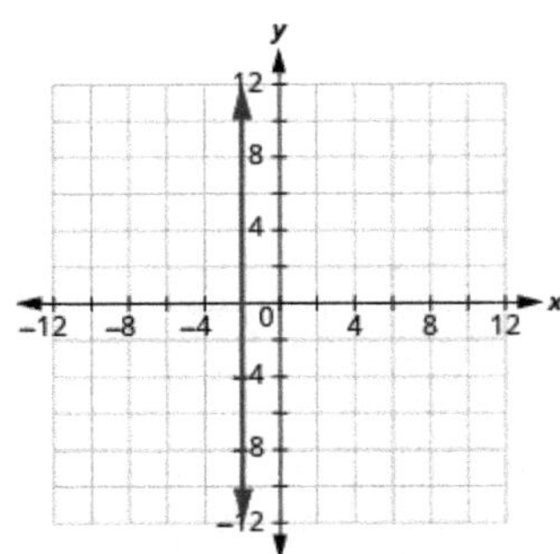

87.

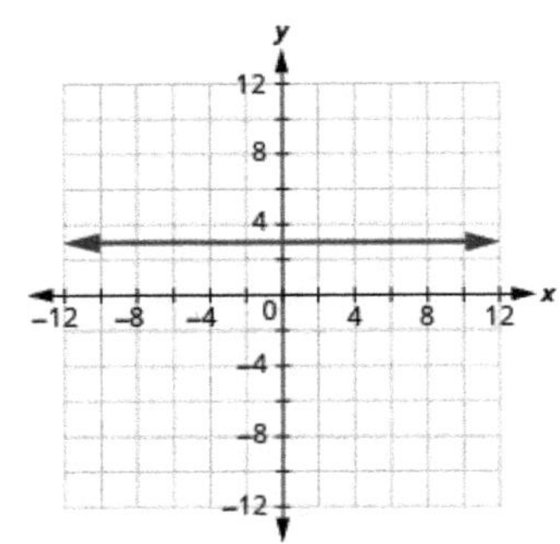

89.

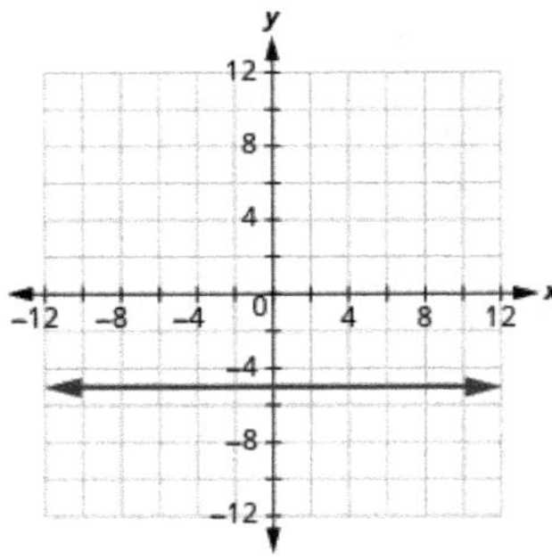

91.

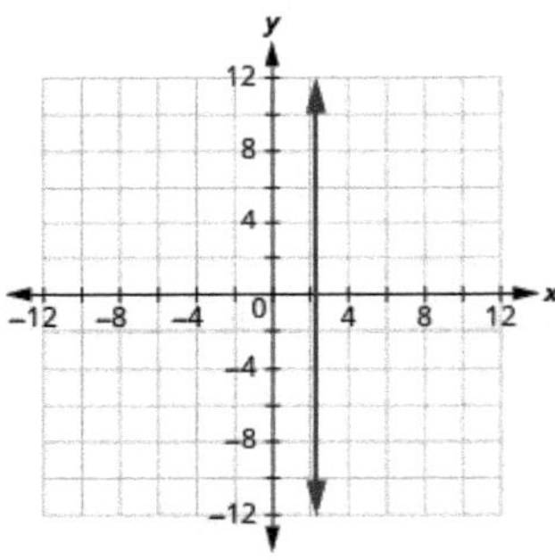

93.

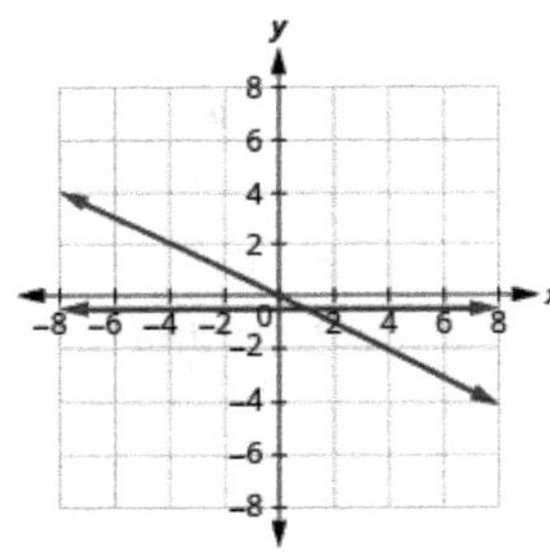

95.

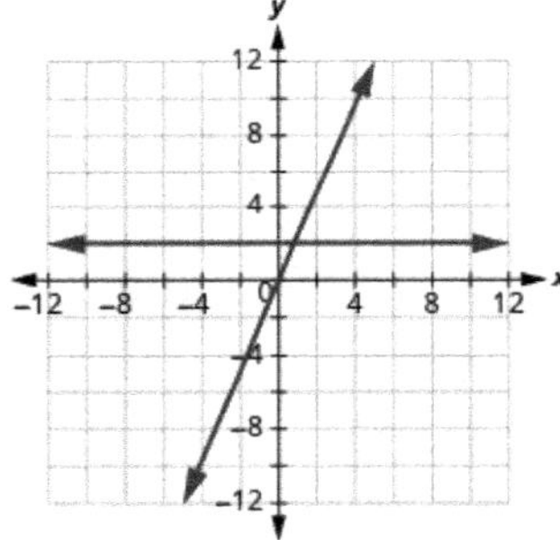

97.

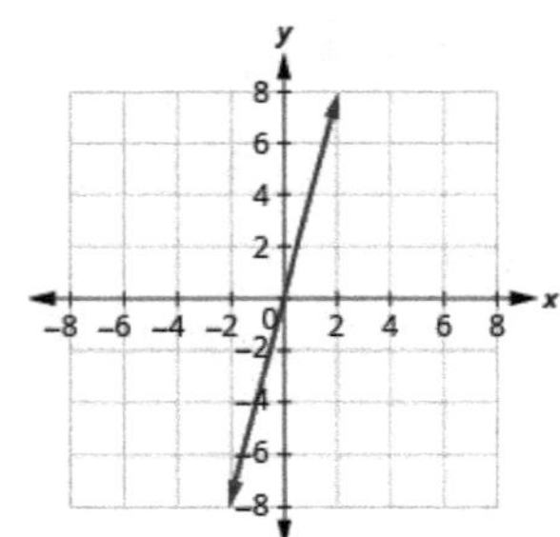

99.

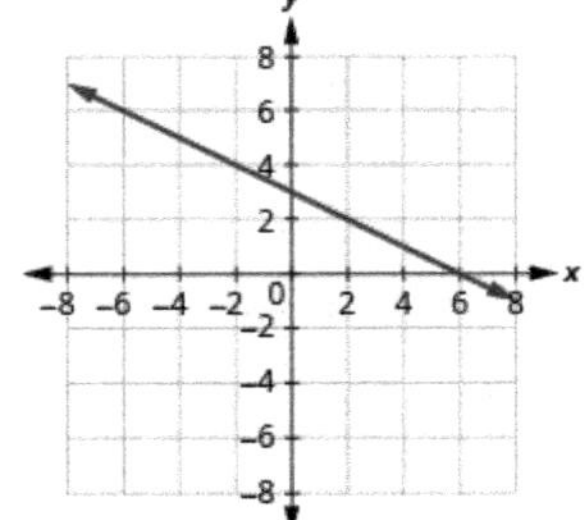

101.

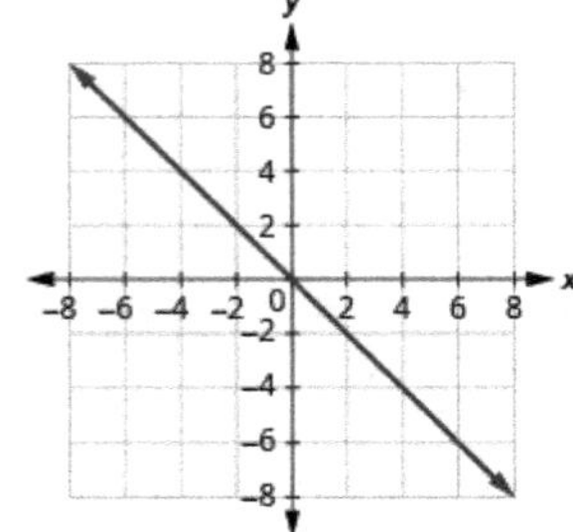

103.

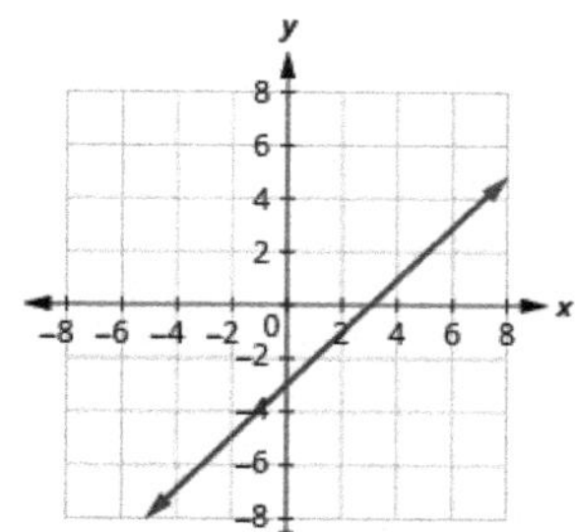

105.

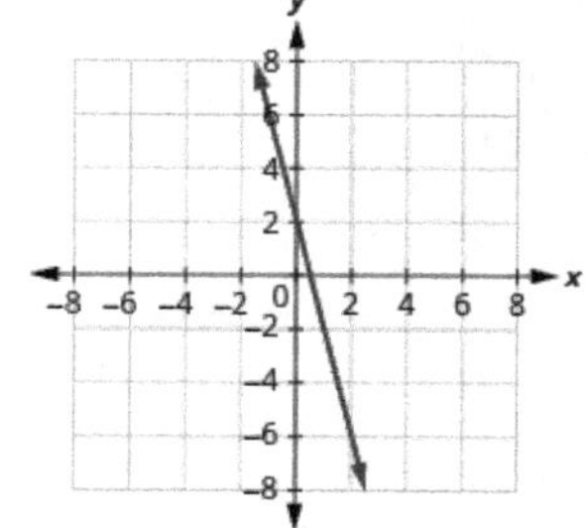

107.

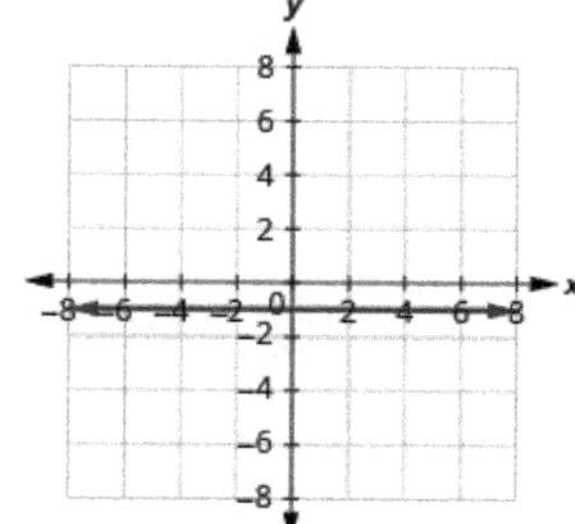

109.

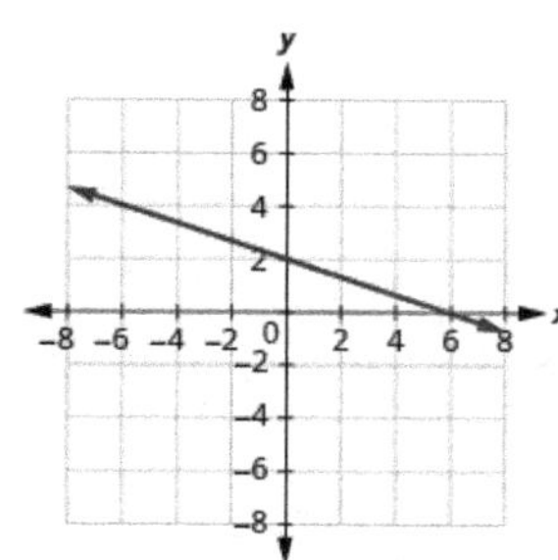

111.

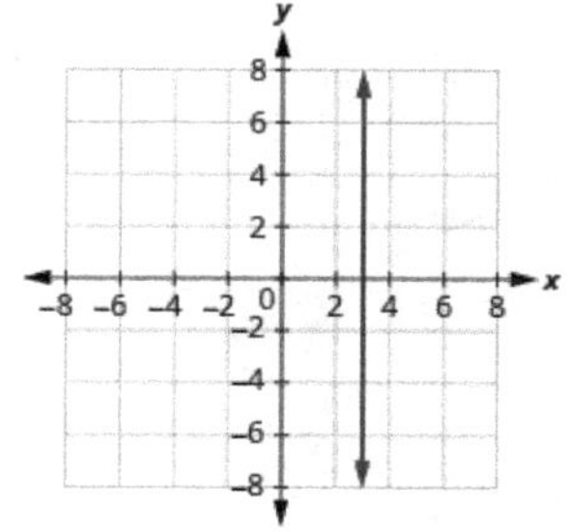

113.

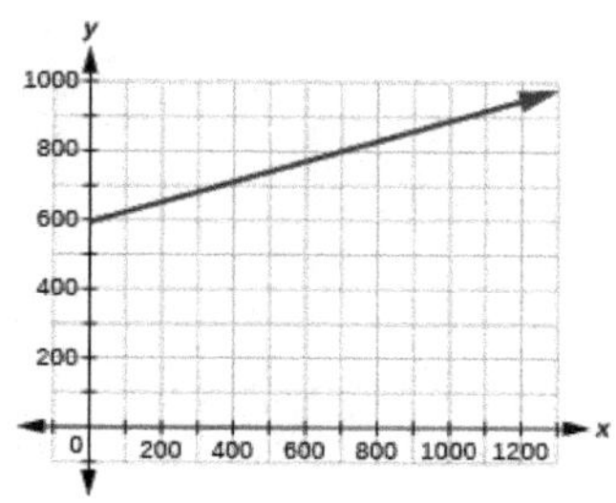

$722, $850, $978

115. Answers will vary.
117. (3,0),(0,3)
119. (5,0),(0,−5)
121. (−2,0),(0,−2)
123. (−1,0),(0,1)
125. (0,0)
127. (4,0),(0,4)
129. (−2,0),(0,−2)
131. (5,0),(0,−5)
133. (−3,0),(0,3)
135. (8,0),(0,4)
137. (2,0),(0,6)
139. (12,0),(0,−4)
141. (2,0),(0,−8)
143. (5,0),(0,2)
145. (4,0),(0,−6)
147. (3,0),(0,−1)
149. (−10,0),(0,2)
151. (0,0)
153. (0,0)
155.

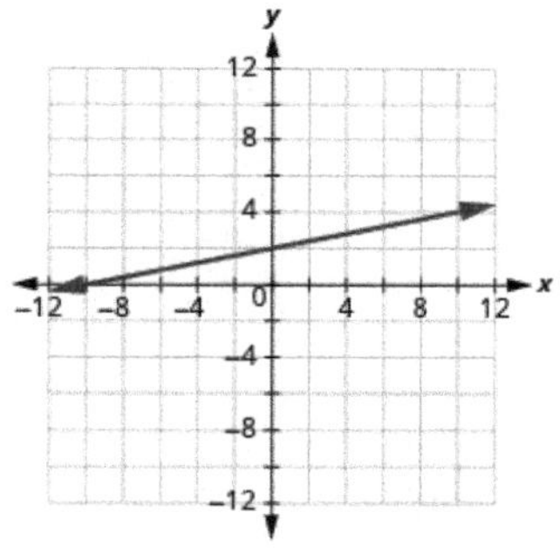

157.

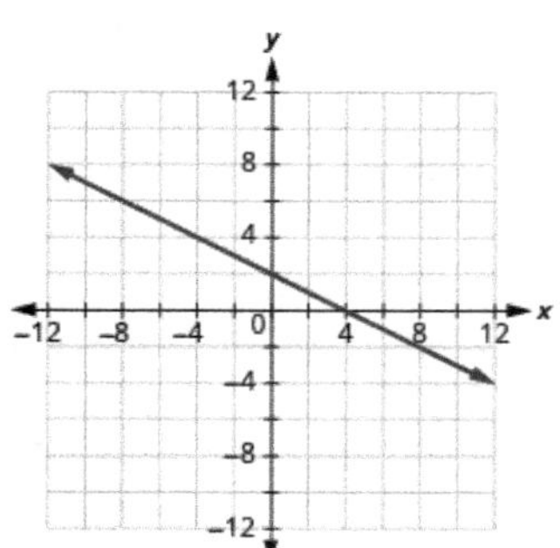

159.

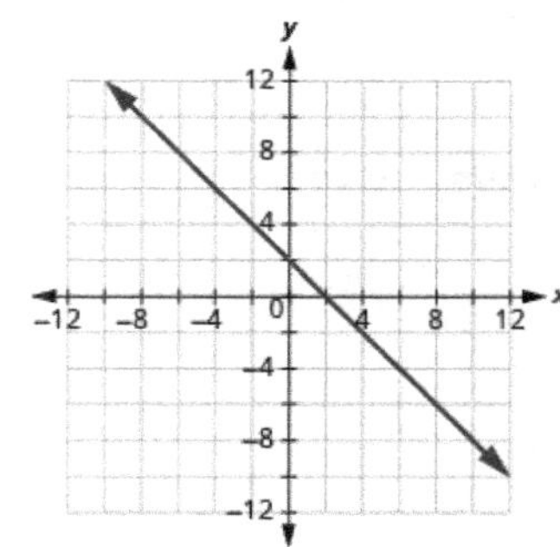

161.

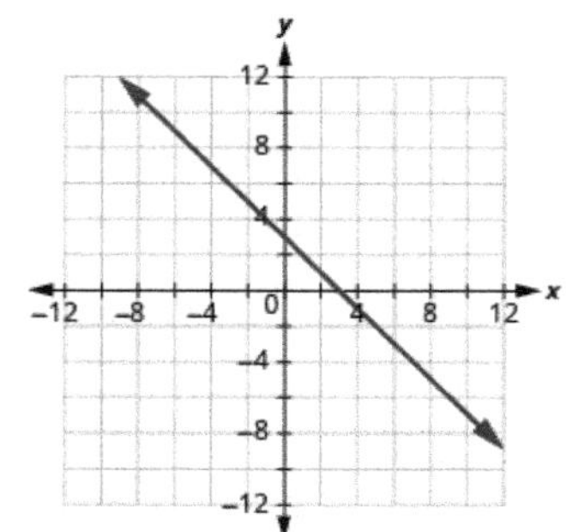

163.

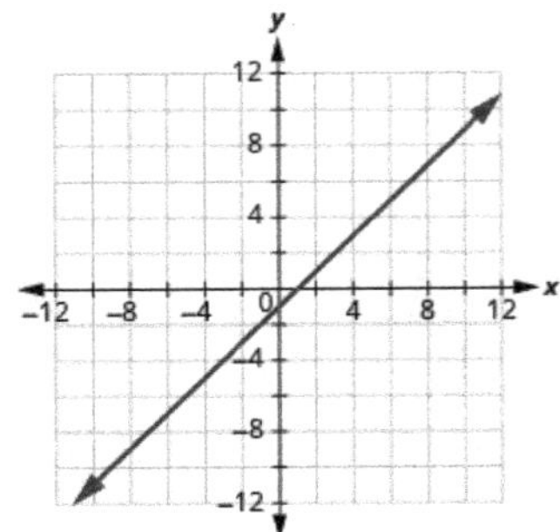

165.

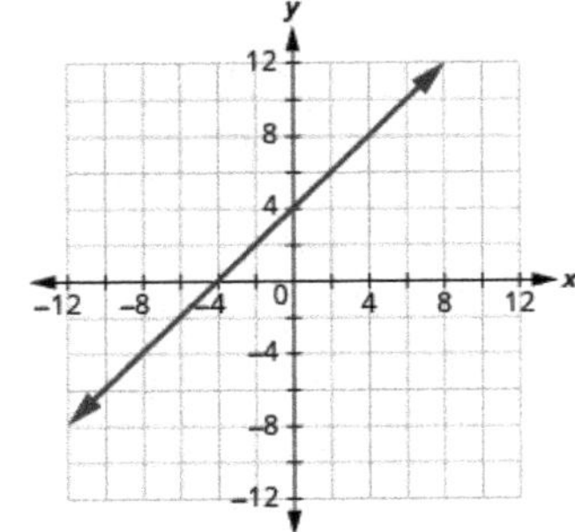

167.

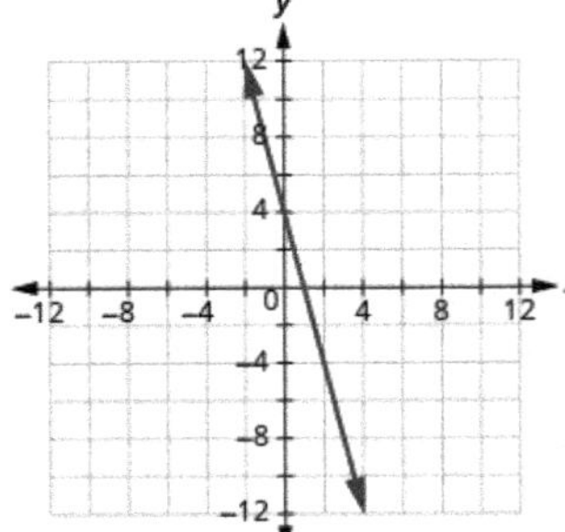

169.

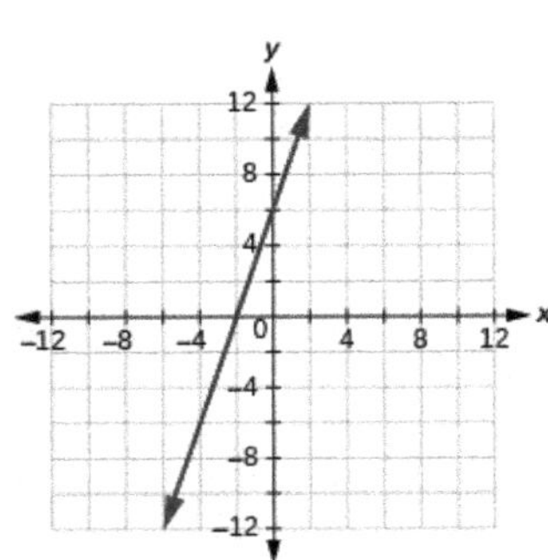

171.

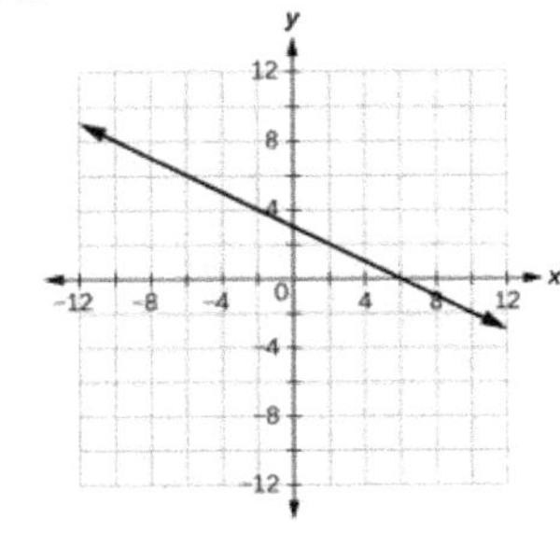

173.

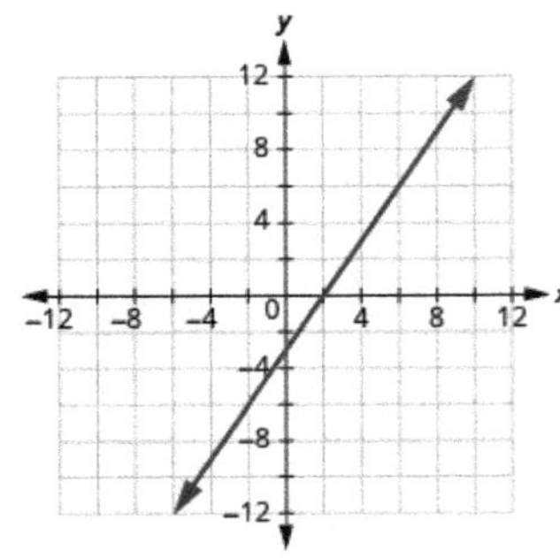

175.

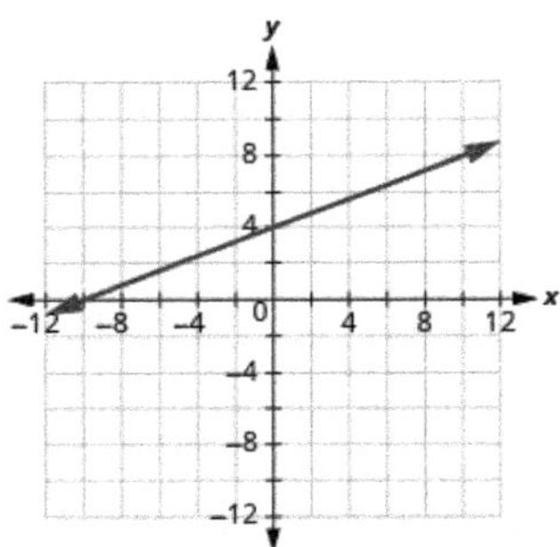

177.

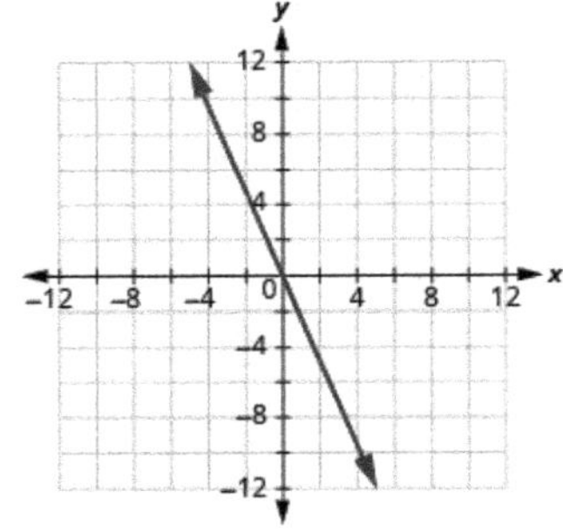

179.

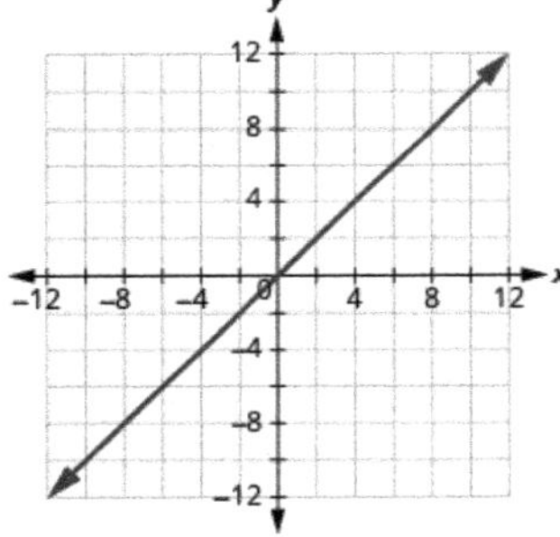

181. vertical line
183. horizontal line
185. plotting points
187. intercepts
189. plotting points
191. horizontal line
193. intercepts
195. plotting points
197. (0,1,000),(15,0). At (0,1,000) he left Chicago 0 hours ago and has 1,000 miles left to drive. At (15,0) he left Chicago 15 hours ago and has 0 miles left to drive.

199. Answers will vary.
201. Answers will vary.
203. –

205. —

207.

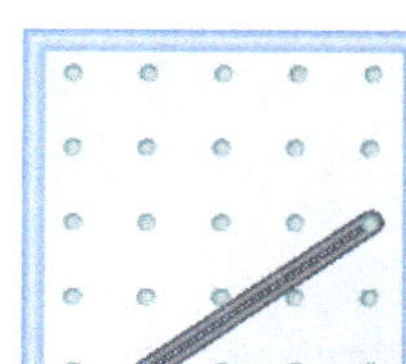

209.

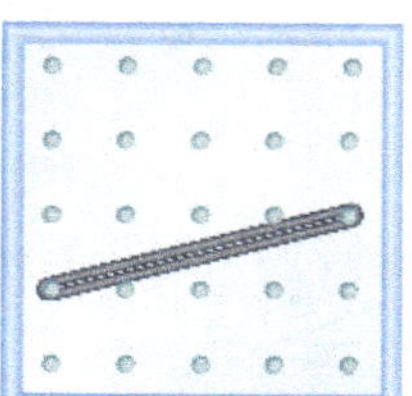

211.

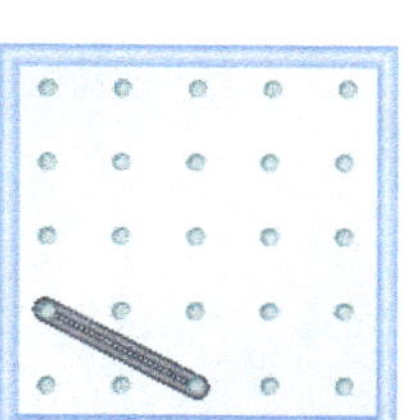

213.

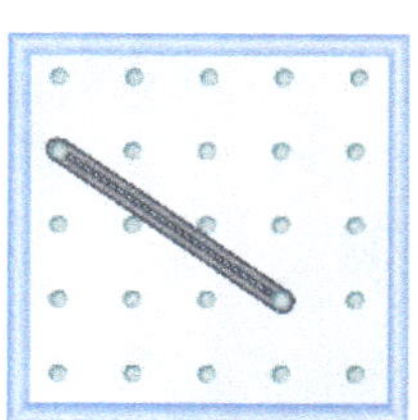

215. —

217. —

219. —

221. —

223. —

225. —

227. —

229. —

231. 0

233. undefined

235. 0

237. undefined

239. —

241. —

243. —

245. —

247. —

249. −1

251.

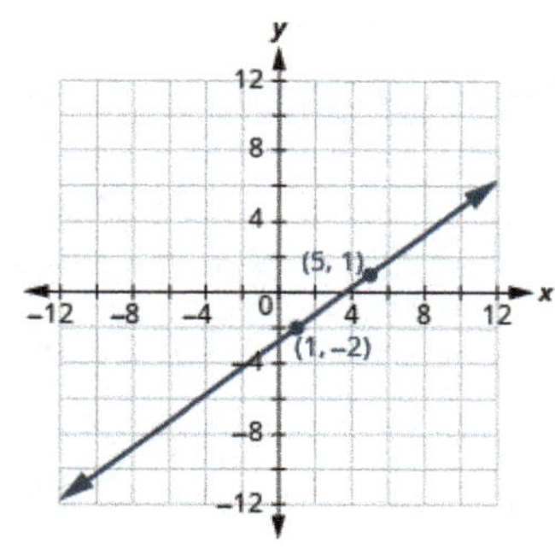

253.

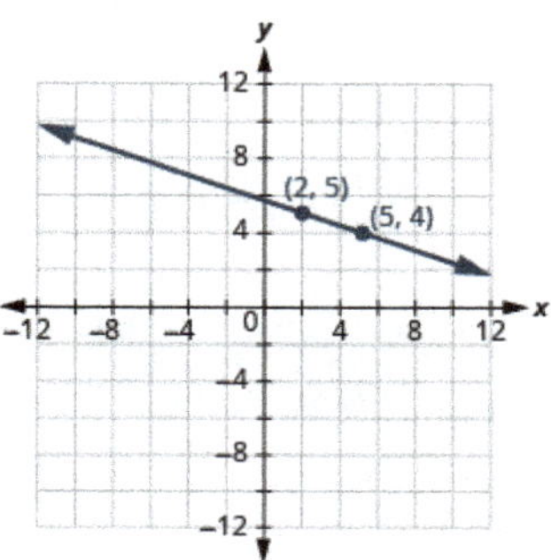

255.

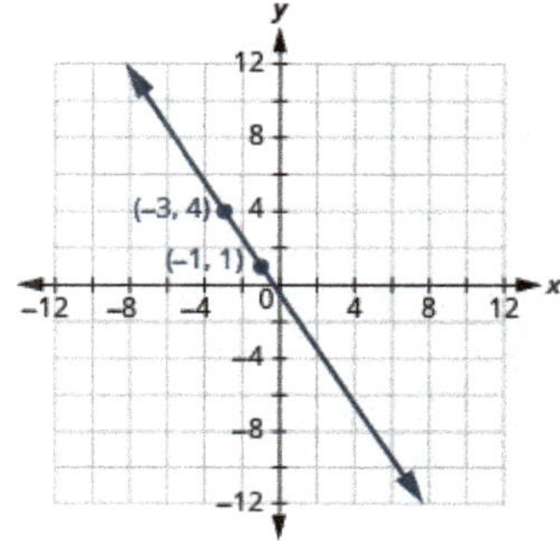

257.

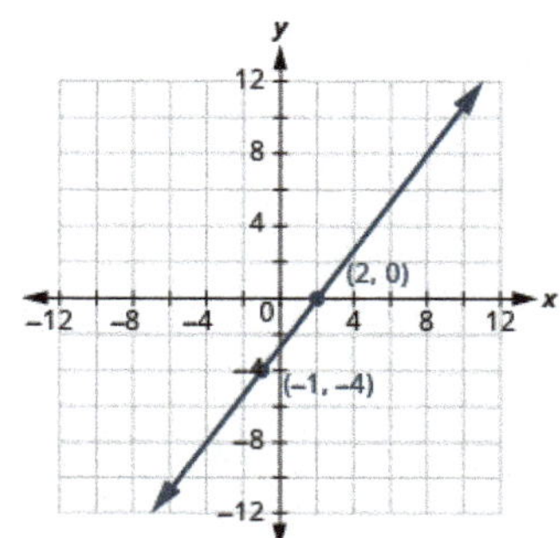

259.

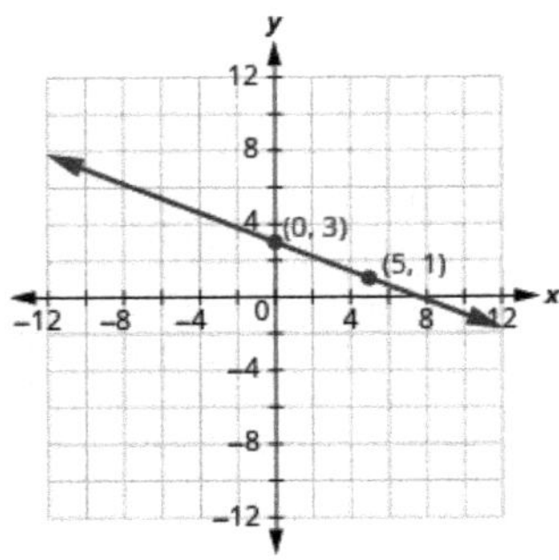

261.

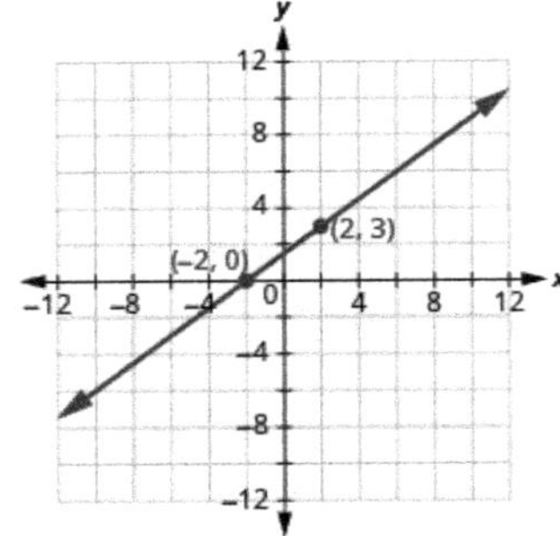

263.

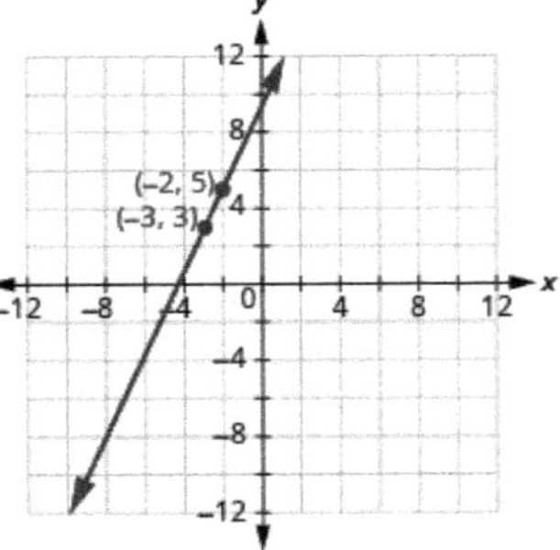

265.

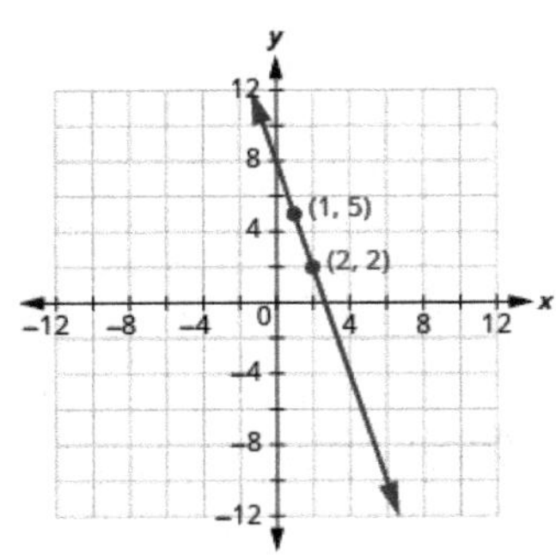

267. –

269. 

271.

288 inches (24 feet)

Models will vary.

273. Answers will vary.

275. Answers will vary.

Review Exercises

277.

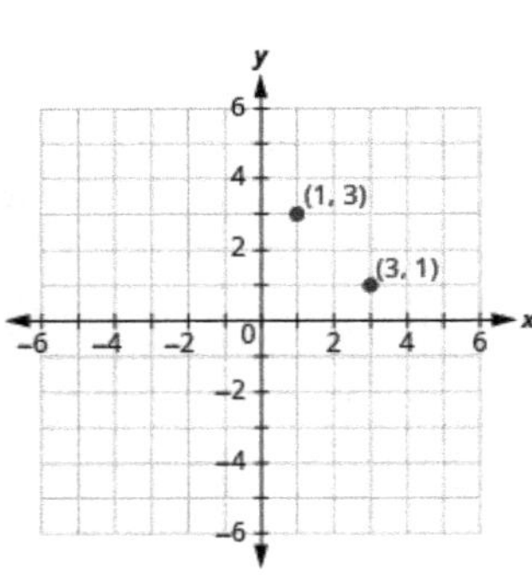

279.

III II IV I

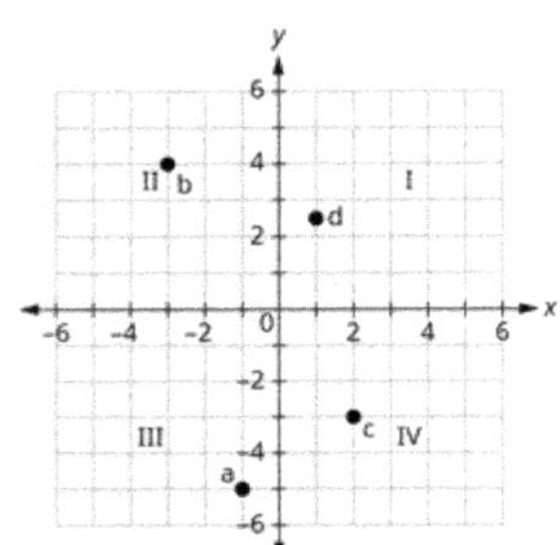

281.

(5,3) (2,−1) (−3,−2)

(−1,4)

283.

(2,0) (0,−5) (−4,0)

(0,3)

285. ,

287.

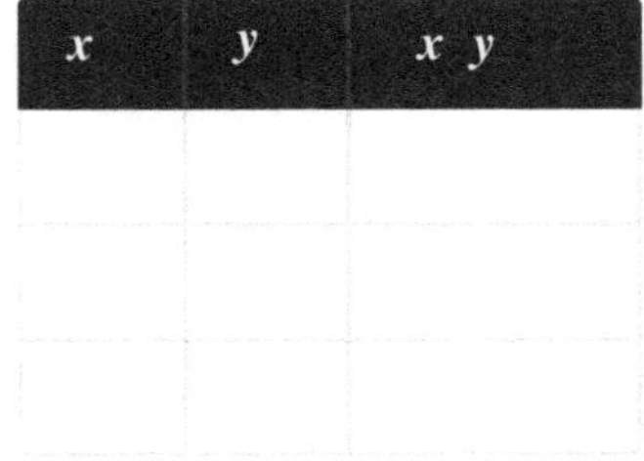

x	y	x y

289.

x	y	(x, y)

291. Answers will vary.

293. Answers will vary.

295.

1. yes yes
2. no no
3. yes yes
4. yes yes

297.

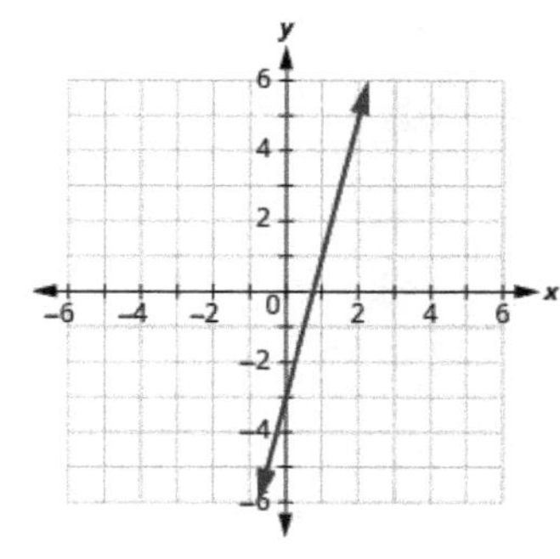

299.

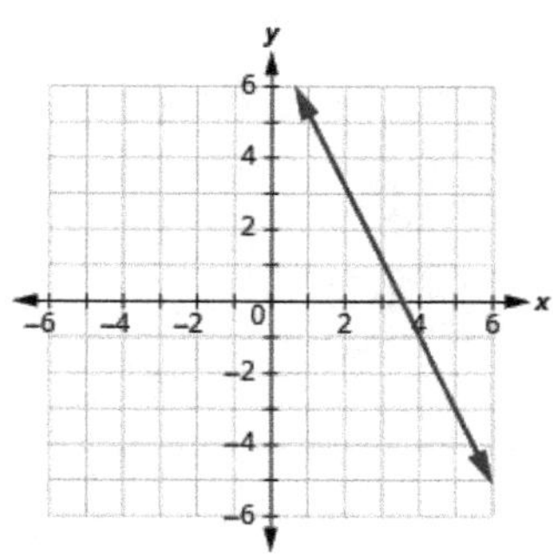

301.

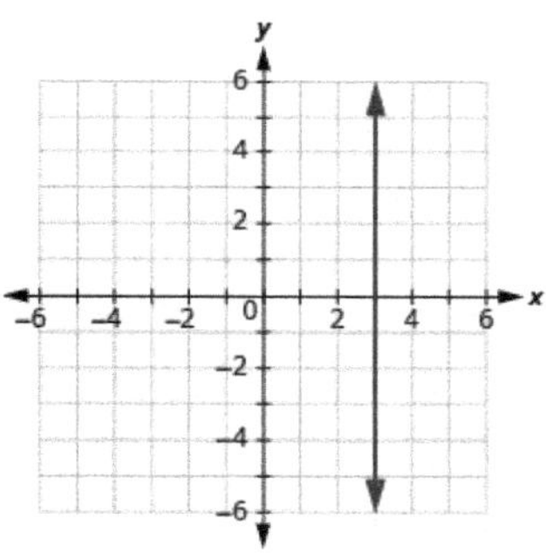

303. (0,3) (3,0)

305. (−1,0) (0,1)

307. (0,0)

309.

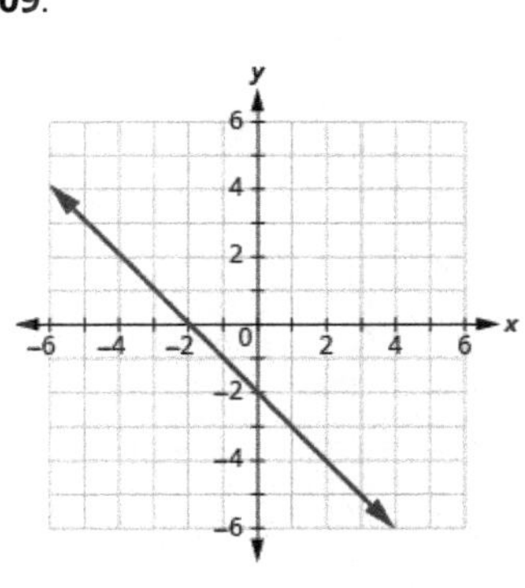

311. horizontal line

313. intercepts

315. plotting points

317. —

319. —

321.

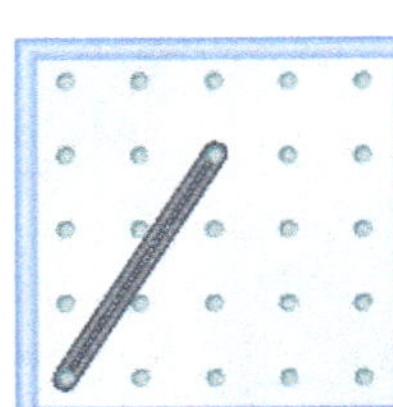

323.

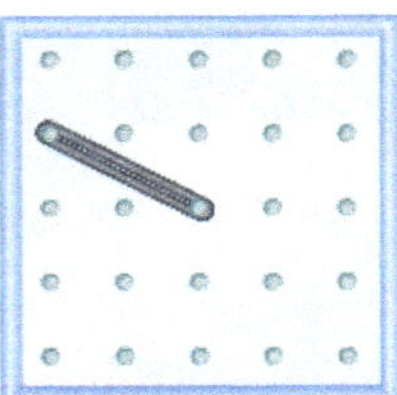

325. 1

327. —

329. undefined

331. 0

333. −4

335. —

337.

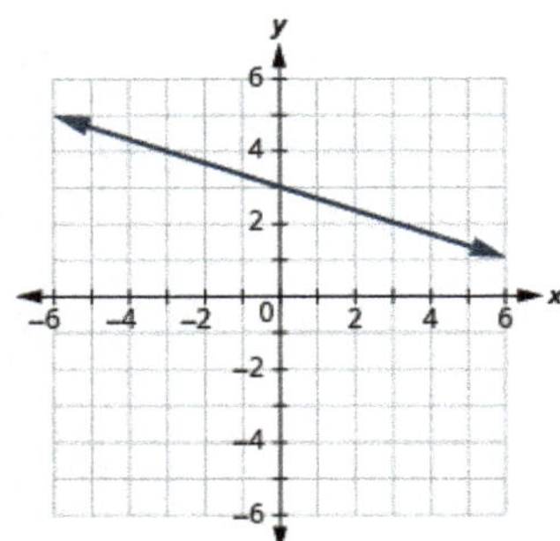

Practice Test

339.

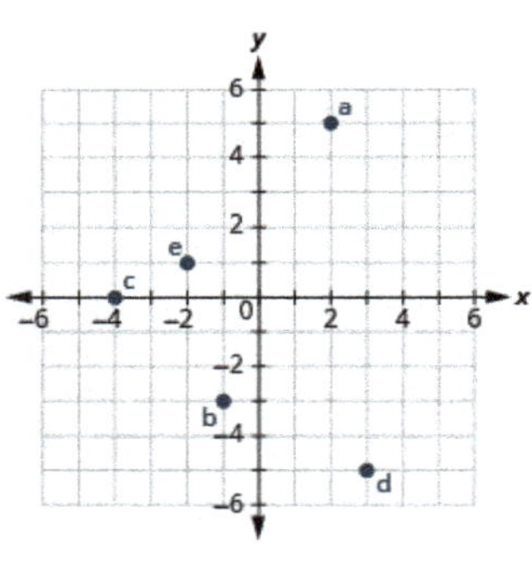

341. (4,0), (0,−2)

343. no; $1 + 4 \cdot 3 \neq 12$

345.

347.

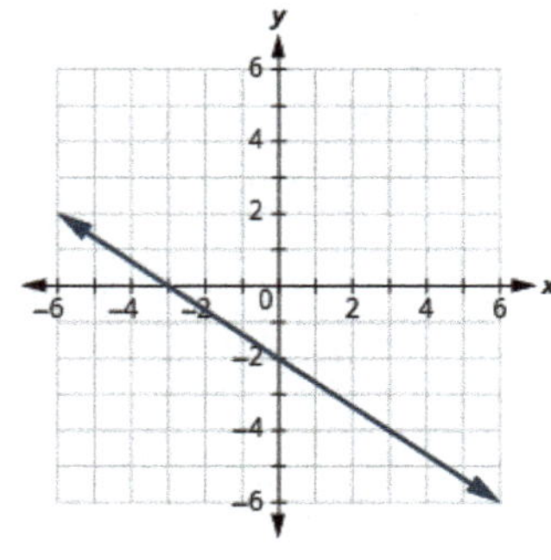

349. —

351. 0

353. —

INDEX

www.ingramcontent.com/pod-product-compliance
Lightning Source LLC
Chambersburg PA
CBHW061327190326
41458CB00011B/3921
* 9 7 8 1 6 8 0 9 2 3 2 6 1 *